Supernovae

Andrei Bykov • Roger Chevalier • John Raymond •
Friedrich-Karl Thielemann • Maurizio Falanga •
Rudolf von Steiger
Editors

Supernovae

Previously published in *Space Science Reviews* in the Topical Collection "Supernovae"

Springer

Editors
Andrei Bykov
High Energy Astrophysics
Ioffe Institute
St. Petersburg, Russia

Roger Chevalier
Department of Astronomy
University of Virginia
Charlottesville, VA, USA

John Raymond
Harvard–Smithsonian
Center for Astrophysics
Cambridge, MA, USA

Friedrich-Karl Thielemann
Department of Physics
University of Basel
Basel, Switzerland

Maurizio Falanga
International Space Science Institute
Bern, Switzerland

Rudolf von Steiger
International Space Science Institute
Bern, Switzerland

ISSN 1385-7525 Space Sciences Series of ISSI
ISBN 978-94-024-1580-3

Library of Congress Control Number: 2018961748

Cover Image: Credit: ESA/Hubble & NASA/D. Milisavljevic (Purdue University)

Printed on acid-free paper

This Springer imprint is published by the registered company Springer Nature B.V.
The registered company address is: Van Godewijckstraat 30, 3311 GX Dordrecht, The Netherlands

Contents

Preface

A.M. Bykov[1,2] · R.A. Chevalier[3] · J.C. Raymond[4] ·
F.-K. Thielemann[5,6] · M. Falanga[7] · R. von Steiger[7]

1 Supernovae in Astrophysics

Being one of the most luminous and spectacular phenomena in the sky, supernovae (SNe) have been actively studied since early Chinese astronomers left records of guest stars, to the most recent high precision, multi-wavelength observations. This intense study has yielded in recent years great insight into the physics of supernovae and their impact on a very wide variety of subjects from cosmology to astrobiology. The unprecedented growth of observational information, the ever increasing computational power for modeling, and the deep multi-disciplinary connections of supernovae to stellar physics, galactic evolution, astrochemistry, and cosmic ray origin has motivated the publication of this topical collection of reviews. Our aim is to provide some insight on current progress and present some open questions for future supernova studies.

We note that relativistic stellar mass objects, i.e., neutron stars and black holes, are the compact remnants of supernovae. Therefore, this "Supernovae" topical collection is generically connected to the topical collection "Jets and Winds in Pulsar Wind Nebulae, Gamma-Ray Bursts and Blazars" published in Space Science Reviews, volume 207, 2017.

✉ A.M. Bykov
byk@astro.ioffe.ru

1 Ioffe Institute, 194021, St. Petersburg, Russian Federation

2 St. Petersburg Polytechnic University, St. Petersburg, Russian Federation

3 Department of Astronomy, University of Virginia, P.O. Box 400325, Charlottesville, VA 22904-4325, USA

4 Center for Astrophysics, 60 Garden St., Cambridge, MA 02138, USA

5 Department of Physics, University of Basel, Basel 4056, Switzerland

6 GSI Darmstadt, Darmstadt 64291, Germany

7 International Space Science Institute, Bern, Switzerland

2 Supernovae in Cosmology

Type Ia supernovae are used to accurately determine cosmological distances and this has helped determine the local expansion rate of the Universe leading to the very important discovery of the accelerated expansion of the Universe (Riess et al. 1998; Perlmutter et al. 1999). This has further led to the dark energy paradigm within general relativity which continues to challenge current concepts in fundamental physics. The assumptions that no selection biases affected the distance measurements, that the corrections applied to nearby Type Ia SNe are valid throughout the whole observed redshift range, and that the corrections for redshift and absorption can be controlled, are discussed in this collection of review papers by Leibundgut and Sullivan. These assumptions are critical for Type Ia cosmological applications. Leibundgut and Sullivan also discuss future studies with the ground-based *Large Synoptic Survey Telescope* (LSST) and the large space-based projects *Euclid* and *WFIRST*. The LSST should provide hundreds of thousands of transient candidates every night and likely find more than 100,000 Type Ia SNe in five years.

3 Modeling of Core-collapse and Thermonuclear Supernovae

Apart from the classical problems of stellar evolution like nucleosynthesis, hydrodynamical transport and instabilities, radiative transfer and magnetic phenomena, the physics of supernovae also includes the accretion and jets of the compact objects with general relativistic effects and neutrinos and astroparticle phenomena. This makes the modeling of supernova explosions and evolution to be extremely complicated, but very exciting. Burrows et al present a review of crucial physical dependencies of the core-collapse supernova mechanism. They emphasize the role of the small changes in the neutrino-matter couplings in the proximity to criticality to produce sizable effects on the core-collapse. Nomoto and Leung review the single degenerate models for Type Ia supernovae. They discuss the evolution and nucleosynthesis yields of the white dwarfs in the binary star systems with non-degenerate companion stars. Röpke and Sim review the possible numerical approaches to model properly the multi-physics and the multi-scale nature of Type Ia supernovae and related astrophysical transients.

4 Neutrinos in Supernovae

Neutrinos play a fundamental role in the collapse of massive stars. Recent 2D and 3D simulations of core-collapse SNe have revealed that even modest changes in the neutrino-matter coupling may have profound effects on the resulting hydrodynamical matter motions associated with the energy release that may power the explosion. A large-scale asymmetric emission of the electron neutrino lepton number (LESA) discovered in 3D simulations may result in neutron-star kicks and affect the nucleosynthesis in the neutrino-driven winds. These processes, as well as the flavor conversions in the stellar envelopes, and the high energy neutrinos from the hidden relativistic outflows associated with the core collapse of massive stars, are reviewed by Tamborra and Murase. Since they are weakly interacting, neutrinos carry unique information on the accretion phase of the core collapse before the first photons can be detected from the supernova. This idea is realized in the *SuperNova Early Warning System* (SNEWS) project which involves an international network of supernova-neutrino-sensitive detectors. Currently, seven neutrino experiments are involved in SNEWS: Super-K

(Japan), LVD (Italy), Ice Cube (South Pole), KamLAND (Japan), Borexino (Italy), Daya Bay (China), and HALO (Canada). SNEWS will alert astronomers of the developing collapse. Detection of the neutrino signal, as well as the associated gravitational waves, will provide important new information to test the complex supernova models. Observational constraints on the neutrino flavor oscillations in the stellar envelope will test the neutrino oscillation physics.

5 Superluminous and Unusual Supernovae

Recent supernova searches have examined open areas of sky, as opposed to nearby galaxies as was done in earlier searches. The searches revealed rare, very luminous events, now called superluminous supernovae (SLSNe). Moriya, Sorokina, and Chevalier review the properties of these events and discuss the possible power sources for the emission, including radioactivity, circumstellar interaction, and magnetar power. Milisavljevic and Margutti discuss peculiar supernovae, starting with the observations that led to the recognition of Type Ib/c supernovae. They go on to describe unusual events, including SN 2014C, which made a transition between supernova types. Analysis of optical observations of unusual events is often aided by multi-wavelength observations, especially at radio and X-ray wavelengths. This is also true for supernovae interacting with dense circumstellar media, which are discussed by P.Chandra. The dense media imply a high rate of mass loss just before the supernova that is not understood in standard stellar evolution theory.

6 Supernova Remnants

Higher and higher spatial and spectral resolution observations of supernova remnants are becoming available at gamma-ray through radio wavelengths. Lopez and Fesen review how the morphology and kinematics of SNRs can be used to determine the supernova type and to explore the nature of the explosion through signatures such as jets and bubbles in the ejecta. The shock waves that make the gas visible are different from shocks in gas of normal astrophysical abundance due to the enormous radiative cooling rates in gas devoid of H and He, and Raymond reviews the physical processes and their consequences for observable spectra.

7 Nuclesynthesis and Dust Formation in Supernovae

While the general picture of the nucleosynthesis in supernovae has been understood for quite a while, some major open questions remains. These are related to the precise amounts of alpha-elements produced in core-collapse SNe, the detailed composition of the Fe-group and the respective role of Type Ia vs. the core-collapse of massive stars. Thielemann, Isern, Perego and von Ballmoos present a detailed review of the status of nucleosynthesis in both Type Ia and core-collapse supernovae and the open problems responsible for the production of the intermediate mass, Fe-group and heavier elements. They discussed the wide range of the observational constraints on the nucleosynthesis in supernovae.

Dust can be produced efficiently in supernovae and supernova remnants. The dust formation scenarios are reviewed in detail by Sarangi, Matsuura and Micelotta. A review of the state of the art results of the dust processing and survival inside supernova remnants, in terms of theoretical modelling and with comparison to observations, is presented by Micelotta et al.

8 Cosmic Rays in Supernovae

The early suggestion by Baade and Zwicky (1934) that cosmic rays (CRs) are produced by supernovae has now a conclusive confirmation in radio, X-ray, and gamma-ray observations of individual supernova remnants (SNRs). It has become clear that SNe are the most likely source of cosmic rays with energies below the so-called "knee" at $\sim 10^{15}$ eV. The origin of CRs with energies beyond the knee is less certain. Bykov, Ellison, Marcowith and Osipov review the possible mechanisms of CR acceleration both below the knee and well beyond PeV energies. The observed energy density and composition of GeV cosmic rays imply that the mechanism of cosmic ray acceleration must be highly efficient with at least 10% of the total kinetic energy of the supernova ejecta going into CRs if they are powered by supernova shocks. If the acceleration efficiency is this high, feedback effects from CR production will influence the shock dynamics. Furthermore, since CRs are connected to the supernova outflows through fluctuating magnetic fields, a high acceleration efficiency is likely to lead to strong magnetic field amplification in fast supernova shocks. Models of Fermi type acceleration in fast supernova shocks with magnetic field amplification, including trans-relativistic SNe, are reviewed.

The circumstellar environment can play a role in high-energy CR acceleration. The role of superluminous interaction powered SNe, and normal core-collapsed SNe in clusters of young massive stars, in efficiently producing CRs to PeV energies is discussed. It is further suggested that galactic sources associated with infrequent but powerful relativistic SNe can accelerate CR protons up to EeV energies (i.e., 10^{18} eV).

Overall, this book provides a timely and comprehensive review of the multi-wavelength observations and the main physical processes involved in the study of supernovae and supernova remnants with applications to nucleosynthesis and cosmic ray physics. It will be particularly useful to postgraduate students and researchers active in various areas of astrophysics and space science.

This volume is based on an international workshop which took place at ISSI Bern during October 3–7 in 2016 where about forty leading experts in physics and astrophysics of supernovae and supernova remnants discussed the issues presented here. The workshop was the second in the series of the three ISSI workshops devoted to physics and astrophysics of the astrophysical objects with extreme energy release. The first book "Jets and Winds in Pulsar Wind Nebulae, Gamma-Ray Bursts and Blazars" ISBN 978-94-024-1291-8 was published by Springer in Space Sciences Series of ISSI.

The Editors would like to thank the participants of the workshop for their participation, and for their contributions to this volume. The scope, the scientific level and the pedagogical aspect of the present reviews testify of the high scientific level of the workshop and of its collaborative spirit. Special thanks are due to the reviewers, who helped improve the balance and the quality of the articles through their constructive reports. Obviously, this workshop could not have taken place without the strong support of the ISSI Bern staff, and we would like to thank most particularly Silvia Wenger, Jennifer Fankhauser, Andrea Fischer, Alexandra Lehmann and Saliba Saliba. Finally we thank the production staff of Space Science Reviews, as well as the production staff of Springer for their competence.

Space Sci Rev (2018) 214:57
https://doi.org/10.1007/s11214-018-0491-8

Type Ia Supernova Cosmology

B. Leibundgut[1,2] · M. Sullivan[3]

Received: 4 December 2017 / Accepted: 28 February 2018 / Published online: 12 March 2018

Abstract The primary agent for Type Ia supernova cosmology is the uniformity of their appearance. We present the current status, achievements and uncertainties. The Hubble constant and the expansion history of the universe are key measurements provided by Type Ia supernovae. They were also instrumental in showing time dilation, which is a direct observational signature of expansion. Connections to explosion physics are made in the context of potential improvements of the quality of Type Ia supernovae as distance indicators. The coming years will see large efforts to use Type Ia supernovae to characterise dark energy.

Keywords Cosmology · Supernovae

1 Background

The monolithic picture of Type Ia supernovae (hereafter SNe Ia) of a decade ago has undergone significant refinements. The observational diversity of SNe Ia includes ranges in luminosity, light-curve shape, colour, spectral evolution, and host galaxy environment (e.g. see Branch and Wheeler 2017, Maguire 2016 and Taubenberger 2016 for summaries on SNe Ia). Although almost certainly thermonuclear explosions of carbon-oxygen white dwarf stars, in detail we now believe that SNe Ia are the result of possibly different explosion mechanisms

Supernovae
Edited by Andrei Bykov, Roger Chevalier, John Raymond, Friedrich-Karl Thielemann, Maurizio Falanga and Rudolf von Steiger

✉ B. Leibundgut
Bruno.Leibundgut@eso.org

M. Sullivan
M.Sullivan@soton.ac.uk

1 European Southern Observatory, Karl-Schwarzschild-Strasse 2, 85748 Garching, Germany

2 Excellence Cluster Universe, Technische Universität München, Boltzmannstrasse 2, 85748 Garching, Germany

3 Department of Physics and Astronomy, University of Southampton, Southampton, Hampshire, SO17 1BJ, UK

after a variety of evolutionary paths towards the explosion. This has opened a rich research field to understand binary stellar evolution and explosion physics. At the same time, the observed diversity has occasionally raised doubts as to whether the cosmological results of SNe Ia are reliable.

SNe Ia are used in three specific areas of cosmology. Historically, they have been employed as an accurate distance indicator to determine the local expansion rate of the Universe (e.g. Leibundgut 2016; Saha and Macri 2016, and references therein), to demonstrate that cosmological redshifts and time dilation are connected as predicted in all theories based on general relativity (Leibundgut et al. 1996; Riess et al. 1997; Goldhaber et al. 2001; Blondin et al. 2008), and to map the expansion history of the universe, which led to the original discovery of the accelerated expansion (Riess et al. 1998; Perlmutter et al. 1999), and more recent studies investigating the nature of dark energy (Betoule et al. 2014, for reviews see Leibundgut 2001, Perlmutter and Schmidt 2003, Leibundgut 2008, Goobar and Leibundgut 2011). The distance determination is critical for the Hubble constant and the expansion history, while the uniformity of the light curves and spectral evolution was used to show how time dilation acts on distant supernovae.

Establishing the uniformity of SNe Ia has been an important factor in their cosmological application. SNe Ia are observed to explode with approximately the same luminosity to within around a factor of two, presumably due to the similarity of the triggering white dwarf mass and therefore the amount of nuclear fuel available to burn. More detailed observations have led to sophisticated treatments and standardisations, involving light curves, spectra, and the star-formation rate and stellar masses of the SN Ia host galaxies. The light-curve shape vs. luminosity relation as originally introduced by Phillips (1993) was the most important improvement to achieve distance accuracies required to detect the acceleration and to bring the uncertainty in the determination of H_0 to less than 5% (see Phillips and Burns 2016, for a review on the different light curve correction methods). Today, the most popular correction methods are SALT2 (Guy et al. 2007), MLCS2k2 (Jha et al. 2007), SNooPy (Burns et al. 2011) and SiFTO (Conley et al. 2008). There are differences in how these methods apply the corrections and in their training sets, but the results appear quite robust and only subtle differences remain.

One critical parameter is how reddening is implemented: this is a free parameter combining colour and absorption in the cosmological interpretation of SALT2 and SiFTO (e.g. Tripp 1998), but follows an astrophysical reddening law in MLCS2k2 (albeit with a non-standard R_V) and SNooPy. While these simple approaches are adequate for the current generation of experiments, they are unlikely to be sufficient as SN Ia measurements become ever more precise. Recent studies have provided evidence that the colour–luminosity parameter (‘β’) may evolve with redshift (e.g. Shariff et al. 2016); evidence that the SN Ia colour corrections can produce biased parameters if the underlying SN colour distribution is not accurately known (e.g. Scolnic and Kessler 2016); evidence for diversity in the extinction laws affecting SNe Ia (Amanullah et al. 2015); and long-standing evidence that conflating SN extinction and intrinsic colour variation into one parameter can (as might be expected) produce biased results (e.g. Mandel et al. 2017). These effects will lead to the handling of SN colour becoming a dominant uncertainty in upcoming analyses.

The most direct and easiest way to show that SNe Ia yield good (relative) distances is the Hubble diagram. Good standard candles delineate a line with a fixed slope, given by the linear expansion in the local universe ($z < 0.1$), when plotting observed magnitude versus the logarithm of the redshift ($m \propto 5\log(z)$; see Fig. 1). The zeropoint of the expansion line is a combination of the Hubble constant and the absolute magnitude of the distance indicator. The scatter around this slope gives an indication on the precision of the standard candle.

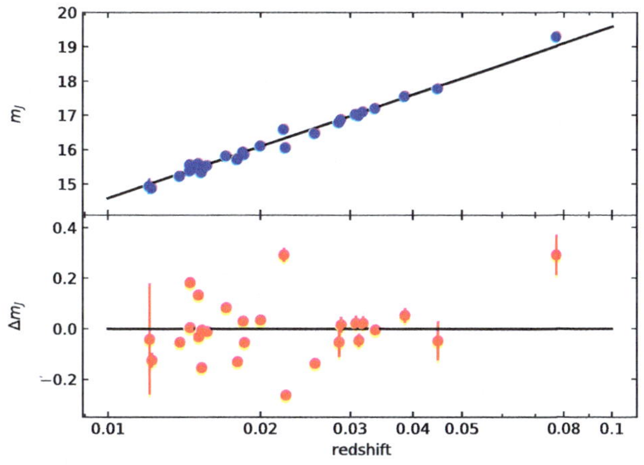

Fig. 1 The Hubble diagram of SNe Ia in the rest-frame J-filter. No correction has been applied to the maximum magnitudes. The slope of the line is fixed by the linear expansion in the local universe. Data from Dhawan et al. (2018).

After correction by the methods mentioned above, SNe Ia provide individual distances of better than 6% (e.g. Riess et al. 2016; Dhawan et al. 2018). This argument is completely independent of any knowledge of the physical nature of SNe Ia and can be regarded as entirely empirical. The Hubble diagram has traditionally been used to establish the quality of distance indicators (see the reviews by Branch 1982, 1998; Branch and Tammann 1992).

Although nearly all published cosmological constraints have been made with distance estimators constructed in the rest-frame optical, redder wavelengths have the potential to provide superior distance estimates. Fig. 1 shows the rest-frame J-band Hubble diagram of nearby SNe Ia (from Dhawan et al. 2018; a discussion of earlier SN Ia infrared Hubble diagrams can be found in Krisciunas 2016). In this case no correction to the J-band peak magnitude has been applied: the SNe Ia have been treated as pure standard candles. The slope of the line has been fixed to the fiducial slope for the linear expansion. The fit is extremely good with a scatter around the Hubble line of 0.12 magnitudes. It is clear from the uncertainties for each supernova that the scatter is intrinsic to the SN Ia sample and a suitable correction might decrease it further. However, there is no clear correlation with other observables (colour, light-curve shape in the optical bands, host galaxy properties; see Dhawan et al. 2018 for a discussion). The scatter should be compared to a standard variation of 0.128 magnitudes for 271 optical light curves with $0.01 < z < 0.15$ used in Riess et al. (2016).

A second method to establish the uniformity of a (normalised) luminosity is by sampling SNe Ia at known distances. If the used distances are absolutely calibrated, then this method is more powerful than the Hubble diagram. Measuring exact distances to SNe Ia has the problem that the volume reachable by reliable distance indicators is rather limited and the low rate of SNe Ia leads to very small samples. The *Hubble Space Telescope* (*HST*) was critical for this method (see Saha and Macri 2016, for a review of these programs) and has by now yielded a sample of 19 SNe Ia with Cepheid distances (Riess et al. 2016). The Cepheids were measured in the near-infrared (1.6 μm) to reduce uncertainties in phase calibrations and absorption. The standard deviation among the absolute luminosities of these 19 SNe Ia with Cepheid distances is 0.17 magnitudes. This is larger than the scatter in the relative distances measured through the Hubble diagram (as already pointed out by Dhawan et al. 2018). Nevertheless, this means that the SNe Ia in the calibrator set show individual distance uncertainties of about 8%.

These two independent methods show that SNe Ia currently provide the most accurate individual distances at cosmological scales. Even though the physics of explosion and radiation transport may not be fully understood, the observed uniformity of the light curves and the empirical normalisation methods yield an excellent consistency in the local universe.

2 SN Ia Cosmology

As already indicated, SNe Ia play decisive roles for the determination of the Hubble constant and the mapping of the expansion history of the universe and hence the contribution of various energy components to the cosmic expansion.

2.1 Hubble Constant

SNe Ia are the secondary distance indicators to step from the gravitationally bound systems, like the Local Group and the Virgo Supercluster, into the Hubble flow, where the cosmic expansion dominates. Once the best-fit expansion line in the Hubble diagram is determined, the Hubble constant can be found from the line zeropoint and an independent measurement of the absolute magnitude of SNe Ia at maximum ($M_{\rm max}$). Attempts to derive the absolute magnitude from models have so far failed due to insufficient knowledge of the explosions and the variety displayed amongst SNe Ia (e.g. Stritzinger and Leibundgut 2005).

The direct way of measuring $M_{\rm max}$ is through Cepheid distances, which themselves are anchored by geometric methods. This has been the classic distance ladder approach (see Jacobi et al. 1992 for a comprehensive description of early methods). The HST key program (Freedman et al. 2001) and the SN Ia programs by Sandage, Tammann and Saha, and $\rm SH_0ES$ (see Saha and Macri 2016, for a complete bibliography of these efforts) have reduced the distance ladder to three steps: a local (geometric) anchor to calibrate the Cepheid period-luminosity (Leavitt) relation, calibrate SNe Ia with Cepheid distances and then use the Hubble diagram to step into the Hubble flow. Most prominent anchors are geometric distances to Milky Way Cepheids, eclipsing binaries in the Large Magellanic Clouds and the water megamasers in NGC 4258 (Humphreys et al. 2013). By reducing the number of steps in the distance ladder the sources of uncertainty have been diminished significantly. Nevertheless, there remain several systematic uncertainties, which will need to be addressed in the future. Among them are the influence of metallicity on the Leavitt relation of the Cepheids and some slight incompatibilities of the different anchors. It is a great achievement that the Hubble constant has been determined to a few percent!

2.2 Time Dilation

Universe models based on General Relativity predict time dilation due to the expansion of the universe. Non-expanding models typically do not predict time dilation for distant objects. The SN Ia light evolution can be used as accurate 'clock' to measure time dilation in distant objects. This had been suggested already shortly after supernovae were discovered as a class (Wilson 1939) and was confirmed only some 50 years later, first through light curves (Leibundgut et al. 1996; Goldhaber et al. 2001) and then in the spectral evolution (Riess et al. 1997; Blondin et al. 2008). Fig. 2 shows a comparison of the observed B light curves of nearby SNe Ia ($z < 0.1$) and observed R light curves of SNe Ia from the JLA sample with $0.45 < z < 0.55$. The B filter is redshifted into R at $z \sim 0.5$ (e.g. Schmidt et al. 1998). Despite considerable diversity in light curve shape that can be seen in the local and the

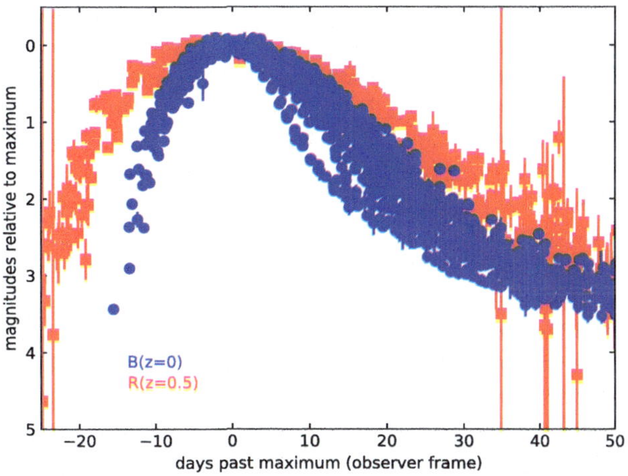

Fig. 2 Observed light curves of SNe Ia in the local universe ($z < 0.1$) in B and at a redshift $0.45 < z < 0.55$ in R. The normalisation is at the time of maximum and the individual light curves have been shifted to the same maximum brightness. The stretching of the observed R light curves relative to the local B light curve is apparent.

distant SN Ia samples, the high-redshift SNe Ia clearly have a wider peak and a significantly larger rise time. These are the signatures of time dilation. Quantitative analyses confirm the expected factor of $1+z$ (Leibundgut et al. 1996; Goldhaber et al. 2001; Blondin et al. 2008). All distant supernova light curves are now routinely corrected for time dilation.

2.3 Cosmic Expansion History

SNe Ia are the most mature and well-exploited probe of cosmic acceleration, and remain a highly-competitive and immediate route to constraining its nature. In a homogeneous and isotropic universe, i.e. using the Robertson-Walker metric, the scale factor changes according to the energy content of the universe. In older formulations, the deceleration parameter was determining the fate of the universe (e.g. Sandage 1961) and cosmology was referred to as the search for two numbers (H_0 and q_0; Sandage 1970). The discovery of the accelerated expansion (Riess et al. 1998; Perlmutter et al. 1999) and the corresponding existence of Dark Energy (in Friedmann–Lemaître models) has been widely documented. The simple argument on the accuracy of the distance indicator used in the local Hubble diagram cannot be applied at large distances, since deviations from the linear expansion are the measurement. Critical assumptions for this application are that the corrections applied to the nearby SNe Ia are valid throughout the observed redshift range, that no selection biases affect the distant samples and that the corrections for redshift and absorption can be controlled. A thorough discussion of many of these fundamental assumptions can be found in Goobar and Leibundgut (2011).

When combined with *Planck* CMB measurements and Baryon Acoustic Oscillation results, current SN Ia samples measure the average dark energy equation-of-state parameter w to around 4–4.5%. The state of the art until recently was the combination of the Sloan Digital Sky Survey-II (SDSS-II) SN program and the first three years of the five-year Canada–France–Hawaii Telescope Supernova Legacy Survey (SNLS), plus various other sources (Betoule et al. 2014). They presented a sample of 740 spectroscopically confirmed SNe Ia from $0.1 < z < 1$ (374 from SDSS, 242 from SNLS). The joint light-curve analysis (JLA) in Betoule et al. (2014) has become the default for cosmological applications of SNe Ia, is most frequently used in comparison with other methods to measure the cosmological parameters (e.g., *Planck*; Planck Collaboration Ade et al. 2016) and is the de facto standard of SN Ia cosmology. The cosmological analysis of the full five-year SNLS and the ESSENCE SN Ia search (Narayan et al. 2016, 213 spectroscopically confirmed SNe Ia) have not yet been presented. Since the systematics of the photometric systems are critical, it is very difficult to combine the SNLS and the ESSENCE data sets for a cosmological analysis. So far, no attempt has been made to combine these data sets. Underlying this problem is the difficulty of absolute calibration of the photometric systems between the different telescopes and instruments (e.g. Scolnic et al. 2014) and the subsequent uncertainties in colour, reddening and K-corrections.

New attempts to combine large SN Ia samples have been made by Scolnic et al. (2017) and Jones et al. (2018). By adding 276 spectroscopically confirmed SNe Ia from the Pan-STARRS1 Medium Deep Survey to the JLA sample, the total number of SNe Ia in the analysis increases to 1049 objects. The conclusions remain unchanged and the uncertainties have decreased slightly. For a constant equation of state parameter ω in a flat universe the uncertainty is now about 4%.

With these large samples, it has been clear for several years that the results are now dominated by experimental systematic uncertainties (Conley et al. 2011), and reducing the statistical error with a larger sample of the same type will only provide limited improvement over the current result. This also means that new data samples based on the same experimental methodology, in particular photometric calibration of the observed optical SN Ia fluxes, will not yield improvements in accuracy.

The SN Ia likelihoods are typically orthogonal to other measurements of cosmological parameters (e.g. Bahcall et al. 1999) because of the lower mean redshift compared to most other methods (see Goobar and Perlmutter 1995, for an illustration of this effect). SNe Ia currently also sample the redshift space out to $z \approx 1.4$ more densely than any other method. While the CMB provides a snapshot at $z = 1000$, Baryonic Acoustic Oscillation, weak lensing and redshift space distortions typically provide measurements at a few redshifts. The latter methods rely on large statistical samples to overcome the considerable individual distance uncertainties. The accuracy to a few percent in distance have allowed the SNe Ia to provide a much denser sampling with a significantly smaller investment of observing time.

3 SN Ia Physics

Taken at face value, the simplicity of the SN Ia technique in cosmological applications – comparing the relative brightnesses of events at different redshifts – suggests the ultimate accuracy of their use may only be limited by the extent to which relative calibrations of their fluxes can be performed. But this is an oversimplification, and ignores the considerable ignorance of the detailed underlying physics governing SN Ia explosions.

The compactness of the progenitor stars of SNe Ia has been demonstrated by the non-detection in deep pre-explosion imaging of the nearest, unobscured SN Ia in decades, SN 2011fe (Nugent et al. 2011; Li et al. 2011). The progenitor star is very likely in a binary system, in which the progenitor white dwarf can be disturbed, or gain mass, and thus explode. But the configuration of this progenitor system prior to explosion remains uncertain.

Searches for compact objects in young SN remnants have so far come up empty handed, indicating that the secondary star is most likely destroyed (e.g. Kerzendorf et al. 2017). There are indications that some SNe Ia appear associated with circumstellar gas (Patat et al. 2007; Sternberg et al. 2011; Maguire et al. 2013; Sternberg et al. 2014; Cikota et al. 2017), which presumably came from a companion to the white dwarf. Sometimes this circumstellar material can be very dense, and interactions with it almost drown out the classical SN Ia signature (e.g., Dilday et al. 2012). But the majority of SNe Ia are consistent with originating from systems of (at least) two white dwarfs in clean environments, with no indication of circumstellar material (Phillips et al. 2013).

The main observational characteristics of SNe Ia are the light curve shapes, which reflect a combination of explosion energy, nickel mass, ejecta mass and composition, and spectral evolution, which traces the ejecta structure and element stratification, if present. The variety of SNe Ia has been described in Maguire (2016), Taubenberger (2016) and Jha (2017).

The majority of SN Ia display very similar characteristics. In particular, the success in applying the Phillips relation for light-curve shape leading to a small scatter in the Hubble diagram, is testimony to the uniformity of a majority of the explosions. There appears to be a continuity in the distribution of light-curve widths and peak luminosity. This has been connected to a possible diversity in the nickel mass produced in the explosion (Nugent et al. 1995; Contardo et al. 2000; Dhawan et al. 2016). A possible connection to the ejecta mass was established later (Stritzinger et al. 2006; Scalzo et al. 2014a,b).

Several distinctions have been introduced among the fast-declining SNe Ia. Taubenberger (2016) provides an exhaustive overview. Many different behaviours have been observed and no clear consensus on a possible physical interpretation of the diversity has emerged.

Several special sub-classes of SNe Ia have been identified. One such are the SNe Iax (Foley et al. 2013; Jha 2017). They have originally been identified through the observations of the peculiar SN 2000cx (Li et al. 2001) and SN 2002cx (Li et al. 2003). SNe Iax are SNe Ia with a faster evolution (fast rise-times and typically also a fast decline-rate after maximum) and low expansion velocities (see Jha 2017, for an overview of SN Iax properties). Most importantly, these SNe Ia show a lower luminosity when compared to most other SNe Ia. Another distinguishing characteristic is the lack of a second maximum in the near infrared light curves. It has been suggested that SNe Iax come from pure deflagration explosions (e.g. Phillips et al. 2007) and it seems that at the low luminosity end of the SN Ia distribution a different explosion mechanism might be in place than for the regular-luminosity objects (Dhawan et al. 2017). Some SNe Iax might not completely incinerate and leave a stripped white dwarf behind (Fink et al. 2014). SNe Iax can be cleanly separated from regular SNe Ia and are eliminated from SN cosmology samples. Due to their lower luminosities they are also discriminated against in distant SN samples.

More problematic for cosmology are 'overluminous' SNe Ia. The first such object identified was SN 1991T (Filippenko et al. 1992; Phillips et al. 1992). While somewhat more luminous than the majority of SNe Ia, it is still considered to be connected to the bulk of SNe Ia. Its light-curve shape fits onto the Phillips relation, and its luminosity can be normalised. A separate class of 'super-Chandrasekhar' mass SNe Ia has been observed. The first example of these objects was SN 2003fg (Howell et al. 2006). The best observed super-Chandrasekhar SNe Ia appear to be SN 2007if and SN 2009dc (see Taubenberger 2016, for a summary of all known objects at the time). These overluminous SNe Ia are characterised by a long rise-time, which leads to estimates of nickel masses in excess of the Chandrasekhar mass (hence the name). They appear UV bright and display blue colours around maximum, but also are more luminous at infrared wavelengths, where most other SNe Ia reach rather similar peak luminosities. Super-Chandrasekhar mass SNe Ia are further characterised by lower ejecta velocities. Simple models predict ejecta masses up to 2 or 3 $M_{\odot}$ (Taubenberger et al. 2011). It is not obvious how such superluminous SNe Ia can be identified in the cosmological samples. For a given redshift they would indicate a larger distance and hence could mimic acceleration, if they would represent a significant fraction of the most distant supernovae. Due to Malmquist bias one expects the fraction of luminous SNe Ia to increase with distance. However, the intrinsic rate of these events is believed to be very small – to date, no clear superluminous SNe Ia have been identified through their slow rise and the blue colours in the distant samples.

While understanding the SN Ia physics is exciting in its own right, it has no *direct* influence on the capability of (normal) SNe Ia to determine accurate distances. Knowing the exact explosion mechanism and the radiation transport involved in these explosions may provide additional refinements in the normalisation methods, like the light-curve shape and colour corrections and the potential dependencies on host galaxy properties. If different explosion mechanisms for subgroups of SNe Ia could be identified, then a cleaner sample might be constructed. However, the identification criteria need to be clear and simple so that they can be applied to noisy data from faint, distant objects.

4 Sources of Uncertainty

Cosmological results from SNe Ia depend on a number of systematic uncertainties. The statistical error has been reduced to rather small levels (about half the total uncertainty; Betoule et al. 2014). Reducing it further with increased samples will not overcome the systematic problems with the distance measurements of SNe Ia. An important limiting factor was the required accurate photometry of the SNe Ia. The latest projects have placed a strong focus on uniform photometry and have mostly used a single telescope/instrument combination to avoid cross-calibrations of different photometric systems. Major fundamental uncertainties are the normalisation of the luminosity through the light curve shapes, colours and host galaxy properties, which are coupled with the unknown absorption in the host galaxy and potential variations of the absorption law in different galaxies. The solution for colour, absorption and K-corrections are intimately connected and are difficult to disentangle.

From Section 3 it is also clear that a better understanding of the progenitor channels and the explosion physics will help in selecting clean SN Ia samples at all redshifts. Any secular evolution, e.g. potential dependencies of the peak luminosity on stellar metallicity, could influence the cosmological measurement. So far, they appear unlikely, but independent tests have been sparse. The main argument against a metallicity dependency is that the local SNe Ia cover a much wider range in metallicities than distant ones, where the average metallicity is presumably smaller. Hence the observed scatter would be largest for nearby SNe Ia. Also, the white dwarf produced in the stellar core by the nuclear burning should be mostly independent of metallicities.

A promising way forward are observations of SNe Ia in the rest frame near-infrared. At these wavelengths the absorption is greatly diminished and SNe Ia behave much more uniformly around maximum light (Krisciunas et al. 2004; Dhawan et al. 2015). The difficulty is to obtain restframe near-infrared light curves of distant SNe as the light is redshifted into wavelength regions with high sky brightness essentially preventing detecting the SNe Ia beyond $z \approx 0.3$ from the ground. The best near-IR Hubble diagrams are from Freedman et al. (2009), Barone-Nugent et al. (2012), Stanishev et al. (2015) and Krisciunas (2016). Extending them to higher redshifts requires space observations. *Euclid* and in particular *WFIRST* will aim to obtain such measurements in the next decade and will significantly reduce the uncertainties due to absorption and colours.

5 Current Surveys

Three distinct phases of SN Ia surveys can be discerned. The first, the early SN searches, were essentially trying to find as many SNe as possible, and establish them as good distance indicators (see Leibundgut 2016, for a historical account). During the 1980s and 1990s SNe Ia were mostly used to measure an accurate value of the Hubble constant (most notably the Calán/Tololo survey; Hamuy et al. 1996). Discovering distant SNe Ia and cosmic acceleration took most of the decade between 1988 and 1998 (see Garnavich 2017, for a summary), followed by another decade to confirm the discovery (Riess 2016, provides a historical overview).

The second phase has been concerned with establishing the machinery required to make the discovery of SNe Ia routine. Surveys such as SNLS, SDSS-SN, and ESSENCE, together with large allocations of spectroscopic follow-up time on 8m-class facilities, made the discovery and confirmation of SNe Ia up to $z = 1$ common place. This phase is responsible for the current state-of-the-art cosmological constraints.

The third phase, which we have just entered, is the establishment of samples of such size that spectroscopic follow-up of all events is no longer practical. The cosmological results from the Pan-STARRS1 Medium Deep Survey have been published by Rest et al. (2014). The full sample is being published now (Scolnic et al. 2017). The results confirm the cosmological parameters of the previous studies (see Section 2.3).

The Dark Energy Survey (DES) will discover about 4000 SNe Ia. The critical host galaxy redshifts are collected in a separate observing program on the AAT, OzDES (Childress et al. 2017). DES should finish with a superior photometric calibration compared to earlier efforts (SNLS, SDSS) and thus should result in reduced uncertainties. DES will also be able to construct Hubble diagrams and perform cosmological analyses for large sub-samples of SNe Ia, equivalent in size and power to SNLS, for the first time, providing a very sensitive test of astrophysical systematic uncertainties.

At lower redshifts, searches for, and studies of, nearby events continue. The SH_0ES project continues to observe Cepheids in nearby galaxies with well-observed SNe Ia. *Gaia* will provide a new calibration of the Cepheid period-luminosity calibration by determining exact distances to many Cepheids in the Milky Way. This will further reduce the systematic uncertainty in the determination of H_0. It appears important to continue to define additional geometric distance indicators to calibrate the SNe Ia.

The Nearby SN Factory has assembled several hundred SNe Ia with detailed observations. The cosmological results are encouraging, although a final publication on H_0 is still outstanding. A correlation between the peak luminosity and the local star formation rate in the region where the SN occurred has been found (Rigault et al. 2015) and a new method by searching for SN Ia 'twins' at different redshifts has been proposed to improve the relative distance measurements (Fakhouri et al. 2015). The dense sampling has also been used to investigate the local flow field (Feindt et al. 2013).

SNe Ia for the purposes of cosmology have not been the main focus of other nearby searches. The Palomar Transient Factory contributed several important observations of SN Ia and how we can understand the physics of these explosions better. The PESSTO/ePESSTO survey (Smartt et al. 2015) has focused on peculiar objects and has given SNe Ia lower priority. The All-Sky Automated Survey for Supernova (ASAS-SN; Holoien et al. 2017a,b,c) is searching for very bright, nearby SNe and emphasises the new physics that can be learned from high-cadence observations.

Another promising approach is to observe SNe Ia in the restframe near-infrared. This has been attempted with *HST* through two programs following SNe Ia detected by Pan-STARRS1 and the DES. Because of the decreased systematics even a small fraction of the NIR light curves can yield cosmological parameters that will be competitive. So far, the results of these programs have not been published.

6 The Future

Supernova cosmology is well-established. SNe Ia remain the foremost secondary distance indicators for the Hubble constant. There is no other distance indicator in the local universe, which can match the accuracy provided by SNe Ia. It is conceivable that the primary distance indicators might be expanded into the Hubble flow but progress has been slow. Megamasers in galaxy cores appeared for a while as the most promising candidate for a precise determination of H_0, but there have been only very few objects with the required geometry. An alternative are Type II supernovae through the expanding photosphere method and related methods (Nugent and Hamuy 2016 provide an overview of the current developments in this area). Extensions to higher redshifts are starting to appear (Gall et al. 2017; de Jaeger et al. 2017a,b).

Searches for distant SNe Ia will be dominated by near whole-sky, high-cadence surveys. The Large Synoptic Survey Telescope (LSST) will find hundreds of thousands of transient candidates every night, and probably in excess of 100,000 SNe Ia after five years. It will provide huge statistical light-curve samples, but spectroscopic follow-up will be limited: a major systematic uncertainty will be the photometric classification (and redshift measurement) of the SN detections. Large multiplex spectrographs, such as 4MOST at the ESO/VISTA telescope, are the most promising solution and will provide redshifts of host galaxies – very similar to OzDES for DES – and spectra of thousands of the SNe providing spectroscopic classifications. But even with 4MOST and other facilities, spectroscopic resources will not be available to target all candidate SNe, and photometric classification techniques are therefore critical for LSST SN Ia cosmological analyses. Plans to harness such statistical samples and improve on systematics are being developed. It remains to be demonstrated that these methods will be able to overcome the above challenges.

The space projects *Euclid* and *WFIRST* will also be major steps in SN Ia cosmology. Astier et al. (2014) chart an extension program for *Euclid* (the Dark Energy Supernova Infra-Red Experiment: DESIRE) to build a densely sampled Hubble diagram to $z < 1.5$ containing nearly 20000 SNe Ia. Part of this sample would be gathered from the ground, while the high-redshift component would be obtained in a 6-month extension of the *Euclid* mission with observations dedicated to distant supernovae. SNe Ia are part of the core program of the *WFIRST* mission. It will provide Hubble diagrams at many different wavelengths and in particular provide the restframe near-infrared.

The goal must be to measure any time dependence of the equation of state parameter. The uncertainties on a constant ω in a flat cold dark matter universe have been reduced to very small values. There are many theories, which allow for a variable ω but constraining any time-dependent $\omega(t)$ has so far been rather inconclusive. The required control on the individual distance measurements are very stringent and have not been achieved so far.

Supernova measurements remain at the core of observational cosmology. They make a unique contribution through their exquisite individual distances. As stellar objects that can be observed across more than half the known universe, they achieve with small samples what needs huge statistics for other objects (weak lensing, redshift surveys, baryonic acoustic oscillations, cosmic microwave background fluctuations) and complement the other methods.

Acknowledgements B.L. acknowledges support for this work by the Deutsche Forschungsgemeinschaft through the TransRegio Project TRR33 "The Dark Universe".

References

R. Amanullah, J. Johansson, A. Goobar et al., Diversity in extinction laws of Type Ia supernovae measured between 0.2 and 2 μm. Mon. Not. R. Astron. Soc. **453**, 3300 (2015)

P. Astier, C. Balland, M. Brescia et al., Extending the supernova Hubble diagram to $z \sim 1.5$ with the Euclid space mission. Astron. Astrophys. **572**, A80 (2014)

N.A. Bahcall, J.P. Ostriker, S. Perlmutter, P.J. Steinhardt, The cosmic triangle: revealing the state of the universe. Science **284**, 1481 (1999)

R.L. Barone-Nugent, C. Lidman, J.S.B. Wyithe et al., Near-infrared observations of Type Ia supernovae: the best known standard candle for cosmology. Mon. Not. R. Astron. Soc. **425**, 1007 (2012)

M. Betoule, R. Kessler, J. Guy et al., Improved cosmological constraints from a joint analysis of the SDSS-II and SNLS supernova samples. Astron. Astrophys. **568**, A22 (2014)

S. Blondin, T.M. Davis, K. Krisciunas et al., Time dilation in Type Ia supernova spectra at high redshift. Astrophys. J. **682**, 724 (2008)

D. Branch, The Hubble diagram for Type I supernovae. Astrophys. J. **258**, 35 (1982)

D. Branch, G.A. Tammann, Type Ia supernovae as standard candles. Annu. Rev. Astron. Astrophys. **30**, 359 (1992)

D. Branch, Type Ia supernovae and the Hubble constant. Annu. Rev. Astron. Astrophys. **36**, 17 (1998)

D. Branch, J.C. Wheeler, *Supernova Explosions: Astron. Astrophys. Library*. (Springer, Germany, 2017). ISBN 978-3-662-55052-6

C.R. Burns, M. Stritzinger, M.M. Phillips et al., The Carnegie supernova project: light-curve fitting with SNooPy. Astron. J. **141**, 19 (2011)

M.J. Childress, C. Lidman, T.M. Davis et al., OzDES multifibre spectroscopy for the Dark Energy Survey: 3-yr results and first data release. Mon. Not. R. Astron. Soc. **472**, 273 (2017)

A. Cikota, F. Patat, S. Cikota, J. Spyromilio, G. Rau, Common continuum polarization properties: a possible link between proto-planetary nebulae and Type Ia supernova progenitors. Mon. Not. R. Astron. Soc. **471**, 2111 (2017)

A. Conley, M. Sullivan, E.Y. Hsiao et al., SiFTO: an empirical method for fitting SN Ia light curves. Astrophys. J. **681**, 482 (2008)

A. Conley, J. Guy, M. Sullivan et al., Supernova constraints and systematic uncertainties from the first three years of the supernova legacy survey. Astrophys. J. Suppl. Ser. **192**, 1 (2011)

G. Contardo, B. Leibundgut, W.D. Vacca, Epochs of maximum light and bolometric light curves of Type Ia supernovae. Astron. Astrophys. **359**, 876 (2000)

T. de Jaeger, L. Galbany, A.V. Filippenko et al., SN 2016jhj at redshift 0.34: extending the Type II supernova Hubble diagram using the standard candle method. Mon. Not. R. Astron. Soc. **472**, 4233 (2017a)

T. de Jaeger, S. González-Gaitán, M. Hamuy et al., A Type II supernova Hubble diagram from the CSP-I, SDSS-II, and SNLS surveys. Astrophys. J. **835**, 166 (2017b)

S. Dhawan, B. Leibundgut, J. Spyromilio, K. Maguire, Near-infrared light curves of Type Ia supernovae: studying properties of the second maximum. Mon. Not. R. Astron. Soc. **448**, 1345 (2015)

S. Dhawan, B. Leibundgut, J. Spyromilio, S. Blondin, A reddening-free method to estimate the ^{56}Ni mass of Type Ia supernovae. Astron. Astrophys. **588**, A84 (2016)

S. Dhawan, B. Leibundgut, J. Spyromilio, S. Blondin, Two classes of fast-declining Type Ia supernovae. Astron. Astrophys. **602**, A118 (2017)

S. Dhawan, S.W. Jha, B. Leibundgut, Measuring the Hubble constant with Type Ia supernovae as near-infrared standard candles. Astron. Astrophys. **609**, A72 (2018)

B. Dilday, D.A. Howell, B. Cenko et al., PTF 11kx: a Type Ia supernova with a symbiotic nova progenitor. Science **237**, 942 (2012)

H.K. Fakhouri, K. Boone, G. Aldering et al., Improving cosmological distance measurements using twin Type Ia supernovae. Astrophys. J. **815**, 58 (2015)

U. Feindt, M. Kerschhaggl, M. Kowalski et al., Measuring cosmic bulk flows with Type Ia supernovae from the Nearby Supernova Factory. Astron. Astrophys. **560**, A90 (2013)

A.V. Filippenko, M.W. Richmond, T. Matheson et al., The peculiar Type Ia SN 1991T—detonation of a white dwarf? Astrophys. J. Lett. **384**, L15 (1992)

M. Fink, M. Kromer, I.R. Seitenzahl et al., Three-dimensional pure deflagration models with nucleosynthesis and synthetic observables for Type Ia supernovae. Mon. Not. R. Astron. Soc. **438**, 1762 (2014)

R.J. Foley, P.J. Challis, R. Chornock et al., Type Iax supernovae: a new class of stellar explosions. Astrophys. J. **767**, 57 (2013)

W.L. Freedman, B.F. Madore, B.K. Gibson et al., Final results from the Hubble Space Telescope key project to measure the Hubble constant. Astrophys. J. **553**, 47 (2001)

W.L. Freedman, C.R. Burns, M.M. Phillips et al., The Carnegie supernova project: first near-infrared Hubble diagram to $z \sim 0.7$. Astrophys. J. **704**, 1036 (2009)

E.E.E. Gall, R. Kotak, B. Leibundgut et al., An updated Type II supernova Hubble diagram (2017). arXiv: 1705.10806
P. Garnavich, Discovery of cosmic acceleration, in *Handbook of Supernovae*, ed. by A.W. Alsabti, P. Murdin (Springer, Berlin, 2017). https://doi.org/10.1007/978-3-319-20794-0_104-1
G. Goldhaber, D.E. Groom, A. Kim et al., Timescale stretch parameterization of Type Ia supernova B-band light curves. Astrophys. J. **558**, 359 (2001)
A. Goobar, S. Perlmutter, Feasibility of measuring the cosmological constant lambda and mass density omega using Type Ia supernovae. Astrophys. J. **450**, 14 (1995)
A. Goobar, B. Leibundgut, Supernova cosmology: legacy and future. Annu. Rev. Nucl. Part. Sci. **61**, 251 (2011)
J. Guy, P. Astier, S. Baumont et al., SALT2: using distant supernovae to improve the use of Type Ia supernovae as distance indicators. Astron. Astrophys. **466**, 11 (2007)
M. Hamuy, M.M. Phillips, N.B. Suntzeff et al., The Hubble diagram of the Calán/Tololo Type Ia supernovae and the value of H_0. Astron. J. **112**, 2398 (1996)
T.W.-S. Holoien, K.Z. Stanek, C.S. Kochanek et al., The ASAS-SN bright supernova catalogue—I. 2013–2014. Mon. Not. R. Astron. Soc. **464**, 2672 (2017a)
T.W.-S. Holoien, J.S. Brown, K.Z. Stanek et al., The ASAS-SN bright supernova catalogue—II. 2015. Mon. Not. R. Astron. Soc. **467**, 1098 (2017b)
T.W.-S. Holoien, J.S. Brown, K.Z. Stanek et al., The ASAS-SN bright supernova catalogue—III. 2016. Mon. Not. R. Astron. Soc. **471**, 4966 (2017c)
D.A. Howell, M. Sullivan, P.E. Nugent et al., The type Ia supernova SNLS-03D3bb from a super-Chandrasekhar-mass white dwarf star. Nature **443**, 308 (2006)
E.M.L. Humphreys, M.J. Reid, J.M. Moran, L.J. Greenhill, A.L. Argon, Toward a new geometric distance to the active galaxy NGC 4258. III. Final results and the Hubble constant. Astrophys. J. **775**, 13 (2013)
S. Jha, A.G. Riess, R.P. Kirshner, Improved distances to Type Ia supernovae with Multicolor Light-Curve Shapes: MLCS2k2. Astrophys. J. **659**, 122 (2007)
S.W. Jha, Type Iax supernovae, in *Handbook of Supernovae*, ed. by A.W. Alsabti, P. Murdin (Springer, Berlin, 2017). https://doi.org/10.1007/978-3-319-20794-0_42-1
D.O. Jones, D.M. Scolnic, A.G. Riess et al., Measuring Dark Energy Properties with Photometrically Classified Pan-STARRS Supernovae. II. Cosmological Parameters (2018). arXiv:1710.00846
W.E. Kerzendorf, G. Strampelli, K.J. Shen et al., A Search for a Surviving White Dwarf Companion in SN 1006 (2017). arXiv:1709.06566
K. Krisciunas, M.M. Phillips, N.B. Suntzeff, Hubble diagrams of Type Ia supernovae in the near-infrared. Astrophys. J. Lett. **602**, L81 (2004)
K. Krisciunas, The infrared Hubble diagram of Type Ia supernovae, in *Handbook of Supernovae*, ed. by A.W. Alsabti, P. Murdin (Springer, Berlin, 2016). https://doi.org/10.1007/978-3-319-20794-0_103-1
B. Leibundgut, Cosmological implications from observations of Type Ia supernovae. Annu. Rev. Astron. Astrophys. **39**, 67 (2001)
B. Leibundgut, Supernovae and cosmology. Gen. Relativ. Gravit. **40**, 221 (2008)
B. Leibundgut, History of supernovae as distance indicators, in *Handbook of Supernovae*, ed. by A.W. Alsabti, P. Murdin (Springer, Berlin, 2016). https://doi.org/10.1007/978-3-319-20794-0_99-1
B. Leibundgut, R. Schommer, M. Phillips et al., Time dilation in the light curve of the distant Type Ia supernova SN 1995K. Astrophys. J. Lett. **466**, L21 (1996)
W. Li, A.V. Filippenko, R. Chornock et al., SN 2002cx: the most peculiar known Type Ia supernova. Publ. Astron. Soc. Pac. **115**, 453 (2003)
W. Li, A.V. Filippenko, E. Gates et al., The unique Type Ia supernova 2000cx in NGC 524. Publ. Astron. Soc. Pac. **113**, 1178 (2001)
W. Li, J.S. Bloom, P. Podsiadlowski et al., Exclusion of a luminous red giant as a companion star to the progenitor of supernova SN 2011fe. Nature **480**, 348 (2011)
K. Maguire, M. Sullivan, F. Patat et al., A statistical analysis of circumstellar material in Type Ia supernovae. Mon. Not. R. Astron. Soc. **436**, 222 (2013)
K. Maguire, Type Ia supernovae, in *Handbook of Supernovae*, ed. by A.W. Alsabti, P. Murdin (Springer, Berlin, 2016). https://doi.org/10.1007/978-3-319-20794-0_36-1
K.S. Mandel, D.M. Scolnic, H. Shariff, R.J. Foley, R.P. Kirshner, Star formation, supernovae, iron, and α: consistent cosmic and galactic histories. Astrophys. J. **842**, 93 (2017)
G. Narayan, A. Rest, B.E. Tucker et al., Light curves of 213 Type Ia supernovae from the ESSENCE survey. Astrophys. J. Suppl. Ser. **224**, 3 (2016)
P. Nugent, M. Phillips, E. Baron, D. Branch, P. Hauschildt, Evidence for a spectroscopic sequence among Type 1a supernovae. Astrophys. J. Lett. **455**, 147 (1995)
P. Nugent, M. Hamuy, Cosmology with Type IIP supernovae, in *Handbook of Supernovae*, ed. by A.W. Alsabti, P. Murdin (Springer, Berlin, 2016). https://doi.org/10.1007/978-3-319-20794-0_108-1

P.E. Nugent, M. Sullivan, S.B. Cenko et al., Supernova SN 2011fe from an exploding carbon–oxygen white dwarf star. Nature **480**, 344 (2011)

F. Patat, P. Chandra, R. Chevalier et al., Detection of circumstellar material in a normal Type Ia supernova. Science **317**, 924 (2007)

S. Perlmutter, G. Aldering, G. Goldhaber et al., Measurements of Ω and Λ from 42 high-redshift supernovae. Astrophys. J. **517**, 565 (1999)

S. Perlmutter, B.P. Schmidt, Measuring cosmology with supernovae, in *Supernovae and Gamma-Ray Bursters*. Lecture Notes in Physics, vol. 598 (2003), p. 195

M.M. Phillips, The absolute magnitudes of Type Ia supernovae. Astrophys. J. Lett. **413**, L105 (1993)

M.M. Phillips, L.A. Wells, N.B. Suntzeff et al., SN 1991T—further evidence of the heterogeneous nature of type Ia supernovae. Astron. J. **103**, 1632 (1992)

M.M. Phillips, W. Li, J.A. Frieman et al., The peculiar SN 2005hk: do some Type Ia supernovae explode as deflagrations? Publ. Astron. Soc. Pac. **119**, 360 (2007)

M.M. Phillips, J.D. Simon, N. Morrell et al., On the source of the dust extinction in Type Ia supernovae and the discovery of anomalously strong Na I absorption. Astrophys. J. **779**, 38 (2013)

M.M. Phillips, C.R. Burns, The peak luminosity–decline rate relationship for Type Ia supernovae, in *Handbook of Supernovae*, ed. by A.W. Alsabti, P. Murdin (Springer, Berlin, 2016). https://doi.org/10.1007/978-3-319-20794-0_100-1

Planck Collaboration, P.A.R. Ade, N. Aghanim et al., Planck 2015 results. XIII. Cosmological parameters. Astron. Astrophys. **594**, A13 (2016)

A. Rest, D. Scolnic, R.J. Foley et al., Cosmological constraints from measurements of Type Ia supernovae discovered during the first 1.5 yr of the Pan-STARRS1 survey. Astrophys. J. **795**, 44 (2014)

A.G. Riess, A.V. Filippenko, D.C. Leonard et al., Time dilation from spectral feature age measurements of Type Ia supernovae. Astron. J. **114**, 722 (1997)

A.G. Riess, A.V. Filippenko, P. Challis et al., Observational evidence from supernovae for an accelerating universe and a cosmological constant. Astron. J. **116**, 1009 (1998)

A.G. Riess, L.M. Macri, S.L. Hoffmann et al., A 2.4% determination of the local value of the Hubble constant. Astrophys. J. **826**, 56 (2016)

A.G. Riess, Confirming cosmic acceleration in the decade that followed from SNe Ia at $z > 1$, in *Handbook of Supernovae*, ed. by A.W. Alsabti, P. Murdin (Springer, Berlin, 2016). https://doi.org/10.1007/978-3-319-20794-0_105-1

M. Rigault, G. Aldering, M. Kowalski et al., Confirmation of a star formation bias in Type Ia supernova distances and its effect on the measurement of the Hubble constant. Astrophys. J. **802**, 20 (2015)

A. Saha, L.M. Macri, The Hubble constant from supernovae, in *Handbook of Supernovae*, ed. by A.W. Alsabti, P. Murdin (Springer, Berlin, 2016). https://doi.org/10.1007/978-3-319-20794-0_102-1

A. Sandage, The ability of the 200-inch telescope to discriminate between selected world models. Astrophys. J. **133**, 355 (1961)

A.R. Sandage, Cosmology: a search for two numbers. Phys. Today **23**, 34 (1970)

R.A. Scalzo, A.J. Ruiter, S.A. Sim, The ejected mass distribution of Type Ia supernovae: a significant rate of non-Chandrasekhar-mass progenitors. Mon. Not. R. Astron. Soc. **445**, 2535 (2014a)

R. Scalzo, G. Aldering, P. Antilogus et al., Type Ia supernova bolometric light curves and ejected mass estimates from the Nearby Supernova Factory. Mon. Not. R. Astron. Soc. **440**, 1498 (2014b)

B.P. Schmidt, N.B. Suntzeff, M.M. Phillips et al., The High-Z supernova search: measuring cosmic deceleration and global curvature of the universe using Type Ia supernovae. Astrophys. J. **507**, 46 (1998)

D. Scolnic, A. Rest, A. Riess et al., Systematic uncertainties associated with the cosmological analysis of the first Pan-STARRS1 Type Ia supernova sample. Astrophys. J. **795**, 45 (2014)

D.M. Scolnic, R. Kessler, Measuring Type Ia supernova populations of stretch and color and predicting distance biases. Astrophys. J. **822**, 35L (2016)

D.M. Scolnic, D.O. Jones, A. Rest et al., The Complete Light-curve Sample of Spectroscopically Confirmed Type Ia Supernovae from Pan-STARRS1 and Cosmological Constraints from The Combined Pantheon Sample (2017). arXiv:1710.00845

H. Shariff, X. Jiao, R. Trotta, D.A. van Dyk, BAHAMAS: new analysis of Type Ia supernovae reveals inconsistencies with standard cosmology. Astrophys. J. **827**, 1 (2016)

S.J. Smartt, S. Valenti, M. Fraser et al., PESSTO: survey description and products from the first data release by the Public ESO Spectroscopic Survey of Transient Objects. Astron. Astrophys. **579**, A40 (2015)

V. Stanishev, A. Goobar, R. Amanullah et al., Type Ia Supernova Cosmology in the Near-Infrared (2015). arXiv:1505.07707

A. Sternberg, A. Gal-Yam, J.D. Simon et al., Circumstellar material in Type Ia supernovae via sodium absorption features. Science **333**, 856 (2011)

A. Sternberg, A. Gal-Yam, J.D. Simon et al., Multi-epoch high-spectral-resolution observations of neutral sodium in 14 Type Ia supernova. Mon. Not. R. Astron. Soc. **443**, 1849 (2014)

M. Stritzinger, B. Leibundgut, Lower limits on the Hubble constant from models of type Ia supernovae. Astron. Astrophys. **431**, 423 (2005)

M. Stritzinger, B. Leibundgut, S. Walch, G. Contardo, Constraints on the progenitor systems of type Ia supernovae. Astron. Astrophys. **450**, 241 (2006)

S. Taubenberger, The extremes of thermonuclear supernovae, in *Handbook of Supernovae*, ed. by A.W. Alsabti, P. Murdin (Springer, Berlin, 2016). https://doi.org/10.1007/978-3-319-20794-0_31-1

S. Taubenberger, S. Benetti, M. Childress et al., High luminosity, slow ejecta and persistent carbon lines: SN 2009dc challenges thermonuclear explosion scenarios. Mon. Not. R. Astron. Soc. **412**, 2735 (2011)

R. Tripp, A two-parameter luminosity correction for Type Ia supernovae. Astron. Astrophys. **331**, 815 (1998)

O.C. Wilson, Possible applications of supernovae to the study of the nebular red shifts. Astrophys. J. **90**, 634 (1939)

Space Sci Rev (2018) 214:33
https://doi.org/10.1007/s11214-017-0450-9

Crucial Physical Dependencies of the Core-Collapse Supernova Mechanism

A. Burrows[1] · D. Vartanyan[1] · J.C. Dolence[2] · M.A. Skinner[3] · D. Radice[4]

Received: 25 October 2017 / Accepted: 19 November 2017 / Published online: 23 January 2018

Abstract We explore with self-consistent 2D FORNAX simulations the dependence of the outcome of collapse on many-body corrections to neutrino-nucleon cross sections, the nucleon-nucleon bremsstrahlung rate, electron capture on heavy nuclei, pre-collapse seed perturbations, and inelastic neutrino-electron and neutrino-nucleon scattering. Importantly, proximity to criticality amplifies the role of even small changes in the neutrino-matter couplings, and such changes can together add to produce outsized effects. When close to the critical condition the cumulative result of a few small effects (including seeds) that individually have only modest consequence can convert an anemic into a robust explosion, or even a dud into a blast. Such sensitivity is not seen in one dimension and may explain the apparent heterogeneity in the outcomes of detailed simulations performed internationally. A natural conclusion is that the different groups collectively are closer to a realistic understanding of the mechanism of core-collapse supernovae than might have seemed apparent.

Supernovae
Edited by Andrei Bykov, Roger Chevalier, John Raymond, Friedrich-Karl Thielemann, Maurizio Falanga and Rudolf von Steiger

✉ A. Burrows
burrows@astro.princeton.edu

D. Vartanyan
dvartany@astro.princeton.edu

J.C. Dolence
jdoelnce@lanl.gov

M.A. Skinner
skinner15@llnl.gov

D. Radice
dradice@astro.princeton.edu

1 Department of Astrophysical Sciences, Princeton University, Princeton, NJ 08544, USA

2 CCS-2, Los Alamos National Laboratory, P.O. Box 1663, Los Alamos, NM 87545, USA

3 Livermore National Laboratory, 7000 East Ave., Livermore, CA 94550-9234, USA

4 Schmidt Fellow, Institute for Advanced Study, 1 Einstein Drive, Princeton, NJ 08540, USA

Keywords Supernova theory · Neutrino physics · Multi-dimensional radiation/hydrodynamics

1 Introduction

A goal of core-collapse supernova theory is to explain the mechanism of explosion. It is an accepted truism of the field that a necessary condition for this is that complicated multi-dimensional simulation codes incorporating the requisite neutrino, nuclear, and gravitational physics reproduce such explosions robustly, yielding at the very least the requisite asymptotic energies, residual neutron star masses, and nucleosynthesis. A simple analytic explanation has not been forthcoming and, given the manifest complexities of the process, would not be deemed credible. It is thought that one litmus test of success would be such a demonstration for a non-rotating model between $\sim$8 and $\sim$20 $M_{\odot}$, the progenitor ZAMS (Zero-Age Main-Sequence) mass regime that must provide the lion's share of core-collapse supernova events.

However, to date the various groups engaged in such efforts around the world have failed to agree on the outcomes of the collapse of otherwise similar progenitor massive star cores. This is despite claims to have embedded the necessary physics and microphysics into the simulations. In fact, the numerical algorithms, resolutions, and input physics all differ, and even when the physics is deemed similar, the implementations and approximations surely differ. The ORNL group (Bruenn et al. 2013, 2016), with their CHIMERA code, uses multi-group flux-limited diffusion (MGFLD) neutrino transport (Bruenn 1985), the VH-1 Newtonian hydrodynamics package, a monopole correction for general relativity (GR) (Marek et al. 2006), but multiple one-dimensional solves for multi-dimensional transport using the so-called "ray-by-ray+" approach (Buras et al. 2003; Burrows et al. 1995). Such a dimensional reduction for the transport, particularly manifest in 2D, ignores lateral, non-radial radiative transport, which has been shown to be of quantitative (Ott et al. 2008; Brandt et al. 2011; Burrows 2013; Dolence et al. 2015; Sumiyoshi et al. 2015) and qualitative (Skinner et al. 2016) importance. To the point, Skinner et al. (2016) have shown that the ray-by-ray anomalies in the angular distribution of the radiation field and corresponding neutrino heating rates can reinforce the axial sloshing motions in 2D and push the shock into explosion.[1] Bruenn et al. (2013, 2016) find explosions in 2D (axially symmetry) for all the progenitors they studied (the 12-, 15-, 20-, and 25-$M_{\odot}$ models of Woosley and Heger 2007), but the models all explode at about the same post-bounce time ($\sim$100 milliseconds) and the shock radii never decrease in value. When performed in 3D for the 15-$M_{\odot}$ model of Woosley and Heger (2007), Lentz et al. (2015) obtain a weaker explosion delayed in its onset by an extra $\sim$100 milliseconds.

Müller et al. (2012a, 2012b), using the conformally-flat CoCoNuT hydrodynamics code in combination with the VERTEX transport solver and the ray-by-ray+ approach, find that models explode in 2D, but explode with lower energies and at post-bounce explosion times that differ significantly from those found by the ORNL group. In addition, their shock radii generally decrease from a peak occurring near a time $\sim$200 ms after bounce, before increasing just prior to explosion hundreds of milliseconds later (and all at different post-bounce times). Using the VERTEX-PROMETHEUS code, the Garching group obtained weak explosions in 2D for a 11.2-$M_{\odot}$ progenitor (Buras et al. 2006) and for a rotating 15-$M_{\odot}$ progenitor (Marek and Janka 2009), though the outcomes when proper correction is made for

[1] However, the general absence in 3D of either an axial effect or the pronounced sloshing seen in 2D may be rendering 3D simulations performed with the ray-by-ray approach less problematic. This has yet to be tested.

the corresponding mantle binding energies were not clear. More recently, the Garching group (Summa et al. 2016, 2018) obtained explosions in 2D for a broad range of Woosley and Heger (2007) progenitors from 11 to 28 $M_{\odot}$, but these calculations as well did not comport with those of the ORNL group in the timescales and energies seen. The Garching group has also used for all their multi-D VERTEX supernova simulations a variant of the problematic ray-by-ray+ dimensionally-reduced transport approach.

However, Murphy et al. (2013) and Couch and Ott (2015), and originally Burrows et al. (1995), highlight the importance of turbulent pressure behind the shock as an aid to explosion, but note that such a pressure is larger (artificially) in 2D than in 3D. Turbulent pressure is also (possibly) more anisotropic in 2D, favoring the radial stress component, further enhancing the prospects (again, artificially) for overcoming the accretion ram pressure. Hence, the current explosions in 2D may in part, or at times, be numerical artifacts.

Importantly, though Hanke et al. (2012) have speculated that the sloshing motion oftimes associated in 2D with the SASI (standing accretion shock instability) may be crucial to explosion, the recent non-rotating default VERTEX-PROMETHEUS calculations in 3D by that same Garching group (that at times manifest the SASI) do not explode (Hanke et al. 2013; Tamborra et al. 2014), even though the corresponding 2D simulations did. As an aside, Burrows et al. (2012) note that such a pronounced axial sloshing motion is rarely in evidence in 3D. However, with altered microphysics, in particular with a change in the axial-vector coupling constant (g_A) due to a speculative enhanced effect on the nucleon spin of the strange quark, Melson et al. (2015) do obtain a weak explosion in 3D of the Woosley and Heger (2007) 20-$M_{\odot}$ in 3D. The difference in outcome is due to the consequent decrease in the neutrino-neutron neutral-current scattering cross section in the neutron-rich envelope of the proto-neutron star bounded by the stalled shock and the resultant increase in the electron-neutrino luminosity and average energy that are instrumental in heating this envelope. However, the magnitude of the strangeness correction employed by Melson et al. may be larger than experiment allows (Ahmed et al. 2012; Green et al. 2017). Furthermore, it is not clear that all these currently published under-powered 3D explosions will actually succeed as an explosion after traversing the entire star and after the mass cut between ejecta and residual core is determined hydrodynamically. An exception is the recent work of Müller et al. (2017), who perform 3D multi-group simulations using their simpler FMT method (Müller and Janka 2015) for an 18-$M_{\odot}$ progenitor model evolved in 3D to collapse. Such a 3D progenitor naturally manifests seed perturbations that Müller et al. (2017) show are instrumental in leading to an explosion with a respectable energy.

The calculations of Burrows et al. (2006a, 2006b, 2007a) and Ott et al. (2008), using the VULCAN/2D code, and Dolence et al. (2015), using the CASTRO code (Zhang et al. 2011, 2013) employed multi-dimensional transport, and not ray-by-ray+, but in neither of these 2D studies did the authors see explosions by the neutrino mechanism. VULCAN/2D did not have all the terms to order v/c in the transport, but CASTRO did, and the results were similar (Dolence et al. 2015). Importantly, neither VULCAN/2D nor CASTRO made corrections for the effects of general relativity, and this could explain in part the different outcome. However, since other self-consistent calculations (summarized previously) that obtained explosions in 2D used the ray-by-ray scheme, while the VULCAN/2D and CASTRO studies did not, one is tempted to suggest that the ray-by-ray approach may in 2D be yielding qualitatively incorrect results.

Suwa et al. (2010) obtain an explosion in 2D of a 13-$M_{\odot}$ progenitor, while Takiwaki et al. (2012) obtain an explosion for an 11.2-$M_{\odot}$ progenitor in both 2D and 3D. Both these efforts, however, neglect ν_{μ} and ν_{τ} neutrinos, which constitute roughly 50% of the total neutrino losses after bounce, and use the IDSA (Liebendörfer et al. 2009) and ray-by-ray approximations for the transport. Suwa et al. (2016) continued their explorations in 2D with a large suite of progenitor simulations, but continue to neglect ν_μ and ν_τ neutrinos and use the IDSA plus ray-by-ray+ transport methods. They highlight a possible role in the explosion systematics of transitions in the mass accretion rate. Ott et al. (2013) perform fully general-relativistic 3D simulations, employing a leakage scheme for all the neutrinos. Their goal was to explore the relative role of the SASI and neutrino-driven convection, and they found for their simulation of the 27-$M_\odot$ progenitor of Woosley et al. (2002) that neutrino-driven convection dominated once it started. Roberts et al. (2016), using full GR hydrodynamics and an M1 transport scheme (Sect. 2) roughly similar to ours (but without the velocity-dependent terms in the transport, inelastic scattering, or many-body effects), emphasize the importance of spatial resolution in determining whether and how the same 27-$M_\odot$ model explodes, as well as the different outcomes for octant and full 4π steradian simulations. Kuroda et al. (2016) have achieved a fully general-relativistic 3D code, also using the M1 transport closure, but have yet to simulate progenitor models beyond a few tens of milliseconds post-bounce. Pan et al. (2016) have constructed a 2D non-relativistic (Newtonian) neutrino radiation/hydrodynamic scheme using the IDSA transport approach for the ν_e and $\bar{\nu}_e$ neutrinos and a leakage scheme for the ν_μ neutrinos. Importantly, they do not use the ray-by-ray+ approach, but follow the transport of ν_e and $\bar{\nu}_e$ neutrinos multi-dimensionally. They obtain explosions for all the progenitor models studied.

Suwa et al. (2010) and Nakamura et al. (2014) found that rotation aided explosion, mostly by rotationally expanding the size of the gain region and increasing the mass it contained. Iwakami et al. (2014) and Takiwaki et al. (2016) highlight the rotational excitation of $m = 1$ spiral-arm modes and find a role for non-axisymmetric rotational instabilities. Iwakami et al. (2014) used a light-bulb neutrino scheme, and neglected "ν_μ" neutrinos. Both Nakamura et al. (2014) and Takiwaki et al. (2016) observed equatorial explosions in the rapidly-rotating context. Recently, Janka et al. (2016) and Summa et al. (2018) published a rapidly-rotating 3D model that exploded and highlighted the potential role of an "$m = 1$" structure in the gain region. Earlier, Fryer and Heger (2000) used SPH for the hydrodynamics and a simplified gray scheme for the neutrino transport to explore the role of rapid rotation. Moreover, in all of these studies the initial rotation rate was not only high in the mantle, but was high in the core. Such rapid initial spins (periods of a few seconds) and final spins (periods of $\sim$2–10 milliseconds) seem not to be consistent with inferred pulsar spin periods at birth (crudely, $\sim 300 \pm 200$ milliseconds; Emmering and Chevalier 1989; Faucher-Giguere and Kaspi 2006; Popov and Turolla 2012; Noutsos et al. 2013) and may be associated only with hypernovae (Burrows et al. 2007c) and/or gamma-ray bursts (MacFadyen and Woosley 1999).

In this paper, we explore, with self-consistent 2D FORNAX (Sect. 2) simulations, the dependence of the outcome of collapse (most notably whether the model explodes) on neutrino-nucleon scattering rates (via modifications of in-medium response corrections due to many-body effects), pre-collapse convective perturbations, inelastic neutrino-electron scattering, and inelastic neutrino-nucleon scattering. We also continue our study, started in Skinner et al. (2016), of the issues raised by the use of the ray-by-ray+ method. What we find is that when the proto-neutron star bounded by a stalled shock is close to the critical condition for explosion (Burrows and Goshy 1993), as it easily can be in the turbulent multi-D context, the sensitivity to explosion of small changes in the physical inputs is amplified. The magnitude of such changes might be only $\sim$20%, but the result can be qualitatively different, in particular whether the model explodes. In the 1D (spherical) case, the core is not sensitive to comparable changes ("Mazurek's Law"), but in the multi-D turbulent (and chaotic) context, small changes in the physics can have a qualitative effect on explodability. This may explain why the various groups around the world simulating core-collapse supernovae can witness very different outcomes, despite the fact that they ostensibly are incorporating almost the same microphysics and similar computational approaches. Small differences are amplified near criticality in this chaotic context. The availability of a new generation of fast, but accurate, simulation codes, such as FORNAX (Sect. 2), and significant supercomputer resources enable rapid multi-parameter investigations in the multi-dimensional (in particular 2D), multi-physics context. Such wide-ranging explorations reveal patterns not easily discerned when one (or only a few) simulations are the focus of a paper and its (or their) results solely are mined.

2 Numerical Method and Computational Setup

We have developed an entirely new multi-dimensional, multi-group radiation/hydrodynamic code, FORNAX, for the study of core-collapse supernovae. This code is described in detail in an upcoming paper (in preparation). For the purposes of this paper, we note that it employs spherical coordinates in one and two spatial dimensions, solves the comoving-frame, multi-group, two-moment, velocity-dependent transport equations to $O(v/c)$, and uses the M1 tensor closure for the second and third moments of the radiation fields (Dubroca and Feugeas 1999; Vaytet et al. 2011). Three species of neutrino (ν_e, $\bar{\nu}_e$, and "ν_μ" [ν_μ, $\bar{\nu}_\mu$, ν_τ, and $\bar{\nu}_\tau$ lumped together]) are followed using an explicit Godunov characteristic method applied to the radiation transport operators, but an implicit solver for the radiation source terms.

The hydrodynamics in FORNAX is based on a directionally-unsplit Godunov-type finite-volume method. Fluxes at cell faces are computed with the fast and accurate HLLC approximate Riemann solver based on left and right states reconstructed from the underlying volume-averaged states. In the interior, to alleviate Courant limits due to converging angular zones, the code can deresolve in angle with decreasing radius, conserving hydrodynamic and radiative fluxes in a manner similar to the method employed in AMR codes at refinement boundaries. Gravity is handled in 2D and 3D with a monopole or a multipole solver (Müller and Steinmetz 1995). When using the latter, we generally set the maximum spherical harmonic order necessary equal to twelve. The monopole gravitational term is altered to accommodate approximate general-relativistic gravity (Marek et al. 2006), and we employ the metric terms, g_{rr} and g_{tt}, derived from this potential to incorporate general relativistic redshift effects in the neutrino transport equations (in the manner of Rampp and Janka 2002). In 2D, rotation and a third component of the velocity vector can be included in the hydrodynamics. We use the SFHo equation of state (EOS) by default (Steiner et al. 2013), but in this paper do compare with results using the DD2 EOS (Banik et al. 2014).[2]

For these simulations, we follow twenty energy (ε_ν) groups for each of the ν_e, $\bar{\nu}_e$, and "ν_μ" species (again, the four species, ν_μ, $\bar{\nu}_\mu$, ν_τ, and $\bar{\nu}_\tau$ lumped together.) For the ν_e types, the neutrino energy ε_ν varies logarithmically from 1 MeV to 300 MeV, while it varies from 1 MeV to 100 MeV for the $\bar{\nu}_e$ s and ν_μ s. We have performed calculations with forty energy groups and found little difference in the results. The radial coordinate, r, runs from 0 to 20,000 kilometers (km) in 608 zones. The radial grid smoothly transitions from uniform spacing with $\Delta r = 0.5$ km in the interior to logarithmic spacing, with a transition radius near $\sim$100 km. The polar angular grid spacing covers the full 180° and varies smoothly from $\approx 0.95°$ at the poles to $\approx 0.65°$ at the equator in 256 zones.

[2] In a earlier version of this paper, we used the K = 220 MeV Lattimer-Swesty equation of state (Lattimer and Swesty 1991) for all our simulations, but since this EOS has been shown to be inconsistent with recent measurements we redid the models and paper using EOSes still consistent with what is known.

A comprehensive set of neutrino-matter interactions are followed in FORNAX, and these are described in Burrows et al. (2006a, 2006b). Inelastic neutrino-nucleon scattering is handled using a modified version of the Thompson et al. (2003) approach (Sect. 4). Our neutrino microphysics is more comprehensively listed in our upcoming code paper.

3 Many-Body Neutrino Response Corrections

Melson et al. (2015) invoked a modification in the axial-vector coupling constant (g_A) due to a possible strangeness contribution to the nucleon spin of $g_A^s = -0.2$. This results in an approximate decrease in the neutrino-nucleon scattering rate of $\sim$20% and in Melson et al. the consequence was an explosion in 3D, even though they did not witness an explosion in 3D when using their default microphysical suite. Curiously, without their strangeness correction, the same model exploded in 2D. However, the value of the correction, g_A^s, they employed in their 3D model is likely too large and g_A^s may be closer to zero (Ahmed et al. 2012; Green et al. 2017).

Many-body corrections to neutral-current and charged-current neutrino-nucleon interactions have been discussed in the past (Burrows and Sawyer 1998, 1999; Hannestad and Raffelt 1998; Reddy et al. 1999; Roberts et al. 2012) in the context of proto-neutron stars and supernovae, and have been known to affect the neutrino-matter reaction rates. Burrows and Sawyer (1998) in particular suggested that many-body corrections to the axial-vector and vector structure factors for neutrino-nucleon scattering could be of a magnitude sufficient to be of relevance to the viability of the neutrino-driven mechanism of core-collapse supernovae, but did not provide a robust estimate of the magnitude of this diminution much below nuclear densities.

The effect of the many-body correction to neutral-current neutrino-nucleon scattering in the supernova context is mostly due to the increase in the ν_μ luminosity occasioned by the decrease in the associated $\nu_\mu + (n, p) \rightarrow \nu_\mu + (n, p)$ scattering cross section for densities above $\sim 10^{12}$ g cm^{-3}; this causes a further compression in the core. Such a compression, similar to the effect of GR, raises the temperatures near the ν_e and $\bar{\nu}_e$ neutrinospheres, thereby raising their associated luminosities and average emergent neutrino energies. These changes increase the neutrino-matter heating rates in the gain region and, hence, facilitate explosion. Since the super-allowed charged-current absorption reactions still dominate the $\nu_e/\bar{\nu}_e$-matter interaction rates, the direct effect of this structure factor correction to the axial-vector term in the neutrino-nucleon scattering rate on the ν_e and $\bar{\nu}_e$ luminosities is slight.

A many-body structure factor (S_A) correction to the axial-vector term in the neutrino-nucleon scattering rate due to the neutrino response to nuclear matter at (low) densities below $\sim 10^{13}$ g cm^{-3} was recently derived by Horowitz et al. (2017) using a virial approach. These authors derive a fit:

$$S_A = \frac{1}{1 + A(1 + Be^{-C})}, \tag{1}$$

where

$$A = A_0 \frac{n(1 - Y_e + Y_e^2)}{T^{1.22}} \tag{2}$$

$$B = \frac{B_0}{T^{0.75}} \tag{3}$$

$$C = C_0 \frac{n Y_e (1 - Y_e)}{T^{0.5}} + D_0 \frac{n^4}{T^6}. \tag{4}$$

In these equations, T is the temperature in MeV, Y_e is the electron fraction, n is the baryon density in fm^{-3}, $A_0 = 920$, $B_0 = 3.05$, $C_0 = 6140$, and $D_0 = 1.5 \times 10^{13}$. Horowitz et al. (2017) join their fit to Burrows and Sawyer (1998) for the higher densities, but were most careful fitting their formula for temperatures between 5 and 10 MeV. Nevertheless, though Horowitz et al. (2017) intended their fit to apply at all densities, temperatures, and Y_e s, (and we use it in this paper at all thermodynamic points), one should be aware that no current approach to this physics is likely to be correct at the highest densities above $\sim 10^{14}\ \text{g}\,\text{cm}^{-3}$. Therefore, use of this formula in supernova and proto-neutron-star cores should be considered provisional. Fortunately, since such densities are too high to affect the evolution during the first second post bounce, and are most relevant during the later proto-neutron-star cooling phase, the values of the Horowitz correction at the highest densities are not germane to the conclusions of this paper.

However, many-body effects in the charged-current sector and on absorption may be comparable (Burrows and Sawyer 1999; Roberts et al. 2012; Fischer 2016), but have not yet been factored in. Therefore, neglected is the effect of final-state nucleon blocking for charged-current absorption reactions. Blocking is important only at high densities, where many-body interaction effects are likely to be even larger. Therefore, we have postponed their inclusion until such corrections, which must for self-consistency be done with the same interaction model that informed the equation of state employed, are available.

The fit derived by Horowitz et al. (2017) to the structure factor applied to g_A^2 translates into a decrease in the neutrino-nucleon scattering cross section in the crucial region at and deeper than the various neutrinospheres of $\sim 5\%$ to $\sim 35\%$. This effect is a function of density, temperature, and electron fraction (Y_e). Horowitz et al. (2017) state that the corresponding structure factor for the vector current is likely greater than one, but since the vector contribution to neutrino-nucleon scattering is small, this is likely to be subdominant. The upshot is a potentially important, and physically plausible, decrease in the neutrino-matter scattering rates that translates into an increase in the driving ν_e and $\bar{\nu}_e$ luminosities and average energies, thereby increasing the heating rate in the gain region.

We note that the corresponding many-body correction for charged-current interactions may be in the same direction (Fischer 2016; Burrows and Sawyer 1999, although see Roberts et al. 2012) and could also be important (Sect. 5.3). The correction would be small at low densities in the gain region, but higher deeper inside, where the neutrinospheres reside. If the rates are suppressed, this could increase the ν_e and $\bar{\nu}_e$ luminosities, while simultaneously not decreasing the heating in the gain region and is not like a uniform correction at all radii. Such behavior, if it obtains, would be near optimal for aiding the explosion.

4 Inelastic Scattering

Neutrino-electron scattering rates and cross sections are much smaller than those for neutrino-nucleon scattering, which themselves are smaller still than those for super-allowed charged-current reactions such as $\nu_e + n \rightarrow p + e^-$. For $\varepsilon_\nu = 10$ MeV, this deficit is approximately two orders of magnitude. However, due to the small mass of the electron, the energy transfer to the matter during a "Compton-like" neutrino-electron scattering is on average quite large, while, due to the large mass of the nucleon, the corresponding average energy transfer during neutrino-nucleon scattering is rather small. Therefore, the large cross section for neutrino-nucleon scattering multiplied by the small associated energy transfer can be comparable to the product of the small neutrino-electron cross section with the large per-interaction energy transfer and depend upon temperature, density, and neutrino energy

(Janka et al. 1996; Thompson et al. 2000). The upshot is that inelastic scattering off both electrons and nucleons can modify thermal profiles and heating rates exterior to the neutrinospheres and in the gain region and contribute to explosion by the neutrino heating mechanism. Moreover, Müller et al. (2012b) make the point that heating by ν_μ-nucleon inelastic energy transfer can boost the temperatures near the ν_e and $\bar{\nu}_e$ neutrinospheres and results in slightly higher ν_e and $\bar{\nu}_e$ energy luminosities, which in turn enhance heating behind the shock correspondingly.

The detailed theory of the structure functions and redistribution kernels for such inelastic scattering, including the effects of final-state blocking and the thermal spectrum of the targets, can be found in Burrows et al. (2006a, 2006b), Reddy et al. (1999), Thompson et al. (2003), and Thompson et al. (2000). Pioneering work on inelastic scattering off electrons in the core-collapse context can be found in Bruenn (1985) and Mezzacappa and Bruenn (1993). However, those latter papers were focussed on the downscattering effect due to inelastic ν_e–e^- scattering on the electron neutrinos produced during infall and the consequent decrease in the trapped lepton fraction. Since a larger trapped lepton fraction could help facilitate immediate post-bounce explosions (Burrows and Lattimer 1983), this quantity was more relevant when the prompt hydrodynamic supernova mechanism still seemed viable. However, with the emergence of the delayed, neutrino-driven mechanism (Bethe and Wilson 1985), and the conclusion that the prompt mechanism could not work due to catastrophic neutrino losses at and around shock breakout, the value of the trapped lepton fraction, and its precise value, receded in significance.

Nevertheless, heating behind the stalled shock due to inelastic energy transfer from neutrinos to both electrons and nucleons, or boosting the ν_e and $\bar{\nu}_e$ luminosities by ν_μ downscattering near their neutrinospheres (Müller et al. 2012b), may help achieve the critical condition for explosion, particularly when in concert with the inclusion of the in-medium neutrino response (Horowitz et al. 2017; Sect. 3) and the slightly net positive influence of GR. Heating by inelastic neutrino-electron scattering is still sub-dominant with respect to that due to charged-current ν_e and $\bar{\nu}_e$ absorption, and in 1D (spherical) simulations there is almost no hydrodynamic consequence of its inclusion (Thompson et al. 2003). The same can be said of inelastic neutrino-nucleon scattering (Janka et al. 1996; Burrows and Sawyer 1998, 1999; Thompson et al. 2000). However, in the realistic multi-D context of the core-collapse phenomenon, the cumulative effect of the addition of a few sub-dominant heating mechanisms, compression due to enhanced ν_μ neutrino losses and consequent ν_e and $\bar{\nu}_e$ neutrinosphere heating,[3] and greater proximity to the critical condition due to multi-D effects[4] amplifies the leverage of even small additions to the gain-region heating power over their effects individually and can convert an anemic explosion into a robust explosion, or even a dud into a blast. One recalls that the original delayed mechanism of Wilson required for explosion only a modest enhancement of $\sim$25% in the neutrino luminosity.[5]

From the work of Thompson et al. (2000) and Tubbs (1979), we find that the crossover energy between upscattering and downscattering is nearer $6k_BT$ (not $3k_BT$, as in Müller and Janka 2015), where k_B is Boltzmann's constant and T is the temperature, around densities of $\sim 10^{11}\ \mathrm{g\,cm^{-3}}$ to $\sim 10^{13}\ \mathrm{g\,cm^{-3}}$ and so our approximate redistribution rate for ν_μ s is

[3] as in neutrino-electron scattering and many-body scattering rate suppression

[4] Examples include the enhancement of the stress behind the shock due to turbulent pressure (Burrows et al. 1995; Murphy et al. 2013; Janka 2012; Burrows 2013) and the modest increase in the dwell time in the gain region of the post-shock matter (Murphy and Burrows 2008; Dolence et al. 2013)

[5] In that case, it was due to "neutron-finger convection," subsequently later shown not to occur (Bruenn and Dineva 1996; Dessart et al. 2006).

proportional to $\kappa_{scat}(\varepsilon_\mu - 6k_BT)/m_nc^2$, where κ_{scat} is the ν_μ scattering opacity and m_n is the neutron mass.

Inelastic scattering off nucleons lowers the ν_μ luminosity, while the corresponding quantities for the ν_e and $\bar{\nu}_e$ are slightly increased (Müller et al. 2012b). The latter responses explain the positive effect on explodability of the inclusion of inelastic scattering off nucleons. The direct effect of such inelasticity is in the core, and the effect in the shocked mantle is indirect.

5 Microphysical Dependences of the 2D Explosion Models

For a set of 4 fiducial models (with progenitor masses of 13, 15, 16, and 20 $M_\odot$), we performed a parameter study of input physics highlighting the role of small changes in promoting (or preventing) explosion. We refer the reader to the upcoming paper by Vartanyan et al. (2018, submitted) for a more detailed description of these simulations. When models explode, the time to explosion can vary significantly with inputs. The relative time of explosion can help one gauge the role of the inputs in question.[6] While there are many combinatoric possibilities, we settled on just a few comparisons to demonstrate the effects we see universally. We employ the notation IES, INS, MB, pert, and rbrp to indicate "inelastic neutrino-electron," "inelastic neutrino-nucleon," "many-body," "perturbations,", and "ray-by-ray+," respectively.

5.1 Inelastic Scattering

Models with only inelastic scattering off electrons (IES), inelastic scattering off both electrons and nucleons (IES_INS), and additionally with the many-body effect (IES_INS_MB) can illuminate the respective roles of each.[7] Figure 1 compares outcomes for the 16-$M_\odot$ progenitor. We toggled the role of inelastic scattering off electrons and nucleons as well as the many-body effect for the 16 $M_\odot$ progenitor of Woosley and Heger (2007), which we found to explode early at $\sim$300 ms using the default setup IES_INS_MB. However, performing the simulation with any of these components removed prevents explosion.

Figure 2 depicts the role of IES, INS, and MB on the emergent spectra at two different times after bounce. One of the central results that can be gleaned is the time of explosion. This quantity can help one gauge the relative role of the inputs in question. The default model (def) here does explode. Prior to explosion, both inelastic scattering off electrons and off nucleons and the many-body correction led to higher spectral fluxes. After the default model explodes, the flux spectrum diminishes relative to that of the non-exploding model 16_def_noMB. The slight increase in the heating in the gain region prior to explosion due to the inclusion of inelastic neutrino-electron and neutrino-nucleon scattering has certainly helped the core achieve the critical explosion condition. Figure 3 compares the emergent

[6]The "time of explosion" is sometimes defined as the time the mean shock radius achieves 400 km. We think this definition somewhat arbitrary, but acknowledge some degree of arbitrariness in any definition, however useful. We here define the time of explosion as the approximate time at which the curve of the mean shock radius versus time is inflected upwards. All our models that inflect in this way explode and all these models have been continued to at least one second after bounce. The mean shock radii of these exploding models all achieve radii beyond 5000 kilometers by the end of the simulation.

[7]In a previous version of this paper, we had employed an opacity file that was compromised by a compiler bug. The net result of this error was a slightly enhanced magnitude of the neutral-current many-body effect. This error has now been fixed.

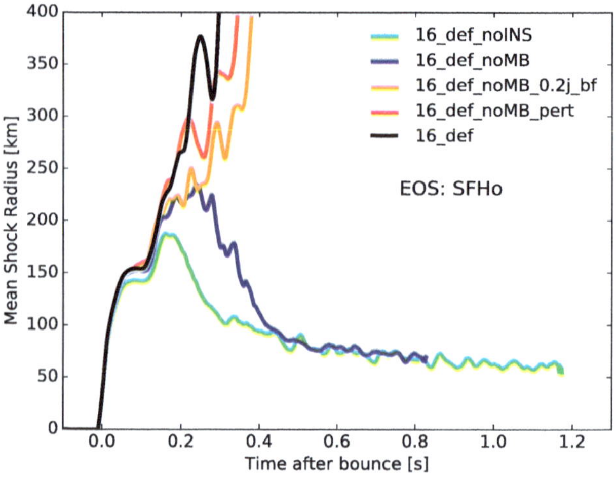

Fig. 1 Shock radii (in km) versus time post-bounce (in s) for variations on the default 16 $M_{\odot}$ progenitor. In all figures, default model "def" refers to inclusion of inelastic scattering off both electrons (IES) and nucleons (INS), as well as the inclusion of the Horowitz et al. (2017) many-body correction (MB). This model explodes at ~250 ms post-bounce. We then remove and add certain inputs, denoted by a subscript with "def". Removing either the many-body correction (*blue*, "16_def_noMB") or inelastic scattering off nucleons (green, "16_def_noINS") leads to a dud. However, even without the many-body correction (noMB), adding either perturbations (red) ("_pert"; Sect. 6) or modifying the opacity table to include Fischer's (2016) correction to the nucleon-nucleon bremsstrahlung rate ("bf") and only 20% of the electron capture rate (Juodagalvis et al. 2010) on heavy nuclei (*orange*, "0.2j"), leads to an explosion ~50 and ~100 ms, respectively, after our default model. This helps illustrate the sensitive dependence of the outcome—explosion or dud—on the microphysical inputs when near criticality

Fig. 2 The role of inelastic scattering off electrons and nucleons and the neutral-current many-body correction on the emergent spectra (in 10^{52} erg/s/MeV) at 500 km at 100 (*solid*) and 400 (*dashed*) ms post-bounce. At early times, prior to explosion, both inelastic scattering off electrons and nucleons and the many-body correction lead to upscattering. At 400 ms, the default model has exploded and hence has a diminished spectrum vis-à-vis the non-exploding model 16_def_noMB

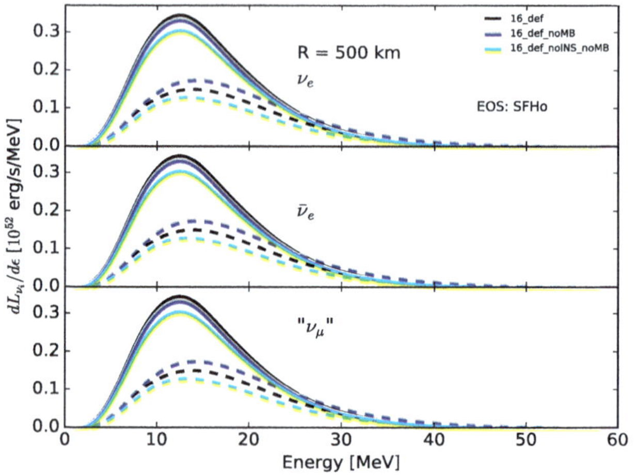

neutrino luminosities (left panel) and root-mean-square (rms) neutrino energies (right) for three models. The boosting in the ν_μ luminosity and rms energy is clearly manifest, as are the corresponding boosts in those quantities for the ν_e and $\bar{\nu}_e$ neutrinos. While one may have speculated that enhanced neutrino losses, particularly due to ν_μ s that are almost ineffectual in heating the shocked mantle, would have had a negative effect on explodability, the converse is true. Greater losses lead to a further compaction of the core with an increase in the

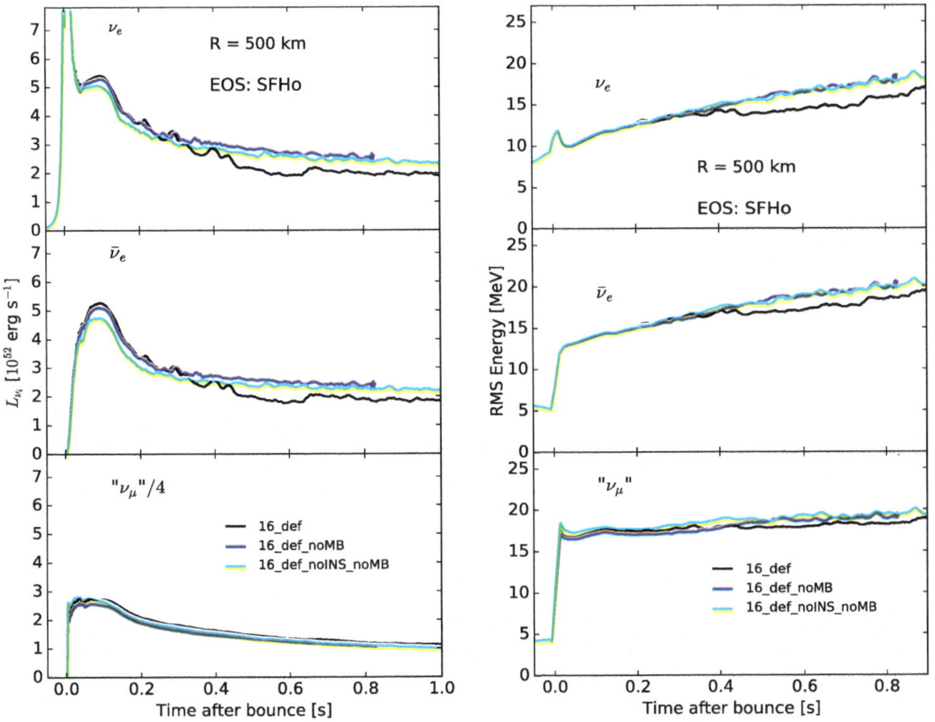

Fig. 3 Modification due to inelastic scattering off electrons and nucleons of the luminosities (*left*) and RMS energies (*right*) of neutrinos at 500 km, redshifted to the lab frame. Including inelastic scattering off nucleons decreases the ν_μ luminosities and RMS energies by $\sim$10%, as in Müller et al. (2012b), while slightly increasing the ν_e and $\bar{\nu}_e$ luminosities. RMS energies of the latter are mostly unaffected by inelastic scattering. The default model (*black*, with many-body corrections and both inelastic scatterings) shows a dip in luminosity and RMS energy after 300 ms post-bounce, the time of its explosion

matter temperatures near the ν_e and $\bar{\nu}_e$ neutrinospheres. The result is similar in effect to that of GR, whereby such core heating enhances the driving ν_e and $\bar{\nu}_e$ luminosities and the average neutrino energies, which in turn enhance the heating power in the gain region. Since it is this power deposition that ultimately drives explosions, the net effect is quite supportive of explosion.

However, the actual magnitude and form of the correction to the axial-vector coupling term in the expression for the neutrino-nucleon scattering rate may be different from that derived by Horowitz et al. (2017) and this still needs to be verified. Moreover, the effects of similar many-body corrections to the absorption rates need to be incorporated, as do those for the vector coupling strengths. One prediction of the consequence of the structure-factor correction we have employed is the enhanced ν_μ luminosities and average energies seen in Fig. 3. It is noteworthy that, as it stands, the effect on the outcome of core-collapse of many-body rate suppressions (Burrows and Sawyer 1998, 1999) might be large.

5.2 Equation of State

Through its control of the pressure for a given temperature, density, and Y_e, the equation of state of hot, lepton-rich nuclear matter will determine the structure of the proto-neutron

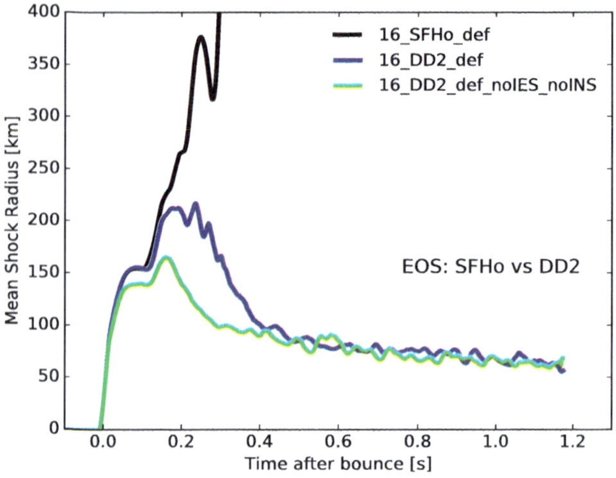

Fig. 4 Shock radii (km) versus time after bounce (s) for the SFHo and DD2 EOS for the 16 $M_\odot$ WH07 progenitor as a function of time after bounce. Only the former (our default model) explodes. We also plot for comparison a model with the DD2 EOS, but without any inelastic scattering off electrons or nucleons. Though neither DD2 model explodes, including inelastic scattering increases the stalled shock radius by ~70 km

star and its evolution after bounce. The stiffer the EOS, the more extended the core. On the one hand, a stiff EOS will resist the quick increase in temperature near the ν_e and $\bar{\nu}_e$ neutrinospheres, and the consequent enhancement of the neutrino heating rate in the gain region, seen both in the comparison of GR and Newtonian models and in the increase in the ν_μ losses due to the many-body effect. On the other, a stiffer EOS will provide a more stable inner platform that won't as easily (by its inward motion with time after bounce) send out weakening rarefactions to the outer bounce shock that could inhibit explosion. Figure 4 indicates (at least for this 16-$M_\odot$ model) that the first effect seems to win, since the DD2 EOS model does not explode, while the SFHo model (the softer of the two at high densities) does. On this figure we also show that, though the DD2 model didn't explode, the behavior of the shock with time was significantly less vigorous when the inelastic scattering effects were turned off, reiterating the conclusions of Sect. 4 and Sect. 5.1. We also note that since the shock in the default DD2 model achieved a large mean shock radius (~200 km) before subsiding, this model was nevertheless very close to exploding.

5.3 Nucleon-Nucleon Bremsstrahlung and Electron-Capture on Heavies

On Fig. 1, we also provide a model that drops the many-body correction to neutrino-nucleon scattering, as derived by Horowitz et al. (2017), but substitutes in the Fischer (2016) correction to the nucleon-nucleon bremsstrahlung rate ("bf") and divides the electron capture rate on heavy nuclei derived for the SFHo EOS by Juodagalvis et al. (2010) by five ("0.2j"). The former effect lowers the bremsstrahlung production rate of ν_μ neutrinos, as well as the inverse bremsstrahlung absorption. The latter effect slightly retards the infall rate, thereby decreasing the mass accretion rate post-bounce at a given time. Decreasing this rate can facilitate explosion. As the figure suggests, these microphysical changes can compensate for ignoring the many-body effect to result in a similarly-aided explosion. Clearly, such modest alterations in the microphysics of relevance, still within the envelope of our ignorance, could play a positive role. In Fig. 5 below, we see a similar positive effect of these potential changes in the default microphysics for a 13-$M_\odot$ model. Together with the model in Fig. 1, this suggests that a more extensive exploration of such physics would be fruitful.

6 Progenitor Perturbations

Performing the last stages of stellar evolution before collapse hydrodynamically in 2D and 3D has been shown to alter, perhaps in significant ways, the compositional, entropy, and density profiles of the core (Meakin et al. 2011; Couch et al. 2015; Chatzopoulos et al. 2016; Müller et al. 2016; Abdikamalov et al. 2016; Müller et al. 2017), and it has long been known that progenitor density profiles have an impact on the outcome of collapse. This is implicit in the critical curve analysis of Burrows and Goshy (1993), where $\dot{M}$ and the accretion ram pressure play central roles. It is also a factor in discussions of the compactness parameter (O'Connor and Ott 2011) and its extensions (Ertl et al. 2015). In this vein, one notes that at least three groups (Kitaura et al. 2006; Burrows et al. 2007b; Radice et al. 2017) have already demonstrated that the 8.8-$M_{\odot}$ "electron-capture" supernova progenitor of Nomoto and Hashimoto (1988), with its extremely steep density ledge, can explode in 1D by the neutrino-driven wind mechanism, though the explosion energy is low ($\sim$1–2 $\times$ 10^{50} ergs).

However, when spherical progenitor models do not readily lead to explosion, the initial perturbation spectrum in the progenitor's convective silicon and oxygen zones could certainly affect the timescales for the generation of turbulence behind the stalled shock and be a factor in the onset of explosion (Couch and Ott 2013; Müller and Janka 2015; Couch et al. 2015; Abdikamalov et al. 2016; Müller et al. 2017). Specifically, the magnitude, character, and spectra of seed perturbations will affect how quickly turbulence reaches the non-linear regime and, perhaps, whether turbulence grows to non-linearity at all during the finite time the accreta are in the unstable gain region.

Therefore, introducing physically-motivated seed perturbations that originate from and reflect the three-dimensional character of the convective core of an actual massive star at its terminal stage of evolution is a topic of some interest. However, most calculations done to date do not start with true 3D convective structures, but with 1D models from the literature, and impose either ad hoc perturbations in density or velocity randomly at the grid level or allow grid asphericities (such as obtained when using a Cartesian grid) or truncation errors to act as seeds. Neither of these approaches is physical, and the resulting initial perturbation spectra lead to early growth rates in the linear regime that reflect not the multi-D progenitor perturbation structure, but the numerical development of convenient artificial power spectra. This will affect how quickly turbulence reaches the non-linear regime and, perhaps, whether turbulence grows to non-linearity at all during the finite time the accreta are in the unstable gain region. In addition, this may have a bearing on the post-bounce delay to a turbulence-aided explosion, with the consequent effect on the time and energy of that explosion.

Therefore, the seed perturbations that arise during oxygen and silicon burning prior to collapse might be key inputs into core-collapse supernova theory, and Couch and Ott (2013), Müller and Janka (2015), Couch et al. (2015), Müller (2016), and Müller et al. (2017) have begun to explore this. However, mixing-length theory, though inadequate as a comprehensive theory, still provides a measure of the magnitude of velocity perturbations at the onset of collapse (Müller et al. 2016), and they are only a few hundred to $\sim$500 km s^{-1}, with Mach numbers bounded by $\sim$0.08 (Woosley and Heger 2007). This is not large.

In this section, we provide a quick glimpse at the possible relative role of significant perturbations on the timing and character of explosion in light of the other physics. To this end, we employ the methodology of Müller and Janka (2015). These authors impose a vector velocity perturbation map on their progenitor that is more realistic than many past approaches and renders a perturbation field that is independent of grid resolution. This (surprisingly) was rarely attempted in the past and provides a specific context for future comparison. We set the maximum perturbation speed on the grid to 1000 km s^{-1}, which as indicated earlier

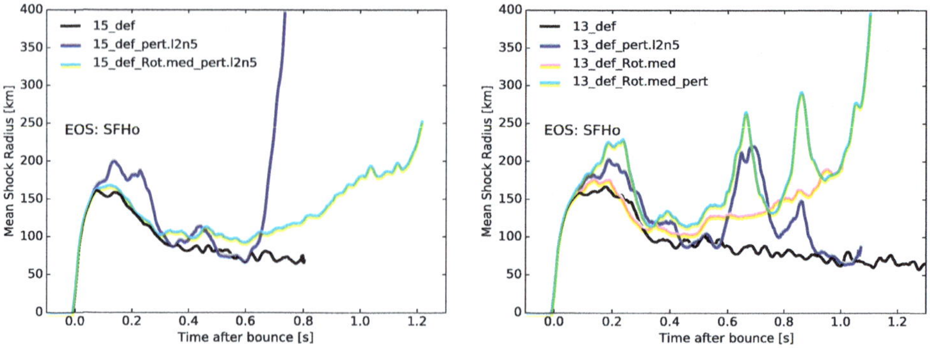

Fig. 5 Shock radius (in km) versus time after bounce (in s) for the 15 (*left*) and 13 (*right*) $M_\odot$ Woosley and Heger (2007) models, showcasing the effect of significant perturbations and moderate rotation. We follow Müller and Janka (2015) prescription in implementing perturbations to radial velocities on infall over three regions with the maximal radial velocity (1000 km s^{-1}), n (number of radial convective cells), and l (number of angular convective cells) as parameters. The interior region spans 1000 to 2000 km, outside the nascent core, the middle region 2100 to 4000 km, and the outer region 4100 to 6000 km, truncated roughly where accretion ends after the first second for our simulations. All regions have $l = 2, n = 5$ and maximum radial velocity of 1000 km s^{-1}. See text for a discussion

may be near or beyond the expected upper end of the range, a spherical harmonic index ℓ of 2, and a radial "quantum number" n of 5. Both ℓ and n are parameters in the Müller and Janka (2015) formulation. With this parameter set, we simulate in 2D the self-consistent multi-group evolution of 13- and 15-$M_\odot$ progenitors and compare the result to a default model for which the initial perturbations are much smaller and arose numerically from grid noise.

Figure 5 portrays the evolution of the mean shock radii, with and without progenitor perturbations. The left panel shows that the imposed perturbations converted failure into success for the 15-$M_\odot$. We also provide on this panel a model including the effect of a modest rotation rate, along with the perturbation. Including the latter resulted in an explosion, though later. Our rotational study has revealed that the effect of rotation is not monotonic with initial internal rotational speed or spatial profile. On the right panel of Fig. 5, we show the corresponding behavior for the 13-$M_\odot$ progenitor, which does not explode in our study for the default microphysics. It does, however, explode when similar perturbations are imposed and when a similar rotational profile is assumed. In this case, rotation promotes explosion, highlighting its non-monotonic effect upon outcome.

We note that the initial perturbations we imposed have an amplitude that is somewhat larger than expected (Müller et al. 2016) and that their character is still rather artificial. The magnitude of the initial perturbations may well be lower, but their character and magnitude will certainly vary from progenitor to progenitor. Therefore, a much more thorough study with 3D progenitors and 3D collapse models is called for.

7 Ray-by-Ray+ Anomalies

As shown by Skinner et al. (2016), the ray-by-ray+ approach to neutrino transport, whereby multi-D transport is replaced by multiple 1D transport calculations with corrections for matter advection, but not lateral transport, can introduce systematic errors in the heating rates along the poles in axisymmetric 2D simulations. Such enhancements, when in proximity to

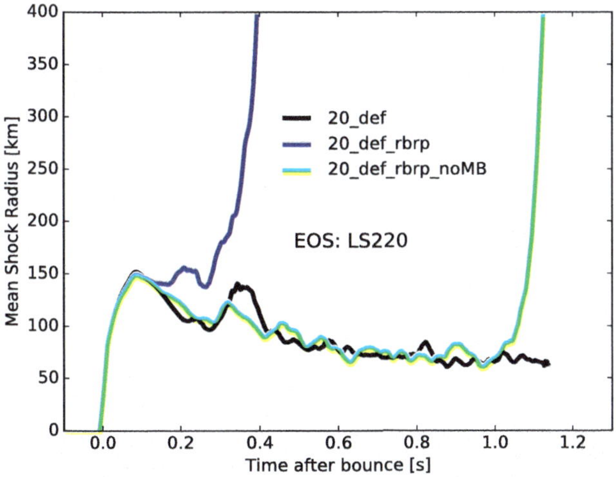

Fig. 6 Shock radius (km) versus time after bounce (s) for the 20 $M_\odot$ WH07 progenitor using the LS220 EOS, with and without the ray-by-ray-plus (rbrp) approximation to neutrino transport. All models include inelastic scattering off electrons and nucleons. We see that including ray-by-ray+ (rbrp) leads to an explosion when otherwise there was none. The model with ray-by-ray+ and without many-body (*green*) explodes as well, though 700 ms after the model with ray-by-ray+ and many-body, suggesting that the ray-by-ray+, though artificial, is more significant to explosion than the physical inclusion of the many-body correction

criticality, may be producing explosions artificially. At the very least, the time to explosion is artificially shortened, perhaps significantly. Since there is little or no accumulation of explosion energy prior to global instability (Burrows et al. 1995),[8] an earlier explosion may make more of the emitted neutrinos available for explosive driving.

Figure 6 compares three 20-$M_\odot$ models, using this time the LS220 EOS. Here, the default model does not explode, but the one employing the ray-by-ray+ simplification does. Note that, for both ray-by-ray+ models, the model without the many-body (MB) correction explodes much later (by 700 ms) than the model with MB. Our results suggest, for this progenitor, that the artificial ray-by-ray+ approximation is more significant to explosion than the physical many-body effect. One is left to speculate whether 2D simulations in the literature that employ ray-by-ray+ would indeed explode if they used more realistic transport. This is all the more relevant in 3D, given that extant published models that do explode in 2D have more difficulty exploding in 3D, a context in which it is not clear that the ray-by-ray+ method introduces as great an artifact as in 2D. It is true that the turbulent pressure spectra in 2D and 3D are different, with the turbulent cascade in 2D resulting in enhanced stresses on larger scales, and that the turbulent-stress boost to explodability may be smaller in 3D. This could also be a factor in the more anemic outcomes in published 3D models. Nevertheless, it may be that the more problematic nature of published 3D models vis-à-vis 2D models is a consequence of some combination of the use of ray-by-ray+ and the reduced turbulent stress in 3D.

We conclude this section by noting that the calculations of Skinner et al. (2016) did not include various physical effects (such as inelasticities and the many-body correction) that we highlight here. In Skinner et al., we obtained explosions using ray-by-ray+ for some models that did not explode otherwise, and when they exploded they did so late.

[8]Only those neutrinos emitted after the onset of the explosion contribute to the asymptotic explosion energy.

8 Compactness

Finally, we conclude that although the compactness parameter (O'Connor and Ott 2011, 2013; Pejcha and Thompson 2015), defined at bounce for a given mass interior (M) at a given radius (R) as $\xi \equiv M/\mathrm{M}_\odot(1000\ \mathrm{km}/R)$, is one measure of the important density structure of the progenitor, it is not necessarily predictive of explosion, at least during the first second after bounce. We find that, depending upon the neutrino physics employed, the temporal order in which models with different compactness parameters explode after bounce varies. If time of explosion is a fit measure of explodability, then this alone would challenge the usefulness of the compactness concept vis à vis explodability. Nevertheless, though the compactness parameter is large for the 21-$\mathrm{M}_\odot$ model and small for the 12-$\mathrm{M}_\odot$, we found that the former can be more explodable. One might have thought that if compactness were the sole predictor the details of the neutrino-matter interaction would have mattered little in this regard. However, we see that the time to explosion does not necessarily correlate well with compactness. Models with dense envelopes have larger accretion and total neutrino luminosities after bounce, and it is these luminosities that drive explosion. In addition, models with dense envelopes and higher compactness have a greater optical depth to neutrinos in the gain region. Since the neutrino power deposition goes approximately as the product of this depth and the luminosities, high-compactness progenitors have an advantage. Therefore, it is feasible that more massive models might be more explosive, and this trend has been seen by other modelers (Summa et al. 2016, 2018; Bruenn et al. 2016). In fact, in the Summa et al. and Bruenn et al. papers for the Woosley and Heger (2007) models, the post-bounce explosion times increase in the sequence 20, 25, 15, and 12 $\mathrm{M}_\odot$, which is clearly not monotonic with compactness. This behavior is the reverse of what might be expected if low compactness signaled greater explodability. Finally, the work of Nakamura et al. (2015) suggests that though they see a weak correlation, it is not monotonic with compactness. In fact, many of their plots versus compactness resemble scatter plots.

A critical issue is whether these more massive models with shallower density profiles that explode early can maintain explosion during the traversal of the shock through the outer stellar mantle. The binding energy penalty of the outer envelope generally increases with progenitor mass and might be too steep a price to pay during subsequent evolution for those more massive cores that explode earlier and, perhaps, more energetically. As previously stated, for most of the relevant mass function, the envelope binding energy exterior to a given interior mass is an increasing function of progenitor mass. It is this "barrier" that may set the limit to the range of massive stars that can explode and leave behind neutron stars, although it cannot be excluded that the progenitor mass range that yields neutron stars, and not black holes, may be discontinuous (Sukhbold et al. 2016). Figure 7 portrays various exterior binding energies and Fig. 8 follows with a depiction of the close correspondence between the compactness and the envelope binding energy exterior to a baryonic mass cut of 1.5 $\mathrm{M}_\odot$. Therefore, we contend that whatever significance there may be to the compactness parameter is likely due to its correspondence with the binding energy of the outer envelope exterior to a given mass cut and that compactness need not correlate with the apparent explodability during the first second after core bounce. Importantly, the ultimate outcome will depend upon the progress of the shock at post-bounce times that generally exceed those to which most core-collapse simulations currently go.

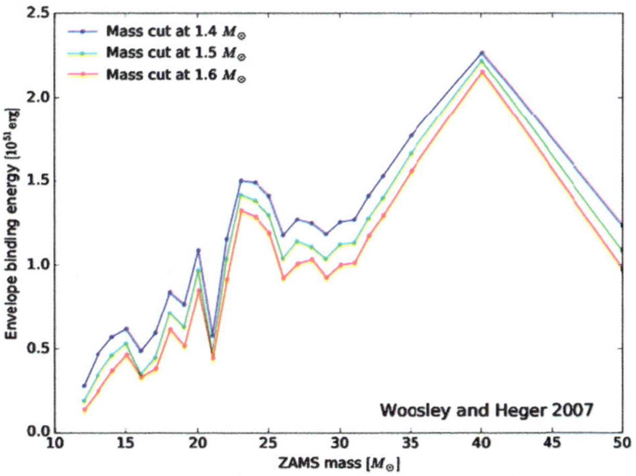

Fig. 7 The outer envelope binding energies (in Bethes [$\equiv 10^{51}$ erg]) for baryon mass cuts of 1.4, 1.5, and 1.6 $M_\odot$ versus ZAMS mass (in $M_\odot$) for the Woosley and Heger (2007) progenitor models

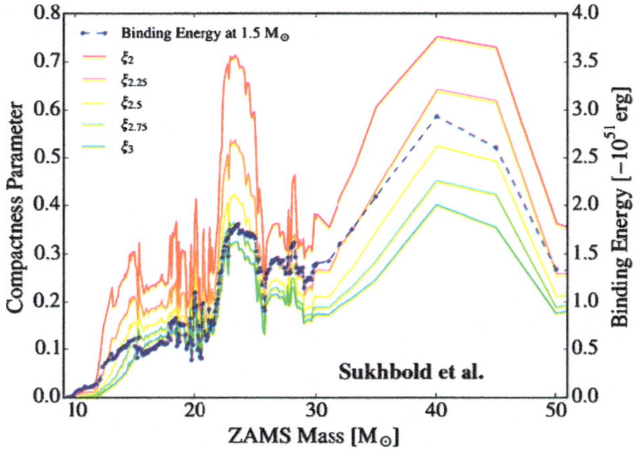

Fig. 8 This plot depicts the dependence of the compactness parameter (O'Connor and Ott 2011, 2013), calculated at various interior masses, versus progenitor ZAMS mass, as well as the corresponding envelope binding energy (*blue dots*; in Bethes [10^{51} ergs]) for a baryon mass cut of 1.5 $M_\odot$ (see Fig. 7). The progenitor models of Sukhbold and Woosley (2014) (also in Sukhbold et al. 2016) are used. As this figure shows, whatever the position at which compactness is defined, it correlates extremely well with envelope binding energy. It is our contention that it is the latter quantity that is more germane to the outcome of core collapse

9 Conclusions

In this paper, we have generated and explored detailed 2D (axisymmetric) approximate GR models of core-collapse supernovae using the new code FORNAX, treating all the relevant physics to determine the dependence of the mechanism of explosion and its timing on the physical and numerical inputs and assumptions. These include inelastic neutrino-electron and neutrino-nucleon scattering, many-body/structure-factor corrections to the neutrino-nucleon scattering rates, nucleon-nucleon bremsstrahlung, electron capture on heavies, and physically-motivated initial perturbations. We have also reexamined in brief the effect of using the ray-by-ray+ simplification to neutrino transport. We found that much of the wide variation between the results obtained by different groups around the world simulating stellar collapse (as well as whether the core explodes at all) might be explained by slight variations at the $\sim$10–30% level in the microphysical inputs when the models are near the critical condition for explosion. In the process, we gauged the relative importance of otherwise sub-dominant neutrino physics processes to the outcome of collapse. Proximity to criticality amplifies the dependence upon small changes in the neutrino sector that translate into slight, but crucial, changes in the emergent luminosities and average neutrino energies and the consequent post-shock heating rates; such sensitivity is not manifest in 1D simulations for which the core is (most often) far from explosive.

Thus, "Mazurek's Law" of severe feedback under variations in neutrino cross sections and rates is overturned due to the proximity to instability possible in the realistic multi-D turbulent context. While alterations in the neutrino coupling rates have little effect in 1D, in multi-D the hydrodynamic response to even small changes and/or corrections to neutrino interaction rates can be more substantial due to greater proximity to the critical curve.

The upshot is that small variations between the methods, microphysics, and resolutions used by groups who ostensibly are incorporating the "same" inputs translates naturally into post-bounce explosion time differences that can range by many hundreds of milliseconds, and in some cases can turn a dud into an explosion (or vice versa). We suggest that this thereby explains in large measure the apparent heterogeneity in the outcomes of detailed simulations performed internationally. A natural conclusion is that, viewed correctly, the different groups are collectively closer to a realistic understanding of the neutrino-driven mechanism of supernova explosion than might have seemed apparent and that a push to rationalize approaches and understand microphysical details in the neutrino-matter interaction sector and the nuclear equation of state could bring a resolution to the decades-long quest for a predictive model of core-collapse supernova explosions.

We have found that many-body corrections to neutrino-matter interaction rates, even at sub-nuclear densities, can have similar effects as general relativity, progenitor perturbations, or inelastic neutrino-electron and neutrino-nucleon scattering. However, we caution that the actual magnitude and functional form of all the various many-body corrections to the neutrino-matter rates (both neutral- and charged-current), however important they seem from our current simulations, still need to be explored and verified.

We performed a test using the ray-by-ray+ approximation to neutrino transport in a manner similar to that employed by Skinner et al. (2016) to gauge its effect on the outcome of collapse when a full physics suite is employed. In Skinner et al. (2016), it was shown that ray-by-ray+ artificially enhanced heating along the poles in synchrony with the axial sloshing seen in 2D simulations and thereby made models more explosive. The shift in the explosion times can be as large as the full range currently witnessed by the various supernova simulation groups for a given progenitor. One can speculate that had groups that use the ray-by-ray+ simplification used real multi-D transport their 2D models would have exploded later, perhaps much later or not at all. Speculating further, one wonders whether the fact that 3D models explode later than the corresponding 2D models (Lentz et al. 2015) or not at all (Melson et al. 2015, if without their strangeness fix) may be connected with their use of ray-by-ray+, since non-rotating 3D simulations seldom manifest the axial sloshing seen in 2D. In short, without ray-by-ray+, it is not clear that 2D would explode much earlier than 3D. However, the differences between the 2D and 3D convective cascades may be equally in play here and the associated simulations still need to be performed to assess this.

The possible role of initial perturbations as seeds to the growth of convective instability in and around the gain region has been a subject of recent focus. Clearly, allowing grid noise, truncation error, or other numerical noise to initiate linear growth to the non-linear phase, or imposing artificial initial perturbations, is less than satisfactory. This is particularly true given that seeds have a finite time to grow after accreting through the shock before leaving the unstable region and given that the time to instability and explosion is germane to which phase of the neutrino light curve is driving explosion. Inaugurating the non-linear convective phase early and maintaining it until explosion may be important, and the magnitude and timing of convective seeding is therefore of interest. Though we have deferred until later a more comprehensive study of this subject, we tested the effect of adding to the progenitor velocity perturbations whose magnitude was informed by mixing-length theory (Müller et al. 2016). In fact, we allowed the maximum perturbation speed to be 1000 $\mathrm{km\,s^{-1}}$, with a Mach number as high as $\sim$0.12, which is a bit larger than found in 1D progenitor models (cf. Woosley and Heger 2007). Indeed, we found that large imposed perturbations of this sort could enable explosion, though the magnitude of the imposed perturbation seems large. Clearly, it will be important to determine their character, and the many 3D-progenitor studies now in process promise to do just that.

Many-body corrections to scattering rates, inelastic scattering, decreases in nucleon-nucleon bremsstrahlung rates and in electron capture rates on heavy nuclei, and initial seed perturbations all boost "explodability" and shorten the time to explosion. In fact, these effects add synergistically and non-linearly to aide explosion, despite the fact that they individually amount to effects at the $\sim$10–30% level. This is due to the proximity of multi-D models to criticality, and is not seen in 1D simulations.

Furthermore, the turbulence behind the shock that has been shown to aid explosion in the realistic multi-D context naturally introduces indeterminacy in detail. Even the same progenitor star, but with different random seed perturbations and rotational structures at collapse, should yield a range of explosion energies, nucleosynthesis, ^{56}Ni yields, pulsar kicks, and explosion morphologies. Therefore, it is expected that Nature provides distribution functions, and not one-to-one maps, in all signatures of explosion. In the long run, theory will need to come to grips with this, but in the short run one should not expect that in the chaotic context of turbulent convection and pre-collapse structures the best models will correspond in detail. This is the physical and natural consequence of turbulence and chaos, and will be paralleled in comparison verification studies.

The next stage is to explore the same issues in three dimensions, and one expects there to be important differences (Takiwaki et al. 2014; Lentz et al. 2015; Müller 2015; Müller et al. 2017). It is only after performing such simulations, and their subsequent verification, that a robust resolution to the core-collapse supernova problem can be claimed. Nevertheless, there has been significant progress of late in unraveling this central mystery in astrophysics, in demonstrating the viability of the neutrino-driven mechanism of explosion, and in illuminating its component physics.

Acknowledgements The authors acknowledge the help of Evan O'Connor with the Lattimer-Swesty equation of state and of Todd Thompson, who was instrumental in developing the inelastic scattering tables and schema. In addition, they thank Chuck Horowitz for early conversations concerning the neutrino response in nuclear matter at low densities, his insights into neutrino-matter interaction physics, and for an advanced look at his recent paper on these topics. Finally, they would like to thank Yukiya Saito and Junichiro Iwasawa for help scrutinizing the various progenitor model sets in the literature and Sean Couch for stimulating conversations on a variety of core-collapse topics. Support was provided by the Max-Planck/Princeton Center (MPPC) for Plasma Physics (NSF PHY-1144374) and NSF grant AST-1714267. The authors employed computational resources provided by the TIGRESS high performance computer center at Princeton University, which is jointly supported by the Princeton Institute for Computational Science and Engineering (PICSciE) and the Princeton University Office of Information Technology, and by the National Energy Research Scientific Computing Center (NERSC), which is supported by the Office of Science of the US Department of Energy (DOE) under contract DE-AC03-76SF00098. The authors express their gratitude to Ted Barnes of the DOE Office of Nuclear Physics for facilitating their use of NERSC. This work was originally part of the "Three Dimensional Modeling of Core-Collapse Supernovae" PRAC allocation support by the National Science Foundation (award number ACI-1440032) and was part of the Blue Waters sustained-petascale computing project, which is supported by the National Science Foundation (awards OCI-0725070 and ACI-1238993) and the state of Illinois. Blue Waters is a joint effort of the University of Illinois at Urbana-Champaign and its National Center for Supercomputing Applications. This paper has been assigned a LANL preprint # LA-UR-16-28849.

References

E. Abdikamalov, A. Zhaksylykov, D. Radice, S. Berdibek, Mon. Not. R. Astron. Soc. **461**, 3864 (2016)
Z. Ahmed et al., Phys. Rev. Lett. **108**, 102001 (2012)
S. Banik, M. Hempel, D. Bandyopadhyay, Astrophys. J. Suppl. Ser. **214**, 22–37 (2014). arXiv:1404.6173
H. Bethe, J.R. Wilson, Astrophys. J. **295**, 14 (1985)
T.D. Brandt, A. Burrows, C.D. Ott, E. Livne, Astrophys. J. **728**, 8 (2011)
S.W. Bruenn, Astrophys. J. Suppl. Ser. **58**, 771 (1985)
S.W. Bruenn, T. Dineva, Astrophys. J. Lett. **458**, L71 (1996)
S.W. Bruenn, A. Mezzacappa, W.R. Hix et al., Astrophys. J. Lett. **767**, L6 (2013)
S.W. Bruenn, E.J. Lentz, W.R. Hix et al., Astrophys. J. **818**, 123 (2016)
R. Buras, M. Rampp, H.-T. Janka, K. Kifonidis, Phys. Rev. Lett. **90**, 241101 (2003)
R. Buras, H.-T. Janka, M. Rampp, K. Kifonidis, Astron. Astrophys. **457**, 281 (2006)
A. Burrows, Rev. Mod. Phys. **85**, 245 (2013)
A. Burrows, J. Goshy, Astrophys. J. Lett. **416**, L75 (1993)
A. Burrows, J.M. Lattimer, Astrophys. J. **270**, 735 (1983)
A. Burrows, R.F. Sawyer, Phys. Rev. C **58**, 554 (1998)
A. Burrows, R.F. Sawyer, Phys. Rev. C **59**, 510 (1999)
A. Burrows, J. Hayes, B.A. Fryxell, Astrophys. J. **450**, 830 (1995)
A. Burrows, S. Reddy, T.A. Thompson, Nucl. Phys. A **777**, 356 (2006a)
A. Burrows, E. Livne, L. Dessart, C.D. Ott, J. Murphy, Astrophys. J. **640**, 878 (2006b)
A. Burrows, L. Dessart, E. Livne, C.D. Ott, J. Murphy, Astrophys. J. **655**, 416 (2007a)
A. Burrows, L. Dessart, E. Livne, The multi-dimensional character and mechanisms of core-collapse supernovae, in *The Proceedings of the Conference "SUPERNOVA 1987A: 20 YEARS AFTER; Supernovae and Gamma-Ray Bursters"*. AIP Proceedings Series, vol. 937 (2007b), pp. 370–380. Held in Aspen, CO, February 19–23
A. Burrows, L. Dessart, E. Livne, C.D. Ott, J. Murphy, Astrophys. J. **664**, 416 (2007c)
A. Burrows, J.C. Dolence, J. Murphy, Astrophys. J. **759**, 5 (2012)
E. Chatzopoulos, S.M. Couch, W.D. Arnett, F.X. Timmes, Astrophys. J. **822**, 61 (2016)
S.M. Couch, C.D. Ott, Astrophys. J. Lett. **778**, L7 (2013)
S.M. Couch, C.D. Ott, Astrophys. J. **799**, 5 (2015). arXiv:1408.1399
S.M. Couch, E. Chatzopoulos, W.D. Arnett, F.X. Timmes, Astrophys. J. Lett. **808**, L21 (2015). arXiv: 1503.02199
L. Dessart, A. Burrows, E. Livne, C.D. Ott, Astrophys. J. **645**, 534 (2006)
J.C. Dolence, A. Burrows, J. Murphy, J. Nordhaus, Astrophys. J. **765**, 110 (2013)
J.C. Dolence, A. Burrows, W. Zhang, Astrophys. J. **800**, 10 (2015). arXiv:1403.6115
B. Dubroca, J.L. Feugeas, C. R. Acad. Sci. **329**, 915 (1999)
R.T. Emmering, R.A. Chevalier, Astrophys. J. **345**, 931 (1989)
T. Ertl, H.T. Janka, S.E. Woosley, T. Sukhbold, M. Ugliano, Astrophys. J. **818**, 124 (2015). arXiv:1503.07522
C.-A. Faucher-Giguere, V. Kaspi, Astrophys. J. **643**, 355 (2006)
T. Fischer, Astron. Astrophys. **593**, 103 (2016). arXiv:1608.05004
C.L. Fryer, A. Heger, Astrophys. J. **541**, 1033 (2000)
J. Green, N. Hasan, S. Meinel, M. Engelhardt, S. Krieg, J. Laeuchli, J. Negele, K. Orginos, A. Pochinsky, S. Syritsyn, Phys. Rev. D **95**, 114502 (2017). arXiv:1703.06703
F. Hanke, A. Marek, B. Müller, H.-T. Janka, Astrophys. J. **755**, 138 (2012)
F. Hanke, B. Müller, A. Wongwathanarat, A. Marek, H.-T. Janka, Astrophys. J. **770**, 66 (2013)
S. Hannestad, G. Raffelt, Astrophys. J. **507**, 339 (1998)
C.J. Horowitz, O.L. Caballero, Z. Lin, E. O'Connor, A. Schwenk, Phys. Rev. C **95**, 025801 (2017). arXiv: 1611.05140

W. Iwakami, H. Nagakura, S. Yamada, Astrophys. J. **793**, 5 (2014). arXiv:1404.2646
H.-T. Janka, Annu. Rev. Nucl. Part. Sci. **62**, 407 (2012)
H.-T. Janka, W. Keil, G. Raffelt, D. Seckel, Phys. Rev. Lett. **76**, 2621 (1996)
H.-T. Janka, T. Melson, A. Summa, Annu. Rev. Astron. Astrophys. **66**, 341 (2016). arXiv:1602.05576
A. Juodagalvis, K. Langanke, W.R. Hix, G. Martínez-Pinedo, J. Sampaio, Nucl. Phys. **848**, 454 (2010)
F.S. Kitaura, H.-T. Janka, W. Hillebrandt, Astron. Astrophys. **450**, 345 (2006)
T. Kuroda, T. Takiwaki, K. Kotake, Astrophys. J. Suppl. Ser. **222**, 20 (2016)
J.M. Lattimer, F.D. Swesty, Nucl. Phys. A **535**, 331 (1991)
E.J. Lentz, S.W. Bruenn, W.R. Hix, A. Mezzacappa, O.E.B. Messer, E. Endeve, J.M. Blondin, J.A. Harris, P. Marronetti, K.N. Yakunin, Astrophys. J. Lett. **807**, 31 (2015). arXiv:1505.05110
M. Liebendörfer, S.C. Whitehouse, T. Fischer, Astrophys. J. **698**, 1174 (2009)
A. MacFadyen, S.E. Woosley, Astrophys. J. **524**, 262 (1999)
A. Marek, H.-T. Janka, Astrophys. J. **694**, 664 (2009)
A. Marek, H. Dimmelmeier, H.-T. Janka, E. Müller, R. Buras, Astron. Astrophys. **445**, 273 (2006)
C.A. Meakin, T. Sukhbold, W.D. Arnett, Astrophys. Space Sci. **336**, 123 (2011)
T. Melson, H.-T. Janka, R. Bollig et al., Astrophys. J. Lett. **808**, 42 (2015). arXiv:1504.07631
A. Mezzacappa, S.W. Bruenn, Astrophys. J. **405**, 637 (1993)
B. Müller, Mon. Not. R. Astron. Soc. **453**, 287 (2015). arXiv:1506.05139
B. Müller, Publ. Astron. Soc. Aust. **33**, 048 (2016). arXiv:1608.03274
B. Müller, H.-T. Janka, Mon. Not. R. Astron. Soc. **448**, 2141 (2015). arXiv:1409.4783
E. Müller, M. Steinmetz, Comput. Phys. Commun. **89**, 45 (1995)
B. Müller, H.-T. Janka, A. Heger, Astrophys. J. **761**, 72 (2012a)
B. Müller, H.-T. Janka, A. Marek, Astrophys. J. **756**, 84 (2012b)
B. Müller, M. Viallet, A. Heger, H.-T. Janka, Astrophys. J. **833**, 124 (2016). arXiv:1605.01393
B. Müller, T. Melson, A. Heger, H.-T. Janka, Mon. Not. R. Astron. Soc. **472**, 491 (2017). arXiv:1705.00620
J. Murphy, A. Burrows, Astrophys. J. **688**, 1159 (2008)
J. Murphy, J.C. Dolence, A. Burrows, Astrophys. J. **771**, 52 (2013)
K. Nakamura, T. Kuroda, T. Takiwaki, K. Kotake, Astrophys. J. **793**, 45 (2014). arXiv:1403.7290
K. Nakamura, T. Takiwaki, T. Kuroda, K. Kotake, Publ. Astron. Soc. Jpn. **67**, 107 (2015). arXiv:1406.2415
K. Nomoto, M. Hashimoto, Phys. Rep. **163**, 13 (1988)
A. Noutsos, D.H.F.M. Schnitzeler, E.F. Keane, M. Kramer, S. Johnston, Mon. Not. R. Astron. Soc. **430**, 2281 (2013)
E. O'Connor, C.D. Ott, Astrophys. J. **730**, 70 (2011)
E. O'Connor, C.D. Ott, Astrophys. J. **762**, 126 (2013)
C.D. Ott, A. Burrows, L. Dessart, E. Livne, Astrophys. J. **685**, 1069 (2008)
C.D. Ott, E. Abdikamalov, P. Mösta, R. Haas, S. Drasco, E. O'Connor, C. Reisswig, C.A. Meakin, E. Schnetter, Astrophys. J. **768**, 115 (2013). arXiv:1210.6674
K.-C. Pan, M. Liebendörfer, M. Hempel, F.-K. Thielemann, Astrophys. J. **817**, 72 (2016)
O. Pejcha, T.A. Thompson, Astrophys. J. **801**, 90 (2015)
S.B. Popov, R. Turolla, Astrophys. Space Sci. **341**, 457 (2012)
D. Radice, A. Burrows, D. Vartanyan, M.A. Skinner, J.C. Dolence, Astrophys. J. **850**, 43 (2017). arXiv:1702.03927
M. Rampp, H.-T. Janka, Astron. Astrophys. **396**, 361 (2002)
S. Reddy, M. Prakash, J.M. Lattimer, J.A. Pons, Phys. Rev. C **59**, 2888 (1999)
L.F. Roberts, S. Reddy, G. Shen, Phys. Rev. C **86**, 065803 (2012). arXiv:1205.4066
L.F. Roberts, C.D. Ott, R. Haas, E. O'Connor, P. Diener, E. Schnetter, Astrophys. J. **831**, 98 (2016)
M.A. Skinner, A. Burrows, J.C. Dolence, Astrophys. J. **831**, 81 (2016). arXiv:1512.00113
A.W. Steiner, M. Hempel, T. Fischer, Astrophys. J. **774**, 17 (2013)
T. Sukhbold, S.E. Woosley, Astrophys. J. **783**, 10 (2014)
T. Sukhbold, T. Ertl, S.E. Woosley, J.M. Brown, H.-T. Janka, Astrophys. J. **821**, 38 (2016)
K. Sumiyoshi, T. Takiwaki, H. Matsufuru, S. Yamada, Astrophys. J. Suppl. Ser. **216**, 37 (2015). arXiv:1403.4476
A. Summa, F. Hanke, H.-T. Janka, T. Melson, A. Marek, B. Müller, Astrophys. J. **825**, 6 (2016). arXiv:1511.07871
A. Summa, H.-T. Janka, T. Melson, A. Marek, Astrophys. J. **852**, 28 (2018). arXiv:1708.04154
Y. Suwa, K. Kotake, T. Takiwaki, S.C. Whitehouse, M. Liebendörfer, K. Sato, Publ. Astron. Soc. Jpn. **62**, L49 (2010)
Y. Suwa, S. Yamada, T. Takiwaki, K. Kotake, Astrophys. J. **816**, 16 (2016). arXiv:1406.6414
T. Takiwaki, K. Kotake, Y. Suwa, Astrophys. J. **749**, 98 (2012). arXiv:1108.3989
T. Takiwaki, K. Kotake, Y. Suwa, Astrophys. J. **786**, 83 (2014). arXiv:1308.5755
T. Takiwaki, K. Kotake, Y. Suwa, Mon. Not. R. Astron. Soc. **461**, L112 (2016). arXiv:1602.06759

I. Tamborra, F. Hanke, H.-T. Janka, B. Mueller, G. Raffelt, A. Marek, Astrophys. J. **792**, 96 (2014). arXiv: 1402.5418
T.A. Thompson, A. Burrows, J.E. Horvath, Phys. Rev. C **62**, 035802 (2000)
T.A. Thompson, A. Burrows, P. Pinto, Astrophys. J. **592**, 434 (2003)
D.L. Tubbs, Astrophys. J. **231**, 846 (1979)
N.M.H. Vaytet, E. Audit, B. Dubroca, F. Delahaye, J. Quant. Spectrosc. Radiat. Transf. **112**, 1323 (2011)
D. Vartanyan, A. Burrows, D. Radice, M.A. Skinner, J. Dolence, Astrophys. J. (2018, submitted)
S.E. Woosley, A. Heger, Phys. Rep. **442**, 269 (2007)
S.E. Woosley, A. Heger, T.A. Weaver, Rev. Mod. Phys. **74**, 1015 (2002)
W. Zhang, L. Howell, A. Almgren, A. Burrows, J. Bell, Astrophys. J. Suppl. Ser. **196**, 20 (2011)
W. Zhang, L. Howell, A. Almgren, A. Burrows, J.C. Dolence, J. Bell, Astrophys. J. Suppl. Ser. **204**, 7 (2013)

Space Sci Rev (2018) 214:67
https://doi.org/10.1007/s11214-018-0499-0

Single Degenerate Models for Type Ia Supernovae: Progenitor's Evolution and Nucleosynthesis Yields

Ken'ichi Nomoto[1] · Shing-Chi Leung[1]

Received: 8 March 2018 / Accepted: 28 March 2018 / Published online: 17 April 2018

Abstract We review how the single degenerate models for Type Ia supernovae (SNe Ia) works. In the binary star system of a white dwarf (WD) and its non-degenerate companion star, the WD accretes either hydrogen-rich matter or helium and undergoes hydrogen and helium shell-burning. We summarize how the stability and non-linear behavior of such shell-burning depend on the accretion rate and the WD mass and how the WD blows strong wind. We identify the following evolutionary routes for the accreting WD to trigger a thermonuclear explosion. Typically, the accretion rate is quite high in the early stage and gradually decreases as a result of mass transfer. With decreasing rate, the WD evolves as follows: (1) At a rapid accretion phase, the WD increase its mass by stable H burning and blows a strong wind to keep its moderate radius. The wind is strong enough to strip a part of the companion star's envelope to control the accretion rate and forms circumstellar matter (CSM). If the WD explodes within CSM, it is observed as an "SN Ia-CSM". (X-rays emitted by the WD are absorbed by CSM.) (2) If the WD continues to accrete at a lower rate, the wind stops and an SN Ia is triggered under steady-stable H shell-burning, which is observed as a super-soft X-ray source: "SN Ia-SSXS". (3) If the accretion continues at a still lower rate, H shell-burning becomes unstable and many flashes recur. The WD undergoes recurrent nova (RN) whose mass ejection is smaller than the accreted matter. Then the WD evolves to an "SN Ia-RN". (4) If the companion is a He star (or a He WD), the accretion of He can trigger He and C double detonations at the sub-Chandrasekhar mass or the WD grows to the Chandrasekhar mass while producing a He-wind: "SN Ia-He CSM". (5) If the accreting WD rotates quite rapidly, the WD mass can exceed the Chandrasekhar mass of the spherical WD, which delays the trigger of an SN Ia. After angular momentum is lost from

Supernovae
Edited by Andrei Bykov, Roger Chevalier, John Raymond, Friedrich-Karl Thielemann, Maurizio Falanga and Rudolf von Steiger

✉ K. Nomoto
nomoto@astron.s.u-tokyo.ac.jp

S.-C. Leung
shingchi.leung@ipmu.jp

1 Kavli Institute for the Physics and Mathematics of the Universe (WPI), The University of Tokyo, Kashiwa, Japan

the WD, the (super-Chandra) WD contracts to become a delayed SN Ia. The companion star has become a He WD and CSM has disappeared: "SN Ia-He WD". We update nucleosynthesis yields of the carbon deflagration model W7, delayed detonation model WDD2, and the sub-Chandrasekhar mass model to provide some constraints on the yields (such as Mn) from the comparison with the observations. We note the important metallicity effects on ^{58}Ni and ^{55}Mn.

Keywords Supernova · Progenitor · White dwarf · Nucleosynthesis

1 Introduction

The thermonuclear explosion of a carbon+oxygen (C+O) white dwarf has successfully explained the basic observed features of Type Ia supernovae (SNe Ia) (e.g., Nomoto and Leung 2017; Leung and Nomoto 2017). Both the Chandrasekhar and the sub-Chandrasekhar mass models have been examined (e.g., Livio 2000). However, no clear observational indication exists as to how the white dwarf mass grows until carbon ignition, i.e., whether the white dwarf accretes H/He-rich matter from its binary companion [single-degenerate (SD) scenario] or whether two C+O white dwarfs merge [double-degenerate (DD) scenario] (e.g., Arnett 1969, 1996; Hillebrandt and Niemeyer 2000; Iben and Tutukov 1984; Ilkov and Soker 2012; Maoz et al. 2014; Nomoto 1982a; Nomoto et al. 1994, 1997, 2000a, 2009; Webbink 1984).

Here we focus on the possible evolutionary paths for the accreting white dwarf to increase its mass to the Chandrasekhar mass in the binary systems.

2 Hydrogen Shell-Burning in Accreting White Dwarfs

2.1 Effects of Mass Accretion on White Dwarfs

Isolated white dwarfs are simply cooling stars that eventually end up as invisible frigid stars. The white dwarf in a close binary system evolves differently because the companion star expands and transfers matter over to the white dwarf at a certain stage of its evolution. The mass accretion can *rejuvenate* the cold white dwarf (e.g., Nomoto and Sugimoto 1977), which could lead to a SNe Ia or accretion-induced collapse (AIC) in some cases.

The scenario that possibly brings a close binary system to a SN Ia or AIC is as follows: Initially the close binary system consists of two intermediate mass stars ($M \lesssim 8M_\odot$). As a result of Roche lobe overflow, the primary star of this system becomes a white dwarf composed of carbon and oxygen (C+O). When the secondary star evolves, it begins to transfer hydrogen-rich matter over to the white dwarf.

The mass accretion onto the white dwarf releases gravitational energy at the white dwarf surface. Most of the released energy is radiated away from the shocked region as UV and does not contribute much to heating the white dwarf interior. The continuing accretion compresses the previously accreted matter and releases gravitational energy in the interior. A part of this energy is transported to the surface and radiated away from the surface (*radiative cooling*) but the rest goes into thermal energy of the interior matter (*compressional heating*). Thus the interior temperature of the white dwarf is determined by the competition between compressional heating and radiative cooling; i.e., the white dwarf is hotter if the mass accretion rate $\dot{M}$ is larger, and vice versa (e.g., Nomoto 1982a,b).

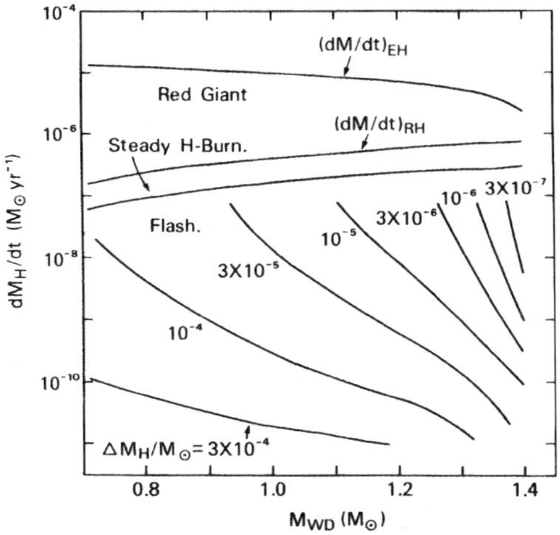

Fig. 1 The properties of accreted hydrogen-rich materials as functions of $M_{\rm WD}$ and $dM_{\rm H}/dt$ (Nomoto 1982a). Hydrogen burning is stable in the region indicated by "Steady H-Burn" between the two lines of $\dot{M}_{\rm stable}$ in Eq. (1) and $\dot{M}_{\rm cr}$ $(= (dM/dt)_{\rm RH})$ in Eq. (2) (Kato et al. 2014). In the region below $\dot{M}_{\rm stable}$, hydrogen shell burning is thermally unstable, and the WD experiences shell flashes. Black solid lines indicate the hydrogen-ignition masses $\Delta M_{\rm H}$, the values of which are shown beside each line. In the region above $(dM/dt)_{\rm RH}$ (and below the Eddington limit $(dM/dt)_{\rm EH}$), the accreted matter is piled up to form a red-giant size envelope as seen in Fig. 2 if no wind is included in the calculation (Nomoto et al. 1979). It has been found that optically thick wind are accelerated in this region, which prevents the formation of a red-giant (Hachisu et al. 1996)

2.2 Hydrogen Shell-Flashes

Hydrogen shell-burning is ignited when the mass of the accumulated hydrogen-rich matter reaches the ignition mass $M_{\rm ig}$ $(= \Delta M_{\rm H})$. When $M_{\rm ig}$ is reached, the compressional heating due to accretion is just balanced with the cooling due to heat conduction (Nomoto et al. 1979; Nomoto 1982a). $M_{\rm ig}$ is presented as contours on the $M_{\rm WD} - \dot{M}$ $(= dM_{\rm H}/dt)$ plane in Fig. 1. For a given $\dot{M}$, $M_{\rm ig}$ is smaller for a larger $M_{\rm WD}$ because of the smaller radius R and thus higher pressure for the same mass of accreted matter (see Eq. (3) below). For a given $M_{\rm WD}$, $M_{\rm ig}$ is smaller for a higher $\dot{M}$ because of the faster compressional heating and thus higher temperature of accreted matter.

The stability of the hydrogen burning shell in the accreting white dwarf is crucial for its evolution. Figure 1 summarizes the properties of hydrogen shell burning (Nomoto 1982a; Nomoto et al. 2007; Kato et al. 2014).

(1) The hydrogen shell burning is unstable to flash in the area below the solid line to show $\dot{M}_{\rm stable}$. This stability line is approximately represented by Kato et al. (2014)

$$\dot{M}_{\rm stable} = 4.17 \times 10^{-7} \left(\frac{M_{\rm WD}}{M_\odot} - 0.53 \right) M_\odot \,{\rm yr}^{-1}. \tag{1}$$

(2) Above the solid line for $\dot{M}_{\rm cr}$ $(= (dM/dt)_{\rm RH})$, the accreted matter is accumulated faster than consumed into He by H-shell burning. This critical accretion rate is represented as Kato et al. (2014)

$$\dot{M}_{\rm cr} = 8.18 \times 10^{-7} \left(\frac{M_{\rm WD}}{M_\odot} - 0.48 \right) M_\odot \,{\rm yr}^{-1}. \tag{2}$$

Fig. 2 Increase in the radius of the accreting white dwarf upon the rapid accretion (Nomoto et al. 1979)

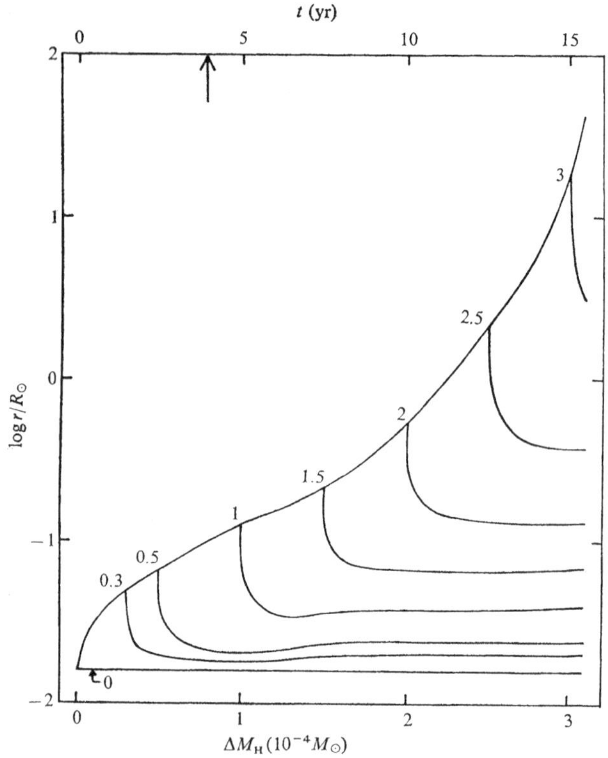

(3) For the region with $\dot{M} > \dot{M}_{cr}$, the accreted matter is piled up to form a red-giant size envelope as seen in Fig. 2 (Nomoto et al. 1979). This could lead to the formation of a common envelope and prevent further mass accretion onto the white dwarf. This problem for has been resolved by the strong optically thick winds (Hachisu et al. 1996, 1999a,b). If the wind is sufficiently strong, the white dwarf radius stays small enough to avoid the formation of a common envelope. Then steady hydrogen burning increases its mass at a rate $\dot{M}_{cr}$ by blowing the extra mass away in a wind.
(4) In the area $\dot{M}_{stable} < \dot{M} < \dot{M}_{cr}$, accreting white dwarfs are thermally stable so that hydrogen burns steadily in the burning shell. Then the white dwarf mass increases at a rate of $\dot{M}$.
(5) For $\dot{M} < \dot{M}_{stable}$, the flash of hydrogen shell burning is stronger (weaker) for lower (higher) $\dot{M}$ and thus for larger (smaller) M_{ig} and larger (smaller) M_{WD}.

The progress and the strength of the flashes are determined by two parameters (P^*, Ω^*),

$$P^* = \frac{G M_{WD} M_{ig}}{4\pi R^4}, \qquad \Omega^* = \frac{GM}{R}, \tag{3}$$

i.e., by the pressure and potential at the burning shell corresponding to completely *flat* configuration (Sugimoto and Fujimoto 1978; Sugimoto et al. 1979). A set of (P^*, Ω^*) can be transformed into (M, M_{ig}) since the radius at the burning shell is well approximated by the WD radius R which is smaller for larger M_{WD}.

Progress of the shell flash can be treated semi-analytically. Initially the temperature at the burning shell increases along $P = P^*$. As the nuclear energy is released, the pressure

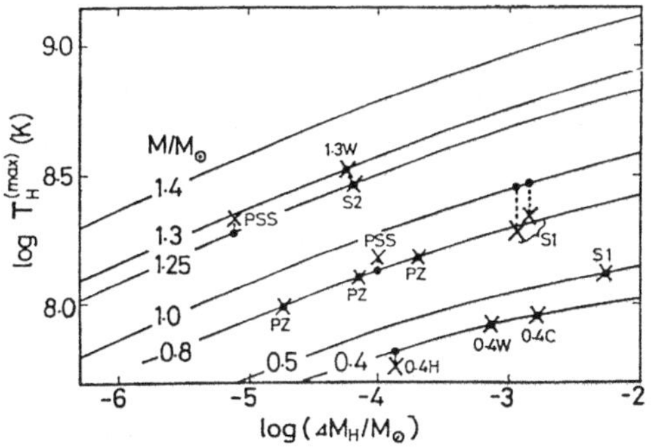

Fig. 3 Maximum temperature attained during the hydrogen shell flash as a function of the accreted mass $\Delta M_{\rm H}$ and the white dwarf mass M. Analytical values (solid curves) are compared with those obtained by hydrodynamical calculations (X-mark) (see Sugimoto et al. 1979, for details)

decreases as a result of expansion, which is described by $P = fP^*$. Here the flatness parameter f is unity for plane-parallel configuration and $f < 1$ for more spherical configuration. This is expressed as

$$\frac{1}{f(V,N)} = \sum_{k=0}^{\infty} b_k, \quad b_0 = 1, \quad b_k = b_{k-1}\frac{k+3}{N+k+1}\frac{N+1}{V}, \quad V = r/H_{\rm p}, \tag{4}$$

where N and $H_{\rm p}$ denote the polytropic index for the convective envelope and the scale height of pressure, respectively (Sugimoto and Fujimoto 1978). As the specific entropy s in the hydrogen-burning shell increases, f decreases because of increasing $H_{\rm p}$, i.e., expansion of the accreted envelope. This corresponds to the change in the configuration of the burning shell from plane parallel to spherical (Sugimoto and Nomoto 1980).

Then the temperature reaches its maximum $T_{\rm H}^{\rm max}$, which is higher for higher P^* and thus for higher M (smaller r) and larger $\Delta M_{\rm H}$ (Eq. (3)). Such a relation between $T_{\rm H}^{\rm max}$ and $\Delta M_{\rm H}$ for several M obtained from Eq. (4) is shown in Fig. 3 (Sugimoto et al. 1979). Results of some hydrodynamical calculations (X-mark) are in excellent agreement with the corresponding analytical predictions (filled circles). (Some discrepancies are likely to be caused by a coarse zoning in numerical calculations.)

The results in Figs. 1 and 3 show that generally smaller M and higher $\dot{M}$ lead to a weaker flash because of the lower pressure at the flashing shell. For $\dot{M} \sim 2 \times 10^{-7}$–$10^{-8}$ $M_\odot$ yr^{-1}, the flash is so weak that the ejected mass is smaller than the mass which hydrogen burning converts into helium. Then the mass of the He layer can increase.

For $\dot{M} \sim 10^{-9}$–$10^{-8} M_\odot$ yr^{-1}, the H-flash is stronger so that the mass of the He layer grows but at much lower rate.

For slow accretion ($\dot{M} \lesssim 1 \times 10^{-9}$ $M_\odot$ yr^{-1}), hydrogen shell flash is strong enough to grow into a *nova* explosion, which leads to the ejection of most of the accreted matter from the white dwarf (e.g., Nariai et al. 1980). Moreover, a part of the original white dwarf matter is dredged up and lost in the outburst wind. Then $M_{\rm WD}$ decreases after the nova outburst. For these cases, the white dwarf does not become a *supernova* since its mass hardly grows. However, if the white dwarfs are close to the Chandrasekhar mass, novae could grow into AIC of SN Ia because the ejected mass from nova explosion is found to be significantly smaller than the accreted mass (Starrfield et al. 1991).

In these cases, the hydrogen flash recurs, and the recurrence period is proportional to $M_{\rm ig}/\dot{M}$, which is shorter for higher $\dot{M}$. If the recurrence period is short, the flashes are observed as *recurrent novae*, which occurs in the upper-right region of Fig. 1.

3 Evolution of Helium Accreting White Dwarfs

3.1 Accretion of Helium

As discussed above, a thin He layer is produced and grows by H-burning for $\dot{M} > 10^{-9} M_{\odot}$ yr^{-1}. The He layer grows also by direct transfer of helium if the companion is a He star (e.g., Iben et al. 1987). Further evolution and final fates depend on the accretion rate of He and the mass of the C+O core $M_{\rm CO}$ as summarized in Fig. 4.

When a certain mass $\Delta M_{\rm He}$ is accumulated, He shell-burning is ignited; the solid lines in Fig. 5 show $\Delta M_{\rm He}$ as a function of $\dot{M}$ and the white dwarf masses M (Kawai et al. 1987). It is seen that $\Delta M_{\rm He}$ is larger for the slower mass-accumulation rate of the He layer $\dot{M}_{\rm He}$.

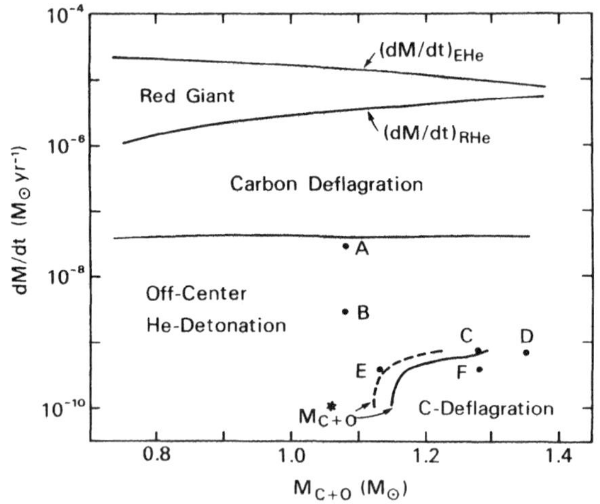

Fig. 4 The properties of He accreting white dwarf as functions of $M_{\rm WD}$ and dM/dt (Nomoto 1982a). In the region above $(dM/dt)_{\rm RHe}$ (and below the Eddington limit $(dM/dt)_{\rm EHe}$), the accreted He envelope is extended to a red-giant size. In the region below $\dot{M}_{\rm RHe}$, He shell burning is thermally unstable, and the WD experiences shell flashes

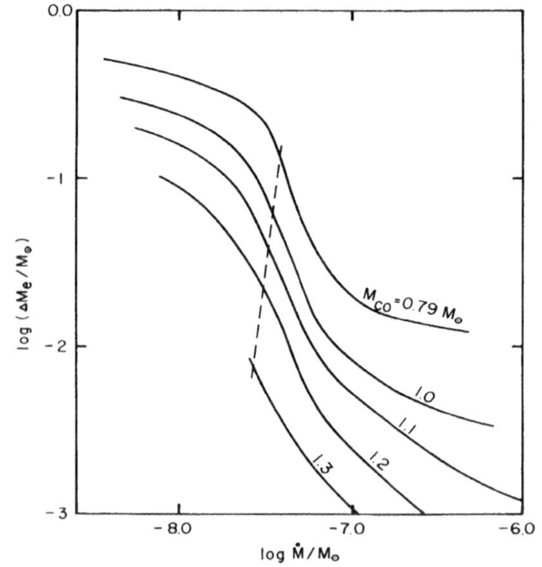

Fig. 5 The mass of accreted He envelope, $\Delta M_{\rm e}$, at the He ignition as a function of $\dot{M}$ and the mass of underlying C+O core, $M_{\rm CO}$ (Kawai et al. 1987)

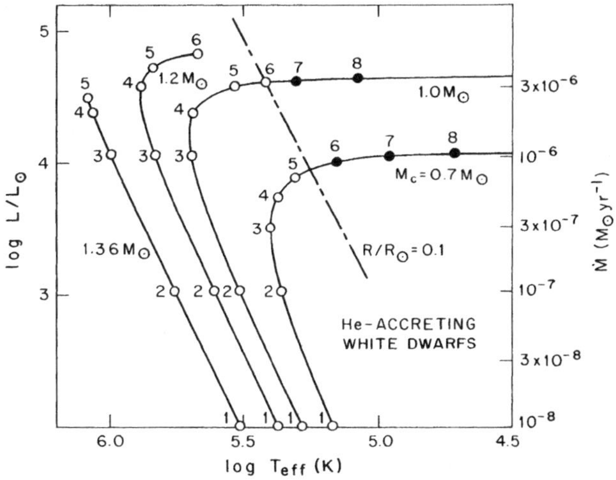

Fig. 6 Locations of He-accreting steady state white dwarf models in the H-R diagram. The accretion rate is indicated on the right vertical axis. The solid lines connect models with the same C+O core mass. The thermally unstable and stable models are indicated by open and filled circles, respectively (Kawai et al. 1988)

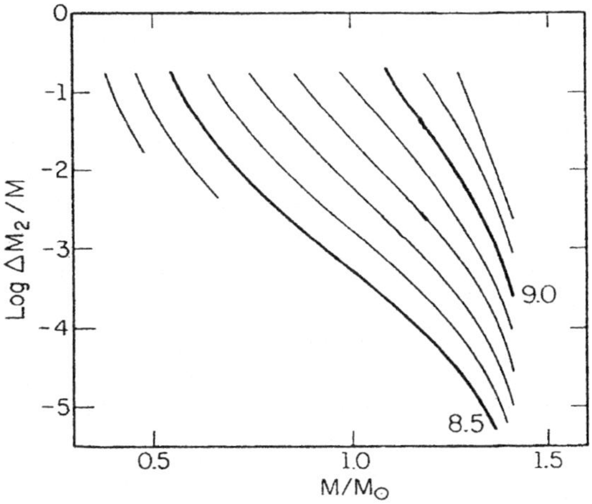

Fig. 7 Maximum temperature log T (K) attained during the He shell flash as a function of the accreted mass ΔM_2 and the white dwarf mass M (Fujimoto and Sugimoto 1982)

If $(dM/dt)_{\rm EHe} \gtrsim \dot{M} \gtrsim (dM/dt)_{\rm RHe}$ as in Fig. 4, the accreted He envelope is extended to a red-giant size as seen in Fig. 6. In the region below $\dot{M}_{\rm RHe}$, He shell burning is thermally unstable, and the WD experiences shell flashes as discussed in the next subsection.

If the accretion of He is as slow as $\dot{M} \lesssim 1 \times 10^{-9}\ M_\odot\ {\rm yr}^{-1}$, the accreted material is too cold to ignite He burning, so that the white dwarf mass increases. An exception is the case with $M_{\rm CO} \lesssim 1.1\ M_\odot$ where pycnonuclear He burning is ignited at high enough densities (Nomoto 1982a,b).

3.2 Helium Shell-Flashes and Detonation

In the early stages of He shell-burning, the He envelope is electron-degenerate and geometrically almost flat. Because of the almost constant pressure at the bottom of the He-burning shell (Eq. (3)), the temperature there increases and makes a He flash. Heated by He burning, the He envelope gradually expands, which decreases the pressure. Then, the temperature attains its maximum and starts decreasing.

The maximum temperature attained during the He shell-flash depends on ΔM_{He} and M (Fig. 7: Fujimoto and Sugimoto 1982) as discussed for hydrogen in Sect. 2.2. The maximum temperature is higher for more massive WD and more massive envelope because of higher pressure. The strength depends on the He envelope mass M_{env}, thus depending mainly on $\dot{M}$ as follows.

3.2.1 He Detonation

For $\dot{M}_{\mathrm{det}} \gtrsim \dot{M} \gtrsim 10^{-9}\ M_{\odot}\,\mathrm{yr}^{-1}$, the He shell-flash is strong enough to initiate an off-center He detonation, which prevents the white dwarf mass from growing (e.g., Nomoto 1982b; Woosley et al. 1986). Here we adopt $\dot{M}_{\mathrm{det}} \sim 1 \times 10^{-8}\ M_{\odot}\,\mathrm{yr}^{-1}$, since the $^{14}\mathrm{N}(e^-, \nu)^{14}\mathrm{C}(\alpha, \gamma)^{18}\mathrm{O}$ (NCO) reaction ignites weak He flashes (Hashimoto et al. 1986; Limongi and Tornambe 1991) if the mass fraction of CNO elements in the accreting material exceeds 0.005. For smaller CNO abundances, the NCO reaction is not effective and thus $\dot{M}_{\mathrm{det}} \sim 4 \times 10^{-8}\ M_{\odot}\,\mathrm{yr}^{-1}$ (Nomoto 1982a).

Two dimensional hydrodynamical simulations after the initiation of He detonation have been performed by several groups (e.g., Livne and Glasner 1991; Nomoto and Leung 2017, and references therein). The outcome is the off-center He-detonation, which develops into double detonation supernovae.

3.2.2 Recurring He Flashes

For intermediate accretion rates ($3 \times 10^{-6}\ M_{\odot}\,\mathrm{yr}^{-1} \gtrsim \dot{M} \gtrsim 1 \times 10^{-8}\ M_{\odot}\,\mathrm{yr}^{-1}$), He shell-flashes are of moderate strength, thereby recurring many times to increase the white dwarf mass (Taam 1980; Fujimoto and Sugimoto 1982; Nomoto 1982b; Kato et al. 2017). When the white dwarf mass becomes close to the Chandrasekhar mass, either thermonuclear explosion or collapse would occur.

Nucleosynthesis in such He shell-flashes has been calculated for various set of (M_{WD}, M_{env}) (Nomoto et al. 2013a; Shen and Bildsten 2007). For higher maximum temperatures, heavier elements, such as $^{28}\mathrm{Si}$ and $^{32}\mathrm{S}$, are synthesized. However, the maximum temperature is not high enough to produce $^{40}\mathrm{Ca}$. After the peak, some amount of He remains unburned in the flash and burns into C+O during the stable He shell-burning. In this way, it is possible that an interesting amount of intermediate mass elements, including Si and S, already exist in the unburned C+O layer at $M_r \geq 1.2 M_{\odot}$.

3.3 Neutron Star Formation

For the C+O white dwarfs, whether they explode or collapse depends not only on $\dot{M}$ but also on the initial mass M_{CO} as summarized in Fig. 8. For $M_{\mathrm{CO}} < 1.2\ M_{\odot}$, substantial heat inflow from the surface layer into the central region ignites carbon at relatively low central density ($\rho_{\mathrm{c}} \sim 3 \times 10^9\ \mathrm{g\,cm^{-3}}$) (Nomoto et al. 1984).

On the other hand, if the white dwarf is sufficiently massive and cold at the onset of accretion, the central region is compressed only adiabatically, thereby being cold (and solid) when carbon is ignited in the center of density as high as $10^{10}\ \mathrm{g\,cm^{-3}}$. C-burning at such high densities is likely to result in a collapse due to rapid electron capture in NSE (nuclear statistical equilibrium).

Neutron star formation is a possible outcome from high C-accretion rates as discussed in the next section on merging C+O WD (as indicated in Fig. 8).

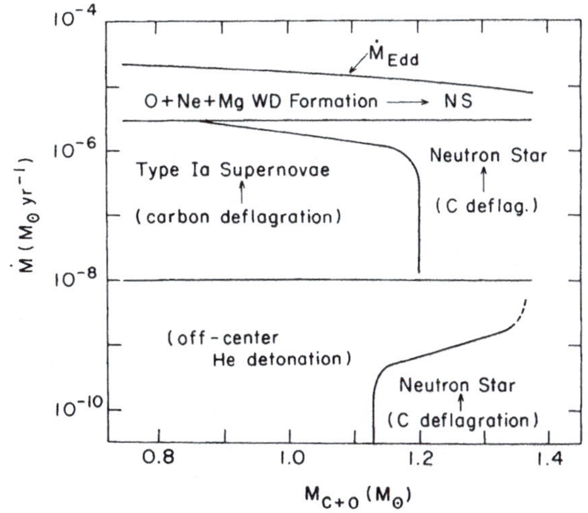

Fig. 8 The final fate of C+O white dwarfs expected for their initial mass and accretion rate $\dot{M}$ of C+O materials (Nomoto and Kondo 1991)

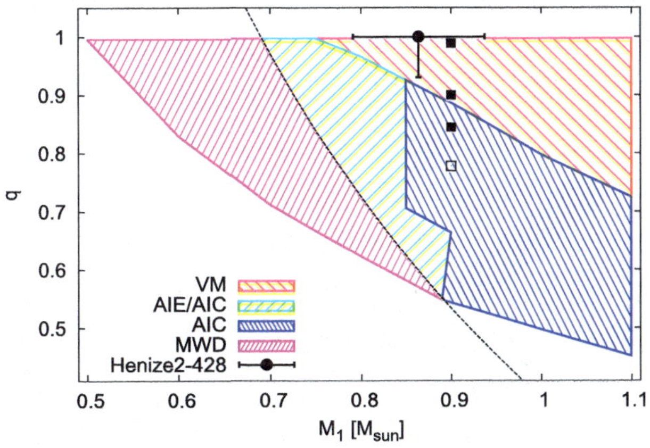

Fig. 9 Final outcomes of our merger simulations in the mass ratio versus total mass diagram, i.e., violent merger, AIC (accretion-induced collapse), and AIE (accretion-induced explosion)/AIC (Sato et al. 2017). The black squares denote the models calculated in Pakmor et al. (2011). Filled squares indicate the models which satisfied the detonation condition of Seitenzahl et al. (2009), while the open square indicates the model that does not. Henize2-428 is the observed double white dwarfs (Sato et al. 2017)

4 Merging of Double C+O White Dwarfs

Merging of double C+O white dwarfs is estimated to take place as frequently as SNe Ia (Iben and Tutukov 1984; Webbink 1984). Sato et al. (2017) performed SPH simulations of merging and summarized the final outcomes of mergers in the primary white dwarf mass (M_1) versus the mass ratio of the two white dwarfs ($q = M_2/M_1$) in Fig. 9 as follows:
(1) Violent Merger (VM): If the temperature of the shock-heated merged region becomes high enough, i.e., the timescale of the temperature rise due to C-burning becomes shorter than the dynamical timescale, C-detonation is generated which then triggers the central C-detonation at a relatively low central density as determined by M_1.

In Fig. 9 (Sato et al. 2017), the black squares denote the models calculated in Pakmor et al. (2011). Filled squares indicate the models which satisfied the detonation condition of Seitenzahl et al. (2009), while the open square indicates the model that does not.
(2) Accretion-induced collapse (AIC): If the temperature at the merged region leads to a stable off-center C-burning, C-flame propagates through the center to convert the C+O white dwarf into the O+Ne+Mg white dwarf, which would eventually collapse due to electron capture.

Such a C-ignition and subsequent flame propagation have been approximately simulated by spherical models. After the smaller mass white dwarf fills its Roche lobe, mass transfer of carbon onto the more massive white dwarf would be very rapid, which ignites off-center C-burning if $\dot{M} \gtrsim 2.7 \times 10^{-6}\ M_\odot\ \mathrm{yr}^{-1}$ (Nomoto and Iben 1985). Subsequent flame propagation coverts the entire C+O into O+Ne+Mg (Saio and Nomoto 2004), which leads to collapse as induced by electron capture.
(3) Accretion-induced explosion (AIE): If the temperature after merging is too low to ignite carbon, the C+O white dwarf would increase its mass toward the Chandrasekhar mass (unless the accretion ignites off-center C burning) and leads to an SN Ia (Yoon et al. 2007).
(4) C+O white dwarf (MWD): If the total mass $M_{\mathrm{tot}} = M_1 + M_2$ of the two white dwarfs does not exceed the Chandrasekhar mass, the above cases (2) and (3) could not occur. Instead, the merger results in the formation of a single C+O white dwarf.

5 Four Cases of Pre-Explosion Configurations of Type Ia Supernovae

Based on the above properties of accretion-induced hydrogen shell burning, the binary system in the SD scenario evolves through stages (a)–(d) below (also shown in Figs. 10(a)–(d)) (Hachisu et al. 2008a).

The more massive (primary) component of a binary evolves to a red giant star (with a helium core) or an AGB star (with a C+O core) and fills its Roche lobe. Mass transfer from the primary to the secondary begins and a common envelope is formed. After the first common envelope evolution, the separation shrinks and the primary component becomes a helium star or a C+O WD. The helium star evolves to a C+O WD after a large part of helium is exhausted by core-helium-burning. We eventually have a close pair of a C+O WD and a main-sequence (MS) star (Fig. 10(a)).

Further evolution of the system depends on the binary parameters. Depending on at which stage SNe Ia are triggered, the SD scenario predicts the following four variations of SNe Ia.

5.1 SNe Ia—Circumstellar Matter (CSM)

After the secondary evolves to fill its Roche lobe, the mass transfer to the WD begins. This mass transfer occurs on a thermal timescale because the secondary mass is more massive than the WD. The mass transfer rate exceeds $\dot{M}_{\mathrm{cr}}$ for the optically thick wind to blow from the WD (Hachisu et al. 1996, 1999a,b) (Fig. 10(b)).

Optically thick winds from the WD collide with the secondary surface and strip off its surface layer. This mass-stripping attenuates the rate of mass transfer from the secondary to the WD, thus preventing the formation of a common envelope for a more massive secondary in the case with than in the case without this effect. Thus the mass-stripping effect widens the donor mass range of SN Ia progenitors (Fig. 10(c)).

Such stripped-off matter forms a massive circumstellar torus on the orbital plane, which may be gradually expanding with an outward velocity of $\sim 10 - 100$ km s^{-1} (Fig. 10(d)), because the escape velocity from the secondary surface to L3 point is $v_{\mathrm{esc}} \sim 100$ km s^{-1}. Subsequent interaction between the fast wind from the WD and the very slowly expanding circumbinary torus forms an hourglass structure (Fig. 10(c)–(d)). When we observe the SN Ia from a high inclination angle such as denoted by "line of sight," circumstellar matter can be detected as absorption lines like in SN 2006X.

This scenario predicts the presence of several types of circumstellar matter around the binary system, which are characterized various wind velocities v_{w}: (1) white dwarf winds

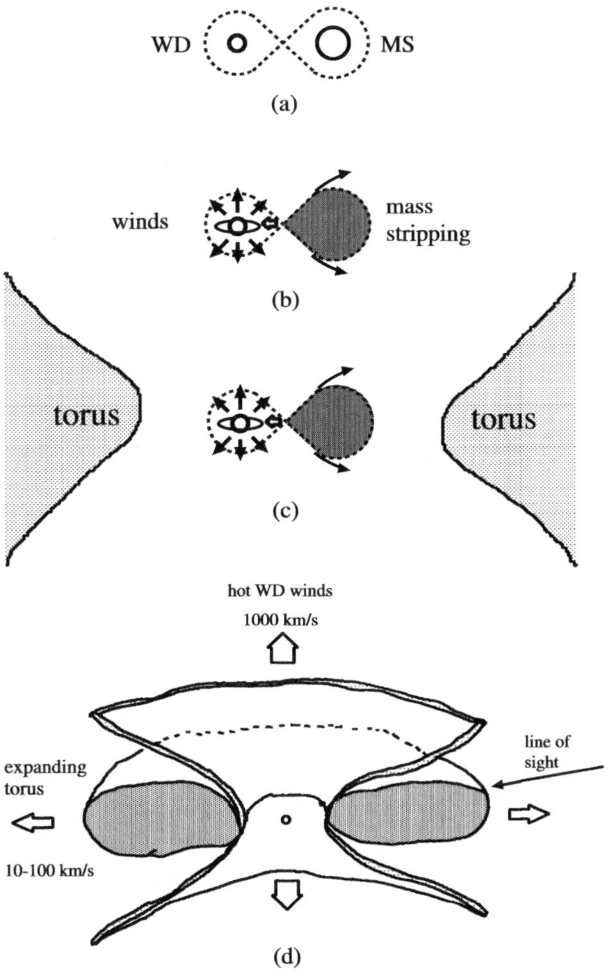

Fig. 10 A schematic configuration of a binary evolution including mass-stripping effect (Hachisu et al. 2008a). (**a**) Here we start a pair of a C+O WD and a more massive main-sequence (MS) star with a separation of several to a few tens of solar radii. (**b**) When the secondary evolves to fill its Roche lobe, mass transfer onto the WD begins. The mass transfer rate exceeds a critical rate M_{cr} for optically thick winds. Strong winds blow from the WD. (**c**) The hot wind from the WD hits the secondary and strips off its surface. (**d**) Such stripped-off material forms a massive circumstellar disk or torus and it gradually expands with an outward velocity of ~ 10–$100\ \mathrm{km\,s^{-1}}$. The interaction between the WD wind and the circumstellar torus forms an hourglass structure. The WD mass increases up to $M_{\mathrm{Ia}} = 1.38\ M_\odot$ and explodes as an SN Ia

with such high velocities as $v_{\mathrm{w}} \sim 1000\ \mathrm{km\,s^{-1}}$, (2) slow dense matter stripped off the companion star by the white dwarf wind, (3) slow wind matter ejected from a red-giant, and (4) moderate wind velocities blown from the main-sequence star.

The above features are supported by observations of the presence of circumstellar matter in some SNe Ia (Patat et al. 2007; Sternberg et al. 2011; Foley et al. 2012), and the detection of H in circumstellar-interaction type supernovae (Ia/IIn) such as SN 2002ic (Hamuy et al. 2003). SN 2002ic shows the typical spectral features of SNe Ia near maximum light, but also apparent hydrogen features that have been absent in ordinary SNe Ia. Its light curve has been reproduced by the model of interaction between the SN Ia ejecta and the H-rich circumstellar medium (Fig. 11) (Nomoto et al. 2005).

5.2 SNe Ia—Supersoft X-Ray Sources

When the mass transfer rate decreases to the following range: $\dot{M}_{\mathrm{stable}} < \dot{M} < \dot{M}_{\mathrm{cr}}$, optically thick winds stop, and the WDs undergo steady H-burning. The WDs are observed as supersoft X-ray sources (SSXSs) until the SN Ia explosion. The stripped-off material forms

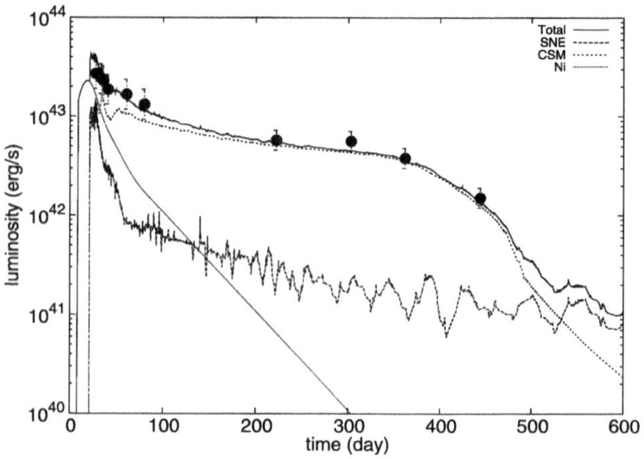

Fig. 11 The observed light curve of SN Ia 2002ic (filled circles) and the calculated light curve with circumstellar interaction (Nomoto et al. 2005)

circumstellar matter (CSM) but it has been dispersed too far to be detected immediately after the SN Ia explosion.

5.3 SNe Ia—Recurrent Novae

When the mass transfer rate from the secondary further decreases below the lowest rate of steady hydrogen burning, i.e., $\dot{M}_{\rm transfer} < \dot{M}_{\rm stable}$, hydrogen shell burning is unstable to trigger a mild flashes, which recur many times in a short period as a recurrent nova (RN) (e.g., Nomoto et al. 2007). Its recurrent period is as short as ~ 1 yr, which can be realized for high M and high $\dot{M}$ as discussed in Sect. 2.2. These flashes burn a large enough fraction of accreted hydrogen to increase M to SNe Ia.

Observationally, PTF11kx (Dilday et al. 2012) provides strong evidences that the accreting white dwarf was a recurrent nova and the companion star was a red supergiant.

5.4 SNe Ia—He White Dwarf Remnants

In the rotating white dwarf scenario, which will be discussed in a later section (e.g., Benvenuto et al. 2015), ignition of central carbon burning is delayed in some cases due to the larger Chandrasekhar mass of the rotating white dwarfs than non-rotating white dwarfs. This delay time after the end of accretion up to the SN Ia explosion depends on the time scale of angular momentum loss from the C+O white dwarfs, and could be long enough for the companion star to evolve into a He white dwarf and for circumstellar materials to disperse. For such a delayed SN Ia, it would be difficult to detect a companion star or circumstellar matter.

5.5 Companion Stars in the SD Scenario

In SD scenario, SNe Ia can occur for a wide range of $\dot{M}$. The progenitor white dwarfs can grow their masses to the Chandrasekhar mass by accreting hydrogen-rich matter at a rate as high as $\dot{M} \gtrsim 10^{-7} - 10^{-6} M_\odot\,{\rm yr}^{-1}$ (e.g., Hachisu et al. 1996, 1999a,b; Li and van den Heuvel 1997; Langer et al. 2000; Han and Podsiadlowski 2004; Nomoto et al. 2000b).

Two types of binary systems can provide such high accretion rates, i.e., (1) a white dwarf and a lobe-filling, more massive (up to $\sim 7 M_\odot$), slightly evolved main-sequence

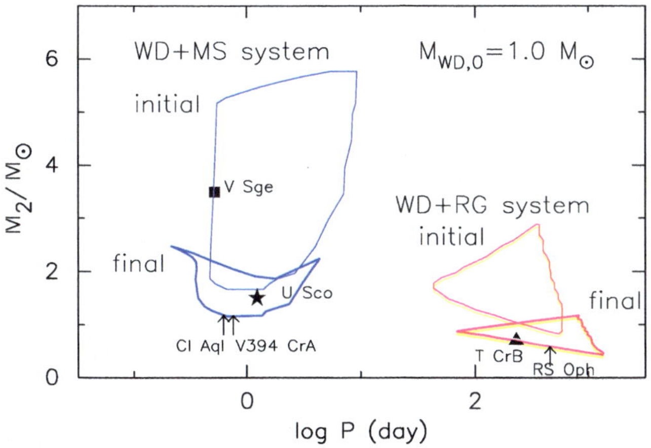

Fig. 12 The regions that produce SNe Ia are plotted in the $\log P - M_2$ (orbital period–secondary mass) plane for the (WD+MS) system (*left*) and the (WD+RG) system (*right*) (Hachisu et al. 2008b). Currently known positions of the recurrent novae and supersoft X-ray sources are indicated by a star mark ($\star$) for U Sco, a triangle for T CrB, a square for V Sge, but by arrows for the other three recurrent novae, V394 CrA, CI Aql, and RS Oph. Two subclasses of the recurrent novae, the U Sco type and the RS Oph type, correspond to the WD + MS channel and the WD + RG channel of SNe Ia, respectively

or sub-giant star (WD+MS), and (2) a white dwarf and a lobe-filling, less massive (typically $\sim 1M_\odot$), red-giant (WD+RG) (Hachisu et al. 1999a,b). Figure 12 shows these two regions of (WD+MS) and (WD+RG) in the $\log P - M_2$ (orbital period–secondary mass) plane (Hachisu et al. 2008b).

Here the metallicity of $Z = 0.02$ and the initial white dwarf mass of $M_{\rm WD,0} = 1.0\,M_\odot$ are assumed. The initial system inside the region encircled by a thin solid line (labeled "initial") increases its WD mass up to the critical mass ($M_{\rm Ia} = 1.38M_\odot$) for the SN Ia explosion, the regions of which are encircled by a thick solid line (labeled "final").

Note that the "initial" region of WD + MS systems extends up to such a massive ($M_{2,0} \sim$ 5–6 $M_\odot$) secondary, which consists of a very young population of SNe Ia with such a short delay time as $t \lesssim 0.1$ Gyr. On the other hand, the WD + RG systems with a less massive RG ($M_{2,0} \sim 0.9 - 1.0\,M_\odot$) consist of a very old population of SNe Ia of $t \gtrsim 10$ Gyr.

The delay time distribution (DTD) of SNe Ia on the basis of the above SD model (Fig. 12) has a featureless power law in good agreement with the observation (Hachisu et al. 2008b). This is because the mass of the secondary star of the SN Ia system ranges from $M_{2,0} \sim 0.9$ to $6\,M_\odot$ due to the effects of the WD winds and the mass stripping. In our model, moreover, the number ratio of SNe Ia between the WD + MS component and the WD + RG component is $r_{\rm MS/RG} = 1.4$. Such almost equal contributions of the two components help to yield a featureless power law.

5.6 Rotating White Dwarf

5.6.1 Uniform Rotation and Delayed Carbon Ignition

In the above sections, some observations that support the SD scenario are given. However, there has been no direct indication of the presence of companions, e.g., the lack of companion stars in images of SN 2011fe (Li et al. 2011) and some Type Ia supernova rem-

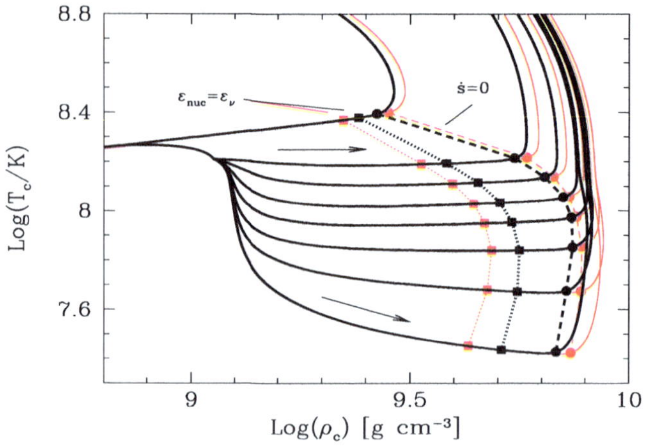

Fig. 13 The evolutionary tracks of the center of the WD up to the onset of the hydrodynamical explosion (Benvenuto et al. 2015). $\varepsilon_{\rm nuc} = \varepsilon_\nu$ indicates the conditions at which neutrino losses equal nuclear energy release while $\dot{S} = 0$ show the stages at which central entropy per baryon begins to increase. Thick black and thin red lines correspond to the treatments of screening given by Kitamura (2000) and Potekhin and Chabrier (2012), respectively. Arrows indicate the sense of the evolution

nants (Schaefer and Pagnotta 2012). The rotating white dwarf scenario solves this missing-companion problem (Justham 2011; Di Stefano et al. 2011; Hachisu et al. 2012a; Benvenuto et al. 2015).

The rotating WD evolves as follows (Hachisu et al. 2012a; Benvenuto et al. 2015; Piersanti et al. 2003).

(1) For certain ranges of binary parameters, the accretion rate ($\dot{M}$) always exceeds $10^{-7} M_\odot\,{\rm yr}^{-1}$ so that the WD increases its mass until it undergoes "prompt" carbon ignition. The mass of the uniformly rotating WD at the carbon ignition, $M_{\rm ig}^{\rm R}$, is larger for smaller $\dot{M}$. For $\dot{M} = 10^{-7}\ M_\odot\,{\rm yr}^{-1}$, $M_{\rm ig}^{\rm R} = 1.43\ M_\odot$, which is the largest mass because nova-like hydrogen flashes prevent the WD mass from growing for the lower $\dot{M}$. Because of the centrifugal force in the rotating WD, $M_{\rm ig}^{\rm R} = 1.43\ M_\odot$ is larger than $M_{\rm ig}^{\rm NR} = 1.38\ M_\odot$ (Nomoto et al. 1984).

(2) For adjacent ranges of binary parameters, the mass of the rotating WD exceeds $M_{\rm ig}^{\rm NR} = 1.38\ M_\odot$ but does not reach $M_{\rm ig}^{\rm R} = 1.43\ M_\odot$ because of the decreasing accretion rate. After the accretion rate falls off, the WD undergoes the angular momentum-loss (J-loss) evolution. The exact mechanism and the time scale of the J-loss are highly uncertain, although the magneto-dipole braking WD is responsible. J-loss induces the contraction of the WD, which leads to the "delayed" carbon ignition after the "delay" time due to neutrino and radiative cooling.

Figure 13 shows the evolution of the center of the WD since before the end of accretion up to the onset of the hydrodynamical stage (Benvenuto et al. 2015). The upper-most line corresponds to the case (1) evolution that leads to the "prompt" carbon ignition. Below that, the lines from upper to lower correspond to the evolutions with increasing J-loss timescale ($\tau_{\rm J} = 1$, 3, 10, 30, 100, 300, and 1000 Myr, respectively), that lead to the "delayed" carbon ignition.

In what binary systems (P and M_2) does a uniformly rotating WD undergo the delayed carbon ignition? The result for the initial WD masses of $0.9 M_\odot$ is shown in Fig. 14. Here the binary systems starting from the "painted" region of the ($P - M_2$) plane reach

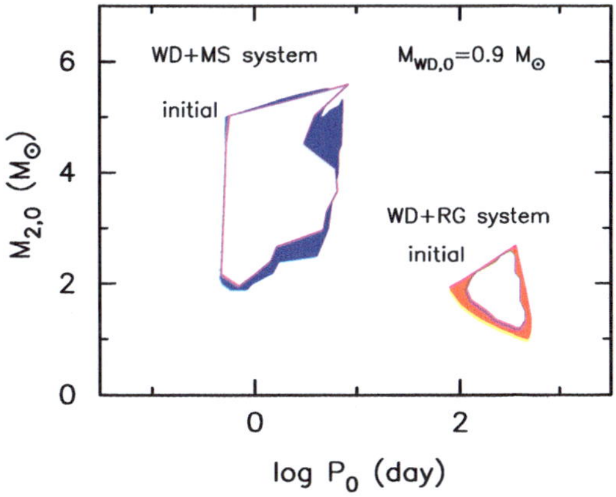

Fig. 14 The outcome of the binary evolution of the WD + companion star systems is shown in the parameter space of the initial orbital period P and the companion mass M_2 for the initial WD mass of $0.9M_\odot$ (Benvenuto et al. 2015). The mass M of the WD starting from the "painted" region reaches $1.38\ M_\odot < M < 1.43\ M_\odot$ (delayed carbon ignition), while the systems starting from the blank region encircled by the solid line reach $M = 1.43\ M_\odot$ (prompt carbon ignition)

$1.38\ M_\odot < M < 1.43\ M_\odot$, while the systems starting from the blank region encircled by the solid line reach $M = 1.43\ M_\odot$. The occurrence frequency of the delayed carbon ignition would roughly be one-third of the total frequency of the carbon ignition.

For the values of τ_J considered here, the WD spends enough time to undergo SN Ia explosion for the donor star to evolve to a structure completely different from the one it had when acted as a donor. For the red-giant donor, its H-rich envelope would be lost as a result of H-shell burning and mass loss so that it would become a He WD in ~ 10 Myr. For the main sequence donor, it would also evolve to become a low mass He WD in 1 Myr, a hot He WD in 10 Myr, and a cold He WD in 1000 Myr (Di Stefano et al. 2011). So, the J-losses should delay the explosion long enough for the former donor to be undetectable. Therefore, this scenario provides a way to account for the failure in detecting companions to SNe Ia.

Such He white dwarf companions would be faint enough not to be seen before or after the Type Ia supernova explosion. This new single-degenerate scenario can explain in a unified manner why no signatures of the companion star are seen in some Type Ia supernovae, whereas some Type Ia supernovae indicate the presence of the companion star.

5.6.2 Differential Rotation and Super-Chandra SNe Ia

If the accretion leads to non-uniform, differentially rotating WDs, carbon ignition occurs at super-Chandrasekhar masses (Hachisu et al. 2012b). The WD mass can increase by accretion up to 2.3 (2.7) $M_\odot$ from the initial value of 1.1 (1.2) $M_\odot$, being consistent with high luminosity SNe Ia such as SN 2003fg, SN 2006gz, SN 2007if, and SN 2009dc (Kamiya et al. 2012). Such very bright super-Chandrasekhar mass SNe Ia are suggested to born in a low metallicity environment.

6 Nucleosynthesis Yields

To distinguish the progenitor scenarios between SD and DD (or the Chandrasekhar mass and sub-Chandrasekhar mass models), comparisons of nucleosynthesis features between these scenarios/models predictions must be useful (e.g., Thielemann et al. 1986; Nomoto et al. 2013b).

To provide the basis to obtain such nucleosynthesis constraints, we recalculate nucleosynthesis yields of 1D explosion models with new nuclear reaction rates and weak interaction rates (Leung and Nomoto 2017; see also Mori et al. 2016).

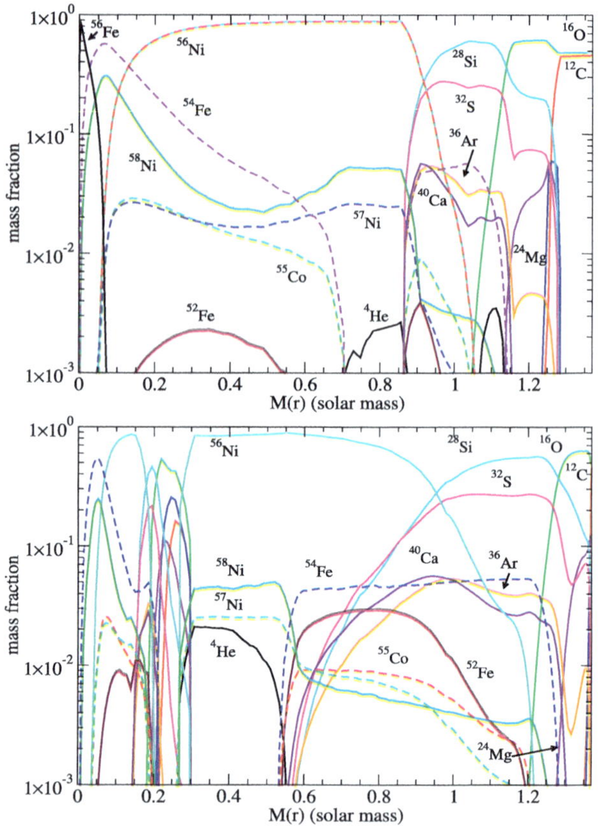

Fig. 15 (Upper panel) Mass fraction against M_r for the W7 model using the new nuclear reaction rates and electron capture rates (Leung and Nomoto 2017). (Lower panel) Same as the upper panel, but for the WDD2 model

6.1 Yields from Chandrasekhar Mass White Dwarfs

We renew the nucleosynthesis yields of 1D models W7 and WDD2 (Nomoto et al. 1984; Iwamoto et al. 1999). In these Chandrasekhar mass models, carbon burning ignited in the central region is unstable to flash because of strong electron degeneracy and release a large amount of nuclear energy explosively. However, the central density is too high and thus the shock wave is too weak to initiate a carbon detonation (because of temperature-insensitive pressure of strongly degenerate electrons).

Then the explosive thermonuclear burning front propagates outward as a convective deflagration wave (subsonic flame) (Nomoto et al. 1976). Rayleigh-Taylor instabilities at the flame front cause the development of turbulent eddies, which increase the flame surface area, enhancing the net burning rate and accelerating the flame. In the 1D convective deflagration model W7 (Nomoto et al. 1984), the flame speed is prescribed by time-dependent mixing-length theory with the mixing length being 0.7 of the pressure scale height. (In the central region, the mixing length is assumed to be equal to the radial distance from the center.)

In some cases the deflagration may undergo "deflagration to detonation transition (DDT)" (Khokhlov 1991). In the 1D DDT model WDD2 (Iwamoto et al. 1999), DDT is assumed to occur when the density at the flame front decreases to 2×10^7 g cm^{-3}. Such a turbulent nature of the flame propagation has been studied in multi-dimensional simulations (e.g., Leung et al. 2015a,b; Nomoto and Leung 2017; Leung and Nomoto 2017).

In Fig. 15 we plot the abundance distributions (mass fractions of main species) of W7 (upper panel) and WDD2 (lower panel). In the central region of these models, the temperature behind the deflagration wave exceeds $\sim 5 \times 10^9$ K, so that the reactions are rapid enough (compared with the expansion timescale) to realize nuclear statistical equilibrium (NSE). The central densities of the WDs are so high ($\sim 3 \times 10^9\ \mathrm{g\,cm^{-3}}$) that electron capture reduces the electron mole fraction, Y_e, that is the number of electrons per baryon (e.g., Fuller et al. 1982; Brachwitz et al. 2000). Then a significant amount of neutron-rich species, such as ^{58}Ni, ^{57}Ni, ^{56}Fe, ^{54}Fe, and ^{55}Co (which decays to ^{55}M), are synthesized.

In W7, 0.65 $M_\odot$ ^{56}Ni are produced. The surrounding layers at $M_r > 0.8\ M_\odot$ gradually expand during the subsonic flame propagation, so that the densities and temperatures get lower. As a result, explosive burning produces the intermediate mass elements ^{28}Si, ^{32}S, ^{36}Ar and ^{40}Ca due to lower peak temperatures than in the central region. Beyond $M_r = 1.2\ M_\odot$ explosive O-burning is slow so that ^{16}O is dominant. The deflagration wave does not reach beyond $M_r = 1.3\ M_\odot$ so that ^{12}C and ^{16}O remain unburned (Nomoto et al. 1984; Thielemann et al. 1986).

In WDD2, the deflagration speed is assumed to be slower than W7 and undergoes the deflagration-detonation-transition (DDT). Then the chemical profile shows a two-layer structure. In the innermost 0.3 $M_\odot$ where the materials are burnt by deflagration, dominant products are similar to W7. After DDT, dominant nucleosynthesis products are somewhat neutron-rich Fe-peak species (e.g., ^{58}Ni, ^{57}Ni, ^{56}Ni) at $M_r < 1.0\ M_\odot$, intermediate mass elements (e.g. ^{28}Si and ^{32}S) at $1.0\ M_\odot < M_r < 1.3\ M_\odot$, and ^{16}O in the outermost layers. There is almost no unburnt ^{12}C, showing the detonation wave can sweep the whole white dwarf before it quenches. The total amount of ^{56}Ni is 0.67 $M_\odot$ (Iwamoto et al. 1999).

Figure 16 shows $[X_i/^{56}\mathrm{Fe}]$ of the stable isotopes from C to Zn for W7 (upper panel) and WDD2 (lower panel). (Here $[X_i/^{56}\mathrm{Fe}] = \log_{10}(X(i)/X(^{56}\mathrm{Fe})) - \log_{10}(X(i)/X(^{56}\mathrm{Fe}))_\odot$ for the mass fraction X of specie X_i.) The detailed abundance ratios with respect to ^{56}Fe depend on the convective flame speed and the central densities, and also the weak reaction rates. In the old W7 and WDD2, electron capture rates by Fuller et al. (1982) were used. In Fig. 16, the most updated weak reaction rates are applied for electron capture Langanke and Martinez-Pinedo (2001). The ratio of [^{58}Ni/^{56}Fe] is reduced from ~ 0.6 in the old W7 to ~ 0.3 in the new W7, although some neutron-rich species are still enhanced relative to ^{56}Fe.

For W7, the metallicity-dependent yields are calculated for the initial mass fraction of ^{22}Ne in the C+O white dwarfs $X(^{22}\mathrm{Ne}) = 0.025$, 0.014, and 0.0025 as given in Yield Table (2018). The metallicity effects are not so large in contrast to the sub-Chandrasekhar mass models (see next section) because synthesis of neutron-rich species is mostly due to electron capture in the NSE region rather than the initial metallicity $X(^{22}\mathrm{Ne})$.

6.2 Yields from Sub-Chandrasekhar Mass White Dwarf Models

There exist possible cases where an explosive carbon ignition occurs in sub-Chandrasekhar mass C+O white dwarfs. In the single degenerate model, an off-center He detonation is induced by relatively slow accretion of He as seen in Fig. 5. In some cases, a resulting shock wave does not induce an immediate off-center C-detonation but propagates toward the central region and increase its strength because of decreasing area at the shock front. Then the shock wave becomes strong enough to ignite explosive carbon burning which forms a detonation wave (e.g., Livne 1990; Woosley and Weaver 1994). In the merging model, a merging of double white dwarfs ignites an off-center carbon flash that produces a shock wave. The shock wave propagates through the center and induces C-detonation (e.g., Pakmor et al. 2011).

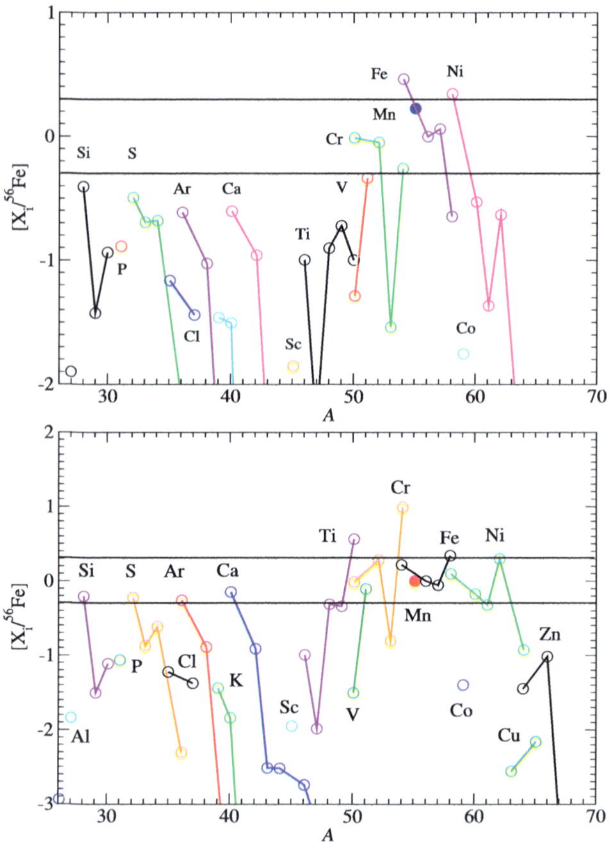

Fig. 16 (Upper panel) $[X_i/^{56}\mathrm{Fe}]$ for the W7 model using the new nuclear reaction rates and electron capture rates (Leung and Nomoto 2017). (Lower panel) Same as the upper panel, but for the WDD2 model. Here $[X_i/^{56}\mathrm{Fe}] = \log_{10}(X(i)/X(^{56}\mathrm{Fe})) - \log_{10}(X(i)/X(^{56}\mathrm{Fe}))_\odot$ for the mass fraction X of specie X_i. The corresponding masses of both stable and radioactive isotopes are given in Tables 1, 2.

Shigeyama et al. (1992) and Nomoto et al. (1994) calculated explosive nucleosynthesis in such sub-Chandrasekhar mass models by artificially inducing a C-detonation at center of the white dwarf. We have calculated similar sub-Chandrasekhar mass models by updating nuclear reaction rates. Figures 17 show the abundance distributions of the $M = 1.0\ M_\odot$ white dwarf models for the solar metallicity $Z = Z_\odot$ (upper) and the lower metallicity $Z = 0.1\ Z_\odot$ (lower). For both models, ^{56}Ni mass is $\sim 0.6\ M_\odot$.

In the Fe-peak region, some neutron-rich species (^{58}Ni and ^{57}Ni) are synthesized because initial CNO elements are converted to neutron-rich ^{22}Ne during He-burning. The distribution of ^{58}Ni is centrally concentrated in the Chandrasekhar mass models, while it is more extended in the sub-Chandrasekhar mass model. Also the amount of ^{58}Ni in the low metallicity sub-Chandrasekhar mass model is much smaller than the solar metallicity model.

Figure 18 shows integrated abundances of $[X_i/^{56}\mathrm{Fe}]$ for both metallicities. Because of the low neutron excess in the central region, there is no overproduction of $[X_i/^{56}\mathrm{Fe}]$, in particular, ^{54}Fe, ^{58}Ni, and ^{64}Cr. In fact, for the solar metallicity model, [^{58}Ni/^{56}Fe] ~ 0. However, large underproduction is seen in [^{55}Mn/^{56}Fe] for both metallicities and [^{58}Ni/^{56}Fe] for $Z = 0.1\ Z_\odot$. These underproductions are important difference between the Chandrasekhar mass and sub-Chandrasekhar mass models when compared with the observations.

For intermediate mass elements, there are almost no metallicity effects on ^{28}Si, ^{32}S, ^{36}Ar and ^{40}Ca. However, underproduction is seen in slightly neutron-rich isotopes such as ^{33}S, ^{38}Ar and ^{42}Ca for $Z = 0.1Z_\odot$. These species have a closer Y_e to ^{22}Ne.

Fig. 17 Abundance profiles of the sub-Chandrasekhar mass model of 1.0 $M_\odot$ as a function of M_r of the white dwarf

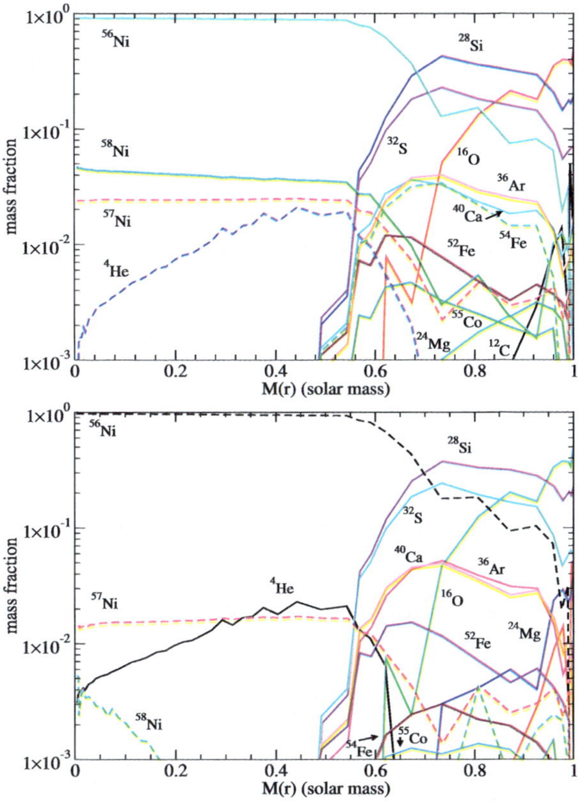

Fig. 18 The ratios of integrated abundances of the sub-Chandrasekhar mass model of 1.0 $M_\odot$ after the decay of unstable nuclei, normalized to ^{56}Fe, relative to solar abundances. Models with $Z = Z_\odot$ (solid line) and $0.1Z_\odot$ (dashed line) are used to contrast the effects of metallicity

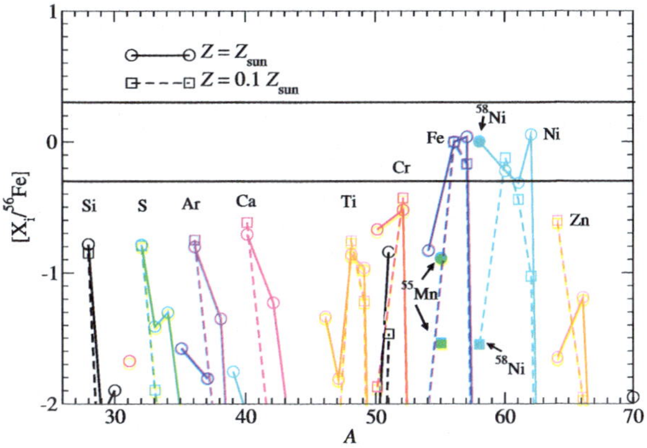

In short, the yields of the sub-Chandrasekhar mass models with the solar metallicity are similar to the Chandrasekhar mass model. The important difference, however, is the Mn yield, which is much smaller in the sub-Chandrasekhar model than the Chandrasekhar mass models. For the smaller metallicity, the differences are larger, so that the comparison with observations of SNe and SNRs in small galaxies is important.

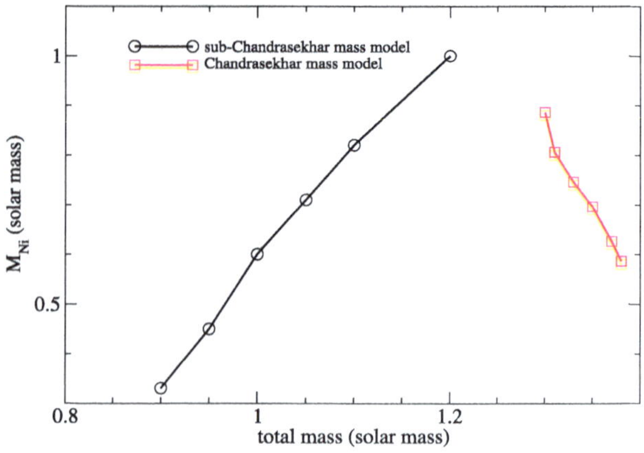

Fig. 19 The ^{56}Ni mass against white dwarf mass for the sub-Chandrasekhar mass models and the near-Chandrasekhar mass models. Solar metallicity is assumed for all models shown here. For the sub-Chandrasekhar mass models, the double detonation model, i.e., the He envelope with the minimal mass required to trigger the carbon detonation is used. For Chandrasekhar mass models, the turbulent deflagration model with deflagration-detonation transition is used (Leung and Nomoto 2017)

6.3 ^{56}Ni Mass as a Function of White Dwarf Mass M

It is interesting to know the maximum $M_{\rm Ni}$ which can be produced in the explosion of WDs. Figure 19 shows the ^{56}Ni mass ($M_{\rm Ni}$) against the white dwarf mass M for the sub-Chandrasekhar mass models with solar metallicity. The double detonation model is used as the explosion mechanism for the sub-Chandrasekhar mass white dwarfs, i.e., we consider carbon detonation model triggered by spherical helium detonation. Due to the extra degree of freedom for the He envelope in the sub-Chandrasekhar model, models with the minimal He envelope mass sufficient for triggering the carbon detonation are used.

By symmetry, the spherical shock wave can propagate into the carbon-oxygen core without significant shock compression in the low density region. As a result, the shock can reach deep in the core where shock convergence can be significant to heat up the core and trigger the carbon detonation.

In the sub-Chandrasekhar mass model sequence, $M_{\rm Ni}$ sharply increases with M for all models. Models with $M \sim 1\ M_\odot$ gives a typical $M_{\rm Ni}$ of 0.6 $M_\odot$.

There exist two factors that could affect the monotonic increase in $M_{\rm Ni}$ with M. One is the asphericity. The aspherical model always produces less ^{56}Ni than the spherical because the off-center carbon detonation allows more expansion before the detonation wave sweeps across the whole white dwarf.

The other is the increasing central density with M. At high central density of the near-Chandrasekhar mass model, the shock wave is too weak to form a detonation, so that a convective deflagration wave is formed as discussed in the earlier subsection.

Figure 19 shows the $M_{\rm Ni}$ vs. M relation for the near-Chandrasekhar mass models. Here the turbulent deflagration model with deflagration-detonation transition (DDT) is used (Leung and Nomoto 2017). There is a monotonic decrease with mass for the Chandrasekhar mass sequence. Since DDT occurs at sufficiently low density, a fraction of the detonation that tends to produce larger $M_{\rm Ni}$ than the deflagration is smaller for larger M. (Even at higher central densities, electron capture suppresses the production of ^{56}Ni.)

As seen in Fig. 19, the maximum $M_{\rm Ni}$ depends on at which M, a deflagration rather than a detonation is formed in the center.

Table 1 The yield table of the stable isotopes for the W7, WDD2 and the 1.0 $M_\odot$ sub-Chandrasekhar mass models with $Z = Z_\odot$ and $Z = 0.1 Z_\odot$. All short-lived isotopes are assumed to have decayed. All masses are in units of solar mass (Yield Table 2018)

Isotope Metallicity	W7 $Z_\odot$	W7 0.1 $Z_\odot$	WDD2 $Z_\odot$	Sub-Chand. $Z_\odot$	Sub-Chand. 0.1 $Z_\odot$
^{12}C	4.75×10^{-2}	6.67×10^{-2}	1.0×10^{-2}	1.1×10^{-3}	1.3×10^{-3}
^{13}C	5.17×10^{-8}	1.28×10^{-12}	2.8×10^{-7}	3.2×10^{-10}	2.53×10^{-11}
^{14}N	1.1×10^{-5}	7.83×10^{-10}	1.54×10^{-7}	1.80×10^{-9}	5.44×10^{-10}
^{15}N	5.46×10^{-8}	1.32×10^{-8}	1.27×10^{-8}	2.50×10^{-10}	4.54×10^{-9}
^{16}O	5.0×10^{-2}	9.95×10^{-2}	9.94×10^{-2}	2.44×10^{-2}	2.29×10^{-2}
^{17}O	4.6×10^{-6}	1.32×10^{-11}	6.88×10^{-8}	8.36×10^{-10}	5.39×10^{-11}
^{18}O	1.43×10^{-7}	7.60×10^{-13}	3.46×10^{-9}	1.85×10^{-11}	1.82×10^{-12}
^{19}F	1.28×10^{-9}	3.46×10^{-12}	4.22×10^{-10}	8.77×10^{-12}	4.75×10^{-12}
^{20}Ne	2.1×10^{-3}	1.56×10^{-3}	1.54×10^{-2}	1.47×10^{-3}	1.60×10^{-3}
^{21}Ne	3.29×10^{-6}	3.12×10^{-9}	2.41×10^{-6}	7.32×10^{-8}	4.85×10^{-9}
^{22}Ne	2.5×10^{-3}	3.4×10^{-4}	1.38×10^{-5}	6.89×10^{-9}	3.41×10^{-9}
^{23}Na	7.2×10^{-5}	2.49×10^{-6}	1.47×10^{-4}	8.56×10^{-6}	4.55×10^{-6}
^{24}Mg	4.61×10^{-3}	8.68×10^{-3}	1.3×10^{-2}	8.70×10^{-4}	1.62×10^{-3}
^{25}Mg	1.19×10^{-4}	2.49×10^{-6}	2.98×10^{-4}	1.27×10^{-5}	5.65×10^{-7}
^{26}Mg	7.64×10^{-5}	1.26×10^{-6}	4.76×10^{-4}	2.2×10^{-5}	1.54×10^{-6}
^{26}Al	2.51×10^{-6}	1.61×10^{-6}	2.84×10^{-5}	2.18×10^{-6}	2.22×10^{-6}
^{27}Al	4.89×10^{-4}	1.58×10^{-4}	1.2×10^{-3}	8.49×10^{-5}	3.76×10^{-5}
^{28}Si	1.63×10^{-1}	1.34×10^{-1}	2.29×10^{-1}	6.27×10^{-2}	5.68×10^{-2}
^{29}Si	8.36×10^{-4}	1.79×10^{-4}	1.31×10^{-3}	1.59×10^{-4}	2.23×10^{-5}
^{30}Si	1.77×10^{-3}	6.29×10^{-5}	1.32×10^{-3}	1.72×10^{-4}	1.22×10^{-5}
^{31}P	4.25×10^{-4}	5.33×10^{-5}	3.4×10^{-4}	6.29×10^{-5}	1.28×10^{-5}
^{32}S	7.90×10^{-2}	7.65×10^{-2}	1.30×10^{-1}	3.69×10^{-2}	3.85×10^{-2}
^{33}S	4.7×10^{-4}	1.18×10^{-4}	2.38×10^{-4}	7.25×10^{-5}	2.51×10^{-5}
^{34}S	2.38×10^{-3}	1.14×10^{-4}	2.46×10^{-3}	5.33×10^{-4}	2.41×10^{-5}
^{36}S	2.86×10^{-7}	6.88×10^{-10}	1.75×10^{-7}	4.48×10^{-8}	1.35×10^{-11}
^{35}Cl	1.35×10^{-4}	1.82×10^{-5}	1.2×10^{-4}	4.89×10^{-5}	7.31×10^{-6}
^{37}Cl	2.46×10^{-5}	8.98×10^{-6}	2.53×10^{-5}	9.88×10^{-6}	3.86×10^{-6}
^{36}Ar	1.25×10^{-2}	1.56×10^{-2}	2.50×10^{-2}	7.47×10^{-3}	8.95×10^{-3}
^{38}Ar	9.45×10^{-4}	5.50×10^{-5}	1.15×10^{-3}	4.19×10^{-4}	2.32×10^{-5}
^{40}Ar	7.94×10^{-9}	1.47×10^{-11}	3.17×10^{-9}	1.17×10^{-9}	7.2×10^{-13}
^{39}K	7.3×10^{-5}	1.24×10^{-5}	6.59×10^{-5}	3.32×10^{-5}	7.89×10^{-6}
^{40}K	8.98×10^{-8}	2.34×10^{-9}	3.19×10^{-8}	2.42×10^{-8}	6.9×10^{-10}
^{41}K	8.27×10^{-9}	4.1×10^{-10}	6.46×10^{-9}	1.62×10^{-9}	1.1×10^{-10}
^{40}Ca	9.67×10^{-3}	1.42×10^{-2}	2.47×10^{-2}	7.0×10^{-3}	9.16×10^{-3}
^{42}Ca	2.88×10^{-5}	1.64×10^{-6}	2.92×10^{-5}	1.46×10^{-5}	7.73×10^{-7}
^{43}Ca	1.46×10^{-7}	5.76×10^{-9}	1.65×10^{-7}	5.48×10^{-7}	2.21×10^{-7}
^{44}Ca	9.9×10^{-7}	1.5×10^{-6}	2.44×10^{-6}	1.30×10^{-6}	1.61×10^{-6}
^{46}Ca	3.50×10^{-10}	4.15×10^{-11}	1.40×10^{-9}	2.76×10^{-11}	2.51×10^{-16}
^{48}Ca	2.20×10^{-12}	1.99×10^{-12}	1.36×10^{-9}	7.91×10^{-16}	4.89×10^{-24}
^{45}Sc	2.93×10^{-7}	9.74×10^{-8}	2.22×10^{-7}	1.10×10^{-7}	6.43×10^{-8}
^{46}Ti	1.38×10^{-5}	9.55×10^{-7}	1.26×10^{-5}	6.1×10^{-6}	5.74×10^{-7}

Table 1 (*Continued*)

Isotope Metallicity	W7 $Z_\odot$	W7 $0.1\ Z_\odot$	WDD2 $Z_\odot$	Sub-Chand. $Z_\odot$	Sub-Chand. $0.1\ Z_\odot$
^{47}Ti	5.38×10^{-7}	7.74×10^{-8}	1.21×10^{-6}	1.88×10^{-6}	7.59×10^{-7}
^{48}Ti	1.70×10^{-4}	2.44×10^{-4}	5.99×10^{-4}	1.73×10^{-4}	2.34×10^{-4}
^{49}Ti	1.95×10^{-5}	1.45×10^{-5}	4.29×10^{-5}	1.4×10^{-5}	6.15×10^{-6}
^{50}Ti	1.2×10^{-5}	9.30×10^{-6}	2.22×10^{-4}	1.58×10^{-10}	1.83×10^{-14}
^{50}V	2.17×10^{-8}	8.55×10^{-9}	1.15×10^{-8}	1.41×10^{-9}	4.33×10^{-12}
^{51}V	9.58×10^{-5}	6.68×10^{-5}	1.41×10^{-4}	2.73×10^{-5}	6.89×10^{-6}
^{50}Cr	4.31×10^{-4}	2.10×10^{-4}	3.99×10^{-4}	8.60×10^{-5}	5.74×10^{-6}
^{52}Cr	8.0×10^{-3}	8.31×10^{-3}	1.54×10^{-2}	2.44×10^{-3}	3.19×10^{-3}
^{53}Cr	3.1×10^{-5}	2.72×10^{-5}	1.28×10^{-4}	2.77×10^{-9}	2.31×10^{-10}
^{54}Cr	1.46×10^{-4}	1.32×10^{-4}	1.77×10^{-3}	2.47×10^{-8}	2.32×10^{-10}
^{55}Mn	1.38×10^{-2}	1.19×10^{-2}	7.59×10^{-3}	9.86×10^{-4}	2.34×10^{-4}
^{54}Fe	1.31×10^{-1}	1.8×10^{-1}	6.98×10^{-2}	6.17×10^{-3}	4.31×10^{-4}
^{56}Fe	7.41×10^{-1}	6.83×10^{-1}	6.54×10^{-1}	6.81×10^{-1}	7.22×10^{-1}
^{57}Fe	2.7×10^{-2}	1.85×10^{-2}	1.34×10^{-2}	1.81×10^{-2}	1.19×10^{-2}
^{58}Fe	6.24×10^{-4}	5.64×10^{-4}	4.70×10^{-3}	6.83×10^{-9}	2.11×10^{-10}
^{60}Fe	1.21×10^{-8}	1.10×10^{-8}	5.73×10^{-7}	5.33×10^{-18}	1.7×10^{-20}
^{59}Co	3.69×10^{-5}	3.34×10^{-5}	6.59×10^{-5}	5.60×10^{-8}	5.88×10^{-9}
^{58}Ni	6.65×10^{-2}	5.95×10^{-2}	3.0×10^{-2}	2.82×10^{-2}	8.40×10^{-4}
^{60}Ni	3.56×10^{-3}	3.21×10^{-3}	6.82×10^{-3}	6.71×10^{-3}	8.94×10^{-3}
^{61}Ni	2.42×10^{-5}	2.17×10^{-5}	2.35×10^{-4}	2.50×10^{-4}	1.97×10^{-4}
^{62}Ni	4.3×10^{-4}	3.64×10^{-4}	3.5×10^{-3}	1.83×10^{-3}	1.60×10^{-4}
^{64}Ni	5.5×10^{-7}	4.56×10^{-7}	1.70×10^{-5}	4.71×10^{-15}	5.33×10^{-16}
^{63}Cu	1.55×10^{-7}	1.60×10^{-7}	8.66×10^{-7}	1.44×10^{-6}	2.68×10^{-6}
^{65}Cu	3.14×10^{-8}	2.81×10^{-8}	1.3×10^{-6}	1.51×10^{-6}	1.15×10^{-6}
^{64}Zn	1.73×10^{-7}	1.55×10^{-7}	1.96×10^{-5}	1.30×10^{-5}	1.52×10^{-4}
^{66}Zn	2.86×10^{-7}	2.59×10^{-7}	3.12×10^{-5}	2.20×10^{-5}	4.1×10^{-6}
^{67}Zn	1.76×10^{-10}	1.64×10^{-10}	1.90×10^{-8}	1.87×10^{-8}	1.73×10^{-8}
^{68}Zn	4.77×10^{-10}	4.32×10^{-10}	1.61×10^{-8}	1.35×10^{-8}	4.44×10^{-7}
^{70}Zn	4.0×10^{-14}	3.61×10^{-14}	1.29×10^{-11}	8.98×10^{-20}	4.51×10^{-22}

7 Summary

In this review, we review how the single degenerate models for Type Ia supernovae (SNe Ia) work. In this binary system, the white dwarf (WD) accretes either hydrogen-rich matter or helium and undergoes hydrogen and helium shell-burning from its non-degenerate companion star. We summarize how the stability and non-linear behavior of such shell-burning depend on the accretion rate and the WD mass and how the WD blows strong wind.

We identify the following evolutionary routes for the accreting WD to trigger a thermonuclear explosion. Typically, the accretion rate is quite high in the early stage and gradually decreases as a result of mass transfer. With decreasing rate, the WD evolves as follows:
(1) At a rapid accretion phase, the WD increase its mass by stable H burning and blows a strong wind to keep its moderate radius. The wind is strong enough to strip a part of the companion star's envelope to control the accretion rate and forms circumstellar matter

Table 2 The yield table of the radioactive isotope for the W7, WDD2 and the 1.0 $M_\odot$ sub-Chandrasekhar mass models with $Z = Z_\odot$ and $Z = 0.1Z_\odot$. All masses are in units of solar mass (Yield Table 2018)

Isotope Metallicity	W7 $Z_\odot$	W7 0.1 $Z_\odot$	WDD2 $Z_\odot$	Sub-Chand. $Z_\odot$	Sub-Chand. 0.1 $Z_\odot$
^{22}Na	6.0×10^{-9}	2.76×10^{-9}	6.20×10^{-8}	5.13×10^{-9}	3.53×10^{-9}
^{26}Al	2.27×10^{-6}	1.61×10^{-6}	2.84×10^{-5}	2.18×10^{-6}	2.22×10^{-6}
^{39}Ar	1.67×10^{-8}	2.54×10^{-10}	6.11×10^{-9}	4.70×10^{-9}	3.86×10^{-11}
^{40}K	8.12×10^{-8}	2.34×10^{-9}	3.19×10^{-8}	2.42×10^{-8}	6.9×10^{-10}
^{41}Ca	3.92×10^{-6}	1.56×10^{-6}	4.99×10^{-6}	2.15×10^{-6}	9.87×10^{-7}
^{44}Ti	5.53×10^{-6}	9.67×10^{-6}	2.21×10^{-5}	1.17×10^{-5}	1.47×10^{-5}
^{48}V	4.66×10^{-8}	8.39×10^{-9}	6.88×10^{-8}	2.30×10^{-8}	2.72×10^{-9}
^{49}V	2.39×10^{-7}	5.56×10^{-8}	1.25×10^{-7}	6.89×10^{-8}	1.67×10^{-9}
^{53}Mn	2.51×10^{-4}	2.32×10^{-4}	1.44×10^{-4}	3.56×10^{-6}	1.1×10^{-7}
^{60}Fe	1.10×10^{-8}	1.10×10^{-8}	5.73×10^{-7}	5.33×10^{-18}	1.7×10^{-20}
^{56}Co	1.13×10^{-4}	1.2×10^{-4}	5.60×10^{-5}	3.2×10^{-6}	3.13×10^{-7}
^{57}Co	7.83×10^{-4}	7.74×10^{-4}	3.48×10^{-4}	1.94×10^{-6}	1.34×10^{-7}
^{60}Co	8.8×10^{-8}	8.8×10^{-8}	3.52×10^{-7}	1.5×10^{-13}	3.93×10^{-15}
^{56}Ni	6.45×10^{-1}	6.59×10^{-1}	6.32×10^{-1}	6.81×10^{-1}	7.22×10^{-1}
^{57}Ni	1.78×10^{-2}	1.77×10^{-2}	1.28×10^{-2}	1.81×10^{-2}	1.19×10^{-2}
^{59}Ni	2.74×10^{-4}	2.72×10^{-4}	1.1×10^{-4}	9.34×10^{-7}	5.7×10^{-8}
^{63}Ni	8.97×10^{-8}	8.98×10^{-8}	8.17×10^{-7}	1.16×10^{-15}	2.13×10^{-16}

(CSM). If the WD explodes within CSM, it is observed as an "SN Ia-CSM". (X-rays emitted by the WD are absorbed by CSM.)
(2) If the WD continues to accrete at a lower rate, the wind stops and an SN Ia is triggered under steady-stable H shell-burning, which is observed as a super-soft X-ray source: "SN Ia-SXSS".
(3) If the accretion continues at a still lower rate, H shell-burning becomes unstable and many flashes recur. The WD undergoes recurrent nova (RN) whose mass ejection is smaller than the accreted matter. Then the WD evolves to an "SN Ia-RN".
(4) If the companion is a He star (or a He WD), the accretion of He can trigger He and C double detonations at the sub-Chandrasekhar mass or the WD grows to the Chandrasekhar mass under a He-wind: "SN Ia-He CSM".
(5) If the accreting WD rotates quite rapidly, the WD mass can exceed the Chandrasekhar mass of the spherical WD, which delays the trigger of an SN Ia. After angular momentum is lost from the WD, the (super-Chandra) WD contracts to become a delayed SN Ia. The companion star has become a He WD and CSM has disappeared: "SN Ia-He WD".

Finally, we update nucleosynthesis yields of the carbon deflagration model W7, delayed detonation model WDD2, and the sub-Chandrasekhar mass model to provide some constraints on the yields (such as Mn) from the comparison with the observations. We note the important metallicity effects on ^{58}Ni and ^{55}Mn.

Acknowledgements This work has been supported by the World Premier International Research Center Initiative (WPI Initiative), MEXT, Japan, and JSPS KAKENHI Grant Numbers JP26400222, JP16H02168, JP17K05382.

References

D. Arnett, A possible model of supernovae: detonation of ^{12}C. Astrophys. Space Sci. **5**, 180 (1969)
D. Arnett, *Supernovae and Nucleosynthesis* (Princeton University Press, Princeton, 1996)
O.G. Benvenuto, J.A. Panei, K. Nomoto, H. Kitamura, I. Hachisu, Final evolution and delayed explosions of spinning white dwarfs in single degenerate models for Type Ia supernovae. Astrophys. J. Lett. **809**, L6 (2015)
F. Brachwitz, D.J. Dean, W.R. Hix et al., The role of electron captures in Chandrasekhar-mass models for Type Ia Supernovae. Astrophys. J. **536**, 934 (2000)
B. Dilday, D.A. Howell, S.B. Cenko et al., PTF 11kx: a Type Ia supernova with a symbiotic nova progenitor. Science **337**, 942 (2012)
R. Di Stefano, R. Voss, J.S.W. Claeys, Spin-up/spin-down models for Type Ia supernovae. Astrophys. J. Lett. **738**, L1 (2011)
R.J. Foley, J.D. Simon, C.R. Burns et al., Linking Type Ia supernova progenitors and their resulting explosions. Astrophys. J. **752**, 101 (2012)
M.-Y. Fujimoto, D. Sugimoto, Helium shell flashes and evolution of accreting white dwarfs. Astrophys. J. **257**, 291 (1982)
G. Fuller, W. Fowler, M. Newman, Stellar weak interaction rates for intermediate mass nuclei. III—Rate tables for the free nucleons and nuclei with $A = 21$ to $A = 60$. Astrophys. J. Suppl. Ser. **48**, 279 (1982)
I. Hachisu, M. Kato, K. Nomoto, A new model for progenitor systems of Type Ia supernovae. Astrophys. J. Lett. **470**, 97 (1996)
I. Hachisu, M. Kato, K. Nomoto, H. Umeda, A new evolutionary path to Type Ia Supernovae: a helium-rich supersoft X-ray source channel. Astrophys. J. **519**, 314 (1999a)
I. Hachisu, M. Kato, K. Nomoto, A wide symbiotic channel to Type Ia Supernovae. Astrophys. J. **522**, 487 (1999b)
I. Hachisu, M. Kato, K. Nomoto, Young and massive binary progenitors of Type Ia supernovae and their circumstellar matter. Astrophys. J. **679**, 1390–1404 (2008a)
I. Hachisu, M. Kato, K. Nomoto, The delay-time distribution of Type Ia supernovae and the single-degenerate model. Astrophys. J. **683**, L27 (2008b)
I. Hachisu, M. Kato, K. Nomoto, Final fates of rotating white dwarfs and their companions in the single degenerate model of Type Ia supernovae. Astrophys. J. Lett. **756**, L4 (2012a)
I. Hachisu, M. Kato, H. Saio, K. Nomoto, A single degenerate progenitor model for Type Ia supernovae highly exceeding the Chandrasekhar mass limit. Astrophys. J. **744**, 69 (2012b)
M. Hamuy, M.M. Phillips, N.B. Suntzeff et al., An asymptotic-giant-branch star in the progenitor system of a Type Ia supernova. Nature **424**, 651 (2003)
Z. Han, Ph. Podsiadlowski, The single-degenerate channel for the progenitors of Type Ia supernovae. Mon. Not. R. Astron. Soc. **350**, 1301 (2004)
M. Hashimoto, K. Nomoto, K. Arai, K. Kaminishi, The (^{14}N) (e^-, ν) (^{14}C) (α, γ) (^{18}O) reaction and helium flashes in accreting helium white dwarfs. Astrophys. J. **307**, 687 (1986)
W. Hillebrandt, J.C. Niemeyer, Type Ia supernova explosion models. Annu. Rev. Astron. Astrophys. **38**, 191 (2000)
I. Iben Jr., A.V. Tutukov, Supernovae of type I as end products of the evolution of binaries with components of moderate initial mass (M not greater than about 9 solar masses). Astrophys. J. Suppl. Ser. **54**, 335 (1984)
I. Iben Jr., K. Nomoto, A. Tornambe, A.V. Tutukov, On interacting helium star-white dwarf pairs as supernova precursors. Astrophys. J. **317**, 717 (1987)
M. Ilkov, N. Soker, Type Ia supernovae from very long delayed explosion of core-white dwarf merger. Mon. Not. R. Astron. Soc. **419**, 1695 (2012)
K. Iwamoto, F. Brachwitz, K. Nomoto et al., Nucleosynthesis in Chandrasekhar mass models for Type Ia supernovae and constraints on progenitor systems and burning-front propagation. Astrophys. J. Suppl. Ser. **125**, 439 (1999)
S. Justham, Single-degenerate Type Ia supernovae without hydrogen contamination. Astrophys. J. Lett. **730**, L34 (2011)
Y. Kamiya, M. Tanaka, K. Nomoto et al., Super-Chandrasekhar-mass light curve models for the highly luminous Type Ia supernova 2009dc. Astrophys. J. **756**, 191 (2012)
M. Kato, H. Saio, I. Hachisu, K. Nomoto, Shortest recurrence periods of novae. Astrophys. J. **793**, 136 (2014)
M. Kato, I. Hachisu, H. Saio, Recurrent novae and long-term evolution of mass-accreting white dwarfs—toward the accurate mass retention efficiency (2017). arXiv:1711.01529
Y. Kawai, H. Saio, K. Nomoto, Off-center ignition of nuclear burning in merging white dwarfs. Astrophys. J. **315**, 229 (1987)

Y. Kawai, H. Saio, K. Nomoto, Steady state models of white dwarfs accreting helium or carbon/oxygen-rich matter. Astrophys. J. **315**, 229 (1988)

A. Khokhlov, Mechanisms for the initiation of detonations in the degenerate matter of supernovae. Astron. Astrophys. **245**, 114 (1991)

H. Kitamura, Pycnonuclear reactions in dense matter near solidification. Astrophys. J. **539**, 888 (2000)

K. Langanke, G. Martinez-Pinedo, Rate tables for the weak processes of pf-SHELL nuclei in stellar environments. At. Data Nucl. Data Tables **79**, 1 (2001)

N. Langer, A. Deutschmann, S. Wellstein, P. Höflich, The evolution of main sequence star + white dwarf binary systems towards Type Ia supernovae. Astron. Astrophys. **362**, 1046 (2000)

S.-C. Leung, K. Nomoto, Dependence of nucleosynthesis on Model Parameters of Type Ia supernovae. Astrophys. J. Suppl. Ser. (2017, submitted). arXiv:1710.04254

S.C. Leung, M.C. Chu, L.M. Lin, A new hydrodynamics code for Type Ia supernovae. Mon. Not. R. Astron. Soc. **454**, 1238 (2015a)

S.C. Leung, M.C. Chu, L.M. Lin, Dark matter admixed Type Ia supernovae. Astrophys. J. **812**, 110 (2015b)

W. Li, J.S. Bloom, P. Podsiadlowski et al., Exclusion of a luminous red giant as a companion star to the progenitor of supernova SN 2011fe. Nature **480**, 348 (2011)

X.-D. Li, E.P.J. van den Heuvel, Evolution of white dwarf binaries: supersoft X-ray sources and progenitors of Type Ia supernovae. Astron. Astrophys. **322**, L9 (1997)

M. Limongi, A. Tornambe, He stars and He-accreting CO white dwarfs. Astrophys. J. **371**, 317 (1991)

M. Livio, The progenitors of Type Ia supernovae, in *Type Ia Supernovae, Theory and Cosmology*, ed. by J.C. Niemeyer, J.W. Truran (Cambridge University Press, Cambridge, 2000), p. 33

E. Livne, Successive detonations in accreting white dwarfs as an alternative mechanism for type I supernovae. Astrophys. J. Lett. **354**, L53 (1990)

E. Livne, A.S. Glasner, Numerical simulations of off-center detonations in helium shells. Astrophys. J. **370**, 272 (1991)

D. Maoz, F. Mannucci, G. Nelemans, Observational clues to the progenitors of Type Ia supernovae. Annu. Rev. Astron. Astrophys. **52**, 107 (2014)

K. Mori et al., Impact of new Gamow-Teller strengths on explosive Type Ia supernova nucleosynthesis. Astrophys. J. **833**, 179 (2016)

K. Nariai, K. Nomoto, D. Sugimoto, Nova explosion of mass-accreting white dwarfs. Publ. Astron. Soc. Jpn. **32**, 473 (1980)

K. Nomoto, Accreting white dwarf models for type I supernovae. I. Presupernova evolution and triggering mechanisms. Astrophys. J. **253**, 798 (1982a)

K. Nomoto, Accreting white dwarf models for type 1 supernovae. II. Off-center detonation supernovae. Astrophys. J. **257**, 780 (1982b)

K. Nomoto, I. Iben Jr., Carbon ignition in a rapidly accreting degenerate dwarf—a clue to the nature of the merging process in close binaries. Astrophys. J. **297**, 531 (1985)

K. Nomoto, Y. Kondo, Conditions for accretion-induced collapse of white dwarfs. Astrophys. J. Lett. **367**, 19 (1991)

K. Nomoto, D. Sugimoto, Rejuvenation of helium white dwarfs by mass accretion. Publ. Astron. Soc. Jpn. **31**, 287 (1977)

K. Nomoto, D. Sugimoto, S. Neo, Carbon deflagration supernova, an alternative to carbon detonation. Astrophys. Space Sci. **39**, L37 (1976)

K. Nomoto, K. Nariai, D. Sugimoto, Rapid mass accretion onto white dwarfs and formation of an extended envelope. Publ. Astron. Soc. Jpn. **31**, 287 (1979)

K. Nomoto, F.-K. Thielemann, K. Yokoi, Accreting white dwarf models of type I supernovae. III. Carbon deflagration supernovae. Astrophys. J. **286**, 644 (1984)

K. Nomoto, H. Yamaoka, T. Shigeyama, S. Kumagai, T. Tsujimoto, Type I supernovae and evolution of interacting binaries, in *Supernovae, Proc. of Session LIV held in Les Houche 1990*, ed. by S. Bludmann et al.. NATO ASI Ser. C (North-Holland, Amsterdam, 1994), p. 199. http://supernova.astron.s.u-tokyo.ac.jp/~nomoto/reference

K. Nomoto, K. Iwamoto, N. Kishimoto, Type Ia supernovae: their origin and possible applications in cosmology. Science **276**, 1378 (1997)

K. Nomoto, H. Umeda, C. Kobayashi et al., Type Ia supernova progenitors, environmental effects, and cosmic supernova rates, in *Type Ia Supernovae, Theory and Cosmology*, ed. by J.C. Niemeyer, J.W. Truran (Cambridge University Press, Cambridge, 2000a), p. 63

K. Nomoto, H. Umeda, C. Kobayashi et al., Type Ia supernovae: progenitors and evolution with redshift, in *Cosmic Explosions AIP Conference Proceedings*, vol. 522 (2000b), p. 35

K. Nomoto, T. Suzuki, J. Deng, T. Uenishi, I. Hachisu, Progenitors of Type Ia supernovae: circumstellar interaction, rotation, and steady hydrogen burning, in *Supernovae as Cosmological Lighthouses 1604–2004*, ed. by Turatto et al.. ASP Conference Series, vol. 342 (2005), p. 105

K. Nomoto, H. Saio, M. Kato, I. Hachisu, Thermal stability of white dwarfs accreting hydrogen-rich matter and progenitors of Type Ia supernovae. Astrophys. J. **663**, 1269 (2007)
K. Nomoto, Y. Kamiya, N. Nakasato et al., Progenitors of Type Ia supernovae: single degenerate and double degenerates. AIPC News **1111**, 267 (2009)
K. Nomoto, M. Kamiya, N. Nakasato, Type Ia supernova models and progenitor scenarios, in *IAU Symposium 281, Binary Paths to Type Ia Supernovae Explosions* (Cambridge University Press, Cambridge, 2013a), p. 253
K. Nomoto, C. Kobayashi, N. Tominaga, Nucleosynthesis in stars and the chemical enrichment of galaxies. Annu. Rev. Astron. Astrophys. **51**, 457 (2013b)
K. Nomoto, S.-C. Leung, Thermonuclear explosions of Chandrasekhar mass white dwarfs, in *Handbook of Supernovae*, vol. 2, ed. by A.W. Alsabti, P. Murdin (Springer) (2017), p. 1275. http://supernova.astron.s.u-tokyo.ac.jp/~nomoto/reference
R. Pakmor, S. Hachinger, F.K. Röpke, W. Hillebrandt, Violent mergers of nearly equal-mass white dwarf as progenitors of subluminous Type Ia supernovae. Astron. Astrophys. **528**, A117 (2011)
F. Patat, P. Chandra, R. Chevalier et al., Detection of circumstellar material in a normal Type Ia supernova. Science **317**, 924 (2007)
L. Piersanti, S. Gagliardi, I. Iben, A. Tornambe, Carbon-oxygen white dwarf accreting CO-rich matter. II. Self-regulating accretion process up to the explosive stage. Astrophys. J. **598**, 1229 (2003)
A.Y. Potekhin, G. Chabrier, Thermonuclear fusion in dense stars. Electron screening, conductive cooling, and magnetic field effects. Astron. Astrophys. **538**, AA115 (2012)
H. Saio, K. Nomoto, Off-center carbon ignition in rapidly rotating, accreting carbon-oxygen white dwarfs. Astrophys. J. **615**, 444 (2004)
Y. Sato, N. Nakasato, A. Tanikawa, K. Nomoto, K. Maedak, I. Hachisu, The critical mass ratio of double white dwarf binaries for violent merger-induced Type Ia supernova explosions. Astrophys. J. **821**, 67 (2017)
B.E. Schaefer, A. Pagnotta, An absence of ex-companion stars in the Type Ia supernova remnant SNR 0509-67.5. Nature **481**, 164 (2012)
I.R. Seitenzahl, C.A. Meakin, D.M. Townsley, D.Q. Lamb, J.W. Truran, Spontaneous initiation of detonations in white dwarf environments: determination of critical sizes. Astrophys. J. **696**, 515 (2009)
K. Shen, L. Bildsten, Thermally stable nuclear burning on accreting white dwarfs. Astrophys. J. **660**, 1444 (2007)
T. Shigeyama, K. Nomoto, H. Yamoka, F.-K. Thielemann, Possible models for the Type Ia supernova 1990N. Astrophys. J. Lett. **386**, L13 (1992)
S. Starrfield, W.M. Sparks, J.W. Truran, G. Shaviv, Neon novae recurrent novae and type-I supernovae, in *Supernovae*, ed. by S.E. Woosley (Springer, Berlin, 1991), p. 602
A. Sternberg, A. Gal-Yam, J.D. Simon et al., Circumstellar material in Type Ia supernovae via sodium absorption features. Science **333**, 856 (2011)
D. Sugimoto, M. Fujimoto, A general theory for thermal pulses of finite amplitude in nuclear shell-burnings. Publ. Astron. Soc. Jpn. **30**, 467 (1978)
D. Sugimoto, M. Fujimoto, K. Nariai, K. Nomoto, in *IAU Colloquium 53, White Dwarfs and Variable Degenerate Stars*, ed. by H.M. Van Horn, V. Weidemann (University of Rochester, Rochester, 1979), p. 280
D. Sugimoto, K. Nomoto, Presupernova models and supernovae. Space Sci. Rev. **25**, 155 (1980)
R.E. Taam, Helium runaways in white dwarfs. Astrophys. J. **237**, 142 (1980)
F.-K. Thielemann, K. Nomoto, K. Yokoi, Explosive nucleosynthesis in carbon deflagration models of type I supernovae. Astron. Astrophys. **158**, 17 (1986)
R.F. Webbink, Double white dwarfs as progenitors of R Coronae Borealis stars and type I supernovae. Astrophys. J. **277**, 355 (1984)
S.E. Woosley, R.E. Taam, T.A. Weaver, Models for type I supernova. I. Detonations in white dwarfs. Astrophys. J. **301**, 601 (1986)
S.E. Woosley, T.A. Weaver, Sub-Chandrasekhar mass models for Type Ia supernovae. Astrophys. J. **423**, 371 (1994)
Yield Table, http://supernova.astron.s.u-tokyo.ac.jp/~nomoto/yields (2018)
S.C. Yoon, P. Podsiadlowski, S. Rosswog, Remnant evolution after a carbon-oxygen white dwarf merger. Mon. Not. R. Astron. Soc. **380**, 933 (2007)

Space Sci Rev (2018) 214:72
https://doi.org/10.1007/s11214-018-0503-8

Models for Type Ia Supernovae and Related Astrophysical Transients

Friedrich K. Röpke[1,2] · **Stuart A. Sim**[3]

Received: 2 February 2018 / Accepted: 30 March 2018 / Published online: 30 April 2018

Abstract We give an overview of recent efforts to model Type Ia supernovae and related astrophysical transients resulting from thermonuclear explosions in white dwarfs. In particular we point out the challenges resulting from the multi-physics multi-scale nature of the problem and discuss possible numerical approaches to meet them in hydrodynamical explosion simulations and radiative transfer modeling. We give examples of how these methods are applied to several explosion scenarios that have been proposed to explain distinct subsets or, in some cases, the majority of the observed events. In case we comment on some of the successes and shortcoming of these scenarios and highlight important outstanding issues.

Keywords Stellar astrophysics · Supernovae · Fluid dynamics · Radiative transfer

1 Introduction

The theoretical description of Type Ia supernovae and related astrophysical transients as thermonuclear explosions of white dwarfs stars has seen rapid development over the past decade. Multidimensional hydrodynamical simulations of the explosion phase were conducted, and the results could be directly used as input for radiative transfer simulations that derive synthetic observables from such models in a consistent way. This allowed to connect modern supernova theory directly to astronomical observations and facilitated a way to validate modeling assumptions by comparison with astronomical data.

Supernovae
Edited by Andrei Bykov, Roger Chevalier, John Raymond, Friedrich-Karl Thielemann, Maurizio Falanga and Rudolf von Steiger

✉ F.K. Röpke
friedrich.roepke@h-its.org

[1] Zentrum für Astronomie der Universität Heidelberg, Philosophenweg 12, 69120 Heidelberg, Germany

[2] Heidelberger Institut für Theoretische Studien, Schloss-Wolfsbrunnenweg 35, 69118 Heidelberg, Germany

[3] School of Mathematics and Physics, Queen's University Belfast, Belfast BT7 1NN, UK

The result of theoretical efforts is a consistent theoretical modeling pipeline for thermonuclear explosions in white dwarf stars. It starts out from a model of the progenitor and extends over multidimensional hydrodynamical simulations of the explosion phase. Nucleosynthesis processes in it are usually determined in a post-processing step. This gives a multidimensional picture of the structure (in particular the density, the velocity and the chemical composition) of the ejecta cloud, that serves as an input to radiative transfer calculations. These, in turn, allow to derive synthetic observables.

In the two parts of this article, we discuss two main ingredients to this modeling pipeline: hydrodynamics simulations of the explosion phase together with nucleosynthesis calculations, and the radiative transfer in the ejecta.

2 Explosion Modeling

2.1 Ansatz and Scale Challenges

The progenitor star of a thermonuclear supernova event is a macroscopic object; densities are high and the typical spatial scales of interest are large. Therefore, the modeling ansatz is based on the equations of fluid dynamics, specifically the Euler equations describing ideal fluids. Viscosity effects are sub-dominant on the scales considered in the model and numerical viscosity in any case outweighs physical effects.

As nuclear burning powers the supernova explosion, reactions have to be taken into account. In addition to the usual fluid dynamics equations describing mass conservation, momentum, and energy balance, a set of equations is necessary to capture species balance. Source terms account for species conversion and the associated energy release. For a complete description of thermonuclear combustion processes, several other effects, such as heat conduction, have to be included in the model (see, e.g., Röpke 2017 for a recent overview). The solution of this system of equations in a numerical supernova simulation, however, is not straightforward because of massive challenges arising from the extremely wide range of relevant scales.

For a numerical treatment, the underlying system of equations is discretized, most commonly in a finite-volume approach. The supernova explosion takes place on the order of the dynamical time scale and therefore time discretization usually follows an explicit scheme. Such schemes are only conditionally stable and consequently the numerical time step has to be restricted according to the Courant-Friedrichs-Lewi (CFL) condition, which, in loose terms, requires numerical time steps to be taken smaller than the sound crossing time over a computational grid cell. The sound crossing time over an entire WD star is on the order of a second; hence the numerical time steps in supernova simulations stay far below a second. The evolution time scale of the progenitor, in contrast, is set by nuclear burning, that lasts many orders of magnitude longer than this. Clearly, this phase is not accessible to multidimensional hydrodynamic simulations. Depending on the propagation mode of the resulting thermonuclear combustion wave, the ignition itself may take centuries or happen dynamically. Thus, ignition is a marginal case that may be addressed in the framework of hydrodynamical simulations, at least as much as time scales are concerned. The explosion itself proceeds in the transonic regime and is certainly accessible to such a numerical treatment.

The time scales on which most observables form are much longer—days, weeks, or months. Because the supernova ejecta are in homologous expansion by then (hydrodynamical effects are frozen out) and the radiation field is dynamically unimportant (at least to zeroth order; see Woosley et al. 2007 for a discussion of the effect of ^{56}Ni decay on the density and velocity profiles), this can be treated in a modeling approach that is separated from the hydrodynamical simulations of the explosion phase and uses their results only to define the background state of the expanding ejecta (see Sect. 3).

The spatial scale problem in thermonuclear supernova models is no less challenging. Due to the extreme temperature sensitivity of the involved nuclear reactions, burning is confined to the hottest regions and propagates in thin fronts. Typically, these have widths far below the millimeter scale. This scale is extremely small compared with that of the exploding white dwarfs (with radii of a few thousand kilometers). Seen from the large global scales, it is well-justified to approximate combustion waves as sharp discontinuities separating the fuel from the ash material. In this discontinuity approximation, jump conditions over the combustion front can be established according to the laws of fluid dynamics. They distinguish between two modes of propagation for the combustion front: subsonic *deflagration* and supersonic *detonation*.

Both deflagrations and detonations are subject to multidimensional hydrodynamic instabilities (for a recent review see Röpke 2017). While for the latter case, it is generally assumed that the effects on the overall explosion process are weak, deflagration burning is most likely dominated—and as a consequence significantly boosted—by the interaction with such instabilities. If ignited near the center of the white dwarf star, a deflagration becomes turbulent. This is an implication of buoyancy instability between the central hot and light ashes and the dense and cold unburnt fuel ahead of the flame. As a result, in the non-linear regime of the Landau-Rayleigh-Taylor instability, bubbles of burning material rise towards the stellar surface (but see Hristov et al. 2017). The flame front is located at their interfaces. Outside of the bubbles, cold unburnt material sinks down towards the center of the white dwarf. This leads to shear motions at the flame. Typical Reynolds numbers are as high as 10^{14} and consequently a *turbulent energy cascade* forms. At the largest scales, kinetic energy is injected by large-scale turbulent eddies, that subsequently decay to smaller scales constituting the inertial range, in which kinetic energy is transported from the large to the small scales without energy loss. Only at the microscopic Kolmogorov scale, the turbulent energy is finally converted to heat by viscous effects.

On a wide sub-range of that turbulent cascade, the deflagration flame interacts with turbulent eddies (see Röpke and Schmidt 2009 and Röpke 2017 for discussions of turbulent deflagrations in SNe Ia). The effect of this interaction depends on whether turbulence corrugates the flame structure only on large scales, or whether it penetrates the internal flame structure and modifies the microphysical transport in it. The first case, which corresponds to the so-called *flamelet regime* of turbulent combustion, applies to most of the explosion period. Here, the flame front is stretched out and wrinkled so that its surface area is greatly enlarged.

Only at the latest times, when the star has expanded significantly and the burning densities are low, the flame structure broadens. With the expansion, turbulence gradually freezes out, but if the prevailing turbulent intensities are still high, a modification of the flame structure is expected. It has been suggested (e.g. Khokhlov et al. 1997; Lisewski et al. 2000a; Röpke et al. 2007; Woosley 2007; Schmidt et al. 2010; Poludnenko et al. 2011) that in this regime transitions of the flame propagation mode from subsonic deflagration to supersonic detonation are possible. Such *deflagration-to-detonation transitions* (*DDTs*) are observed in terrestrial chemical combustion, but there they are mostly associated with obstacles or walls of the combustion vessel. The existence of unconfined DDTs, as would be required in the astrophysical context, remains unproven. Sufficiently strong turbulent mixing inside a broad flame structure is proposed to lead to conditions in which a detonation wave can form via the Zel'dovich gradient mechanism (Zel'dovich et al. 1970).

In addition to these uncertainties in the flame propagation mechanism, the problem of the initial conditions poses a fundamental challenge to modeling thermonuclear supernova explosions. As to now, progenitor systems of Type Ia supernovae are not observationally established. Although the astronomical identification of a progenitor would help to constrain potential scenarios, it would not completely solve all problems of initial conditions for explosion simulations. The progenitor structure and the ignition process are not directly accessible to observations and have to be modeled. As discussed above, the timescales dominating the pre-ignition evolution phases cannot easily be addressed in multidimensional simulations. The resulting uncertainty in the initial conditions is a fundamental obstacle to explosion modeling. The equations of hydrodynamics forming the basis for the description of the explosion processes are hyperbolic partial differential equations. Thus they pose initial value problems. The choice of the initial conditions therefore has a strong impact on the numerical solution (or even determines it). One should thus avoid to draw conclusions from thermonuclear supernova simulations that are dominated by an arbitrary of the initial conditions.

2.2 Numerical Implementation

Several approaches have been taken by different groups to meet the challenges laid out above and perform simulations of thermonuclear supernova explosions. An overview of modeling the combustion physics is given in Röpke (2017). Here, we will focus on one particular choice.

The impracticality to resolve the tiny internal structure of combustion waves in full-star supernova explosion simulations requires to model their propagation in parametrized approaches. The physical structure is either artificially broadened so that it can be represented on the computational grid (Khokhlov 1995; Vladimirova et al. 2006; Calder et al. 2007), or it is completely ignored and the combustion front is treated as a sharp discontinuity separating the fuel from the ashes. An appropriate technique to achieve this (at least up to the spread in hydrodynamical quantities introduced by the numerical Riemann solver) is the so-called level-set scheme (Osher and Sethian 1988; Reinecke et al. 1999). In this front-tracking method, the combustion wave is associated with the zero level-set of a signed distance function G. Its motion is due to advection of the G-field and due to burning. This is captured by an appropriate "level-set equation". While the advection part can be determined from the underlying hydrodynamics scheme, the advancement due to burning is not consistently treated in the discontinuity approximation. It is a parameter of the model that has to be determined externally. For laminar deflagration flames, for instance, it can be derived from resolved one-dimensional simulations (e.g. Timmes and Woosley 1992). For detonations, the Chapman-Jouguet case is a reasonable approximation at low fuel densities. At higher densities, however, nuclear statistical equilibrium establishes behind the detonation front and reactions are partially endothermic. This gives rise to detonations of "pathological" type, that have to be studied in off-line simulations (Sharpe 1999).

The fundamental importance of hydrodynamical instabilities for the propagation of deflagrations requires special modeling approaches. As discussed in Sect. 2.1, the interaction of the flame front with self-generated turbulence boosts the burning efficiency. Because only the largest scales of the turbulent cascade are resolved, the effect of flame surface enlargement due to interaction with turbulent eddies on smaller scales has to be compensated by imposing an effective *turbulent burning speed* on the scale of numerical resolution. This effective turbulent flame propagation velocity replaces the laminar flame speed in the level-set equation. According to Damköhler (1940), it scales with the turbulent velocity fluctuations on the considered length scale. Because of numerical dissipation, these are difficult to determine close to the resolution of the computational grid, and therefore turbulent subgrid-scale models are employed to determine them (see e.g. Niemeyer and Hillebrandt 1995; Schmidt et al. 2006; Röpke and Schmidt 2009; Ciaraldi-Schoolmann et al. 2009; Hicks and Rosner 2013; Jackson et al. 2014; Röpke 2017 for a discussion of approaches used in SN Ia explosion models). It is one of the important achievements of multidimensional simulations to capture the effect of turbulent flame acceleration in a self-consistent way.

Another challenge is the modeling of nuclear reactions that take place in and behind the combustion wave. Two major obstacles have to be overcome in this context. The first is that many reactions are involved in the burning and an extended nuclear network is necessary to predict the synthesis of all involved isotopes. Solving the full network concurrently with the hydrodynamic simulation requires substantial computational effort, in particular in three-dimensional setups. While this is a practical challenge, the second is more fundamental. If combustion waves are represented as discontinuities, their internal structure and details of the reactions are not captured. Artificially broadened combustion waves face the problem that the length scales on which the species conversion and energy release proceed physically are not resolved. They are also challenged by the numerical effort of an extended nuclear network. For this reason, reduced nuclear networks are usually employed in the hydrodynamic explosion simulations, that follow only a few representative species (accounting, for instance, for unburnt fuel material, intermediate mass elements, and nuclear statistical equilibrium compositions). The primary goal of the description of nuclear reactions in the hydrodynamic explosion simulations is to model the energy release driving the dynamics. With reduced networks and artificially broadened combustion waves, it is possible to approximate the energy release to a sufficient accuracy. In models with very few representative species and/or discontinuity descriptions of the combustion waves, the energy release cannot be consistently reproduced and has to be calibrated. This is either done on the basis of one-dimensional resolved flame simulations or in an iterative procedure involving a sequence of explosion models and nucleosynthesis post-processing step. Such post-processing is also necessary to achieve the secondary goal of modeling the burning processes: the determination of detailed nucleosynthetic yields and the chemical structure of ejected material in thermonuclear supernova explosions. The key idea is to place virtual particles (so-called tracers) in the material of the exploding white dwarf star so that each represents a certain fraction of the total mass. These tracer particles are then passively advected with the flow of the exploding material and record the thermodynamic trajectories representative for the fraction of mass they follow. This data is then used as input to a post-processing step that reconstructs the details of the nuclear reactions based on an extended nuclear reaction network (see, e.g., Travaglio et al. 2004).

The detailed hydrodynamic and chemical structure of the ejected material is part of a modeling pipeline that follows the supernova event from the progenitor structure over hydrodynamical explosion simulations and nucleosynthetic post-processing to the formation of observables that can then be compared to astronomical data. It is input to multidimensional radiative transfer calculations that will be discussed in Sect. 3.

2.3 Requirements for a Viable Explosion Scenario

A fundamental goal of modeling thermonuclear explosions in white dwarfs is to reproduce the characteristic spectral features of Type Ia supernovae. The lack of hydrogen and helium is characteristic for this class of objects. Moreover, spectral features indicate the presence of substantial amounts of iron group and intermediate-mass elements. This is prototypical for burning carbon-oxygen white dwarf matter. Irrespective of the combustion wave being a deflagration or a detonation, the released energy and the composition of the ash depends on the fuel density ahead of the front.

At the highest densities, as encountered in the cores of massive white dwarf stars, the ash temperatures become high enough to establish nuclear statistical equilibrium (NSE) conditions. Freeze-out from NSE occurs when the ejecta expand and iron group nuclei are formed. At lower fuel densities, burning is incomplete and intermediate-mass elements (Si, S, Ca, etc.) are synthesized. At even lower densities, carbon burns to oxygen, and below a certain threshold, burning ceases and unprocessed carbon-oxygen white dwarf material is left behind.

The fact that intermediate-mass elements are seen in the spectra implies that a substantial amount of the stellar material must be processed at sufficiently low densities ($\rho_\mathrm{fuel} \lesssim 10^7\ \mathrm{g\,cm^{-3}}$) to enable incomplete burning. The burning front therefore must either (1) pre-expand a Chandrasekhar-mass WD, which requires a sub-sonic flame propagation mode, (2) proceed as a detonation in a pre-expanded Chandrasekhar-mass WD in a delayed detonation scenario, or (3) form a detonation in a sub-Chandrasekhar mass WD. We will discuss these possibilities in Sect. 4.

3 Prediction of Observables

3.1 Radiative Transfer Considerations

As outlined above, hydrodynamical explosion models can simulate the dynamics and nucleosynthesis in thermonuclear supernovae from the point of ignition until the ejecta reach near-homologous expansion. However, to assess the validity of such models, the explosion model output must be mapped onto the space of observable quantities that can be compared to data. In general, this requires additional calculations that yield predicted light curves, spectra and/or spectropolarimetry. Fortunately, as noted above, such calculations can usually be performed as a post-processing step on the ejecta that have been dynamically simulated into the homologous phase. There are important exceptions, however, most notably for scenarios that involve ongoing dynamics as the explosion ejecta interact with a dense environment (see, e.g., Fryer et al. 2010; Blinnikov and Sorokina 2010; Noebauer et al. 2016).

Thermonuclear supernovae have ejecta that are rich in heavy elements: in models of normal SNe Ia, around half of the ejected mass is composed of iron-peak elements, with around half of the remainder being the so-called intermediate-mass elements (Si, S, Ca etc.). The relative complexity of the atomic structure of the first few ions of these elements (compared to e.g. H or He, which are more usually dominant) means that large numbers of bound-bound transitions need to be taken into account when simulating radiative transfer. Moreover, the large expansion velocities blend the transitions together meaning that it is usually very difficult to define a simple continuum or separate out individual lines. As a consequence, the net contribution of bound-bound transitions tends to dominate over any continuum thermalization opacity (see, e.g. Pinto and Eastman 2000) and fluorescent frequency redistribution becomes key to understanding the overall spectral energy distribution.

The large expansion velocities (and associated velocity gradients) have a central role in shaping the spectral features of supernovae and need to be considered in any method that aims to predict synthetic spectra. Several contemporary radiative transfer approaches embed the assumption of high-velocity gradients in the form of the Sobolev approximation (see e.g. Sobolev 1960; Lamers and Casinelli 1999). This approach makes it relatively easy to take into account very large numbers of bound-bound transitions at modest computational cost (either directly, e.g. Mazzali and Lucy 1993, or via an expansion opacity formalism, e.g. Karp et al. 1977; Blinnikov et al. 1998). However, the Sobolev approximation does have limitations, particularly in relation to the treatment of overlapping lines (e.g. Baron et al. 1996), which becomes increasingly common at short wavelengths. Consequently, the most sophisticated radiative transfer codes avoid this approximation and treat individual line profiles in detail.

Reasonably accurate radiative transfer simulations also depend strongly on the calculation of the temperatures and ionization/excitation conditions in the ejecta. Local thermodynamic equilibrium (LTE) is often adopted as a first estimate but departures from equilibrium have important consequences and non-LTE effects become increasingly important with time as the ejecta expand. As illustrated by Dessart et al. (2014), accurate synthetic observables depend on describing a range of complicated microphysics, whose role evolves with time. By the latest epochs commonly observed for thermonuclear supernovae ($\sim$ 200–300 days post explosion), the ejecta are very far from LTE: at these "nebular" epochs, ionization and heating controlled by the ongoing injection of non-thermal particles (from radioactive decay) and cooling dominated by multiplets of forbidden lines, predominantly of the iron group elements (for a recent review of the modeling of nebular spectra, see Jerkstrand 2017). Consequently, modeling of spectra at these epochs is critically dependent on the microphysics and quality/quantity of the atomic data (radiative and collisional) available for the necessary ions.

3.2 Implementation and Application to Modern Explosion Models

Given the competing demands on implementations and computational resources (i.e. need to address complicated/time-dependent microphysics in expanding, inhomogeneous 3D ejecta models), most published studies to date have made several necessary approximations. Currently, perhaps the most important trade-off made in relation to the study of hydrodynamical explosion models is that between simplified microphysics and multi-dimensional effects in the ejecta. For example, while several of the Monte Carlo radiative transfer codes (e.g. SEDONA, Kasen et al. 2006, or ARTIS, Kromer and Sim 2009) can compute orientation-dependent synthetic observables for fully 3D ejecta models, these codes use the Sobolev approximation and either use LTE or relatively simple nLTE approximations. In contrast, sophisticated 1D codes can avoid the Sobolev approximation and treat many levels of many ions in full nLTE for SNe Ia explosion models (see, e.g., Höflich et al. 1998; Baron et al. 2006; Blondin et al. 2013).

4 Scenarios and Simulations

To meet the requirements for a viable explosion scenario discussed in Sect. 2.3, different modes of ignition and flame propagation are necessary, depending on whether the exploding star is close to the Chandrasekhar-mass or below that limit. The first case constitutes so-called Chandrasekhar-mass explosion models and the second sub-Chandrasekhar mass models. We explore these model classes below. They are addressed with our modeling pipeline shown in Fig. 1.

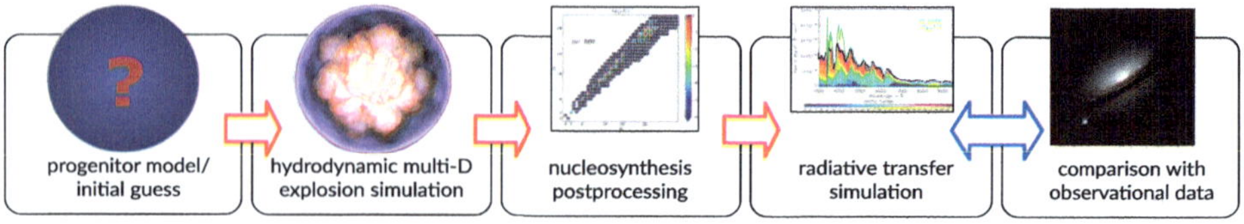

Fig. 1 3D modeling pipeline

4.1 Chandrasekhar-Mass White Dwarf Explosion Models

Approaching the Chandrasekhar mass, the density in the core of a white dwarf increases steadily. This will lead to the ignition of carbon fusion in the so-called intermediate thermopycnonuclear regime (Gasques et al. 2005), i.e. under conditions where the density has an appreciable effect on the reaction rate. Initially, energy losses due to neutrinos formed in plasmon decays and electron-nucleus bremsstrahlung cool the stellar center. Ignition occurs when the central density reaches high enough values so that neutrino losses become insufficient to balance the energy production due to carbon burning.

This does, however, not yet trigger the explosion process. It rather leads to a century of "simmering", in which convective motions efficiently transport the energy generated in the stellar center outward. The fluid motions are highly turbulent. On this background, a hotspot finally develops, out of which a deflagration wave is formed by thermonuclear runaway. Simulating the entire simmering phase is virtually impossible, because a century cannot be bridged and the spatial resolution is far from resolving turbulence. Nonetheless, ignition simulations have been performed (Höflich and Stein 2002; Kuhlen et al. 2006; Zingale et al. 2009; Nonaka et al. 2012). A three-dimensional simulation following the last hours until the first thermonuclear runaway occurs at a radius of ~ 50 km off-center is presented by Nonaka et al. (2012). The results indicate that a second runaway at a different location shortly after the first is unlikely. Thus, to current best knowledge, the deflagration will form in a single region off-center of the WD star.

4.1.1 Pure Deflagrations

The pure deflagration scenario follows possibility (1) described in Sect. 2.3. After formation of the deflagration wave near the center, it propagates towards the surface, subject to buoyancy instability. Multidimensional simulations clearly show the formation of "mushroom-shaped" bubbles giving rise to a complex morphology of the flame front (see Fig. 2 for an illustration). The flame strongly accelerates due to the interaction with turbulence. The scaling behavior of turbulent motions in this situation was unclear for a long time. Based on highly resolved three-dimensional simulations, however, it could be shown that at small length scales and for most of the burning turbulence is isotropic and follows Kolmogorov scaling (Zingale et al. 2005; Ciaraldi-Schoolmann et al. 2009).

The strength of the deflagration and the overall outcome of the explosion phase fundamentally depend on the initial conditions. Several parameters are expected to vary in nature from event to event, such as the central density and the chemical composition of the exploding white dwarf star. Other parameters are unknown or subject to uncertainties in the numerical modeling. A parameter of models of the explosion phase is the ignition geometry. The length scales of the actual flame formation cannot be resolved in multidimensional simulations. Therefore, the effect is usually mimicked by placing a number of flame kernels near

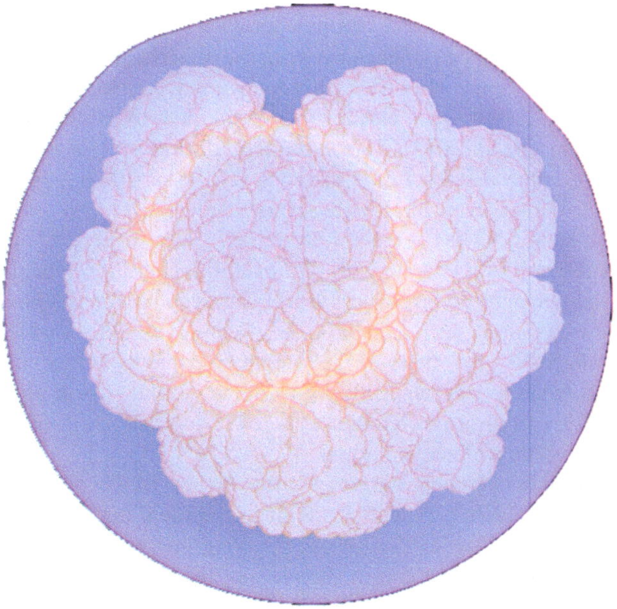

Fig. 2 Simulation of a deflagration (orange/white contour) in a Chandrasekhar-mass white dwarf star (blue color)

the stellar center. Although this does not necessarily capture the ignition physics, it is a way of defining a well-posed initial setup. For a single sphere, the Rayleigh-Taylor instability is seeded by random numerical noise and imprinting a certain spectrum of resolved perturbations on the flame geometry ensures convergence of the model. The simulations of Fink et al. (2014) showed that the initial flame shape has a tremendous impact on the strength of the burning. In this study, a sequence of models was presented with varying numbers of ignition sparks, that were randomly placed in the central region of the white dwarf star. Sparse ignitions naturally lead to aspherical flame geometry evolution, whereas an on average isotropic flame propagation is only possible with dense ignitions. In the former case, only a small fraction of the star is burned. Due to buoyancy, the flame quickly rises towards the surface leaving the far side of the star unaffected. The energy liberated in this process leads to the ejection of parts of the stellar material, and a bound remnant is left behind. In contrast, a complete disruption of the white dwarf star is possible for dense isotropic ignition configurations. In all cases, however, the ejecta structure is well-mixed on large scales due to the flame instabilities. Even with dense ignition configurations, the production of ^{56}Ni hardly exceeds $0.35 M_{\odot}$.

The relatively low ^{56}Ni production means that pure deflagration models generally fail to account for the observed brightness of the majority of SNe Ia (the predicted peak luminosity of such models is too low). However, the ^{56}Ni masses are consistent with those required for some low-luminosity events, of which several sub-classes have now been identified (Taubenberger 2017). In particular, the range of ^{56}Ni masses predicted in pure deflagration models (Jordan et al. 2012b; Fink et al. 2014) is roughly consistent with the observed range of brightness spanned by the Type Iax supernovae (Foley et al. 2013).

The potential identification of Type Iax supernovae with pure deflagrations is supported by comparison of synthetic spectra and light curves to observations. For example, Kromer et al. (2013) compared model predictions for one of the pure deflagration models of Fink et al. (2014) for a range of photospheric-phase epochs to SN2005hk (Phillips et al. 2007), a well-observed member of the SNe Iax class. They found fair agreement in both the strengths and shapes of spectral features across a range of phases in the optical and infrared, and also good correspondence between the model and observed colors around maximum lights. However, some important discrepancies do remain. In particular, the model light curves evolve too quickly, most notably in the post-maximum decline phase in the redder bands (Kromer et al. 2013). This systematic discrepancy is also apparent in the comparison of a different model from the Fink et al. (2014) sample to a fainter member of the SNe Ia class, 2015H (Magee et al. 2016). Several studies have also drawn attention to evidence that the ejecta of SNe Iax are not fully mixed (Stritzinger et al. 2015; Barna et al. 2017), which is difficult to reconcile with a turbulent deflagration model. In addition, it remains unclear whether pure deflagration models can account for the lowest luminosity members of the SNe Iax class, such as SN2008ha (Foley et al. 2009), which requires less than $0.01 M_\odot$ of ^{56}Ni. Such a low mass of ^{56}Ni might be achieved under conditions whereby burning in the deflagration is occurs only in a limited central region of the WD, for example due to an exotic composition (Kromer et al. 2015)—however, it remains to be demonstrated whether this can be realized in nature.

One outstanding, but noteworthy, feature of the comparison of SNe Iax and pure deflagration models is potential role of the residual material from the WD, that was still bound at the end of the explosion phase (in e.g. the models compared to observations mentioned above, $\sim 1 M_\odot$ or more of the mass of the initial WD remains bound at the end of the explosion simulation). Some of the ^{56}Ni synthesized in the explosion remains in this material (Kromer et al. 2013; Fink et al. 2014), meaning that it will experience ongoing energy injection which will plausible drive further expulsion of mass (Foley et al. 2016). Further study, both of physical conditions in the residual material (Shen et al. 2017) and of late-phase observations of SNe Iax (Foley et al. 2016) are needed to explore this topic in more detail.

4.1.2 Delayed Detonations

Enhancing the ^{56}Ni production and the explosion energy to values necessary to reproduce normal SNe Ia is not possible by simply increasing the number of ignition kernels or tuning the initial parameters of the exploding white dwarf. A fundamental change in the burning mode is required—a transition from the initial deflagration, that is necessary to pre-expand the material, to a subsequent detonation. This scenario follows possibility (2) described in Sect. 2.3 This is the idea of the class of *delayed detonation models*. The key question in these is obviously if and how a detonation is triggered in a late burning stage.

Several mechanisms have been proposed for initiating detonations in delayed detonation models. A spontaneous deflagration-to-detonation transition (DDT) may be caused by intrinsic processes in the burning wave. Whether or not such DDTs occur in thermonuclear supernova explosions remains uncertain. Some necessary conditions were laid out in the studies of Lisewski et al. (2000b), Woosley (2007), Woosley et al. (2009, 2011). Two other mechanisms, the gravitationally confined detonation (GCD, Plewa et al. 2004) and the pulsational delayed detonation (PDD) mechanisms (e.g., Bravo and García-Senz 2006), rely on weak initial deflagration stages that fail to gravitationally unbind the white dwarf star.

The mechanism for GCD assumes that a one-sided ignition leads to an asymmetric deflagration that is too weak to gravitationally unbind the WD star. The deflagration ash breaks out of the star's surface and sweeps around it to collide in the antipode. Clearly, a successful ignition of a detonation in this collision favors stronger impact which in turn implies a weak deflagration phase. The resulting detonation then burns the bound core of the object. With weak deflagrations, it will not be very expanded and thus particularly bright events with high masses of synthesized ^{56}Ni are expected. A bound white dwarf resulting from a weak deflagration will pulsate. These pulsations may aid the formation of a detonation (Jordan et al. 2012a). Both the GCD and the PDD scenarios share the characteristic feature that the products of high-density deflagration will be located in the outer part of the ejected material at high velocities. This is in conflict with observations (Seitenzahl et al. 2016).

Also for the classical (DDT) delayed detonation scenario, several problems persist. It has been suggested as a model for the bulk of normal SNe Ia. This requires them to reproduce individual supernova observations. Studies based on 1D DDT models have generally been fairly successful in this regard (e.g. Höflich et al. 1998; Blondin et al. 2013). Radiative transfer simulations based on multi-dimensional simulations of DDTs (e.g., in 2D Kasen et al. 2009; Blondin et al. 2011, or in 3D Seitenzahl et al. 2013; Sim et al. 2013) have also generally found that some DDT models can provide a fairly good (albeit far from perfect) match to many properties of the light curves spectra, and indeed spectropolarimetry (Bulla et al. 2016) of individual SNe Ia.

In addition, if DDT models are responsible for the full population, observed trends between characteristic features should be reproduced. The model explosions should be able to cover the range of brightnesses observed from normal SNe Ia. This requires a nickel mass production in the range from below $0.4\,M_\odot$ to $0.8\,M_\odot$, see e.g. Scalzo et al. (2014). Delayed detonations face a fundamental challenge here. Generally, stronger deflagrations lead to increased expansion before the detonation phase sets in Röpke and Niemeyer (2007), Mazzali et al. (2007). Consequently, the detonation runs over lower-density material and produces less ^{56}Ni. Therefore, the faintest models are expected to be those with the strongest deflagration. This was tested in multi-spot ignition setups that allow to vary the deflagration strength significantly. The strongest deflagrations produce $\gtrsim 0.3\,M_\odot$ of ^{56}Ni and in the subsequent detonation little is added to this amount. This means that multi-spot ignitions with many, on average isotropically distributed kernels are required to reach the fainter end of the distribution of normal SNe Ia. These, however seem unlikely to be realized in nature (Nonaka et al. 2012). Furthermore, when such models are invoked, they appear to fail to fully produce the relatively rapid light curve evolution that is observed to coincide with low luminosity (i.e. the light curve width-luminosity relation: see e.g. Sim et al. 2013). The second problem in this context is that the brightness distribution of normal events is observed to peak at explosions producing $\sim 0.6\,M_\odot$ of ^{56}Ni. It is not obvious why the initial parameters such as flame ignition geometry, central density and chemical composition of the progenitor white dwarf star, or the initiation mechanism of the detonation, should favor a configuration producing this amount of radioactive nickel.

We note that, although the 3D simulations of Sim et al. (2013) have difficulties reproducing the observed width-luminosity relation with a faster decline of the B-band light curve for fainter events, it may be possible to construct such a relation in Chandrasekhar-mass white dwarf star explosions (Kasen and Woosley 2007). Recent studies (Blondin et al. 2017; Goldstein and Kasen 2018), however, increasingly indicate that sub-Chandrasekhar mass explosions are required to capture the full range of the observed relation.

4.2 Sub-Chandrasekhar Mass White Dwarf Explosions

The alternative to Chandrasekhar-mass white dwarf explosions as models for SNe Ia are detonations in sub-Chandrasekhar mass objects following possibility (3) of Sect. 2.3. The capability of the scenario to reproduce basic characteristics of observed SNe Ia was demonstrated by sequences of toy models. Here, if white dwarfs of varying mass are set up and a central detonation ignited artificially, the resulting ejecta structures provide a relatively simple sequence of models that do a fair job of reproducing many of the observed characteristics of SNe Ia (e.g. Shigeyama et al. 1992; Sim et al. 2010; Shen et al. 2017). A very attractive feature of this class of model is that a single simple physical parameter can be recognized as responsible for driving many of the differences between different explosions: the mass of the exploding WD. In particular, a wide range of nickel mass can be produced from only a modest range of WD masses: an initial WD mass of around $1 M_\odot$ is required to produce an explosion with brightness characteristic of the most common SNe Ia (see Sim et al. 2010; Shen et al. 2017). In addition, the low densities in sub-Chandrasekhar mass white dwarfs means that their detonation yields significant quantities of intermediate mass elements, allowing for a relatively good match to observed spectra across a broad range of explosion luminosity. One important challenge for this class of simple (toy) sub-Chandrasekhar mass models is that they struggle to sustain sufficiently slow light curve evolution to account for the brighter end of the SNe Ia distribution. However, the variation in ejecta mass amongst sub-Chandrasekhar mass models does naturally suggest a link between brightness and light curve evolution and simulations have favored sub-Chandrasekhar models for relatively faint explosions (see, e.g., Blondin et al. 2017).

The problem with sub-Chandrasekhar mass explosion models of the sort described above is that the physical ignition of detonations in such objects does not intrinsically arise from an evolutionary process (such as the accretion of mass towards the Chandrasekhar limit). It has to be caused by some vigorous event. Two possibilities are commonly discussed.

One way to ignite a detonation in a carbon-oxygen white dwarf star is that it accretes helium-rich material from a companion. Due to instabilities in the accretion process or once the accreted shell has grown massive enough, a detonation triggers in the He material. It propagates the carbon-oxygen core and drives a shock wave into it. This shock wave may trigger a secondary detonation in carbon-oxygen rich material—constituting the so-called *double detonation explosion scenario* (see Fink et al. 2010; Moll and Woosley 2013 for recent multidimensional supernova simulations following this paradigm). It is conceivable that this occurs when the shock hits the outer edge of the core ("edge-lit double detonation") or due to a spherical collimation of the inwards propagating oblique shock wave near its center. The latter case was shown to robustly lead to detonations of the core by a geometric amplification effect (Fink et al. 2007). It has to be emphasized, however, that many of the published models simply *assume* a primary detonation of the He shell. The ignition process is very hard to resolve numerically and the success of the scenario hinges on it to occur in reality. Synthetic spectra have been computed from double detonation models for a range of conditions (e.g. Nugent et al. 1997; Kromer et al. 2010; Woosley and Kasen 2011), with results that depend significantly on the assumed structure and mass of the helium layer at explosion. Indeed, it generally appears to be the case that the influence of the outer ejecta layers (rich in helium and/or helium-detonation ash) is mostly detrimental to the agreement of the models with observations: if substantial helium shells are invoked this leads to effectively suppressing the characteristic features of intermediate mass elements and to very dramatic line blanketing effects associated with heavy elements that are synthesized in the helium detonation. Thus, for such double detonation models to be viable, very low mass helium shells must be invoked: whether such low masses of helium can really be ignited and/or sustain a detonation is a topic of active study (e.g. Shen and Bildsten 2007, 2014).

An alternative scenario is that of *violent mergers* (Pakmor et al. 2010, 2011, 2012). In contrast to the classical merger paradigm, the explosion happens before the two white dwarfs have formed a single object. In the inspiral process, the lighter of the pair is disrupted and its material plunges into the more massive primary that is only weakly affected by tidal forces. This impact may trigger a detonation of the primary—a sub-Chandrasekhar mass white dwarf. A recent update of the model (Pakmor et al. 2013) suggests that the ignition of the detonation is triggered even earlier in the inspiral process when He rich material (that always exists in low quantities on top of a carbon-oxygen white dwarf) is accreted from the secondary to the primary. This rapid accretion process leads to hydrodynamical instabilities in the He-layer on the primary and triggers a detonation in this shell. Similar to the double detonation scenario, the actual supernovae results from a secondary detonation of the core material. In contrast to that scenario, however, in the violent merger case the He shell is less massive and less dense so that its imprint on the predicted observables is much reduced.

It is remarkable that population synthesis studies predict a peak of the distribution of white dwarf mergers at primary masses that produce $\sim 0.6\,M_\odot$ of ^{56}Ni (Ruiter et al. 2013). Moreover, also the temporal evolution of the luminosity function resulting from sub-Chandrasekhar mass WD detonations seems to match the observations (Shen et al. 2017).

5 Conclusions

Thanks to advances in computing resources and numerical methods in recent years, we are now able to perform meaningful fully 3D explosion simulations for a range of progenitor scenarios that have been proposed for Type Ia supernovae. Combined with radiative transfer post-processing, which allows predictions to made that can be directly compared to observations, such simulations are not playing a key role in driving our understanding of the nature and physics of thermonuclear supernovae. However, the state of the art remains incomplete and far from unsatisfactory—numerous limitations persist. These include clearly posing initial conditions for explosion simulations in the context of particular progenitor modeling, proper representation of the dynamics and instabilities during the thermonuclear combustion in full star models, and adequate description of the complex radiation processes responsible for spectrum formation in the evolving ejecta. The last decade has demonstrated that such multi-dimensional simulations are possible. The goal for the future will be their development towards a level of predictive power than allows for ever-improving quantitative testing by comparison to the increasing wealth of observational data.

Acknowledgements The work of FKR is supported by the Klaus Tschira Foundation and by the Collaborative Research Center SFB 881 "The Milky Way System" (subproject A10) of the German Research Foundation (DFG).

References

B. Barna, T. Szalai, M. Kromer, W.E. Kerzendorf, J. Vinkó, J.M. Silverman, G.H. Marion, J.C. Wheeler, Abundance tomography of Type Iax SN 2011ay with tardis. Mon. Not. R. Astron. Soc. **471**, 4865–4877 (2017).

E. Baron, P.H. Hauschildt, A. Mezzacappa, Radiative transfer in the comoving frame. Mon. Not. R. Astron. Soc. **278**, 763–772 (1996)

E. Baron, S. Bongard, D. Branch, P.H. Hauschildt, Spectral modeling of SNe Ia near maximum light: probing the characteristics of hydrodynamical models. Astrophys. J. **645**, 480–487 (2006).

S.I. Blinnikov, E.I. Sorokina, Supernova explosions inside carbon-oxygen circumstellar shells. ArXiv e-prints (2010)

S.I. Blinnikov, R. Eastman, O.S. Bartunov, V.A. Popolitov, S.E. Woosley, A comparative modeling of supernova 1993J. Astrophys. J. **496**, 454–472 (1998).

S. Blondin, D. Kasen, F.K. Röpke, R.P. Kirshner, K.S. Mandel, Confronting 2d delayed-detonation models with light curves and spectra of Type Ia supernovae. Mon. Not. R. Astron. Soc. **417**, 1280–1302 (2011).

S. Blondin, L. Dessart, D.J. Hillier, A.M. Khokhlov, One-dimensional delayed-detonation models of Type Ia supernovae: confrontation to observations at bolometric maximum. Mon. Not. R. Astron. Soc. **429**, 2127–2142 (2013).

S. Blondin, L. Dessart, D.J. Hillier, A.M. Khokhlov, Evidence for sub-Chandrasekhar-mass progenitors of Type Ia supernovae at the faint end of the width-luminosity relation. Mon. Not. R. Astron. Soc. **470**, 157–165 (2017).

E. Bravo, D. García-Senz, Beyond the bubble catastrophe of Type Ia supernovae: Pulsating reverse detonation models. Astrophys. J. **642**, 157–160 (2006).

M. Bulla, S.A. Sim, M. Kromer, I.R. Seitenzahl, M. Fink, F. Ciaraldi-Schoolmann, F.K. Röpke, W. Hillebrandt, R. Pakmor, A.J. Ruiter, S. Taubenberger, Predicting polarization signatures for double-detonation and delayed-detonation models of Type Ia supernovae. Mon. Not. R. Astron. Soc. **462**, 1039–1056 (2016).

A.C. Calder, D.M. Townsley, I.R. Seitenzahl, F. Peng, O.E.B. Messer, N. Vladimirova, E.F. Brown, J.W. Truran, D.Q. Lamb, Capturing the fire: flame energetics and neutronization for Type Ia supernova simulations. Astrophys. J. **656**, 313–332 (2007).

F. Ciaraldi-Schoolmann, W. Schmidt, J.C. Niemeyer, F.K. Röpke, W. Hillebrandt, Turbulence in a three-dimensional deflagration model for Type Ia supernovae. I. Scaling properties. Astrophys. J. **696**, 1491–1497 (2009).

G. Damköhler, Der Einflußder Turbulenz auf die Flammengeschwindigkeit in Gasgemischen. Z. Elektrochem. **46**(11), 601–652 (1940)

L. Dessart, D.J. Hillier, S. Blondin, A. Khokhlov, Critical ingredients of Type Ia supernova radiative-transfer modelling. Mon. Not. R. Astron. Soc. **441**, 3249–3270 (2014).

M. Fink, W. Hillebrandt, F.K. Röpke, Double-detonation supernovae of sub-Chandrasekhar mass white dwarfs. Astron. Astrophys. **476**, 1133–1143 (2007).

M. Fink, F.K. Röpke, W. Hillebrandt, I.R. Seitenzahl, S.A. Sim, M. Kromer, Double-detonation sub-Chandrasekhar supernovae: can minimum helium shell masses detonate the core? Astron. Astrophys. **514**, 53 (2010).

M. Fink, M. Kromer, I.R. Seitenzahl, F. Ciaraldi-Schoolmann, F.K. Röpke, S.A. Sim, R. Pakmor, A.J. Ruiter, W. Hillebrandt, Three-dimensional pure deflagration models with nucleosynthesis and synthetic observables for Type Ia supernovae. Mon. Not. R. Astron. Soc. **438**, 1762–1783 (2014).

R.J. Foley, R. Chornock, A.V. Filippenko, M. Ganeshalingam, R.P. Kirshner, W. Li, S.B. Cenko, P.J. Challis, A.S. Friedman, M. Modjaz, J.M. Silverman, W.M. Wood-Vasey, SN 2008ha: an extremely low luminosity and exceptionally low energy supernova. Astron. J. **138**, 376–391 (2009).

R.J. Foley, P.J. Challis, R. Chornock, M. Ganeshalingam, W. Li, G.H. Marion, N.I. Morrell, G. Pignata, M.D. Stritzinger, J.M. Silverman, X. Wang, J.P. Anderson, A.V. Filippenko, W.L. Freedman, M. Hamuy, S.W. Jha, R.P. Kirshner, C. McCully, S.E. Persson, M.M. Phillips, D.E. Reichart, A.M. Soderberg, Type Iax supernovae: a new class of stellar explosion. Astrophys. J. **767**, 57 (2013).

R.J. Foley, S.W. Jha, Y.-C. Pan, W.K. Zheng, L. Bildsten, A.V. Filippenko, D. Kasen, Late-time spectroscopy of Type Iax supernovae. Mon. Not. R. Astron. Soc. **461**, 433–457 (2016).

C.L. Fryer, A.J. Ruiter, K. Belczynski, P.J. Brown, F. Bufano, S. Diehl, C.J. Fontes, L.H. Frey, S.T. Holland, A.L. Hungerford, S. Immler, P. Mazzali, C. Meakin, P.A. Milne, C. Raskin, F.X. Timmes, Spectra of Type Ia supernovae from double degenerate mergers. Astrophys. J. **725**, 296–308 (2010).

L.R. Gasques, A.V. Afanasjev, E.F. Aguilera, M. Beard, L.C. Chamon, P. Ring, M. Wiescher, D.G. Yakovlev, Nuclear fusion in dense matter: reaction rate and carbon burning. Phys. Rev. C **72**(2), 025806 (2005).

D.A. Goldstein, D. Kasen, Evidence for sub-Chandrasekhar mass Type Ia supernovae from an extensive survey of radiative transfer models. Astrophys. J. **852**, 33 (2018).

E.P. Hicks, R. Rosner, Gravitationally unstable flames: Rayleigh-Taylor stretching versus turbulent wrinkling. Astrophys. J. **771**, 135 (2013).

P. Höflich, J. Stein, On the thermonuclear runaway in Type Ia supernovae: how to run away? Astrophys. J. **568**, 779–790 (2002).

P. Höflich, J.C. Wheeler, F.K. Thielemann, Type Ia supernovae: influence of the initial composition on the nucleosynthesis, light curves, and spectra and consequences for the determination of Ω_M and Λ. Astrophys. J. **495**, 617–629 (1998).

B. Hristov, D.C. Collins, P. Hoeflich, C.A. Weatherford, T.R. Diamond, Magneto-hydrodynamical effects on nuclear deflagration fronts in Type Ia supernovae. ArXiv e-prints (2017)

A.P. Jackson, D.M. Townsley, A.C. Calder, Power-law wrinkling turbulence-flame interaction model for astrophysical flames. Astrophys. J. **784**, 174 (2014).

A. Jerkstrand, Spectra of supernovae in the nebular phase, in *Handbook of Supernovae*, ed. by A. Alsabti, P. Murdin (Springer, Berlin, 2017), pp. 795–842

G.C. Jordan IV, C. Graziani, R.T. Fisher, D.M. Townsley, C. Meakin, K. Weide, L.B. Reid, J. Norris, R. Hudson, D.Q. Lamb, The detonation mechanism of the pulsationally assisted gravitationally confined detonation model of type Ia supernovae. Astrophys. J. **759**, 53 (2012a).

G.C. Jordan IV, H.B. Perets, R.T. Fisher, D.R. van Rossum, Failed-detonation Supernovae: subluminous low-velocity Ia supernovae and their kicked remnant white dwarfs with iron-rich cores. Astrophys. J. **761**, 23 (2012b).

A.H. Karp, G. Lasher, K.L. Chan, E.E. Salpeter, The opacity of expanding media—the effect of spectral lines. Astrophys. J. **214**, 161–178 (1977).

D. Kasen, S.E. Woosley, On the origin of the Type Ia supernova width-luminosity relation. Astrophys. J. **656**, 661–665 (2007).

D. Kasen, R.C. Thomas, P. Nugent, Time-dependent Monte Carlo radiative transfer calculations for three-dimensional supernova spectra, light curves, and polarization. Astrophys. J. **651**, 366–380 (2006).

D. Kasen, F.K. Röpke, S.E. Woosley, The diversity of type Ia supernovae from broken symmetries. Nature **460**, 869–872 (2009).

A.M. Khokhlov, Propagation of turbulent flames in supernovae. Astrophys. J. **449**, 695–713 (1995).

A.M. Khokhlov, E.S. Oran, J.C. Wheeler, Deflagration-to-detonation transition in thermonuclear supernovae. Astrophys. J. **478**, 678–688 (1997).

M. Kromer, S.A. Sim, Time-dependent three-dimensional spectrum synthesis for Type Ia supernovae. Mon. Not. R. Astron. Soc. **398**, 1809–1826 (2009).

M. Kromer, S.A. Sim, M. Fink, F.K. Röpke, I.R. Seitenzahl, W. Hillebrandt, Double-detonation sub-Chandrasekhar supernovae: synthetic observables for minimum helium shell mass models. Astrophys. J. **719**, 1067–1082 (2010).

M. Kromer, M. Fink, V. Stanishev, S. Taubenberger, F. Ciaraldi-Schoolman, R. Pakmor, F.K. Röpke, A.J. Ruiter, I.R. Seitenzahl, S.A. Sim, G. Blanc, N. Elias-Rosa, W. Hillebrandt, 3D deflagration simulations leaving bound remnants: a model for 2002cx-like Type Ia supernovae. Mon. Not. R. Astron. Soc. **429**, 2287–2297 (2013).

M. Kromer, S.T. Ohlmann, R. Pakmor, A.J. Ruiter, W. Hillebrandt, K.S. Marquardt, F.K. Röpke, I.R. Seitenzahl, S.A. Sim, S. Taubenberger, Deflagrations in hybrid CONe white dwarfs: a route to explain the faint Type Iax supernova 2008ha. Mon. Not. R. Astron. Soc. **450**, 3045–3053 (2015).

M. Kuhlen, S.E. Woosley, G.A. Glatzmaier, Carbon ignition in Type Ia supernovae. II. A three-dimensional numerical model. Astrophys. J. **640**, 407–416 (2006).

H.J.G.L.M. Lamers, J.P. Casinelli, *Introduction to Stellar Winds* (Cambridge University Press, Cambridge, 1999)

A.M. Lisewski, W. Hillebrandt, S.E. Woosley, Constraints on the delayed transition to detonation in Type Ia supernovae. Astrophys. J. **538**, 831–836 (2000a).

A.M. Lisewski, W. Hillebrandt, S.E. Woosley, J.C. Niemeyer, A.R. Kerstein, Distributed burning in Type Ia supernovae: a statistical approach. Astrophys. J. **537**, 405–413 (2000b).

M.R. Magee, R. Kotak, S.A. Sim, M. Kromer, D. Rabinowitz, S.J. Smartt, C. Baltay, H.C. Campbell, T.-W. Chen, M. Fink, A. Gal-Yam, L. Galbany, W. Hillebrandt, C. Inserra, E. Kankare, L. Le Guillou, J.D. Lyman, K. Maguire, R. Pakmor, F.K. Röpke, A.J. Ruiter, I.R. Seitenzahl, M. Sullivan, S. Valenti, D.R. Young, The type Iax supernova, SN 2015H. A white dwarf deflagration candidate. Astron. Astrophys. **589**, 89 (2016).

P.A. Mazzali, L.B. Lucy, The application of Monte Carlo methods to the synthesis of early-time supernovae spectra. Astron. Astrophys. **279**, 447–456 (1993).

P.A. Mazzali, F.K. Röpke, S. Benetti, W. Hillebrandt, A common explosion mechanism for Type Ia supernovae. Science **315**, 825–828 (2007).

R. Moll, S.E. Woosley, Multi-dimensional models for double detonation in sub-Chandrasekhar mass white dwarfs. Astrophys. J. **774**, 137 (2013).

J.C. Niemeyer, W. Hillebrandt, Turbulent nuclear flames in type Ia supernovae. Astrophys. J. **452**, 769–778 (1995).

U.M. Noebauer, S. Taubenberger, S. Blinnikov, E. Sorokina, W. Hillebrandt, Type Ia supernovae within dense carbon- and oxygen-rich envelopes: a model for 'super-Chandrasekhar' explosions? Mon. Not. R. Astron. Soc. **463**, 2972–2985 (2016).

A. Nonaka, A.J. Aspden, M. Zingale, A.S. Almgren, J.B. Bell, S.E. Woosley, High-resolution simulations of convection preceding ignition in Type Ia supernovae using adaptive mesh refinement. Astrophys. J. **745**, 73 (2012).

P. Nugent, E. Baron, D. Branch, A. Fisher, P.H. Hauschildt, Synthetic spectra of hydrodynamic models of Type IA supernovae. Astrophys. J. **485**, 812–819 (1997).

S. Osher, J.A. Sethian, Fronts propagating with curvature-dependent speed: algorithms based on Hamilton–Jacobi formulations. J. Comput. Phys. **79**, 12–49 (1988)

R. Pakmor, M. Kromer, F.K. Röpke, S.A. Sim, A.J. Ruiter, W. Hillebrandt, Sub-luminous Type IA supernovae from the mergers of equal-mass white dwarfs with mass $\sim 0.9 m_{\odot}$. Nature **463**, 61–64 (2010).

R. Pakmor, S. Hachinger, F.K. Röpke, W. Hillebrandt, Violent mergers of nearly equal-mass white dwarf as progenitors of subluminous Type Ia supernovae. Astron. Astrophys. **528**, 117 (2011).

R. Pakmor, M. Kromer, S. Taubenberger, S.A. Sim, F.K. Röpke, W. Hillebrandt, Normal Type Ia supernovae from violent mergers of white dwarf binaries. Astrophys. J. **747**, 10 (2012).
R. Pakmor, M. Kromer, S. Taubenberger, V. Springel, Helium-ignited violent mergers as a unified model for normal and rapidly declining type Ia supernovae. Astrophys. J. **770**, 8 (2013).
M.M. Phillips, W. Li, J.A. Frieman, S.I. Blinnikov, D. DePoy, J.L. Prieto, P. Milne, C. Contreras, G. Folatelli, N. Morrell, M. Hamuy, N.B. Suntzeff, M. Roth, S. González, W. Krzeminski, A.V. Filippenko, W.L. Freedman, R. Chornock, S. Jha, B.F. Madore, S.E. Persson, C.R. Burns, P. Wyatt, D. Murphy, R.J. Foley, M. Ganeshalingam, F.J.D. Serduke, K. Krisciunas, B. Bassett, A. Becker, B. Dilday, J. Eastman, P.M. Garnavich, J. Holtzman, R. Kessler, H. Lampeitl, J. Marriner, S. Frank, J.L. Marshall, G. Miknaitis, M. Sako, D.P. Schneider, K. van der Heyden, N. Yasuda, The peculiar SN 2005hk: do some Type Ia supernovae explode as deflagrations? Publ. Astron. Soc. Pac. **119**, 360–387 (2007).
P.A. Pinto, R.G. Eastman, The physics of Type Ia supernova light curves. II. Opacity and diffusion. Astrophys. J. **530**, 757–776 (2000).
T. Plewa, A.C. Calder, D.Q. Lamb, Type Ia supernova explosion: gravitationally confined detonation. Astrophys. J. **612**, 37–40 (2004).
A.Y. Poludnenko, T.A. Gardiner, E.S. Oran, Spontaneous transition of turbulent flames to detonations in unconfined media. Phys. Rev. Lett. **107**(5), 054501 (2011).
M. Reinecke, W. Hillebrandt, J.C. Niemeyer, R. Klein, A. Gröbl, A new model for deflagration fronts in reactive fluids. Astron. Astrophys. **347**, 724–733 (1999)
F.K. Röpke, Combustion in thermonuclear supernova explosions, in *Handbook of Supernovae*, ed. by A. Alsabti, P. Murdin (Springer, Berlin, 2017), pp. 1185–1209
F.K. Röpke, J.C. Niemeyer, Delayed detonations in full-star models of type Ia supernova explosions. Astron. Astrophys. **464**, 683–686 (2007).
F.K. Röpke, W. Schmidt, Turbulent combustion in thermonuclear supernovae, in *Interdisciplinary Aspects of Turbulence*, ed. by W. Hillebrandt, F. Kupka Lecture Notes in Physics (Springer, Berlin, 2009), pp. 255–289.
F.K. Röpke, S.E. Woosley, W. Hillebrandt, Off-center ignition in Type Ia supernovae. I. Initial evolution and implications for delayed detonation. Astrophys. J. **660**, 1344–1356 (2007).
A.J. Ruiter, S.A. Sim, R. Pakmor, M. Kromer, I.R. Seitenzahl, K. Belczynski, M. Fink, M. Herzog, W. Hillebrandt, F.K. Röpke, S. Taubenberger, On the brightness distribution of Type Ia supernovae from violent white dwarf mergers. Mon. Not. R. Astron. Soc. **429**, 1425–1436 (2013).
R. Scalzo, G. Aldering, P. Antilogus, C. Aragon, S. Bailey, C. Baltay, S. Bongard, C. Buton, F. Cellier-Holzem, M. Childress, N. Chotard, Y. Copin, H.K. Fakhouri, E. Gangler, J. Guy, A.G. Kim, M. Kowalski, M. Kromer, J. Nordin, P. Nugent, K. Paech, R. Pain, E. Pecontal, R. Pereira, S. Perlmutter, D. Rabinowitz, M. Rigault, K. Runge, C. Saunders, S.A. Sim, G. Smadja, C. Tao, S. Taubenberger, R.C. Thomas, B.A. Weaver (Nearby Supernova Factory), Type Ia supernova bolometric light curves and ejected mass estimates from the Nearby Supernova Factory. Mon. Not. R. Astron. Soc. **440**, 1498–1518 (2014).
W. Schmidt, J.C. Niemeyer, W. Hillebrandt, F.K. Röpke, A localised subgrid scale model for fluid dynamical simulations in astrophysics. II. Application to type Ia supernovae. Astron. Astrophys. **450**, 283–294 (2006).
W. Schmidt, F. Ciaraldi-Schoolmann, J.C. Niemeyer, F.K. Röpke, W. Hillebrandt, Turbulence in a three-dimensional deflagration model for Type Ia supernovae. II. Intermittency and the deflagration-to-detonation transition probability. Astrophys. J. **710**, 1683–1693 (2010).
I.R. Seitenzahl, F. Ciaraldi-Schoolmann, F.K. Röpke, M. Fink, W. Hillebrandt, M. Kromer, R. Pakmor, A.J. Ruiter, S.A. Sim, S. Taubenberger, Three-dimensional delayed-detonation models with nucleosynthesis for Type Ia supernovae. Mon. Not. R. Astron. Soc. **429**, 1156–1172 (2013).
I.R. Seitenzahl, M. Kromer, S.T. Ohlmann, F. Ciaraldi-Schoolmann, K. Marquardt, M. Fink, W. Hillebrandt, R. Pakmor, F.K. Röpke, A.J. Ruiter, S.A. Sim, S. Taubenberger, Three-dimensional simulations of gravitationally confined detonations compared to observations of SN 1991T. Astron. Astrophys. **592**, 57 (2016).
G.J. Sharpe, The structure of steady detonation waves in Type Ia supernovae: pathological detonations in C-O cores. Mon. Not. R. Astron. Soc. **310**, 1039–1052 (1999).
K.J. Shen, L. Bildsten, Thermally stable nuclear burning on accreting white dwarfs. Astrophys. J. **660**, 1444–1450 (2007).
K.J. Shen, L. Bildsten, The ignition of carbon detonations via converging shock waves in white dwarfs. Astrophys. J. **785**, 61 (2014).
K.J. Shen, S. Toonen, O. Graur, The evolution of the Type Ia supernova luminosity function. Astrophys. J. **851**, 50 (2017).
T. Shigeyama, K. Nomoto, H. Yamaoka, F. Thielemann, Possible models for the Type IA supernova 1990N. Astrophys. J. **386**, 13–16 (1992).

S.A. Sim, F.K. Röpke, W. Hillebrandt, M. Kromer, R. Pakmor, M. Fink, A.J. Ruiter, I.R. Seitenzahl, Detonations in sub-Chandrasekhar-mass C + O white dwarfs. Astrophys. J. **714**, 52–57 (2010).
S.A. Sim, I.R. Seitenzahl, M. Kromer, F. Ciaraldi-Schoolmann, F.K. Röpke, M. Fink, W. Hillebrandt, R. Pakmor, A.J. Ruiter, S. Taubenberger, Synthetic light curves and spectra for three-dimensional delayed-detonation models of Type Ia supernovae. Mon. Not. R. Astron. Soc. **436**, 333–347 (2013).
V.V. Sobolev, *Moving Envelopes of Stars* (Harvard University Press, Cambridge, 1960)
M.D. Stritzinger, S. Valenti, P. Hoeflich, E. Baron, M.M. Phillips, F. Taddia, R.J. Foley, E.Y. Hsiao, S.W. Jha, C. McCully, V. Pandya, J.D. Simon, S. Benetti, P.J. Brown, C.R. Burns, A. Campillay, C. Contreras, F. Förster, S. Holmbo, G.H. Marion, N. Morrell, G. Pignata, Comprehensive observations of the bright and energetic Type Iax SN 2012Z: interpretation as a Chandrasekhar mass white dwarf explosion. Astron. Astrophys. **573**, 2 (2015).
S. Taubenberger, The extremes of thermonuclear supernovae, in *Handbook of Supernovae*, ed. by A. Alsabti, P. Murdin (Springer, Cham, 2017), pp. 317–373
F.X. Timmes, S.E. Woosley, The conductive propagation of nuclear flames. I. Degenerate C + O and O + Ne + Mg white dwarfs. Astrophys. J. **396**, 649–667 (1992).
C. Travaglio, W. Hillebrandt, M. Reinecke, F.-K. Thielemann, Nucleosynthesis in multi-dimensional SN Ia explosions. Astron. Astrophys. **425**, 1029–1040 (2004).
N. Vladimirova, G.V. Weirs, L. Ryzhik, Flame capturing with an advection-reaction-diffusion model. Combust. Theory Model. **10**(5), 727–747 (2006).
S.E. Woosley, Type Ia supernovae: burning and detonation in the distributed regime. Astrophys. J. **668**, 1109–1117 (2007).
S.E. Woosley, D. Kasen, Sub-Chandrasekhar mass models for supernovae. Astrophys. J. **734**, 38 (2011).
S.E. Woosley, D. Kasen, S. Blinnikov, E. Sorokina, Type Ia supernova light curves. Astrophys. J. **662**, 487–503 (2007).
S.E. Woosley, A.R. Kerstein, V. Sankaran, A.J. Aspden, F.K. Röpke, Type Ia supernovae: calculations of turbulent flames using the Linear Eddy Model. Astrophys. J. **704**, 255–273 (2009).
S.E. Woosley, A.R. Kerstein, A.J. Aspden, Flames in type Ia supernova: deflagration-detonation transition in the oxygen-burning flame. Astrophys. J. **734**, 37 (2011).
Y.B. Zel'dovich, V.B. Librovich, G.M. Makhviladze, G.I. Sivashinskii, On the onset of detonation in a nonuniformly heated gas. J. Appl. Mech. Tech. Phys. **11**, 264–270 (1970).
M. Zingale, S.E. Woosley, C.A. Rendleman, M.S. Day, J.B. Bell, Three-dimensional numerical simulations of Rayleigh-Taylor unstable flames in Type Ia supernovae. Astrophys. J. **632**, 1021–1034 (2005).
M. Zingale, A.S. Almgren, J.B. Bell, A. Nonaka, S.E. Woosley, Low Mach number modeling of Type IA supernovae. IV. White dwarf convection. Astrophys. J. **704**, 196–210 (2009).

Space Sci Rev (2018) 214:31
https://doi.org/10.1007/s11214-018-0468-7

Neutrinos from Supernovae

Irene Tamborra[1] · **Kohta Murase**[2]

Received: 22 November 2017 / Accepted: 6 January 2018 / Published online: 23 January 2018

Abstract Neutrinos are fundamental particles in the collapse of massive stars. Because of their weakly interacting nature, neutrinos can travel undisturbed through the stellar core and be direct probes of the still uncertain and fascinating supernova mechanism. Intriguing recent developments on the role of neutrinos during the stellar collapse are reviewed, as well as our current understanding of the flavor conversions in the stellar envelope. The detection perspectives of the next burst and of the diffuse supernova background will be also outlined. High-energy neutrinos in the GeV-PeV range can follow the MeV neutrino emission. Various scenarios concerning the production of high-energy neutrinos are discussed.

Keywords Core-collapse supernova · Neutrinos · Flavor oscillations · Neutrino telescopes

1 Introduction

Three decades ago, the SN 1987A was detected in neutrinos. This has been the first and, up to now, the only supernova (SN) seen through neutrinos. Although very few neutrino events were detected at that time, these events allowed us to directly prove the stellar collapse theory and constrain properties of particles beyond the ones foreseen by the Standard Model of Particle Physics (Raffelt 1990). Since then, substantial progress has been made as for our understanding of the physics leading to the stellar explosion (Janka 2012; Janka et al.

Supernovae
Edited by Andrei Bykov, Roger Chevalier, John Raymond, Friedrich-Karl Thielemann, Maurizio Falanga and Rudolf von Steiger

✉ I. Tamborra
tamborra@nbi.ku.dk

K. Murase
murase@psu.edu

1 Niels Bohr Institute, Blegdamsvej 17, 2100 Copenhagen, Denmark

2 Department of Physics; Department of Astronomy & Astrophysics; Center for Particle and Gravitational Astrophysics, The Pennsylvania State University, University Park, Pennsylvania 16802, USA

2016; Janka 2017), the role of neutrinos in the star, and the flavor conversion physics in neutrino-dense media (Mirizzi et al. 2016; Chakraborty et al. 2016a). Several large scale detectors are working (or they will be soon) to detect the next galactic burst (Mirizzi et al. 2016; Scholberg 2012; Lang et al. 2016). Nevertheless, we are still far from fully grasping the core collapse physics and the neutrino oscillation phenomenology. In this sense, the detection of the next galactic burst will provide us with a precious test of our understanding of the stellar dynamics with high event statistics.

Besides the single SN burst, the detection of the diffuse SN neutrino background (DSNB), i.e. the neutrino flux emitted from all SNe exploding somewhere in the Universe, is approaching with the upcoming enrichment of with gadolinium of Super-Kamiokande (Fernández 2016) and the JUNO experiment (An et al. 2016). The DSNB will allow us to learn about the stellar population, other than provide with an independent test of the SN rate (Mirizzi et al. 2016; Beacom 2010; Lunardini 2016). In what follows, we will outline some of the open issues in SN and neutrino astrophysics, as well as future directions.

After the MeV neutrino burst, in the later phase, higher-energy neutrinos can be produced non-thermally. High-energy neutrinos can be produced in outflows (jets or winds) launched from a proto-neutron star. One of the possibilities is caused by the entrainment of neutrons (Bahcall and Mészáros 2000). GeV-TeV neutrinos can be produced via inelastic neutron-proton collisions if there is a relativistic outflow embedded in the SN ejecta. The second possibility involves the acceleration of cosmic rays, although the Fermi acceleration in dense and/or radiation-rich environments is not guaranteed in general (Murase and Ioka 2013). Such SNe accompanied by relativistic outflows may be rare, but the investigation of that population is important to reveal the connection between gamma-ray bursts (GRBs) and SNe (Mészáros and Waxman 2001).

Another more guaranteed neutrino production occurs by interaction between SN ejecta and ambient circumstellar material (CSM). Recent observations of optical and infrared transients have revealed that the violent eruption of progenitor masses or the inflation of the progenitor envelope is commonly found in the SN data (see Smith 2014; Yaron et al. 2017, and references therein), which leads to new prospects for detecting high-energy neutrino signals from nearby SNe (Murase 2017).

2 Features of the Neutrino Signal from the Core-Collapse Burst

A core-collapse SN explosion originates from the death of stars with $M > 8\ M_\odot$ (Janka 2017). The iron core is surrounded by outer layers with light elements deposited in the outer shells; once the Chandrasekhar limit is reached, the core collapses and the explosion is triggered. 99% of the explosion energy is released in neutrinos with average energies of $\mathcal{O}(10)$ MeV. The neutrino signal is expected to last for about 10 s.

The neutrino signal from the core-collapse can be divided in three time windows, as shown in Fig. 1. The *neutronization burst* is marked by a large peak in the ν_e luminosity (generated because of the rapid electron capture by nuclei and free protons, as the shock wave crosses the iron core dissociating its nuclei); the *accretion phase* is characterized by large differences among the fluxes of different flavors, especially between the electron and non-electron flavors; the *cooling phase* where the neutrino emission properties of the different flavors become very similar and the luminosity progressively decreases. Neutrino flavor conversions could have the largest impact on the neutrino energy distributions during the neutronization phase and accretion phase, where the electron and non-electron neutrino fluxes are very different between themselves.

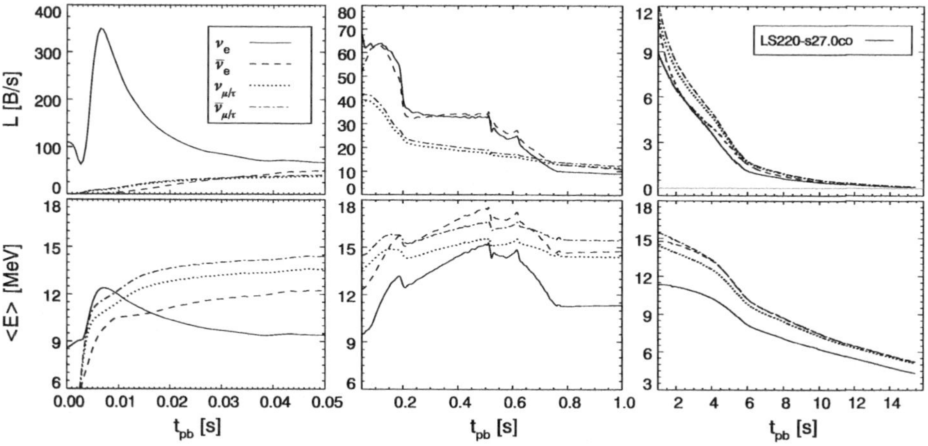

Fig. 1 Neutrino luminosities (on the *top*) and mean energies (on the *bottom*) for the three flavors [*continue line* stands for ν_e's, *dashed line* for $\bar{\nu}_e$'s, *dotted* (*dashed-dotted*) *line* for $\nu_{\mu,\tau}$ ($\bar{\nu}_{\mu,\tau}$)] as a function of the post-bounce time for a 27 $M_\odot$ SN progenitor. The panels on the left refer to the neutronization phase, the middle panels to the accretion phase, and the panels on the right to the cooling phase. Figure adapted from Mirizzi et al. (2016)

3 Intriguing Neutrino Features in Three-Dimensional Supernova Simulations

Hydrodynamical simulations of the core-collapse have recently reached the three-dimensional modelling front (Janka et al. 2016; Janka 2017). The first successful explosions have been obtained in 3D by several groups (Melson et al. 2015; Lentz et al. 2015; Takiwaki et al. 2016; Roberts et al. 2016) with different degrees of sophistication. Although the SN modeling still needs to be refined and long-term 3D simulations of the core-collapse are not yet available, the available hydro simulations already proved to be a precious benchmark to test the neutrino mechanism.

The current paradigm adopted to explain the SN explosion is the delayed neutrino explosion mechanism (Bethe and James 1985), according to which neutrinos provide new energy to revive the stalled shock wave and trigger the explosion. During the shock revival phase, hydrodynamical instabilities occur, such as convective overturns and the standing accretion shock instability (SASI). Those instabilities contribute to enhance the efficiency of the energy transfer between the neutrinos and the stalled shock wave.

Recent three-dimensional SN simulations suggest that the neutrino signal carries clear imprints of such instabilities (Tamborra et al. 2013, 2014b). In fact the SASI activity is responsible for large-amplitude sinusoidal modulations in the neutrino signal clearly detectable with neutrino telescopes such as IceCube and Hyper-Kamiokande, as shown in the top panel of Fig. 2. The event rate shown in Fig. 2 has been derived for a 27 $M_\odot$ progenitor; this progenitor exhibits two SASI phases interspersed by one convective phase. The frequency of the SASI sloshing motions can be extrapolated from the power spectrum of the neutrino signal plotted in the bottom panel of Fig. 2. Indeed a clear peak is visible in the power spectrum around 85 Hz which corresponds to the SASI modulation frequency.

The gravitational wave signal from the same 3D progenitor models has also been investigated (Andresen et al. 2017). An enhanced activity below 250 Hz has been observed in correspondence of the SASI episodes. This means that we will be able to probe our current

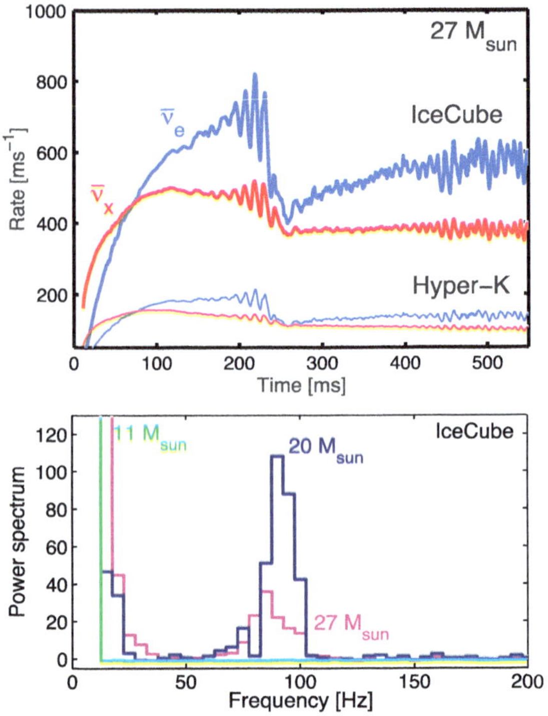

Fig. 2 *Top panel*: Detection rate in IceCube and Hyper-Kamiokande as a function of the post-bounce time for a 27 $M_{\odot}$ SN progenitor for an observer at 10 kpc located along a direction where the neutrino signal shows strong SASI modulations. *Bottom panel*: Power spectrum of the neutrino signal seen in the IceCube detector for 27, 20 and 11 $M_{\odot}$ SN progenitors. A peak in the power spectrum is visible at about 85 Hz corresponding to the SASI frequency. Figure adapted from Tamborra et al. (2013, 2014b)

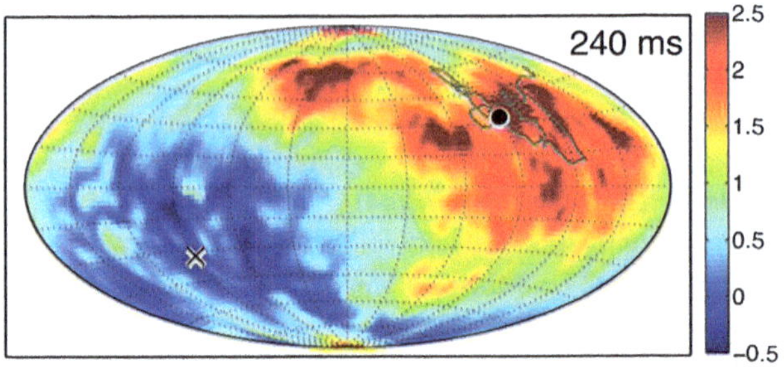

Fig. 3 Neutrino lepton number flux ($\nu_e - \bar{\nu}_e$) normalized by its average value for a 11.2 $M_{\odot}$ progenitor at $t_{p.b.} = 240$ ms. Figure adapted from Tamborra et al. (2014a)

conjecture on the explosion mechanism by employing the joint observation of the signal from gravitational waves and neutrinos.

The advent of the 3D SN simulations has also led to the discovery of a new hydrodynamical instability: The lepton emission self-sustained asymmetry, LESA (Janka et al. 2016; Tamborra et al. 2014a). LESA is the first instability completely driven by neutrinos and it consists of an asymmetric emission of the ν_e number flux with respect to the $\bar{\nu}_e$ one as shown in Fig. 3. Such an asymmetry can reach up to the 20–30% with respect to the average. LESA is characterized by a large scale dipolar character, independent of the numerical grid of the simulation and of the SASI sloshing motions.

LESA is responsible for a strong directional dependence of the neutrino fluxes. The protoneutron-star convection is stronger in the hemisphere of maximum lepton number flux and the electron abundance emission is also asymmetric as a deleptonized shell in the hemisphere of minimal number flux forms. LESA may have a possible impact on SN nucleosynthesis, flavor oscillation physics and neutron star kicks.

Another recent development of interest for the neutrino detection is related to black-hole forming SNe. These bursts mainly occur when a SN collapses directly into a black hole; the neutrino signal abruptly ends after a few hundreds of ms (e.g., Sumiyoshi et al. 2006). In the past we thought that only progenitors with mass larger than 40 $M_{\odot}$ could evolve in black-hole forming SNe. Recent work shows that even low mass progenitors may collapse into black holes. Black-hole forming SN progenitors can reach up to the 20–40% of the total SN population (e.g., O'Connor and Ott 2011; Horiuchi et al. 2014; Ertl et al. 2016; Sukhbold et al. 2016). Black hole forming SNe may considerably enhance the expected diffuse SN neutrino background signal because of the different neutrino properties (Nakazato 2013; Mirizzi et al. 2016). A fraction of these bursts can be associated with the delayed collapse into a black hole, which may be accompanied by optical transients as well as neutrinos (Blum and Kushnir 2016; Kashiyama and Quataert 2015; Kashiyama et al. 2017). If the angular momentum is so large that the massive accretion disk is formed, the failed SNe could be associated with powerful transients such as hypernovae (MacFadyen and Woosley 1999), which can also be potential targets for multi-messenger facilities (e.g., Kobayashi and Meszaros 2003; Suwa and Murase 2009).

4 Neutrino Flavor Evolution in Supernovae

While neutrinos propagate in the stellar envelope, they interact among themselves rather than with the matter background. The flavor evolution of neutrinos in neutrino-dense media is described by matrices of densities whose elements along the main diagonal represent the neutrino survival probabilities. The equations describing the neutrino flavor evolution are

$$\partial_t + \mathbf{v} \cdot \nabla_p \rho(t, \mathbf{x}, \mathbf{p}) = -i\left[H(t, \mathbf{x}, \mathbf{p}), \rho(t, \mathbf{x}, \mathbf{p})\right] + \mathcal{C}\left[\rho(t, \mathbf{x}, \mathbf{p})\right], \tag{1}$$

where the Hamiltonian matrix is defined for each energy mode E and it contains vacuum, matter, and neutrino–neutrino terms: $H_E = H_E^{\text{vac}} + H_E^{\text{m}} + H_{E,\vartheta}^{\nu\nu}$; in particular, $H^{\nu\nu}$ depends on the angle between the momenta of the colliding neutrinos for each energy mode E (Mirizzi et al. 2016; Chakraborty et al. 2016a; Duan et al. 2010). The ν–ν interaction term makes the equations non-linear and it also depends on the angle between the momenta of the colliding neutrinos. Solving the full neutrino equations of motion means that we need to deal with a 7D problem involving quantities changing on different time scales. Simplifications are therefore required.

The resonant flavor conversion due to the interactions of the neutrinos with the matter background in the stellar envelope (MSW effect) is a well understood phenomenon (Mikheev and Smirnov 1986; Dighe and Smirnov 2000). As neutrinos stream from the decoupling region out in the SN envelope, the first MSW resonance is due to the atmospheric neutrino mass difference and it occurs at about 10^3 km from the neutrino sphere in the neutrino (antineutrino) channel for normal (inverted) mass ordering. The second MSW conversion takes place at larger radii and it is due to the smaller solar neutrino mass difference.

Neutrino self-interactions, on the other hand, occur at smaller radii, closer to the SN core. Neutrino–neutrino interactions are responsible for making the equations of motion non-linear and inducing exponentially growing instabilities in the flavor space. As our understanding of this phenomenon becomes more robust, growing attention is dedicated to

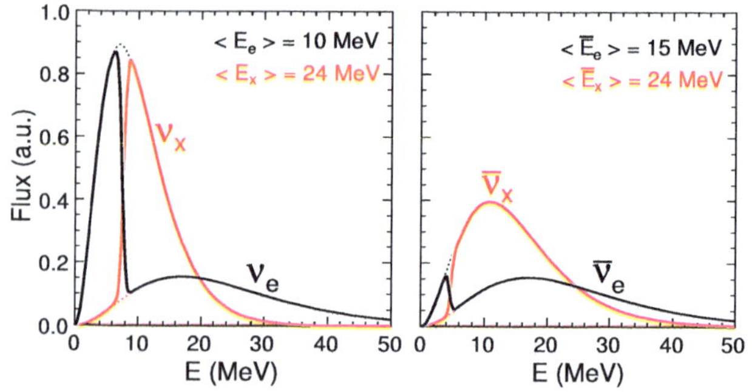

Fig. 4 Neutrino (on the left) and antineutrino fluxes (on the right) as a function of the energy after ν–ν interactions assuming inverted mass ordering. The fluxes at the neutrino sphere are shown as *dotted lines*. Figure taken from Fogli et al. (2008)

the role played by the neutrino angular distributions in triggering flavor conversions. In fact neutrino–neutrino interactions do not only depend on the neutrino energy spectra, but also on the angle between the momenta of the colliding neutrinos. Both variables are important in defining the final distribution of the neutrino fluxes.

The modeling of neutrino–neutrino interactions has been first developed by assuming a stationary and spherically symmetric SN within the so-called *neutrino bulb model* (Duan et al. 2006). According to this model, neutrinos of all flavors are emitted from each point of the neutrino-sphere in the forward solid angle uniformly and isotropically. Under this approximation, general features were found such as the *spectral split*: A complete swap of the neutrino energy spectra for certain energy ranges and according to the neutrino mass hierarchy, as shown in Fig. 4. Within this model, it was proved that the high matter density profile, typical of early post-bounce times (i.e., during the SN accretion phase), locks the neutrino modes inhibiting multi-angle flavor conversion effects ($N_e \geq N_\nu$) (Esteban-Pretel et al. 2008). On the other hand, during the cooling phase, where the fluxes of the neutrinos of different flavors are more similar to each other, multiple spectral splits were expected to occur (Fogli et al. 2009).

On the other hand, more recently, it has been pointed out that new instabilities may arise in the flavor space by perturbing the stationary, spherically symmetric SN bulb model. For example, within a simplified setup, it has been shown as breaking the axial symmetry (Raffelt et al. 2013), the spatial and directional symmetry (Duan and Shalgar 2015), or by introducing temporal instabilities (Abbar and Duan 2015), flavor conversions could be induced (i.e., flavor instabilities can be determined because of the non-homogeneous or non-stationary conditions occurring within the stellar envelope). The same should be expected by considering a neutrino angular distribution not limited to the outward direction, as well as in the presence of large 3D effects that make the system inhomogeneous, non-stationary and anisotropic (Tamborra et al. 2013, 2014a).

Recently, efforts have been devoted to understand the role of fast pairwise conversions in SNe (Sawyer 2005, 2016; Chakraborty et al. 2016b; Dasgupta et al. 2017; Izaguirre et al. 2017). Such conversions are different than ordinary flavor oscillations in the sense that it is not necessary to have a non-null vacuum term in the Hamiltonian in order for them to occur. While MSW conversions are responsible for a lepton flavor violation by mass and mixing, fast pairwise conversions are responsible for no lepton flavor change. Such pairwise conversions develop on timescales much shorter than ordinary flavor conversions as their typical frequency is given by the ν–ν interaction potential.

It has been pointed out that favorable conditions for fast conversions may be present already in the neutrino decoupling region (Sawyer 2005, 2016). This means that fast conversions, if they occur, could be potentially harmful to the SN dynamics. Note that up to

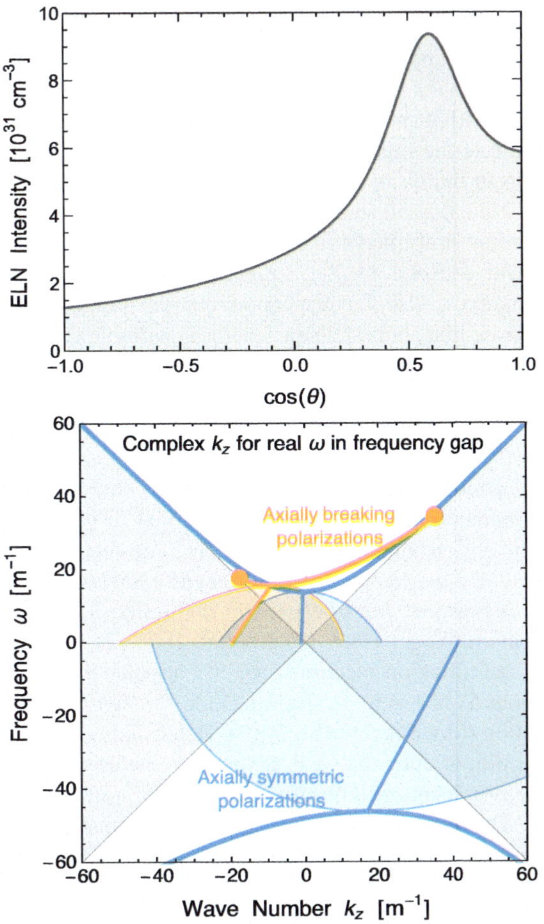

Fig. 5 *Top panel*: Electron neutrino lepton number angular distribution of a 15 $M_\odot$ SN simulation at 280 ms post bounce and at 37 km. No crossings in the electron neutrino lepton number are visible. *Bottom panel*: Dispersion relations for $\mathbf{k} = (0, 0, k_z)$ for the electron neutrino lepton number distribution shown in the left panel of this figure. The axially symmetric polarizations are plotted in *blue*, while the two degenerate axially breaking polarizations are drawn in *orange*. The filled regions mark complex solutions. Zone of avoidance are plotted in *gray*. Figures adapted from Izaguirre et al. (2017)

now, neutrino flavor conversions have not been considered relevant for the stellar dynamics as they were thought to occur at large radii, see e.g. Dasgupta et al. (2012).

A new approach to deal with fast pairwise conversions has been proposed (Izaguirre et al. 2017). The latter is based on a dispersion relation for the frequency and wavenumber $(\Omega, \mathbf{K})$ of disturbances in the local mean field of $\nu_e \nu_x$ coherence. By applying the dispersion relation approach, one finds that the local ν_e–$\bar{\nu}_e$ angular distribution is one crucial ingredient behind such conversions. Run-away solutions may occur when one has crossings in the neutrino electron lepton number angle distributions, giving origin to forbidden gaps in the k space. Otherwise, if we have a non-negligible backward flux, this can induce forbidden ω gaps. In the latter case, one needs to define appropriate boundary conditions for the problem in such a way to properly include a non-negligible flux of neutrinos streaming backwards, currently not consistently taken into account in the solution of the neutrino equations of motion.

As shown in the top panel of Fig. 5, while a non-negligible backward neutrino flux seems to be present in the neutrino decoupling region, crossings in the electron neutrino lepton number are not observed in existing 1D SN simulations (Tamborra et al. 2017), but they may occur, e.g. in the presence of LESA (Tamborra et al. 2014a; Izaguirre et al. 2017).

The bottom panel of Fig. 5 shows an example of dispersion relation for a mode $\mathbf{k} = (0, 0, k_z)$ applied to the angular distribution of the electron neutrino lepton number as from a 1D simulation of a 15 $M_\odot$ SN simulation at 280 ms post bounce and at 37 km (Izaguirre et al. 2017) shown in the left panel of the same figure. The blue and orange shaded regions indicate the imaginary part of k_z: k_z in the frequency gap has a real part (semi-thick line) plus or minus an imaginary part (edge of the blob).

Note that all studies developed up to now beyond the SN bulb model, should be considered as simplified cases of study. More work is needed to finally understand whether fast flavor conversions are responsible for inducing conversions that may affect the explosion dynamics. Also, the dispersion relation alone is not sufficient to prove the occurrence of fast conversions, but it tells us whether favorable conditions for them to occur exist.

5 What Could We Learn from the Next Supernova Burst?

Neutrinos reach the Earth earlier than photons, therefore they can be used to inform astronomers that a core-collapse is about to go off. To this purpose, a network of neutrino detectors around the world (SNEWS) has been established (Antonioli et al. 2004). SNEWS will alert astronomers in the event of a SN burst.

Neutrinos could be used to determine the angular location of a SN in the sky (Beacom and Vogel 1999; Tomàs et al. 2003). By employing the directionality of the scattering of neutrinos on electrons, the SN angular location could be determined with an error of about 5° in e.g. Super-Kamiokande. Triangulation would be useful to determine the SN location (Muehlbeier et al. 2013). Determining the SN location with neutrinos will be crucial for dim or optically weak SNe, where neutrinos could be the only detected particles. These measurements will be also important for multi-messenger searches.

The neutronization phase of the neutrino signal is independent of the progenitor mass and the nuclear equation of state. Therefore, in this time window, the SN neutrino signal can be used as a "standard candle." As a consequence, the neutronization phase could be exploited to determine the SN distance (Kachelriess et al. 2005). By reconstructing the neutrino light-curve in this time window, one can extrapolate the distance at which the SN is located.

The neutronization phase could also be of help to constrain the oscillation physics occurring within the stellar envelope. Since the slope of the $\bar{\nu}_e$ and $\bar{\nu}_x$ light-curves is different, the observed neutrino event rate will be sensitive to the neutrino mass ordering in, e.g., Cherenkov telescopes (Serpico et al. 2012). The same holds for the neutrino channel; if we consider the ν_e event rate as seen in, e.g., a liquid argon detector such as the upcoming DUNE, the absence (presence) of the peak of the neutronization burst will hint towards a normal (inverted) mass ordering (Wallace et al. 2016).

During the accretion phase, the neutrino signal is instead sensitive to the stellar dynamics. As discussed in the previous section, in this time window the neutrino signal carries characteristic signatures of the SASI motions and convective overturns, clearly detectable in, e.g., Cherenkov telescopes (Tamborra et al. 2013, 2014b). Those signatures are complementary to the ones coming from gravitational waves. Moreover, by combining the neutrino signal with the one coming from gravitational waves, one could also probe the core bounce time (Halzen and Raffelt 2009).

The cooling phase signal is strongly sensitive to the nuclear equation of state as well as to the SN progenitor mass. The exact composition in neutrinos of different flavors is responsible for determining the nucleosynthesis outcome in the neutrino driven wind (Duan et al. 2011; Pllumbi et al. 2015). One of the hottest unsolved problems in nuclear astrophysics

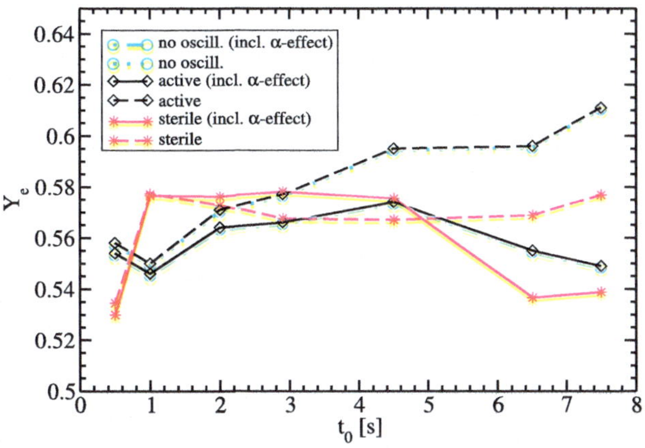

Fig. 6 Asymptotic electron fraction (Y_e) as a function of the post-bounce time (t_0) in the presence of only active states, 2 active+1 sterile states, and with and without oscillations. The electron abundance Y_e should be lower than 0.5 in order to create favorable conditions for the r-process. Figure adapted from Pllumbi et al. (2015)

is to pinpoint the location of the r-process, i.e., to find the sites where elements heavier than iron originate. The neutrino-driven wind of a core-collapse SN may be a natural site, however it is important to consider the exact relative ratio among neutrino fluxes of different flavors as the electron abundance is extremely sensitive to the ν_e and $\bar{\nu}_e$ exact energy spectra. Recent first attempts of coupling the oscillation codes to nucleosynthesis networks suggest that, even by taking into account the existence of an extra light sterile neutrino family, it is difficult to create a n-rich environment in the SN neutrino driven wind (i.e., $Y_e < 0.5$) and to activate the r-process (Pllumbi et al. 2015), see e.g. Fig. 6. However, the role of oscillations in the production of heavy elements remains to be clarified.

In synthesis, each of the three different time windows of the SN neutrino signal offers different opportunities to learn about the stellar collapse or neutrino properties. The neutronization burst is not sensitive to the nuclear equation of state and the progenitor mass and it allows us to learn about the distance from Earth at which a SN burst occurred. This will be especially helpful in the case of SNe dim or dark in photons. The accretion phase is strongly sensitive to the SN dynamics and its detection will be instrumental to test our conjectures on the SN explosion mechanism. The cooling phase of the neutrino signal is sensitive to the nuclear equation of state and to the progenitor mass. This third phase of the neutrino signal will be helpful to learn about the nuclear physics occurring in the star, the transition from the SN to the neutron star phases and the synthesis of new elements occurring in the neutrino driven wind.

6 Diffuse Supernova Neutrino Background

We expect two or three SNe to occur in our galaxy per century on average. However, a SN explodes every second in the Universe and we could detect the cumulative neutrino flux, the so-called diffuse SN neutrino background (DSNB) (e.g., Beacom 2010; Lunardini 2016; Mirizzi et al. 2016; Horiuchi et al. 2017). The DSNB should be clearly detectable in the region around 20–30 MeV, where it is expected to be above the reactor and atmospheric backgrounds.

Figure 7 shows the expected DSNB energy spectrum for ν_e (top panels) and $\bar{\nu}_e$ (bottom panels) in normal and inverted hierarchy (dashed and solid lines respectively), obtained by including MSW effects as well as neutrino self-interactions. While ν–ν interactions do not leave any energy dependent signature on the observable DSNB as it gets smeared out by the

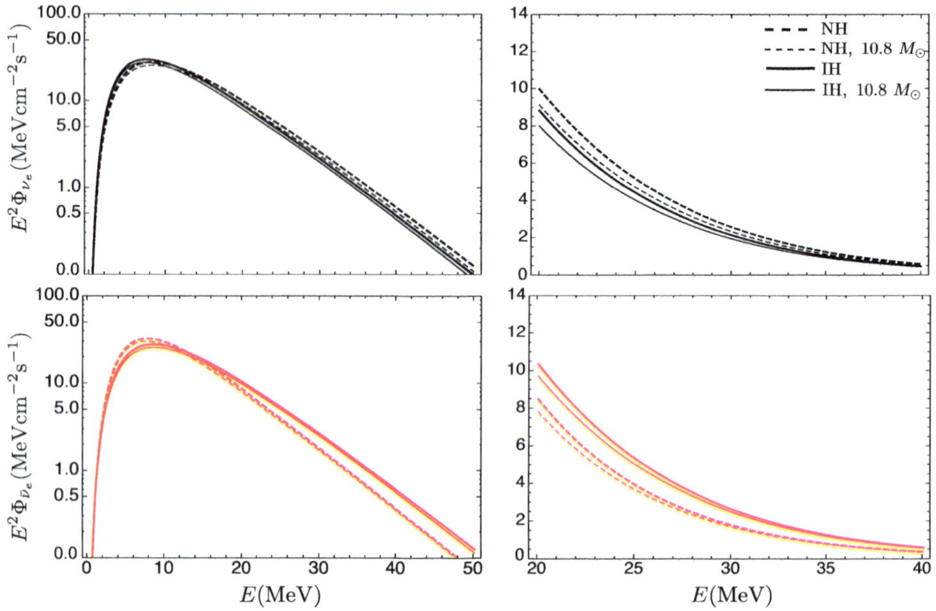

Fig. 7 *Left panel*: DSNB as a function of the energy, MSW effects and ν–ν interactions are included. *Right panel*: DSNB as a function of the energy, zoom in the region of interest of Super-Kamiokande. The ν_e flux is plotted in *black* (top panels) while the $\bar{\nu}_e$ flux is plotted in *red* (bottom panels). The *solid lines* refer to the DSNB in IH, while the *dashed one* represent the NH case. Figure taken from Lunardini and Tamborra (2012)

Fig. 8 Event rates as a function of the detected energy based on a estimation of the DSNB done by using a 27 $M_\odot$ progenitor as input, MSW effects under normal and inverted hierarchy assumptions are included. The rates for a 560-kton water Cherenkov detector are shown in *blue*, the rates for a 20-kton liquid scintillator are shown in *green* and the ones for a 40-kton liquid Argon detector are shown in *red*. Figure taken from Mirizzi et al. (2016)

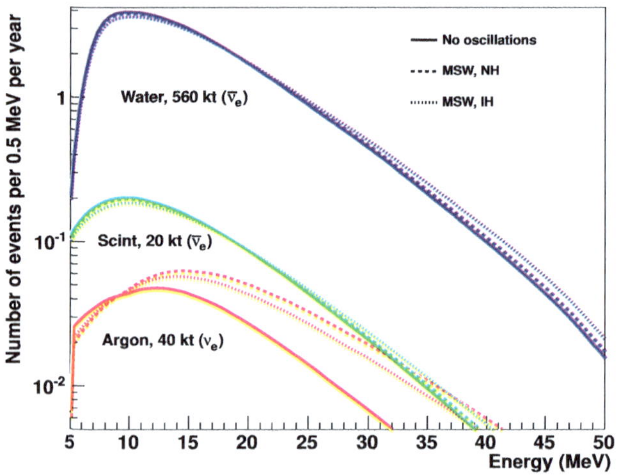

redshift integration, it is still clearly visible a difference between normal and inverted mass hierarchy.

The possibility of detecting the DSNB will be improved in the next future with the planned JUNO scintillator detector as well as with the approval of the Gd-project for the Super-K detector (Mirizzi et al. 2016). Figure 8 shows the expected event rate as a function of the detected energy for a 560-kton water Cherenkov detector, a 20 kton liquid scintillator and a 40-kton liquid Argon detector.

The detection of the DSNB will be instrumental to constrain the stellar population and it will also provide us with an independent test of the SN rate. Black-hole forming progenitors will contribute to the DNSB by creating a higher tail of the energy spectrum at high energies. Therefore, by reconstructing the measured DSNB, one can constrain the fraction of black-hole forming progenitors by using neutrinos. At the same time, the DSNB detection could help us to constrain the neutrino emission properties as well as exotic physics scenarios.

7 GeV-PeV Neutrinos from Hidden Relativistic Outflows

After the core-collapse, a proto-neutron star is left and cools via MeV neutrino emission. During the Kelvin-Helmholtz time scale, a fraction of these thermal neutrinos should deposit their energy in the atmosphere, and a neutrino-driven wind leads to significant mass losses (Qian and Woosley 1996). If the proto-neutron star is rotating and magnetized, the wind changes from nonrelativistic and thermally driven to relativistic and Poynting dominated winds, which could explain the diversity of SNe, including super-luminous SNe, hypernovae, and GRBs (Thompson et al. 2004). The wind may be entrained with neutrons, which will be composed of neutrons, protons, and nuclei. During the wind acceleration, the neutrons become relativistic due to collisional coupling with the ions. The ions will form a termination shock that is mediated by Coulomb collisions or plasma instabilities. However, the relativistic neutrons eventually undergo inelastic collisions around the termination shock inside the stellar material, producing GeV-TeV neutrinos (Murase et al. 2014a). The typical energy of the quasi-thermal neutrinos is given by

$$E_\nu \approx 0.05 \Gamma_n m_n c^2 \sim 0.5 \text{ GeV } (\Gamma_n/10), \qquad (2)$$

where Γ_n is the bulk Lorentz factor of the neutrons. The emission of GeV-TeV neutrinos is naturally expected for core-collapse SNe leaving rapidly rotating neutron starts, and detecting the GeV-TeV neutrino signals is possible with future neutrino detectors such as Hyper-K, for a Galactic SN event (Murase et al. 2014a). Possibly, if ions can be accelerated to ultra-high energies in the wind zone, higher-energy neutrinos could be produced although the acceleration mechanism is uncertain (Gaisser and Stanev 1987; Beall and Bednarek 2002; Murase et al. 2009; Fang et al. 2014).

More powerful relativistic outflows are gamma-ray burst (GRB) jets. The GRB jets are expected to have a large Lorentz factor, $\Gamma \sim 100$–1000. Ions and neutrons are tightly coupled in the early phase (Derishev et al. 1999; Bahcall and Mészáros 2000). If the decouple occurs during the acceleration phase, only the ion-component can be accelerated by the radiation pressure. Then, a large relative velocity between ion and neutron flows leads to inelastic np collisions and associated neutrino emission at a few GeV energies from pion and muon decays (Bahcall and Mészáros 2000; Meszaros and Rees 2000). For example, when a slower neutron outflow is overtaken by a faster ion-dominated or compound flow, we have (Murase et al. 2013; Bartos et al. 2013)

$$E_\nu \approx 0.05 \Gamma_r m_p c^2 \sim 150 \text{ GeV } (\Gamma/500), \qquad (3)$$

where Γ_r is the bulk Lorentz factor of the faster outflow and we have used the relative Lorentz factor between faster and slower outflows is ~ 3. If the decouple occurs in the coasting phase, internal shocks occur (Rees and Meszaros 1994) and GeV-TeV neutrinos

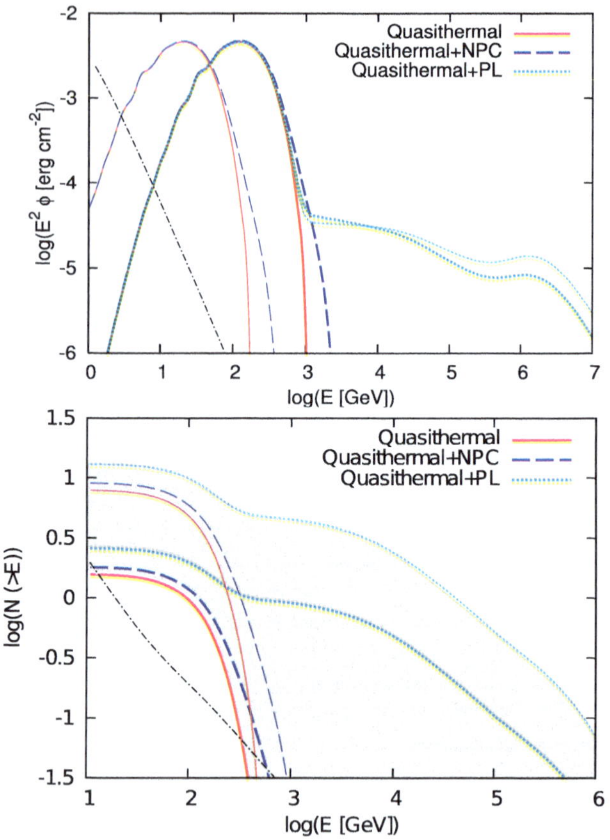

Fig. 9 *Top panel*: The energy fluence of muon neutrinos from a high-luminosity GRB at $z = 0.1$. The atmospheric neutrino background in 30 s is shown. *Bottom panel*: The expected number of muon neutrino events, which can be detected by coincident 20 yr observations with DeepCore+IceCube. The *dot-dashed curve* represents the atmospheric neutrino background. Figure taken from Murase et al. (2013)

are still produced by the inelastic dissipation of neutrons (Meszaros and Rees 2000; Murase et al. 2013).

The compound-flow model has been studied especially in the literature of prompt emission from GRBs. In the photospheric emission model, gamma-rays mainly originate from thermal or Comptonized emission rather than synchrotron radiation (Rees and Meszaros 2005). It has been shown that the spectrum of prompt emission is explained by Coulomb or other slow heating processes via the hadronic injection of secondary electron-positron pairs (Beloborodov 2010; Vurm et al. 2011; Murase et al. 2012). The slow heating model via inelastic *np* collisions (Beloborodov 2010; Vurm et al. 2011) inevitably predicts quasi-thermal neutrino emission in the 10–100 GeV range (see also Bykov and Meszaros 1996; Asano and Terasawa 2009; Murase et al. 2012, for the dissipation via internal shocks). As shown in Fig. 9, although detecting neutrinos from a single GRB is challenging, the signal can be detected by stacking analyses for ~ 10 yr data (Murase et al. 2013).

The quasi-thermal neutrino emission naturally occurs when neutrons are loaded in the outflow. If particle acceleration occurs in the jet, even higher-energy neutrinos in the TeV-PeV range can be produced. Sufficiently high-energy cosmic rays accelerated in the jet must be depleted inside the SN ejecta, so the system is calorimetric in the sense that the non-thermal energy is used for neutrino and gamma-ray production. Theoretically, it is natural to expect that a significant fraction of the GRB jets may fail to break through the stellar envelope or circumstellar material (Bromberg et al. 2011; Mizuta and Ioka 2013). This is

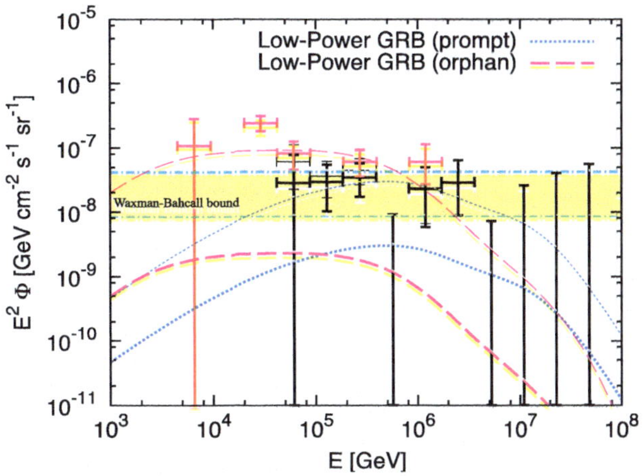

Fig. 10 The diffuse neutrino intensity from low-power GRBs. "Orphan" neutrinos correspond to the possible contribution from choked jets (Murase and Ioka 2013). For comparison, early calculations for prompt emission from low-luminosity GRBs are shown as "prompt" (Murase et al. 2006)

especially when the jets are less powerful or the progenitors are extended (as expected in blue supergiants and the first generation of stars) (Iocco et al. 2008; Murase and Ioka 2013; Matsumoto et al. 2015). Then optical or other electromagnetic emissions are blocked, and only neutrinos can escape. These high-energy neutrinos can be used as precious probes of not only relativistic jets embedded in the SN ejecta but also the properties of their progenitors, and would be particularly relevant to understand the physical connection between GRBs and SNe (Mészáros and Waxman 2001; Razzaque et al. 2003, 2004; Ando and Beacom 2005; Enberg et al. 2009; Bartos et al. 2012; Murase and Ioka 2013; Tamborra and Ando 2016; Senno et al. 2016).

However, the assumption of the cosmic-ray acceleration, which was often made *ad hoc* (Mészáros and Waxman 2001; Razzaque et al. 2003, 2004; Ando and Beacom 2005; Enberg et al. 2009; Bartos et al. 2012), is unrealistic for powerful GRB jets. The high radiation energy density deforms the shock structure (Levinson and Bromberg 2008; Murase and Ioka 2013), and the shock transition layer becomes thicker than the collisionless mean free path of particles. In such radiation-mediated shocks, low-energy particles cannot be efficiently accelerated to very high energies via the diffusive shock acceleration mechanism. Murase and Ioka (2013) considered radiation constraints on high-energy neutrino production, taking into account effects of the jet collimation. While high-power GRB jets are unlikely to be efficient neutrino emitters, low-power GRB jets are found to be more promising sources of TeV-PeV neutrinos. Since lower-power jets or more extended progenitors are more difficult to penetrate the star, it implies that "choked jets" or "failed GRBs" may be better neutrino sources (Murase and Ioka 2013; Senno et al. 2016). If the fraction of the choked GRBs is high, the choked jets may significantly contribute to the diffuse neutrino flux observed in IceCube (Murase and Ioka 2013; Bhattacharya et al. 2015; Tamborra and Ando 2016; Senno et al. 2016). See Fig. 10.

Finally, we note that another type of the Fermi acceleration mechanism may operate even in radiation-mediated shocks. When compound flows form internal shocks due to the inhomogeneity, cold neutrons can easily cross the shock transition layer and can be converted into protons in the downstream. Then, the protons are rapidly isotropized by magnetic fields, and some of the nucleons can go back to the upstream as neutrons. This neutron-proton-converter acceleration mechanism (Derishev et al. 2003) may efficiently work in high-luminosity GRB jets as the dominant nucleon acceleration mechanism at subphotospheres (Murase et al. 2013). Kashiyama et al. (2013) first performed numerical simulations, and showed that a good fraction of the incoming neutron energy is converted into high-energy nucleons, which can enhance the detectability of high-energy neutrinos.

8 GeV-PeV Neutrinos from Supernovae After Shock Breakout

SN remnants have been believed to be the primary sources of Galactic cosmic rays. The amount of energy of cosmic rays per SN event is thought to be $\mathcal{E}_{\rm CR} \sim 10^{50}$ erg, which corresponds to $\sim 10\%$ of the SN kinetic energy. However, the energy fraction carried by cosmic rays accelerated in the early phase (~ 0.1–10 months after the core collapse) had been expected to be negligibly small. This is simply because, in the free expansion phase, most of the energy is still in a form of the kinetic energy. The dissipated energy becomes significant only around the deceleration time, i.e. the Sedov time at a few hundred years after the explosion.

A SN shock propagates inside its progenitor star, and the shock eventually breaks out from the star (Waxman and Katz 2016). During the shock propagation in the star, the shock is radiation-mediated or even collisional, in which efficient Fermi acceleration does not occur. After the shock breakout, the SN ejecta forms a shock by interaction with the circumstellar material or inflated envelope (Chevalier and Fransson 2016). (For convenience, we use CSM for both cases of the stellar material.) In ordinary Type II-P, II-L, IIb, and Ibc SNe, the CSM density is low enough for the shock to be collisionless and radiation-unmediated. Thus, by analogy with SN remnants, we expect that the diffusive shock acceleration mechanism operates, but only a fraction of the kinetic energy is converted into non-thermal particles for typical mass-loss rates expected by the stellar evolution theory.

However, the situation has changed recently. The recent observations of transients in the optical and infrared bands have revealed that a dense material, which can be either an CSM or an inflated envelope, commonly exists in the immediate environment of a SN progenitor (see Smith 2014; Morozova et al. 2017; Yaron et al. 2017). Based on these SN measurements, Murase (2017) quantitatively predicted high-energy neutrino emission by developing a new time-dependent model based on the self-similar solutions (Chevalier 1982) for the SN dynamics. Surprisingly, for a Galactic SN, Murase (2017) found that Gton neutrino detectors like IceCube and KM3Net can detect ~ 10–1000 neutrinos from Type II-P, II-L, and IIb SNe that occur in our Galaxy. Note that these SNe are Type II SNe which form an ordinary, dominant class of core-collapse SNe. The time-dependent model also allows us to optimize the time window that is crucial for experimental searches for high-energy neutrinos. The targeted time window is ~ 0.1–1 day after the MeV neutrino detection for these Type II-P, II-L, and IIb SNe. The neutrino light curves and muon neutrino detection rates for IceCube-like detectors are shown in Fig. 11.

Detecting high-energy neutrinos will give us various implications. A significant mass loss before the SN explosion has not been well-understood with the conventional stellar evolution theory, and high-energy neutrinos give us clues to the origin of the dense CSM without depending on the properties of the SN ejecta. Another appealing point is that high-energy neutrinos enable us to investigate cosmic-ray ion acceleration in "real time". The neutrino signals from a Galactic SN may be detected even at multi-energies. Observing the highest-energy neutrinos in the PeV range would indicate the maximum energy of cosmic rays. Theoretically, it ha been expected that the maximum energy of cosmic rays may be higher in the presence of the dense CSM. Indeed, Type IIn and IIb SNe have been suggested as the possible origin of cosmic rays around the knee energy (e.g., Sveshnikova 2003; Murase et al. 2014b), which can be probed by the PeV neutrinos. In addition, a Galactic SN should be one

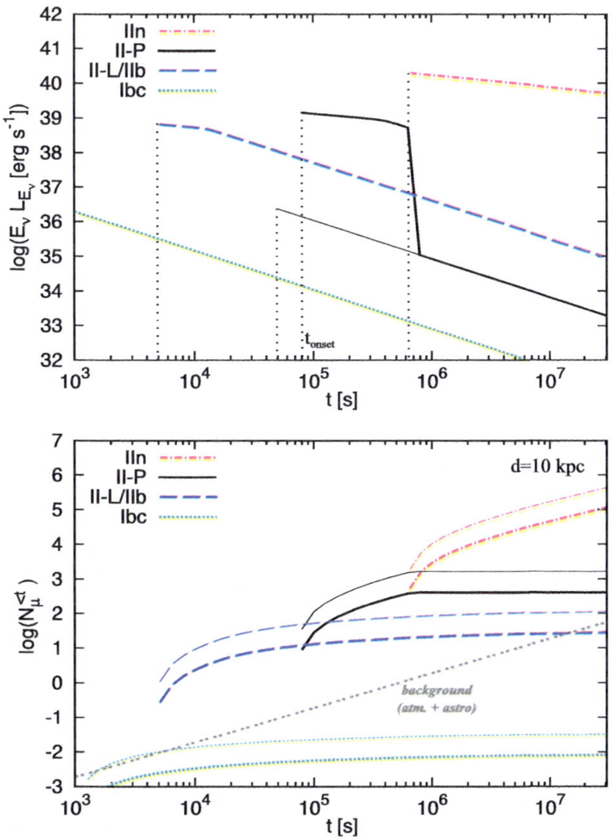

Fig. 11 *Top panel*: Predictions of TeV neutrino light curves for various types of SNe. *Bottom panel*: Expected numbers of through-going muon tracks in an IceCube-like detector, for a Galactic SN. For the signals, we consider $s = 2.2$ (*thick curves*) and $s = 2.0$ (*thin curves*), respectively. Figure taken from Murase (2017)

of the ideal targets for multi-messenger astrophysics. As discussed in the previous sections, not only MeV neutrinos but also gravitational waves may be detected. Furthermore, thermal radiation in the optical and X-ray bands will coincide with high-energy neutrinos. Hadronic gamma rays should also be generated from the production and decay of neutral pions. If the system is transparent to gamma rays (although the validity of this assumption depends on details), imaging atmospheric Cherenkov telescopes, *Fermi*, and HAWC can detect the Galactic SN.

Detections of high-energy neutrinos from Type II-P, II-L, and IIb, which are ordinary SNe II, with IceCube and KM3Net are promising for nearby SN events including Galactic SNe. However, it seems difficult to detect for distant extragalactic events. In this sense, the more extreme class of Type II SNe, i.e. Type IIn SNe accompanied by violent pre-explosion mass losses, is of interest. Optical and infrared observations have revealed that some Type IIn SN have a CSM mass of $M_{\rm cs} \sim 1$–$10\ M_\odot$, in which most of the SN kinetic energy may be dissipated by shock interactions with the CSM (Chevalier and Irwin 2011). Then, if $\sim 10\%$ of the ejecta energy is used for cosmic rays, the cosmic-ray energy may reach $\mathcal{E}_{\rm CR} \sim 10^{49}$–$10^{50}$ erg (Murase et al. 2011; Katz et al. 2011). Murase et al. (2011) proposed that such interaction-powered SNe may be detected as a new population of high-energy neutrino sources, and the results for two representative cases are shown in Fig. 12. For Model A, the SN 2006gy-like is considered, in which the CSM shell is opaque for Compton scattering and strong thermalization is expected. For Model B, the SN 2008iy-like case is considered,

Fig. 12 Energy fluences of muon neutrinos from a Type IIn SN at 10 Mpc. Thick and thin curves represent SN 2016gy-like and SN 2008iy-like cases, respectively. *Dot-dashed curves* represent the atmospheric neutrino background at different time windows, 10^7 s (*thick curve*) and $10^{7.8}$ s (*thin curve*), respectively. Figure taken from Murase et al. (2011)

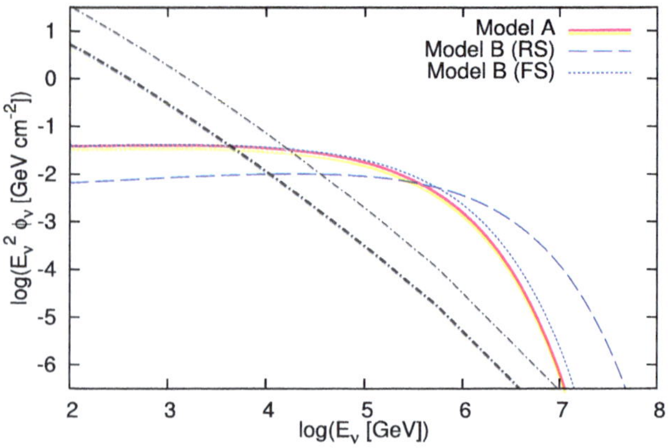

in which the CSM shell is optically thin for Compton scattering. The Type IIn SNe are also considered as contributors to the diffuse neutrino flux observed in IceCube (Zirakashvili and Ptuskin 2016). However, one should keep in mind that the physics of Type IIn SNe may be more complicated than ordinary Type II SNe. First, the mass eruption is likely to be unsteady, and the CSM is inhomogeneous and anisotropic. The CSM mass may also be comparable to the ejecta mass. As a result, the self-similar solutions may not be applicable to Type IIn SNe, where numerical calculations may be required. Second, since the CSM density is too high, the shock is initially radiation-mediated, where the diffusive shock acceleration is highly suppressed. Third, other cooing processes such as Coulomb collisions and neutral-ion collisions can affect the cosmic-ray acceleration, and resulting predictions for high-energy neutrinos are influenced. In other words, the detection of high-energy neutrinos from Type IIn SNe is important, since it will give us new information about these physical processes, independent of gamma rays that can be severely attenuated due to the two-photon annihilation and photon-matter interactions.

9 Conclusions

Neutrinos play a fundamental role in the physics of a core-collapse SN. Very recently, three-dimensional hydrodynamical SN simulations with sophisticate neutrino transport became available. However, despite the great advancement in the modeling of the SN microphysics, we are still far from fully grasping the role of neutrinos and of their flavor conversions in the stellar envelope.

The first successful SN simulations in 3D revealed unexpected and fascinating phenomena related to neutrinos. Indeed, it has been shown as the detection of the accretion phase of the neutrino signal from core-collapse SNe will be instrumental to test the explosion mechanism. Moreover, the first neutrino-driven hydrodynamical instability has been discovered: LESA. LESA is a large-scale asymmetric emission of the electron neutrino lepton number that can have major consequences for neutron-star kicks, nucleosynthesis and flavor conversions.

Since SNe are neutrino-dense environments, ν–ν interactions cannot be neglected. A careful modeling of the neutrino flavor conversions phenomenology is compulsory and it is still incomplete at the moment, despite the intense theoretical activity in this direction. Neutrino self-interactions induce a non-linear behavior of the flavor field. It has been shown that, by releasing some of the symmetry assumptions traditionally adopted in the neutrino-bulb model, instabilities in the flavor space could arise. Moreover, favorable conditions for fast pairwise flavor conversions exist in the neutrino decoupling region. Such conversions may affect the SN dynamics and may require a self-consistent treatment of the flavor evolution with the SN dynamics. However, we are still far from fully grasping the phenomenology of fast conversions and new methods need to be developed to take them properly into account.

Each phase of the neutrino signal emitted during the core collapse could offer different opportunities to learn about the SN physics and the synthesis of the new elements. In particular, neutrinos reach the Earth earlier than photons; SNEWS, a network of neutrino detectors, will alert astronomers of the upcoming collapse. Neutrinos could also help us to determine the SN distance and its location in the sky. Together with gravitational waves, neutrinos could help us to test the our theories on the explosion mechanism. The neutrino signal is also affected by flavor oscillations and the detection of the neutrino signal on Earth could help us to constrain the oscillation physics occurring in the stellar envelope. Moreover, neutrinos also play a fundamental role on the synthesis of new elements in the neutrino-driven wind.

Independently from the detection of the single SN burst, the detection of the DSNB is expected within the next decade and it will offer us with a chance to constrain the stellar population as well as to independently test the SN rate. We could use the DSNB signal to define the percentage of black-hole forming SNe present in the whole SN population and to validate non-standard physics scenarios. The detection of the DSNB is excitingly very close given the enrichment of Super-Kamiokande with gadolinium and the upcoming JUNO detector.

We also discussed high-energy neutrino emission in the later phase. For rare classes of core-collapse SNe that may leave a rapidly rotating and magnetized neutron star as a compact remnant, quasi-thermal GeV-TeV neutrinos or non-thermal TeV-PeV neutrinos from relativistic outflows can be produced $\sim$ 10–1000 s after the MeV neutrino burst. In addition, non-thermal GeV-PeV neutrinos are produced $\sim$ 1–100 day after the MeV neutrino burst and the shock breakout in the optical or ultraviolet band, which are promising targets for IceCube and KM3Net, and $\gtrsim$ 10 events can be detected for the next Galactic SN.

Acknowledgements The authors thank the organizers of the ISSI SN Workshop held in October 2016 for their kind invitation and warm hospitality. I.T. acknowledges support from the Knud Højgaard Foundation, the Villum Foundation (Project No. 13164) and the Danish National Research Foundation (DNRF91). The work of K.M. is supported by Alfred P. Sloan Foundation and NSF Grant No. PHY-1620777.

References

S. Abbar, H. Duan, Neutrino flavor instabilities in a time-dependent supernova model. Phys. Lett. B **751**, 43–47 (2015)

F. An et al., Neutrino physics with JUNO. J. Phys. G **43**(3), 030401 (2016)

S. Ando, J.F. Beacom, Revealing the supernova-gamma-ray burst connection with TeV neutrinos. Phys. Rev. Lett. **95**, 061103 (2005). astro-ph/0502521

H. Andresen, B. Mueller, E. Mueller, H.T. Janka, Gravitational wave signals from 3D neutrino hydrodynamics simulations of core-collapse supernovae. Mon. Not. R. Astron. Soc. **468**(2), 2032–2051 (2017)

P. Antonioli et al., SNEWS: the supernova early warning system. New J. Phys. **6**, 114 (2004)

K. Asano, T. Terasawa, Slow heating model of gamma-ray burst: photon spectrum and delayed emission. Astrophys. J. **705**, 1714–1720 (2009). 0905.1392

J.N. Bahcall, P. Mészáros, 5-GeV to 10-GeV neutrinos from gamma-ray burst fireballs. Phys. Rev. Lett. **85**, 1362–1365 (2000). hep-ph/0004019

I. Bartos, B. Dasgupta, S. Marka, Probing the structure of jet driven core-collapse supernova and long gamma ray burst progenitors with high energy neutrinos. Phys. Rev. D **86**, 083007 (2012). 1206.0764
I. Bartos, A.M. Beloborodov, K. Hurley, S. Márka, Detection prospects for GeV neutrinos from collisionally heated gamma-ray bursts with IceCube/DeepCore. Phys. Rev. Lett. **110**(24), 241101 (2013). 1301.4232
J.F. Beacom, The diffuse supernova neutrino background. Annu. Rev. Nucl. Part. Sci. **60**, 439–462 (2010)
J.F. Beacom, P. Vogel, Can a supernova be located by its neutrinos? Phys. Rev. D **60**, 033007 (1999)
J.H. Beall, W. Bednarek, Neutrinos from early phase, pulsar driven supernovae. Astrophys. J. **569**, 343–348 (2002). astro-ph/0108447
A.M. Beloborodov, Collisional mechanism for GRB emission. Mon. Not. R. Astron. Soc. **407**, 1033 (2010). 0907.0732
H.A. Bethe, W.R. James, Revival of a stalled supernova shock by neutrino heating. Astrophys. J. **295**, 14–23 (1985)
A. Bhattacharya, R. Enberg, M.H. Reno, I. Sarcevic, Charm decay in slow-jet supernovae as the origin of the IceCube ultra-high energy neutrino events. J. Cosmol. Astropart. Phys. **1506**(06), 034 (2015). 1407.2985
K. Blum, D. Kushnir, Neutrino signal of collapse-induced thermonuclear supernovae: the case for prompt black hole formation in SN1987A. Astrophys. J. **828**(1), 31 (2016). 1601.03422
O. Bromberg, E. Nakar, T. Piran, R. Sari, The propagation of relativistic jets in external media. Astrophys. J. **740**, 100 (2011). 1107.1326
A.M. Bykov, P. Meszaros, Electron acceleration and efficiency in nonthermal gamma-ray sources. Astrophys. J. **461**, L37–L40 (1996). astro-ph/9602016
S. Chakraborty, R. Hansen, I. Izaguirre, G. Raffelt, Collective neutrino flavor conversion: recent developments. Nucl. Phys. B **908**, 366–381 (2016a)
S. Chakraborty, R.S. Hansen, I. Izaguirre, G. Raffelt, Self-induced neutrino flavor conversion without flavor mixing. J. Cosmol. Astropart. Phys. **1603**(03), 042 (2016b)
R.A. Chevalier, Self-similar solutions for the interaction of stellar ejecta with an external medium. Astrophys. J. **258**, 790–797 (1982)
R.A. Chevalier, C. Fransson, Thermal and non-thermal emission from circumstellar interaction (2016). 1612.07459
R.A. Chevalier, C.M. Irwin, Shock breakout in dense mass loss: luminous supernovae. Astrophys. J. **729**, L6 (2011). 1101.1111
B. Dasgupta, E.P. O'Connor, C.D. Ott, The role of collective neutrino flavor oscillations in core-collapse supernova shock revival. Phys. Rev. D **85**, 065008 (2012)
B. Dasgupta, A. Mirizzi, M. Sen, Fast neutrino flavor conversions near the supernova core with realistic flavor-dependent angular distributions. J. Cosmol. Astropart. Phys. **1702**(02), 019 (2017)
E.V. Derishev, V.V. Kocharovsky, V.V. Kocharovsky, The neutron component in fireballs of gamma-ray bursts: dynamics and observable imprints. Astrophys. J. **521**, 640–649 (1999)
E.V. Derishev, F.A. Aharonian, V.V. Kocharovsky, V.V. Kocharovsky, Particle acceleration through multiple conversions from charged into neutral state and back. Phys. Rev. D **68**, 043003 (2003). astro-ph/0301263
A.S. Dighe, A.Yu. Smirnov, Identifying the neutrino mass spectrum from the neutrino burst from a supernova. Phys. Rev. D **62**, 033007 (2000)
H. Duan, S. Shalgar, Flavor instabilities in the neutrino line model. Phys. Lett. B **747**, 139–143 (2015)
H. Duan, G.M. Fuller, J. Carlson, Y.Z. Qian, Simulation of coherent non-linear neutrino flavor transformation in the supernova environment. 1. Correlated neutrino trajectories. Phys. Rev. D **74**, 105014 (2006)
H. Duan, G.M. Fuller, Y.Z. Qian, Collective neutrino oscillations. Annu. Rev. Nucl. Part. Sci. **60**, 569–594 (2010)
H. Duan, A. Friedland, G.C. McLaughlin, R. Surman, The influence of collective neutrino oscillations on a supernova r-process. J. Phys. G **38**, 035201 (2011)
R. Enberg, M.H. Reno, I. Sarcevic, High energy neutrinos from charm in astrophysical sources. Phys. Rev. D **79**, 053006 (2009). 0808.2807
T. Ertl, H.T. Janka, S.E. Woosley, T. Sukhbold, M. Ugliano, A two-parameter criterion for classifying the explodability of massive stars by the neutrino-driven mechanism. Astrophys. J. **818**(2), 124 (2016)
A. Esteban-Pretel, A. Mirizzi, S. Pastor, R. Tomàs, G.G. Raffelt, P.D. Serpico, G. Sigl, Role of dense matter in collective supernova neutrino transformations. Phys. Rev. D **78**, 085012 (2008)
K. Fang, K. Kotera, K. Murase, A.V. Olinto, Testing the newborn pulsar origin of ultrahigh energy cosmic rays with EeV neutrinos. Phys. Rev. D **90**(10), 103005 (2014). 1311.2044
P. Fernández, GADZOOKS! (SuperK-Gd): status and physics potential (2016). PoS ICRC2015:1131
G. Fogli, E. Lisi, A. Marrone, I. Tamborra, Supernova neutrinos and antineutrinos: ternary luminosity diagram and spectral split patterns. J. Cosmol. Astropart. Phys. **0910**, 002 (2009)

G.L. Fogli, E. Lisi, A. Marrone, A. Mirizzi, I. Tamborra, Low-energy spectral features of supernova (anti)neutrinos in inverted hierarchy. Phys. Rev. D **78**, 097301 (2008)

T.K. Gaisser, T. Stanev, Energetic (>GeV) neutrinos as a probe of acceleration in the new supernova. Phys. Rev. Lett. **58**, 1695 (1987). Erratum: Phys. Rev. Lett. **59**, 844(E) (1987)

F. Halzen, G.G. Raffelt, Reconstructing the supernova bounce time with neutrinos in IceCube. Phys. Rev. D **80**, 087301 (2009)

S. Horiuchi, K. Nakamura, T. Takiwaki, K. Kotake, M. Tanaka, The red supergiant and supernova rate problems: implications for core-collapse supernova physics. Mon. Not. R. Astron. Soc. **445**, L99 (2014). 1409.0006

S. Horiuchi, K. Sumiyoshi, K. Nakamura, T. Fischer, A. Summa, T. Takiwaki, H.T. Janka, K. Kotake, Diffuse supernova neutrino background from extensive core-collapse simulations of 8–100 $M_\odot$ progenitors (2017). 1709.06567

F. Iocco, K. Murase, S. Nagataki, P.D. Serpico, High energy neutrino signals from the epoch of reionization. Astrophys. J. **675**, 937–945 (2008). 0707.0515

I. Izaguirre, G. Raffelt, I. Tamborra, Fast pairwise conversion of supernova neutrinos: a dispersion-relation approach. Phys. Rev. Lett. **118**(2), 021101 (2017)

H.T. Janka, Explosion mechanisms of core-collapse supernovae. Annu. Rev. Nucl. Part. Sci. **62**, 407–451 (2012)

H.T. Janka, Neutrino emission from supernovae (2017). 1702.08713

H.T. Janka, T. Melson, A. Summa, Physics of core-collapse supernovae in three dimensions: a sneak preview. Annu. Rev. Nucl. Part. Sci. **66**, 341–375 (2016)

M. Kachelriess, R. Tomàs, R. Buras, H.T. Janka, A. Marek, M. Rampp, Exploiting the neutronization burst of a galactic supernova. Phys. Rev. D **71**, 063003 (2005)

K. Kashiyama, E. Quataert, Fast luminous blue transients from newborn black holes. Mon. Not. R. Astron. Soc. **451**(3), 2656–2662 (2015). 1504.05582

K. Kashiyama, K. Murase, P. Mészáros, Neutron-proton-converter acceleration mechanism at subphotospheres of relativistic outflows. Phys. Rev. Lett. **111**, 131103 (2013). 1304.1945

K. Kashiyama, K. Hotokezaka, K. Murase, Radio transients from newborn black holes (2017). 1710.10765

B. Katz, N. Sapir, E. Waxman, X-rays, gamma-rays and neutrinos from collisionless shocks in supernova wind breakouts (2011). 1106.1898

S. Kobayashi, P. Meszaros, Gravitational radiation from gamma-ray burst progenitors. Astrophys. J. **589**, 861–870 (2003). astro-ph/0210211

R.F. Lang, C. McCabe, S. Reichard, M. Selvi, I. Tamborra, Supernova neutrino physics with xenon dark matter detectors: a timely perspective. Phys. Rev. D **94**(10), 103009 (2016)

E.J. Lentz, S.W. Bruenn, W.R. Hix, A. Mezzacappa, O.E.B. Messer, E. Endeve, J.M. Blondin, J.A. Harris, P. Marronetti, K.N. Yakunin, Three-dimensional core-collapse supernova simulated using a 15 $M_\odot$ progenitor. Astrophys. J. **807**(2), L31 (2015)

A. Levinson, O. Bromberg, Relativistic photon mediated shocks. Phys. Rev. Lett. **100**, 131101 (2008). 0711.3281

C. Lunardini, Diffuse supernova neutrinos at underground laboratories. Astropart. Phys. **79**, 49–77 (2016)

C. Lunardini, I. Tamborra, Diffuse supernova neutrinos: oscillation effects, stellar cooling and progenitor mass dependence. J. Cosmol. Astropart. Phys. **1207**, 012 (2012)

A. MacFadyen, S.E. Woosley, Collapsars: gamma-ray bursts and explosions in 'failed supernovae'. Astrophys. J. **524**, 262 (1999). astro-ph/9810274

T. Matsumoto, D. Nakauchi, K. Ioka, A. Heger, T. Nakamura, Can direct collapse black holes launch gamma-ray bursts and grow to supermassive black holes? Astrophys. J. **810**(1), 64 (2015). 1506.05802

T. Melson, H.T. Janka, R. Bollig, F. Hanke, A. Marek, B. Mueller, Neutrino-driven explosion of a 20 solar-mass star in three dimensions enabled by strange-quark contributions to neutrino-nucleon scattering. Astrophys. J. **808**(2), L42 (2015)

P. Meszaros, M.J. Rees, Multi GeV neutrinos from internal dissipation in GRB fireballs. Astrophys. J. **541**, L5–L8 (2000). astro-ph/0007102

P. Mészáros, E. Waxman, TeV neutrinos from successful and choked gamma-ray bursts. Phys. Rev. Lett. **87**, 171102 (2001). astro-ph/0103275

S.P. Mikheev, A.Yu. Smirnov, Neutrino oscillations in a variable density medium and neutrino bursts due to the gravitational collapse of stars. Sov. Phys. JETP **64**, 4–7 (1986). Zh. Eksp. Teor. Fiz. **91**, 7 (1986). 0706.0454

A. Mirizzi, I. Tamborra, H.T. Janka, N. Saviano, K. Scholberg, R. Bollig, L. Huedepohl, S. Chakraborty, Supernova neutrinos: production, oscillations and detection. Riv. Nuovo Cimento **39**(1–2), 1–112 (2016)

A. Mizuta, K. Ioka, Opening angles of collapsar jets. Astrophys. J. **777**, 162 (2013). 1304.0163

V. Morozova, A.L. Piro, S. Valenti, Unifying type II supernova light curves with dense circumstellar material. Astrophys. J. **838**(1), 28 (2017). 1610.08054

T. Muehlbeier, H. Nunokawa, R. Zukanovich Funchal, Revisiting the triangulation method for pointing to supernova and failed supernova with neutrinos. Phys. Rev. D **88**, 085010 (2013)
K. Murase, Detecting high-energy neutrinos from the next galactic supernova (2017). 1705.04750
K. Murase, K. Ioka, TeV-PeV neutrinos from low-power gamma-ray burst jets inside stars. Phys. Rev. Lett. **111**(12), 121102 (2013). 1306.2274
K. Murase, K. Ioka, S. Nagataki, T. Nakamura, High energy neutrinos and cosmic-rays from low-luminosity gamma-ray bursts? Astrophys. J. **651**, L5–L8 (2006). astro-ph/0607104
K. Murase, P. Mészáros, B. Zhang, Probing the birth of fast rotating magnetars through high-energy neutrinos. Phys. Rev. D **79**, 103001 (2009). 0904.2509
K. Murase, T.A. Thompson, B.C. Lacki, J.F. Beacom, New class of high-energy transients from crashes of supernova ejecta with massive circumstellar material shells. Phys. Rev. D **84**, 043003 (2011). 1012.2834
K. Murase, K. Asano, T. Terasawa, P. Meszaros, The role of stochastic acceleration in the prompt emission of gamma-ray bursts: application to hadronic injection. Astrophys. J. **746**, 164 (2012). 1107.5575
K. Murase, K. Kashiyama, P. Mészáros, Subphotospheric neutrinos from gamma-ray bursts: the role of neutrons. Phys. Rev. Lett. **111**, 131102 (2013). 1301.4236
K. Murase, B. Dasgupta, T.A. Thompson, Quasithermal neutrinos from rotating protoneutron stars born during core collapse of massive stars. Phys. Rev. D **89**(4), 043012 (2014a). 1303.2612
K. Murase, T.A. Thompson, E.O. Ofek, Probing cosmic-ray ion acceleration with radio-submm and gamma-ray emission from interaction-powered supernovae. Mon. Not. R. Astron. Soc. **440**(3), 2528–2543 (2014b). 1311.6778
K. Nakazato, Imprint of explosion mechanism on supernova relic neutrinos. Phys. Rev. D **88**(8), 083012 (2013). 1306.4526
E. O'Connor, C.D. Ott, Black hole formation in failing core-collapse supernovae. Astrophys. J. **730**, 70 (2011)
E. Pllumbi, I. Tamborra, S. Wanajo, H.T. Janka, L. Huedepohl, Impact of neutrino flavor oscillations on the neutrino-driven wind nucleosynthesis of an electron-capture supernova. Astrophys. J. **808**(2), 188 (2015)
Y.Z. Qian, S.E. Woosley, Nucleosynthesis in neutrino driven winds: 1. The physical conditions. Astrophys. J. **471**, 331–351 (1996). astro-ph/9611094
G. Raffelt, S. Sarikas, D. de Sousa Seixas, Axial symmetry breaking in self-induced flavor conversion of supernova neutrino fluxes. Phys. Rev. Lett. **111**(9), 091101 (2013)
G.G. Raffelt, What have we learned from SN 1987A? Mod. Phys. Lett. A **5**, 2581–2592 (1990)
S. Razzaque, P. Meszaros, E. Waxman, Neutrino tomography of gamma-ray bursts and massive stellar collapses. Phys. Rev. D **68**, 083001 (2003). astro-ph/0303505
S. Razzaque, P. Meszaros, E. Waxman, TeV neutrinos from core collapse supernovae and hypernovae. Phys. Rev. Lett. **93**, 181101 (2004). Erratum: Phys. Rev. Lett. **94**, 109903 (2005). astro-ph/0407064
M.J. Rees, P. Meszaros, Unsteady outflow models for cosmological gamma-ray bursts. Astrophys. J. **430**, L93–L96 (1994). astro-ph/9404038
M.J. Rees, P. Meszaros, Dissipative photosphere models of gamma-ray bursts and x-ray flashes. Astrophys. J. **628**, 847–852 (2005). astro-ph/0412702
L.F. Roberts, C.D. Ott, R. Haas, E.P. O'Connor, P. Diener, E. Schnetter, General-relativistic three-dimensional multi-group neutrino radiation-hydrodynamics simulations of core-collapse supernovae. Astrophys. J. **831**, 98 (2016). 1604.07848
R.F. Sawyer, Speed-up of neutrino transformations in a supernova environment. Phys. Rev. D **72**, 045003 (2005)
R.F. Sawyer, Neutrino cloud instabilities just above the neutrino sphere of a supernova. Phys. Rev. Lett. **116**(8), 081101 (2016)
K. Scholberg, Supernova neutrino detection. Annu. Rev. Nucl. Part. Sci. **62**, 81–103 (2012)
N. Senno, K. Murase, P. Mészáros, Choked jets and low-luminosity gamma-ray bursts as hidden neutrino sources. Phys. Rev. D **93**(8), 083003 (2016). 1512.08513
P.D. Serpico, S. Chakraborty, T. Fischer, L. Huedepohl, H.T. Janka, A. Mirizzi, Probing the neutrino mass hierarchy with the rise time of a supernova burst. Phys. Rev. D **85**, 085031 (2012)
N. Smith, Mass loss: its effect on the evolution and fate of high-mass stars. Annu. Rev. Astron. Astrophys. **52**, 487–528 (2014). 1402.1237
T. Sukhbold, T. Ertl, S.E. Woosley, J.M. Brown, H.T. Janka, Core-collapse supernovae from 9 to 120 solar masses based on neutrino-powered explosions. Astrophys. J. **821**(1), 38 (2016)
K. Sumiyoshi, S. Yamada, H. Suzuki, S. Chiba, Neutrino signals from the formation of black hole: a probe of equation of state of dense matter. Phys. Rev. Lett. **97**, 091101 (2006). astro-ph/0608509
Y. Suwa, K. Murase, Probing the central engine of long gamma-ray bursts and hypernovae with gravitational waves. Phys. Rev. D **80**, 123008 (2009). 0906.3833

L.G. Sveshnikova, The knee in galactic cosmic ray spectrum and variety in supernovae. Astron. Astrophys. **409**, 799–808 (2003). astro-ph/0303159

T. Takiwaki, K. Kotake, Y. Suwa, Three-dimensional simulations of rapidly rotating core-collapse supernovae: finding a neutrino-powered explosion aided by non-axisymmetric flows. Mon. Not. R. Astron. Soc. **461**(1), L112–L116 (2016)

I. Tamborra, S. Ando, Inspecting the supernova-gamma-ray-burst connection with high-energy neutrinos. Phys. Rev. D **93**(5), 053010 (2016). 1512.01559

I. Tamborra, F. Hanke, B. Mueller, H.T. Janka, G. Raffelt, Neutrino signature of supernova hydrodynamical instabilities in three dimensions. Phys. Rev. Lett. **111**(12), 121104 (2013)

I. Tamborra, F. Hanke, H.T. Janka, B. Mueller, G.G. Raffelt, A. Marek, Self-sustained asymmetry of lepton-number emission: a new phenomenon during the supernova shock-accretion phase in three dimensions. Astrophys. J. **792**(2), 96 (2014a)

I. Tamborra, G. Raffelt, F. Hanke, H.T. Janka, B. Mueller, Neutrino emission characteristics and detection opportunities based on three-dimensional supernova simulations. Phys. Rev. D **90**(4), 045032 (2014b)

I. Tamborra, L. Huedepohl, G. Raffelt, H.T. Janka, Flavor-dependent neutrino angular distribution in core-collapse supernovae. Astrophys. J. **839**, 132 (2017)

T.A. Thompson, P. Chang, E. Quataert, Magnetar spindown, hyper-energetic supernovae, and gamma ray bursts. Astrophys. J. **611**, 380–393 (2004). astro-ph/0401555

R. Tomàs, D. Semikoz, G.G. Raffelt, M. Kachelriess, A.S. Dighe, Supernova pointing with low-energy and high-energy neutrino detectors. Phys. Rev. D **68**, 093013 (2003)

I. Vurm, A.M. Beloborodov, J. Poutanen, Gamma-ray bursts from magnetized collisionally-heated jets. Astrophys. J. **738**, 77 (2011). 1104.0394

J. Wallace, A. Burrows, J.C. Dolence, Detecting the supernova breakout burst in terrestrial neutrino detectors. Astrophys. J. **817**(2), 182 (2016)

E. Waxman, B. Katz, Shock breakout theory (2016). 1607.01293

O. Yaron et al., Confined dense circumstellar material surrounding a regular type II supernova: the unique flash-spectroscopy event of SN 2013fs. Nat. Phys. **13**, 510 (2017). 1701.02596

V.N. Zirakashvili, V.S. Ptuskin, Type IIn supernovae as sources of high energy astrophysical neutrinos. Astropart. Phys. **78**, 28–34 (2016). 1510.08387

Space Sci Rev (2018) 214:59
https://doi.org/10.1007/s11214-018-0493-6

Superluminous Supernovae

Takashi J. Moriya[1] · **Elena I. Sorokina**[2] ·
Roger A. Chevalier[3]

Received: 25 December 2017 / Accepted: 5 March 2018 / Published online: 13 March 2018

Abstract Superluminous supernovae are a new class of supernovae that were recognized about a decade ago. Both observational and theoretical progress has been significant in the last decade. In this review, we first briefly summarize the observational properties of superluminous supernovae. We then introduce the three major suggested luminosity sources to explain the huge luminosities of superluminous supernovae, i.e., the nuclear decay of ^{56}Ni, the interaction between supernova ejecta and dense circumstellar media, and the spin down of magnetars. We compare these models and discuss their strengths and weaknesses.

Keywords Supernovae · Superluminous supernovae · Massive stars

1 Introduction

Superluminous supernovae (SLSNe) are supernovae (SNe) that become more luminous than ~ -21 mag in optical. They are more than 1 mag more luminous than broad-line Type Ic

T.J. Moriya is a NAOJ Fellow.

Supernovae
Edited by Andrei Bykov, Roger Chevalier, John Raymond, Friedrich-Karl Thielemann, Maurizio Falanga and Rudolf von Steiger

✉ T.J. Moriya
takashi.moriya@nao.ac.jp

E.I. Sorokina
sorokina@sai.msu.su

R.A. Chevalier
rac5x@virginia.edu

[1] Division of Theoretical Astronomy, National Astronomical Observatory of Japan, National Institutes of Natural Sciences, 2-21-1 Osawa, Mitaka, Tokyo 181-8588, Japan

[2] Sternberg Astronomical Institute, M.V. Lomonosov Moscow State University, Universitetski pr. 13, 119234 Moscow, Russia

[3] Department of Astronomy, University of Virginia, P.O. Box 400325, Charlottesville, VA 22904-4325, USA

SNe, or the so-called "hypernovae," which have kinetic energy of more than $\sim 10^{52}$ erg and are the most luminous among the classical core-collapse SNe. The first glimpse of their existence was in SN 1999as (Knop et al. 1999). Other SLSNe, such as SCP06F6 (Barbary et al. 2009), were subsequently discovered. However, their nature was not clarified until SLSNe started to be discovered regularly with unbiased transient surveys such as that conducted by the ROTSE-IIIb telescope (Quimby 2006). Nowadays, SLSNe are discovered even at redshifts around 2 and beyond (Cooke et al. 2012; Pan et al. 2017; Smith et al. 2018; Moriya et al. 2018; Curtin et al. 2018), and they are becoming an important probe of, e.g., massive star formation in the high-redshift universe. They may be eventually applicable to cosmology like Type Ia SNe (Quimby et al. 2013; Inserra and Smartt 2014; Inserra et al. 2018a; Scovacricchi et al. 2016; Wei et al. 2015). They can also be a back light source to study interstellar media (Berger et al. 2012).

This review mainly focuses on theoretical aspects of the energy sources of SLSNe. Three energy sources are mainly suggested to account for the huge luminosities of SLSNe, i.e., ^{56}Ni decay (Sect. 3), interaction between ejecta and dense circumstellar media (CSM, Sect. 4), and magnetar spin-down (Sect. 5). We start this review with a brief summary of the observational properties in Sect. 2 and then discuss the theoretical aspects of the power sources. We compare three models in Sect. 6. For a review of observational aspects of SLSNe, see, e.g., Gal-Yam (2012) and Howell (2017). Observational data on SLSNe are now available at https://slsn.info/, which is maintained by T.-W. Chen.

2 Observational Properties

Broadly speaking, SLSNe can be classified in two different spectral types; those with narrow hydrogen emission lines (Type IIn, e.g., Smith et al. 2010) and those without (e.g., Quimby et al. 2011). Most of SLSNe without the narrow hydrogen emission lines are classified as Type Ic and we simply call them SLSNe Ic. However, some SLSNe show broad hydrogen lines (e.g., Inserra et al. 2018b) and there exist some SLSNe that are initially Type Ic but start to show hydrogen emissions about 1 year after the luminosity peak (Yan et al. 2015, 2017a). We introduce Type IIn SLSNe in Sect. 2.1 and then the other SLSNe in Sect. 2.2. We discuss the event rates of SLSNe in Sect. 2.3 and their environment in Sect. 2.4.

2.1 Type IIn SLSNe

SLSNe IIn are characterized by narrow emission lines (Fig. 1). Narrow hydrogen emission lines are particularly strong. The spectral characteristics of SLSNe IIn are similar to those of less luminous SNe IIn. Therefore, they are believed to be extreme cases of SNe IIn where their luminosities are mainly powered by the interaction between ejecta and dense CSM (Sect. 4). SN 2006gy is the closest and the best observed SN in this class (e.g., Ofek et al. 2007; Smith et al. 2007, 2008b, 2010; Kawabata et al. 2009; Agnoletto et al. 2009; Miller et al. 2010).

Although the light curves (LCs) of SLSNe IIn share the characteristics that they become extremely luminous, their LC shapes are diverse (Fig. 1). SN 2008fz (Drake et al. 2010) and SN 2006gy (Smith et al. 2007) have broad round LCs lasting for more than 100 days while SN 2003ma (Rest et al. 2011) has a short round phase followed by a long phase of constant luminosity. The LCs of less luminous Type IIn SNe like SN 2006tf (Smith et al. 2008a) and SN 2010jl (Zhang et al. 2012; Fransson et al. 2014) often decline with a power law. These

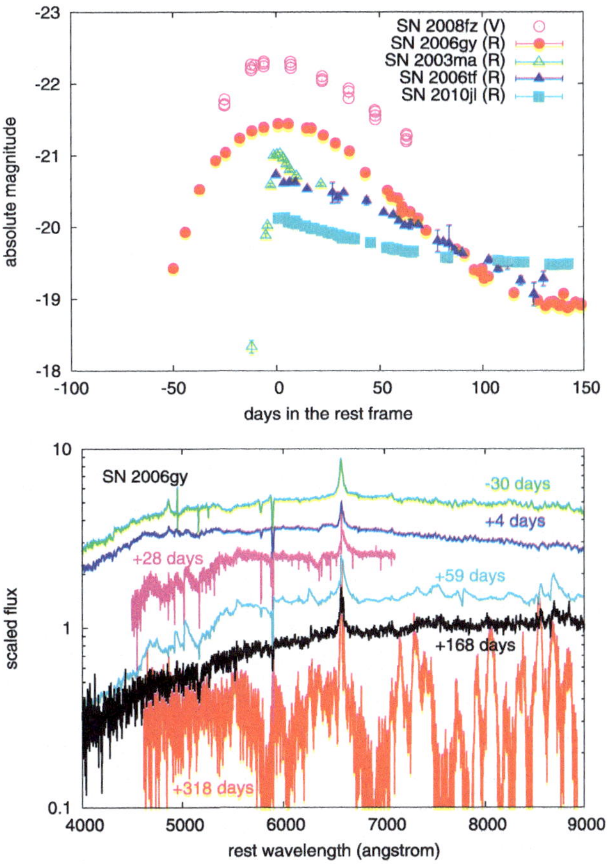

Fig. 1 *Top:* LC diversities in Type IIn SLSNe. The SN names and observed bands are indicated in the figure. The data sources are: Drake et al. (2010) (SN 2008fz), Smith et al. (2007) (SN 2006gy), Rest et al. (2011) (SN 2003ma), Smith et al. (2008a) (SN 2006tf), and Zhang et al. (2012) (SN 2010jl). *Bottom:* Spectroscopic evolution of SLSN IIn 2006gy (e.g., Smith et al. 2010; Kawabata et al. 2009). The dates are relative to the luminosity peak in the *R* band in the rest frame

diversities are likely to originate from the diversity in the CSM related to the diversity in the mass loss processes of the progenitors.

It is not yet clear if SLSNe IIn are just the most luminous end of a SN IIn luminosity function or SLSNe IIn make a separate population of their own (e.g., Richardson et al. 2014). There exist several SNe IIn whose peak luminosity is between ~ -20 mag and ~ -21 mag. These luminous populations can be observed more easily but there do not exist many SNe IIn with which we can make a volume-limited sample because of their small event rate. The most luminous SLSN IIn reported so far is CSS100217:102913+404220 reaching -23 mag in the optical at the LC peak (Drake et al. 2011, see also Kankare et al. 2017). Although CSS100217:102913+404220 and other similar objects are suggested to originate from stellar explosions, they appear at the centers of galaxies with active galactic nuclei and their stellar explosion origin is doubted (e.g., Blanchard et al. 2017; Moriya et al. 2017b). The most luminous published SLSN IIn without a clear association with an active galactic nucleus is SN 2008fz (Drake et al. 2010).

2.2 Type Ic SLSNe

SLSNe Ic, which are often simply called SLSNe I, are SLSNe without hydrogen spectral features around the time of the luminosity peak (e.g., Quimby et al. 2011). Figure 2 shows

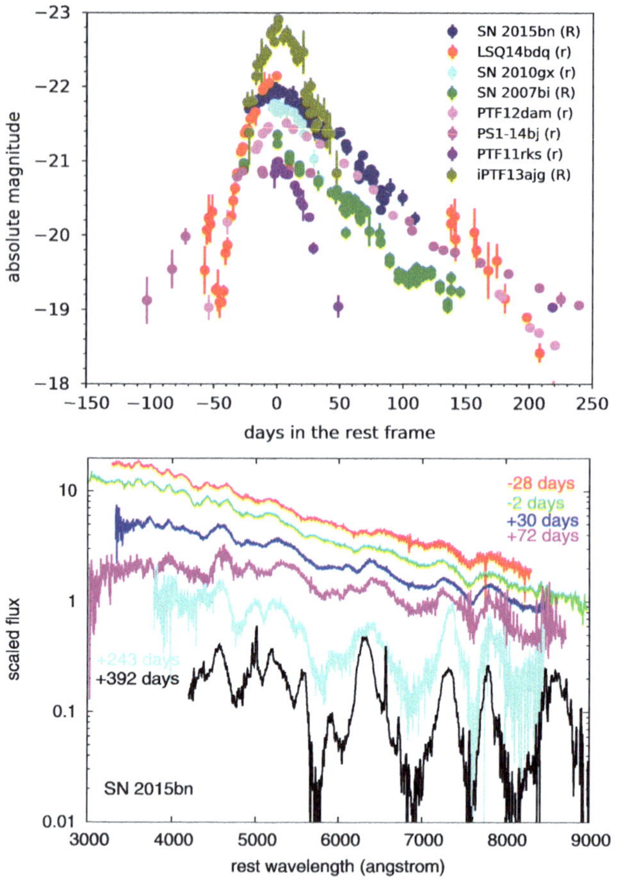

Fig. 2 *Top:* LC diversities in Type Ic SLSNe. The SN names and observed bands are indicated in the figure. The data sources are: Nicholl et al. (2016b) (SN 2015bn), Nicholl et al. (2015b) (LSQ14bdq), Pastorello et al. (2010) (SN 2010gx), Gal-Yam et al. (2009) (SN 2007bi), Lunnan et al. (2016) (PS1-14bj), Inserra et al. (2013) (PTF11rks), and Vreeswijk et al. (2014) (iPTF13ajg). *Bottom:* Spectroscopic evolution of SLSN Ic 2015bn (Nicholl et al. 2016a,b). The dates are relative to the optical luminosity peak in the rest frame

LCs and spectra of SLSNe Ic. The peak optical magnitudes of SLSNe Ic are typically between ∼ −21 mag and −22 mag (e.g., Nicholl et al. 2015a; De Cia et al. 2017; Lunnan et al. 2018). The distribution of their peak bolometric magnitudes derived from optical LCs peaks at −20.7 mag with a small dispersion of 0.5 mag (Nicholl et al. 2015a). The luminosity function of Type Ibc SNe peaks at around −18 mag with a dispersion of more than 0.5 mag (Drout et al. 2011; Taddia et al. 2015; Lyman et al. 2016). Therefore, there may exist a gap in the hydrogen-poor SN luminosity function between ∼ −20 mag and ∼ −21 mag, where a small number of Type Ic BL supernovae and some gap transients exist (Arcavi et al. 2016; Roy et al. 2016). The rise times of SLSNe Ic have a diversity, ranging from ≃ 20 days (PTF11rks, Inserra et al. 2013) to more than 125 days (PS1-14bj, Lunnan et al. 2016). However, their LC decline rates may have two distinct classes: rapidly-declining ones and slowly-declining ones (e.g., Nicholl et al. 2015a). The LC decline rates of slowly-declining SLSNe Ic are often consistent with the nuclear decay rate of ^{56}Co (e.g., Gal-Yam et al. 2009; Inserra et al. 2017) and they are sometimes referred as SLSN R, with "R" for radioactive decay (Gal-Yam 2012). However, whether they are powered by nuclear decay or not cannot be judged solely based on the LCs (e.g., Moriya et al. 2017a). The slowly-declining SLSNe Ic may tend to be intrinsically more luminous than rapidly-declining SLSNe Ic, which may enable us to use SLSNe Ic as a distance indicator like Type Ia SNe (Inserra and Smartt 2014, but see also Quimby et al. 2013).

The LCs of SLSNe Ic have a distinguishing feature at the very beginning. They often show a precursor before the main luminosity increase (e.g., Nicholl et al. 2015b; Smith et al. 2016; Vreeswijk et al. 2017). The possible existence of the precursor was first found in SN 2006oz (Leloudas et al. 2012) and it was systematically studied by Nicholl and Smartt (2016). The precursor lasts for about 5–10 days and its magnitude is around −20 mag. Its blackbody temperature is around 30,000 K (Nicholl and Smartt 2016). The precursor is followed by a fading lasting for about 5 days and the luminosity is about 1 mag fainter than the precursor luminosity in this period. Then, the LC starts to rise again to be a SLSN.

LC behaviors of slowly-declining SLSNe Ic during the fading phase after the LC peak have been intensively studied by Inserra et al. (2017). The LCs show fluctuations that may originate from the interaction with the CSM. The LC decline rates are consistent with that of the ^{56}Co decay at first, but tend to decline faster after a few hundred days after the LC peak.

The early-time spectra of SLSNe Ic at around the luminosity peak are characterized by a series of broad carbon and oxygen features (Quimby et al. 2011, 2018; Chomiuk et al. 2011; Howell et al. 2013; Nicholl et al. 2014, 2016b; Yan et al. 2017c,b; Liu et al. 2017b). The blackbody temperature is around 15,000 K at these epochs, but non-thermal excitations play essential roles in forming the spectral features (e.g., Mazzali et al. 2016). Photospheric velocities measured by using Fe II λ5169 are typically around 12,000 km s^{-1} and they do not evolve fast (e.g., Nicholl et al. 2015a). It is suggested that the existence of helium is required for the line formation (Mazzali et al. 2016) and a helium line may have been observed in SN 2012il (Inserra et al. 2013). At later phases, the spectra often start to be similar to those of broad-line Type Ic SNe (Pastorello et al. 2010; Nicholl et al. 2016a). In addition, there are several SLSNe Ic that do not show hydrogen features near the LC peak but hydrogen emission lines start to appear about 1 year after the LC peak (Yan et al. 2015, 2017a). The late-phase hydrogen emissions have been suggested to originate from a detached hydrogen-rich circumstellar shell (Yan et al. 2015, 2017a) or matter stripped from the progenitor's hydrogen-rich companion star (Moriya et al. 2015).

There are a few SLSNe in which broad hydrogen features are observed (Gezari et al. 2009; Miller et al. 2009; Inserra et al. 2018b). They usually do not show narrow features in spectra like SLSNe IIn at early times. Therefore, their powering source may be related to SLSNe Ic. However, their spectra start to show possible signatures of the interaction in the late phase and some luminosity contributions from the interaction may exist after around 1 year since the LC peak. The SLSN CSS121015:004244+132827 showed both broad and narrow hydrogen features from early times (Benetti et al. 2014).

Polarization of SLSNe Ic was measured in three objects: LSQ14mo, SN 2015bn, and SN 2017egm. The first attempt was made for LSQ14mo. Leloudas et al. (2015a) measured photometric polarization in the V band for 5 epochs and they did not detect clear polarization signatures. SN 2015bn appeared closer and both imaging and spectroscopic polarizations were measured for several epochs (Leloudas et al. 2017; Inserra et al. 2016). Aspherical shapes may be required to explain the polarization features. SN 2017egm is the closest SLSN Ic and a spectropolarimetry observation was reported (Bose et al. 2018). Polarization of about 0.5% without a strong dependence on wavelength is observed, indicating a modest departure from spherical symmetry. More observations are needed to obtain a systematic view of the asphericity in SLSNe Ic. Polarization of one SLSN II with broad hydrogen features is observed but a meaningful constraint on its shape could not be made (Inserra et al. 2018b).

SLSNe Ic are also actively observed in X-rays. SCP06F6 is the first SLSN Ic with a possible detection of X-rays (Gänsicke et al. 2009; Levan et al. 2013). It was detected in X-rays

(0.1–2 keV) by the *XMM-Newton* satellite. The X-ray luminosity reached $\sim 10^{45}$ erg s^{-1} in one epoch and then it was not detected any more. Another case is PTF12dam (Margutti et al. 2017). It was detected by *Chandra* at 0.5–2 keV in a similar epoch to SCP06F6 but its luminosity is $\sim 10^{40}$ erg s^{-1} and is 5 orders of magnitude below that of SCP06F6. The X-ray luminosity at the location of PTF12dam is consistent with that expected from the underlying star-formation activities at the SN location and the X-rays may not originate from PTF12dam itself. X-ray observations in many other SLSNe Ic have been performed but SCP06F6 and PTF12dam are the only cases with X-ray detections (Margutti et al. 2017). No radio emission is observed in SLSNe so far (Coppejans et al. 2017).

Gamma-ray bursts (GRBs) are often associated with Type Ic-BL SNe. It is interesting to note that SLSNe Ic have spectra similar to SNe Ic-BL at late phases (e.g., Pastorello et al. 2010; Nicholl et al. 2016a). In addition, SN 2011kl, which was associated with an ultra-long GRB111209A, had a LC peak at -20 mag and it was much more luminous than SNe usually associated with GRBs (Greiner et al. 2015). Although SN 2011kl was not as luminous as SLSNe and its spectroscopic features, which are overall featureless, do not appear to be similar to SLSNe Ic, some GRBs may be related to SLSNe. Yu and Li (2017) speculated that LC modulations in SN 2015bn may be related to flare activity of GRBs. Similarities in GRBs and SLSNe are also suggested in their host environment which will be discussed in a following section.

The most luminous SLSN Ic reported so far is ASASSN-15lh, which reached about -23.5 mag (Dong et al. 2016). However, it appeared at the center of the host galaxy and its origin is questioned (Leloudas et al. 2016). Its host galaxy is massive and metal-rich and, therefore, is peculiar if it is actually a SLSN Ic (e.g., Leloudas et al. 2016). Except for ASASSN-15lh, the most luminous SLSNe Ic reach $-22.5\sim-23$ mag in optical (e.g., PTF13ajg, Vreeswijk et al. 2014).

2.3 Event Rates

Quimby et al. (2013) calculate the rates of SLSNe at $z \sim 0.2$. They find a total SLSN rate of 199^{+137}_{-86} Gpc^{-3} yr^{-1} h_{71}^3, where h_{71} is the Hubble constant scaled with 71 km s^{-1} Mpc^{-1}. They estimate a SLSN II rate of 151^{+151}_{-82} Gpc^{-3} yr^{-1} h_{71}^3. Their SLSN II sample consists of SLSNe IIn, with the exception of SN 2008es. The rate of SLSNe Ic is estimated to be 32^{+77}_{-26} Gpc^{-3} yr^{-1} h_{71}^3. Because the core-collapse SN rate at $z \sim 0.2$ is $\sim 10^5$ Gpc^{-3} yr^{-1} (Madau and Dickinson 2014), the fraction of SLSNe to core-collapse SNe are estimated to be $\sim 10^{-3}$. The total SLSN rate is similar to the long GRB rate (e.g., Soderberg et al. 2006) and the SLSN Ic rate may be about 10% of the long GRB rate.

There are two works deriving the SLSN Ic rate at $z \sim 1$. The first work is by McCrum et al. (2015) who use a SN sample from the Pan-STARRS survey to derive a relative fraction of SLSNe Ic to core-collapse SNe. They found only $\sim 10^{-5}$ of core-collapse SNe are SLSNe Ic at $z \sim 1$, which is roughly a factor of 10 smaller than the rate at $z \sim 0.2$. The subsequent work by Prajs et al. (2017) measured the SLSN Ic rate at $z \sim 1$ based on the SLSN Ic samples from the Supernova Legacy Survey. The survey has two spectroscopically confirmed SLSNe Ic at $z = 1.50$ and 1.59 and one additional SLSN Ic candidate at $z = 0.76$ which is found by fitting the LCs found in the survey. The three SLSNe Ic in the survey give a SLSN Ic rate of 91^{+76}_{-36} Gpc^{-3} yr^{-1} h_{70}^3 at $z \sim 1$. This rate is equivalent to $\sim 10^{-4}$ of the core-collapse rate at $z \sim 1$ and an order of magnitude higher than that estimated by McCrum et al. (2015).

Cooke et al. (2012) reported two SLSNe at $z = 2.05$ and $z = 3.90$. Although their spectral types are not clear, the detection of two SLSNe in their survey very roughly corresponds to the SLSN rate of $\sim 400\ \mathrm{Gpc}^{-3}\,\mathrm{yr}^{-1}\,h_{70}^{3}$ at $z = 2$–4. The increase in the SLSN rate is still consistent with that expected from the increase in the cosmic star-formation rate and it is still not clear if the intrinsic SLSN frequency increases at high redshifts (see also a recent estimate by Moriya et al. 2018).

2.4 Environments

The host environments of SLSNe have been studied extensively. SLSNe IIn are known to come from a broad range of metallicities (Neill et al. 2011; Perley et al. 2016; Schulze et al. 2018; Angus et al. 2016), while SLSNe Ic only come from metal-poor environments (e.g., Schulze et al. 2018; Chen et al. 2017c, 2015, 2013; Leloudas et al. 2015b; Perley et al. 2016; Vreeswijk et al. 2014; Lunnan et al. 2016). There is likely a suppression of SLSNe Ic above around the half-solar metallicity environment (Perley et al. 2016; Chen et al. 2017c; Schulze et al. 2018). SLSNe Ic also tend to come from star-bursting galaxies with high specific star-formation rates (e.g., Schulze et al. 2018). This tendency can be explained if their progenitors are very massive stars (e.g., Thöne et al. 2015; Leloudas et al. 2015b), but this interpretation is also questioned (e.g., Perley et al. 2016). SLSNe Ic are often found in interacting galaxies where the star formation can be triggered by the interaction (e.g., Chen et al. 2017a; Cikota et al. 2017). SLSNe Ic may prefer metal-poorer environments than long GRBs (e.g., Schulze et al. 2018). However, the differences between SLSNe Ic and long GRB host environments are not yet clear (e.g., Lunnan et al. 2014, 2015; Angus et al. 2016; Japelj et al. 2016).

A SLSN Ic SN 2017egm has appeared in a nearby spiral galaxy that has a high metallicity (Bose et al. 2018; Nicholl et al. 2017a) but the SN location itself may have low metallicity (Izzo et al. 2018, but see also Chen et al. 2017b).

An interesting relation between the host metallicities and the spin periods of magnetars required to fit the LCs of SLSNe Ic has been suggested by Chen et al. (2017c). This relation may simply come from the relation between the host metallicities and total radiated energy in SLSNe Ic, because the spin periods in the magnetar model represent the total available energy to power SLSNe (Sect. 5). The suggested relation is still based on a small number of SLSNe Ic and it needs to be studied more. Recent follow-up studies based on larger SLSN samples infer that the relation may not be as strong as originally suggested (Nicholl et al. 2017b; De Cia et al. 2017).

3 ^{56}Ni-Powered Models

The standard source of early luminosity in SNe is the nuclear decay of ^{56}Ni. The simplest approach to explain the huge luminosities of SLSNe is to increase the amount of ^{56}Ni. ^{56}Ni decays to ^{56}Co with a decay time of 8.76 ± 0.01 days (da Cruz et al. 1992) and then ^{56}Co decays to ^{56}Fe with a decay time of 111.42 ± 0.04 days (Funck et al. 1992). The total available energy from the nuclear decay is

$$\left[6.48\exp\left(-\frac{t}{8.76\ \mathrm{days}}\right) + 1.44\exp\left(-\frac{t}{111.42\ \mathrm{days}}\right)\right]\frac{M_{^{56}\mathrm{Ni}}}{M_\odot}10^{43}\ \mathrm{erg\,s^{-1}}, \tag{1}$$

where $M_{^{56}\mathrm{Ni}}$ is the total mass of ^{56}Ni available at the beginning (Nadyozhin 1994, but updated based on http://www.nndc.bnl.gov/chart). However, all the available energy from the decay is not necessarily absorbed by the SN ejecta to heat them up. All the decay energy of ^{56}Ni and about 97% of the decay energy of ^{56}Co are released in the form of γ-rays. The released γ-rays travel in the ejecta and some fraction is absorbed by the ejecta and heats them up. The effective γ-ray opacity is 0.027 cm^2 g^{-1} (Axelrod 1980). The remaining 3% of the ^{56}Co decay energy is released as positrons that can be all absorbed *in situ* by the ejecta to heat them up. During the early phases when the ejecta are dense, most γ-rays can be absorbed. However, in the nebular phases when the ejecta are optically thin, a proper treatment of the γ-ray transport is needed to estimate the expected luminosity. Because heating by the ^{56}Ni decay is a standard way to power SNe, we refer to a standard text such as Arnett (1996) for further details of the SN properties powered by ^{56}Ni decay.

The required ^{56}Ni masses to explain the peak luminosity of SLSNe are 5–20 $M_\odot$ (e.g., Nicholl et al. 2015a). There are several suggested ways to make such a large amount of ^{56}Ni. One is the very energetic explosions of massive stars. Umeda and Nomoto (2008) investigated the explosive nucleosynthesis of very massive stars up to the zero-age main-sequence (ZAMS) mass of 100 $M_\odot$ with $Z_\odot/200$. They show that more than 5 $M_\odot$ of ^{56}Ni can be produced in massive cores if the explosion energy exceeds 10^{52} erg. To produce more than 10 $M_\odot$ of ^{56}Ni, which is often required to reproduce the SLSN LCs, an unrealistically high explosion energy of 10^{53} erg is required. Relatively faint SLSNe requiring ~ 5 $M_\odot$ of ^{56}Ni, like SN 2007bi, may be explained by the energetic core-collapse explosions of very massive stars (Young et al. 2010; Moriya et al. 2010).

A promising mechanism to synthesize more than 10 $M_\odot$ of ^{56}Ni in massive stars is through pair-instability (e.g., Rakavy and Shaviv 1967; Barkat et al. 1967). If a star has a core that is massive enough, photons in the stellar core can be efficiently converted to electron and positron pairs and the center of the star can suffer a strong reduction in radiative pressure supporting the massive core. The reduction of the pressure support triggers the contraction of the massive core by making the adiabatic index below 4/3 and leads to explosive carbon and oxygen burning. If the energy released by the explosive nuclear burning is enough to unbind the whole star, it can explode without leaving any compact remnant. This kind of explosion is called pair-instability SN (PISN). The amount of ^{56}Ni produced in PISNe strongly depends on the core mass of the progenitor and it ranges from almost zero to ~ 50 $M_\odot$ (e.g., Heger and Woosley 2002; Takahashi et al. 2016). PISNe can occur if helium cores of massive stars are between ~ 65 $M_\odot$ and ~ 135 $M_\odot$ which corresponds to ZAMS masses between ~ 150 $M_\odot$ and ~ 250 $M_\odot$ at zero metallicity (e.g., Heger and Woosley 2002). The ZAMS mass can be as low as 65 $M_\odot$ if massive stars rotate rapidly (e.g., Chatzopoulos and Wheeler 2012; Yusof et al. 2013). The rapid rotation also helps PISN progenitors to remove their hydrogen through quasi-chemically homogeneous evolution (e.g., Yoon et al. 2012). It is interesting to note that most PISN candidates are hydrogen-free SLSNe (but see Terreran et al. 2017), although hydrogen-rich PISNe can also be very luminous (e.g., Kasen et al. 2011). The surface instability of red supergiant PISN progenitors can also help reduce the amount of hydrogen in their explosions (Moriya and Langer 2015). PISNe are considered to exist in environments with metallicities less than $\sim Z_\odot/3$ (e.g., Langer et al. 2007), but they may exist even in a solar-metallicity environment if magnetic fields of massive stars can suppress their mass loss (Georgy et al. 2017).

Predicted LCs of PISNe (e.g., Scannapieco et al. 2005; Kasen et al. 2011; Dessart et al. 2013; Whalen et al. 2014; Kozyreva et al. 2014; Chatzopoulos et al. 2015) are roughly consistent with those of slowly-declining SLSNe I whose late-phase decline rates are consistent with the ^{56}Co nuclear decay rate. PISNe tend to have large rise times compared to SLSNe I but some PISN models are consistent with the relatively short rise times of SLSNe Ic (e.g., Kozyreva et al. 2017). The internal mixing of PISNe can help reduce their rise times as well (e.g., Kozyreva and Blinnikov 2015) but multidimensional simulations of PISN explosions generally do not find strong mixing in them (Joggerst and Whalen 2011; Chen et al.

2014; Gilmer et al. 2017). Crucial issues of PISN models for SLSNe Ic are in their spectroscopic properties. Although PISNe produce a huge amount of ^{56}Ni, it is not sufficient to heat the massive core as much as observed (e.g., Dessart et al. 2012). The production of a large amount of ^{56}Ni also leads to the existence of a large amount of Fe-group elements in the ejecta. The absorptions and emissions by the Fe-group elements results in much redder spectra than those observed in SLSN Ic (e.g., Dessart et al. 2012; Jerkstrand et al. 2016).

Finally, we summarize pros and cons of the ^{56}Ni-powered models.

Pros:

- ^{56}Ni power can naturally explain the decay rate of slowly declining SLSNe Ic.

Cons:

- ^{56}Ni power cannot explain the rapidly declining SLSNe Ic. A different model must be invoked for the rapidly declining SLSNe Ic.
- Spectra of SLSNe Ic do not match those predicted for ^{56}Ni-powered SNe.

4 Interaction-Powered Models

Interaction of the gas ejected during a SN explosion with an ambient medium is a process that happens quite often at different stages of SN evolution. Interaction leads to shock wave formation, and shocks very efficiently transform the kinetic energy to thermal energy, which can be then radiated at different wavelengths (Chevalier and Fransson 2003). The stage of SN evolution at which significant interaction starts depends on the density of the CSM. If the ejecta expand into low density material, the interaction becomes important at the SN remnant stage, a few tens, hundreds, or even thousands of years after the SN explosion. Up to this age, the ejecta are cold and faint, but the shock interaction with the surrounding material heats it up again (Reynolds 2016). In the case of a dense CSM, the interaction occurs already on the SN stage (e.g., for SNe of Type IIn, see Chandra 2018). The SN becomes more luminous than a normal one exploded in lower density surroundings.

In the case of SLSNe, the radiation energy emitted during the first few months after the explosion is of the order of 10^{50}–10^{51} ergs (Quimby et al. 2011). The kinetic energy of a standard SN explosion is also $\sim 10^{51}$ ergs. Naked SNe preserve almost all their kinetic energy until the remnant stage, when a sufficient amount of ambient mass (comparable to the ejecta mass) is swept up, and the interaction becomes important. If the ambient mass is concentrated close to the exploding star, the interaction starts from the very beginning and there is a good chance to transform the kinetic energy of the ejecta to radiation during the first months, exactly as it is needed to produce a SLSN event.

In this Section we first make crude estimates of some parameters of the explosion in order to understand what kind of structure can produce such a powerful event as a SLSN in the CSM interaction scenario (Sect. 4.1). Then we briefly discuss how the shock wave propagates through the dense extended CSM (Sect. 4.2). After that we describe analytical and numerical models for the SLSN light curves that originate from ejecta–CSM interaction and discuss the progenitors that are able to provide the required CSM structure (Sect. 4.3). Finally, we list pros and cons of the CSM interaction scenario for SLSNe (Sect. 4.4).

4.1 Estimates of Model Parameters

Mass Ratio m_{ej}/m_{CSM} Needed for SLSNe In the CSM interaction scenario for SLSNe, the expanding ejecta collide with the CS envelope that can be static or expanding. Let us consider for simplicity an inelastic collision of two blobs of matter. The first one, which represents the ejecta, has mass m_1 and momentum $\mathbf{p}_1$. Its kinetic energy is

$$E_{\text{init}} = \frac{\mathbf{p}_1^2}{2m_1}. \tag{2}$$

It collides with another blob with mass m_2 and, for now, we suppose its momentum to be zero. The final energy of two merged blobs in a fully inelastic collision is

$$E_{\text{fin}} = \frac{\mathbf{p}_1^2}{2(m_1 + m_2)}. \tag{3}$$

The momentum is conserved, but the energy $E_{\text{init}} - E_{\text{fin}}$ is radiated away, since $E_{\text{fin}} < E_{\text{init}}$. If $m_2 \ll m_1$ then only a tiny fraction of E_{init} is radiated, but if $m_2 \gg m_1$ then $E_{\text{fin}} \ll E_{\text{init}}$ and almost all initial kinetic energy of the system is radiated away.

The situation is a bit more complicated when the second blob, which represents the CSM envelope, is not static at the time of the explosion, but expands with a momentum $\mathbf{p}_2$. This is most probable for the case of SLSNe I, because velocities higher than 10,000 km/s are observed for several months starting from the maximum (Liu et al. 2017b). The energy before the interaction in this case is the sum of ejecta and CSM energies. In the prescription of two blobs, that is the sum of the blobs' energies:

$$E_{\text{init}} = \frac{\mathbf{p}_1^2}{2m_1} + \frac{\mathbf{p}_2^2}{2m_2}. \tag{4}$$

After the inelastic interaction the kinetic energy of the system becomes

$$E_{\text{fin}} = \frac{(\mathbf{p}_1 + \mathbf{p}_2)^2}{2(m_1 + m_2)}. \tag{5}$$

So an amount of energy that can be thermalized during the interaction is

$$\Delta E = E_{\text{init}} - E_{\text{fin}} = \frac{(\mathbf{p}_1 m_2 - \mathbf{p}_2 m_1)^2}{2m_1 m_2 (m_1 + m_2)} = \frac{m_1 m_2}{2(m_1 + m_2)} (\mathbf{v}_1 - \mathbf{v}_2)^2. \tag{6}$$

Here $\mathbf{v}_1$ and $\mathbf{v}_2$ are velocities of the colliding blobs that represent, respectively, the ejecta and the CSM envelope. Again, the important question is how much of the total kinetic energy of the system can be lost to radiation, because for the superluminous event this energy transformation must be very effective. If we define $\mu = m_1/m_2$ and $\beta = v_2/v_1$ then we get

$$\frac{\Delta E}{E_{\text{init}}} = \frac{\mu(1-\beta)^2}{(\mu + 1)(\mu + \beta^2)}. \tag{7}$$

It is clear that $v_1 > v_2$ ($\beta < 1$). Otherwise, the collision will never happen. This means that the larger the difference between v_1 and v_2, the larger the energy that can be radiated. For a few fixed values of β, the dependence of $\Delta E / E_{\text{init}}$ on the mass ratio μ is shown in Fig. 3. When the CS envelope is expanding ($v_2 > 0$) the maximum losses happen not at $m_1 \sim 0$, but at some finite mass. Still, m_1 has to be noticeably lower than m_2 for the effective energy loss to happen. This means that the ejecta must be less massive than the CS envelope for a SLSN event.

Fig. 3 Energy losses relative to the initial kinetic energy of the system of two inelastically colliding blobs

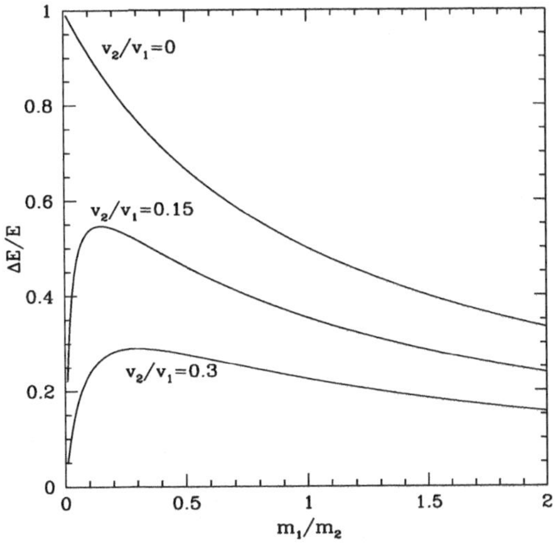

Typical Radius A crude estimate yields the photospheric radius R_{ph}, and, therefore, the typical size of the dense circumstellar envelope that can provide the observed luminosity L of the SLSNe. For simplicity, assume that the SLSN emits like a blackbody with a temperature T_{bb}. Then the photospheric radius can be obtained from $L = 4\pi R_{ph}^2 \sigma T_{bb}^4$. The typical temperature measured for SLSNe is $T_{bb} \simeq 10^4$ K. The order of magnitude estimate of the luminosity is $L \simeq 10^{44}$ erg/s. Hence, the radius is roughly $R_{ph} = 4 \times 10^{15}$ cm, which is about an order of magnitude larger than typical photospheric radii of standard SNe near maximum light.

Ginzburg and Balberg (2012) in their simple numerical hydrodynamic diffusion model got a similar value for the CSM radius (of order 10^{15} cm), and also stated that the masses of the ejecta and the CSM involved into the interaction must be comparable for the effective transformation of energy into radiation. As we see from the Fig. 3, a smaller ejecta-to-CSM mass ratio is even better.

4.2 Hydrodynamical Properties

Shock Propagation Through the Optically Thick Extended CSM The shock waves produced at the ejecta-CSM interaction are initially radiation-dominated. The photons behind the shock bear a large amount of energy compared to the gas energy. The photons are locked within the CSM and heat it up making the shock front wider. Precursor heating extends to the optical depth $\tau = c/v_s$ from the shock front, where v_s is a shock velocity. When it reaches the outer edge of the CSM, the photons that were heated on the shock start to gradually leave the system. The shock begins to break out. Due to the large size of the system the breakout stage lasts much longer than it does for standard SN Ib/c. In the latter case, $R < 10\ R_\odot$ and the shock breaks out for less than $R/v_s \sim 200$ s, while in the case of CSM-interacting SLSN with typical $R \sim 10^5\ R_\odot$, the shock breakout time is proportionally larger and lasts a few months. On the whole the light curve of the SLSNe in the CSM-interaction scenario can be considered as a prolonged shock breakout. At some stage the shock can become strongly radiative. The temperature jump at the shock almost disappears, and the shock becomes almost isothermal. In this case, the density can jump up by a few orders of magnitude, and a rather dense and thin layer forms behind the shock. A very detailed study

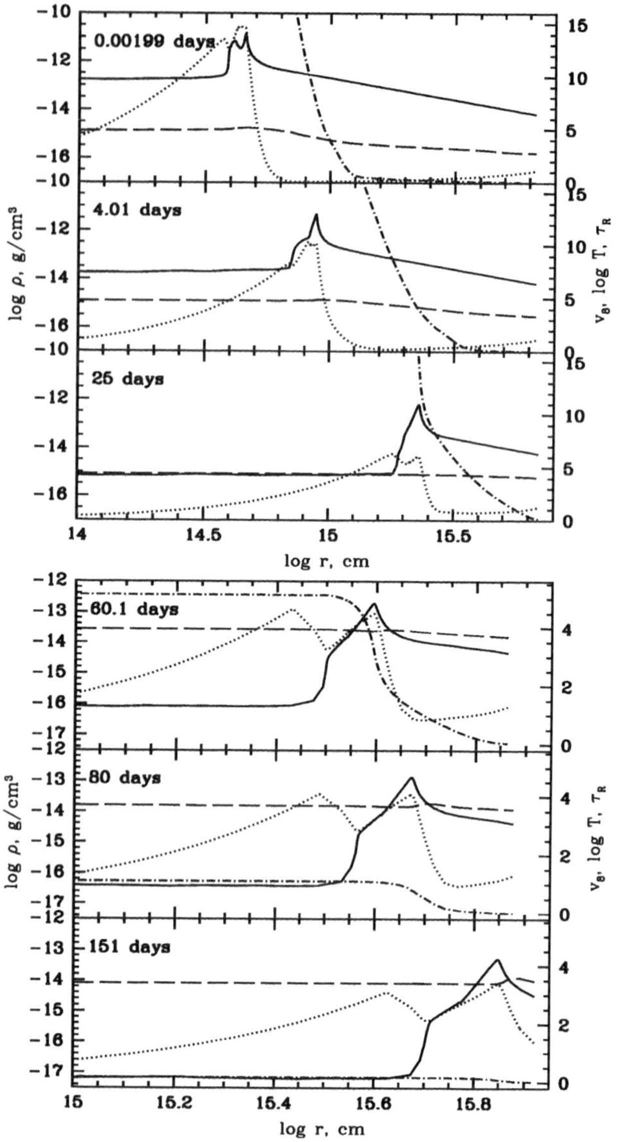

Fig. 4 Evolution of radial profiles of the density (*solid lines*), velocity (in 10^8 cm s^{-1}, *dots*), gas temperature (*dashes*), and Rosseland optical depth (*dash-dots*) for the model N0 from Sorokina et al. (2016). The scale for the density is on the left Y axis, for all other quantities, on the right Y axis. Upper panel: pre-maximum hydrodynamical structure for three moments, very soon after the explosion and at days 4 and 25. Lower panel: the same parameters, but after maximum, at days 60, 80, and 151. Note that different scales for the axes are used on the upper and lower panels

of the structure of the SN shocks is provided by Weaver (1976). More details are presented in Tolstov et al. (2015). The basic formulas can be found in Blinnikov (2016).

Figure 4 shows the results of a numerical calculation of the ejecta-CSM interaction for one of the SLSN models that demonstrates how the radiative shock propagates throughout the dense extended CSM and what happens with the Rosseland optical depth (and therefore, photospheric radius) at the same time. While the CS envelope is cool before the SN explosion, it is transparent to the radiation, but when it becomes warm due to the pre-heating by the photons from the shock, it quickly becomes optically thick and strongly diffusive. The photosphere ($\tau_R \sim 1$) shifts to the outer edge of the CSM. As a result, the shock is left deep inside the photosphere for a few months. Conditions are not appropriate for hard emission at the early stages despite the high velocity of the shock. But an ultraviolet flash, or perhaps even an X-ray flash, can appear later, when the shock becomes weaker, closer to the outer edge of the CS envelope. This can happen in the fading stage of the light curve evolution. SCP 06F6 is a possible example of such a SLSN originating from interaction due to the X-ray flash observed about 2 months after the maximum (Gänsicke et al. 2009). At the earlier stage, near maximum, the temperature conditions are appropriate for strong emission in the near UV range. Almost all SLSNe are relatively blue near maximum, and the shock heating in the CSM interaction scenario reproduces the color better then other mechanisms (Tolstov et al. 2017b).

4.3 Light Curves

Analytical Models Analytical methods to obtain rough estimates for the interaction-powered SNe are available. In this review, we focus more on numerical studies and we only briefly mention the analytic studies.

The dense CSM required to explain the large luminosity of SLSNe by the interaction is optically thick and we need to take large optical depths of the dense CSM into account in treating them analytically unlike the case of less luminous Type IIn SNe (e.g., Chevalier and Fransson 2003; Moriya et al. 2013b). When the CSM optical depth is large enough, the shock breakout itself can occur in the dense CSM. Chevalier and Irwin (2011) provide a simple analytic way to estimate the rise time and peak luminosities followed by the shock breakout in the dense CSM. They show that the LC features differ depending on where the shock breakout occurs in the dense CSM. This method is extended to the case of non-steady mass loss by Moriya and Tominaga (2012).

Chatzopoulos et al. (2012, 2013) develop a semi-analytic approach to estimate the overall LC properties of the interaction-powered SNe that are commonly used recently. Although it is a simple powerful method to estimate the CSM properties, it is important to keep its limitation in mind. The semi-analytic method is based on the formalism of Arnett (1980) that assumes that the heating source is at the center of SNe. However, the energy source of the interaction model is not located at the center. It moves outwards as the shock progresses. Moreover, it can be situated within a very thin layer, since gas between the forward and the reverse shocks can cool down strongly. In this case, it is almost impossible to spatially separate the emission from the two shocks, while the analytic solution by Chatzopoulos et al. (2012, 2013) takes the radiation of the forward and the reverse shocks into account separately, which leads to an inaccurate luminosity prediction. Indeed, the CSM configuration obtained by the semi-analytic model has been shown to be inconsistent with those found by numerical approaches (Moriya et al. 2013b; Sorokina et al. 2016). Chatzopoulos et al. (2013) study how to find the CSM configuration from the semi-analytic model that better matches the numerical results.

Numerical Modeling Numerical modeling allows taking more physical processes into account and simulating them in detail, so the results should be more robust.

The interaction scenario is challenging for numerical modeling. A numerical code must be able to resolve optically thick shocks. This is a reason why the number of SLSN interaction models calculated up to now is not very large. The very first numerical modeling of the SN ejecta interaction with the dense extended medium has been done by Grasberg et al. (1971), Falk and Arnett (1977), Grasberg and Nadyozhin (1987). More recently, most calculations were produced with the radiation hydrodynamic code STELLA (Blinnikov et al. 1998, 2006).

SLSNe II Hydrogen-rich SLSNe are spectroscopically similar to SNe IIn, and therefore, their origin from the interaction of SN ejecta with slowly expanding dense CSM is the most probable scenario. To provide the high luminosity of SLSN, the CSM is expected to be sufficiently more massive than for SN II case.

Woosley et al. (2007) suggest a pulsational pair instability (PPI) scenario for the last stage of the pre-SN evolution, which leads to the formation of the required density structure. PPI can happen for stars with an initial main-sequence mass 95–130 $M_\odot$, or, more certain, with a helium-core mass 40–60 $M_\odot$. Near the end of their lives, when the central temperature exceeds 10^9 K, such stars lose stability through electron-positron pair creation. Explosive oxygen burning in the cores of these stars cannot unbind the whole star, but leads to the ejection of a very massive envelope with large kinetic energy. The remaining stellar core contracts, and some time later, when the central temperature again becomes high enough for pair creation, another ejection happens. The process repeats until the stellar mass becomes too low for the pair instability, and the core finally collapses due to iron dissociation.

As a result of the PPI scenario, the final core collapse happens inside the system of the massive expanding shells. Collision of these shells, as well as of the final ejecta with shells, can lead to a very bright event, as observed for the SLSNe.

Woosley (2017) presents a wider study of the models of the stellar evolution exploding as PPISNe. He concludes that no solar metallicity star ends up as PPISN since it loses too much mass through the wind, while stars with the metallicity $Z/Z_\odot < 0.1$ can lead to this end. For SLSNe II, whose host galaxies have metallicities in a wide range, including high-metallicity objects (Perley et al. 2016), this means that the PPI explosion mechanism is not so likely to explain all these events, though more study of PPI is still required. At the same time, PPI may be promising to explain SLSNe I, since they prefer low-metallicity galaxies and their photospheric velocities are much larger than what more standard stellar mass loss mechanism can provide.

Woosley et al. (2007) apply the PPI scenario to one of the first SLSN II, namely, SN 2006gy. They found that the collisions of two subsequent mass ejections of about 25 $M_\odot$ and 5 $M_\odot$, separated in time by less than 7 years, with a total energy of about 3×10^{51} ergs, provides a good fit to the observed light curve of SN 2006gy.

Moriya et al. (2013b) studied a large number of toy models for the colliding ejections and found slightly different parameters for the best-fit model for SN 2006gy. Both ejections contain a mass of about 15 $M_\odot$ (the second one could be lighter), and their kinetic energy is 4×10^{51} ergs, which is only a little larger than the energy of normal SNe. The influence of the density distribution on the light curve is also studied, which showed that the first ejection could not be formed by a steady wind.

Despite a slightly different estimate of the ejection parameters, both Woosley et al. (2007) and Moriya et al. (2013b) converge on the conclusion that two subsequent ejections of a mass of a few tens solar mass with an energy close to the typical one for SN explosions within a few years can easily lead to the very high luminosities typical for SLSNe. Similar results are obtained by Dessart et al. (2015) and Vlasis et al. (2016), who also investigate aspherical CSM configurations.

SLSNe I In the modeling of hydrogen-poor SLSNe, one more parameter appears in addition to those that are important for hydrogen-rich events. It is the chemical composition of the ejecta and the CSM. There are several observational hints for choosing the correct composition for numerical modeling of SLSN I light curves and spectra. Observationally, strong broad oxygen lines are typical for SLSNe I from the very beginning, but very few SLSNe I have helium lines in their nebular spectra. So the observations favor a CO composition for the ejecta and CSM. Numerical modeling of the multicolor broad-band light

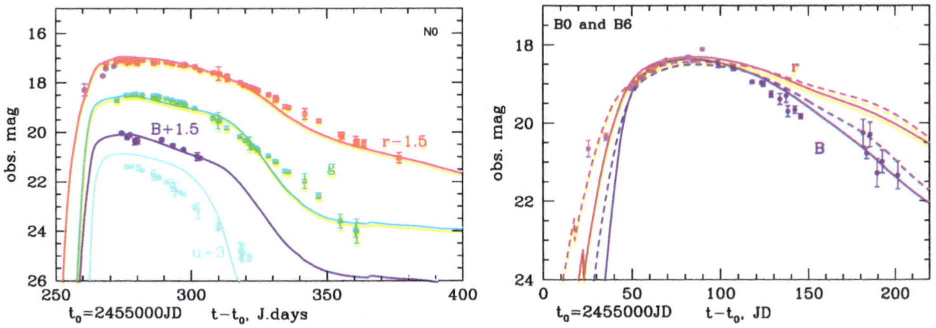

Fig. 5 The narrow light curve of SN 2010gx (*left*, Pastorello et al. 2010) and the broad one of PTF09cnd (*right*, Quimby et al. 2011) in different filters. Observations are shown with *dots*, calculations (Sorokina et al. 2016) with *lines*. *Solid* and *dashed* lines in the right plot mean different composition:*solid lines* correspond to the model B0 with the outer envelope containing 90% carbon and 10% oxygen, *dashed lines* correspond to the model B6 similar to B0, but roughly half of C and O in the outer envelope is replaced with He

curves (Sorokina et al. 2016) is in agreement with the observations from this point of view. The presence of helium leads the light curve to rise much longer than observed, while a CO composition results in a faster rise. This difference is explained by different opacities for CO and He. CO material requires a lower temperature for excitation and ionization, so the carbon-oxygen CSM becomes opaque soon after the shock starts to heat the CSM gas in front of it. Helium needs a higher temperature to become opaque, and it takes a longer time. Some amount of helium can be mixed with the CO gas. The light curve is not much changed when carbon and oxygen are the main absorbers. The photometric colors near maximum help determine the proportions of C to O, since these elements have different spectral line distributions. Oxygen is more absorptive in the blue, and it shifts the larger part of the emission to this wavelength range. The modeling by Sorokina et al. (2016) shows that models with a C to O ratio from 0.7 to 0.9 show the best fit to the observations of the narrow and the broad SLSN I light curves, respectively. The light curves are shown in Fig. 5, and demonstrate how the pure ejecta–CSM interaction scenario, without any radioactive material, can reproduce the emission of SLSNe I. The masses and energies of the models are similar to those needed for SLSNe II. A fast evolving SLSN I can be reproduced by the interaction with about 10 $M_{\odot}$ CS envelope, while for slowly fading events, a few tens $M_{\odot}$ of hydrogen-poor material is needed. The explosion energy in any case is a few 10^{51} erg.

Bumpy Light Curves Light curves of some SLSNe I show some undulations on the fading part (Inserra et al. 2017). The CSM interaction scenario explains this in the most natural way: each bump on the light curve can denote the collision of CSM layers or the SN ejecta with a CSM layer. If the progenitor star experiences several episodes of massive ejection before the final collapse as it can happen in the PPI mechanism, then several collisions of CSM layers could be later observed as bumps on the SLSN light curve. In the spectra of the SLSNe I with the most undulating light curves, hydrogen lines appear at a late stage (Yan et al. 2017a). This also can be considered as a proof for the CSM interaction scenario for some SLSNe I. Initially the luminosity of the SLSN is explained by the collision of the innermost hydrogen-free layers, while the outer layers that can contain some hydrogen are cold and transparent. Later, the inner hydrogen-poor layers, that are faster, reach the hydrogen-rich one, heat it, and the hydrogen lines appear in the spectrum.

The origin of the early pre-maximum luminosity excess that is observed in many SLSNe I is more questionable. Moriya and Maeda (2012) qualitatively explain the possible origin of this excess by heating of a detached CS envelope. Its opacity is enhanced after ionization leading to a temporary decrease in luminosity before a rise to the main maximum. More detailed numerical modeling is needed for this mechanism. Some other explanations of the pre-maximum excess are described in Sect. 6.

4.4 Pros and Cons for the Interaction Scenario

Summarizing the above material, we point out the strengths and weaknesses of the CSM interaction model for SLSNe.

The interaction scenario for SLSNe is favored due to the following reasons:

- SLSNe II are spectrally very similar to SNe IIn that are explained by CSM-ejecta interaction. For a more massive CSM, a more luminous (and superluminous) SN is expected.
- SLSN I light curves with a wide range of rising and fading timescales can quite easily be modeled by the interaction with a CSM of different mass. The larger the light curve timescale, the more massive CSM is needed. Only the widest light curves are problematic for modeling with interaction.
- The undulation on the fading part of the light curve is most naturally explained by a sequence of collisions of several CSM envelopes or clouds.
- Measuring masses of different elements from the nebular spectra of SLSNe (Jerkstrand et al. 2017), energetics of these events, and other observational properties show that SLSNe originate from massive stars. Mass loss is typical for these objects (Vink 2015). In this sense, some kind of CSM interaction is unavoidable after the SN explosion. However, the amount of such material must be very large to provide the strong interaction that is required to explain the SLSN phenomenon.

The last item from the list above takes us to the list of difficulties that still remain for the interaction mechanism.

- At least a few solar masses must be lost by a SLSN progenitor during a few months before the explosion to explain the brightness and the duration of the SLSN events. Although this is not impossible from the stellar evolution point of view, there is no detailed understanding of the very last stages of stellar life yet, neither theoretical, nor observational. The required CSM mass is comparable to observed mass-loss rates only from luminous blue variables (LBVs) in outburst (Smith et al. 2011). For the metal-poor stars that are able to preserve considerable mass without losing it through a steady wind till the end of their evolution, eruptive mass loss with the ejection of several solar masses of gas prior to the explosion could be very important (Smith and Owocki 2006). Anyway, a detailed mechanism for the LBV eruptions is still unknown, as well as the details of the mass loss by the PPI mechanism. It is not, of course, an argument against the origin of SLSNe from CSM interaction, but this topic still requires both theoretical and observational efforts.
- High photospheric velocities (10,000–20,000 km/s) are measured in SLSNe I for about a month from maximum. Existing numerical interaction models do not reproduce these velocities. But an explosion energy only a little larger than standard has been considered in modeling. Enhancing the explosion energy up to a few 10^{52} erg and the kinetic energy of the CSM expansion to about 10^{51} erg can improve the situation. However, this energetic model would bring another problem: such energetic explosions must be separated in time longer than needed for getting a strong interaction because the stronger an eruptive mass ejection is, the longer time is needed for the star to reach the instability conditions for the next explosion or eruption. For example, an estimate of the time intervals between the eruptions for the PPI model can be found in Woosley (2017).

- The slowest (i.e., widest) observed SLSN I light curves together with high velocities are extremely difficult to explain solely by CSM interaction. In this case a combination of energy sources, like CSM interaction for the maximum plus radioactivity for the tail similar to what has been done in Tolstov et al. (2017a,b), is more promising.

Detailed numerical modeling for an extended set of explosion parameters is still needed to understand the applicability of the CSM interaction model to the SLSNe, especially for the hydrogen-poor ones. Not only the light curves must be fitted, but also the spectra. PPI seems to be the best mechanism for making a CSM density and velocity structure that would lead to a SLSN event after the interaction with the SN ejecta. But PPI itself must be better studied in order to understand the limitations it places on the CSM structure.

5 Magnetar-Powered Models

The view that pulsars could play a role in powering supernovae was proposed soon after the discovery of pulsars. Ostriker and Gunn (1971) proposed that the spin-down of a central pulsar could power both the supernova explosion and the light from the supernova. In this model, the pulsar creates a bubble of relativistic fluid that sweeps up and then accelerates the surrounding star. However, this model was not found to be in accord with supernova observations. The current prevailing model is that the power for the supernova is injected over a short period of time (seconds) and the pulsar bubble then evolves in the freely expanding ejecta set up by the initial explosion. This model was proposed for the Crab Nebula and other young pulsar wind nebulae (Chevalier 1977) and for the light from supernovae (Gaffet 1977). Generally, the light from core collapse supernovae does not require power input from a pulsar, and can be explained by a combination of initial explosion energy, radioactivity, and possible circumstellar interaction. However, the supernova SN 2005bf showed two broad peaks, implying the operation of some additional mechanism in addition to the usual. Maeda et al. (2007) suggested that power from pulsar spin-down could be the source of the additional power. The observed luminosity and timescale required a highly magnetized pulsar ($> 10^{14}$ G), or magnetar, with an initial spin period in the ms range. Kasen and Bildsten (2010) and Woosley (2010) found that similar parameters could explain the timescale and luminosity of SLSNe, and this has become one of the leading contenders for the light from SLSNe. Here we discuss the basic magnetar model, giving some of the uncertainties of the model. We primarily deal with simplified models that illustrate the physical principles.

The initial central explosion is assumed to occur on a timescale of seconds, followed by shock traversal of the star and the ejecta transition to free expansion on a timescale of the doubling of the radius. At that point, the velocity profile is given by $v = r/t$. The density profile for the expanding gas can be approximated as a steep power law with radius at large radii and a flat power law at small radii (Chevalier and Soker 1989; Matzner and McKee 1999). We have the inner power law $\rho_{\rm in} \propto t^{-3}(r/t)^{-\delta}$ and outer $\rho_{\rm out} \propto t^{-3}(r/t)^{-n}$. Here, we take $\delta = 0$. The transition velocity between the inner and outer density profiles is

$$v_{\rm tr} = \left[\frac{10(n-5)}{3(n-3)}\frac{E_s}{M_{\rm ej}}\right]^{1/2}. \tag{8}$$

For $n = 7$, we have $v_{\rm tr} = 4080 E_{51}^{1/2} M_5^{-1/2}$ km s^{-1}, where E_{51} is E_s in units of 10^{51} ergs and M_5 is $M_{\rm ej}$ in units of 5 $M_\odot$, where E_s is the supernova energy and $M_{\rm ej}$ is the ejecta mass. The value of E_s is generally taken to be 1×10^{51} ergs, which is a value that is obtained for cases where the energy is well determined.

The newly formed rapidly rotating neutron star is assumed to drive a supersonic/superAlfvenic wind, powered by the spin-down of the neutron star. The details of how the transition is made from an accreting neutron star to a pulsar wind are not known. A plausible estimate for the spin-down power is the result for a force-free magnetic field (Spitkovsky 2006)

$$\dot{E} \approx \frac{\mu^2 \Omega^4}{c^3}\left(1+\sin^2\alpha\right), \tag{9}$$

where $\mu = BR^3$ is the magnetic dipole moment, $\Omega = 2\pi/P$ is the spin frequency, α is the angle between the magnetic and rotation axes, B is the dipole field at the neutron star surface, and R is the neutron star radius. Here, R is taken to be 10 km. The neutron star rotational energy is $E = \frac{1}{2}I\Omega^2$ where I is the neutron star moment of inertia, taken to be 1×10^{45} g cm^2. The initial spin-down time is

$$t_{\rm sd} = \frac{E_0}{\dot{E}_0} = 0.14\left(\frac{P_0}{\rm ms}\right)^2\left(\frac{B_p}{10^{14}\ {\rm G}}\right)^{-2}\ {\rm day}, \tag{10}$$

where the subscript 0 refers to the time $t = 0$. The evolution of the spin-down power is given by

$$\dot{E} = \dot{E}_0\left(1+\frac{t}{t_{\rm sd}}\right)^{-2}. \tag{11}$$

The structure of pulsar magnetospheres is not a fully solved problem; there are other possibilities for the spin-down. For the vacuum magnetic dipole case

$$\dot{E}_{\rm vac} = \frac{2\mu^2\Omega^4}{3c^3}\sin^2\alpha. \tag{12}$$

In the force free case, there is spin-down for an aligned rotator, but not in the vacuum case.

More generally, the evolution of the spin of a pulsar is often taken to be of the form

$$\dot{\Omega} \propto -\Omega^m \tag{13}$$

where m is the braking index. The value of m for the magnetic dipole case is 3 and the constant of proportionality in that case can be related to the neutron star moment of inertia and magnetic field, as discussed above. The value of m can be determined from observations by $m = \ddot{\Omega}\Omega/\dot{\Omega}^2$. The power from the spin-down with constant m is

$$\dot{E} = \dot{E}_0\left(1+\frac{t}{\tau}\right)^{-(m+1)/(m-1)}, \tag{14}$$

where τ is a spindown timescale that cannot now be written simply in terms of the magnetic field. Measured values of m for ordinary pulsars are < 3. For PSR J1119–6127, $m = 2.91$, close to the magnetic dipole value. Other values are lower; the long term value is $m = 2.34$ for the Crab Pulsar and 1.7 for the Vela pulsar (Espinoza et al. 2017). Possible reasons for deviations from the magnetic dipole value are evolution of the neutron star moment of inertia or magnetic field, or a more complex magnetic configuration than a dipole. These considerations show that pulsars generally do not follow the magnetic dipole spin-down and the late evolution of $\dot{E}$ (at $t \gg \tau$) is steeper than in the force-free dipole case; for the case of the Vela pulsar, $\dot{E} \propto t^{-3.9}$ at late times. The parameters for the rapidly rotating magnetars of interest here are quite different from the ordinary pulsars, but these objects show that pulsar spin-down can deviate from the force-free dipole case.

We next consider the evolution of the pulsar bubble on the assumption of adiabatic motion. In the case of a pulsar wind nebula, the shocked pulsar wind of relativistic particles and magnetic field can be approximated by a $\gamma = 4/3$ fluid, where γ is the adiabatic index. For the relatively young SLSN case, the assumption is that the pulsar power thermalizes so the radiation field is the dominant energy density. Again, an initial value $\gamma = 4/3$ is appropriate. If spherical symmetry is assumed, the expansion of the pulsar bubble can be found in the thin shell approximation. Even if the outer shock wave is not radiative, the swept up shell is thin. If the pulsar provides a steady power $\dot{E}$, the radius of the swept up shell is (Chevalier 1977; Blondin et al. 2001)

$$R_s = 1.50\left(\frac{n}{n-5}\right)^{1/5}\left(\frac{n-5}{n-3}\right)^{1/2}\left(\frac{E_s^3\dot{E}^2}{M_{\rm ej}^5}\right)^{1/10} t^{6/5}, \tag{15}$$

where the thin shell approximation has been made. We have

$$t_{\rm tr} = \frac{66(n-5)}{25n}\frac{E_s}{\dot{E}}, \tag{16}$$

where $t_{\rm tr}$ is the time for the shell to reach the inflection point in the supernova density profile. Approximating $\dot{E} = E/t_{\rm sd}$ for a time $t_{\rm sd}$, we have

$$t_{\rm tr} = 0.19 B_{14}^{-2} P_{ms}^4 E_{51} \text{ day}. \tag{17}$$

Once the shell has traveled through the flat part of the supernova density profile, it comes into the steep power law part of the density profile. The mass added from the outer ejecta is negligible compared to that in the shell and, for 1-dimensional (1-D) expansion, the evolution of the shell radius goes to $\propto t^{1.5}$ as in the solution of Ostriker and Gunn (1971). This is another self-similar solution.

The acceleration of the shell by the hot bubble is subject to Rayleigh-Taylor instability (RTI). The action of the RTI while the shell is in the flat part of the ejecta density profile has been well studied (Porth et al. 2014; Bucciantini et al. 2004). Despite the growth of instabilities, the instability saturates and the flow remains self-similar; the outer shock front has a radius that is 7% larger than in the 1-D solution (Blondin and Chevalier 2017). Most of the swept up mass is in the shell, although there is also some matter that is in fingers that extend in towards the neutron star. The RT fingers have relatively little effect on the termination shock of the pulsar wind. Global simulations including a magnetic field show qualitatively similar properties, although the Kelvin-Helmholtz instability is suppressed on small scales.

Once the shell reaches the transition point in the density profile, there is a qualitative change in the evolution. The preshock ejecta are unable to contain the pressure of the shocked wind bubble (Chevalier 2005) and the bubble matter blows out through the shell (Chen et al. 2016; Suzuki and Maeda 2017; Blondin and Chevalier 2017). Channels form in the swept up ejecta through which matter from the inner wind bubble flows. Suzuki and Maeda (2017) used a special relativistic code, allowing for a relativistic pulsar wind that shocked and slowed on passing through the termination shock. When the gas was able to escape through channels, relativistic velocities were again attained.

The power provided by the magnetar initially comes out in a relativistic wind of particles and Poynting flux. Assumptions must be made about the fraction of the wind power that goes into thermal radiation field. There might also be some fraction that remains in nonthermal form. The mechanisms giving rise to thermalization are not well understood.

Standard models for calculating the light curves of SLSNe I (Chatzopoulos et al. 2012; Inserra et al. 2013; Nicholl et al. 2017b) have been proposed based on the formalism developed by Arnett (1980, 1982) for ordinary supernovae. The model assumes that the supernova gas is homologously expanding and is characterized by a constant opacity κ for the thermal radiation. The emitted luminosity is

$$L(t) = 2e^{-(t/t_d)^2} \int_0^t e^{(t'/t_d)^2} \frac{t'}{t_d} \dot{E}(t') \frac{dt'}{t_d}, \tag{18}$$

where

$$t_d = \left(\frac{10\kappa M_{\rm ej}}{3\beta v_{\rm ej} c} \right)^{1/2} \tag{19}$$

is a diffusion time that gives a characteristic timescale for the light curve, $v_{\rm ej}$ is the velocity at the outer edge of the ejecta, and c is the speed of light. Here, β is a constant that depends on the density distribution of the freely expanding ejecta. We have $\rho_{\rm ej} \propto \exp(1.723Ax)$, where A is a constant and $x = r/v_{\rm ej}t$. For uniform density ejecta ($A = 0$), $\beta = 13.8$ and for $A = -1$, $\beta = 13.4$; $\beta = 13.8$ is typically chosen in magnetar models. As discussed above, a fairly flat density distribution is a reasonable approximation for the freely expanding ejecta in a core collapse supernova. However, in the present case, the density distribution is strongly affected by the magnetar wind nebula, causing some uncertainty in the model fits.

There is the assumption in equation (18) that the initial stellar radius is much smaller than the supernova radius so that terms involving the initial radius and thermal energy do not appear. The initial explosion energy is lost to adiabatic expansion. Another assumption in equation (18) is that all the pulsar power goes into the thermal radiation, but it may only be some fraction f that is thermalized. There have been various suggestions on how nonthermal effects can be included. Kotera et al. (2013) suggested that a constant fraction of the pulsar power goes into the thermal radiation. However, Wang et al. (2015) noted that models using equation (18) fit the observations at early times, but fall below the observations at late times. They suggested that

$$f = 1 - e^{Bt^{-2}}, \tag{20}$$

where

$$B = \frac{3\kappa_\gamma M_{\rm ej}}{4\pi v_{\rm ej}^2} \tag{21}$$

and κ_γ is the opacity for high energy photons. This expression has the property that at early times, nonthermal photons are completely thermalized, and at late times there is optically thin absorption of photons, assuming that the absorbing material is uniform.

This model, or some variation on it, has been widely used in modeling the light curves of SLSNe I. Different researchers make different assumption, hindering the comparison of results. Nicholl et al. (2017b) have analyzed the light curves of 38 SLSNe I in a consistent way, using the radiative transfer approximation of Arnett and the energetic photon escape of Wang et al. (2015). They assumed the vacuum magnetic dipole spindown for the pulsar power and made a blackbody assumption to obtain multi-color light curves for the supernovae. A result of their model fitting is shown in Fig. 6. The models fit the cases with a decline that mimics that of ^{56}Co, like SN 2007bi. They deduce the following ranges (1-sigma)

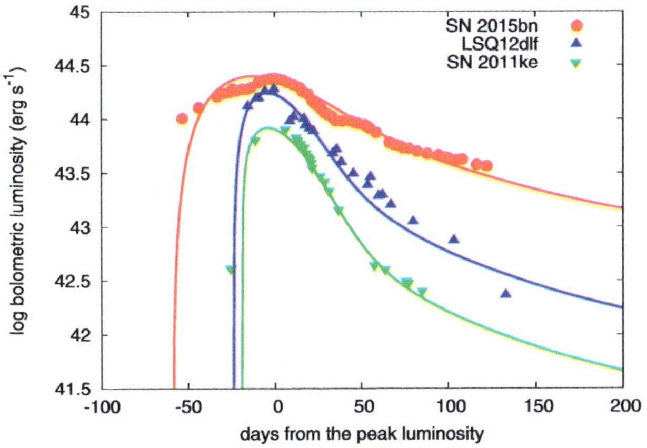

Fig. 6 Examples of magnetar-powered LCs fitted to SLSNe with different LC evolution. The SLSN bolometric LCs are obtained from Nicholl et al. (2016b). The model LCs are calculated with a semi-analytic way based on Arnett (1982) as in Inserra et al. (2013). The magnetar spin down energy is assumed to be thermalized with 100% efficiency. The magnetic field strengths and the spin periods in the models are: 10^{14} G and 2 ms (SN 2015bn), 3.5×10^{14} G and 2.5 ms (LSQ12dlf), and 7×10^{14} G and 2.5 ms (SN 2011ke)

of parameters from their models: initial period 1.2–4 ms, $B = (0.2\text{–}1.8) \times 10^{14}$ G, and $M_{ej} = (2.2\text{–}12.9)\ M_{\odot}$. Using similar methods, Liu et al. (2017a) had previously modeled 19 supernovae and obtained good fits for 14 objects. They found parameter ranges: initial period 1.2–8.3 ms, $B = (0.2\text{–}8.8) \times 10^{14}$ G, and $M_{ej} = 1\text{–}27.6\ M_{\odot}$. These results give some idea of the range of parameters in present modeling. Yu et al. (2017) also modeled 31 SLSNe similarly and found parameter ranges of initial period 1.4–12 ms, $B = (0.5\text{–}4.8) \times 10^{14}$ G, and $M_{ej} = 0.5\text{–}17.9\ M_{\odot}$.

Most modeling efforts have concentrated on the supernova light curves, but there has also been attention to the spectra of SLSNe. Dessart et al. (2012) calculated steady-state, 1-D, non-LTE spectral models with power input from a magnetar. They found that the magnetar has important effects when the energy deposited by the magnetar becomes comparable to the supernova energy. In this case, the spectrum is blue, in accord with observations, and does not have the line-blanketing that occurs when the power is produced by radioactivity. Broad lines can be produced, as observed, if the ejecta mass is relatively small.

Bersten et al. (2016) also went beyond the simple models in their light curve models, using a 1-D, special relativistic, radiation hydrodynamics code. They found that hydrodynamical modeling is especially important when the deposited magnetar energy exceeds that of the initial supernova, consistent with the arguments given above.

Although the magnetar model has had considerable success in reproducing the light curves and spectra of SLSNe I, these results follow from a certain power supplied over a certain time and do not provide a "smoking gun" for the magnetar model. There has been interest in whether some of the detailed features of SLSN I observations require a magnetar. As noted in Sect. 2.2, some SLSNe I show a precursor in the light curve before the main peak. Kasen et al. (2016) suggested that the precursor is emission from the breakout of the shock wave driven by the pulsar wind bubble. However, the power generated by the shell sweeping into the freely expanding ejecta is small. In order to obtain a significant precursor, Kasen et al. (2016) suggest that only a fraction of the initial pulsar power is thermalized, reducing the main peak relative to the shock breakout. Other suggestions for the precursor emission include circumstellar interaction (Moriya and Maeda 2012; Piro 2015) or a highly energetic supernova (Nicholl et al. 2015b).

Metzger et al. (2014) have discussed the breakout of an ionization front through the swept up shell. They assume that a substantial fraction of the pulsar power goes into an ionizing spectrum radiated by energetic particles. The ionization front is initially trapped in the shell but it is able to penetrate the shell as the column density of the shell declines due to expansion. This breakout can occur within months of optical maximum under certain conditions. Accounting for the Rayleigh-Taylor instability of the shell would result in an earlier breakout if the pulsar wind evolves to this phase. The ionization breakout theory was proposed in part to explain the luminous X-ray emission observed from SCP 06F6 (Levan et al. 2013). The complete ionization of the shell allows the X-rays to escape. However, after the detection with XMM, an observation with Chandra set a low upper limit on X-rays from SCP 06F6, and X-ray observations of other SLSN have only yielded upper limits (Margutti et al. 2017).

There may be a connection between SLSNe and long gamma-ray bursts (LGRBs). Both classes of objects occur in small, low metallicity galaxies. An ultra-long GRB, GRB 111209A, has recently been observed along with a bright supernova, SN 2011kl (Greiner et al. 2015). Margalit et al. (2018) suggest that the two classes of objects have similar progenitors (rapidly rotating neutron stars) and that the alignment between the spin axis and the magnetic axis is what determines the relative amount of jet vs thermal emission. In this view, the aligned rotator case is expected to give strong jets. The importance of jets in SLSNe is also discussed in, e.g., Soker and Gilkis (2017), Gilkis et al. (2016).

Although the magnetar model has attractive features, the case is not yet settled. The multidimensional hydrodynamic models show that a complex 3-D structure results from power input from an energetic magnetar (Chen et al. 2016; Suzuki and Maeda 2017; Blondin and Chevalier 2017). Models that account for that structure are ultimately needed.

We summarize pros and cons of the magnetar-powered models.

Pros:

- The magnetar models can explain both the slowly declining and the rapidly declining light curves of SLSNe Ic.
- The observed velocities and temperatures are roughly reproduced.

Cons:

- The magnetar models have a number of free parameters. There is not a "smoking gun" for the presence of magnetar power.
- The magnetar models do not naturally explain the bumps observed in the light curves of some SLSNe.

6 Comparison

Here we compare the three major suggested mechanisms discussed so far. SLSNe IIn show clear signatures of CSM interaction and their LCs and spectra can be explained by the interaction-powered model (Sect. 4). We here focus on SLSNe Ic whose luminosity sources are actively discussed now.

6.1 Light Curves

Both interaction-powered and magnetar-powered models can provide reasonable fits to most SLSN Ic LCs with a proper choice of the model parameters. The ^{56}Ni powered models fail to explain the rapidly declining SLSN Ic LCs as their late-phase luminosity evolution is governed by the nuclear decay rate of ^{56}Co which is rather slow.

Some SLSN Ic LCs show complicated structures that may prefer the interaction model. For example, the LCs of iPTF15esb (Yan et al. 2017a) have two major peaks that are hard to explain with a single luminosity source at the center and prefer the existence of a complicated dense CSM structure. The "spiky" bolometric LC of SN 2017egm is also suggests an interaction model (Wheeler et al. 2017). The LC modulations observed in some SLSNe Ic are also likely to indicate the existence of an outer energy source like a dense CSM. Even if there is a major central power source like a magnetar powering a SLSN Ic LC, some dense CSM that has a minor contribution to the luminosity may also exist. A combination of central and outer powering sources could better explain the general SLSN Ic LCs.

Multi-frequency studies of SLSNe Ic are helpful in distinguishing the different powering mechanisms. For example, the magnetar model predicts a late-phase increase in X-rays at the "ionization breakout" (Metzger et al. 2014), while X-rays in the interaction model are expected in some cases after the shock breakout stage that lasts a few months. Margutti et al. (2017) report their observational efforts to detect SLSNe in X-rays but only a few detections are obtained so far and it is still hard to distinguish the powering mechanisms in this way. Tolstov et al. (2017b) recently suggested that the ultraviolet luminosity could be better explained by the interaction model. It is not totally clear yet if the ultraviolet luminosities are the decisive distinguisher among the models, given the uncertainties in how magnetars transform their spin energy to the surrounding ejecta. More studies are required in this regard. No γ-ray detections from SLSNe could indicate that the magnetic field strengths in the compact remnants of SLSNe are much less than required to power the LCs (Renault-Tinacci et al. 2017).

The precursors found in many SLSNe Ic are hard to explain with any model. If the precursor is from the shock breakout, the progenitor radii need to be more than 500 $R_\odot$ with a reasonable assumption of the explosion energy (Nicholl et al. 2015b; Piro 2015). These huge radii are unexpected in hydrogen-free progenitors and some mechanisms to make an extended envelope are required. The ^{56}Ni-powered model only has the central heating source without much hydrodynamical effect and such extended envelopes are necessary to explain the precursor.

The CSM interaction and magnetar models are able to explain the precursor without invoking the extremely extended progenitor for SLSNe Ic. For the CSM interaction model, Moriya and Maeda (2012) illustrate that the precursor bump can be naturally made by the ionization change in the massive CSM to power the major LCs in SLSNe Ic. Regardless of the luminosity source of the precursor, the precursor LC bump can be made by the massive CSM required to explain the main peak of SLSN Ic LCs. In the magnetar-powered model, so-called "magnetar shock breakout" can occur in the ejecta, making a precursor on some occasions (Kasen et al. 2016, see also Sect. 5). When a huge magnetar rotational energy is released near the center of an exploding star, another shock wave can be launched. If the shock becomes strong enough to be a radiation-dominated shock, another shock breakout can occur when the shock reaches the surface of the ejecta resulting in a precursor bump observed in SLSNe Ic. However, in 1D LC simulations, the shock breakout bump is usually not clearly seen as observed unless the central heating efficiency is artificially reduced (Kasen et al. 2016; Moriya et al. 2016). More studies are needed to see if the magnetar shock breakout can generally explain the precursors.

6.2 Spectra

Spectra of SLSNe Ic provides plenty of information to distinguish the powering mechanisms. The ^{56}Ni-power model is generally less favored recently because of the lack of strong Fe absorptions and emissions in SLSNe Ic (e.g., Jerkstrand et al. 2017, see also Sect. 3). The models with a central engine like magnetars are generally found to match the spectra well (e.g., Mazzali et al. 2016; Dessart et al. 2012). SLSNe Ic do not show narrow features that are found in less luminous Type Ib (e.g., Pastorello et al. 2016) and Type Ic (e.g., Ben-Ami et al. 2014) SNe with strong CSM interactions. There are some suggested mechanisms to hide the possible narrow lines in the interacting SNe (e.g., Chevalier and Irwin 2011; Moriya and Tominaga 2012), but they may not generally occur. Spectral modeling is required to judge if the lack of the narrow emissions in SLSNe Ic is consistent with the interaction model.

Chen et al. (2017a) performed a detailed spectral modeling of the Type Ic SLSN LSQ14mo. They show that its line features did not change for a while but the continuum significantly changed in the same period. The temperature deduced from the line features does not match that deduced from the continuum for some period. In their interpretation, the continuum may be affected by an external heating source like CSM interaction in addition to a central heating source like a magnetar that is determining the temperature of the line forming region in the SN ejecta. This indicates that more than a single energy source is responsible for the LCs for all the periods. Different powering sources could be contributing at the same time to power SLSNe. Another example of this kind is Gaia16apd, which is suggested to be powered by a combination of CSM interaction and ^{56}Ni heating (Tolstov et al. 2017b). ^{56}Ni heating may contribute to SLSNe IIn in addition to CSM interaction if a significant amount of ^{56}Ni is produced, although CSM interaction by itself can explain SN 2006gy (Miller et al. 2010).

7 Other Models

Several other ideas have been raised as the power sources of SLSNe. For example, the fallback accretion onto the central compact remnant of the exploding stars can potentially release a sufficient energy to power SLSNe (Dexter and Kasen 2013). It is also suggested that a latent heat released by the phase transition from neutron stars to quark stars is sufficient to power SLSNe (e.g., Ouyed et al. 2012).

It is possible that not a single energy source is playing a role in powering SLSNe. Two or more energy sources can provide heat sources at the same time.

We also need to keep in mind the possibility that none of the energy sources currently proposed are correct – a power source that is completely missed so far may still exist.

Acknowledgements This review made use of the Weizmann interactive supernova data repository—http://wiserep.weizmann.ac.il (Yaron and Gal-Yam 2012) and the Open Supernova Catalog—https://sne.space/ (Guillochon et al. 2017). TJM is supported by the Grants-in-Aid for Scientific Research of the Japan Society for the Promotion of Science (16H07413, 17H02864) and the Munich Institute for Astro- and Particle Physics (MIAPP) of the DFG cluster of excellence "Origin and Structure of the Universe." The work of ES (interaction models) is supported by the Russian Scientific Foundation grant 16–12–10519. RAC was supported in part by NASA grant NNX12AF90G.

References

I. Agnoletto, S. Benetti, E. Cappellaro, L. Zampieri, M. Turatto, P. Mazzali, A. Pastorello, M. Della Valle, F. Bufano, A. Harutyunyan, H. Navasardyan, N. Elias-Rosa, S. Taubenberger, S. Spiro, S. Valenti,

SN 2006gy: was it really extraordinary? Astrophys. J. **691**, 1348–1359 (2009). 0810.0635. https://doi.org/10.1088/0004-637X/691/2/1348

C.R. Angus, A.J. Levan, D.A. Perley, N.R. Tanvir, J.D. Lyman, E.R. Stanway, A.S. Fruchter, A Hubble Space Telescope survey of the host galaxies of Superluminous Supernovae. Mon. Not. R. Astron. Soc. **458**, 84–104 (2016). 1601.01874. https://doi.org/10.1093/mnras/stw063

I. Arcavi, W.M. Wolf, D.A. Howell, L. Bildsten, G. Leloudas, D. Hardin, S. Prajs, D.A. Perley, G. Svirski, A. Gal-Yam, B. Katz, C. McCully, S.B. Cenko, C. Lidman, M. Sullivan, S. Valenti, P. Astier, C. Balland, R.G. Carlberg, A. Conley, D. Fouchez, J. Guy, R. Pain, N. Palanque-Delabrouille, K. Perrett, C.J. Pritchet, N. Regnault, J. Rich, V. Ruhlmann-Kleider, Rapidly rising transients in the supernova–superluminous supernova gap. Astrophys. J. **819**, 35 (2016). 1511.00704. https://doi.org/10.3847/0004-637X/819/1/35

W.D. Arnett, Analytic solutions for light curves of supernovae of Type II. Astrophys. J. **237**, 541–549 (1980). https://doi.org/10.1086/157898

W.D. Arnett, Type I supernovae. I—Analytic solutions for the early part of the light curve. Astrophys. J. **253**, 785–797 (1982). https://doi.org/10.1086/159681

D. Arnett, *Supernovae and Nucleosynthesis: An Investigation of the History of Matter from the Big Bang to the Present* (Princeton University Press, Princeton, 1996)

T.S. Axelrod, Late time optical spectra from the Ni-56 model for Type 1 supernovae. PhD thesis, California University, Santa Cruz (1980)

K. Barbary, K.S. Dawson, K. Tokita, G. Aldering, R. Amanullah, N.V. Connolly, M. Doi, L. Faccioli, V. Fadeyev, A.S. Fruchter, G. Goldhaber, A. Goobar, A. Gude, X. Huang, Y. Ihara, K. Konishi, M. Kowalski, C. Lidman, J. Meyers, T. Morokuma, P. Nugent, S. Perlmutter, D. Rubin, D. Schlegel, A.L. Spadafora, N. Suzuki, H.K. Swift, N. Takanashi, R.C. Thomas, N. Yasuda, Discovery of an unusual optical transient with the Hubble Space Telescope. Astrophys. J. **690**, 1358–1362 (2009). 0809.1648. https://doi.org/10.1088/0004-637X/690/2/1358

Z. Barkat, G. Rakavy, N. Sack, Dynamics of supernova explosion resulting from pair formation. Phys. Rev. Lett. **18**, 379–381 (1967). https://doi.org/10.1103/PhysRevLett.18.379

S. Ben-Ami, A. Gal-Yam, P.A. Mazzali, O. Gnat, M. Modjaz, I. Rabinak, M. Sullivan, L. Bildsten, D. Poznanski, O. Yaron, I. Arcavi, J.S. Bloom, A. Horesh, M.M. Kasliwal, S.R. Kulkarni, P.E. Nugent, E.O. Ofek, D. Perley, R. Quimby, D. Xu, SN 2010mb: direct evidence for a supernova interacting with a large amount of hydrogen-free circumstellar material. Astrophys. J. **785**, 37 (2014). 1309.6496. https://doi.org/10.1088/0004-637X/785/1/37

S. Benetti, M. Nicholl, E. Cappellaro, A. Pastorello, S.J. Smartt, N. Elias-Rosa, A.J. Drake, L. Tomasella, M. Turatto, A. Harutyunyan, S. Taubenberger, S. Hachinger, A. Morales-Garoffolo, T.W. Chen, S.G. Djorgovski, M. Fraser, A. Gal-Yam, C. Inserra, P. Mazzali, M.L. Pumo, J. Sollerman, S. Valenti, D.R. Young, M. Dennefeld, L. Le Guillou, M. Fleury, P.F. Léget, The supernova CSS121015:004244+132827: a clue for understanding superluminous supernovae. Mon. Not. R. Astron. Soc. **441**, 289–303 (2014). arXiv: 1310.1311. https://doi.org/10.1093/mnras/stu538

E. Berger, R. Chornock, R. Lunnan, R. Foley, I. Czekala, A. Rest, C. Leibler, A.M. Soderberg, K. Roth, G. Narayan, M.E. Huber, D. Milisavljevic, N.E. Sanders, M. Drout, R. Margutti, R.P. Kirshner, G.H. Marion, P.J. Challis, A.G. Riess, S.J. Smartt, W.S. Burgett, K.W. Hodapp, J.N. Heasley, N. Kaiser, R.P. Kudritzki, E.A. Magnier, M. McCrum, P.A. Price, K. Smith, J.L. Tonry, R.J. Wainscoat, Ultraluminous supernovae as a new probe of the interstellar medium in distant galaxies. Astrophys. J. Lett. **755**, L29 (2012). 1206.4050. https://doi.org/10.1088/2041-8205/755/2/L29

M.C. Bersten, O.G. Benvenuto, M. Orellana, K. Nomoto, The unusual super-luminous supernovae SN 2011kl and ASASSN-15lh. Astrophys. J. Lett. **817**, L8 (2016). 1601.01021. https://doi.org/10.3847/2041-8205/817/1/L8

P.K. Blanchard, M. Nicholl, E. Berger, J. Guillochon, R. Margutti, R. Chornock, K.D. Alexander, J. Leja, M.R. Drout, PS16dtm: a tidal disruption event in a narrow-line Seyfert 1 galaxy. Astrophys. J. **843**, 106 (2017). 1703.07816. https://doi.org/10.3847/1538-4357/aa77f7

S. Blinnikov, Radiative shock waves and their role in solving puzzles of Superluminous Supernovae. ArXiv e-prints. 1611.00513 (2016)

S.I. Blinnikov, R. Eastman, O.S. Bartunov, V.A. Popolitov, S.E. Woosley, A comparative modeling of supernova 1993J. Astrophys. J. **496**, 454–472 (1998). astro-ph/9711055. https://doi.org/10.1086/305375

S.I. Blinnikov, F.K. Röpke, E.I. Sorokina, M. Gieseler, M. Reinecke, C. Travaglio, W. Hillebrandt, M. Stritzinger, Theoretical light curves for deflagration models of type Ia supernova. Astron. Astrophys. **453**, 229–240 (2006). astro-ph/0603036. https://doi.org/10.1051/0004-6361:20054594

J.M. Blondin, R.A. Chevalier, Pulsar wind bubble blowout from a supernova. Astrophys. J. **845**, 139 (2017). 1707.07021. https://doi.org/10.3847/1538-4357/aa8267

J.M. Blondin, R.A. Chevalier, D.M. Frierson, Pulsar wind nebulae in evolved supernova remnants. Astrophys. J. **563**, 806–815 (2001). astro-ph/0107076. https://doi.org/10.1086/324042

S. Bose, S. Dong, A. Pastorello, A.V. Filippenko, C.S. Kochanek, J. Mauerhan, C. Romero-Cañizales, T.G. Brink, P. Chen, J.L. Prieto, R. Post, C. Ashall, D. Grupe, L. Tomasella, S. Benetti, B.J. Shappee, K.Z. Stanek, Z. Cai, E. Falco, P. Lundqvist, S. Mattila, R. Mutel, P. Ochner, D. Pooley, M.D. Stritzinger, S. Villanueva Jr., W. Zheng, R.J. Beswick, P.J. Brown, E. Cappellaro, S. Davis, M. Fraser, T. de Jaeger, N. Elias-Rosa, C. Gall, B.S. Gaudi, G.J. Herczeg, J. Hestenes, T.W.S. Holoien, G. Hosseinzadeh, E.Y. Hsiao, S. Hu, S. Jaejin, B. Jeffers, R.A. Koff, S. Kumar, A. Kurtenkov, M.W. Lau, S. Prentice, T. Reynolds, R.J. Rudy, M. Shahbandeh, A. Somero, K.G. Stassun, T.A. Thompson, S. Valenti, J.H. Woo, S. Yunus, Gaia17biu/SN 2017egm in NGC 3191: the closest hydrogen-poor superluminous supernova to date is in a normal, massive, metal-rich spiral galaxy. Astrophys. J. **853**, 57 (2018). 1708.00864. https://doi.org/10.3847/1538-4357/aaa298

N. Bucciantini, E. Amato, R. Bandiera, J.M. Blondin, L. Del Zanna, Magnetic Rayleigh-Taylor instability for Pulsar Wind Nebulae in expanding Supernova Remnants. Astron. Astrophys. **423**, 253–265 (2004). astro-ph/0405276. https://doi.org/10.1051/0004-6361:20040360

P. Chandra, Circumstellar interaction in supernovae in dense environments—an observational perspective. Space Sci. Rev. **214**, 27 (2018). 1712.07405. https://doi.org/10.1007/s11214-017-0461-6

E. Chatzopoulos, J.C. Wheeler, Effects of rotation on the minimum mass of primordial progenitors of pair-instability supernovae. Astrophys. J. **748**, 42 (2012). 1201.1328. https://doi.org/10.1088/0004-637X/748/1/42

E. Chatzopoulos, J.C. Wheeler, J. Vinko, Generalized semi-analytical models of supernova light curves. Astrophys. J. **746**, 121 (2012). 1111.5237. https://doi.org/10.1088/0004-637X/746/2/121

E. Chatzopoulos, J.C. Wheeler, J. Vinko, Z.L. Horvath, A. Nagy, Analytical light curve models of superluminous supernovae: χ^2-minimization of parameter fits. Astrophys. J. **773**, 76 (2013). 1306.3447. https://doi.org/10.1088/0004-637X/773/1/76

E. Chatzopoulos, D.R. van Rossum, W.J. Craig, D.J. Whalen, J. Smidt, B. Wiggins, Emission from pair-instability supernovae with rotation. Astrophys. J. **799**, 18 (2015). 1410.0039. https://doi.org/10.1088/0004-637X/799/1/18

T.W. Chen, S.J. Smartt, F. Bresolin, A. Pastorello, R.P. Kudritzki, R. Kotak, M. McCrum, M. Fraser, S. Valenti, The host galaxy of the super-luminous SN 2010gx and limits on explosive ^{56}Ni production. Astrophys. J. Lett. **763**, L28 (2013). 1210.4027. https://doi.org/10.1088/2041-8205/763/2/L28

K.J. Chen, A. Heger, S. Woosley, A. Almgren, D.J. Whalen, Pair instability supernovae of very massive Population III stars. Astrophys. J. **792**, 44 (2014). 1402.5960. https://doi.org/10.1088/0004-637X/792/1/44

T.W. Chen, S.J. Smartt, A. Jerkstrand, M. Nicholl, F. Bresolin, R. Kotak, J. Polshaw, A. Rest, R. Kudritzki, Z. Zheng, N. Elias-Rosa, K. Smith, C. Inserra, D. Wright, E. Kankare, T. Kangas, M. Fraser, The host galaxy and late-time evolution of the superluminous supernova PTF12dam. Mon. Not. R. Astron. Soc. **452**, 1567–1586 (2015). 1409.7728. https://doi.org/10.1093/mnras/stv1360

K.J. Chen, S.E. Woosley, T. Sukhbold, Magnetar-powered supernovae in two dimensions. I. Superluminous supernovae. Astrophys. J. **832**, 73 (2016). 1604.07989. https://doi.org/10.3847/0004-637X/832/1/73

T.W. Chen, M. Nicholl, S.J. Smartt, P.A. Mazzali, R.M. Yates, T.J. Moriya, C. Inserra, N. Langer, T. Krühler, Y.C. Pan, R. Kotak, L. Galbany, P. Schady, P. Wiseman, J. Greiner, S. Schulze, A.W.S. Man, A. Jerkstrand, K.W. Smith, M. Dennefeld, C. Baltay, J. Bolmer, E. Kankare, F. Knust, K. Maguire, D. Rabinowitz, S. Rostami, M. Sullivan, D.R. Young, The evolution of superluminous supernova LSQ14mo and its interacting host galaxy system. Astron. Astrophys. **602**, A9 (2017a). 1611.09910. https://doi.org/10.1051/0004-6361/201630163

T.W. Chen, P. Schady, L. Xiao, J.J. Eldridge, T. Schweyer, C.H. Lee, P.C. Yu, S.J. Smartt, C. Inserra, Spatially resolved MaNGA observations of the host galaxy of superluminous supernova 2017egm. Astrophys. J. Lett. **849**, L4 (2017b). 1708.04618. https://doi.org/10.3847/2041-8213/aa8f40

T.W. Chen, S.J. Smartt, R.M. Yates, M. Nicholl, T. Krühler, P. Schady, M. Dennefeld, C. Inserra, Superluminous supernova progenitors have a half-solar metallicity threshold. Mon. Not. R. Astron. Soc. **470**, 3566–3573 (2017c). 1605.04925. https://doi.org/10.1093/mnras/stx1428

R.A. Chevalier, Was SN 1054 a Type II supernova? in *Supernovae*. Astrophysics and Space Science Library, vol. 66, ed. by D.N. Schramm (Springer, Berlin, 1977), p. 53. https://doi.org/10.1007/978-94-010-1229-4_5

R.A. Chevalier, Young core-collapse supernova remnants and their supernovae. Astrophys. J. **619**, 839–855 (2005). astro-ph/0409013. https://doi.org/10.1086/426584

R.A. Chevalier, C. Fransson, Supernova interaction with a circumstellar medium, in *Supernovae and Gamma-Ray Bursters*, ed. by K. Weiler. Lecture Notes in Physics, vol. 598 (Springer, Berlin, 2003), pp. 171–194. astro-ph/0110060. https://doi.org/10.1007/3-540-45863-8_10

R.A. Chevalier, C.M. Irwin, Shock breakout in dense mass loss: luminous supernovae. Astrophys. J. Lett. **729**, L6 (2011). 1101.1111. https://doi.org/10.1088/2041-8205/729/1/L6

R.A. Chevalier, N. Soker, Asymmetric envelope expansion of supernova 1987A. Astrophys. J. **341**, 867–882 (1989). https://doi.org/10.1086/167545

L. Chomiuk, R. Chornock, A.M. Soderberg, E. Berger, R.A. Chevalier, R.J. Foley, M.E. Huber, G. Narayan, A. Rest, S. Gezari, R.P. Kirshner, A. Riess, S.A. Rodney, S.J. Smartt, C.W. Stubbs, J.L. Tonry, W.M. Wood-Vasey, W.S. Burgett, K.C. Chambers, I. Czekala, H. Flewelling, K. Forster, N. Kaiser, R.P. Kudritzki, E.A. Magnier, D.C. Martin, J.S. Morgan, J.D. Neill, P.A. Price, K.C. Roth, N.E. Sanders, R.J. Wainscoat, Pan-STARRS1 discovery of two ultraluminous supernovae at $z \simeq 0.9$. Astrophys. J. **743**, 114 (2011). 1107.3552. https://doi.org/10.1088/0004-637X/743/2/114

A. Cikota, A. De Cia, S. Schulze, P.M. Vreeswijk, G. Leloudas, A. Gal-Yam, D.A. Perley, S. Cikota, S. Kim, F. Patat, R. Lunnan, R. Quimby, O. Yaron, L. Yan, P.A. Mazzali, Spatially resolved analysis of superluminous supernovae PTF 11hrq and PTF 12dam host galaxies. Mon. Not. R. Astron. Soc. **469**, 4705–4717 (2017). 1705.01948. https://doi.org/10.1093/mnras/stx1110

J. Cooke, M. Sullivan, A. Gal-Yam, E.J. Barton, R.G. Carlberg, E.V. Ryan-Weber, C. Horst, Y. Omori, C.G. Díaz, Superluminous supernovae at redshifts of 2.05 and 3.90. Nature **491**, 228–231 (2012). 1211.2003. https://doi.org/10.1038/nature11521

D.L. Coppejans, R. Margutti, C. Guidorzi, L. Chomiuk, K.D. Alexander, E. Berger, M.F. Bietenholz, P.K. Blanchard, P. Challis, R. Chornock, M. Drout, W. Fong, A. Mac Fadyen, G. Migliori, D. Milisavljevic, M. Nicholl, J.T. Parrent, G. Terreran, B.A. Zauderer, Jets in hydrogen-poor super-luminous supernovae: constraints from a comprehensive analysis of radio observations. ArXiv e-prints. 1711.03428 (2017)

C. Curtin, J. Cooke, T.J. Moriya, S.R. Bernard, L. Galbany, J. Ja, C.H. Lee, K. Maeda, T. Morokuma, K. Nomoto, G. Pignata, T. Pritchard, R.M. Quimby, N. Suzuki, I. Takahashi, M. Tanaka, M. Tanaka, N. Tominaga, M. Yamaguchi, N. Yasuda, First release of high-redshift superluminous supernovae from the Subaru HIgh-Z sUpernova CAmpaign (SHIZUCA). II. Spectroscopic properties. ArXiv e-prints. 1801.08241 (2018)

M.T. da Cruz, Y. Chan, R.M. Larimer, K.T. Lesko, E.B. Norman, R.G. Stokstad, F.E. Wietfeldt, I. Žlimen, Half-life of ^{56}Ni. Phys. Rev. C **46**, 1132–1135 (1992). https://doi.org/10.1103/PhysRevC.46.1132

A. De Cia, A. Gal-Yam, A. Rubin, G. Leloudas, P. Vreeswijk, D.A. Perley, R. Quimby, L. Yan, M. Sullivan, A. Flörs, J. Sollerman, D. Bersier, S.B. Cenko, M. Gal-Yam, K. Maguire, E.O. Ofek, S. Prentice, S. Schulze, J. Spyromilio, S. Valenti, I. Arcavi, A. Corsi, A. Howell, P. Mazzali, M.M. Kasliwal, F. Taddia, O. Yaron, Light curves of hydrogen-poor Superluminous Supernovae from the Palomar Transient Factory. ArXiv e-prints. 1708.01623 (2017)

L. Dessart, D.J. Hillier, R. Waldman, E. Livne, S. Blondin, Superluminous supernovae: ^{56}Ni power versus magnetar radiation. Mon. Not. R. Astron. Soc. **426**, L76–L80 (2012). 1208.1214. https://doi.org/10.1111/j.1745-3933.2012.01329.x

L. Dessart, R. Waldman, E. Livne, D.J. Hillier, S. Blondin, Radiative properties of pair-instability supernova explosions. Mon. Not. R. Astron. Soc. **428**, 3227–3251 (2013). 1210.6163. https://doi.org/10.1093/mnras/sts269

L. Dessart, E. Audit, D.J. Hillier, Numerical simulations of superluminous supernovae of Type IIn. Mon. Not. R. Astron. Soc. **449**, 4304–4325 (2015). 1503.05463. https://doi.org/10.1093/mnras/stv609

J. Dexter, D. Kasen, Supernova light curves powered by fallback accretion. Astrophys. J. **772**, 30 (2013). 1210.7240. https://doi.org/10.1088/0004-637X/772/1/30

S. Dong, B.J. Shappee, J.L. Prieto, S.W. Jha, K.Z. Stanek, T.W.S. Holoien, C.S. Kochanek, T.A. Thompson, N. Morrell, I.B. Thompson, U. Basu, J.F. Beacom, D. Bersier, J. Brimacombe, J.S. Brown, F. Bufano, P. Chen, E. Conseil, A.B. Danilet, E. Falco, D. Grupe, S. Kiyota, G. Masi, B. Nicholls, E.F. Olivares, G. Pignata, G. Pojmanski, G.V. Simonian, D.M. Szczygiel, P.R. Woźniak, ASASSN-15lh: a highly super-luminous supernova. Science **351**, 257–260 (2016). 1507.03010. https://doi.org/10.1126/science.aac9613

A.J. Drake, S.G. Djorgovski, J.L. Prieto, A. Mahabal, D. Balam, R. Williams, M.J. Graham, M. Catelan, E. Beshore, S. Larson, Discovery of the extremely energetic supernova 2008fz. Astrophys. J. Lett. **718**, L127–L131 (2010). 0908.1990. https://doi.org/10.1088/2041-8205/718/2/L127

A.J. Drake, S.G. Djorgovski, A. Mahabal, J. Anderson, R. Roy, V. Mohan, S. Ravindranath, D. Frail, S. Gezari, J.D. Neill, L.C. Ho, J.L. Prieto, D. Thompson, J. Thorstensen, M. Wagner, R. Kowalski, J. Chiang, J.E. Grove, F.K. Schinzel, D.L. Wood, L. Carrasco, E. Recillas, L. Kewley, K.N. Archana, A. Basu, Y. Wadadekar, B. Kumar, A.D. Myers, E.S. Phinney, R. Williams, M.J. Graham, M. Catelan, E. Beshore, S. Larson, E. Christensen, The discovery and nature of the optical transient CSS100217:102913+404220. Astrophys. J. **735**, 106 (2011). 1103.5514. https://doi.org/10.1088/0004-637X/735/2/106

M.R. Drout, A.M. Soderberg, A. Gal-Yam, S.B. Cenko, D.B. Fox, D.C. Leonard, D.J. Sand, D.S. Moon, I. Arcavi, Y. Green, The first systematic study of Type Ibc supernova multi-band light curves. Astrophys. J. **741**, 97 (2011). 1011.4959. https://doi.org/10.1088/0004-637X/741/2/97

C.M. Espinoza, A.G. Lyne, B.W. Stappers, New long-term braking index measurements for glitching pulsars using a glitch-template method. Mon. Not. R. Astron. Soc. **466**, 147–162 (2017). 1611.08314. https://doi.org/10.1093/mnras/stw3081

S.W. Falk, W.D. Arnett, Radiation dynamics, envelope ejection, and supernova light curves. Astrophys. J. Suppl. Ser. **33**, 515 (1977). https://doi.org/10.1086/190440
C. Fransson, M. Ergon, P.J. Challis, R.A. Chevalier, K. France, R.P. Kirshner, G.H. Marion, D. Milisavljevic, N. Smith, F. Bufano, A.S. Friedman, T. Kangas, J. Larsson, S. Mattila, S. Benetti, R. Chornock, I. Czekala, A. Soderberg, J. Sollerman, High-density circumstellar interaction in the luminous Type IIn SN 2010jl: the first 1100 days. Astrophys. J. **797**, 118 (2014). 1312.6617. https://doi.org/10.1088/0004-637X/797/2/118
E. Funck, U. Schötzig, M.J. Woods, J.P. Sephton, A.S. Munster, J.C.J. Dean, P. Blanchis, B. Chauvenet, ^{56}Co standardization and half-life. Nucl. Instrum. Methods Phys. Res., Sect. A, Accel. Spectrom. Detect. Assoc. Equip. **312**, 334–338 (1992). https://doi.org/10.1016/0168-9002(92)90177-6
B. Gaffet, Pulsar theory of supernova light curves. I—Dynamical effect and thermalization of the pulsar strong waves. Astrophys. J. **216**, 565–577 (1977). https://doi.org/10.1086/155498
A. Gal-Yam, Luminous supernovae. Science **337**, 927 (2012). 1208.3217. https://doi.org/10.1126/science.1203601
A. Gal-Yam, P. Mazzali, E.O. Ofek, P.E. Nugent, S.R. Kulkarni, M.M. Kasliwal, R.M. Quimby, A.V. Filippenko, S.B. Cenko, R. Chornock, R. Waldman, D. Kasen, M. Sullivan, E.C. Beshore, A.J. Drake, R.C. Thomas, J.S. Bloom, D. Poznanski, A.A. Miller, R.J. Foley, J.M. Silverman, I. Arcavi, R.S. Ellis, J. Deng, Supernova 2007bi as a pair-instability explosion. Nature **462**, 624–627 (2009). 1001.1156. https://doi.org/10.1038/nature08579
B.T. Gänsicke, A.J. Levan, T.R. Marsh, P.J. Wheatley, SCP 06F6: a carbon-rich extragalactic transient at redshift $z \sim 0.14$? Astrophys. J. Lett. **697**, L129–L132 (2009). 0809.2562. https://doi.org/10.1088/0004-637X/697/2/L129
C. Georgy, G. Meynet, S. Ekström, G.A. Wade, V. Petit, Z. Keszthelyi, R. Hirschi, Possible pair-instability supernovae at solar metallicity from magnetic stellar progenitors. Astron. Astrophys. **599**, L5 (2017). 1702.02340. https://doi.org/10.1051/0004-6361/201730401
S. Gezari, J.P. Halpern, D. Grupe, F. Yuan, R. Quimby, T. McKay, D. Chamarro, M.D. Sisson, C. Akerlof, J.C. Wheeler, P.J. Brown, S.B. Cenko, A. Rau, J.O. Djordjevic, D.M. Terndrup, Discovery of the ultra-bright Type II-L supernova 2008es. Astrophys. J. **690**, 1313–1321 (2009). 0808.2812. https://doi.org/10.1088/0004-637X/690/2/1313
A. Gilkis, N. Soker, O. Papish, Explaining the most energetic supernovae with an inefficient jet-feedback mechanism. Astrophys. J. **826**, 178 (2016). arXiv:1511.01471. https://doi.org/10.3847/0004-637X/826/2/178
M.S. Gilmer, A. Kozyreva, R. Hirschi, C. Fröhlich, N. Yusof, Pair-instability supernova simulations: progenitor evolution, explosion, and light curves. Astrophys. J. **846**, 100 (2017). 1706.07454. https://doi.org/10.3847/1538-4357/aa8461
S. Ginzburg, S. Balberg, Superluminous light curves from supernovae exploding in a dense wind. Astrophys. J. **757**, 178 (2012). 1205.3455. https://doi.org/10.1088/0004-637X/757/2/178
E.K. Grasberg, D.K. Nadyozhin, The onset of a type II supernova outburst in a circumstellar envelope created by the intense stellar wind from the presupernova. Astron. Zh. **64**, 1199–1209 (1987)
E.K. Grasberg, V.S. Imshenik, D.K. Nadyozhin, On the theory of the light curves of supernovate. Astrophys. Space Sci. **10**, 3 (1971) (in Russian). https://doi.org/10.1007/BF00654603
J. Greiner, P.A. Mazzali, D.A. Kann, T. Krühler, E. Pian, S. Prentice, E.F. Olivares, A. Rossi, S. Klose, S. Taubenberger, F. Knust, P.M.J. Afonso, C. Ashall, J. Bolmer, C. Delvaux, R. Diehl, J. Elliott, R. Filgas, J.P.U. Fynbo, J.F. Graham, A.N. Guelbenzu, S. Kobayashi, G. Leloudas, S. Savaglio, P. Schady, S. Schmidl, T. Schweyer, V. Sudilovsky, M. Tanga, A.C. Updike, H. van Eerten, K. Varela, A very luminous magnetar-powered supernova associated with an ultra-long γ-ray burst. Nature **523**, 189–192 (2015). 1509.03279. https://doi.org/10.1038/nature14579
J. Guillochon, J. Parrent, L.Z. Kelley, R. Margutti, An open catalog for supernova data. Astrophys. J. **835**, 64 (2017). 1605.01054. https://doi.org/10.3847/1538-4357/835/1/64
A. Heger, S.E. Woosley, The nucleosynthetic signature of population III. Astrophys. J. **567**, 532–543 (2002). astro-ph/0107037. https://doi.org/10.1086/338487
D.A. Howell, *Superluminous Supernovae* (Springer, Cham, 2017), pp. 1–29. https://doi.org/10.1007/978-3-319-20794-0_41-1
D.A. Howell, D. Kasen, C. Lidman, M. Sullivan, A. Conley, P. Astier, C. Balland, R.G. Carlberg, D. Fouchez, J. Guy, D. Hardin, R. Pain, N. Palanque-Delabrouille, K. Perrett, C.J. Pritchet, N. Regnault, J. Rich, V. Ruhlmann-Kleider, Two superluminous supernovae from the early universe discovered by the Supernova Legacy Survey. Astrophys. J. **779**, 98 (2013). 1310.0470. https://doi.org/10.1088/0004-637X/779/2/98
C. Inserra, S.J. Smartt, Superluminous supernovae as standardizable candles and high-redshift distance probes. Astrophys. J. **796**, 87 (2014). 1409.4429. https://doi.org/10.1088/0004-637X/796/2/87

C. Inserra, S.J. Smartt, A. Jerkstrand, S. Valenti, M. Fraser, D. Wright, K. Smith, T.W. Chen, R. Kotak, A. Pastorello, M. Nicholl, F. Bresolin, R.P. Kudritzki, S. Benetti, M.T. Botticella, W.S. Burgett, K.C. Chambers, M. Ergon, H. Flewelling, J.P.U. Fynbo, S. Geier, K.W. Hodapp, D.A. Howell, M. Huber, N. Kaiser, G. Leloudas, L. Magill, E.A. Magnier, M.G. McCrum, N. Metcalfe, P.A. Price, A. Rest, J. Sollerman, W. Sweeney, F. Taddia, S. Taubenberger, J.L. Tonry, R.J. Wainscoat, C. Waters, D. Young, Super-luminous Type Ic supernovae: catching a magnetar by the tail. Astrophys. J. **770**, 128 (2013). 1304.3320. https://doi.org/10.1088/0004-637X/770/2/128

C. Inserra, M. Bulla, S.A. Sim, S.J. Smartt, Spectropolarimetry of superluminous supernovae: insight into their geometry. Astrophys. J. **831**, 79 (2016). 1607.02353. https://doi.org/10.3847/0004-637X/831/1/79

C. Inserra, M. Nicholl, T.W. Chen, A. Jerkstrand, S.J. Smartt, T. Krühler, J.P. Anderson, C. Baltay, M. Della Valle, M. Fraser, A. Gal-Yam, L. Galbany, E. Kankare, K. Maguire, D. Rabinowitz, K. Smith, S. Valenti, D.R. Young, Complexity in the light curves and spectra of slow-evolving superluminous supernovae. Mon. Not. R. Astron. Soc. **468**, 4642–4662 (2017). 1701.00941. https://doi.org/10.1093/mnras/stx834

C. Inserra, R.C. Nichol, D. Scovacricchi, J. Amiaux, M. Brescia, C. Burigana, E. Cappellaro, C.S. Carvalho, S. Cavuoti, V. Conforti, J.C. Cuillandre, A. da Silva, A. De Rosa, M. Della Valle, J. Dinis, E. Franceschi, I. Hook, P. Hudelot, K. Jahnke, T. Kitching, H. Kurki-Suonio, I. Lloro, G. Longo, E. Maiorano, M. Maris, J.D. Rhodes, R. Scaramella, S.J. Smartt, M. Sullivan, C. Tao, R. Toledo-Moreo, I. Tereno, M. Trifoglio, L. Valenziano, Euclid: superluminous supernovae in the deep survey. Astron. Astrophys. **609**, A83 (2018a). 1710.09585. https://doi.org/10.1051/0004-6361/201731758

C. Inserra, S.J. Smartt, E.E.E. Gall, G. Leloudas, T.W. Chen, S. Schulze, A. Jerkstrand, M. Nicholl, J.P. Anderson, I. Arcavi, S. Benetti, R.A. Cartier, M. Childress, M. Della Valle, H. Flewelling, M. Fraser, A. Gal-Yam, C.P. Gutiérrez, G. Hosseinzadeh, D.A. Howell, M. Huber, E. Kankare, T. Krühler, E.A. Magnier, K. Maguire, C. McCully, S. Prajs, N. Primak, R. Scalzo, B.P. Schmidt, M. Smith, K.W. Smith, B.E. Tucker, S. Valenti, M. Wilman, D.R. Young, F. Yuan, On the nature of hydrogen-rich superluminous supernovae. Mon. Not. R. Astron. Soc. **475**, 1046–1072 (2018b). 1604.01226. https://doi.org/10.1093/mnras/stx3179

L. Izzo, C.C. Thöne, R. García-Benito, A. de Ugarte Postigo, Z. Cano, D.A. Kann, K. Bensch, M. Della Valle, D. Galadí-Enríquez, R.P. Hedrosa, The host of the Type I SLSN 2017egm. A young, sub-solar metallicity environment in a massive spiral galaxy. Astron. Astrophys. **610**, A11 (2018). 1708.03856. https://doi.org/10.1051/0004-6361/201731766

J. Japelj, S.D. Vergani, R. Salvaterra, L.K. Hunt, F. Mannucci, Taking stock of superluminous supernovae and long gamma-ray burst host galaxy comparison using a complete sample of LGRBs. Astron. Astrophys. **593**, A115 (2016). 1607.01045. https://doi.org/10.1051/0004-6361/201628603

A. Jerkstrand, S.J. Smartt, A. Heger, Nebular spectra of pair-instability supernovae. Mon. Not. R. Astron. Soc. **455**, 3207–3229 (2016). 1510.02698. https://doi.org/10.1093/mnras/stv2369

A. Jerkstrand, S.J. Smartt, C. Inserra, M. Nicholl, T.W. Chen, T. Krühler, J. Sollerman, S. Taubenberger, A. Gal-Yam, E. Kankare, K. Maguire, M. Fraser, S. Valenti, M. Sullivan, R. Cartier, D.R. Young, Long-duration superluminous supernovae at late times. Astrophys. J. **835**, 13 (2017). 1608.02994. https://doi.org/10.3847/1538-4357/835/1/13

C.C. Joggerst, D.J. Whalen, The early evolution of primordial pair-instability supernovae. Astrophys. J. **728**, 129 (2011). 1010.4360. https://doi.org/10.1088/0004-637X/728/2/129

E. Kankare, R. Kotak, S. Mattila, P. Lundqvist, M.J. Ward, M. Fraser, A. Lawrence, S.J. Smartt, W.P.S. Meikle, A. Bruce, J. Harmanen, S.J. Hutton, C. Inserra, T. Kangas, A. Pastorello, T. Reynolds, C. Romero-Cañizales, K.W. Smith, S. Valenti, K.C. Chambers, K.W. Hodapp, M.E. Huber, N. Kaiser, R.P. Kudritzki, E.A. Magnier, J.L. Tonry, R.J. Wainscoat, C. Waters, A population of highly energetic transient events in the centres of active galaxies. Nat. Astron. **1**, 865–871 (2017). 1711.04577. https://doi.org/10.1038/s41550-017-0290-2

D. Kasen, L. Bildsten, Supernova light curves powered by young magnetars. Astrophys. J. **717**, 245–249 (2010). 0911.0680. https://doi.org/10.1088/0004-637X/717/1/245

D. Kasen, S.E. Woosley, A. Heger, Pair instability supernovae: light curves, spectra, and shock breakout. Astrophys. J. **734**, 102 (2011). 1101.3336. https://doi.org/10.1088/0004-637X/734/2/102

D. Kasen, B.D. Metzger, L. Bildsten, Magnetar-driven shock breakout and double-peaked supernova light curves. Astrophys. J. **821**, 36 (2016). 1507.03645. https://doi.org/10.3847/0004-637X/821/1/36

K.S. Kawabata, M. Tanaka, K. Maeda, T. Hattori, K. Nomoto, N. Tominaga, M. Yamanaka, Extremely luminous supernova 2006gy at late phase: detection of optical emission from supernova. Astrophys. J. **697**, 747–757 (2009). 0902.1440. https://doi.org/10.1088/0004-637X/697/1/747

R. Knop, G. Aldering, S. Deustua, G. Goldhaber, M. Kim, P. Nugent, E. Helin, S. Pravdo, D. Rabinowitz, K. Lawrence, Supernovae 1999as and 1999at in anonymous galaxies. IAU Circ. **7128** (1999)

K. Kotera, E.S. Phinney, A.V. Olinto, Signatures of pulsars in the light curves of newly formed supernova remnants. Mon. Not. R. Astron. Soc. **432**, 3228–3236 (2013). 1304.5326. https://doi.org/10.1093/mnras/stt680

A. Kozyreva, S. Blinnikov, Can pair-instability supernova models match the observations of superluminous supernovae? Mon. Not. R. Astron. Soc. **454**, 4357–4365 (2015). 1510.00439. https://doi.org/10.1093/mnras/stv2287

A. Kozyreva, S. Blinnikov, N. Langer, S.C. Yoon, Observational properties of low-redshift pair instability supernovae. Astron. Astrophys. **565**, A70 (2014). 1403.5212. https://doi.org/10.1051/0004-6361/201423447

A. Kozyreva, M. Gilmer, R. Hirschi, C. Fröhlich, S. Blinnikov, R.T. Wollaeger, U.M. Noebauer, D.R. van Rossum, A. Heger, W.P. Even, R. Waldman, A. Tolstov, E. Chatzopoulos, E. Sorokina, Fast evolving pair-instability supernova models: evolution, explosion, light curves. Mon. Not. R. Astron. Soc. **464**, 2854–2865 (2017). 1610.01086. https://doi.org/10.1093/mnras/stw2562

N. Langer, C.A. de Norman, A. Koter, J.S. Vink, M. Cantiello, S.C. Yoon, Pair creation supernovae at low and high redshift. Astron. Astrophys. **475**, L19–L23 (2007). 0708.1970. https://doi.org/10.1051/0004-6361:20078482

G. Leloudas, E. Chatzopoulos, B. Dilday, J. Gorosabel, J. Vinko, A. Gallazzi, J.C. Wheeler, B. Bassett, J.A. Fischer, J.A. Frieman, J.P.U. Fynbo, A. Goobar, M. Jelínek, D. Malesani, R.C. Nichol, J. Nordin, L. Östman, M. Sako, D.P. Schneider, M. Smith, J. Sollerman, M.D. Stritzinger, C.C. Thöne, A. de Ugarte Postigo, SN 2006oz: rise of a super-luminous supernova observed by the SDSS-II SN Survey. Astron. Astrophys. **541**, A129 (2012). 1201.5393. https://doi.org/10.1051/0004-6361/201118498

G. Leloudas, F. Patat, J.R. Maund, E. Hsiao, D. Malesani, S. Schulze, C. Contreras, A. de Ugarte Postigo, J. Sollerman, M.D. Stritzinger, F. Taddia, J.C. Wheeler, J. Gorosabel, Polarimetry of the superluminous supernova LSQ14mo: no evidence for significant deviations from spherical symmetry. Astrophys. J. Lett. **815**, L10 (2015a). 1511.04522. https://doi.org/10.1088/2041-8205/815/1/L10

G. Leloudas, S. Schulze, T. Krühler, J. Gorosabel, L. Christensen, A. Mehner, A. de Ugarte Postigo, R. Amorín, C.C. Thöne, J.P. Anderson, F.E. Bauer, A. Gallazzi, K.G. Hełminiak, J. Hjorth, E. Ibar, D. Malesani, N. Morell, J. Vinko, J.C. Wheeler, Spectroscopy of superluminous supernova host galaxies. A preference of hydrogen-poor events for extreme emission line galaxies. Mon. Not. R. Astron. Soc. **449**, 917–932 (2015b). 1409.8331. https://doi.org/10.1093/mnras/stv320

G. Leloudas, M. Fraser, N.C. Stone, S. van Velzen, P.G. Jonker, I. Arcavi, C. Fremling, J.R. Maund, S.J. Smartt, T. Krìhler, J.C.A. Miller-Jones, P.M. Vreeswijk, A. Gal-Yam, P.A. Mazzali, A. De Cia, D.A. Howell, C. Inserra, F. de Patat, A. Ugarte Postigo, O. Yaron, C. Ashall, I. Bar, H. Campbell, T.W. Chen, M. Childress, N. Elias-Rosa, J. Harmanen, G. Hosseinzadeh, J. Johansson, T. Kangas, E. Kankare, S. Kim, H. Kuncarayakti, J. Lyman, M.R. Magee, K. Maguire, D. Malesani, S. Mattila, C.V. McCully, M. Nicholl, S. Prentice, C. Romero-Cañizales, S. Schulze, K.W. Smith, J. Sollerman, M. Sullivan, B.E. Tucker, S. Valenti, J.C. Wheeler, D.R. Young, The superluminous transient ASASSN-15lh as a tidal disruption event from a Kerr black hole. Nat. Astron. **1**, 0002 (2016). 1609.02927. https://doi.org/10.1038/s41550-016-0002

G. Leloudas, J.R. Maund, A. Gal-Yam, T. Pursimo, E. Hsiao, D. Malesani, F. Patat, A. de Ugarte Postigo, J. Sollerman, M.D. Stritzinger, J.C. Wheeler, Time-resolved polarimetry of the superluminous SN 2015bn with the Nordic Optical Telescope. Astrophys. J. Lett. **837**, L14 (2017). 1702.05494. https://doi.org/10.3847/2041-8213/aa6157

A.J. Levan, A.M. Read, B.D. Metzger, P.J. Wheatley, N.R. Tanvir, Superluminous X-rays from a superluminous supernova. Astrophys. J. **771**, 136 (2013). 1304.1173. https://doi.org/10.1088/0004-637X/771/2/136

L.D. Liu, S.Q. Wang, L.J. Wang, Z.G. Dai, H. Yu, Z.K. Peng, A Monte Carlo approach to magnetar-powered transients. I. Hydrogen-deficient superluminous supernovae. Astrophys. J. **842**, 26 (2017a). 1705.06047. https://doi.org/10.3847/1538-4357/aa73d9

Y.Q. Liu, M. Modjaz, F.B. Bianco, Analyzing the largest spectroscopic data set of hydrogen-poor superluminous supernovae. Astrophys. J. **845**, 85 (2017b). 1612.07321. https://doi.org/10.3847/1538-4357/aa7f74

R. Lunnan, R. Chornock, E. Berger, T. Laskar, W. Fong, A. Rest, N.E. Sanders, P.M. Challis, M.R. Drout, R.J. Foley, M.E. Huber, R.P. Kirshner, C. Leibler, G.H. Marion, M. McCrum, D. Milisavljevic, G. Narayan, D. Scolnic, S.J. Smartt, K.W. Smith, A.M. Soderberg, J.L. Tonry, W.S. Burgett, K.C. Chambers, H. Flewelling, K.W. Hodapp, N. Kaiser, E.A. Magnier, P.A. Price, R.J. Wainscoat, Hydrogen-poor superluminous supernovae and long-duration gamma-ray bursts have similar host galaxies. Astrophys. J. **787**, 138 (2014). 1311.0026. https://doi.org/10.1088/0004-637X/787/2/138

R. Lunnan, R. Chornock, E. Berger, A. Rest, W. Fong, D. Scolnic, D.O. Jones, A.M. Soderberg, P.M. Challis, M.R. Drout, R.J. Foley, M.E. Huber, R.P. Kirshner, C. Leibler, G.H. Marion, M. McCrum, D. Milisavljevic, G. Narayan, N.E. Sanders, S.J. Smartt, K.W. Smith, J.L. Tonry, W.S. Burgett, K.C. Chambers, H. Flewelling, R.P. Kudritzki, R.J. Wainscoat, C. Waters, Zooming in on the progenitors of superluminous supernovae with the HST. Astrophys. J. **804**, 90 (2015). 1411.1060. https://doi.org/10.1088/0004-637X/804/2/90

R. Lunnan, R. Chornock, E. Berger, D. Milisavljevic, D.O. Jones, A. Rest, W. Fong, C. Fransson, R. Margutti, M.R. Drout, P.K. Blanchard, P. Challis, P.S. Cowperthwaite, R.J. Foley, R.P. Kirshner, N. Morrell, A.G. Riess, K.C. Roth, D. Scolnic, S.J. Smartt, K.W. Smith, V.A. Villar, K.C. Chambers, P.W. Draper, M.E. Huber, N. Kaiser, R.P. Kudritzki, E.A. Magnier, N. Metcalfe, C. Waters, PS1-14bj: a hydrogen-poor superluminous supernova with a long rise and slow decay. Astrophys. J. **831**, 144 (2016). 1605.05235. https://doi.org/10.3847/0004-637X/831/2/144

R. Lunnan, R. Chornock, E. Berger, D.O. Jones, A. Rest, I. Czekala, J. Dittmann, M.R. Drout, R.J. Foley, W. Fong, R.P. Kirshner, T. Laskar, C.N. Leibler, R. Margutti, D. Milisavljevic, G. Narayan, Y.C. Pan, A.G. Riess, K.C. Roth, N.E. Sanders, D. Scolnic, S.J. Smartt, K.W. Smith, K.C. Chambers, P.W. Draper, H. Flewelling, M.E. Huber, N. Kaiser, R.P. Kudritzki, E.A. Magnier, N. Metcalfe, R.J. Wainscoat, C. Waters, M. Willman, Hydrogen-poor superluminous supernovae from the Pan-STARRS1 Medium Deep Survey. Astrophys. J. **852**, 81 (2018). 1708.01619. https://doi.org/10.3847/1538-4357/aa9f1a

J.D. Lyman, D. Bersier, P.A. James, P.A. Mazzali, J.J. Eldridge, M. Fraser, E. Pian, Bolometric light curves and explosion parameters of 38 stripped-envelope core-collapse supernovae. Mon. Not. R. Astron. Soc. **457**, 328–350 (2016). 1406.3667. https://doi.org/10.1093/mnras/stv2983

P. Madau, M. Dickinson, Cosmic star-formation history. Annu. Rev. Astron. Astrophys. **52**, 415–486 (2014). 1403.0007. https://doi.org/10.1146/annurev-astro-081811-125615

K. Maeda, M. Tanaka, K. Nomoto, N. Tominaga, K. Kawabata, P.A. Mazzali, H. Umeda, T. Suzuki, T. Hattori, The unique Type Ib supernova 2005bf at nebular phases: a possible birth event of a strongly magnetized neutron star. Astrophys. J. **666**, 1069–1082 (2007). 0705.2713. https://doi.org/10.1086/520054

B. Margalit, B.D. Metzger, T.A. Thompson, M. Nicholl, T. Sukhbold, The GRB-SLSN connection: misaligned magnetars, weak jet emergence, and observational signatures. Mon. Not. R. Astron. Soc. **475**, 2659–2674 (2018). 1705.01103. https://doi.org/10.1093/mnras/sty013

R. Margutti, R. Chornock, B.D. Metzger, D.L. Coppejans, C. Guidorzi, G. Migliori, D. Milisavljevic, E. Berger, M. Nicholl, B.A. Zauderer, R. Lunnan, A. Kamble, M. Drout, M. Modjaz, Results from a systematic survey of X-ray emission from Hydrogen-poor Superluminous Supernovae. ArXiv e-prints. 1704.05865 (2017)

C.D. Matzner, C.F. McKee, The expulsion of stellar envelopes in core-collapse supernovae. Astrophys. J. **510**, 379–403 (1999). astro-ph/9807046. https://doi.org/10.1086/306571

P.A. Mazzali, M. Sullivan, E. Pian, J. Greiner, D.A. Kann, Spectrum formation in superluminous supernovae (Type I). Mon. Not. R. Astron. Soc. **458**, 3455–3465 (2016). 1603.00388. https://doi.org/10.1093/mnras/stw512

M. McCrum, S.J. Smartt, A. Rest, K. Smith, R. Kotak, S.A. Rodney, D.R. Young, R. Chornock, E. Berger, R.J. Foley, M. Fraser, D. Wright, D. Scolnic, J.L. Tonry, Y. Urata, K. Huang, A. Pastorello, M.T. Botticella, S. Valenti, S. Mattila, E. Kankare, D.J. Farrow, M.E. Huber, C.W. Stubbs, R.P. Kirshner, F. Bresolin, W.S. Burgett, K.C. Chambers, P.W. Draper, H. Flewelling, R. Jedicke, N. Kaiser, E.A. Magnier, N. Metcalfe, J.S. Morgan, P.A. Price, W. Sweeney, R.J. Wainscoat, C. Waters, Selecting superluminous supernovae in faint galaxies from the first year of the Pan-STARRS1 Medium Deep Survey. Mon. Not. R. Astron. Soc. **448**, 1206–1231 (2015). 1402.1631. https://doi.org/10.1093/mnras/stv034

B.D. Metzger, I. Vurm, R. Hascoët, A.M. Beloborodov, Ionization break-out from millisecond pulsar wind nebulae: an X-ray probe of the origin of superluminous supernovae. Mon. Not. R. Astron. Soc. **437**, 703–720 (2014). 1307.8115. https://doi.org/10.1093/mnras/stt1922

A.A. Miller, R. Chornock, D.A. Perley, M. Ganeshalingam, W. Li, N.R. Butler, J.S. Bloom, N. Smith, M. Modjaz, D. Poznanski, A.V. Filippenko, C.V. Griffith, J.H. Shiode, J.M. Silverman, The exceptionally luminous Type II-linear supernova 2008es. Astrophys. J. **690**, 1303–1312 (2009). 0808.2193. https://doi.org/10.1088/0004-637X/690/2/1303

A.A. Miller, N. Smith, W. Li, J.S. Bloom, R. Chornock, A.V. Filippenko, J.X. Prochaska, New observations of the very luminous supernova 2006gy: evidence for echoes. Astron. J. **139**, 2218–2229 (2010). 0906.2201. https://doi.org/10.1088/0004-6256/139/6/2218

T.J. Moriya, N. Langer, Pulsations of red supergiant pair-instability supernova progenitors leading to extreme mass loss. Astron. Astrophys. **573**, A18 (2015). 1410.4557. https://doi.org/10.1051/0004-6361/201424957

T.J. Moriya, K. Maeda, A dip after the early emission of superluminous supernovae: a signature of shock breakout within dense circumstellar media. Astrophys. J. Lett. **756**, L22 (2012). 1203.1451. https://doi.org/10.1088/2041-8205/756/1/L22

T.J. Moriya, N. Tominaga, Diversity of luminous supernovae from non-steady mass loss. Astrophys. J. **747**, 118 (2012). 1110.3807. https://doi.org/10.1088/0004-637X/747/2/118

T. Moriya, N. Tominaga, M. Tanaka, K. Maeda, K. Nomoto, A core-collapse supernova model for the extremely luminous Type Ic supernova 2007bi: an alternative to the pair-instability supernova model. Astrophys. J. Lett. **717**, L83–L86 (2010). 1004.2967. https://doi.org/10.1088/2041-8205/717/2/L83

T.J. Moriya, S.I. Blinnikov, N. Tominaga, N. Yoshida, M. Tanaka, K. Maeda, K. Nomoto, Light-curve modelling of superluminous supernova 2006gy: collision between supernova ejecta and a dense circumstellar medium. Mon. Not. R. Astron. Soc. **428**, 1020–1035 (2013b). 1204.6109. https://doi.org/10.1093/mnras/sts075

T.J. Moriya, K. Maeda, F. Taddia, J. Sollerman, S.I. Blinnikov, E.I. Sorokina, An analytic bolometric light curve model of interaction-powered supernovae and its application to Type IIn supernovae. Mon. Not. R. Astron. Soc. **435**, 1520–1535 (2013b). 1307.2644. https://doi.org/10.1093/mnras/stt1392

T.J. Moriya, Z.W. Liu, J. Mackey, T.W. Chen, N. Langer, Revealing the binary origin of Type Ic superluminous supernovae through nebular hydrogen emission. Astron. Astrophys. **584**, L5 (2015). 1510.01621. https://doi.org/10.1051/0004-6361/201527515

T.J. Moriya, B.D. Metzger, S.I. Blinnikov, Supernovae powered by magnetars that transform into black holes. Astrophys. J. **833**, 64 (2016). 1606.09316. https://doi.org/10.3847/1538-4357/833/1/64

T.J. Moriya, T.W. Chen, N. Langer, Properties of magnetars mimicking ^{56}Ni-powered light curves in Type IC superluminous supernovae. Astrophys. J. **835**, 177 (2017a). 1612.06917. https://doi.org/10.3847/1538-4357/835/2/177

T.J. Moriya, M. Tanaka, T. Morokuma, K. Ohsuga, Superluminous transients at AGN centers from interaction between black hole disk winds and broad-line region clouds. Astrophys. J. Lett. **843**, L19 (2017b). 1706.06855. https://doi.org/10.3847/2041-8213/aa7af3

T.J. Moriya, M. Tanaka, N. Yasuda, J. Ja, C.H. Lee, K. Maeda, T. Morokuma, K. Nomoto, R.M. Quimby, N. Suzuki, I. Takahashi, M. Tanaka, N. Tominaga, M. Yamaguchi, S.R. Bernard, J. Cooke, C. Curtin, L. Galbany, S. Gonzalez-Gaitan, G. Pignata, T. Pritchard, Y. Komiyama, R.H. Lupton, First release of high-redshift superluminous supernovae from the Subaru HIgh-Z sUpernova CAmpaign (SHIZUCA). I. Photometric properties. ArXiv e-prints. 1801.08240 (2018)

D.K. Nadyozhin, The properties of Ni to Co to Fe decay. Astrophys. J. Suppl. Ser. **92**, 527–531 (1994). https://doi.org/10.1086/192008

J.D. Neill, M. Sullivan, A. Gal-Yam, R. Quimby, E. Ofek, T.K. Wyder, D.A. Howell, P. Nugent, M. Seibert, D.C. Martin, R. Overzier, T.A. Barlow, K. Foster, P.G. Friedman, P. Morrissey, S.G. Neff, D. Schiminovich, L. Bianchi, J. Donas, T.M. Heckman, Y.W. Lee, B.F. Madore, B. Milliard, R.M. Rich, A.S. Szalay, The extreme hosts of extreme supernovae. Astrophys. J. **727**, 15 (2011). 1011.3512. https://doi.org/10.1088/0004-637X/727/1/15

M. Nicholl, S.J. Smartt, Seeing double: the frequency and detectability of double-peaked superluminous supernova light curves. Mon. Not. R. Astron. Soc. **457**, L79–L83 (2016). 1511.03740. https://doi.org/10.1093/mnrasl/slv210

M. Nicholl, S.J. Smartt, A. Jerkstrand, C. Inserra, J.P. Anderson, C. Baltay, S. Benetti, T.W. Chen, N. Elias-Rosa, U. Feindt, M. Fraser, A. Gal-Yam, E. Hadjiyska, D.A. Howell, R. Kotak, A. Lawrence, G. Leloudas, S. Margheim, S. Mattila, M. McCrum, R. McKinnon, A. Mead, P. Nugent, D. Rabinowitz, A. Rest, K.W. Smith, J. Sollerman, M. Sullivan, F. Taddia, S. Valenti, E.S. Walker, D.R. Young, Superluminous supernovae from PESSTO. Mon. Not. R. Astron. Soc. **444**, 2096–2113 (2014). 1405.1325. https://doi.org/10.1093/mnras/stu1579

M. Nicholl, S.J. Smartt, A. Jerkstrand, C. Inserra, S.A. Sim, T.W. Chen, S. Benetti, M. Fraser, A. Gal-Yam, E. Kankare, K. Maguire, K. Smith, M. Sullivan, S. Valenti, D.R. Young, C. Baltay, F.E. Bauer, S. Baumont, D. Bersier, M.T. Botticella, M. Childress, M. Dennefeld, M. Della Valle, N. Elias-Rosa, U. Feindt, L. Galbany, E. Hadjiyska, L. Le Guillou, G. Leloudas, P. Mazzali, R. McKinnon, J. Polshaw, D. Rabinowitz, S. Rostami, R. Scalzo, B.P. Schmidt, S. Schulze, J. Sollerman, F. Taddia, F. Yuan, On the diversity of superluminous supernovae: ejected mass as the dominant factor. Mon. Not. R. Astron. Soc. **452**, 3869–3893 (2015a). 1503.03310. https://doi.org/10.1093/mnras/stv1522

M. Nicholl, S.J. Smartt, A. Jerkstrand, S.A. Sim, C. Inserra, J.P. Anderson, C. Baltay, S. Benetti, K. Chambers, T.W. Chen, N. Elias-Rosa, U. Feindt, H.A. Flewelling, M. Fraser, A. Gal-Yam, L. Galbany, M.E. Huber, T. Kangas, E. Kankare, R. Kotak, T. Krühler, K. Maguire, R. McKinnon, D. Rabinowitz, S. Rostami, S. Schulze, K.W. Smith, M. Sullivan, J.L. Tonry, S. Valenti, D.R. Young, LSQ14bdq: a Type Ic super-luminous supernova with a double-peaked light curve. Astrophys. J. Lett. **807**, L18 (2015b). 1505.01078. https://doi.org/10.1088/2041-8205/807/1/L18

M. Nicholl, E. Berger, R. Margutti, R. Chornock, P.K. Blanchard, A. Jerkstrand, S.J. Smartt, I. Arcavi, P. Challis, K.C. Chambers, T.W. Chen, P.S. Cowperthwaite, A. Gal-Yam, G. Hosseinzadeh, D.A. Howell, C. Inserra, E. Kankare, E.A. Magnier, K. Maguire, P.A. Mazzali, C. McCully, D. Milisavljevic, K.W. Smith, S. Taubenberger, S. Valenti, R.J. Wainscoat, O. Yaron, D.R. Young, Superluminous supernova SN 2015bn in the nebular phase: evidence for the engine-powered explosion of a stripped massive star. Astrophys. J. Lett. **828**, L18 (2016a). 1608.02995. https://doi.org/10.3847/2041-8205/828/2/L18

M. Nicholl, E. Berger, S.J. Smartt, R. Margutti, A. Kamble, K.D. Alexander, T.W. Chen, C. Inserra, I. Arcavi, P.K. Blanchard, R. Cartier, K.C. Chambers, M.J. Childress, R. Chornock, P.S. Cowperthwaite, M. Drout, H.A. Flewelling, M. Fraser, A. Gal-Yam, L. Galbany, J. Harmanen, T.W.S. Holoien, G. Hosseinzadeh,

D.A. Howell, M.E. Huber, A. Jerkstrand, E. Kankare, C.S. Kochanek, Z.Y. Lin, R. Lunnan, E.A. Magnier, K. Maguire, C. McCully, M. McDonald, B.D. Metzger, D. Milisavljevic, A. Mitra, T. Reynolds, J. Saario, B.J. Shappee, K.W. Smith, S. Valenti, V.A. Villar, C. Waters, D.R. Young, SN 2015BN: a detailed multi-wavelength view of a nearby superluminous supernova. Astrophys. J. **826**, 39 (2016b). 1603.04748. https://doi.org/10.3847/0004-637X/826/1/39

M. Nicholl, E. Berger, R. Margutti, P.K. Blanchard, J. Guillochon, J. Leja, R. Chornock, The superluminous supernova SN 2017egm in the nearby galaxy NGC 3191: a metal-rich environment can support a typical SLSN evolution. Astrophys. J. Lett. **845**, L8 (2017a). 1706.08517. https://doi.org/10.3847/2041-8213/aa82b1

M. Nicholl, J. Guillochon, E. Berger, The magnetar model for Type I superluminous supernovae. I. Bayesian analysis of the full multicolor light-curve sample with MOSFiT. Astrophys. J. **850**, 55 (2017b). 1706.00825. https://doi.org/10.3847/1538-4357/aa9334

E.O. Ofek, P.B. Cameron, M.M. Kasliwal, A. Gal-Yam, A. Rau, S.R. Kulkarni, D.A. Frail, P. Chandra, S.B. Cenko, A.M. Soderberg, S. Immler, SN 2006gy: an extremely luminous supernova in the galaxy NGC 1260. Astrophys. J. Lett. **659**, L13–L16 (2007). astro-ph/0612408. https://doi.org/10.1086/516749

J.P. Ostriker, J.E. Gunn, Do pulsars make supernovae? Astrophys. J. Lett. **164**, L95 (1971). https://doi.org/10.1086/180699

R. Ouyed, M. Kostka, N. Koning, D.A. Leahy, W. Steffen, Quark nova imprint in the extreme supernova explosion SN 2006gy. Mon. Not. R. Astron. Soc. **423**, 1652–1662 (2012). 1010.5530. https://doi.org/10.1111/j.1365-2966.2012.20986.x

Y.C. Pan, R.J. Foley, M. Smith, L. Galbany, C.B. D'Andrea, S. González-Gaitán, M.J. Jarvis, R. Kessler, E. Kovacs, C. Lidman R.C. Nichol, A. Papadopoulos, M. Sako, M. Sullivan, T.M.C. Abbott, F.B. Abdalla, J. Annis, K. Bechtol, A. Benoit-Lévy, D. Brooks, E. Buckley-Geer, D.L. Burke, A. Carnero Rosell, M. Carrasco Kind, J. Carretero, F.J. Castander, C.E. Cunha, L.N. da Costa, S. Desai, H.T. Diehl, P. Doel, T.F. Eifler, D.A. Finley, B. Flaugher, J. Frieman, J. García-Bellido, D.A. Goldstein, D. Gruen, R.A. Gruendl, J. Gschwend, G. Gutierrez, D.J. James, A.G. Kim, E. Krause, K. Kuehn, N. Kuropatkin, O. Lahav, M. Lima, M.A.G. Maia, M. March, J.L. Marshall, P. Martini, R. Miquel, P. Nugent, A.A. Plazas, A.K. Romer, E. Sanchez, V. Scarpine, M. Schubnell, I. Sevilla-Noarbe, R.C. Smith, F. Sobreira, E. Suchyta, M.E.C. Swanson, R.C. Thomas, A.R. Walker (DES Collaboration), DES15E2mlf: a spectroscopically confirmed superluminous supernova that exploded 3.5 Gyr after the Big Bang. Mon. Not. R. Astron. Soc. **470**, 4241–4250 (2017). 1707.06649. https://doi.org/10.1093/mnras/stx1467

A. Pastorello, S.J. Smartt, M.T. Botticella, K. Maguire, M. Fraser, K. Smith, R. Kotak, L. Magill, S. Valenti, D.R. Young, S. Gezari, F. Bresolin, R. Kudritzki, D.A. Howell, A. Rest, N. Metcalfe, S. Mattila, E. Kankare, K.Y. Huang, Y. Urata, W.S. Burgett, K.C. Chambers, T. Dombeck, H. Flewelling, T. Grav, J.N. Heasley, K.W. Hodapp, N. Kaiser, G.A. Luppino, R.H. Lupton, E.A. Magnier, D.G. Monet, J.S. Morgan, P.M. Onaka, P.A. Price, P.H. Rhoads, W.A. Siegmund, C.W. Stubbs, W.E. Sweeney, J.L. Tonry, R.J. Wainscoat, M.F. Waterson, C. Waters, C.G. Wynn-Williams, Ultra-bright optical transients are linked with Type Ic supernovae. Astrophys. J. Lett. **724**, L16–L21 (2010). 1008.2674. https://doi.org/10.1088/2041-8205/724/1/L16

A. Pastorello, X.F. Wang, F. Ciabattari, D. Bersier, P.A. Mazzali, X. Gao, Z. Xu, J.J. Zhang, S. Tokuoka, S. Benetti, E. Cappellaro, N. Elias-Rosa, A. Harutyunyan, F. Huang, M. Miluzio, J. Mo, P. Ochner, L. Tartaglia, G. Terreran, L. Tomasella, M. Turatto, Massive stars exploding in a He-rich circumstellar medium—IX. SN 2014av, and characterization of Type Ibn SNe. Mon. Not. R. Astron. Soc. **456**, 853–869 (2016). 1509.09069. https://doi.org/10.1093/mnras/stv2634

D.A. Perley, R.M. Quimby, L. Yan, P.M. Vreeswijk, A. De Cia, R. Lunnan, A. Gal-Yam, O. Yaron, A.V. Filippenko, M.L. Graham, R. Laher, P.E. Nugent, Host-galaxy properties of 32 low-redshift superluminous supernovae from the Palomar Transient Factory. Astrophys. J. **830**, 13 (2016). 1604.08207. https://doi.org/10.3847/0004-637X/830/1/13

A.L. Piro, Using double-peaked supernova light curves to study extended material. Astrophys. J. Lett. **808**, L51 (2015). 1505.07103. https://doi.org/10.1088/2041-8205/808/2/L51

O. Porth, S.S. Komissarov, R. Keppens, Rayleigh-Taylor instability in magnetohydrodynamic simulations of the Crab nebula. Mon. Not. R. Astron. Soc. **443**, 547–558 (2014). 1405.4029. https://doi.org/10.1093/mnras/stu1082

S. Prajs, M. Sullivan, M. Smith, A. Levan, N.V. Karpenka, T.D.P. Edwards, C.R. Walker, W.M. Wolf, C. Balland, R. Carlberg, D.A. Howell, C. Lidman, R. Pain, C. Pritchet, V. Ruhlmann-Kleider, The volumetric rate of superluminous supernovae at $z \sim 1$. Mon. Not. R. Astron. Soc. **464**, 3568–3579 (2017). 1605.05250. https://doi.org/10.1093/mnras/stw1942

R.M. Quimby, The Texas Supernova Search. PhD thesis, The University of Texas at Austin (2006)

R.M. Quimby, S.R. Kulkarni, M.M. Kasliwal, A. Gal-Yam, I. Arcavi, M. Sullivan, P. Nugent, R. Thomas, D.A. Howell, E. Nakar, L. Bildsten, C. Theissen, N.M. Law, R. Dekany, G. Rahmer, D. Hale, R. Smith, E.O. Ofek, J. Zolkower, V. Velur, R. Walters, J. Henning, K. Bui, D. McKenna, D. Poznanski, S.B. Cenko, D. Levitan, Hydrogen-poor superluminous stellar explosions. Nature **474**, 487–489 (2011). 0910.0059. https://doi.org/10.1038/nature10095

R.M. Quimby, F. Yuan, C. Akerlof, J.C. Wheeler, Rates of superluminous supernovae at $z \sim 0.2$. Mon. Not. R. Astron. Soc. **431**, 912–922 (2013). 1302.0911. https://doi.org/10.1093/mnras/stt213

R.M. Quimby, A. De Cia, A. Gal-Yam, G. Leloudas, R. Lunnan, D.A. Perley, P.M. Vreeswijk, L. Yan, J.S. Bloom, S.B. Cenko, J. Cooke, R. Ellis, A.V. Filippenko, M.M. Kasliwal, I.K.W. Kleiser, S.R. Kulkarni, T. Matheson, P.E. Nugent, Y.C. Pan, J.M. Silverman, A. Sternberg, M. Sullivan, O. Yaron, Spectra of hydrogen-poor superluminous supernovae from the Palomar Transient Factory. ArXiv e-prints. 1802.07820 (2018)

G. Rakavy, G. Shaviv, Instabilities in highly evolved stellar models. Astrophys. J. **148**, 803 (1967). https://doi.org/10.1086/149204

N. Renault-Tinacci, K. Kotera, A. Neronov, S. Ando, Search for gamma-ray emission from super-luminous supernovae with the Fermi-LAT. ArXiv e-prints. 1708.08971 (2017)

A. Rest, R.J. Foley, S. Gezari, G. Narayan, B. Draine, K. Olsen, M.E. Huber, T. Matheson, A. Garg, D.L. Welch, A.C. Becker, P. Challis, A. Clocchiatti, K.H. Cook, G. Damke, M. Meixner, G. Miknaitis, D. Minniti, L. Morelli, S. Nikolaev, G. Pignata, J.L. Prieto, R.C. Smith, C. Stubbs, N.B. Suntzeff, A.R. Walker, W.M. Wood-Vasey, A. Zenteno, L. Wyrzykowski, A. Udalski, M.K. Szymański, M. Kubiak, G. Pietrzyński, I. Soszyński, O. Szewczyk, K. Ulaczyk, R. Poleski, Pushing the boundaries of conventional core-collapse supernovae: the extremely energetic supernova SN 2003ma. Astrophys. J. **729**, 88 (2011). 0911.2002. https://doi.org/10.1088/0004-637X/729/2/88

S.P. Reynolds, *Dynamical Evolution and Radiative Processes of Supernova Remnants* (Springer, Cham, 2016), pp. 1–24. https://doi.org/10.1007/978-3-319-20794-0_89-1

D. Richardson, R.L. Jenkins III, J. Wright, L. Maddox, Absolute-magnitude distributions of supernovae. Astron. J. **147**, 118 (2014). 1403.5755. https://doi.org/10.1088/0004-6256/147/5/118

R. Roy, J. Sollerman, J.M. Silverman, A. Pastorello, C. Fransson, A. Drake, F. Taddia, C. Fremling, E. Kankare, B. Kumar, E. Cappellaro, S. Bose, S. Benetti, A.V. Filippenko, S. Valenti, A. Nyholm, M. Ergon, F. Sutaria, B. Kumar, S.B. Pandey, M. Nicholl, D. Garcia-Álvarez, L. Tomasella, E. Karamehmetoglu, K. Migotto, SN 2012aa: a transient between Type Ibc core-collapse and superluminous supernovae. Astron. Astrophys. **596**, A67 (2016). 1607.00924. https://doi.org/10.1051/0004-6361/201527947

E. Scannapieco, P. Madau, S. Woosley, A. Heger, A. Ferrara, The detectability of pair-production supernovae at $z \lesssim 6$. Astrophys. J. **633**, 1031–1041 (2005). astro-ph/0507182. https://doi.org/10.1086/444450

S. Schulze, T. Krühler, G. Leloudas, J. Gorosabel, A. Mehner, J. Buchner, S. Kim, E. Ibar, R. Amorín, R. Herrero-Illana, J.P. Anderson, F.E. Bauer, L. Christensen, M. de Pasquale, A. de Ugarte Postigo, A. Gallazzi, J. Hjorth, N. Morrell, D. Malesani, M. Sparre, B. Stalder, A.A. Stark, C.C. Thöne, J.C. Wheeler, Cosmic evolution and metal aversion in superluminous supernova host galaxies. Mon. Not. R. Astron. Soc. **473**, 1258–1285 (2018). 1612.05978. https://doi.org/10.1093/mnras/stx2352

D. Scovacricchi, R.C. Nichol, D. Bacon, M. Sullivan, S. Prajs, Cosmology with superluminous supernovae. Mon. Not. R. Astron. Soc. **456**, 1700–1707 (2016). 1511.06670. https://doi.org/10.1093/mnras/stv2752

N. Smith, S.P. Owocki, On the role of continuum-driven eruptions in the evolution of very massive stars and Population III stars. Astrophys. J. Lett. **645**, L45–L48 (2006). astro-ph/0606174. https://doi.org/10.1086/506523

N. Smith, W. Li, R.J. Foley, J.C. Wheeler, D. Pooley, R. Chornock, A.V. Filippenko, J.M. Silverman, R. Quimby, J.S. Bloom, C. Hansen, SN 2006gy: discovery of the most luminous supernova ever recorded, powered by the death of an extremely massive star like η Carinae. Astrophys. J. **666**, 1116–1128 (2007). astro-ph/0612617. https://doi.org/10.1086/519949

N. Smith, R. Chornock, W. Li, M. Ganeshalingam, J.M. Silverman, R.J. Foley, A.V. Filippenko, A.J. Barth, SN 2006tf: precursor eruptions and the optically thick regime of extremely luminous Type IIn supernovae. Astrophys. J. **686**, 467–484 (2008a). 0804.0042. https://doi.org/10.1086/591021

N. Smith, R.J. Foley, J.S. Bloom, W. Li, A.V. Filippenko, R. Gavazzi, A. Ghez, Q. Konopacky, M.A. Malkan, P.J. Marshall, D. Pooley, T. Treu, J.H. Woo, Late-time observations of SN 2006gy: still going strong. Astrophys. J. **686**, 485–491 (2008b). 0802.1743. https://doi.org/10.1086/590141

N. Smith, R. Chornock, J.M. Silverman, A.V. Filippenko, R.J. Foley, Spectral evolution of the extraordinary Type IIn supernova 2006gy. Astrophys. J. **709**, 856–883 (2010). 0906.2200. https://doi.org/10.1088/0004-637X/709/2/856

N. Smith, W. Li, J.M. Silverman, M. Ganeshalingam, A.V. Filippenko, Luminous blue variable eruptions and related transients: diversity of progenitors and outburst properties. Mon. Not. R. Astron. Soc. **415**, 773–810 (2011). 1010.3718. https://doi.org/10.1111/j.1365-2966.2011.18763.x

M. Smith, M. Sullivan, C.B. D'Andrea, F.J. Castander, R. Casas, S. Prajs, A. Papadopoulos, R.C. Nichol, N.V. Karpenka, S.R. Bernard, P. Brown, R. Cartier, J. Cooke, C. Curtin, T.M. Davis, D.A. Finley, R.J. Foley, A. Gal-Yam, D.A. Goldstein, S. González-Gaitán, R.R. Gupta, D.A. Howell, C. Inserra, R. Kessler,

C. Lidman, J. Marriner, P. Nugent, T.A. Pritchard, M. Sako, S. Smartt, R.C. Smith, H. Spinka, R.C. Thomas, R.C. Wolf, A. Zenteno, T.M.C. Abbott, A. Benoit-Lévy, E. Bertin, D. Brooks, E. Buckley-Geer, A. Carnero Rosell, M. Carrasco Kind, J. Carretero, M. Crocce, C.E. Cunha, L.N. da Costa, S. Desai, H.T. Diehl, P. Doel, J. Estrada, A.E. Evrard, B. Flaugher, P. Fosalba, J. Frieman, D.W. Gerdes, D. Gruen, R.A. Gruendl, D.J. James, K. Kuehn, N. Kuropatkin, O. Lahav, T.S. Li, J.L. Marshall, P. Martini, C.J. Miller, R. Miquel, B. Nord, R. Ogando, A.A. Plazas, K. Reil, A.K. Romer, A. Roodman, E.S. Rykoff, E. Sanchez, V. Scarpine, M. Schubnell, I. Sevilla-Noarbe, M. Soares-Santos, F. Sobreira, E. Suchyta, M.E.C. Swanson, G. Tarle, A.R. Walker, W. Wester (DES Collaboration), DES14X3taz: a Type I superluminous supernova showing a luminous, rapidly cooling initial pre-peak bump. Astrophys. J. Lett. **818**, L8 (2016). 1512.06043. https://doi.org/10.3847/2041-8205/818/1/L8

M. Smith, M. Sullivan, R.C. Nichol, L. Galbany, C.B. D'Andrea, C. Inserra, C. Lidman, A. Rest, M. Schirmer, A.V. Filippenko, W. Zheng, S.B. Cenko, C.R. Angus, P.J. Brown, T.M. Davis, D.A. Finley, R.J. Foley, S. González-Gaitán, C.P. Gutiérrez, R. Kessler, S. Kuhlmann, J. Marriner, A. Möller, P.E. Nugent, S. Prajs, R. Thomas, R. Wolf, A. Zenteno, T.M.C. Abbott, F.B. Abdalla, S. Allam, J. Annis, K. Bechtol, A. Benoit-Lévy, E. Bertin, D. Brooks, D.L. Burke, A. Carnero Rosell, M. Carrasco Kind, J. Carretero, F.J. Castander, M. Crocce, C.E. Cunha, L.N. da Costa, C. Davis, S. Desai, H.T. Diehl, P. Doel, T.F. Eifler, B. Flaugher, P. Fosalba, J. Frieman, J. García-Bellido, E. Gaztanaga, D.W. Gerdes, D.A. Goldstein, D. Gruen, R.A. Gruendl, J. Gschwend, G. Gutierrez, K. Honscheid, D.J. James, M.W.G. Johnson, K. Kuehn, N. Kuropatkin, T.S. Li, M. Lima, M.A.G. Maia, J.L. Marshall, P. Martini, F. Menanteau, C.J. Miller, R. Miquel, R.L.C. Ogando, D. Petravick, A.A. Plazas, A.K. Romer, E.S. Rykoff, M. Sako, E. Sanchez, V. Scarpine, R. Schindler, M. Schubnell, I. Sevilla-Noarbe, R.C. Smith, M. Soares-Santos, F. Sobreira, E. Suchyta, M.E.C. Swanson, G. Tarle, A.R. Walker, The DES Collaboration, Studying the ultraviolet spectrum of the first spectroscopically confirmed supernova at redshift two. Astrophys. J. **854**, 37 (2018). arXiv:1712.04535. https://doi.org/10.3847/1538-4357/aaa126

A.M. Soderberg, S.R. Kulkarni, E. Nakar, E. Berger, P.B. Cameron, D.B. Fox, D. Frail, A. Gal-Yam, R. Sari, S.B. Cenko, M. Kasliwal, R.A. Chevalier, T. Piran, P.A. Price, B.P. Schmidt, G. Pooley, D.S. Moon, B.E. Penprase, E. Ofek, A. Rau, N. Gehrels, J.A. Nousek, D.N. Burrows, S.E. Persson, P.J. McCarthy, Relativistic ejecta from X-ray flash XRF 060218 and the rate of cosmic explosions. Nature **442**, 1014–1017 (2006). arXiv:astro-ph/0604389. https://doi.org/10.1038/nature05087

N. Soker, A. Gilkis, Magnetar-powered superluminous supernovae must first be exploded by jets. Astrophys. J. **851**, 95 (2017). arXiv:1708.08356. https://doi.org/10.3847/1538-4357/aa9c83

E. Sorokina, S. Blinnikov, K. Nomoto, R. Quimby, A. Tolstov, Type I superluminous supernovae as explosions inside non-hydrogen circumstellar envelopes. Astrophys. J. **829**, 17 (2016). 1510.00834. https://doi.org/10.3847/0004-637X/829/1/17

A. Spitkovsky, Time-dependent force-free pulsar magnetospheres: axisymmetric and oblique rotators. Astrophys. J. Lett. **648**, L51–L54 (2006). astro-ph/0603147. https://doi.org/10.1086/507518

A. Suzuki, K. Maeda, Supernova ejecta with a relativistic wind from a central compact object: a unified picture for extraordinary supernovae. Mon. Not. R. Astron. Soc. **466**, 2633–2657 (2017). 1612.03911. https://doi.org/10.1093/mnras/stw3259

F. Taddia, J. Sollerman, G. Leloudas, M.D. Stritzinger, S. Valenti, L. Galbany, R. Kessler, D.P. Schneider, J.C. Wheeler, Early-time light curves of Type Ib/c supernovae from the SDSS-II Supernova Survey. Astron. Astrophys. **574**, A60 (2015). 1408.4084. https://doi.org/10.1051/0004-6361/201423915

K. Takahashi, T. Yoshida, H. Umeda, K. Sumiyoshi, S. Yamada, Exact and approximate expressions of energy generation rates and their impact on the explosion properties of pair instability supernovae. Mon. Not. R. Astron. Soc. **456**, 1320–1331 (2016). 1511.03040. https://doi.org/10.1093/mnras/stv2649

G. Terreran, M.L. Pumo, T.W. Chen, T.J. Moriya, F. Taddia, L. Dessart, L. Zampieri, S.J. Smartt, S. Benetti, C. Inserra, E. Cappellaro, M. Nicholl, M. Fraser, Ł. Wyrzykowski, A. Udalski, D.A. Howell, C. McCully, S. Valenti, G. Dimitriadis, K. Maguire, M. Sullivan, K.W. Smith, O. Yaron, D.R. Young, J.P. Anderson, M. Della Valle, N. Elias-Rosa, A. Gal-Yam, A. Jerkstrand, E. Kankare, A. Pastorello, J. Sollerman, M. Turatto, Z. Kostrzewa-Rutkowska, S. Kozłowski, P. Mróz, M. Pawlak, P. Pietrukowicz, R. Poleski, D. Skowron, J. Skowron, I. Soszyński, M.K. Szymański, K. Ulaczyk, Hydrogen-rich supernovae beyond the neutrino-driven core-collapse paradigm. Nat. Astron. **1**, 713–720 (2017). 1709.10475. https://doi.org/10.1038/s41550-017-0228-8

C.C. Thöne, A. de Ugarte Postigo, R. García-Benito, G. Leloudas, S. Schulze, R. Amorín, A young stellar environment for the superluminous supernova PTF12dam. Mon. Not. R. Astron. Soc. **451**, L65–L69 (2015). 1411.1104. https://doi.org/10.1093/mnrasl/slv051

A. Tolstov, S. Blinnikov, S. Nagataki, K. Nomoto, Shock wave structure in astrophysical flows with an account of photon transfer. Astrophys. J. **811**, 47 (2015). 1412.1434. https://doi.org/10.1088/0004-637X/811/1/47

A. Tolstov, K. Nomoto, S. Blinnikov, E. Sorokina, R. Quimby, P. Baklanov, Pulsational pair-instability model for superluminous supernova PTF12dam: interaction and radioactive decay. Astrophys. J. **835**, 266 (2017a). 1612.01634. https://doi.org/10.3847/1538-4357/835/2/266

A. Tolstov, A. Zhiglo, K. Nomoto, E. Sorokina, A. Kozyreva, S. Blinnikov, Ultraviolet light curves of Gaia16apd in superluminous supernova models. Astrophys. J. Lett. **845**, L2 (2017b). 1707.05746. https://doi.org/10.3847/2041-8213/aa808e

H. Umeda, K. Nomoto, How much ^{56}Ni can be produced in core-collapse supernovae? Evolution and explosions of 30–100$M_\odot$ stars. Astrophys. J. **673**, 1014–1022 (2008). 0707.2598. https://doi.org/10.1086/524767

J.S. Vink (ed.), *Very Massive Stars in the Local Universe*. Astrophysics and Space Science Library, vol. 412 (Springer, Berlin, 2015). 1406.4836. https://doi.org/10.1007/978-3-319-09596-7

A. Vlasis, L. Dessart, E. Audit, Two-dimensional radiation hydrodynamics simulations of superluminous interacting supernovae of Type IIn. Mon. Not. R. Astron. Soc. **458**, 1253–1266 (2016). 1602.05864. https://doi.org/10.1093/mnras/stw410

P.M. Vreeswijk, S. Savaglio, A. Gal-Yam, A. De Cia, R.M. Quimby, M. Sullivan, S.B. Cenko, D.A. Perley, A.V. Filippenko, K.I. Clubb, F. Taddia, J. Sollerman, G. Leloudas, I. Arcavi, A. Rubin, M.M. Kasliwal, Y. Cao, O. Yaron, D. Tal, E.O. Ofek, J. Capone, A.S. Kutyrev, V. Toy, P.E. Nugent, R. Laher, J. Surace, S.R. Kulkarni, The hydrogen-poor superluminous supernova iPTF 13ajg and its host galaxy in absorption and emission. Astrophys. J. **797**, 24 (2014). 1409.8287. https://doi.org/10.1088/0004-637X/797/1/24

P.M. Vreeswijk, G. Leloudas, A. Gal-Yam, A. De Cia, D.A. Perley, R.M. Quimby, R. Waldman, M. Sullivan, L. Yan, E.O. Ofek, C. Fremling, F. Taddia, J. Sollerman, S. Valenti, I. Arcavi, D.A. Howell, A.V. Filippenko, S.B. Cenko, O. Yaron, M.M. Kasliwal, Y. Cao, S. Ben-Ami, A. Horesh, A. Rubin, R. Lunnan, P.E. Nugent, R. Laher, U.D. Rebbapragada, P. Woźniak, S.R. Kulkarni, On the early-time excess emission in hydrogen-poor superluminous supernovae. Astrophys. J. **835**, 58 (2017). 1609.08145. https://doi.org/10.3847/1538-4357/835/1/58

S.Q. Wang, L.J. Wang, Z.G. Dai, X.F. Wu, Superluminous supernovae powered by magnetars: late-time light curves and hard emission leakage. Astrophys. J. **799**, 107 (2015). 1411.5126. https://doi.org/10.1088/0004-637X/799/1/107

T.A. Weaver, The structure of supernova shock waves. Astrophys. J. Suppl. Ser. **32**, 233–282 (1976). https://doi.org/10.1086/190398

J.J. Wei, X.F. Wu, F. Melia, Testing cosmological models with Type Ic super luminous supernovae. Astron. J. **149**, 165 (2015). arXiv:1503.06378. https://doi.org/10.1088/0004-6256/149/5/165

D.J. Whalen, J. Smidt, A. Heger, R. Hirschi, N. Yusof, W. Even, C.L. Fryer, M. Stiavelli, K.J. Chen, C.C. Joggerst, Pair-instability supernovae in the local universe. Astrophys. J. **797**, 9 (2014). 1312.5360. https://doi.org/10.1088/0004-637X/797/1/9

J.C. Wheeler, E. Chatzopoulos, J. Vinkó, R. Tuminello, Circumstellar interaction models for the bolometric light curve of Type I superluminous SN 2017egm. Astrophys. J. Lett. **851**, L14 (2017). 1710.04994. https://doi.org/10.3847/2041-8213/aa9d84

S.E. Woosley, Bright supernovae from magnetar birth. Astrophys. J. Lett. **719**, L204–L207 (2010). 0911.0698. https://doi.org/10.1088/2041-8205/719/2/L204

S.E. Woosley, Pulsational pair-instability supernovae. Astrophys. J. **836**, 244 (2017). 1608.08939. https://doi.org/10.3847/1538-4357/836/2/244

S.E. Woosley, S. Blinnikov, A. Heger, Pulsational pair instability as an explanation for the most luminous supernovae. Nature **450**, 390–392 (2007). 0710.3314. https://doi.org/10.1038/nature06333

L. Yan, R. Quimby, E. Ofek, A. Gal-Yam, P. Mazzali, D. Perley, P.M. Vreeswijk, G. Leloudas, A. de Cia, F. Masci, S.B. Cenko, Y. Cao, S.R. Kulkarni, P.E. Nugent, U.D. Rebbapragada, P.R. Woźniak, O. Yaron, Detection of broad Hα emission lines in the late-time spectra of a hydrogen-poor superluminous supernova. Astrophys. J. **814**, 108 (2015). 1508.04420. https://doi.org/10.1088/0004-637X/814/2/108

L. Yan, R. Lunnan, D.A. Perley, A. Gal-Yam, O. Yaron, R. Roy, R. Quimby, J. Sollerman, C. Fremling, G. Leloudas, S.B. Cenko, P. Vreeswijk, M.L. Graham, D.A. Howell, A. De Cia, E.O. Ofek, P. Nugent, S.R. Kulkarni, G. Hosseinzadeh, F. Masci, C. McCully, U.D. Rebbapragada, P. Woźniak, Hydrogen-poor superluminous supernovae with late-time Hα emission: three events from the intermediate Palomar Transient Factory. Astrophys. J. **848**, 6 (2017a). 1704.05061. https://doi.org/10.3847/1538-4357/aa8993

L. Yan, D.A. Perley, A. De Cia, R. Quimby, R. Lunnan, K.H.R. Rubin, P.J. Brown, Far-UV HST spectroscopy of an unusual hydrogen poor superluminous supernova: SN2017egm. ArXiv e-prints. 1711.01534 (2017b)

L. Yan, R. Quimby, A. Gal-Yam, P. Brown, N. Blagorodnova, E.O. Ofek, R. Lunnan, J. Cooke, S.B. Cenko, J. Jencson, M. Kasliwal, Far-ultraviolet to near-infrared spectroscopy of a nearby hydrogen-poor superluminous supernova Gaia16apd. Astrophys. J. **840**, 57 (2017c). 1611.02782. https://doi.org/10.3847/1538-4357/aa6b02

O. Yaron, A. Gal-Yam, WISeREP—an interactive supernova data repository. Publ. Astron. Soc. Pac. **124**, 668 (2012). 1204.1891. https://doi.org/10.1086/666656

S.C. Yoon, A. Dierks, N. Langer, Evolution of massive Population III stars with rotation and magnetic fields. Astron. Astrophys. **542**, A113 (2012). 1201.2364. https://doi.org/10.1051/0004-6361/201117769

D.R. Young, S.J. Smartt, S. Valenti, A. Pastorello, S. Benetti, C.R. Benn, D. Bersier, M.T. Botticella, R.L.M. Corradi, A.H. Harutyunyan, M. Hrudkova, I. Hunter, S. Mattila, E.J.W. de Mooij, H. Navasardyan, I.A.G. Snellen, N.R. Tanvir, L. Zampieri, Two type Ic supernovae in low-metallicity, dwarf galaxies: diversity of explosions. Astron. Astrophys. **512**, A70 (2010). 0910.2248. https://doi.org/10.1051/0004-6361/200913004

Y.W. Yu, S.Z. Li, A possible relation between flare activity in superluminous supernovae and gamma-ray bursts. Mon. Not. R. Astron. Soc. **470**, 197–201 (2017). arXiv:1607.00626. https://doi.org/10.1093/mnras/stx1028

Y.W. Yu, J.P. Zhu, S.Z. Li, H.J. Lü, Y.C. Zou, A statistical study of superluminous supernovae using the magnetar engine model and implications for their connection with gamma-ray bursts and hypernovae. Astrophys. J. **840**, 12 (2017). arXiv:1704.01682. https://doi.org/10.3847/1538-4357/aa6c27

N. Yusof, R. Hirschi, G. Meynet, P.A. Crowther, S. Ekström, U. Frischknecht, C. Georgy, H. Abu Kassim, O. Schnurr, Evolution and fate of very massive stars. Mon. Not. R. Astron. Soc. **433**, 1114–1132 (2013). 1305.2099. https://doi.org/10.1093/mnras/stt794

T. Zhang, X. Wang, C. Wu, J. Chen, J. Chen, Q. Liu, F. Huang, J. Liang, X. Zhao, L. Lin, M. Wang, M. Dennefeld, J. Zhang, M. Zhai, H. Wu, Z. Fan, H. Zou, X. Zhou, J. Ma, Type IIn Supernova SN 2010jl: optical observations for over 500 days after explosion. Astron. J. **144**, 131 (2012). 1208.6078. https://doi.org/10.1088/0004-6256/144/5/131

Space Sci Rev (2018) 214:68
https://doi.org/10.1007/s11214-018-0500-y

Peculiar Supernovae

Dan Milisavljevic[1] · Raffaella Margutti[2]

Received: 16 March 2018 / Accepted: 28 March 2018 / Published online: 17 April 2018

Abstract What makes a supernova truly "peculiar?" In this review we attempt to address this question by tracing the history of the use of "peculiar" as a descriptor of non-standard supernovae back to the original binary spectroscopic classification of Type I vs. Type II proposed by Minkowski (Publ. Astron. Soc. Pac., 53:224, 1941). A handful of noteworthy examples are highlighted to illustrate a general theme: classes of supernovae that were once thought to be peculiar are later seen as logical branches of standard events. This is not always the case, however, and we discuss ASASSN-15lh as an example of a transient with an origin that remains contentious. We remark on how late-time observations at all wavelengths (radio-through-X-ray) that probe 1) the kinematic and chemical properties of the supernova ejecta and 2) the progenitor star system's mass loss in the terminal phases preceding the explosion, have often been critical in understanding the nature of seemingly unusual events.

Keywords Supernovae: general · Stars: mass-loss · X-rays: general

1 Introduction

Modern transient surveys have uncovered ever-increasing diversity in the observational properties of supernovae (SNe). Well-known surveys include the Palomar Transient Factory (PTF; Rau et al. 2009), the Panoramic Survey Telescope and Rapid Response System

Supernovae
Edited by Andrei Bykov, Roger Chevalier, John Raymond, Friedrich-Karl Thielemann, Maurizio Falanga and Rudolf von Steiger

✉ D. Milisavljevic
dmilisav@purdue.edu

R. Margutti
rafmargutti@gmail.com

1 Department of Physics and Astronomy, Purdue University, 525 Northwestern Avenue, West Lafayette, IN 47907, USA

2 Center for Interdisciplinary Exploration and Research in Astrophysics (CIERA) and Department of Physics and Astrophysics, Northwestern University, Evanston, IL 60208, USA

(Pan-STARRS; Chambers et al. 2016), La Silla-QUEST (LSQ; Baltay et al. 2013), the Lick Observatory Supernova Search (LOSS; Li et al. 2000), the Catalina Real-Time Transient Survey (CRTS; Djorgovski et al. 2011), the All-Sky Automated Survey for Supernovae (ASAS-SN; Shappee et al. 2014), and the Texas Supernova Search now operating as the ROTSE Supernova Verification Project (RSVP; Quimby 2006). The original classification system of Type I and Type II proposed by Minkowski (1941) has branched considerably from its binary roots and is continually being updated in the face of new objects that bridge subtypes and extend luminosity ranges.

Given the large range of parameter space that a SN progenitor system can occupy, it should not be surprising that a zoo of diverse events will be observed (Heger and Woosley 2002; Woosley et al. 2002; Sukhbold et al. 2016). Mass, metallicity, rotation, single vs. binary evolution (and if binary, the mass ratio between the companion stars and their initial separation) make for a broad range of possibilities (see, e.g., Podsiadlowski et al. 1992, Yoon et al. 2010, Claeys et al. 2011, and Eldridge et al. 2017). And yet, "peculiar" has a tradition of use in the SN community, and each time non-standard SNe are encountered there is often an element of surprise in the reported analysis. In this review we seek to understand why this might be and explore patterns of the use of peculiar as a descriptor of SNe.

We include in our investigation the many synonyms of peculiar that can be used to describe unexpected phenomena. This includes, but is not limited to, e.g., "unusual" (Bersten et al. 2016), "unique" (Tominaga et al. 2005; Maeda et al. 2007), or "Perplexing, Troublesome, and Possibly Misleading" (Foley et al. 2010). Peculiar is typically characterized through non-standard bolometric luminosity, either in peak value (e.g., Smith et al. 2007) or longevity (e.g., Arcavi et al. 2017), or it is associated with non-standard spectroscopic emission (e.g., Jha et al. 2006). These outstanding properties may be attributable to explosion dynamics or, more often, characteristics related to mass loss environment sculpted by the progenitor star system. Peculiar is sometimes used interchangeably with rare. Most certainly there are psychological motivations behind use of the word: use of peculiar (or a synonym) evokes mystery, intrigue, and curiosity more than something that is normal and familiar. Fundamentally, use of the word reflects *incomplete knowledge*.

Here we focus on peculiar supernovae that at the time of discovery did not fit neatly into any single observational classification. We will discuss how time has shown that some events that were once regarded as highly unusual are in hindsight understood as being normal and/or logical extensions of an already well characterized phenomenon. Our review is by no means a complete census of "extreme" events and the reader is directed to excellent reviews of core-collapse and thermonuclear supernovae by Gal-Yam (2016), Parrent et al. (2014), and Taubenberger (2017).

2 Type Ib—The "First" Peculiar Supernova

The two main classes of supernovae Type I and II (hereafter SNe I and II) were established by Minkowski (1941). Type II have conspicuous hydrogen features and Type I do not. Today we further divide Type I into those that show Si and S lines (Type Ia) and those that do not (Type Ib and Ic). Narrow emission lines of SNe IIn are associated with interaction with circumstellar environments.

Generally, SNe Ia are presumed to be from white dwarf progenitor systems and the remaining classes from massive stars (see discussion and references in Milisavljevic 2013). The classifications have remained chiefly spectroscopic, although some information about

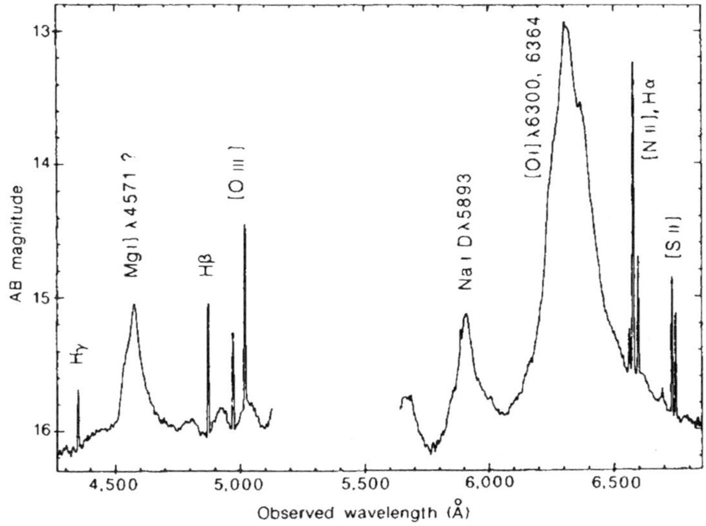

Fig. 1 Late-time spectrum of SN 1985F. These "peculiar" emissions were in stark contrast with any previously published supernova spectrum. We now know that these broad [O I], Mg I], and Na I lines are standard emissions observed from metal-rich ejecta of stripped-envelope supernovae observed at nebular epochs. Reproduced with permission from Filippenko and Sargent (1985)

light curve shape (e.g., IIP and IIL; Barbon et al. 1979) and bolometric luminosity (e.g., superluminous supernovae; Gal-Yam 2012) have been incorporated.

It is challenging to accurately pin down the original use of "peculiar" to describe supernovae in the Minkowski (1941) classification scheme. McLaughlin (1963) reported observations of SN 1954A that were conspicuously different from SNe I and II. However, more notable are the observations of SN 1962I in NGC 1073 by Bertola (1964). The spectral evolution of SN 1962I was distinctly different from normal Type I, in particular lacking the Si II 6115 Å absorption feature and having an absolute luminosity lower by 1.4 mag compared with normal Type I.

Two decades later a fusillade of papers published rapidly between 1985-1987 on "peculiar Type I supernovae" zeroed in on what we now refer to as Type Ib (Wheeler and Levreault 1985; Harkness et al. 1987; Porter and Filippenko 1987; Wheeler et al. 1987). These investigations were spurred on by multiple nearby supernovae (including SN 1984L in NGC 991 and SN 1983N in M83) that also lacked the 6115 Å feature normally associated with Type I maximum-light spectra and exhibited absolute luminosities down by $\approx$1.5 mag compared with normal Type I, despite the shape of the optical light curves being similar. The suggestion that these be called "Type III" supernovae was made (Chevalier 1986), but the term was not adopted.

Perhaps most notable of these reports was on SN 1985F, deemed "A peculiar supernova in the spiral galaxy NGC4618" by Filippenko and Sargent (1985). SN 1985F exhibited an optical spectrum dominated by very strong, broad emission lines of [O I] 6300, 6364, [Ca II] 7291, 7324, the Ca II near-IR triplet, Mg I] 4571, and Na I D (Fig. 1). The complete absence of hydrogen led to a formal classification of SN I. However, at that time no known spectra of SNe I, which generally exhibit a complex blend of P-Cyg profiles, resembled that of SN 1985F. Many properties of SN 1985F suggested that the progenitor star was massive and had lost its hydrogen envelope prior to exploding (Filippenko and Sargent 1986; Begelman and Sarazin 1986), which had been anticipated as a plausible SN progenitor system for the supernova remnant Cassiopeia A a decade previously by Chevalier (1976).

The question of whether or not SN 1985F was distinct from SNe Ia and SNe Ib was resolved when Gaskell et al. (1986) published a spectrum of the Type Ib SN 1983N obtained eight months past maximum. The two spectra were very similar, and formed a crucial link between SN 1985F and other peculiar SN I (now known as Type Ib). These data demonstrated that SN 1985F was a SN Ib discovered long after maximum and, more generally, made it clear that SNe Ib transition into a nebular phase dominated by forbidden transitions that is conspicuously different from SNe Ia. The findings confirmed a previous suggestion by Chugai (1986) that SN 1985F might be a SN Ib discovered long after maximum. The recognition of He I lines in early-time spectra of SNe Ib by Harkness et al. (1987), provided additional evidence that SNe Ib are a physically separate subclass of SNe I initiated by the core collapse of a massive star (Wheeler and Levreault 1985). Wolf–Rayet stars were quickly hypothesized as natural progenitor systems (Begelman and Sarazin 1986; Gaskell et al. 1986). Late-time observations of Type Ib SN 1984L (Schlegel and Kirshner 1989) and detailed analysis of the nebular spectrum of SN 1985F by Fransson and Chevalier (1989) further supported this notion. For additional discussion the reader is directed to Filippenko (1997) and Branch and Wheeler (2017).

As time passed additional heterogeneity in the Type I family beyond the Ia vs. Ib division was recognized. Filippenko et al. (1990) reported on a He-poor Type Ib supernova and adopted the nomenclature Type Ic. New Type Ia classifications emerged with SN 1991T (Filippenko et al. 1992), which was more luminous by $\sim$0.5 mag than regular SNe Ia, and SN 1991bg (Filippenko et al. 1992a), which was less luminous by $R \sim 1$ mag compared to normal SNe Ia at peak. Diversity continues to be recognized to this day; e.g., the class is SN 2002cx-like Type Ia supernovae (also known as SNe Iax) represent one of the most numerous peculiar SN classes (Foley et al. 2013). SNe Iax differ from normal SNe Ia by having a wide range of fainter peak magnitudes, faster decline rates and lower photospheric velocities. We refer the reader to Parrent et al. (2014) for overview of SNe Ia spectra, especially for evolution into the nebular phase. Taubenberger (2017) is another excellent source about extremes of SNe Ia.

The evolving interpretation of the Filippenko and Sargent (1985) observations of SN 1985F is illustrative. In hindsight this event was not so peculiar after all and was no more than spectra of an otherwise normal SN Ib observed several months after explosion. It was, however, perceived as peculiar because observations broke into an unexplored phase space (several months after explosion in this case). In the decades that have followed, larger aperture telescopes along with improvements to detector and instrument efficiencies have made it possible to routinely follow supernovae from explosion to nebular epochs (Fig. 2). Dozens of high-quality spectra of supernovae at epochs $t > 200$ days post explosion are now available (Matheson et al. 2001; Modjaz et al. 2008; Maeda et al. 2008; Taubenberger et al. 2009; Black et al. 2016; Graham et al. 2017), and in some cases (e.g., SNe 1957D, 1970G, 1979C, 1980K, 1993J) supernovae have even been monitored several years to decades after explosion at optical wavelengths (Fesen and Becker 1988; Long et al. 1989; Turatto et al. 1989; Milisavljevic et al. 2012). These observations show that spectra of SNe Ia and SNe Ib/c evolve in very different manners: SNe Ia consist of many forbidden transitions of Fe and Co, whereas SNe Ib/Ic are dominated by O, Mg, Ca, a mix of Na and He, and other intermediate mass elements. The divergence is a valuable method of spectroscopic fingerprinting supernova progenitor systems in cases when photospheric spectra dominated by P-Cyg absorptions provide ambiguous identifications (see Sect. 4).

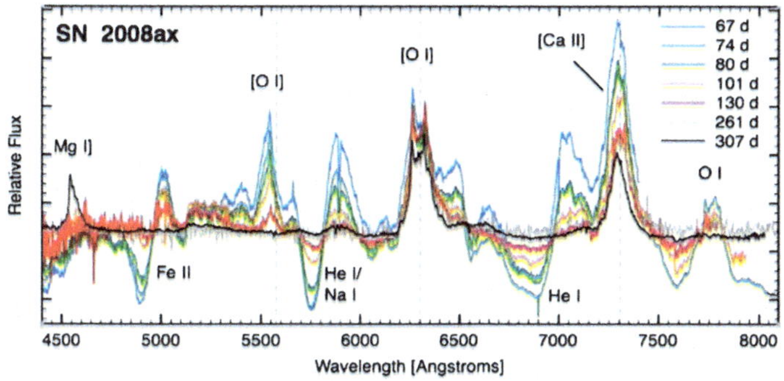

Fig. 2 Representative evolution of a stripped-envelope supernova from photospheric to nebular epochs. As the supernova expands the P-Cyg profiles originating from the thin expanding photosphere subside to broad emission from forbidden lines originating from the metal-rich ejecta being heated by ^{56}Co decay. Data are for the Type IIb SN 2008ax from Milisavljevic et al. (2010)

3 Late-time Observations: Optical, Radio, and X-Ray

Data outside of optical wavelengths have played crucial roles in constraining the nature of peculiar objects. Ultraviolet, optical, and near-infrared wavelengths trace the bulk of the stellar ejecta, while radio and X-ray wavelengths trace emission from the SN's interaction with the local circumstellar environment. Considerable insight into the mass loss environment of the original peculiar Type I (i.e., Type Ib) objects came from observations at radio frequencies; e.g. SN 1983N (Sramek et al. 1984) and SN 1984L (Panagia et al. 1986). Asymmetries in the progenitor star's mass loss can be a major factor contributing to its multi-wavelength emissions and overall degree of "peculiarity" (e.g., Anderson et al. 2017, Bilinski et al. 2017). Notably, many of the aforementioned evolved SNe still detectable at optical wavelengths (e.g., SN 1993J) are also associated with long-lived X-ray and radio emission from interaction between the SN and circumstellar material (CSM) (Weiler et al. 2007). We discuss additional examples of such SN-CSM interaction in Sect. 3.1.

Thermal X-ray emission is one of the clearest indications of circumstellar interaction in supernovae (Chevalier and Fransson 2016). The intensity of the emission depends on the square of the density, and thus is a good estimator of the density of the ambient medium, as long as the emission is not absorbed by the medium. So far, over 60 SNe have been detected in X-rays. All of them until recently had been core-collapse SNe, where the CSM is formed by mass loss from the progenitor star. Deep searches have been conducted for Type Ia explosions with null detections and strong constraints both from radio and X-ray observations (Margutti et al. 2012, 2014b; Chomiuk et al. 2016), suggesting that these explosions occur in "clean" environments ($n < 3\ \mathrm{cm}^{-3}$ at $R \sim 10^{16}$ cm).

Recently, there has been a successful detection of a special case of Type Ia explosion understood to be interacting with H-rich environments. These are known as SN 2002ic-like or Type Ia-CSM events (Wang et al. 2004; Silverman et al. 2013). Late-time (500–800 days after discovery) X-ray detections of SN 2012ca in *Chandra* observations were reported by Bochenek et al. (2018), who favor an asymmetric medium with a high-density component ($n > 10^8\ \mathrm{cm}^{-3}$). This finding provides critical insight into the debate as to whether all SNe Ia-CSM are thermonuclear explosions (Fox et al. 2015; Inserra et al. 2016).

3.1 Supernova Metamorphosis: *SN 2014C*

The remarkable SN 2014C is a closely studied, remarkable example of how late-time SN-CSM interaction can probe fundamental aspects of supernova progenitor systems. SN 2014C was discovered in the nearby spiral galaxy NGC 7331 (D $\sim$ 15.8 Mpc) on 5 January 2014 and identified as a comparatively normal Type Ib. However, when observations resumed in May 2014 after SN 2014C had emerged from behind the Sun, its multi-wavelength properties had become notably different from those of normal SNe Ib. Optical spectra showed that conspicuous emission centered around the Hα line with an overall full-width-half-maximum (FWHM) velocity dispersion of approximately 1400 $\mathrm{km\,s^{-1}}$ had emerged (Fig. 3). This is a feature associated with radiative shocks in dense clumps of CSM normally seen only in Type IIn SNe.

Subsequent observations have demonstrated that this metamorphosis is consistent with a *delayed* interaction between an H-poor star's supernova explosion (SN 2014C) and a massive H-rich envelope that had been stripped decades to centuries before core collapse. Coordinated observations with *Swift*, *Chandra*, *NuSTAR*, the *Hubble Space Telescope*, and the Karl G. Jansky Very Large Array (VLA) captured the evolution in detail and revealed the presence of a massive shell of $\sim$1 $M_{\odot}$ of hydrogen-rich material at $\sim$6 $\times$ 10^{16} cm from the explosion site (Margutti et al. 2017a; see Fig. 4). The IR luminosity of SN 2014C stayed constant over 800 days, possibly due to strong CSM interaction with the H-rich shell, which is rare among stripped-envelope SNe, and suggests that the CSM shell originated from an LBV-like eruption roughly 100 years pre-explosion (Tinyanont et al. 2016). The radio light curve exhibited a double bump morphology indicating two distinct phases of mass-loss from the progenitor star with the transition between density regimes occurring at 100–200 d (Anderson et al. 2017). This reinforces the interpretation that SN 2014C exploded in a low-density region before encountering a dense hydrogen-rich shell of circumstellar material that was likely ejected by the progenitor prior to the explosion.

SN 2014C is a variant of Type IIn and Ibn SNe that exhibit strong shock interaction between their ejecta and pre-existing, slower-moving CSM (Foley et al. 2007; Pastorello et al. 2007; Smith 2016), and part of larger family of supernovae from H-poor progenitor stars that interact with H-rich CSM months to years after explosion. This includes SN 1986J (Rupen et al. 1987; Milisavljevic et al. 2008), the "Wild cousin of SN 1987A" SN 1996cr (Bauer et al. 2008), and SN 2001em (Chugai and Chevalier 2006), which was at first speculated to be a γ-ray burst seen off-axis (Soderberg et al. 2004; Granot and Ramirez-Ruiz 2004). Observations of the majority of these supernovae bridging classifications largely missed the transition from one SN type to another. SN 2014C was unique in the level of details achievable in multi-wavelength follow up across the electromagnetic spectrum.

The metamorphosis emphasizes how SN-CSM interaction can dramatically change SN emission features. Despite originally being an H-poor Type Ib, after only one year SN 2014C rapidly evolved to resemble the H-rich Type II SN 1980K (Fig. 5) at three decades of age. The overall phenomenon poses significant challenges to current theories of massive star evolution, as it requires a physical mechanism responsible for the ejection of the deepest hydrogen layer of H-poor SN progenitors to be synchronized with the onset of stellar collapse. Although the brief timescales between eruption and supernova explosion have been anticipated in special cases of very massive stars (Quataert and Shiode 2012), a growing number of systems are being discovered that fall outside most theoretical regimes (Margutti et al. 2014; Smith and Arnett 2014). Binary interactions and/or instabilities during the last stages of nuclear burning extending all the way to the core C-burning phase in massive stars are believed to be potential triggers of precursor mass loss (Meakin 2006, Meakin and Arnett 2007, Arnett and Meakin 2011,b; see also discussion in Margutti et al. 2017a).

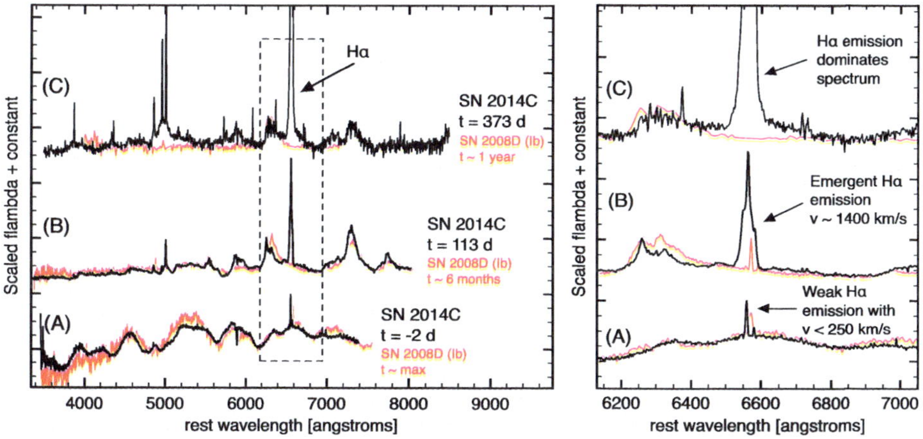

Fig. 3 Real-time monitoring of the delayed interaction between an H-poor star's supernova explosion (SN 2014C) and its previously stripped H-rich envelope. (**A**) A spectrum of SN 2014C obtained around the time of maximum light compared to the H-poor type Ib SN 2008D. (**B**) Months later, SN 2014C exhibits typical type Ib emissions originating from O- and Ca-rich inner ejecta, but also unexpected new and extended (FWHM $\approx$1400 km/s) Hα emission normally only seen in strongly interacting SNe. (**C**) One year after explosion, Hα emission dominates the spectrum. Also seen are low and high ionization emissions of narrow to broad widths originating from many regions (see Fig. 3 for enlargement). SN 2014C represents the first time a type Ib SN has been seen to slowly evolve into a strongly interacting Type IIn. Adapted from Milisavljevic et al. (2015a)

4 Peculiar Supernovae of Debated Origin

4.1 The SN Ib,c–GRB–SLSN Connection: *SN 2012au*

Twentieth century samples of well-studied SNe were gathered primarily through surveys that targeted known, bright, nearby galaxies. Most of the SNe discovered this way had absolute peak magnitudes fainter than about ≈ -20 (see e.g., results in Jha et al. 2006, Richardson et al. 2006). The latest generation of searches, however, have begun gathering SNe via untargeted surveys that are not biased in this way. One surprising result is that a significant percentage of SNe found in dwarf galaxies peak at -21 magnitude or brighter. These superluminous supernovae (SLSNe) are a factor of 10–100 times brighter than normal core-collapse supernovae. The energy emitted in optical light alone rivals the total explosion energy available to typical core-collapse supernovae ($>10^{51}$ erg). SLSNe are broadly differentiated as hydrogen-rich SLSN-II (e.g., Smith et al. 2007), and hydrogen-poor SLSN-I (e.g., Quimby et al. 2011). A further classification for radioactively powered events SLSN-R has also been suggested (Gal-Yam 2012).

Though some SLSNe may be powered mainly by radioactive decay of ^{56}Ni, many reach peak luminosities far too great to be reasonably explained this way. Interactions between SN ejecta with pre-SN winds is a natural explanation for some systems (Smith et al. 2007), but others show no obvious signs of such ongoing interactions or are unclear (Inserra et al. 2013; Nicholl et al. 2013; Lunnan et al. 2016). Fundamentally different engines may be powering these observationally distinct events. A leading candidate is energy injection from a rapidly rotating neutron star (i.e., magnetar) that gets transferred to the kinetic energy of the SN (Kasen and Bildsten 2010; Woosley 2010).

Numerous connections have been established between SLSNe and other classifications. The similarity between the spectra of SN 2010gx (SLSN-I) and those of broad-lined Type

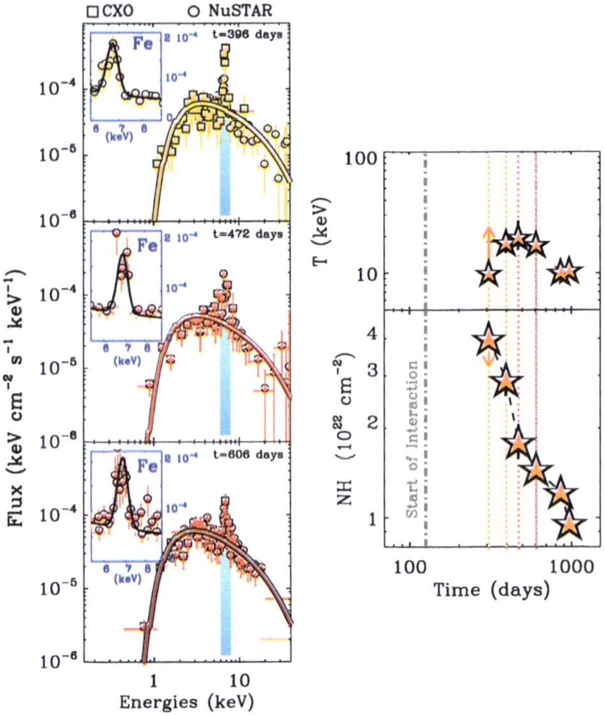

Fig. 4 SN2014C is the first young extragalactic SN for which soft X-rays (*Chandra*—CXO) and hard X-rays (with *NuSTAR*) emission were detected and monitored. These observations reveal a thermal spectrum (left panels) with evolving temperature (T, right upper panel) and absorption (NH, lower right panel). Updated from Margutti et al. (2017a), with the most recent *Chandra-NuSTAR* observations obtained ~1000 days since explosion

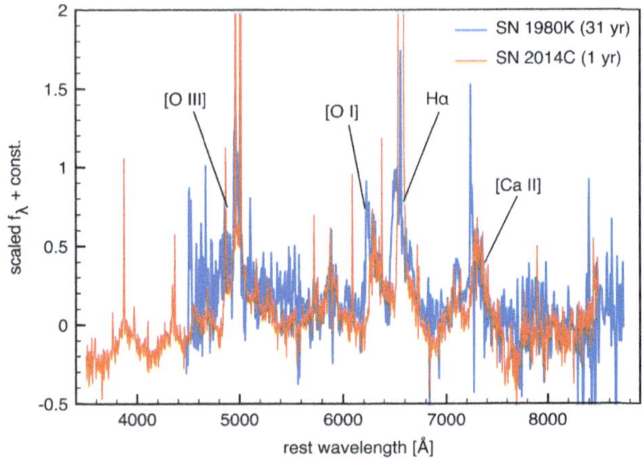

Fig. 5 Optical spectra of SN 2014C (Milisavljevic et al. 2015a) compared to SN 1980K (Milisavljevic et al. 2012). Interaction with a dense H-rich CSM shell accelerated the SN-to-SNR (c.f. Milisavljevic and Fesen 2017) evolution of SN 2014C such that at one year its reverse shock had developed strong enough to excite metal-rich material. Shown here is a comparison to the Type II SN 1980K at 31 yr

Ic SNe first hinted that the two classes potentially had similar progenitor star systems (Pastorello et al. 2010). This connection was further strengthened and extended to lower-luminosity counterparts with discovery of SN 2012au, an energetic explosion having a rarely observed combination of late-time properties linking subsets of energetic and H-poor SNe with SLSNe. Initial spectroscopic observations of SN 2012au showed prominent helium absorption features of an otherwise ordinary Type Ib supernovae. However, continued monitoring through to nebular stages ($t > 250$ days) revealed extraordinary emission properties

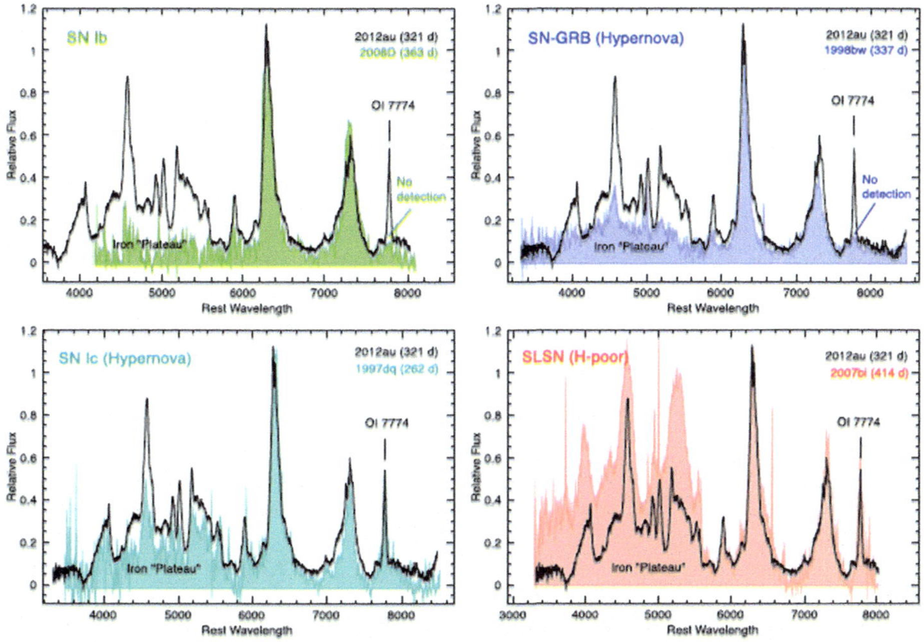

Fig. 6 Spectral fingerprinting SN 2012au. The top panels show the SN Ib SN 2008D (Tanaka et al. 2009) and the hypernova SN 1998bw (Patat et al. 2001), where spectroscopic evolution is normal and O I λ7774 is not detected. These SNe are representative of late-time emissions from the majority of SNe Ib/c. The bottom panels show the hypernova SN 1997dq (Matheson et al. 2001) and the SLSN SN 2007bi (Gal-Yam et al. 2009), where spectroscopic evolution is slow, strong O I λ7774 emission is detected, and emission from Fe lines forms an emission plateau between 4000 and 5600 Å. This connection between SNe Ib/c and SLSNe initially made in Milisavljevic et al. (2013b) was strengthened with late-time observations of the SLSN SN 2015bn (Nicholl et al. 2016). Adapted from Milisavljevic et al. (2013b)

in its optical and near-infrared spectra not unlike those observed in SLSNe (Milisavljevic et al. 2013b).

SN 2012au had a large explosion kinetic energy of $\sim 10^{52}$ erg and ^{56}Ni mass of $\approx 0.3\ M_\odot$ on par with SN 1998bw (Takaki et al. 2013). Its intriguing late-time properties included persistent P-Cyg absorptions attributable to Fe II at $<2000\ \mathrm{km\,s^{-1}}$, and unusually strong emissions from Ca II H&K, Na I D, and O I 7774 Å (see Fig. 6). Persistent P-Cyg absorptions and asymmetries between elements and their ions in the emission line profiles are consistent with expectations of a moderately aspherical and potentially jetted explosion that was most likely initiated by the core collapse of a massive progenitor star. Aside from some differences in the strength and velocity widths of the Fe and Mg emissions, above $\approx$5600 Å emissions from SN 2012au are almost indistinguishable from those of hypernovae SN 1997dq and SN 1997ef, and the superluminous SN 2007bi. Like SN 2012au, these objects exhibited slow spectroscopic evolution and slowly declining light curves.

Extensive radio and X-ray observations of SN 2012au closely follow models of synchrotron emission from a CSM with wind density profile $\rho \propto r^{-2}$ and mass-loss rate of the progenitor of $\dot{M} = 3.6 \times 10^{-6}\ M_\odot\ \mathrm{yr}^{-1}$ (Kamble et al. 2014). Together with the large explosion energy and the large ^{56}Ni mass, the progenitor of SN 2012au likely had a main-sequence mass $>20\ M_\odot$, for which the outer hydrogen envelope had been stripped away but the helium layer still remained. Milisavljevic et al. (2013b) concluded that a single framework involving the core collapse of a massive progenitor and a subsequent asymmetric explosion could unify subsets of SNe and SLSNe that span $-21 < M_B < -17$ mag.

Observations of other extraordinary events have corroborated the notion that connections between some seemingly ordinary Type Ib,c, SLSNe, and SNe associated with long duration gamma-ray bursts (GRBs) may exist. SN 2011kl was associated with the ultra-long-duration gamma-ray burst, GRB 111209A, at a redshift z of 0.677 (Greiner et al. 2015). Its light curve was significantly overluminous compared to other GRB-associated SNe and suggested a link between SN-GRB and SLSNe. Also significant in this regard is nebular-phase spectroscopy of the SLSN-I SN 2015bn spanning +250–400 days after maximum light. These spectra (among the latest ever obtained for a SLSN) are virtually identical to SN 2012au and other energetic SNe Ic (Nicholl et al. 2016), and supported the Milisavljevic et al. (2013b) interpretation that relatively narrow O I λ7774 line may be the signature of a central engine.

4.2 Calcium-rich Transients: *iPTF15eqv*

The class of "Ca-rich transients" defined by unusually strong calcium line emissions that develop in optical spectra months after explosion has garnered considerable attention in the last decade. They were first reported by Filippenko et al. (2003). Later, thorough investigations were provided by Perets et al. (2010) of SN 2005E, and Kasliwal et al. (2012) who discovered several examples in PTF survey data.

Ca-rich transients have distinct He lines in their spectra during the first couple months of evolution and fall within the supernova Type Ib classification. However, Ca-rich transients are less luminous than normal SNe Ib and reach nebular stages much more rapidly (within two months post-explosion). Perhaps most intriguing is that Ca-rich transients are often located at large projected distances (as high as 150 kpc) from the centers of their host galaxies, implying that progenitors have traveled significant distances before exploding (Lyman et al. 2014; Foley 2015). Many occur in early-type galaxies lacking obvious massive star populations (Kasliwal et al. 2012), and examinations of explosion sites with deep, high resolution images have thus far failed to uncover any sign of in situ star formation (Lyman et al. 2013, 2014, 2016; Lunnan et al. 2017).

The progenitor systems of Ca-rich transients have been a source of contention. A variety of spectroscopic properties of Ca-rich transients, including the Type Ib classification, are most naturally understood as originating from explosions of massive star progenitors (Kawabata et al. 2010). However, the prevailing view is that Ca-rich transients are somehow related to WD progenitor systems. The relatively large delay-time distribution required to travel large distances favors an older white dwarf (WD) population that also contributes to SNe Ia. Many explosion channels have been proposed involving helium detonations occurring on a helium-accreting WD (Shen and Bildsten 2009; Perets et al. 2010; Woosley and Kasen 2011). Attempts to understand Ca-rich transients using models involving the tidal detonation of a low-mass WD by a neutron star or intermediate-/stellar-mass black hole have also been suggested (Rosswog et al. 2008; MacLeod et al. 2014; Sell et al. 2015).

The discovery of iPTF15eqv—a "Peculiar Ca-rich Transient"—complicated the WD progenitor paradigm and corroborated suspicions that the observational classification may encompass multiple progenitor channels (Milisavljevic et al. 2017). The close proximity and relatively slow decline rate of iPTF15eqv enabled spectroscopic observations of high signal-to-noise ratio at epochs >200 days after explosion, which is among the latest epochs ever observed for a Ca-rich transient. The transient was originally discovered by K. Itagaki, and then independently discovered and classified by the iPTF survey (Cao et al. 2015). Initial spectra led Cao et al. (2015) to classify iPTF15eqv as an SN IIb/Ib in the nebular phase.

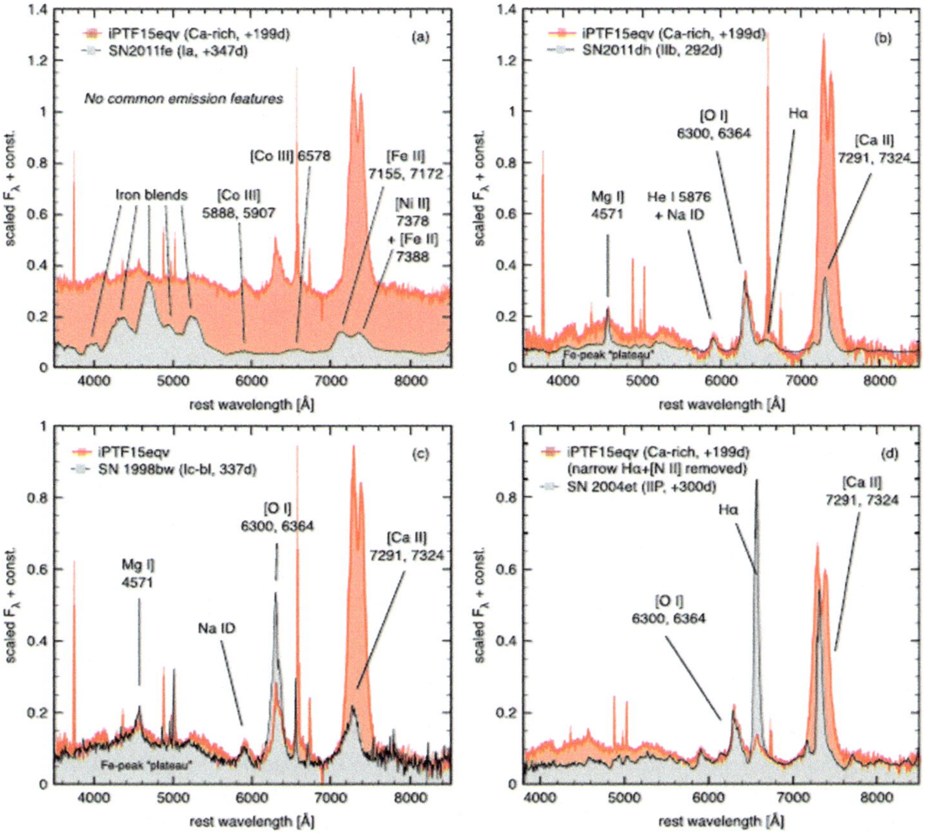

Fig. 7 Spectral fingerprinting iPTF15eqv. Spectra have been arbitrarily scaled to compare and contrast the relative strengths of specific emission features between nebular epoch spectra of iPTF15eqv and other supernovae (both thermonuclear and core collapse). The spectral features of iPTF15eqv best match those of a core-collapse explosion and have little similarity to those observed in thermonuclear WD explosions. Data were originally published in Mazzali et al. (2015), Ergon et al. (2015), Patat et al. (2001), and Sahu et al. (2006). Adapted from Milisavljevic et al. (2017)

However, a multi-wavelength follow-up campaign supported by radio observations with the VLA and X-ray observations with *Chandra* showed a combination of properties that bridge those observed in Ca-rich transients and SNe Ib/c (Milisavljevic et al. 2017).

Perhaps most revealing about iPTF15eqv is its conspicuous Type Ib/c and Type II spectroscopic signatures in late-time optical and NIR data (Fig. 7). iPTF15eqv exhibits [O I], [Ca II], a blend of He I + Na I, and a "plateau? of emission spanning 3800–5600 Å associated with Fe-peak elements. To varying degrees these are all observed in Type IIb SN 2011dh (Ergon et al. 2015), the GRB-SN SN 1998bw, and the Type IIP SN 2004et (Sahu et al. 2006). A broad emission feature redward of [O I] that is likely Hα but may include contribution from [N II] and/or Ca I] λ6572, is also shared in some cases. In sharp contrast, iPTF15eqv does not share any conspicuous features with normal Type Ia that are dominated by blended forbidden and permitted Fe-peak lines between 4000–5500 Å and 7000–7600 Å.

On the one hand, the late-time spectra of iPTF15eqv are dominated by strong calcium emission lines, which is a defining characteristic of Ca-rich transients. Its [CaII]/[OI] $\approx$ 10 emission line ratio is among the highest encountered among Ca-rich events (Fig. 8).

Fig. 8 Emission line ratio of [Ca II] λλ7291, 7324 to [O I] λλ6300, 6364 for Type II, Type IIb/Ib/Ic, and Ca-rich transients. At all epochs where both lines are visible and the conditions can be reasonably assessed to be nebular, all Ca-rich transients have [Ca II]/[O I]>2, and all Type Ib/c are <2. Solid lines connect different epochs of the same object. iPTF15eqv stands out with the highest [Ca II]/[O I] ratio of ≈10. The gray silhouette of iPTF15eqv connected by dashed lines reflects the uncertainty in the explosion date. Adapted from Milisavljevic et al. (2017)

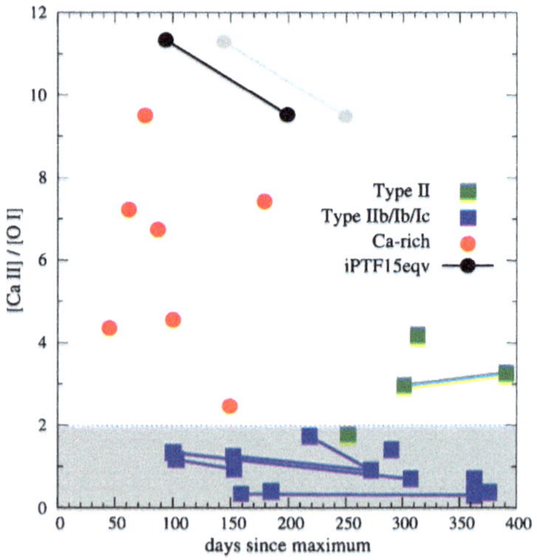

On the other hand, iPTF15eqv differs from other Ca-rich transients in terms of its light curve evolution. iPTF15eqv was a slower evolving and potentially much more luminous object than Ca-rich transients (estimated to potentially be up to ~2 mag brighter than other examples; Kawahara et al. 2018) and exhibited a light curve that resembles SNe Ib/c.

Analysis of the chemical abundances associated with observed line emission from iPTF15eqv is consistent with the supernova explosion of a $<10\ M_{\odot}$ star that was stripped of its H-rich envelope via binary interaction. This result challenges the notion that spectroscopically classified Ca-rich transients only originate from WD progenitor systems. Milisavljevic et al. (2017) conclude that the relatively long delay-time distribution and distinguishable abundance patterns of electron capture supernovae from binary systems make them an attractive progenitor system for at least some Ca-rich transients.

4.3 Unusual Longevity: *iPTF14hls*

iPTF14hls (Arcavi et al. 2017) commanded recent attention for exhibiting both extremely normal and extremely peculiar characteristics simultaneously. iPTF14hls showed spectra that are identical to normal Type II-P explosions, yet persisted nearly unchanged for ~600 days while the light-curve experienced multiple re-brigthenings (Fig. 9). The absorption lines showed negligible decreases in velocity throughout the five peaks of the luminous light curve. A pre-explosion outburst in 1954 was also reported.

The spectral and temporal evolution of iPTF14hls is unprecedented. Possible frameworks to understand its nature include magnetar central engines and pulsational pair-instability supernovae (Dessart 2018; Woosley 2018), or common envelope jets (Soker and Gilkis 2018). High-energy γ-ray emission may have been detected (Yuan et al. 2017), which suggests very fast particle acceleration by the supernova explosion.

A significant breakthrough in constraining the nature of iPFT14hls came with an abrupt change in its spectral emissions observed in data obtained by Andrews and Smith (2017) three years after discovery. A double-peaked intermediate-width Hα line indicative of expansion speeds around 1000 $\mathrm{km\,s^{-1}}$ was observed. A similar profile was also observed in the [O I] 6300, 6364 lines. Andrews and Smith (2017) interpreted this as clear evidence of interaction between the SN and dense CSM having a disc-like geometry, which is an important clue for understanding the nature of the explosion. They conclude that interaction

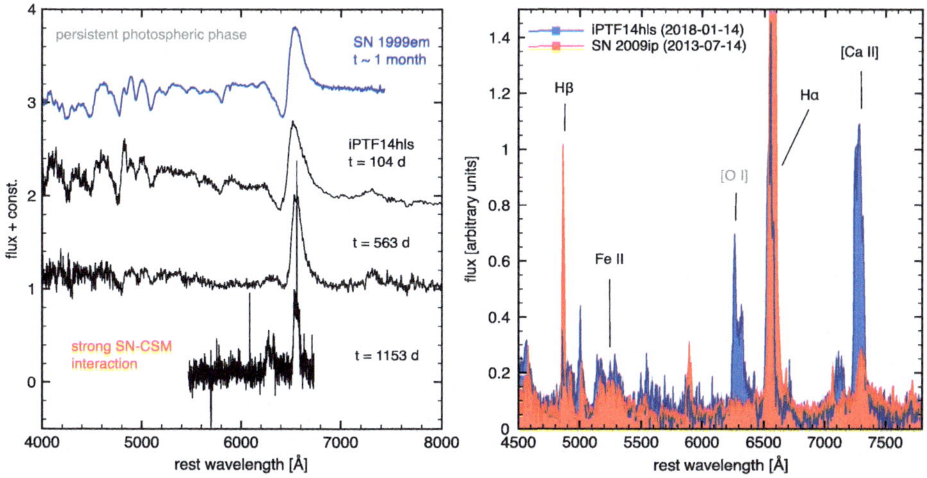

Fig. 9 Left: Spectral sequence of iPTF14hls. iPTF14hls exhibited spectral features identical to normal Type II-P explosions (shown here is SN 1999em for comparison), yet persisted nearly unchanged for ~600 days while the light-curve experienced multiple re-brightenings. Data are from Arcavi et al. (2017) and downloaded from the Weizmann Interactive Supernova data Repository (Yaron and Gal-Yam 2012). Andrews and Smith (2017) published a spectrum obtained on day 1153 showing dramatic transformation consistent with strong SN-CSM interaction. Right: Keck optical spectrum of iPTF14hls obtained 1210 days after discovery (Terreran et al., in preparation) compared to the strongly interacting Type IIn SN 2009ip (Margutti et al. 2014). This spectrum, which spans a wider wavelength window compared to the Andrews and Smith (2017) data and shows a combination of broad emission lines associated with shocked ejecta and narrow emission lines associated with nearby photoionized material, is consistent with their interpretation that the dramatic change is the consequence of SN-CSM interaction. Notably, the broad [O I] 6300, 6364 emission line feature, not observed (and possibly obscured) in SN 2009ip, signifies considerable metal-rich O-rich ejecta in iPTF14hls that is interacting with a disk-like environment

with variations in the density structure of the CSM may be adequate to explain the peculiar evolution of iPTF14hls. A later epoch Keck spectrum obtained by Terreran et al. (in preparation) shows a combination of narrow and broad emission lines seen in strongly interacting Type IIn supernovae consistent with the Andrews and Smith (2017) SN-CSM interpretation (Fig. 9). Further monitoring across wavelengths will hopefully continue to shed additional light on the nature of this exciting transient.

5 Special Case: ASASSN-15lh

The transient ASASSN-15lh is an excellent example of how some phenomena fail to fit naturally into any known classification. Indeed, the properties of ASASSN-15lh are so peculiar that, at the time of writing, it is not even clear if ASASSN-15lh is connected with a stellar explosion and might be the first (and only member so far) of its own class.

ASASSN-15lh was discovered (Dong et al. 2016) by the All-Sky Automated Survey for Supernovae (ASAS-SN) at $z = 0.2326$ ($d = 1171$ Mpc for standard *Planck* cosmology). The location of ASASSN-15ls is astrometrically consistent with he nucleus of its massive early type host galaxy (Dong et al. 2016). The transient exhibited an extremely large peak luminosity $L_{\rm pk} \sim 2 \times 10^{45}$ erg s^{-1} and a blue, almost featureless spectrum with no apparent sign of H or He, leading Dong et al. (2016) and Godoy-Rivera et al. (2017) to suggest it to be the most luminous SLSN ever detected (Fig. 10). However, the spectral and temporal

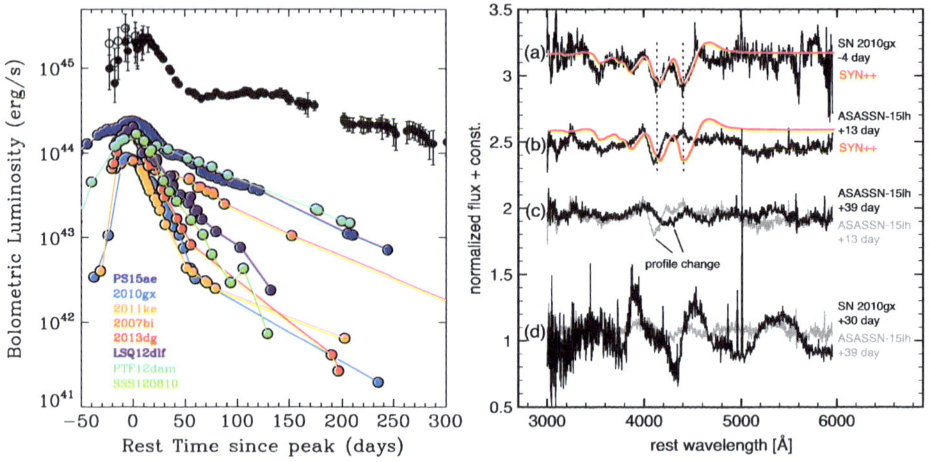

Fig. 10 Left panel: Bolometric luminosity emission from ASASSN-15lh (black dots from Dong et al. 2016, Margutti et al. 2017b) compared to a sample of H-stripped SLSNe (from Inserra et al. 2013, Nicholl et al. 2016). ASASSN-15lh is >10 times more luminous then the most luminous SLSNe known at peak and shows an unusually very mild decay at later times. Right panel: early-times spectral evolution of ASASSN-15lh in context with SLSNe (here we use the SLSN SN2010gx, which shows the typical features of the class). The O II ion "W-shaped" feature is typical of SLSNe (vertical dashed lines), and here we show how using the simple assumptions of SYN++ and a photospheric velocity of 19,000 km s^{-1}, the -4 day spectrum of the SLSN 2010gx can be reproduced. By contrast, we cannot reproduce +13 day spectrum of ASASSN-15lh. Even more importantly, ASASSN-15lh progressively evolves towards a featureless spectrum with time, which is the opposite evolution of SNe. Adapted from Margutti et al. (2017b)

evolution of the transient, as well as its host galaxy, are unprecedented among SLSNe and other explanations have been proposed, including a tidal disruption event (TDE) by a supermassive BH (Leloudas et al. 2016).

The observed properties of ASASSN-15lh challenge any classification scheme and can be summarized as follows:

- The peak luminosity of ASASSN-15lh and its total radiated energy ($E_{\rm rad} \sim 2 \times 10^{52}$ erg) are extreme even among SLSNe, and require sources of energy that are different from the standard radioactive decay of ^{56}Ni that powers normal H-stripped SNe in the local Universe (Dong et al. 2016, Chatzopoulos et al. 2016, Kozyreva et al. 2016, van Putten and Della Valle 2016).
- ASASSN-15lh experienced a UV rebrightening beginning at $t \sim 90$-d (observer frame) after the primary peak and was followed by a ~120-d long plateau in the bolometric luminosity (Fig. 10), before starting to fade again. Throughout its initial decline, subsequent rebrightening and renewed decline, the spectra did not show evidence of interactions between the ejecta and CSM such as narrow emission lines. There are hints of weak Hα emission at late-times, but Margutti et al. (2017b) have shown that it is narrow line emission consistent with star formation in the host nucleus.
- Differently from SNe, ASASSN-15lh evolved towards a progressively featureless spectrum (Fig. 10).
- The host galaxy of ASASSN-15lh is old, massive $M_* \sim 2 \times 10^{11}\ M_\odot$ and with limited star formation (Dong et al. 2016). These properties are very different from galaxies that host core-collapse SNe and SLSNe, which are typically younger star forming galaxies with significantly lower stellar mass (Milisavljevic et al. 2015b, Lunnan et al. 2015, Perley

et al. 2016, Leloudas et al. 2016). In the context of TDEs, the host galaxy properties are also unprecedented, and ASASSN-15lh would be the most luminous TDE ever observed, associated with a SMBH with mass $M_\bullet \sim 10^{8.6}\ M_\odot$ significantly larger than the average TDE SMBHs (e.g., Komossa 2015). The large SMBH mass motivated speculation of a fast rotating SMBH, in order to disrupt the star outside the innermost stable stellar orbit (Leloudas et al. 2016; Margutti et al. 2017b). Coughlin and Armitage (2017) argue that its features, including its anomalous rebrightening at $\sim$100 days after detection, are consistent with the tidal disruption of a star by a supermassive black hole in a binary system.

Debate has continued on whether ASASSN-15lh is an H-poor superluminous supernova or a tidal disruption event. It is clear that the properties of ASASSN-15lh and of its host galaxy are unprecedented in both scenarios. It is possible that a clue to the real nature of ASASSN-15lh has already been provided by its emission outside the optical range. A faint X-ray source has been detected at the location consistent with ASASSN-15lh (Margutti et al. 2017b). The very soft spectrum of the source and its persistent emission are not consistent with SNe, and suggest instead an origin connected with the nucleus of the host galaxy. If ASASSN-15lh is a TDE, the expectation is fading of the X-ray source in the next few years.

Future multi-wavelength follow-up of ASASSN-15lh, and a larger sample of similar events, are the only way to advance our understanding of the physics of this extremely peculiar transient.

6 Concluding Remarks: What Will be "Peculiar" in the Future?

In this review we have reviewed several examples of peculiar supernovae in an effort to understand what sets them apart from normal events. A general pattern emerged that classes of supernovae that were originally interpreted to have unusual properties are later seen as logical branches of phenomena that are already well understood. In some cases, the nature of a peculiar transient can remain debated, and we discussed the special case of ASASSN-15lh as an example of this contentious variety. The shift in interpretation from extraordinary to mundane has often been brought about by improvements in telescope and detector technology, and by observations that push into new phase spaces of luminosity, time scale, and wavelength.

What will be identified as peculiar in the future? Observations of supernovae in the first hours to days following explosion have been limited thus far but provide enticing glimpses at the wealth of information that can be extracted at extremely early phases (Nugent et al. 2011; Milisavljevic et al. 2013a; Gal-Yam et al. 2014; Nicholl et al. 2015; Garnavich et al. 2016). The Pan-STARRS, PTF, and $D < 40$ Mpc (DLT40) surveys have made commendable progress in this regard (Drout et al. 2014; Yaron et al. 2017; Tartaglia et al. 2018), and the upcoming surveys Zwicky Transient Facility[1] and Large Synoptic Survey Telescope[2] (LSST) will allow even greater exploration of these very early phases. Complementary to these advances will be explorations at very late epochs. This will be a particular strength of LSST, which will provide a deep, all-sky survey sampled in a consistent rapid cadence, and next-generation telescopes such as the Giant Magellan Telescope,[3] the Thirty Meter

[1] https://www.ptf.caltech.edu/ztf.

[2] https://www.lsst.org/.

[3] https://www.gmto.org/.

Telescope,[4] and the European Extremely Large Telescope[5] that will revolutionize late-time investigations of supernovae both in the volume of space that can be sampled (and hence the number of objects to study) and the quality of data to be obtained. Particularly exciting will be the ability to carefully monitor the evolution of emission line widths in spectra, which can distinguish between sources of energy in extreme supernovae (Milisavljevic et al. 2012). As these and other facilities, now operating in the multi-messenger era (Abbott et al. 2017), move into uncharted frontiers of transient phase space, they are destined to continue finding new examples of supernovae that defy expectations and shape our understanding of the terminal stages of stellar evolution.

Acknowledgements D. M. & R. M. thank ISSI organizers for their kind invitation to the Supernova Workshop in Bern. D. M. acknowledges support from NASA through grant number GO-14202 from the Space Telescope Science Institute, which is operated by AURA, Inc., under NASA contract NAS 5-26555. R. M. acknowledges partial support to her group from NASA through NuSTAR grants NNX17AI13G and NNX17AG80G. Partial support for this work was provided by the National Aeronautics and Space Administration through Chandra Award Number GO5-16064A and GO6-17054A issued by the Chandra X-ray Center, which is operated by the Smithsonian Astrophysical Observatory for and on behalf of the National Aeronautics Space Administration under contract NAS8-03060. We especially thank Roger Chevalier for his help and patience preparing the manuscript. We thank J. Andrews and N. Smith for sharing their iPTF14hls spectrum before publication. We also thank G. Terreran for sharing Keck observations of iPTF14hls. This work is based in part on observations from the Low Resolution Imaging Spectrometer at the Keck-1 telescope. We are grateful to the staff at the Keck Observatory for their assistance, and we extend special thanks to those of Hawaiian ancestry on whose sacred mountain we are privileged to be guests. The W. M. Keck Observatory is operated as a scientific partnership among the California Institute of Technology, the University of California, and NASA; it was made possible by the generous financial support of the W. M. Keck Foundation.

References

B.P. Abbott, R. Abbott, T.D. Abbott, F. Acernese, K. Ackley, C. Adams, T. Adams, P. Addesso, R.X. Adhikari, V.B. Adya et al., Multi-messenger observations of a binary neutron star merger. Astrophys. J. **848**, L12 (2017). https://doi.org/10.3847/2041-8213/aa91c9. arXiv:1710.05833

G.E. Anderson, A. Horesh, K.P. Mooley, A.P. Rushton, R.P. Fender, T.D. Staley, M.K. Argo, R.J. Beswick, P.J. Hancock, M.A. Pérez-Torres, Y.C. Perrott, R.M. Plotkin, M.L. Pretorius, C. Rumsey, D.J. Titterington, The peculiar mass-loss history of SN 2014C as revealed through AMI radio observations. Mon. Not. R. Astron. Soc. **466**, 3648–3662 (2017). https://doi.org/10.1093/mnras/stw3310. arXiv:1612.06059

J.E. Andrews, N. Smith, Strong late-time circumstellar interaction in the not-so-impossible supernova iPTF14hls (2017). ArXiv e-prints, arXiv:1712.00514

I. Arcavi, D.A. Howell, D. Kasen, L. Bildsten, G. Hosseinzadeh, C. McCully, Z.C. Wong, S.R. Katz, A. Gal-Yam, J. Sollerman, F. Taddia, G. Leloudas, C. Fremling, P.E. Nugent, A. Horesh, K. Mooley, C. Rumsey, S.B. Cenko, M.L. Graham, D.A. Perley, E. Nakar, N.J. Shaviv, O. Bromberg, K.J. Shen, E.O. Ofek, Y. Cao, X. Wang, F. Huang, L. Rui, T. Zhang, W. Li, Z. Li, J. Zhang, S. Valenti, D. Guevel, B. Shappee, C.S. Kochanek, T.W.S. Holoien, A.V. Filippenko, R. Fender, A. Nyholm, O. Yaron, M.M. Kasliwal, M. Sullivan, N. Blagorodnova, R.S. Walters, R. Lunnan, D. Khazov, I. Andreoni, R.R. Laher, N. Konidaris, P. Wozniak, B. Bue, Energetic eruptions leading to a peculiar hydrogen-rich explosion of a massive star. Nature **551**, 210–213 (2017). https://doi.org/10.1038/nature24030. arXiv:1711.02671

W.D. Arnett, C. Meakin, Toward realistic progenitors of core-collapse supernovae. Astrophys. J. **733**, 78 (2011). https://doi.org/10.1088/0004-637X/733/2/78. arXiv:1101.5646

W.D. Arnett, C. Meakin, Turbulent cells in stars: fluctuations in kinetic energy and luminosity. Astrophys. J. **741**, 33 (2011b). arXiv:1012.1848. https://doi.org/10.1088/0004-637X/741/1/33

C. Baltay, D. Rabinowitz, E. Hadjiyska, E.S. Walker, P. Nugent, P. Coppi, N. Ellman, U. Feindt, R. McKinnon, B. Horowitz, A. Effron, The La Silla-QUEST low redshift supernova survey. Publ. Astron. Soc. Pac. **125**, 683 (2013). https://doi.org/10.1086/671198

[4] https://www.tmt.org/.

[5] https://www.eso.org/sci/facilities/eelt/.

R. Barbon, F. Ciatti, L. Rosino, Photometric properties of type II supernovae. Astron. Astrophys. **72**, 287–292 (1979)

F.E. Bauer, V.V. Dwarkadas, W.N. Brandt, S. Immler, S. Smartt, N. Bartel, M.F. Bietenholz, Supernova 1996cr: SN 1987A's wild cousin? Astrophys. J. **688**, 1210–1234 (2008). https://doi.org/10.1086/589761. arXiv:0804.3597

M.C. Begelman, C.L. Sarazin, SN 1985f—death of a Wolf–Rayet star. Astrophys. J. **302**, L59–L62 (1986). https://doi.org/10.1086/184637

M.C. Bersten, O.G. Benvenuto, M. Orellana, K. Nomoto, The unusual super-luminous supernovae SN 2011kl and ASASSN-15lh. Astrophys. J. **817**, L8 (2016). https://doi.org/10.3847/2041-8205/817/1/L8. arXiv:1601.01021

F. Bertola, The supernovae in NGC 1073 and NGC 1058. Ann. Astrophys. **27**, 319 (1964)

C. Bilinski, N. Smith, G.G. Williams, P. Smith, W. Zheng, M.L. Graham, J.C. Mauerhan, J.E. Andrews, A.V. Filippenko, C. Akerlof, E. Chatzopoulos, J.L. Hoffman, L. Huk, D.C. Leonard, G.H. Marion, P. Milne, R.M. Quimby, J.M. Silverman, J. Vinkó, J.C. Wheeler, F. Yuan, SN2012ab: A Peculiar Type IIn Supernova with Aspherical Circumstellar Material (2017). ArXiv e-prints, arXiv:1712.03370

C.S. Black, R.A. Fesen, J.T. Parrent, Progressive redshifts in the late-time spectra of Type Ia supernovae. Mon. Not. R. Astron. Soc. **462**, 649–662 (2016). https://doi.org/10.1093/mnras/stw1680. arXiv:1604.01044

C.D. Bochenek, V.V. Dwarkadas, J.M. Silverman, O.D. Fox, R.A. Chevalier, N. Smith, A.V. Filippenko, X-ray emission from SN 2012ca: a type Ia-CSM supernova explosion in a dense surrounding medium. Mon. Not. R. Astron. Soc. **473**, 336–344 (2018). https://doi.org/10.1093/mnras/stx2029. arXiv:1708.07181

D. Branch, J.C. Wheeler, *Supernova Explosions* (Springer, Berlin, 2017). https://doi.org/10.1007/978-3-662-55054-0

Y. Cao, S.R. Kulkarni, D.A. Howell, A. Gal-Yam, M.M. Kasliwal, S. Valenti, J. Johansson, R. Amanullah, A. Goobar, J. Sollerman, F. Taddia, A. Horesh, I. Sagiv, S.B. Cenko, P.E. Nugent, I. Arcavi, J. Surace, P.R. Woźniak, D.I. Moody, U.D. Rebbapragada, B.D. Bue, N. Gehrels, A strong ultraviolet pulse from a newborn type Ia supernova. Nature **521**, 328–331 (2015). https://doi.org/10.1038/nature14440. arXiv:1505.05158

K.C. Chambers, E.A. Magnier, N. Metcalfe, H.A. Flewelling, M.E. Huber, C.Z. Waters, L. Denneau, P.W. Draper, D. Farrow, D.P. Finkbeiner, C. Holmberg, J. Koppenhoefer, P.A. Price, R.P. Saglia, E.F. Schlafly, S.J. Smartt, W. Sweeney, R.J. Wainscoat, W.S. Burgett, T. Grav, J.N. Heasley, K.W. Hodapp, R. Jedicke, N. Kaiser, R.P. Kudritzki, G.A. Luppino, R.H. Lupton, D.G. Monet, J.S. Morgan, P.M. Onaka, C.W. Stubbs, J.L. Tonry, E. Banados, E.F. Bell, R. Bender, E.J. Bernard, M.T. Botticella, S. Casertano, S. Chastel, W.P. Chen, X. Chen, S. Cole, N. Deacon, C. Frenk, A. Fitzsimmons, S. Gezari, C. Goessl, T. Goggia, B. Goldman, E.K. Grebel, N.C. Hambly, G. Hasinger, A.F. Heavens, T.M. Heckman, R. Henderson, T. Henning, M. Holman, U. Hopp, W.H. Ip, S. Isani, C.D. Keyes, A. Koekemoer, R. Kotak, K.S. Long, J.R. Lucey, M. Liu, N.F. Martin, B. McLean, E. Morganson, D.N.A. Murphy, M.A. Nieto-Santisteban, P. Norberg, J.A. Peacock, E.A. Pier, M. Postman, N. Primak, C. Rae, A. Rest, A. Riess, A. Riffeser, H.W. Rix, S. Roser, E. Schilbach, A.S.B. Schultz, D. Scolnic, A. Szalay, S. Seitz, B. Shiao, E. Small, K.W. Smith, D. Soderblom, A.N. Taylor, A.R. Thakar, J. Thiel, D. Thilker, Y. Urata, J. Valenti, F. Walter, S.P. Watters, S. Werner, R. White, W.M. Wood-Vasey, R. Wyse, The Pan-STARRS1 Surveys (2016). ArXiv e-prints, arXiv:1612.05560

E. Chatzopoulos, J.C. Wheeler, J. Vinko, A.P. Nagy, B.K. Wiggins, W.P. Even, Extreme supernova models for the super-luminous transient ASASSN-15lh. Astrophys. J. **828**, 94 (2016). https://doi.org/10.3847/0004-637X/828/2/94. arXiv:1603.06926

R.A. Chevalier, Cassiopeia A, faint supernovae, and heavy-element ejection by supernovae. Astrophys. J. **208**, 826–828 (1976). https://doi.org/10.1086/154669

R.A. Chevalier, Supernovae and stellar mass loss. Highlights Astron. **7**, 599–609 (1986)

R.A. Chevalier, C. Fransson, Thermal and non-thermal emission from circumstellar interaction (2016). ArXiv e-prints, arXiv:1612.07459

L. Chomiuk, A.M. Soderberg, R.A. Chevalier, S. Bruzewski, R.J. Foley, J. Parrent, J. Strader, C. Badenes, C. Fransson, A. Kamble, R. Margutti, M.P. Rupen, J.D. Simon, A deep search for prompt radio emission from thermonuclear supernovae with the very large array. Astrophys. J. **821**, 119 (2016). https://doi.org/10.3847/0004-637X/821/2/119. arXiv:1510.07662

N.N. Chugai, Normal and peculiar type-I supernovae—why do they differ. Sov. Astron. Lett. **12**, 192–195 (1986)

N.N. Chugai, R.A. Chevalier, Late emission from the type Ib/C SN 2001em: overtaking the hydrogen envelope. Astrophys. J. **641**, 1051–1059 (2006). https://doi.org/10.1086/500539. arXiv:astro-ph/0510362

J.S.W. Claeys, S.E. de Mink, O.R. Pols, J.J. Eldridge, M. Baes, Binary progenitor models of type IIb supernovae. Astron. Astrophys. **528**, A131 (2011). https://doi.org/10.1051/0004-6361/201015410. arXiv:1102.1732

E.R. Coughlin, P.J. Armitage, Tidal disruption by extreme mass ratio binaries and application to ASASSN-15lh (2017). ArXiv e-prints, arXiv:1705.04689

L. Dessart, A magnetar model for the hydrogen-rich super-luminous supernova iPTF14hls (2018). ArXiv e-prints, arXiv:1801.05340

S.G. Djorgovski, A.J. Drake, A.A. Mahabal, M.J. Graham, C. Donalek, R. Williams, E.C. Beshore, S.M. Larson, J. Prieto, M. Catelan, E. Christensen, R.H. McNaught, The Catalina Real-Time Transient Survey (CRTS) (2011). ArXiv e-prints. arXiv:1102.5004

S. Dong, B.J. Shappee, J.L. Prieto, S.W. Jha, K.Z. Stanek, T.W.S. Holoien, C.S. Kochanek, T.A. Thompson, N. Morrell, I.B. Thompson, U. Basu, J.F. Beacom, D. Bersier, J. Brimacombe, J.S. Brown, F. Bufano, P. Chen, E. Conseil, A.B. Danilet, E. Falco, D. Grupe, S. Kiyota, G. Masi, B. Nicholls, E.F. Olivares, G. Pignata, G. Pojmanski, G.V. Simonian, D.M. Szczygiel, P.R. Woźniak, ASASSN-15lh: a highly super-luminous supernova. Science **351**, 257–260 (2016). https://doi.org/10.1126/science.aac9613. arXiv:1507.03010

M.R. Drout, R. Chornock, A.M. Soderberg, N.E. Sanders, R. McKinnon, A. Rest, R.J. Foley, D. Milisavljevic, R. Margutti, E. Berger, M. Calkins, W. Fong, S. Gezari, M.E. Huber, E. Kankare, R.P. Kirshner, C. Leibler, R. Lunnan, S. Mattila, G.H. Marion, G. Narayan, A.G. Riess, K.C. Roth, D. Scolnic, S.J. Smartt, J.L. Tonry, W.S. Burgett, K.C. Chambers, K.W. Hodapp, R. Jedicke, N. Kaiser, E.A. Magnier, N. Metcalfe, J.S. Morgan, P.A. Price, C. Waters, Rapidly evolving and luminous transients from Pan-STARRS1. Astrophys. J. **794**, 23 (2014). https://doi.org/10.1088/0004-637X/794/1/23. arXiv:1405.3668

J.J. Eldridge, E.R. Stanway, L. Xiao, L.A.S. McClelland, G. Taylor, M. Ng, S.M.L. Greis, J.C. Bray, Binary population and spectral synthesis version 2.1: construction, observational verification, and new results. Publ. Astron. Soc. Aust. **34**, e058 (2017). https://doi.org/10.1017/pasa.2017.51. arXiv:1710.02154

M. Ergon, A. Jerkstrand, J. Sollerman, N. Elias-Rosa, C. Fransson, M. Fraser, A. Pastorello, R. Kotak, S. Taubenberger, L. Tomasella, S. Valenti, S. Benetti, G. Helou, M.M. Kasliwal, J. Maund, S.J. Smartt, J. Spyromilio, The Type IIb SN 2011dh: two years of observations and modelling of the lightcurves. Astron. Astrophys. **580**, A142 (2015). https://doi.org/10.1051/0004-6361/201424592. arXiv:1408.0731

R.A. Fesen, R.H. Becker, Optical detection of the remnant of SN 1980K in NGC 6946, in *Bulletin of the American Astronomical Society*, vol. 20 (1988), p. 962

A.V. Filippenko, Optical spectra of supernovae. Annu. Rev. Astron. Astrophys. **35**, 309–355 (1997). https://doi.org/10.1146/annurev.astro.35.1.309

A.V. Filippenko, W.L.W. Sargent, A peculiar supernova in the spiral galaxy NGC4618. Nature **316**, 407–412 (1985). https://doi.org/10.1038/316407a0

A.V. Filippenko, W.L.W. Sargent, The unique supernova (1985f) in NGC 4618. Astron. J. **91**, 691–696 (1986). https://doi.org/10.1086/114051

A.V. Filippenko, A.C. Porter, W.L.W. Sargent, The type IC (helium-poor Ib) supernova 1987M—transition to the supernebular phase. Astron. J. **100**, 1575–1587 (1990). https://doi.org/10.1086/115618

A.V. Filippenko, M.W. Richmond, D. Branch, M. Gaskell, W. Herbst, C.H. Ford, R.R. Treffers, T. Matheson, L.C. Ho, A. Dey, W.L.W. Sargent, T.A. Small, W.J.M. van Breugel, The subluminous, spectroscopically peculiar type IA supernova 1991bg in the elliptical galaxy NGC 4374. Astron. J. **104**, 1543–1556 (1992a). https://doi.org/10.1086/116339

A.V. Filippenko, M.W. Richmond, T. Matheson, J.C. Shields, E.M. Burbidge, R.D. Cohen, M. Dickinson, M.A. Malkan, B. Nelson, J. Pietz, D. Schlegel, P. Schmeer, H. Spinrad, C.C. Steidel, H.D. Tran, W. Wren, The peculiar Type IA SN 1991T—detonation of a white dwarf? Astrophys. J. **384**, L15–L18 (1992). https://doi.org/10.1086/186252

A.V. Filippenko, R. Chornock, B. Swift, M. Modjaz, R. Simcoe, M. Rauch, Supernovae 2001co, 2003H, 2003dg, and 2003dr. IAU Circ. 8159 (2003)

R.J. Foley, Kinematics and host-galaxy properties suggest a nuclear origin for calcium-rich supernova progenitors. Mon. Not. R. Astron. Soc. **452**, 2463–2478 (2015). https://doi.org/10.1093/mnras/stv789. arXiv:1501.07607

R.J. Foley, N. Smith, M. Ganeshalingam, W. Li, R. Chornock, A.V. Filippenko, SN 2006jc: a Wolf–Rayet star exploding in a dense He-rich circumstellar medium Astrophys. J. **657**, L105–L108 (2007). https://doi.org/10.1086/513145. arXiv:astro-ph/0612711

R.J. Foley, G. Narayan, P.J. Challis, A.V. Filippenko, R.P. Kirshner, J.M. Silverman, T.N. Steele, SN 2006bt: a perplexing, troublesome, and possibly misleading Type Ia supernova. Astrophys. J. **708**, 1748–1759 (2010). https://doi.org/10.1088/0004-637X/708/2/1748. arXiv:0912.0263

R.J. Foley, P.J. Challis, R. Chornock, M. Ganeshalingam, W. Li, G.H. Marion, N.I. Morrell, G. Pignata, M.D. Stritzinger, J.M. Silverman, X. Wang, J.P. Anderson, A.V. Filippenko, W.L. Freedman, M. Hamuy, S.W. Jha, R.P. Kirshner, C. McCully, S.E. Persson, M.M. Phillips, D.E. Reichart, A.M. Soderberg, Type Iax supernovae: a new class of stellar explosion. Astrophys. J. **767**, 57 (2013). https://doi.org/10.1088/0004-637X/767/1/57. arXiv:1212.2209

O.D. Fox, J.M. Silverman, A.V. Filippenko, J. Mauerhan, J. Becker, H.J. Borish, S.B. Cenko, K.I. Clubb, M. Graham, E. Hsiao, P.L. Kelly, W.H. Lee, G.H. Marion, D. Milisavljevic, J. Parrent, I. Shivvers, M. Skrutskie, N. Smith, J. Wilson, W. Zheng, On the nature of Type IIn/Ia-CSM supernovae: optical and near-infrared spectra of SN 2012ca and SN 2013dn. Mon. Not. R. Astron. Soc. **447**, 772–785 (2015). https://doi.org/10.1093/mnras/stu2435. arXiv:1408.6239

C. Fransson, R.A. Chevalier, Late emission from supernovae—a window on stellar nucleosynthesis. Astrophys. J. **343**, 323–342 (1989). https://doi.org/10.1086/167707

A. Gal-Yam, Luminous supernovae. Science **337**, 927 (2012). https://doi.org/10.1126/science.1203601. arXiv:1208.3217

A. Gal-Yam, Observational and Physical Classification of Supernovae (2016). ArXiv e-prints, arXiv:1611.09353

A. Gal-Yam, P. Mazzali, E.O. Ofek, P.E. Nugent, S.R. Kulkarni, M.M. Kasliwal, R.M. Quimby, A.V. Filippenko, S.B. Cenko, R. Chornock, R. Waldman, D. Kasen, M. Sullivan, E.C. Beshore, A.J. Drake, R.C. Thomas, J.S. Bloom, D. Poznanski, A.A. Miller, R.J. Foley, J.M. Silverman, I. Arcavi, R.S. Ellis, J. Deng, Supernova 2007bi as a pair-instability explosion. Nature **462**, 624–627 (2009). https://doi.org/10.1038/nature08579. arXiv:1001.1156

A. Gal-Yam, I. Arcavi, E.O. Ofek, S. Ben-Ami, S.B. Cenko, M.M. Kasliwal, Y. Cao, O. Yaron, D. Tal, J.M. Silverman, A. Horesh, A. De Cia, F. Taddia, J. Sollerman, D. Perley, P.M. Vreeswijk, S.R. Kulkarni, P.E. Nugent, A.V. Filippenko, J.C. Wheeler, A Wolf–Rayet-like progenitor of SN 2013cu from spectral observations of a stellar wind. Nature **509**, 471–474 (2014). https://doi.org/10.1038/nature13304. arXiv:1406.7640

P.M. Garnavich, B.E. Tucker, A. Rest, E.J. Shaya, R.P. Olling, D. Kasen, A. Villar, Shock breakout and early light curves of Type II-P supernovae observed with Kepler. Astrophys. J. **820**, 23 (2016). https://doi.org/10.3847/0004-637X/820/1/23. arXiv:1603.05657

C.M. Gaskell, E. Cappellaro, H.L. Dinerstein, D.R. Garnett, R.P. Harkness, J.C. Wheeler, Type Ib supernovae 1983n and 1985f—oxygen-rich late time spectra. Astrophys. J. **306**, L77–L80 (1986). https://doi.org/10.1086/184709

D. Godoy-Rivera, K.Z. Stanek, C.S. Kochanek, P. Chen, S. Dong, J.L. Prieto, B.J. Shappee, S.W. Jha, R.J. Foley, Y.C. Pan, T.W.S. Holoien, T.A. Thompson, D. Grupe, J.F. Beacom, The unexpected, long-lasting, UV rebrightening of the superluminous supernova ASASSN-15lh. Mon. Not. R. Astron. Soc. **466**, 1428–1443 (2017). https://doi.org/10.1093/mnras/stw3237. arXiv:1605.00645

M.L. Graham, S. Kumar, G. Hosseinzadeh, D. Hiramatsu, I. Arcavi, D.A. Howell, S. Valenti, D.J. Sand, J.T. Parrent, C. McCully, A.V. Filippenko, Nebular-phase spectra of nearby Type Ia supernovae. Mon. Not. R. Astron. Soc. **472**, 3437–3454 (2017). https://doi.org/10.1093/mnras/stx2224. arXiv:1708.07799

J. Granot, E. Ramirez-Ruiz, The case for a misaligned relativistic jet from SN 2001em. Astrophys. J. **609**, L9–L12 (2004). https://doi.org/10.1086/422516. arXiv:astro-ph/0403421

J. Greiner, P.A. Mazzali, D.A. Kann, T. Krühler, E. Pian, S. Prentice, E.F. Olivares, A. Rossi, S. Klose, S. Taubenberger, F. Knust, P.M.J. Afonso, C. Ashall, J. Bolmer, C. Delvaux, R. Diehl, J. Elliott, R. Filgas, J.P.U. Fynbo, J.F. Graham, A.N. Guelbenzu, S. Kobayashi, G. Leloudas, S. Savaglio, P. Schady, S. Schmidl, T. Schweyer, V. Sudilovsky, M. Tanga, A.C. Updike, H. van Eerten, K. Varela, A very luminous magnetar-powered supernova associated with an ultra-long γ-ray burst. Nature **523**, 189–192 (2015). https://doi.org/10.1038/nature14579. arXiv:1509.03279

R.P. Harkness, J.C. Wheeler, B. Margon, R.A. Downes, R.P. Kirshner, A. Uomoto, E.S. Barker, A.L. Cochran, H.L. Dinerstein, D.R. Garnett, R.M. Levreault, The early spectral phase of type Ib supernovae—evidence for helium. Astrophys. J. **317**, 355–367 (1987). https://doi.org/10.1086/165283

A. Heger, S.E. Woosley, The nucleosynthetic signature of population III. Astrophys. J. **567**, 532–543 (2002). https://doi.org/10.1086/338487. arXiv:astro-ph/0107037

C. Inserra, S.J. Smartt, A. Jerkstrand, S. Valenti, M. Fraser, D. Wright, K. Smith, T.W. Chen, R. Kotak, A. Pastorello, M. Nicholl, F. Bresolin, R.P. Kudritzki, S. Benetti, M.T. Botticella, W.S. Burgett, K.C. Chambers, M. Ergon, H. Flewelling, J.P.U. Fynbo, S. Geier, K.W. Hodapp, D.A. Howell, M. Huber, N. Kaiser, G. Leloudas, L. Magill, E.A. Magnier, M.G. McCrum, N. Metcalfe, P.A. Price, A. Rest, J. Sollerman, W. Sweeney, F. Taddia, S. Taubenberger, J.L. Tonry, R.J. Wainscoat, C. Waters, D. Young, Super-luminous Type Ic supernovae: catching a magnetar by the tail. Astrophys. J. **770**, 128 (2013). https://doi.org/10.1088/0004-637X/770/2/128. arXiv:1304.3320

C. Inserra, M. Fraser, S.J. Smartt, S. Benetti, T.W. Chen, M. Childress, A. Gal-Yam, D.A. Howell, T. Kangas, G. Pignata, J. Polshaw, M. Sullivan, K.W. Smith, S. Valenti, D.R. Young, S. Parker, T. Seccull, M. McCrum, On Type IIn/Ia-CSM supernovae as exemplified by SN 2012ca*. Mon. Not. R. Astron. Soc. **459**, 2721–2740 (2016). https://doi.org/10.1093/mnras/stw825. arXiv:1510.01109

S. Jha, D. Branch, R. Chornock, R.J. Foley, W. Li, B.J. Swift, D. Casebeer, A.V. Filippenko, Late-time spectroscopy of SN 2002cx: the prototype of a new subclass of Type Ia supernovae. Astron. J. **132**, 189–196 (2006). https://doi.org/10.1086/504599. arXiv:astro-ph/0602250

A. Kamble, A.M. Soderberg, L. Chomiuk, R. Margutti, M. Medvedev, D. Milisavljevic, S. Chakraborti, R. Chevalier, N. Chugai, J. Dittmann, M. Drout, C. Fransson, E. Nakar, N. Sanders, Radio observations reveal a smooth circumstellar environment around the extraordinary Type Ib supernova 2012au. Astrophys. J. **797**, 2 (2014). https://doi.org/10.1088/0004-637X/797/1/2

D. Kasen, L. Bildsten, Supernova light curves powered by Young magnetars. Astrophys. J. **717**, 245–249 (2010). https://doi.org/10.1088/0004-637X/717/1/245. arXiv:0911.0680

M.M. Kasliwal, S.R. Kulkarni, A. Gal-Yam, P.E. Nugent, M. Sullivan, L. Bildsten, O. Yaron, H.B. Perets, I. Arcavi, S. Ben-Ami, V.B. Bhalerao, J.S. Bloom, S.B. Cenko, A.V. Filippenko, D.A. Frail, M. Ganeshalingam, A. Horesh, D.A. Howell, N.M. Law, D.C. Leonard, W. Li, E.O. Ofek, D. Polishook, D. Poznanski, R.M. Quimby, J.M. Silverman, A. Sternberg, D. Xu, Calcium-rich gap transients in the remote outskirts of galaxies. Astrophys. J. **755**, 161 (2012). https://doi.org/10.1088/0004-637X/755/2/161. arXiv:1111.6109

K.S. Kawabata, K. Maeda, K. Nomoto, S. Taubenberger, M. Tanaka, J. Deng, E. Pian, T. Hattori, K. Itagaki, A massive star origin for an unusual helium-rich supernova in an elliptical galaxy. Nature **465**, 326–328 (2010). https://doi.org/10.1038/nature09055. arXiv:0906.2811

N. Kawahara, M. Yamanaka, K. Kawabata, T. Nakaoka, M. Kawabata, K. Maeda, K. Takaki, H. Akitaya, R. Itoh, Y. Moritani, M. Uemura, M. Yoshida, Optical and near-infrared study of the Ca-rich transient iPTF15eqv in the early phase, in *American Astronomical Society Meeting Abstracts*, vol. 231 (2018). 446.03

S. Komossa, Tidal disruption of stars by supermassive black holes: status of observations. J. High Energy Astrophys. **7**, 148–157 (2015). https://doi.org/10.1016/j.jheap.2015.04.006. arXiv:1505.01093

A. Kozyreva, R. Hirschi, S. Blinnikov, J. den Hartogh, How much radioactive nickel does ASASSN-15lh require? Mon. Not. R. Astron. Soc. **459**, L21–L25 (2016). https://doi.org/10.1093/mnrasl/slw036. arXiv:1603.00335

G. Leloudas, M. Fraser, N.C. Stone, S. van Velzen, P.G. Jonker, I. Arcavi, C. Fremling, J.R. Maund, S.J. Smartt, T. Krìhler, J.C.A. Miller-Jones, P.M. Vreeswijk, A. Gal-Yam, P.A. Mazzali, A. De Cia, D.A. Howell, C. Inserra, F. Patat, A. de Ugarte Postigo, O. Yaron, C. Ashall, I. Bar, H. Campbell, T.W. Chen, M. Childress, N. Elias-Rosa, J. Harmanen, G. Hosseinzadeh, J. Johansson, T. Kangas, E. Kankare, S. Kim, H. Kuncarayakti, J. Lyman, M.R. Magee, K. Maguire, D. Malesani, S. Mattila, C.V. McCully, M. Nicholl, S. Prentice, C. Romero-Cañizales, S. Schulze, K.W. Smith, J. Sollerman, M. Sullivan, B.E. Tucker, S. Valenti, J.C. Wheeler, D.R. Young, The superluminous transient ASASSN-15lh as a tidal disruption event from a Kerr black hole. Nat. Astron. **1**, 0002 (2016). https://doi.org/10.1038/s41550-016-0002. arXiv:1609.02927

W.D. Li, A.V. Filippenko, R.R. Treffers, A. Friedman, E. Halderson, R.A. Johnson, J.Y. King, M. Modjaz, M. Papenkova, Y. Sato, T. Shefler, The Lick Observatory Supernova Search, in *American Institute of Physics Conference Series*, ed. by S.S. Holt, W.W. Zhang, vol. 522 (2000), pp. 103–106. https://doi.org/10.1063/1.1291702. arXiv:astro-ph/9912336

K.S. Long, W.P. Blair, W. Krzeminski, Discovery of optical emission from the remnant of SN 1957D in M83. Astrophys. J. **340**, L25–L28 (1989). https://doi.org/10.1086/185430

R. Lunnan, R. Chornock, E. Berger, A. Rest, W. Fong, D. Scolnic, D.O. Jones, A.M. Soderberg, P.M. Challis, M.R. Drout, R.J. Foley, M.E. Huber, R.P. Kirshner, C. Leibler, G.H. Marion, M. McCrum, D. Milisavljevic, G. Narayan, N.E. Sanders, S.J. Smartt, K.W. Smith, J.L. Tonry, W.S. Burgett, K.C. Chambers, H. Flewelling, R.P. Kudritzki, R.J. Wainscoat, C. Waters, Zooming in on the progenitors of superluminous supernovae with the HST. Astrophys. J. **804**, 90 (2015). https://doi.org/10.1088/0004-637X/804/2/90. arXiv:1411.1060

R. Lunnan, R. Chornock, E. Berger, D. Milisavljevic, D.O. Jones, A. Rest, W. Fong, C. Fransson, R. Margutti, M.R. Drout, P.K. Blanchard, P. Challis, P.S. Cowperthwaite, R.J. Foley, R.P. Kirshner, N. Morrell, A.G. Riess, K.C. Roth, D. Scolnic, S.J. Smartt, K.W. Smith, V.A. Villar, K.C. Chambers, P.W. Draper, M.E. Huber, N. Kaiser, R.P. Kudritzki, E.A. Magnier, N. Metcalfe, C. Waters, PS1-14bj: a hydrogen-poor superluminous supernova with a long rise and slow decay. Astrophys. J. **831**, 144 (2016). https://doi.org/10.3847/0004-637X/831/2/144. arXiv:1605.05235

R. Lunnan, M.M. Kasliwal, Y. Cao, L. Hangard, O. Yaron, J.T. Parrent, C. McCully, A. Gal-Yam, J.S. Mulchaey, S. Ben-Ami, A.V. Filippenko, C. Fremling, A.S. Fruchter, D.A. Howell, J. Koda, T. Kupfer, S.R. Kulkarni, R. Laher, F. Masci, P.E. Nugent, E.O. Ofek, M. Yagi, L. Yan, Two new calcium-rich gap transients in group and cluster environments. Astrophys. J. **836**, 60 (2017). https://doi.org/10.3847/1538-4357/836/1/60. arXiv:1612.00454

J.D. Lyman, P.A. James, H.B. Perets, J.P. Anderson, A. Gal-Yam, P. Mazzali, S.M. Percival, Environment-derived constraints on the progenitors of low-luminosity Type I supernovae. Mon. Not. R. Astron. Soc. **434**, 527–541 (2013). https://doi.org/10.1093/mnras/stt1038. arXiv:1306.2474

J.D. Lyman, A.J. Levan, R.P. Church, M.B. Davies, N.R. Tanvir, The progenitors of calcium-rich transients are not formed in situ*. Mon. Not. R. Astron. Soc. **444**, 2157–2166 (2014). https://doi.org/10.1093/mnras/stu1574. arXiv:1408.1424

J.D. Lyman, A.J. Levan, P.A. James, C.R. Angus, R.P. Church, M.B. Davies, N.R. Tanvir, Hubble Space Telescope observations of the host galaxies and environments of calcium-rich supernovae. Mon. Not. R. Astron. Soc. **458**, 1768–1777 (2016). https://doi.org/10.1093/mnras/stw477. arXiv:1602.08098

M. MacLeod, J. Goldstein, E. Ramirez-Ruiz, J. Guillochon, J. Samsing, Illuminating massive black holes with white dwarfs: orbital dynamics and high-energy transients from tidal interactions. Astrophys. J. **794**, 9 (2014). https://doi.org/10.1088/0004-637X/794/1/9. arXiv:1405.1426

K. Maeda, M. Tanaka, K. Nomoto, N. Tominaga, K. Kawabata, P.A. Mazzali, H. Umeda, T. Suzuki, T. Hattori, The unique Type Ib supernova 2005bf at nebular phases: a possible birth event of a strongly magnetized neutron star. Astrophys. J. **666**, 1069–1082 (2007). https://doi.org/10.1086/520054. arXiv:0705.2713

K. Maeda, K. Kawabata, P.A. Mazzali, M. Tanaka, S. Valenti, K. Nomoto, T. Hattori, J. Deng, E. Pian, S. Taubenberger, M. Iye, T. Matheson, A.V. Filippenko, K. Aoki, G. Kosugi, Y. Ohyama, T. Sasaki, T. Takata, Asphericity in supernova explosions from late-time spectroscopy. Science **319**, 1220 (2008). https://doi.org/10.1126/science.1149437. arXiv:0801.1100

R. Margutti, A.M. Soderberg, L. Chomiuk, R. Chevalier, K. Hurley, D. Milisavljevic, R.J. Foley, J.P. Hughes, P. Slane, C. Fransson, M. Moe, S. Barthelmy, W. Boynton, M. Briggs, V. Connaughton, E. Costa, J. Cummings, E. Del Monte, H. Enos, C. Fellows, M. Feroci, Y. Fukazawa, N. Gehrels, J. Goldsten, D. Golovin, Y. Hanabata, K. Harshman, H. Krimm, M.L. Litvak, K. Makishima, M. Marisaldi, I.G. Mitrofanov, T. Murakami, M. Ohno, D.M. Palmer, A.B. Sanin, R. Starr, D. Svinkin, T. Takahashi, M. Tashiro, Y. Terada, K. Yamaoka, Inverse compton X-ray emission from supernovae with compact progenitors: application to SN2011fe. Astrophys. J. **751**, 134 (2012). https://doi.org/10.1088/0004-637X/751/2/134. arXiv:1202.0741

R. Margutti, D. Milisavljevic, A.M. Soderberg, R. Chornock, B.A. Zauderer, K. Murase, C. Guidorzi, N.E. Sanders, P. Kuin, C. Fransson, E.M. Levesque, P. Chandra, E. Berger, F.B. Bianco, P.J. Brown, P. Challis, E. Chatzopoulos, C.C. Cheung, C. Choi, L. Chomiuk, N. Chugai, C. Contreras, M.R. Drout, R. Fesen, R.J. Foley, W. Fong, A.S. Friedman, C. Gall, N. Gehrels, J. Hjorth, E. Hsiao, R. Kirshner, M. Im, G. Leloudas, R. Lunnan, G.H. Marion, J. Martin, N. Morrell, K.F. Neugent, N. Omodei, M.M. Phillips, A. Rest, J.M. Silverman, J. Strader, M.D. Stritzinger, T. Szalai, N.B. Utterback, J. Vinko, J.C. Wheeler, D. Arnett, S. Campana, R. Chevalier, A. Ginsburg, A. Kamble, P.W.A. Roming, T. Pritchard, G. Stringfellow, A panchromatic view of the restless SN 2009ip reveals the explosive ejection of a massive star envelope. Astrophys. J. **780**, 21 (2014). https://doi.org/10.1088/0004-637X/780/1/21. arXiv:1306.0038

R. Margutti, J. Parrent, A. Kamble, A.M. Soderberg, R.J. Foley, D. Milisavljevic, M.R. Drout, R. Kirshner, No X-rays from the very nearby type Ia SN 2014J: constraints on its environment. Astrophys. J. **790**, 52 (2014b). https://doi.org/10.1088/0004-637X/790/1/52. arXiv:1405.1488

R. Margutti, A. Kamble, D. Milisavljevic, E Zapartas, S.E. de Mink, M. Drout, R. Chornock, G. Risaliti, B.A. Zauderer, M. Bietenholz, M. Cantiello, S. Chakraborti, L. Chomiuk, W. Fong, B. Grefenstette, C. Guidorzi, R. Kirshner, J.T. Parrent, D. Patnaude, A.M. Soderberg, N.C. Gehrels, Ejection of the massive hydrogen-rich envelope timed with the collapse of the stripped SN 2014C. Astrophys. J. **835**, 140 (2017a). https://doi.org/10.3847/1538-4357/835/2/140. arXiv:1601.06806

R. Margutti, B.D. Metzger, R. Chornock, D. Milisavljevic, E. Berger, P.K. Blanchard, C. Guidorzi, G. Migliori, A. Kamble, R. Lunnan, M. Nicholl, D.L. Coppejans, S. Dall'Osso, M.R. Drout, R. Perna, B. Sbarufatti, X-rays from the location of the double-humped transient ASASSN-15lh. Astrophys. J. **836**, 25 (2017b). https://doi.org/10.3847/1538-4357/836/1/25. arXiv:1610.01632

T. Matheson, A.V. Filippenko, W. Li, D.C. Leonard, J.C. Shields, Optical spectroscopy of type IB/C supernovae. Astron. J. **121**, 1648–1675 (2001). https://doi.org/10.1086/319390. arXiv:astro-ph/0101119

P.A. Mazzali, M. Sullivan, A.V. Filippenko, P.M. Garnavich, K.I. Clubb, K. Maguire, Y.-C. Pan, B. Shappee, J.M. Silverman, S. Benetti, S. Hachinger, K. Nomoto, E. Pian, Nebular spectra and abundance tomography of the Type Ia supernova SN 2011fe: a normal SN Ia with a stable Fe core. Mon. Not. R. Astron. Soc. **450**, 2631–2643 (2015). https://doi.org/10.1093/mnras/stv761. arXiv:1504.04857

D.B. McLaughlin, The spectrum of the supernova of 1954 in NGC 4214. Publ. Astron. Soc. Pac. **75**, 133 (1963). https://doi.org/10.1086/127916

C.A. Meakin, Hydrodynamic modeling of massive star interiors. PhD thesis, The University of Arizona, Arizona, USA (2006)

C.A. Meakin, D. Arnett, Turbulent convection in stellar interiors. I. Hydrodynamic simulation. Astrophys. J. **667**, 448–475 (2007). https://doi.org/10.1086/520318. arXiv:astro-ph/0611315

D. Milisavljevic, The progenitor systems and explosion mechanisms of supernovae, in *New Horizons in Astronomy*, BASH 2013 (2013), p. 9

D. Milisavljevic, R.A. Fesen, The supernova—supernova remnant connection (2017). ArXiv e-prints, arXiv:1701.00891. http://adsabs.harvard.edu/abs/2017arXiv170100891M. Provided by the SAO/NASA Astrophysics Data System

D. Milisavljevic, R.A. Fesen, B. Leibundgut, R.P. Kirshner, The evolution of late-time optical emission from SN 1986J. Astrophys. J. **684**, 1170–1173 (2008). https://doi.org/10.1086/590426. arXiv:0804.1545

D. Milisavljevic, R.A. Fesen, C.L. Gerardy, R.P. Kirshner, P. Challis, Doublets and double peaks: late-time [O I] λλ6300, 6364 line profiles of stripped-envelope, core-collapse supernovae. Astrophys. J. **709**, 1343–1355 (2010). https://doi.org/10.1088/0004-637X/709/2/1343. arXiv:0904.4256

D. Milisavljevic, R.A. Fesen, R.A. Chevalier, R.P. Kirshner, P. Challis, M. Turatto, Late-time optical emission from core-collapse supernovae. Astrophys. J. **751**, 25 (2012). https://doi.org/10.1088/0004-637X/751/1/25. arXiv:1203.0006

D. Milisavljevic, R. Margutti, A.M. Soderberg, G. Pignata, L. Chomiuk, R.A. Fesen, F. Bufano, N.E. Sanders, J.T. Parrent, S. Parker, P. Mazzali, E. Pian, T. Pickering, D.A.H. Buckley, S.M. Crawford, A.A.S. Gulbis, C. Hettlage, E. Hooper, K.H. Nordsieck, D. O'Donoghue, T.O. Husser, S. Potter, A. Kniazev, P. Kotze, E. Romero-Colmenero, P. Vaisanen, M. Wolf, M.F. Bietenholz, N. Bartel, C. Fransson, E.S. Walker, A. Brunthaler, S. Chakraborti, E.M. Levesque, A. MacFadyen, C. Drescher, G. Bock, P. Marples, J.P. Anderson, S. Benetti, D. Reichart, K. Ivarsen, Multi-wavelength observations of supernova 2011ei: time-dependent classification of Type IIb and Ib supernovae and implications for their progenitors. Astrophys. J. **767**, 71 (2013a). https://doi.org/10.1088/0004-637X/767/1/71. arXiv:1207.2152

D. Milisavljevic, A.M. Soderberg, R. Margutti, M.R. Drout, G. Howie Marion, N.E. Sanders, E.Y. Hsiao, R. Lunnan, R. Chornock, R.A. Fesen, J.T. Parrent, E.M. Levesque, E. Berger, R.J. Foley, P. Challis, R.P. Kirshner, J. Dittmann, A. Bieryla, A. Kamble, S. Chakraborti, G. De Rosa, M. Fausnaugh, K.N. Hainline, C.T. Chen, R.C. Hickox, N. Morrell, M.M. Phillips, M. Stritzinger, SN 2012au: a golden link between superluminous supernovae and their lower-luminosity counterparts. Astrophys. J. **770**, L38 (2013b). https://doi.org/10.1088/2041-8205/770/2/L38. arXiv:1304.0095

D. Milisavljevic, R. Margutti, A. Kamble, D.J. Patnaude, J.C. Raymond, J.J. Eldridge, W. Fong, M. Bietenholz, P. Challis, R. Chornock, M.R. Drout, C. Fransson, R.A. Fesen, J.E. Grindlay, R.P. Kirshner, R. Lunnan, J. Mackey, G.F. Miller, J.T. Parrent, N.E. Sanders, A.M. Soderberg, B.A. Zauderer, Metamorphosis of SN 2014C: delayed interaction between a hydrogen poor core-collapse supernova and a nearby circumstellar shell. Astrophys. J. **815**, 120 (2015a). https://doi.org/10.1088/0004-637X/815/2/120. arXiv:1511.01907

D. Milisavljevic, R. Margutti, J.T. Parrent, A.M. Soderberg, R.A. Fesen, P. Mazzali, K. Maeda, N.E. Sanders, S.B. Cenko, J.M. Silverman, A.V. Filippenko, A. Kamble, S. Chakraborti, M.R. Drout, R.P. Kirshner, T.E. Pickering, K. Kawabata, T. Hattori, E.Y. Hsiao, M.D. Stritzinger, G.H. Marion, J. Vinko, J.C. Wheeler, The broad-lined type Ic SN 2012ap and the nature of relativistic supernovae lacking a gamma-ray burst detection. Astrophys. J. **799**, 51 (2015b). https://doi.org/10.1088/0004-637X/799/1/51. arXiv:1408.1606

D. Milisavljevic, D.J. Patnaude, J.C. Raymond, M.R. Drout, R. Margutti, A. Kamble, R. Chornock, J. Guillochon, N.E. Sanders, J.T. Parrent, L. Lovisari, I.V. Chilingarian, P. Challis, R.P. Kirshner, M.T. Penny, K. Itagaki, J.J. Eldridge, T.J. Moriya, iPTF15eqv: multiwavelength exposé of a peculiar calcium-rich transient. Astrophys. J. **846**, 50 (2017). https://doi.org/10.3847/1538-4357/aa7d9f. arXiv:1706.01887

R. Minkowski, Spectra of supernovae. Publ. Astron. Soc. Pac. **53**, 224 (1941). https://doi.org/10.1086/125315

M. Modjaz, R.P. Kirshner, S. Blondin, P. Challis, T. Matheson, Double-peaked oxygen lines are not rare in nebular spectra of core-collapse supernovae. Astrophys. J. **687**, L9 (2008). https://doi.org/10.1086/593135. arXiv:0801.0221

M. Nicholl, S.J. Smartt, A. Jerkstrand, C. Inserra, M. McCrum, R. Kotak, M. Fraser, D. Wright, T.W. Chen, K. Smith, D.R. Young, S.A. Sim, S. Valenti, D.A. Howell, F. Bresolin, R.P. Kudritzki, J.L. Tonry, M.E. Huber, A. Rest, A. Pastorello, L. Tomasella, E. Cappellaro, S. Benetti, S. Mattila, E. Kankare, T. Kangas, G. Leloudas, J. Sollerman, F. Taddia, E. Berger, R. Chornock, G. Narayan, C.W. Stubbs, R.J. Foley, R. Lunnan, A. Soderberg, N. Sanders, D. Milisavljevic, R. Margutti, R.P. Kirshner, N. Elias-Rosa, A. Morales-Garoffolo, S. Taubenberger, M.T. Botticella, S. Gezari, Y. Urata, S. Rodney, A.G. Riess, D. Scolnic, W.M. Wood-Vasey, W.S. Burgett, K. Chambers, H.A. Flewelling, E.A. Magnier, N. Kaiser, N. Metcalfe, J. Morgan, P.A. Price, W. Sweeney, C. Waters, Slowly fading super-luminous supernovae that are not pair-instability explosions. Nature **502**, 346–349 (2013). https://doi.org/10.1038/nature12569. arXiv:1310.4446

M. Nicholl, S.J. Smartt, A. Jerkstrand, S.A. Sim, C. Inserra, J.P. Anderson, C. Baltay, S. Benetti, K. Chambers, T.W. Chen, N. Elias-Rosa, U. Feindt, H.A. Flewelling, M. Fraser, A. Gal-Yam, L. Galbany, M.E. Huber, T. Kangas, E. Kankare, R. Kotak, T. Krühler, K. Maguire, R. McKinnon, D. Rabinowitz,

S. Rostami, S. Schulze, K.W. Smith, M. Sullivan, J.L. Tonry, S. Valenti, D.R. Young, LSQ14bdq: a type Ic super-luminous supernova with a double-peaked light curve. Astrophys. J. **807**, L18 (2015). https://doi.org/10.1088/2041-8205/807/1/L18. arXiv:1505.01078

M. Nicholl, E. Berger, R. Margutti, R. Chornock, P.K. Blanchard, A. Jerkstrand, S.J. Smartt, I. Arcavi, P. Challis, K.C. Chambers, T.W. Chen, P.S. Cowperthwaite, A. Gal-Yam, G. Hosseinzadeh, D.A. Howell, C. Inserra, E. Kankare, E.A. Magnier, K. Maguire, P.A. Mazzali, C. McCully, D. Milisavljevic, K.W. Smith, S. Taubenberger, S. Valenti, R.J. Wainscoat, O. Yaron, D.R. Young, Superluminous supernova SN 2015bn in the nebular phase: evidence for the engine-powered explosion of a stripped massive star. Astrophys. J. **828**, L18 (2016). https://doi.org/10.3847/2041-8205/828/2/L18. arXiv:1608.02995

P.E. Nugent, M. Sullivan, S.B. Cenko, R.C. Thomas, D. Kasen, D.A. Howell, D. Bersier, J.S. Bloom, S.R. Kulkarni, M.T. Kandrashoff, A.V. Filippenko, J.M. Silverman, G.W. Marcy, A.W. Howard, H.T. Isaacson, K. Maguire, N. Suzuki, J.E. Tarlton, Y.C. Pan, L. Bildsten, B.J. Fulton, J.T. Parrent, D. Sand, P. Podsiadlowski, F.B. Bianco, B. Dilday, M.L. Graham, J. Lyman, P. James, M.M. Kasliwal, N.M. Law, R.M. Quimby, I.M. Hook, E.S. Walker, P. Mazzali, E. Pian, E.O. Ofek, A. Gal-Yam, D. Poznanski, Supernova SN 2011fe from an exploding carbon-oxygen white dwarf star. Nature **480**, 344–347 (2011). https://doi.org/10.1038/nature10644. arXiv:1110.6201

N. Panagia, R.A. Sramek, K.W. Weiler, Subluminous, radio emitting Type I supernovae. Astrophys. J. **300**, L55–L58 (1986). https://doi.org/10.1086/184602

J. Parrent, B. Friesen, M. Parthasarathy, A review of type Ia supernova spectra. Astrophys. Space Sci. **351**, 1–52 (2014). https://doi.org/10.1007/s10509-014-1830-1. arXiv:1402.6337

A. Pastorello, S.J. Smartt, S. Mattila, J.J. Eldridge, D. Young, K. Itagaki, H. Yamaoka, H. Navasardyan, S. Valenti, F. Patat, I. Agnoletto, T. Augusteijn, S. Benetti, E. Cappellaro, T. Boles, J.M. Bonnet-Bidaud, M.T. Botticella, F. Bufano, C. Cao, J. Deng, M. Dennefeld, N. Elias-Rosa, A. Harutyunyan, F.P. Keenan, T. Iijima, V. Lorenzi, P.A. Mazzali, X. Meng, S. Nakano, T.B. Nielsen, J.V. Smoker, V. Stanishev, M. Turatto, D. Xu, L. Zampieri, A giant outburst two years before the core-collapse of a massive star. Nature **447**, 829–832 (2007). https://doi.org/10.1038/nature05825. arXiv:astro-ph/0703663

A. Pastorello, S.J. Smartt, M.T. Botticella, K. Maguire, M. Fraser, K. Smith, R. Kotak, L. Magill, S. Valenti, D.R. Young, S. Gezari, F. Bresolin, R. Kudritzki, D.A. Howell, A. Rest, N. Metcalfe, S. Mattila, E. Kankare, K.Y. Huang, Y. Urata, W.S. Burgett, K.C. Chambers, T. Dombeck, H. Flewelling, T. Grav, J.N. Heasley, K.W. Hodapp, N. Kaiser, G.A. Luppino, R.H. Lupton, E.A. Magnier, D.G. Monet, J.S. Morgan, P.M. Onaka, P.A. Price, P.H. Rhoads, W.A. Siegmund, C.W. Stubbs, W.E. Sweeney, J.L. Tonry, R.J. Wainscoat, M.F. Waterson, C. Waters, C.G. Wynn-Williams, Ultra-bright optical transients are linked with Type Ic supernovae. Astrophys. J. **724**, L16–L21 (2010). https://doi.org/10.1088/2041-8205/724/1/L16. arXiv:1008.2674

F. Patat, E. Cappellaro, J. Danziger, P.A. Mazzali, J. Sollerman, T. Augusteijn, J. Brewer, V. Doublier, J.F. Gonzalez, O. Hainaut, C. Lidman, B. Leibundgut, K. Nomoto, T. Nakamura, J. Spyromilio, L. Rizzi, M. Turatto, J. Walsh, T.J. Galama, J. van Paradijs, C. Kouveliotou, P.M. Vreeswijk, F. Frontera, N. Masetti, E. Palazzi, E. Pian, The metamorphosis of SN 1998bw. Astrophys. J. **555**, 900–917 (2001). https://doi.org/10.1086/321526. arXiv:astro-ph/0103111

H.B. Perets, A. Gal-Yam, P.A. Mazzali, D. Arnett, D. Kagan, A.V. Filippenko, W. Li, I. Arcavi, S.B. Cenko, D.B. Fox, D.C. Leonard, D.S. Moon, D.J. Sand, A.M. Soderberg, J.P. Anderson, P.A. James, R.J. Foley, M. Ganeshalingam, E.O. Ofek, L. Bildsten, G. Nelemans, K.J. Shen, N.N. Weinberg, B.D. Metzger, A.L. Piro, E. Quataert, M. Kiewe, D. Poznanski, A faint type of supernova from a white dwarf with a helium-rich companion. Nature **465**, 322–325 (2010). https://doi.org/10.1038/nature09056. arXiv:0906.2003

D.A. Perley, R. Quimby, L. Yan, P. Vreeswijk, A. De Cia, R. Lunnan, A. Gal-Yam, O. Yaron, A.V. Filippenko, M.L. Graham, P.E. Nugent, Host-Galaxy Properties of 32 Low-Redshift Superluminous Supernovae from the Palomar Transient Factory (2016). ArXiv e-prints, arXiv:1604.08207

P. Podsiadlowski, P.C. Joss, J.J.L. Hsu, Presupernova evolution in massive interacting binaries. Astrophys. J. **391**, 246–264 (1992). https://doi.org/10.1086/171341

A.C. Porter, A.V. Filippenko, The observational properties of Type Ib supernovae. Astron. J. **93**, 1372–1380 (1987). https://doi.org/10.1086/114420

E. Quataert, J. Shiode, Wave-driven mass loss in the last year of stellar evolution: setting the stage for the most luminous core-collapse supernovae. Mon. Not. R. Astron. Soc. **423**, L92–L96 (2012). https://doi.org/10.1111/j.1745-3933.2012.01264.x. arXiv:1202.5036

R.M. Quimby, The Texas Supernova Search. PhD thesis, The University of Texas at Austin (2006)

R.M. Quimby, S.R. Kulkarni, M.M. Kasliwal, A. Gal-Yam, I. Arcavi, M. Sullivan, P. Nugent, R. Thomas, D.A. Howell, E. Nakar, L. Bildsten, C. Theissen, N.M. Law, R. Dekany, G. Rahmer, D. Hale, R. Smith, E.O. Ofek, J. Zolkower, V. Velur, R. Walters, J. Henning, K. Bui, D. McKenna, D. Poznanski, S.B. Cenko, D. Levitan, Hydrogen-poor superluminous stellar explosions. Nature **474**, 487–489 (2011). https://doi.org/10.1038/nature10095. arXiv:0910.0059

A. Rau, S.R. Kulkarni, N.M. Law, J.S. Bloom, D. Ciardi, G.S. Djorgovski, D.B. Fox, A. Gal-Yam, C.C. Grillmair, M.M. Kasliwal, P.E. Nugent, E.O. Ofek, R.M. Quimby, W.T. Reach M. Shara, L. Bildsten, S.B. Cenko, A.J. Drake, A.V. Filippenko, D.J. Helfand, G. Helou, D.A. Howell, D. Poznanski, M. Sullivan, Exploring the optical transient sky with the palomar transient factory. Publ. Astron. Soc. Pac. **121**, 1334 (2009). https://doi.org/10.1086/605911. arXiv:0906.5355

D. Richardson, D. Branch, E. Baron, Absolute magnitude distributions and light curves of stripped-envelope supernovae. Astron. J. **131**, 2233–2244 (2006). https://doi.org/10.1086/500578. arXiv: astro-ph/0601136

S. Rosswog, E. Ramirez-Ruiz, W.R. Hix, Atypical thermonuclear supernovae from tidally crushed white dwarfs. Astrophys. J. **679**, 1385–1389 (2008). https://doi.org/10.1086/528738. arXiv:0712.2513

M.P. Rupen, J.H. van Gorkom, G.R. Knapp, J.E. Gunn, D.P. Schneider, Observations of SN 1986J in NGC 891. Astron. J. **94**, 61–70 (1987). https://doi.org/10.1086/114447

D.K. Sahu, G.C. Anupama, S. Srividya, S. Muneer, Photometric and spectroscopic evolution of the Type IIP supernova SN 2004et. Mon. Not. R. Astron. Soc. **372**, 1315–1324 (2006). https://doi.org/10.1111/j.1365-2966.2006.10937.x. arXiv:astro-ph/0608432

E.M. Schlegel, R.P. Kirshner, The type Ib supernova 1984L in NGC 991. Astron. J. **98**, 577–589 (1989). https://doi.org/10.1086/115158

P.H. Sell, T.J. Maccarone, R. Kotak, C. Knigge, D.J. Sand, Calcium-rich gap transients: tidal detonations of white dwarfs? Mon. Not. R. Astron. Soc. **450**, 4198–4206 (2015). https://doi.org/10.1093/mnras/stv902

B.J. Shappee, J.L. Prieto, D. Grupe, C.S. Kochanek, K.Z. Stanek, G. De Rosa, S. Mathur, Y. Zu, B.M. Peterson, R.W. Pogge, S. Komossa, M. Im, J. Jencson, T.W.S. Holoien, U. Basu, J.F. Beacom, D.M. Szczygieł, J. Brimacombe, S. Adams, A. Campillay, C. Choi, C. Contreras, M. Dietrich, M. Dubberley, M. Elphick, S. Foale, M. Giustini, C. Gonzalez, E. Hawkins, D.A. Howell, E.Y. Hsiao, M. Koss, K.M. Leighly, N. Morrell, D. Mudd, D. Mullins, J.M. Nugent, J. Parrent, M.M. Phillips, G. Pojmanski, W. Rosing, R. Ross, D. Sand, D.M. Terndrup, S. Valenti, Z. Walker, Y. Yoon, The man behind the curtain: X-rays drive the UV through NIR variability in the 2013 active galactic nucleus outburst in NGC 2617. Astrophys. J. **788**, 48 (2014). https://doi.org/10.1088/0004-637X/788/1/48. arXiv:1310.2241

K.J. Shen, L. Bildsten, Unstable helium shell burning on accreting white dwarfs. Astrophys. J. **699**, 1365–1373 (2009). https://doi.org/10.1088/0004-637X/699/2/1365. arXiv:0903.0654

J.M. Silverman, P.E. Nugent, A. Gal-Yam, M. Sullivan, D.A. Howell, A.V. Filippenko, I. Arcavi, S. Ben-Ami, J.S. Bloom, S.B. Cenko, Y. Cao, R. Chornock, K.I. Clubb, A.L. Coil, R.J. Foley, M.L. Graham, C.V. Griffith, A. Horesh, M.M. Kasliwal, S.R. Kulkarni, D.C. Leonard, W. Li, T. Matheson, A.A. Miller, M. Modjaz, E.O. Ofek, Y.C. Pan, D.A. Perley, D. Poznanski, R.M. Quimby, T.N. Steele, A. Sternberg, D. Xu, O. Yaron, Type Ia supernovae strongly interacting with their circumstellar medium. Astrophys. J. Suppl. Ser. **207**, 3 (2013). https://doi.org/10.1088/0067-0049/207/1/3. arXiv:1304.0763

N. Smith, Interacting Supernovae: Types IIn and Ibn (2016). ArXiv e-prints, arXiv:1612.02006

N. Smith, W.D. Arnett, Preparing for an explosion: hydrodynamic instabilities and turbulence in presupernovae. Astrophys. J. **785**, 82 (2014). https://doi.org/10.1088/0004-637X/785/2/82. arXiv:1307.5035

N. Smith, W. Li, R.J. Foley, J.C. Wheeler, D. Pooley, R. Chornock, A.V. Filippenko, J.M. Silverman, R. Quimby, J.S. Bloom, C. Hansen, SN 2006gy: discovery of the most luminous supernova ever recorded, powered by the death of an extremely massive star like η Carinae. Astrophys. J. **666**, 1116–1128 (2007). https://doi.org/10.1086/519949. arXiv:astro-ph/0612617

A.M. Soderberg, A. Gal-Yam, S.R. Kulkarni, Type Ic SN2001em (off-axis GRB jet?), optical spectrum. GRB Coordinates Network 2586, 1 (2004)

N. Soker, A. Gilkis, Explaining iPTF14hls as a common-envelope jets supernova. Mon. Not. R. Astron. Soc. **475**, 1198–1202 (2018). https://doi.org/10.1093/mnras/stx3287. arXiv:1711.05180

R.A. Sramek, N. Panagia, K.W. Weiler, Radio emission from a type I supernova—SN 1983.51 IN NGC 5236. Astrophys. J. **285**, L59–L62 (1984). https://doi.org/10.1086/184365

T. Sukhbold, T. Ertl, S.E. Woosley, J.M. Brown, H.T. Janka, Core-collapse supernovae from 9 to 120 solar masses based on neutrino-powered explosions. Astrophys. J. **821**, 38 (2016). https://doi.org/10.3847/0004-637X/821/1/38. arXiv:1510.04643

K. Takaki, K.S. Kawabata, M. Yamanaka, K. Maeda, M. Tanaka, H. Akitaya, Y. Fukazawa, R. Itoh, K. Kinugasa, Y. Moritani, T. Ohsugi, M. Sasada, M. Uemura, I. Ueno, T. Ui, T. Urano, M. Yoshida, K. Nomoto, A luminous and fast-expanding Type Ib supernova SN 2012au. Astrophys. J. **772**, L17 (2013). https://doi.org/10.1088/2041-8205/772/2/L17. arXiv:1306.5490

M. Tanaka, M. Yamanaka, K. Maeda, K.S. Kawabata, T. Hattori, T. Minezaki, S. Valenti, M. Della Valle, D.K. Sahu, G.C. Anupama, N. Tominaga, K. Nomoto, P.A. Mazzali, E. Pian, Nebular phase observations of the Type Ib supernova 2008D/X-ray transient 080109: side-viewed bipolar explosion. Astrophys. J. **700**, 1680–1685 (2009). https://doi.org/10.1088/0004-637X/700/2/1680. arXiv:0906.1061

L. Tartaglia, D.J. Sand, S. Valenti, S. Wyatt, J.P. Anderson, I. Arcavi, C. Ashall, M.T. Botticella, R. Cartier, T.W. Chen, A. Cikota, D. Coulter, M. Della Valle, R.J. Foley, A. Gal-Yam, L. Galbany, C. Gall,

J.B. Haislip, J. Harmanen, G. Hosseinzadeh, D.A. Howell, E.Y. Hsiao, C. Inserra, S.W. Jha, E. Kankare, C.D. Kilpatrick, V.V. Kouprianov, H. Kuncarayakti, T.J. Maccarone, K. Maguire, S. Mattila, P.A. Mazzali, C. McCully, A. Melandri, N. Morrell, M.M. Phillips, G. Pignata, A.L. Piro, S. Prentice, D.E. Reichart, C. Rojas-Bravo, S.J. Smartt, K.W. Smith, J. Sollerman, M.D. Stritzinger, M. Sullivan, F. Taddia, D.R. Young, The early detection and follow-up of the highly obscured Type II supernova 2016ija/DLT16am. Astrophys. J. **853**, 62 (2018). https://doi.org/10.3847/1538-4357/aaa014. arXiv:1711.03940

S. Taubenberger, The Extremes of Thermonuclear Supernovae (2017). ArXiv e-prints, arXiv:1703.00528

S. Taubenberger, S. Valenti, S. Benetti, E. Cappellaro, M. Della Valle, N. Elias-Rosa, S. Hachinger, W. Hillebrandt, K. Maeda, P.A. Mazzali, A. Pastorello, F. Patat, S.A. Sim, M. Turatto, Nebular emission-line profiles of Type Ib/c supernovae—probing the ejecta asphericity. Mon. Not. R. Astron. Soc. **397**, 677–694 (2009). https://doi.org/10.1111/j.1365-2966.2009.15003.x. arXiv:0904.4632

S. Tinyanont, M.M. Kasliwal, O.D. Fox, R. Lau, N. Smith, R. Williams, J. Jencson, D. Perley, D. Dykhoff, R. Gehrz, J. Johansson, S.D. Van Dyk, F. Masci, A.M. Cody, T. Prince, A systematic study of mid-infrared emission from core-collapse supernovae with SPIRITS. Astrophys. J. **833**, 231 (2016). https://doi.org/10.3847/1538-4357/833/2/231. arXiv:1601.03440

N. Tominaga, M. Tanaka, K. Nomoto, P.A. Mazzali, J. Deng, K. Maeda, H. Umeda, M. Modjaz, M. Hicken, P. Challis, R.P. Kirshner, W.M. Wood-Vasey, C.H. Blake, J.S. Bloom, M.F. Skrutskie, A. Szentgyorgyi, E.E. Falco, N. Inada, T. Minezaki, Y. Yoshii, K. Kawabata, M. Iye, G.C. Anupama, D.K. Sahu, T.P. Prabhu, The unique Type Ib supernova 2005bf: a WN star explosion model for peculiar light curves and spectra. Astrophys. J. **633**, L97–L100 (2005). https://doi.org/10.1086/498570. arXiv:astro-ph/0509557

M. Turatto, E. Cappellaro, I.J. Danziger, The remnant of SN 1957d in M83. Messenger **56**, 36–37 (1989)

M.H.P.M. van Putten, M. Della Valle, On extreme transient events from rotating black holes and their gravitational wave emission (2016). ArXiv e-prints, arXiv:1610.00535

L. Wang, D. Baade, P. Höflich, J.C. Wheeler, K. Kawabata, K. Nomoto, On the hydrogen emission from the Type Ia supernova SN 2002ic. Astrophys. J. **604**, L53–L56 (2004). https://doi.org/10.1086/383411. arXiv:astro-ph/0312508

K.W. Weiler, C.L. Williams, N. Panagia, C.J. Stockdale, M.T. Kelley, R.A. Sramek, S.D. Van Dyk, J.M. Marcaide, Long-term radio monitoring of SN 1993J. Astrophys. J. **671**, 1959–1980 (2007). https://doi.org/10.1086/523258. arXiv:0709.1136

J.C. Wheeler, R. Levreault, The peculiar Type I supernova in NGC 991. Astrophys. J. **294**, L17–L20 (1985). https://doi.org/10.1086/184500

J.C. Wheeler, R.P. Harkness, E.S. Barker, A.L. Cochran, D. Wills, Supernovae 1983I and 1983v—evidence for abundance variations in type Ib supernovae. Astrophys. J. **313**, L69–L73 (1987). https://doi.org/10.1086/184833

S.E. Woosley, Bright supernovae from magnetar birth. Astrophys. J. **719**, L204–L207 (2010). https://doi.org/10.1088/2041-8205/719/2/L204. arXiv:0911.0698

S.E. Woosley, Models for the Unusual Supernova iPTF14hls (2018). ArXiv e-prints, arXiv:1801.08666

S.E. Woosley, D. Kasen, Sub-Chandrasekhar mass models for supernovae. Astrophys. J. **734**, 38 (2011). https://doi.org/10.1088/0004-637X/734/1/38. arXiv:1010.5292

S.E. Woosley, A. Heger, T.A. Weaver, The evolution and explosion of massive stars. Rev. Mod. Phys. **74**, 1015–1071 (2002). https://doi.org/10.1103/RevModPhys.74.1015

O. Yaron, A. Gal-Yam, WISeREP—an interactive supernova data repository. Publ. Astron. Soc. Pac. **124**, 668–681 (2012). https://doi.org/10.1086/666656. arXiv:1204.1891

O. Yaron, D.A. Perley, A. Gal-Yam, J.H. Groh, A. Horesh, E.O. Ofek, S.R. Kulkarni, J. Sollerman, C. Fransson, A. Rubin, P. Szabo, N. Sapir, F. Taddia, S.B. Cenko, S. Valenti, I. Arcavi, D.A. Howell, M.M. Kasliwal, P.M. Vreeswijk, D. Khazov, O.D. Fox, Y. Cao, O. Gnat, P.L. Kelly, P.E. Nugent, A.V. Filippenko, R.R. Laher, P.R. Wozniak, W.H. Lee, U.D. Rebbapragada, K. Maguire, M. Sullivan, M.T. Soumagnac, Confined dense circumstellar material surrounding a regular type II supernova. Nat. Phys. **13**, 510–517 (2017). https://doi.org/10.1038/nphys4025. arXiv:1701.02596

S.C. Yoon, S.E. Woosley, N. Langer, Type Ib/c supernovae in binary systems. I. Evolution and properties of the progenitor stars. Astrophys. J. **725**, 940–954 (2010). https://doi.org/10.1088/0004-637X/725/1/940. arXiv:1004.0843

Q. Yuan, N.H. Liao, Y.L. Xin, Y. Li, Y.Z. Fan, B. Zhang, H.B. Hu, X.J. Bi, Fermi Large Area Telescope detection of gamma-ray emission from the direction of supernova iPTF14hls (2017). ArXiv e-prints, arXiv:1712.01043

Space Sci Rev (2018) 214:27
https://doi.org/10.1007/s11214-017-0461-6

Circumstellar Interaction in Supernovae in Dense Environments—An Observational Perspective

Poonam Chandra[1,2]

Received: 24 October 2017 / Accepted: 15 December 2017 / Published online: 9 January 2018

Abstract In a supernova explosion, the ejecta interacting with the surrounding circumstellar medium (CSM) give rise to variety of radiation. Since CSM is created from the mass loss from the progenitor, it carries footprints of the late time evolution of the star. This is one of the unique ways to get a handle on the nature of the progenitor system. Here, I will focus mainly on the supernovae (SNe) exploding in dense environments, a.k.a. Type IIn SNe. Radio and X-ray emission from this class of SNe have revealed important modifications in their radiation properties, due to the presence of high density CSM. Forward shock dominance in the X-ray emission, internal free-free absorption of the radio emission, episodic or non-steady mass loss rate, and asymmetry in the explosion seem to be common properties of this class of SNe.

Keywords Radiation mechanisms: general · Radiative transfer · Stars: mass-loss · Supernovae: general · X-rays: general · Radio continuum: general

1 Introduction

A massive star (mass $M > 8\mathrm{M}_{\odot}$) evolves for millions of years, burning nuclear fuel, keeping the star stable with resulting outward radiation pressure. However, when the nuclear fuel in the star is exhausted, the equilibrium between gravity and radiation pressure ceases to exist and the star collapses under its own gravity within a fraction of a second. At some point the implosion turns into an explosion, and a core-collapse supernova (SN) is born, leaving behind a relic which is either a neutron star or a black hole. This is the most simplistic

Supernovae
Edited by Andrei Bykov, Roger Chevalier, John Raymond, Friedrich-Karl Thielemann, Maurizio Falanga and Rudolf von Steiger

✉ P. Chandra
poonam@ncra.tifr.res.in

[1] National Centre for Radio Astrophysics, Tata Institute of Fundamental Research, Pune University Campus, Ganeshkhind, Pune 411 007, India

[2] Department of Astronomy, Stockholm University, AlbaNova, 106 91 Stockholm, Sweden

picture of a SN explosion. The classic paper by Brown and Bethe (1985) remains one of the best resources to get an overall qualitative picture of the SN explosion. Understanding the detailed and more quantitative characteristics of the explosion requires complex neutrino physics, general relativity and magnetohydrodynamics. Other kinds of supernovae (SNe), thermonuclear SNe, arising from the detonation of a white dwarf residing in a binary system are not the subject of this chapter.

Stars lose mass from their least gravitationally bound outermost layers. The mass-loss rate $\dot{M}$ from a massive star can be expressed as $\dot{M} = 4\pi r_\infty^2 \rho(r_\infty) v_\infty$, where $\rho(r_\infty)$ is the average mass density of the star at a distance r_∞ from the centre where the wind speed $v_{\rm wind}$ has reached terminal velocity v_∞ (Smith 2014). While the Sun is losing mass at a rate of $\dot{M} \sim 10^{-14}\,{\rm M}_\odot\,{\rm yr}^{-1}$, the massive stars lose mass more profusely owing to their much larger radii. The mass lost manifests itself in the form of dense winds moving with velocities of the order 10–1000 $\rm km\,s^{-1}$, creating a high density medium around the star, a.k.a. circumstellar medium (CSM). The wind speed is generally proportional to the star's escape velocity, thus the outflows from the yellow super giants (YSGs), and red supergiants (RSGs) with larger radii are normally slower ($\sim$ 10–20 $\rm km\,s^{-1}$), whereas luminous blue variable (LBV) and blue supergiants (BSGs) with smaller radii have faster winds ($\sim$ 100 $\rm km\,s^{-1}$). In reality, though, high radiation from the SN can potentially accelerate the CSM winds to higher speeds (Smith 2014).

In normal core-collapse SNe, the spectra and energetics are mainly governed by the explosion dynamics, and do not have imprints from the CSM. However, in a class of core-collapse SNe exploding in dense environments, termed type IIn SNe (hereafter SNe IIn), the high CSM densities reveal their presence in the form of narrow emission lines in their optical spectra. While the radioactive decay mainly powers the optical light curves in normal core collapse SNe, in SNe IIn, the energetics have a significant contribution from the SN explosion ejecta interacting with the dense CSM, resulting in high bolometric and H-α luminosities (Chugai 1990). Due to this extra source of energy, they are detectable up to cosmological distances. The farthest detected Type IIn SN is at a redshift, $z = 2.36$ (Cooke et al. 2009). SNe IIn are the focus of this chapter.

2 Progenitors of SNe IIn

The progenitor mass loss rate required to explain the extreme densities of SNe IIn are quite high, e.g. 10^{-3}–$10^{-1}\,{\rm M}_\odot\,{\rm yr}^{-1}$. Chugai et al. (2004) derived mass loss rate for SN 1994W to be $\sim 0.2\,{\rm M}_\odot\,{\rm yr}^{-1}$, whereas the mass loss rate of SN 1995G was found to be $\sim 0.1\,{\rm M}_\odot\,{\rm yr}^{-1}$ (Chugai and Danziger 2003). In case of SN 1997ab, Salamanca et al. (1998) argued that narrow P Cygni profile superimposed on the broad emission lines implied a mass loss rate of $\sim 10^{-2}\,{\rm M}_\odot\,{\rm yr}^{-1}$, for a presupernova wind velocity of 90 $\rm km\,s^{-1}$. Chandra et al. (2012a) and Chandra et al. (2015) found $\dot{M} \sim 10^{-3}\,{\rm M}_\odot\,{\rm yr}^{-1}$ and $\dot{M} \sim 10^{-1}\,{\rm M}_\odot\,{\rm yr}^{-1}$ for SN 2006jd and SN 2010jl, respectively. This indicates that the mass loss rates in SNe IIn are extremely high and not explained in standard stellar evolution theories of normal class of evolved stars, such as Wolf-Rayet (WR) stars, LBVs, YSGs, RSGs etc (Fullerton et al. 2006). One possibility is that such high mass loss rates may be related to an explosive event a few years before the SN outburst (Chugai and Danziger 2003; Pastorello et al. 2007). Radiation production close to the Eddington limit can make the star unstable, which can lead to eruptive mass loss in a dramatically short time. However, the only stars which could incorporate enhanced mass loss shortly before explosion are super-asymptotic giant branch stars (AGB; mass 8–10 ${\rm M}_\odot$), massive RSGs (mass 17–25 $M_\odot$) and LBV giant eruptions (mass $> 35\,M_\odot$) (Fullerton et al. 2006).

While erupting LBV progenitors are favorite models for bright SNe IIn because of their extreme mass loss rates (Humphreys et al. 1999), they are a transient phase between O-type and WR stars (Humphreys and Davidson 1994), and are not supposed to explode at this phase. Even if they explode, the fine tuning of the explosion at the core coinciding with the enhanced episodic mass loss from the outer most layer is not understood. Yet archival data in some SNe IIn seem to suggest LBVs as progenitors, in particular SN 2005gl (Gal-Yam et al. 2007), SN 2009ip (Mauerhan et al. 2013), and SN 2010jl (Smith et al. 2011). The pre-explosion Hubble Space Telescope (HST) observations suggested an LBV progenitor for SN 2005gl (Gal-Yam et al. 2007). In addition, the inferred pre-shock wind speed of 420 $\mathrm{km\,s^{-1}}$ deduced from narrow hydrogen lines was also found to be consistent with the speeds of LBV eruptive winds (Smith et al. 2010). In SN 2009ip, multiple pre-SN eruptions were observed before the star finally thought to explode as SN IIn in 2012, strengthening the connection with an LBV like progenitor (Mauerhan et al. 2013). However, whether the 2012 event was a true supernova is still debated (Fraser et al. 2013). In SN 2010jl, the pre-explosion archival HST data revealed a luminous, blue point source at the position of the SN, suggesting $M > 30\,M_\odot$ LBV progenitor (Smith et al. 2011). But recently Fox et al. (2017) presented HST Wide-Field Camera (WFC3) imaging of the SN 2010jl field obtained 4–6 years after the SN explosion and show that the SN is $0.61''$ offset from the blue source previously considered the progenitor LBV star. The pre-SN eruptions were also looked for in a sample of ~ 16 SNe IIn from the Palomar Transient Factory archival data, and were found in roughly half the SNe (Ofek et al. 2014a). However, these arguments were contradicted by Bilinski et al. (2015), who analysed 12 years of Katzman Automatic Imaging Telescope archival data having five known SNe IIn locations and found no precursors. Anderson et al. (2012) used H-α emission and near-ultraviolet (UV) emission as tracers of ongoing (< 16 Myr old) and recent (16–100 Myr old) star formation, respectively. He used pixel statistics to build distributions of associations of different types of SNe with host galaxy star formation. His sample of 19 SNe IIn was found to trace the recent star formation but not the ongoing star formation. Under the assumption that more massive a star is, shorter its life time is, this implied that contrary to the general belief, the majority of SNe IIn do not arise from the most massive stars. This was also suggested by Chugai and Danziger (1994) based on the low ejecta mass $M_{\mathrm{ej}} < 1\,M_\odot$ in SN 1988Z. In addition, Sanders et al. (2013) reported the detection of another interactive kind (Type Ibn) of SN, PS1-12sk, in a giant elliptical galaxy CGCG 208-042 ($z = 0.054$), with no evidence for ongoing star formation at the explosion site. This further challenges the idea of interacting SNe arising from very massive progenitor stars, and puts a question mark on the LBV progenitors.

Questions have been raised on the true nature of LBVs. Smith and Tombleson (2015) challenged the current understanding of LBVs being a transient phase between O-type and WR stars. They found that the LBVs are not spatially located in O-type clusters and are statistically more isolated than O-type and WR stars. Based on the relatively isolated environments of LBVs, they suggested that LBVs are the product of binary evolution. The explained LBVs to be evolved massive blue stragglers. Smith et al. (2016) took late time HST observations of the SN 2009ip site to search for evidence of recent star formation in the local environment, and found none within a few kpc. If LBVs are indeed a transient phase between O-type and WR stars, they should be found in the massive star forming regions. They proposed that the SN 2009ip progenitor may have been the product of a merger or binary mass transfer, rejuvenated after 4–5 Myr.

Even though LBVs remain the most favoured and exotic progenitors of SNe IIn, alternative models too have been suggested. Chugai and Danziger (1994), Chevalier (2012) proposed a binary model where common envelope evolution can give rise to SNe IIn. In this model, the SN is triggered by the inspiral of a compact neutron star or black hole to the central core of the binary star. Pulsational pair instability models too have been suggested, especially in Type IIn SLSNe, e.g. SN 2006gy (Woosley et al. 2007). Quataert and Shiode (2012) proposed a gravity-wave driven mass loss model for such SNe IIn. According to this model, the fusion luminosity post C-burning stage can reach super-Eddington values and generate gravity waves in massive stars. As the gravity waves propagate towards stellar surface, they can convert into sound waves and dissipation of the sound waves can trigger a sudden enhanced mass loss.

Overall, there is a large discrepancy in the progenitor models of SNe IIn. The heart of the problem is that SNe IIn classification is an external one, where any explosion surrounded by very high density can mimic a type IIn SN. Thus SNe IIn are governed by external factors rather than by internal explosion dynamics. Recently some superluminous supernovae (SLSNe) have also been explained in the extreme ejecta-CS interaction scenario by Chevalier and Irwin (2011). These are expected to be the extreme version of SNe IIn (Quimby et al. 2013). Thus SNe IIn certainly encompass objects of different stellar evolution and mass loss history. For example, SN 1994W had a very fast decline after about $>$ 125 days, (Sollerman et al. 1998), contrary to SN 1988Z which remained bright for a very long time (Turatto et al. 1993). SN 1998S was initially classified as a bright Type IIL SN, however, at a very early stage ($<$ 10 days) it started to show characteristics of type IIn SN (Li et al. 1998). A special class of SNe IIn is characterized by spectral similarities to Type Ia SNe at peak light, but later shows SN IIn properties, e.g., SNe 2002ic, 2005gj, PTF 11kx. While an explosion mechanism is not determined in the first two SNe, PTF 11kx was clearly considered to be a thermonuclear event (Dilday et al. 2012). Some SNe IIn also appear to be related to Type Ib/c SNe. For example, SNe 2001em and 2014C were initially classified as type Ib SNe, however, at late epochs, signatures of the dense CSM interaction were found and these SNe were reclassified as SNe IIn (Chugai and Chevalier 2006; Milisavljevic et al. 2015). These are not the only SNe Ib/c to show late time CSM interaction signatures. While analysing a sample of 183 SNe Ib/c, Margutti et al. (2017) found these late rebrightenings in SNe 2003gk, 2007bg, and PTF11qcj as well.

A well sampled long term observational follow up of ejecta-CSM interaction is the key to unravel the true nature of the SNe IIn progenitors through their footprints in their light curve and spectra.

3 Circumstellar Interaction

After a SN explosion, the shock wave propagates through the star and after shock breakout the expanding ejecta start to interact with the surrounding medium. While the inner part of the ejecta density profile is flatter, it is less important for the ejecta-CSM interaction dynamics. The outer ejecta density profile, more relevant part for the interaction, can be described as $\rho_{\rm ej} = \rho_o (t/t_o)^{-3} (V_o t/r)^n$ (where for a mass element m_i, ejecta velocity $V(m_i)$, radius $r(m_i)$ and density $\rho(m_i)$ have time dependence of $r(m_i) = V(m_i)t$ and $\rho(m_i) = \rho_o(m_i)(t_o/t)^3$, Chevalier and Fransson 2016). The factor t^{-3} comes due to free expansion of the gas.

The ejecta shock moving through stellar layers is radiation dominated, mediated by photons. However, eventually the photons will diffuse out, leading to the disappearance of the radiation dominated shock and acceleration of the outer gas. The velocity of the accelerated outer gas drops with radius because of the decreasing flux, hence the inner gas would catch

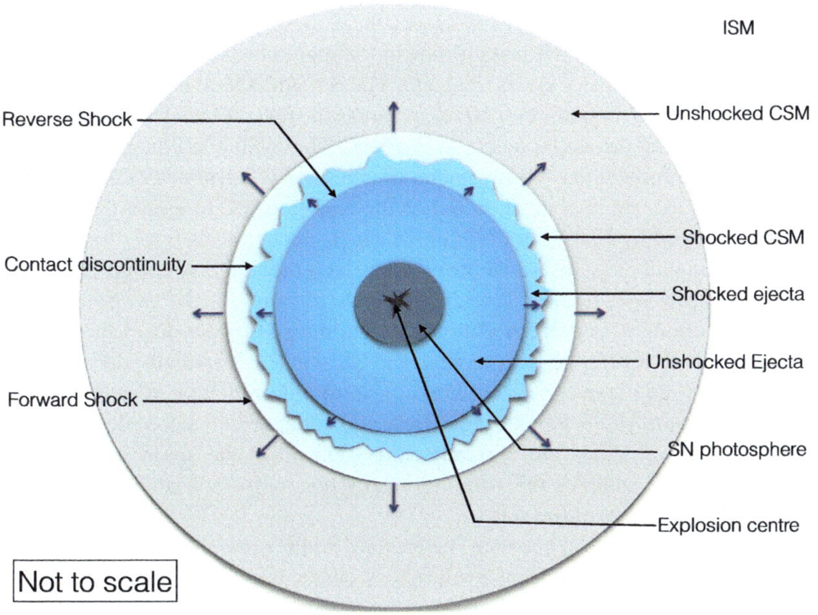

Fig. 1 Schematic diagram of the SN ejecta interacting with the surrounding CSM. This creates a hot 10^9 K forward shock, a 10^7 K reverse shock and a contact discontinuity in between. Hot forward and reverse shock produce X-ray emission, whereas the electrons accelerated in the forward shock in the presence of the enhanced magnetic field produce synchrotron radio emission. Reprocessed X-rays can come out in optical and UV radiation

up with the outer gas. Since the ejecta velocity is still supersonic, a viscous shock will be formed, which will commence the interaction with the surrounding medium.

Diffusion of the photons will happen when the optical depth is smaller than the ratio of speed of light, c, and ejecta velocity v, i.e. $\leq c/v$ and the photon diffusion time scale becomes comparable to that of age of the SN (Chevalier and Fransson 2016). In normal core-collapse SNe, the diffusion happens at the progenitor radius, in the form of shock breakout and a huge amount of energy (10^{46}–10^{48} ergs, Matzner and McKee 1999; Chevalier and Fransson 2016) is released, mainly in extreme ultraviolet (EUV) to soft X-rays. The shock breakout time scale is small ($\sim$ 1 hr), and is followed by the optical peak due to radioactivity. For most of the SNe, shock breakout happens at the photosphere of the star. However, in some SNe IIn (especially superluminous SNe IIn), the density of the medium surrounding the exploding star is so high that it becomes optically thick $\tau_{\rm wind} > c/V_{\rm ej}$, the shock breakout happens in the CSM, over a longer period of time and powers the bolometric light curve. Radiation from shock breakout can ionize and accelerate the immediate medium around it. Svirski et al. (2012) predicted the generation of late time hard X-ray emission in these prolongated shock breakouts, which has been detected in a few SNe (Ofek et al. 2013).

The structure of the unshocked CSM is important for the further ejecta-CSM interaction post shock-breakout. The structure of the CSM density profile $\rho_{\rm wind}$ can be written as $\rho_{\rm wind}(r) = \dot{M}/4\pi r^{-s} v_{\rm wind}$ (where $s = 2$ for a steady wind). Thus if one has information of the $\rho_{\rm wind}$, the mass loss rate of the progenitor star can be estimated, provided the wind velocity is known. The fast moving ejecta collide with the CSM leading to the formation of a 'forward' shock moving in the CSM and a 'reverse' shock propagating back into the stellar envelope *relative* to the *expanding* stellar ejecta (Fig. 1 and Chevalier 1981, 1982a).

In between the forward and the reverse shocks there exists a contact discontinuity, which is anything but smooth due to the instabilities in the region caused by low density shocked CSM decelerating high density shocked ejecta. The forward shock has a velocity of the order 10,000 $\mathrm{km\,s^{-1}}$ and temperature $\sim 10^9$ K, whereas the reverse shock moves with velocity $\sim 1000\ \mathrm{km\,s^{-1}}$, heating the stellar ejecta to $\sim 10^7$ K. The main consequences of the ejecta-CSM interaction is production of X-ray and radio emission (Chevalier 1982b; Chevalier and Fransson 2003, 2016). An analytical model of the emission due to ejecta-CSM interaction in SNe was developed by Chevalier (1982b), in which ejecta moves freely for several years in the ejecta dominated phase (mass of the ejecta is larger than the swept-up CSM mass). In normal core-collapse SNe, it takes several thousands of years for the swept-up CSM mass to be significant enough (as compared to the ejecta mass) to slow down the free moving ejecta. However, in SNe IIn, the extremely dense CSM can substantially decelerate the SN ejecta very quickly, and converts ejecta kinetic energy into radiation very efficiently. This also causes the forward shock to cool radiatively and collapse into a cool dense shell.

Since the shock velocities are typically 100–1000 times the speed of the progenitor winds, the shock wave samples the wind lost many hundreds–thousands of years ago and thus probes the past history of the star.

When the interaction region between the forward and reverse shocks is treated in a thin-shell approximation, the shock front evolution is characterized by a self-similar solution (Chevalier 1981, 1982a,b). By balancing the ram pressure from the CSM and the ejecta, one can derive the shock radii R_s expansion, which is a power law in time as $R_s \propto t^m$ ($m = (n-3)/(n-2)$). These solutions are valid for $n > 5$ to meet the finite energy requirements (Matzner and McKee 1999).

Chevalier and Fransson (2016) have shown how to obtain various shock parameters. Please refer to it for details. Here I write down the main equations needed to interpret the observational data. Throughout the chapter, the subscript 'CS' refers to circumstellar shock and 'rev' to reverse shock.

The swept up masses behind the two shocks can be written as

$$\begin{aligned} M_{\mathrm{CS}} &= \frac{\dot{M} R_{\mathrm{s}}}{v_{\mathrm{wind}}} \\ M_{\mathrm{rev}} &= 4\pi \int_{R_s}^{\infty} \rho(r) r^2 dr \end{aligned} \tag{1}$$

Since $\rho_{\mathrm{rev}}/\rho_{\mathrm{CS}} = \rho_{\mathrm{ej}}/\rho_{\mathrm{wind}}$, the ratios of the masses and the density of the shocked shell are related as

$$\begin{aligned} M_{\mathrm{rev}} &= \frac{(n-4)}{2} M_{\mathrm{CS}} \\ \rho_{\mathrm{rev}} &= \frac{(n-4)(n-3)}{2} \rho_{\mathrm{CS}} \end{aligned} \tag{2}$$

The maximum ejecta velocity V_{ej}, circumstellar shock velocity at the contact discontinuity V_{s}, and reverse shock velocity V_{rev} can be written as

$$\begin{aligned} V_{\mathrm{ej}} &= R_s/t \propto t^{-1/(n-2)} \\ V_{\mathrm{s}} &= dR_s/dt = (n-3)/(n-2) V_{\mathrm{ej}} \\ V_{\mathrm{rev}} &= V_{\mathrm{ej}} - V_{\mathrm{s}} = V_{\mathrm{ej}}/(n-2) \end{aligned} \tag{3}$$

The temperature of the forward CSM shock $T_{\rm CS}$ (assuming cosmic abundances), and that of the reverse shock $T_{\rm rev}$ are

$$
\begin{aligned}
kT_{\rm CS} &= 117\left(\frac{n-3}{n-2}\right)^2\left(\frac{V_{\rm ej}}{10^4\ {\rm km\,s^{-1}}}\right)^2\ {\rm keV} \\
kT_{\rm rev} &= \frac{T_{\rm CSM}}{(n-3)^2} = 1.2\left(\frac{10}{n-2}\right)^2\left(\frac{V_{\rm ej}}{10^4\ {\rm km\,s^{-1}}}\right)^2\ {\rm keV}
\end{aligned}
\tag{4}
$$

The above equations assume energy equipartition between ions and electrons, for which the characteristic time is

$$
t_{\rm eq} = 2.5\times 10^7\left(\frac{T_e}{10^9\ {\rm K}}\right)^{1.5}\left(\frac{n_e}{10^7\ {\rm cm^{-3}}}\right)^{-1}\ {\rm s} \tag{5}
$$

The ratio of the electron temperatures of the shocked CSM and shocked ejecta is $(n-3)^2$. Due to the low temperature and higher density, the condition of equipartition between ions and electrons may be valid in the reverse shock, but it is highly unlikely for the forward shock to obtain electron ion equilibrium. However, in SNe IIn, it is possible to meet this condition for both reverse as well as forward shocks owing to extreme densities.

3.1 Radiation in Ejecta CSM Interaction

During ejecta CSM interaction in SNe, radiation can come in multiple frequency bands from the shocked as well as unshocked regions via multiple channels.

In the shocked CSM, the photospheric photons may undergo inverse-Compton (IC) scattering by the energetic electrons and may emit in UV and X-rays. For an electron scattering optical depth $\tau_{\rm e}$ with $\tau_{\rm e} < 1$ behind the CSM, a fraction $\tau_{\rm e}^N$ electrons will scatter N times and boost their energies to UV and X-ray with a power-law spectrum of spectral index between -1 to -3 (Chevalier and Fransson 2003).

Forward shock and reverse shocks are very hot (Eq. (4)) and can emit soft and hard X-rays by free-free bremsstrahlung radiation. The total X-ray luminosity in free-free emission depends upon the medium density ρ_i (where subscript i refers to 'CS' and 'rev' for circumstellar and reverse shocks, respectively), and emitting volume $\propto R_s^3$. If Λ is the cooling function, then X-ray luminosity can be written as $L_i \propto \rho_i^2 R_s^3 \Lambda$ (Fransson et al. 1996), where $\Lambda \sim T^{0.5}$ (Chevalier and Fransson 1994). Chevalier and Fransson (2003) derive the expression for the free-free X-ray luminosity from forward and reverse shocks to be

$$
L_{\rm i} \approx 3.0\times 10^{39}\overline{g}_{\rm ff} C_{\rm n}\left(\frac{\dot M}{10^{-5}\ {\rm M_\odot\,yr^{-1}}}\right)^2\left(\frac{v_{\rm wind}}{10\ {\rm km\,s^{-1}}}\right)^{-2}\left(\frac{t}{10}\right)^{-1}\ {\rm erg\,s^{-1}} \tag{6}
$$

Here C_n depends on the ejecta density index. C_n is 1 for the forward shock and is $(n-3)(n-4)^2/4(n-2)$ for the reverse shock. $\overline{g}_{\rm ff}$ is the free-free Gaunt factor. A caveat here is that this equation is valid only under electron-ion equipartition which is questionable for the forward shock.

Eq. (6) suggests that the free-free X-ray luminosity goes as t^{-1}. However this is valid only for the constant mass loss medium ($s = 2$). For the general case Fransson et al. (1996) derive

$$
L_i \propto t^{-(12-7s+2ns-3n)/(n-s)}
$$

An important point is that this formula assumes the X-ray emission covers the X-ray band, whereas the X-ray satellites measure only in a very narrow band, e.g. *Chandra* and *XMM-*Newton cover 0.2–10 keV range. This means unless the luminosity ratio in the narrow observing band is same as that in the total X-ray band, one will not see the above evolution of the X-ray luminosity. As shown in Fransson et al. (1996), the luminosity in a given energy band with $E < kT_i$ can be written as $L_i \propto t^{-(6-5s+2ns-3n)/(n-s)}$, which has a much flatter time dependence. For a steady wind ($s = 2$), $L_i \propto t^{-(n-4)/(n-2)}$.

The above treatment is valid for adiabatic shocks with $T > 2 \times 10^7$ K. For high mass loss rates and slow moving shocks, the radiative cooling may be important as the cooling function will depend on temperature as $\Lambda \propto T^{-0.6}$ (Chevalier and Fransson 1994). If the temperature of the shocks goes below 2×10^7 K, then the efficient radiative cooling will lower the ejecta temperature to 10^4 K, where photoelectric heating from the shocks balances the cooling. Since the ram pressure is maintained, a cool dense shell (CDS) will form. The column density of the CDS can be estimated from $N_{\rm cool} = M_{\rm rev}/(4\pi R_{\rm s}^2 m_p)$ (m_p is mass of a proton; Chevalier and Fransson 2003), which can be simplified to

$$N_{\rm cool}(\text{RS}) \sim 10^{21}(n-4)\left(\frac{\dot{M}}{10^{-5}\,\text{M}_\odot\,\text{yr}^{-1}}\right)\left(\frac{v_{\rm wind}}{10\,\text{km}\,\text{s}^{-1}}\right)^{-1}\left(\frac{V_{\rm ej}}{10^4\,\text{km}\,\text{s}^{-1}}\right)^{-1} \times \left(\frac{t}{100\,\text{days}}\right)^{-1}\,\text{cm}^{-2} \tag{7}$$

The effect of the CDS is that it will absorb the soft X-ray emission from the reverse shock and re-radiate into UV and optical emission and will contribute significantly to the bolometric luminosity (Fransson 1984). In the CDS, the atoms recombine and produce mainly H-α emission lines of intermediate line widths ~ 1000 km s^{-1}.

Even though owing to high density, the intrinsic X-ray luminosity is high from the reverse shock, due to absorption of the emission (Nymark et al. 2006), the X-ray emission may be dominated by the forward shock, especially in SNe IIn (Chandra et al. 2012a, 2015). An additional factor towards dominance of the forward shock is that once the reverse shock becomes radiative, its luminosity rises proportional to density while the luminosity from the forward shock continues to grow as density squared. While the cooling time of the forward shock is long, if the mass loss rate is very high, even the forward shock may become radiative (e.g. SN 2010jl, Chandra et al. 2015).

While the radiation due to ejecta-CSM interaction in SNe is affected by complicated hydrodynamics, in a simple approach the radiation luminosity of the shocks is the total kinetic luminosity times the radiation efficiency $\eta = t/(t + t_c)$, where t_c is the cooling time of the shocked shell at the age t. In normal core-collapse SNe, the total radiated energy is only a few per cent of its kinetic energy, but could reach up to 50% in SNe IIn due to efficient cooling (Smith 2016). In case of radiative forward and reverse shocks, the kinetic luminosity can be written as (Chevalier and Fransson 2016)

$$L_i = \pi R_{\rm s}^2\left(1/2\rho_i V_i^3\right) = c_i\frac{(n-3)}{(n-2)} \times \frac{1}{2}\frac{\dot{M} V_{\rm ej}^3}{v_{\rm wind}} \tag{8}$$

Here $c_i = (n-4)/2(n-2)^2$ for the reverse shock and $c_i = (n-3)^2/(n-2)^2$ for the forward shock. Thus radiative luminosity goes as $L_i \propto t^{-3/(n-2)}$ for a steady wind. However, for the more general case, it will evolve as $L_i \propto t^{-(15-6s+ns-2n)/(n-s)}$.

While the X-ray emission discussed above is from the shocked shells, the unshocked ejecta and CSM can also contribute to significant radiation. During the shock breakout, the unshocked CSM can be ionized and emit in optical and UV by subsequent recombination, mainly in H-α and Lyman-α and C III, C IV, N V, and Si IV, whose line widths give information about the wind velocity. The unshocked CSM can also be excited by the X-rays from the forward and reverse shocked shells and can produce coronal emission lines, e.g., Fe VII and Fe X lines. The outermost layers of the unshocked ejecta, ionized and heated by radiation can contribute towards broad emission and absorption lines, such as C III, C IV, N V, and Si IV. In dense CSM, electron scattering can be the dominant scattering and broaden the width of the above discussed emission lines.

The above treatment assumes spherical symmetry of the system. However, if the CSM is asymmetric and/or clumpy, there are strong observational consequences for the CSM interaction with the SN ejecta (Chugai and Danziger 1994). In a clumpy CSM, the predominance of soft radiation is likely to come out from the clumps, since due to high density the velocity of the shock in the clump will be much smaller than the $V_c = V_s(\rho_s/\rho_c)^{1/2}$. The higher density and lower velocity of the clump can lead to radiative cooling of the clump shock. Indeed, if the bulk of the CSM were in clumps, radiation from clumps becomes the dominant source of the radiation in the soft XUV band.

The ejecta diagnostics based on the CSM interaction cannot constrain the ejecta mass and energy uniquely because the same density versus velocity distribution in the ejecta outer layers can be produced by a different combination of mass and energy. Yet the observed interaction luminosity and the final velocity V_f of the decelerated shell constrain the energy of the outer ejecta from the condition $V > V_f$. For the power-law index n, the obvious relations, such as the ejecta density turnover velocity $V_0 \propto (E/M)^{1/2}$, $\rho(V) \propto \rho_0(V_0/V)^n$, and $\rho_0 \propto M/V^3$ result in the energy-mass scaling $E \propto M^{(n-5)/(n-3)}$. This scaling, when combined with the requirement that the velocity V_f should be larger than V_0, provides us with the lowest plausible values of E and M.

An important consequence of ejecta-CSM interaction is the radio emission. In the fast moving forward shock, the particles can be accelerated to relativistic energies, and the presence of magnetic field give rise to non-thermal synchrotron radio emission, which we discuss below.

3.1.1 Radio Emission

In the shock front, the charged particles are accelerated most likely via diffusive Fermi shock acceleration (Blandford and Ostriker 1978; Bell 1978). The magnetic field is also considered to be a seed CSM magnetic field, compressed and enhanced in the shocked region. Rayleigh-Taylor instabilities at the contact discontinuity can further enhance the magnetic field (Chevalier and Blondin 1995). If the energy density of the magnetic field as well as the relativistic particles are considered to be proportional to the thermal energy density, then one can easily obtain the radio emission formulae.

Radio emission arises primarily from the higher temperature forward shock, where it is easier to accelerate the particles to relativistic energies. The electron energy distribution is assumed to be a power law $N(E) = N_o E^{-p}$, where E is the electron energy, N_o is the normalization of the distribution and p is the electron energy index, which is related to spectral index α (in $F_\nu \propto \nu^{-\alpha}$, where F_ν is the flux density at frequency ν) as $\alpha = (p-1)/2$.

The radio emission is affected by either the external free-free absorption (FFA) process by the surrounding ionized wind, or by the internal synchrotron self absorption (SSA) by the same electrons responsible for the emission. The dominant absorption mechanism depends upon the mass loss rate, magnetic field in the shocked shells, shock velocity and density of the ejecta. One can distinguish between the two processes from the optically thick part of the light curve or spectrum.

If SSA is the dominant absorption mechanism, then for SSA optical depth $\tau_\nu^{\rm SSA}$, one can use Chevalier (1998) formulation to derive the radio flux density evolution.

$$
\begin{aligned}
F_\nu^{\rm SSA} &= \frac{\pi R_s^2}{D^2}\frac{c_5}{c_6}B^{-1/2}\left(\frac{\nu}{2c_1}\right)^{5/2}\left[1-\exp\left(-\tau_\nu^{\rm SSA}\right)\right] \\
\tau_\nu^{\rm SSA} &= \int_0^s \kappa(\nu)ds \approx \kappa(\nu)s = \left(\frac{\nu}{2c_1}\right)^{-\frac{p+4}{2}}\left(\frac{4fR_s c_6 N_o B^{\frac{p+2}{2}}}{3}\right)
\end{aligned}
\tag{9}
$$

where c_1, c_5 and c_6 are the constants defined in Pacholczyk (1970), D, and f, and B are the distance to the SN from observer, filling factor and magnetic field strength, respectively.

One can determine N_o from $\int_{E_l}^{\infty} N(E)EdE$, assuming electron rest mass energy to be the lower energy limit, i.e. $E_l = 0.51$ MeV, then

$$
N_o = \frac{aB^2(p-2)E_l^{p-2}}{8\pi}
$$

where a is the equipartition factor. The above equation can be written in terms of optically thick $\tau > 1$ and optically thin $\tau \le 1$ limits as

$$
\begin{aligned}
F_\nu^{\rm SSA}|_{\tau\le 1} &= \frac{4\pi f R^3}{3D^2}c_5 N_o B^{\frac{p+1}{2}}\left(\frac{\nu}{2c_1}\right)^{-\frac{p-1}{2}} \\
F_\nu^{\rm SSA}|_{\tau>1} &= \frac{\pi R^2}{D^2}\frac{c_5}{c_6}B^{-1/2}\left(\frac{\nu}{2c_1}\right)^{5/2}
\end{aligned}
\tag{10}
$$

From these equations, one can estimate the R_p and B_p at the peak of the spectrum, when $F_\nu^{\rm SSA}$ is F_p. They will depend on various parameters as

$$
\begin{aligned}
R_p &\propto a^{-\frac{1}{2p+13}} f^{-\frac{1}{2p+13}}(p-2)^{-\frac{1}{2p+13}} F_p^{\frac{p+6}{2p+13}} D^{\frac{2p+12}{2p+13}} \nu^{-1} \\
B_p &\propto a^{-\frac{4}{2p+13}} f^{-\frac{4}{2p+13}}(p-2)^{-\frac{4}{2p+13}} F_p^{-\frac{2}{2p+13}} D^{-\frac{4}{2p+13}} \nu
\end{aligned}
\tag{11}
$$

Chevalier (1998) has given these equations for $p=3$ to be:

$$
\begin{aligned}
R_p &= 8.8\times 10^{15} a^{-\frac{1}{19}}\left(\frac{f}{0.5}\right)^{-\frac{1}{19}}\left(\frac{F_p}{\rm Jy}\right)^{\frac{9}{19}}\left(\frac{D}{\rm Mpc}\right)^{\frac{18}{19}}\left(\frac{\nu}{5\ {\rm GHz}}\right)^{-1}\ {\rm cm} \\
B_p &= 0.58a^{-\frac{4}{19}}\left(\frac{f}{0.5}\right)^{-\frac{4}{19}}\left(\frac{F_p}{\rm Jy}\right)^{-\frac{2}{19}}\left(\frac{D}{\rm Mpc}\right)^{-\frac{4}{19}}\left(\frac{\nu}{5\ {\rm GHz}}\right)\ {\rm G}
\end{aligned}
\tag{12}
$$

In Eq. (10), if R can be measured in some independent way (e.g. VLBI), then both the magnetic field and the column density of the relativistic electrons can be determined. Another possibility to independently estimate the magnetic field is from synchrotron cooling effects; if it is observationally seen in the supernova spectrum, one can derive the exact equipartition fraction (e.g. SN 1993J, Chandra et al. 2004).

By relating the post-shock magnetic energy density to the shock ram pressure, i.e. $B^2/8\pi = \varsigma\rho_{\rm wind}V_s^2$, where ς ($\varsigma \le 1$) is a numerical constant, one can obtain the mass loss rate to be

$$\dot{M} = \frac{6\times10^{-7}}{m_H^2\varsigma}\left(\frac{B}{1\ \mathrm{G}}\right)^2\left(\frac{t}{100\ \mathrm{d}}\right)^2\left(\frac{v_{\mathrm{wind}}}{10\ \mathrm{km\,s^{-1}}}\right)\mathrm{M_\odot\,yr^{-1}} \tag{13}$$

While SSA seems to dominate in high velocity shocks, in the case of high mass loss rate, external free-free absorption may be quite dominant. The optical depth of the free-free absorption can be defined as

$$\tau_\nu^{\mathrm{FFA}} = \int_{R_S}^{\infty} \kappa_\nu^{\mathrm{FFA}} n_e n_i ds$$

Here $\kappa_\nu^{\mathrm{FFA}}$ is the absorption coefficient, which depends upon n_e and n_i, the electron and ion number densities, respectively. Here $n_e = \dot{M}/4\pi R_s^2 v_{\mathrm{wind}}\mu_e m_H$, and $n_i = \overline{Z}n_e$, where $\overline{Z} = \sum X_j Z_j^2/\sum X_j Z_j$. Thus

$$\tau_\nu^{\mathrm{FFA}} = \frac{\dot{M}\overline{Z}\kappa_\nu^{\mathrm{FFA}}}{3(4\pi)^2 R^3 m_H^2 \mu_e^2 v_{\mathrm{wind}}^2} \tag{14}$$

Panagia and Felli (1975), Weiler et al. (2002) derived $\kappa_\nu^{\mathrm{FFA}}$ to be

$$\kappa_\nu^{\mathrm{FFA}} = 4.74\times10^{-27}\left(\frac{\nu}{1\ \mathrm{GHz}}\right)^{-2.1}\left(\frac{T_e}{10^5\ \mathrm{K}}\right)^{-1.35} \tag{15}$$

FFA will cause the radio emission arising out of the (mainly) forward shock to be attenuated by $\exp(-\tau_\nu^{\mathrm{FFA}})$.

The above equation for a fully ionized wind, and assuming the wind to be singly ionized, i.e. $\overline{Z} = 1$ and $\mu_e = 1.3$, gives a mass loss rate

$$\dot{M} = 4.76\times10^{-5}(\tau_\nu^{\mathrm{FFA}})^{0.5}\left(\frac{V_{\mathrm{ej}}}{10^4\ \mathrm{km\,s^{-1}}}\right)^{1.5}\left(\frac{t}{100\ \mathrm{d}}\right)^{1.5}\left(\frac{T_e}{10^5\ \mathrm{K}}\right)^{0.675}\left(\frac{v_{\mathrm{wind}}}{10\ \mathrm{km\,s^{-1}}}\right)\mathrm{M_\odot\,yr^{-1}} \tag{16}$$

Here it is important to mention that the above formula will overestimate the mass loss rate if the wind if clumpy. Puls et al. (2008) has shown that in a clumpy wind, the mass loss rate can change (to a lower value) by as much as a factor of 3.

Weiler et al. (1990) proposed a model in which thermal absorbing gas is mixed into the synchrotron emitting gas. This can be a likely case with SNe IIn, in which a CDS is formed. A fraction of the cool gas (10^4–10^5 K) from the CDS mixed in the forward shocked region can also give rise to internal FFA. This will attenuate the radio emission by a factor $(1-\exp(-\tau_\nu^{\mathrm{intFFA}}))$, where $\tau_\nu^{\mathrm{intFFA}}$ is the internal FFA optical depth.

To fit the observational data, one can put the above formulae in simpler form, and derive time and frequency dependence of various parameters. When SSA is the dominant absorption mechanism, the data can be fit with the following model,

$$\begin{aligned} F_\nu(t) &= K_1\nu^{5/2}t^a\left(1-\exp\left(-\tau_\nu^{\mathrm{SSA}}\right)\right) \\ \tau_\nu^{\mathrm{SSA}} &= K_2\nu^{-(p+4)/2}t^{-(a+b)} \end{aligned} \tag{17}$$

Here a gives the time evolution of the radio flux density in the optically thick phase ($F \propto t^a$) and b in the optically thin phase ($F \propto t^{-b}$). Under the assumption that the energy density in the particles and the fields is proportional to the postshock energy density, these quantities are related with expansion parameter m and electron energy index p as $a = 2m + 0.5$, and $b = (p + 5 - 6m)/2$.

When FFA is the dominant absorption mechanism, the evolution of the radio flux density can be fit with

$$F_\nu(t) = K_1 \nu^{-\alpha} t^{-\beta} \exp\left(-\tau_\nu^{\mathrm{FFA}}(t)\right)$$
$$\tau_\nu^{\mathrm{FFA}}(t) = K_2 \nu^{-2.1} t^{-\delta}, \tag{18}$$

where α is the frequency spectral index, which relates to the electron energy index as $p = 2\alpha + 1$. Here K_1 is the radio flux density normalization parameter and K_2 is the FFA optical depth normalization parameter. The parameter δ is related to the expansion parameter m in $R_s \propto t^m$ as $\delta \approx 3m$. The parameter β is the time dependence, which under the assumption that the energy density in the particles and the fields is proportional to the postshock energy density leads to $\beta = (p + 5 - 6m)/2$ (Chevalier 1982a,b).

When the mixing of cool gas in the synchrotron emitting region is doing much of the absorption, in such a case the flux takes the form

$$F_\nu(t) = K_1 \nu^{\alpha} t^{\beta} \frac{(1 - \exp(-\tau_\nu^{\mathrm{intFFA}}))}{\tau_\nu^{\mathrm{intFFA}}}$$
$$\tau_\nu^{\mathrm{intFFA}} = K_2 \nu^{-2.1} t^{\delta'}, \tag{19}$$

where δ' is time evolution of internal-FFA optical depth.

The above discussion is based on the spherical geometry of the SN-CSM system. However, there is evidence that, at least in some cases, the ejecta and the CSM may be far from spherical and very complex. Smith et al. (2009) found anisotropic and clumpy CSM medium in the RSG VY CMa, whereas the structure of η Carinae was found to be bipolar (Smith 2010). The deviation from spherical geometry can have strong observational consequences in studying the SN-CSM interaction.

From this section onwards, I will concentrate on the observational aspects of SNe IIn, mainly in X-ray and radio bands.

4 X-ray Observations of SNe IIn

Since a high CSM density is a prerequisite for radio and X-ray emission, the ejecta CSM interaction is expected to emit copiously in these bands in SNe IIn. However, the statistics are contradictory. While more than ~ 400 SNe IIn are known, only 12 are known to emit in the X-ray bands (Fig. 2). These include SNe 1978K, 1986J, 1988Z, 1995N, 1994W, 1996cr, 1998S, 2005ip, 2005kd, 2006gy, 2006jd and 2010jl (Ross 2017). In Fig. 2, we plot X-ray luminosities of SNe IIn in 0.3–8 keV and 0.5–2 keV bands. In 0.3–8 keV band, SNe IIn have high luminosities spread within an order of magnitude, with the exception of SNe 1978K and 1998S which are much weaker X-ray emitters. Other than SN 2010jl, most of the X-ray detected SNe IIn have X-ray emission after around a year. This could have a contribution from observational biases due to the lack of early observations. However, recently, in SN 2017gas, a nearby ($d = 43$ Mpc) SN IIn, early observations covering 20–60 days after the discovery resulted in a non-detection (Chandra and Chevalier 2017). In soft X-ray bands (lower panel of Fig. 2), SN 2006gy is a weak X-ray emitter. However, it was a SLSN and SLSNe are generally known to be weak X-ray emitters. Here the most unique SN is SN 1996cr, which was detected after 1000 days and since then it's luminosity continued to rise.

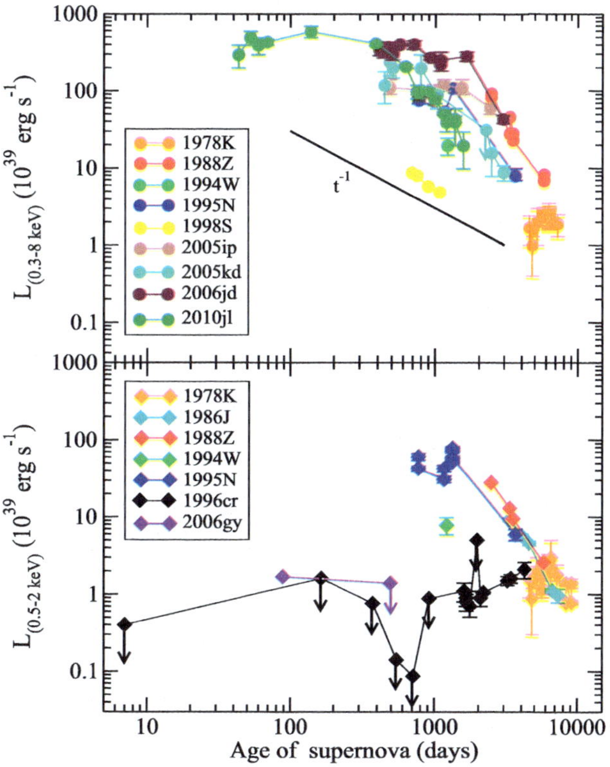

Fig. 2 The 0.3–8 keV (*upper panel*) and 0.5–2 keV (*lower panel*) X-ray luminosities of SNe IIn (Ross 2017). In the *upper panel* the X-ray luminosities of SN 1998S and SN 2006jd are in 0.2–10 keV and in the *lower panel*, SN 1986J and SN 1995N X-ray luminosities are in the range 0.5–2.4 keV

Due to higher density of the reverse shock, the thermal X-ray emission from core collapse SNe, in general, has a dominant contribution from the reverse shock. In SNe IIn, the radiative cooling shell (CDS) formed between the reverse and forward shock absorbs most of the X-rays emanating from the reverse shock and observable X-rays are from the forward shock. This effect has observationally seen in many SNe IIn. In SN 2005ip, the X-rays were fit by temperature $T \geq 7$ keV, for at least up to 6 years after the explosion (Katsuda et al. 2014). In SN 2005kd, the early emission (up to ≤ 2 years) was dominated by hard X-rays (Dwarkadas et al. 2016). Chandra et al. (2012a) have shown that the *Chandra* and *XMM-Newton* observations in SN 2006jd are best fit with a thermal plasma with electron temperature ≥ 20 keV. While this implies X-ray emission coming from the forward shock, these telescopes did not have the sensitivity to better constrain the temperature since both work below 10 keV range. In SN 2010jl, using the joint *NuSTAR* and *XMM-Newton* observations, Ofek et al. (2014b), Chandra et al. (2015) derived the X-ray spectrum in the energy range 0.3–80 keV, and constrained the X-ray emitting shock temperature to be 19 keV, thus confirming that the dominant X-ray emission in the SN is indeed coming from the hotter forward shock (Fig. 3). This is the first time the (forward) shock temperature was observationally accurately measured in a SN.

As discussed in section, Sect. 3, the thermal X-ray luminosity is expected to decline as $\sim 1/t$ for a steady wind. However, observations of several SNe IIn have shown to not follow this evolution (see Fig. 2 as well as Dwarkadas and Gruszko 2012; Dwarkadas et al. 2016). While this deviation could be attributed to effects like inverse-Compton scattering etc. at early times, the late time deviation seen in many SNe IIn cannot be explained by it. A more

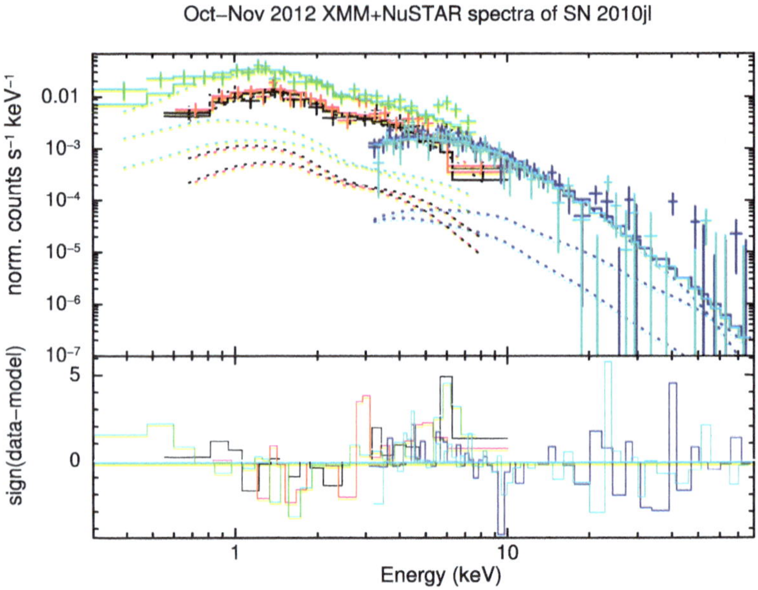

Fig. 3 Joint *NuSTAR* and *XMM-Newton* spectrum of the SN 2010jl covering energy range 0.3–80 keV. Chandra et al. (reproduced from, 2015)

general reason for deviation could be due to the complex nature of the progenitor star of SNe IIn. If the winds are not steady (deviation from $\rho_{\rm wind} \propto 1/R^2$), and have clumpy, asymmetric structure, the luminosity evolution will be far from $L_{\rm Xray} \propto 1/t$. In the energy range 0.2–10 keV, where the current telescopes *Chandra*, *XMM-Newton* and *Swift-XRT* are most sensitive, SN 2005kd followed an X-ray luminosity evolution of $L_{\rm Xray} \propto t^{-1.6}$ (Dwarkadas et al. 2016). Contrary to this, SN 2006jd followed a flatter decline of $t^{-0.2}$ for up to 4.5 years (Chandra et al. 2012a), followed by a much steeper decline 8 years later (Katsuda et al. 2016). The X-ray luminosity for SN 2010jl was found to be roughly constant for the first ~ 200 days with a power-law index of $t^{0.13\pm0.08}$, followed by a much faster decay with a power-law index of $\sim -2.12 \pm 0.13$ after day 400. Unfortunately there are no data between 200 and 400 days to see the luminosity evolution in this range. Figure 4 shows the plot of X-ray and bolometric luminosity light curves of SN 2010jl and SN 2006jd. The deviation from t^{-1} is quite evident here. The luminosity decline is much flatter in SN 2006jd than that of SN 2010jl. Around the same epoch, the SN 2006jd X-ray light curve declines as $t^{-0.24}$, while SN 2010jl declines as $t^{-2.12}$. The relative flatness of the bolometric luminosity of SN 2006jd for a longer duration indicates that the CSM interaction powered the light curve for a much longer time than SN 2010jl. This indicates that the duration of mass ejection in SN 2006jd may have been longer in this case than for SN 2010jl, though in both cases it occurred shortly before the explosion. This suggests a different nature of the progenitors for the two SNe.

The narrow energy ranges of present X-ray instruments pose a problem because one measures spectral X-ray luminosity, whereas the $\sim 1/t$ dependence is valid for total X-ray luminosity (Fransson et al. 1996). Observations have indicated that the temperatures of the X-ray emitting regions in many SNe IIn are higher than can be measured in the 0.2–10 keV bandpass of the X-ray detectors, thus we may be missing the peak of the X-ray emission. A measurable effect of this instrument bias is that, as shock sweeps up more and more mate-

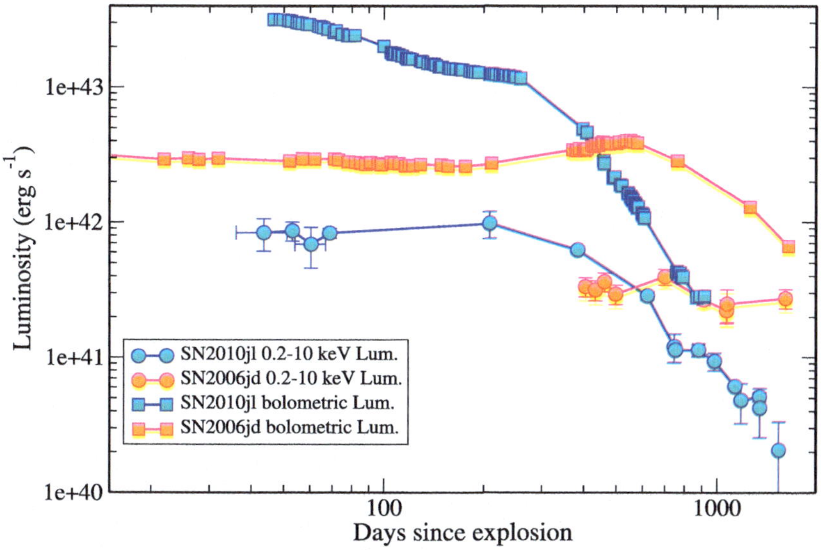

Fig. 4 The X-ray (*filled circles*) and bolometric luminosity (*filled squares*) evolution of SN 2006jd and SN 2010jl. The *blue curves* are for SN 2010jl and *orange curves* indicate SN 2006jd. None of them follow $\sim 1/t$ dependence. Chandra et al. (reproduced from, 2015)

rial with time, the shock temperature decreases, shifting the X-ray emission to progressively lower temperatures at later epochs. Thus one should see flattening or even increase in the soft X-ray emission flux with time. This was seen in SN 1978K, where 2–10 keV flux showed a decline, but 0.5–2 keV soft X-ray emission remained constant for a long time without any signs of decline (Schlegel et al. 2004). However, there are exceptions like SN 1986J and SN 1988Z where the soft X-ray luminosity was found to evolve as $L_{\mathrm{Xray}} \propto t^{-3}$ (Temple et al. 2005) and $L_{\mathrm{Xray}} \propto t^{-2.6}$ (Schlegel and Petre 2006), respectively. This may suggest that more physical reasons (like complex progenitor, unsteady mass loss rates etc.) are responsible for this behaviour. In this respect, SN 1996cr was a unique SN which demonstrated a factor of > 30 increase in X-ray flux between 1997 and 2000 (Bauer et al. 2008). The observations implied that the progenitor of SN 1996cr exploded in a cavity and freely expanded for 1–2 before striking the dense CSM. The above observational study shows that It is crucial to have wide band X-ray coverage to disentangle the instrumental issues from the physical reasons in order to truly understand the nature of the X-ray emission.

4.1 Evolution of Column Density

In some SNe IIn with well sampled X-ray observations, the X-ray column density seems to evolve with time, indicating the cause of absorption as CSM as opposed to ISM. Katsuda et al. (2014) found a column density of $N_H \sim 5 \times 10^{22}$ cm^{-2} for the first few years in SN 2005ip, gradually decreasing to $N_H \sim 4 \times 10^{20}$ cm^{-2}, consistent with Galactic absorption. The fact that the spectra was fit well with a thermal emission model with $kT > 7$ keV implied a forward shock origin of the X-rays, hence absorbed by the CSM which was evolving with time.

SN 2010jl is the best studied X-ray supernova, where X-ray observations covered a period up to day 1500 with *Chandra*, *XMM-Newton*, *NuSTAR*, and *Swift* X-ray Telescope. This

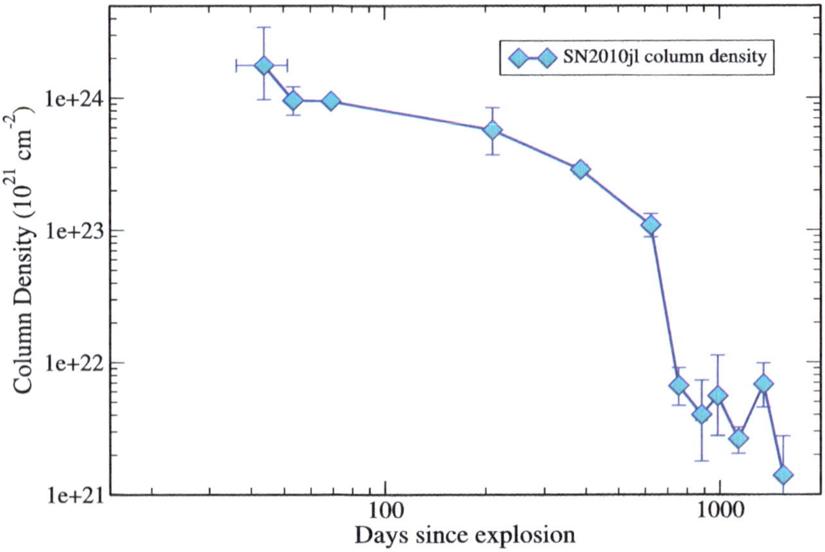

Fig. 5 The column density evolution in SN 2010jl. The figure is reproduced from Chandra et al. (2015)

SN provided the best studied evolution of column density from 40 to 1500 days (Fig. 5 and Chandra et al. 2012b, 2015). During this time period, three orders of magnitude change in column density was witnessed. At the first epoch the column density associated with the SN, $N_H = 10^{24}$ cm^{-2}, 3000 which was times higher than the Galactic column density, which declined slowly by an order of magnitude up to day ~ 650. Followed by a sharp decline, it settled to 10 times the Galactic value. The higher column density observed in the X-ray observations for SN 2010jl was not found to be associated with the high host galaxy extinction, indicating that the higher column density is due to the mass loss near the forward shock and thus is arising from the CSM. This was the first time that external circumstellar X-ray absorption had been clearly observed in a SN and enabled one to trace the precise mass loss evolution history of the star.

As discussed in previous section, due to high density of the CSM, much of the X-ray is reprocessed to visible bands. Thus one would expect less efficient reprocessing as material becomes more transparent with time. In the X-ray and bolometric luminosities plot of SN 2006jd and SN 2010jl (Fig. 4), the bolometric luminosities are at least an order of magnitude higher for first several hundred days. However, one can see that the difference between the two luminosities is decreasing with time. This indicates that with decreasing density, reprocessing of X-rays is becoming less efficient. Long term follow up of the X-ray and bolometric luminosities will be very informative here.

5 Radio Emission of SNe IIn

Radio emission in SNe IIn is very intriguing. Out of ~ 400, less than half SNe IIn (~ 154) have been looked at the radio bands (mainly with the Very Large Array (VLA)) and $\sim 10\%$ have been detected. However, there is a very interesting trend in radio and X-ray detected SNe IIn. While X-ray luminosities are towards the high end in SNe IIn, the radio luminosities are diverse. In Fig. 6, we plot peak X-ray and radio luminosities of radio and X-ray

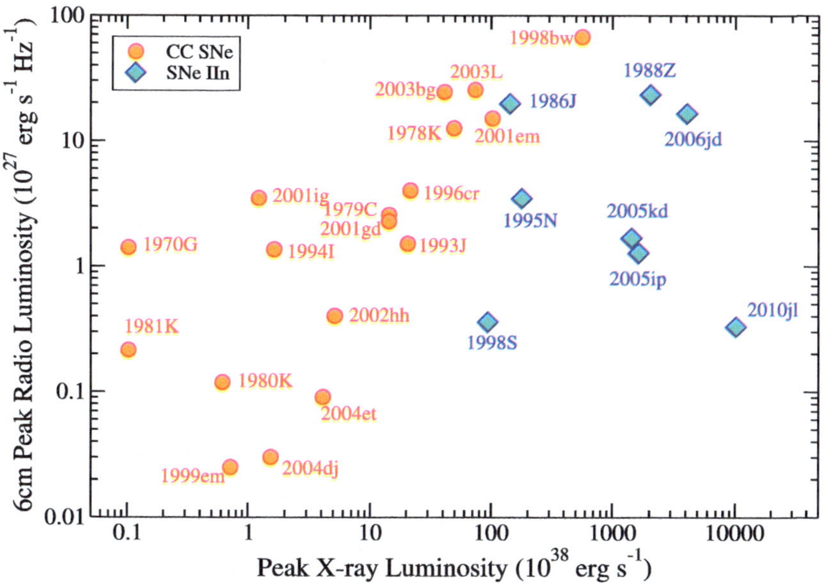

Fig. 6 Radio and X-ray peak luminosities of SNe IIn compared with some well observed core collapse SNe. While in radio bands, SNe IIn do not stand out, they occupy higher values in the X-ray luminosity space. The X-ray data are taken from Immler (2007) and radio data mainly from Weiler et al. (2002)

detected SNe IIn and compare them with some well sampled core collapse SNe. While the X-ray luminosities of SNe IIn are systematically higher than their counterparts, radio luminosities do not stand out and occupy four orders of magnitude space.

This probably could be explained as a side effect of high density. While high density of the CSM means efficient production of synchrotron radio emission, it also translates into higher absorption. Hence, by the time emission reaches optical depth of unity, the strength of synchrotron emission is already weakened. This can explain that radio light curves are extremely diverse; the diversity translate to diversity in the CSM density.

The support for this argument also comes from the fact that most of the SNe IIn are late radio emitters (Fig. 7), indicating higher absorption early on (Chandra et al., to be submitted). Some of the radio emitting SNe IIn were either discovered late or classified late. The prototypical SNe IIn SN 1986J, SN 1988Z and SN 1978K were observed years after their explosions. Most of the SNe IIn discovered soon after the explosion, did not have early radio observations within a month, except for SN 2009ip (Margutti et al. 2014). SN 2009ip was detected in radio bands early on but faded below detection within a few tens of days. However, SN 2009ip was a peculiar SN with a poorly understood explosion mechanism. In contrast the observations of SN 2010jl started by day 45 but the first detection happened only after 500 days (Chandra et al. 2015), clearly showcasing the efficient absorption of radio emission at early epochs.

In addition to high absorption, observational biases could also be partially responsible. The late radio turn on may also explain the observed overall low detection rates of SNe IIn. If a SN is not bright in radio bands in a first few epochs, the observational campaign for that SN is usually over. So even if a SN IIn was a potential radio emitter at later epochs, it would be missed in radio bands. To understand this observational bias, van Dyk et al. (1996) observed 10 SNe IIn at an age of a few hundred days and still did not detect any; they set upper

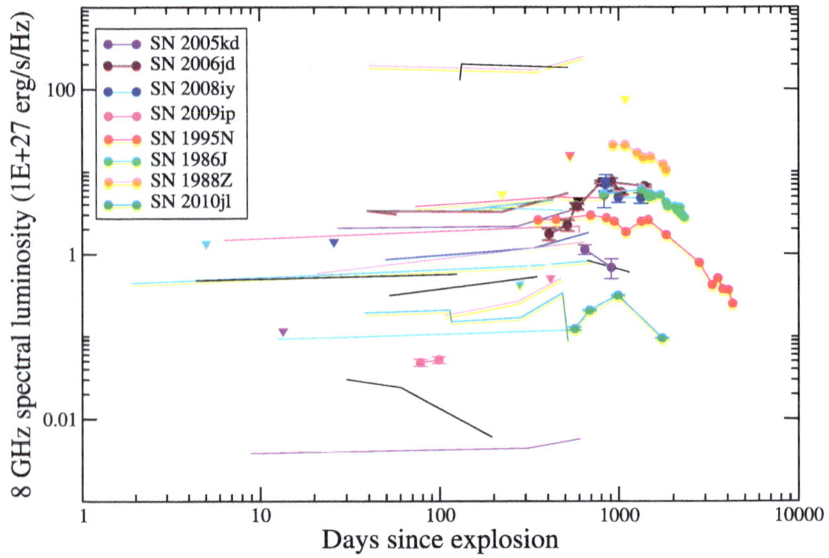

Fig. 7 8 GHz radio spectral luminosity of some of the well observed SNe IIn (Chandra et al. to be submitted). Here *circles* are detections. The *lines* indicate upper limits at observations at many epochs. Single *triangles* indicate upper limits at individual epochs

limits of 150–250 μJy. However, the current telescopes are able to reach at least an order of magnitude better sensitivity. In addition, the current upgraded VLA is extremely sensitive at higher frequencies where the radio absorption effects vanish much sooner ($\tau_\nu \propto \nu^{-2}$). Strategic study of SNe IIn at higher frequencies at early times to low frequencies at late epochs is likely to result in the most complete sample of radio supernovae (Perez-Torres et al. 2015; Wang et al. 2015).

5.1 Absorption of Radio Emission

In classic core collapse SNe, the radio emission is absorbed either by the ionized CS medium (FFA) or by the in situ relativistic electrons (SSA). Observations when the SN is rising in the light curve evolution can disentangle the underlying absorption mechanism. The role of absorption mechanisms is different for different SNe (Chevalier and Fransson 2003). SNe IIn are expected to have external FFA due to their high density. However, in SNe IIn due to high density, a radiative cooling shell is formed in the shocked region. The mixing of the cool gas in the forward shocked shell may cause some SNe IIn to undergo internal FFA. This indeed has been seen in SN 2006jd (Fig. 8 and Chandra et al. 2012a), SN 1986J (Chandra et al. 2012a) and SN 1988Z (van Dyk et al. 1993; Williams et al. 2002).

In SN 2006jd, Chandra et al. (2012a) estimated the mass of the mixed absorbing gas by assuming that the absorbing low temperature (10^4–10^5 K) gas is in pressure equilibrium with the X-ray emitting gas. They showed that a modest amount of $\sim 10^4$ K cool gas, i.e. Mass $\sim 10^{-8}\,M_\odot$ could explain internal absorption in SN 2006jd (Fig. 8).

Irrespective of the fact whether absorption is internal or external, it implies very high mass loss rates. However, the mass loss rate constraints will become less severe by a factor of 3 if the wind is clumpy (Puls et al. 2008). In addition, calculations of mass loss rate are usually done assuming solar metallicity. However, metallicity will lower mass loss rate as

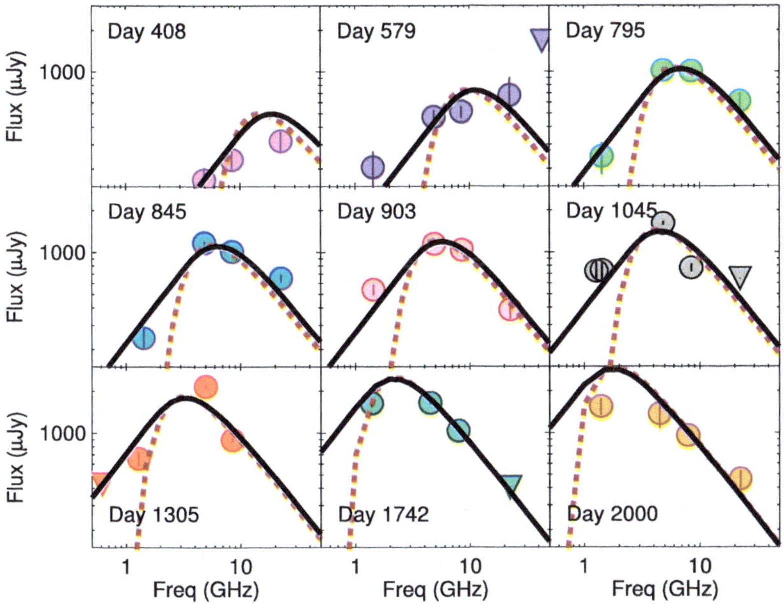

Fig. 8 Spectra of SN 2006jd reproduced from Chandra et al. (2012a). Here internal free-free absorption (*black solid lines*) best fits the observed absorption

$\dot{M} \sim Z^{0.69}$ (Vink et al. 2001) Interestingly, recent observational findings demonstrated that erratic mass-loss behavior preceding core-collapse extends to H-poor progenitors as well.

6 Episodic Mass Loss?

Multiwaveband observations of SNe IIn have shown that the mass loss rate from some SNe IIn may not be steady. Unfortunately it is difficult to disentangle this effect because SNe IIn are likely to remain highly absorbed for a long duration. By the time they become optically thin in radio bands, the strength of synchrotron emission is too weak to see any modulations due to changing mass loss. In addition, one needs a long term follow up to see the episodic mass loss. In X-ray observations, due to the lack of well sampled data, such trends are easy to miss. Yet, it seems that some SNe IIn seem to suggest episodic mass loss rate.

In Fig. 9, X-ray and radio light curves are plotted for a Type IIn SN 1995N (Chandra et al. 2009). The X-ray bands were chosen to mimic the *ROSAT* and *ASCA* bands (Chandra et al. 2005). The radio data is at the four representative VLA frequencies. Around day ~ 1300 onwards, bumps in the X-ray and radio light curves are seen. The long gap in the X-ray light curve does not show the evolution of this bump, however, in the radio data one can clearly see this effect. Due to the sensitivity limitation of the telescopes and lack of long term follow up, its not possible to explore the further bumps in time, if any.

Hints of bumps near simultaneously in both X-ray and radio bands light curves can be possibly seen in SN 2006jd as well (Chandra et al. 2012a). The radio emission is too weak in SN 2010jl to note any such bumps. However, X-ray data seems to suggest a steady decline followed by a sudden decline settling down to steady decline again at ~ 1000 days (Chandra et al. 2015). These trends could be suggestive of the episodic mass loss rates in some SNe IIn

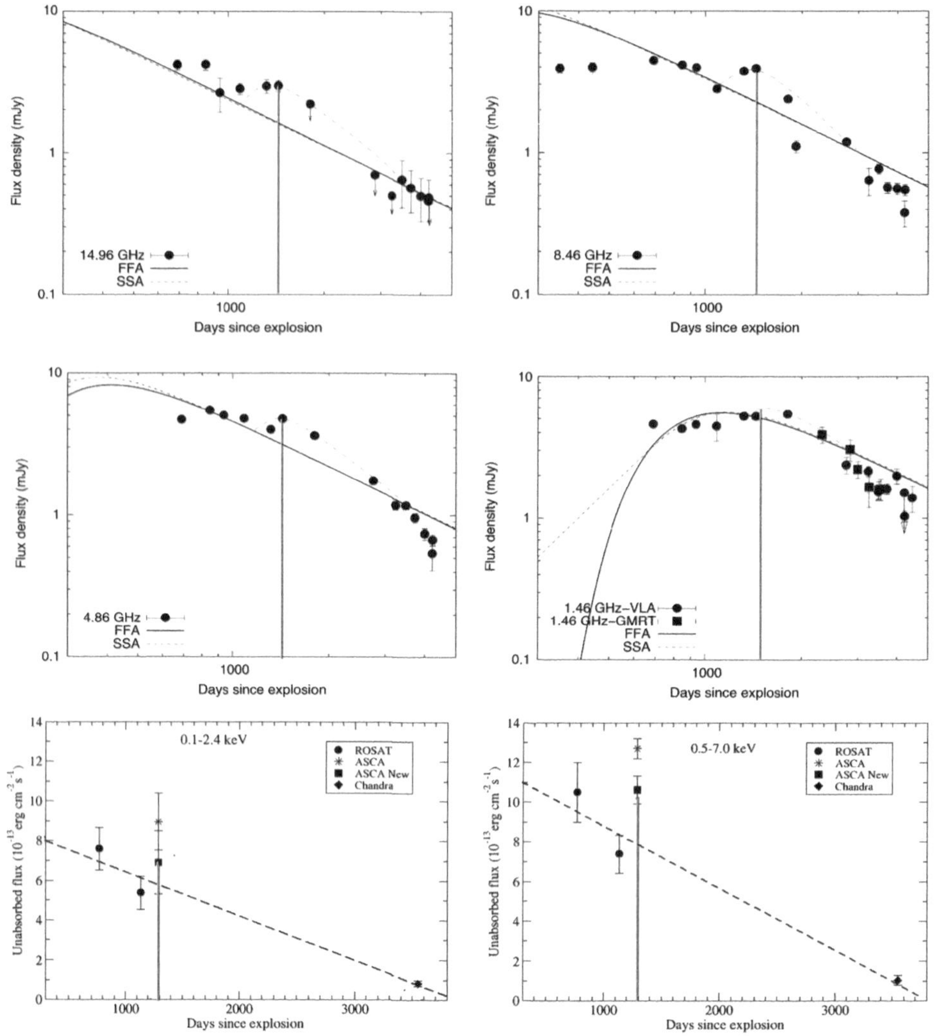

Fig. 9 Radio and X-ray light curves of SN 1995N. A hint of enhanced emission bump in radio and X-ray bands of a Type IIn SN 1995N around $\sim$ 1300 days. The figure is reproduced from Chandra et al. (2005, 2009)

and may support the LBV scenario. However, further study and much longer follow ups are needed. Episodic mass-loss rates have been seen in many type Ib/Ic/IIb SNe too.

7 Asymmetry in the Explosion

Although in normal core collapse SNe circumstellar interaction can be successfully described in terms of spherical models, there are indications of a more complex situation in SNe exploding in dense environments. In optical bands, there have been various pieces of evidence suggesting asymmetry in the explosion via polarization observations. The optical spectropolarimetry observations with Keck telescope showed a high degree of linear polarization in SN 1998S, implying significant asphericity for its CSM environment (Leonard et al. 2000). In SN 1997eg, spectropolarimetric observations indicated the presence of a flattened disk-like CSM surrounding aspherical ejecta, which Hoffman et al. (2008) interpreted as supporting the LBV progenitor scenario. Patat et al. (2011) found significant polarization in SN 2010jl two weeks after the discovery, suggesting similarity with SN 1998S and SN 1997eg. Many of the models explaining SNe IIn require a bipolar wind or disk like geometry (see Smith 2016 for a review).

In some SNe IIn, interpretation of the radio and X-ray data too seem to suggest asymmetry. In case of SN 2006jd, it was difficult to reconcile the X-ray and radio data with spherical models (Chandra et al. 2012a). The column density of the matter absorbing the X-ray emission was found to be ~ 50 times smaller than what was needed to produce the X-ray luminosity. In addition, the implied column density from radio observations were also found to be low. Chandra et al. (2012a) attempted to explain it with the clumpiness of the CSM clouds, since the absorption column depends on the CSM wind density as $\propto \rho_{\rm wind}$ and the X-ray luminosity as $\propto \rho^2_{\rm wind}$. However, it lead to an inconsistent shock velocity derived from the observed X-ray spectrum and this model was ruled out. They explained it in a scenario in which the CSM clouds interact primarily with the CDS. This suggested that there is a global asymmetry in the CSM gas distribution that allowed a low column density in one direction, unaffected by the dense interaction taking place over much of the rest of the solid angle as viewed from the SN. This would also explain the low column density to the radio emission, because both radio and X-ray are from the same region. The lack of external FFA seen in this SN could also be explained in this model.

SN 2010jl early optical spectra showed the presence of broad emission lines (Smith et al. 2012; Fransson et al. 2014; Ofek et al. 2014b). This could be explained by an electron scattering optical depth > 1–3, i.e., a column density $N_H \geq 3 \times 10^{24}$ cm^{-2}. This was comparable with the column density seen in the first X-ray observations (Chandra et al. 2012b). However, at later epochs ($t > 70$ day), the X-rays column density declined to $N_H < 3 \times 10^{24}$ cm^{-2}. These constraints are difficult to reconcile with a spherical model of column density 3×10^{24} cm^{-2}. This may suggest that the X-rays escape the interaction region, avoiding the high column density CSM. This situation can be reconciled well in a scenario for the CSM having a bipolar geometry.

If LBVs indeed are the progenitor of some SNe IIn, they are most likely asymmetric as known Galactic LBVs show a complex CSM structure. In addition, binary evolution can also lead to aspherical structure (e.g., McCray and Fransson 2016).

8 Summary and Open Problems

In this chapter, I have discussed the observational aspects of the ejecta-CSM interaction mainly in SNe exploding in dense environments. The chapter mainly concentrates on the radio and X-ray emission, with the aim of understanding the nature of intriguing progenitor system of SNe IIn.

Observations have indicated that in SNe IIn show high X-ray luminosities dominated by high temperatures for years, implying the forward shock is responsible for the dominant X-ray emission. Long term observations of X-ray emission have indicated evolving CSM in some SNe IIn, directly confirming the evidence that medium surrounding stars is modified via the progenitor star, the properties of which can be studied by the detailed X-ray light curves and spectra. However, one needs to make observations in wide X-ray bands, to avoid biases coming from the narrow X-ray range, and so to disentangle the observational effects versus the ones coming due to the nature of progenitor wind.

In radio bands, SNe IIn are mostly late emitters. Radio detected SNe IIn do not tend to be necessarily bright, most likely due to absorption. The absorption mechanism could have a dominant contribution from the cool gas mixed in the radio emitting region. Only 10% SNe IIn show radio emission. More systematic studies are needed to remove observational biases and pinpoint physical causes of the low detection statistics. The property of late time turn on may be causing us to miss many of them due to lack of sufficiently late observations. With the wealth of new and refurbished sensitive telescopes, a carefully planned sensitive study is needed. A good strategy is to observe at high frequencies for at least a year before discarding it as a radio non-emitter.

The nature of progenitors of SNe IIn remains the biggest challenge. The mass loss rates needed to explain SNe IIn are extremely high. Some SNe IIn have indicated episodic mass loss rates. There are hints of eruptive events towards the end stages of evolution. These effects need to be included in the current generations of stellar evolution models. However, it is important to note here that the mass loss rate constraints will become less severe by a factor of 3 if the wind is clumpy.

Many of these stars are expected to live in binary systems; their influence in the stellar evolution needs to be explored. The observations have indicated low progenitor masses in some SNe IIn. This has challenged the idea of interacting SNe arising from very massive progenitor stars, and puts a question mark on the LBV progenitors. In fact, the very nature of the LBV has been questioned and much work is needed in this direction. A very sensitive observation of the Galactic OB regions and LBV stars could throw light on this.

Acknowledgements I thank referee for useful comments, which helped improve this chapter. I acknowledge the support from the Department of Science and Technology via SwaranaJayanti Fellowship award (file no. DST/SJF/PSA-01/2014-15). The chapter has benefited immensely from discussions with Roger Chevalier, Claes Fransson, and Nikolai Chugai at various stages. I am highly indebted to Roger Chevalier for proof reading the manuscript.

References

J.P. Anderson, S.M. Habergham, P.A. James, M. Hamuy, Progenitor mass constraints for core-collapse supernovae from correlations with host galaxy star formation. Mon. Not. R. Astron. Soc. **424**, 1372–1391 (2012). arXiv:1205.3802. https://doi.org/10.1111/j.1365-2966.2012.21324.x

F.E. Bauer, V.V. Dwarkadas, W.N. Brandt, S. Immler, S. Smartt, N. Bartel, M.F. Bietenholz, Supernova 1996cr: SN 1987A's wild cousin? Astrophys. J. **688**, 1210–1234 (2008). arXiv:0804.3597. https://doi.org/10.1086/589761

A.R. Bell, The acceleration of cosmic rays in shock fronts. II. Mon. Not. R. Astron. Soc. **182**, 443–455 (1978). https://doi.org/10.1093/mnras/182.3.443

C. Bilinski, N. Smith, W. Li, G.G. Williams, W. Zheng, A.V. Filippenko, Constraints on type IIn supernova progenitor outbursts from the Lick Observatory Supernova Search. Mon. Not. R. Astron. Soc. **450**, 246–265 (2015). arXiv:1503.04252. https://doi.org/10.1093/mnras/stv566

R.D. Blandford, J.P. Ostriker, Particle acceleration by astrophysical shocks. Astrophys. J. Lett. **221**, L29–L32 (1978). https://doi.org/10.1086/182658

G. Brown, H.A. Bethe, How a supernova explodes. Sci. Am. **252**, 60–68 (1985). https://doi.org/10.1038/scientificamerican0585-60

P. Chandra, R.A. Chevalier, Swift-XRT observations of type IIn supernova ASASSN-17kr a.k.a. SN 2017gas. The Astronomer's Telegram 10705 (2017)

P. Chandra, A. Ray, S. Bhatnagar, Synchrotron aging and the radio spectrum of SN 1993J. Astrophys. J. Lett. **604**, L97–L100 (2004). arXiv:astro-ph/0402391. https://doi.org/10.1086/383615

P. Chandra, A. Ray, E.M. Schlegel, F.K. Sutaria, W. Pietsch, Chandra's Tryst with SN 1995N. Astrophys. J. **629**, 933–943 (2005). arXiv:astro-ph/0505051. https://doi.org/10.1086/431573

P. Chandra, C.J. Stockdale, R.A. Chevalier, S.D. Van Dyk, A. Ray, M.T. Kelley, K.W. Weiler, N. Panagia, R.A. Sramek, Eleven years of radio monitoring of the type IIn supernova SN 1995N. Astrophys. J. **690**, 1839–1846 (2009). arXiv:0809.2810. https://doi.org/10.1088/0004-637X/690/2/1839

P. Chandra, R.A. Chevalier, N. Chugai, C. Fransson, C.M. Irwin, A.M. Soderberg, S. Chakraborti, S. Immler, Radio and X-ray observations of SN 2006jd: another strongly interacting type IIn supernova. Astrophys. J. **755**, 110 (2012a). arXiv:1205.0250. https://doi.org/10.1088/0004-637X/755/2/110

P. Chandra, R.A. Chevalier, C.M. Irwin, N. Chugai, C. Fransson, A.M. Soderberg, Strong evolution of X-ray absorption in the type IIn supernova SN 2010jl. Astrophys. J. Lett. **750**, L2 (2012b). arXiv:1203.1614. https://doi.org/10.1088/2041-8205/750/1/L2

P. Chandra, R.A. Chevalier, N. Chugai, C. Fransson, A.M. Soderberg, X-ray and radio emission from type IIn supernova SN 2010jl. Astrophys. J. **810**, 32 (2015). arXiv:1507.06059. https://doi.org/10.1088/0004-637X/810/1/32

R.A. Chevalier, The interaction of the radiation from a type II supernova with a circumstellar shell. Astrophys. J. **251**, 259–265 (1981). https://doi.org/10.1086/159460

R.A. Chevalier, Self-similar solutions for the interaction of stellar ejecta with an external medium. Astrophys. J. **258**, 790–797 (1982a). https://doi.org/10.1086/160126

R.A. Chevalier, The radio and X-ray emission from type II supernovae. Astrophys. J. **259**, 302–310 (1982b). https://doi.org/10.1086/160167

R.A. Chevalier, Synchrotron self-absorption in radio supernovae. Astrophys. J. **499**, 810–819 (1998). https://doi.org/10.1086/305676

R.A. Chevalier, Common envelope evolution leading to supernovae with dense interaction. Astrophys. J. Lett. **752**, L2 (2012). arXiv:1204.3300. https://doi.org/10.1088/2041-8205/752/1/L2

R. Chevalier, J.M. Blondin, Hydrodynamic instabilities in supernova remnants: early radiative cooling. Astrophys. J. **444**, 312–317 (1995). https://doi.org/10.1086/175606

R.A. Chevalier, C. Fransson, Emission from circumstellar interaction in normal type II supernovae. Astrophys. J. **420**, 268–285 (1994). https://doi.org/10.1086/173557

R.A. Chevalier, C. Fransson, Supernova interaction with a circumstellar medium, in *Supernovae and Gamma-Ray Bursters*, ed. by K. Weiler. Lecture Notes in Physics, vol. 598 (Springer, Berlin, 2003), pp. 171–194. arXiv:astro-ph/0110060

R.A. Chevalier, C. Fransson, Thermal and non-thermal emission from circumstellar interaction. arXiv:1612.07459 (2016)

R.A. Chevalier, C.M. Irwin, Shock breakout in dense mass loss: luminous supernovae. Astrophys. J. Lett. **729**, L6 (2011). arXiv:1101.1111. https://doi.org/10.1088/2041-8205/729/1/L6

N.N. Chugai, Late stage radiation source in type-II supernovae—radioactivity or shock heating. Sov. Astron. Lett. **16**, 457 (1990)

N.N. Chugai, R.A. Chevalier, Late emission from the Type Ib/c SN 2001em: overtaking the hydrogen envelope. Astrophys. J. **641**, 1051–1059 (2006). arXiv:astro-ph/0510362. https://doi.org/10.1086/500539

N.N. Chugai, I.J. Danziger, Supernova 1988Z—low-mass ejecta colliding with the clumpy wind. Mon. Not. R. Astron. Soc. **268**, 173 (1994). https://doi.org/10.1093/mnras/268.1.173

N.N. Chugai, I.J. Danziger, A massive circumstellar envelope around the Type-IIn supernova SN 1995G. Astron. Lett. **29**, 649–657 (2003). arXiv:astro-ph/0306330. https://doi.org/10.1134/1.1615333

N.N. Chugai, S.I. Blinnikov, R.J. Cumming, P. Lundqvist, A. Bragaglia, A.V. Filippenko, D.C. Leonard, T. Matheson, J. Sollerman, The Type IIn supernova 1994W: evidence for the explosive ejection of a circumstellar envelope. Mon. Not. R. Astron. Soc. **352**, 1213–1231 (2004). arXiv:astro-ph/0405369. https://doi.org/10.1111/j.1365-2966.2004.08011.x

J. Cooke, M. Sullivan, E.J. Barton, J.S. Bullock, R.G. Carlberg, A. Gal-Yam, E. Tollerud, Type IIn supernovae at redshift $z \sim 2$ from archival data. Nature **460**, 237–239 (2009). arXiv:0907.1928. https://doi.org/10.1038/nature08082

B. Dilday, D.A. Howell, S.B. Cenko, J.M. Silverman, P.E. Nugent, M. Sullivan, S. Ben-Ami, L. Bildsten, M. Bolte, M. Endl, A.V. Filippenko, O. Gnat, A. Horesh, E. Hsiao, M.M. Kasliwal, D. Kirkman, K. Maguire, G.W. Marcy, K. Moore, Y. Pan, J.T. Parrent, P. Podsiadlowski, R.M. Quimby, A. Sternberg, N. Suzuki, D.R. Tytler, D. Xu, J.S. Bloom, A. Gal-Yam, I.M. Hook, S.R. Kulkarni, N.M. Law, E.O. Ofek, D. Polishook, D. Poznanski, PTF 11kx: a type Ia supernova with a symbiotic nova progenitor. Science **337**, 942 (2012). arXiv:1207.1306. https://doi.org/10.1126/science.1219164

V.V. Dwarkadas, J. Gruszko, What are published X-ray light curves telling us about young supernova expansion? Mon. Not. R. Astron. Soc. **419**, 1515–1524 (2012). arXiv:1109.2616. https://doi.org/10.1111/j.1365-2966.2011.19808.x

V.V. Dwarkadas, C. Romero-Cañizales, R. Reddy, F.E. Bauer, X-ray and radio emission from the luminous supernova 2005kd. Mon. Not. R. Astron. Soc. **462**, 1101–1110 (2016). arXiv:1607.06104. https://doi.org/10.1093/mnras/stw1717

O.D. Fox, S.D. Van Dyk, E. Dwek, N. Smith, A.V. Filippenko, J. Andrews, R.G. Arendt, R.J. Foley, P.L. Kelly, A.A. Miller, I. Shivvers, The candidate progenitor of the type IIn SN 2010jl is not an optically luminous star. Astrophys. J. **836**, 222 (2017). arXiv:1611.00369. https://doi.org/10.3847/1538-4357/836/2/222

C. Fransson, Comptonization and UV emission lines from Type II supernovae. Astron. Astrophys. **133**, 264–284 (1984)

C. Fransson, P. Lundqvist, R.A. Chevalier, Circumstellar interaction in SN 1993J. Astrophys. J. **461**, 993 (1996). https://doi.org/10.1086/177119

C. Fransson, M. Ergon, P.J. Challis, R.A. Chevalier, K. France, R.P. Kirshner, G.H. Marion, D. Milisavljevic, N. Smith, F. Bufano, A.S. Friedman, T. Kangas, J. Larsson, S. Mattila, S. Benetti, R. Chornock, I. Czekala, A. Soderberg, J. Sollerman, High-density circumstellar interaction in the luminous type IIn SN 2010jl: the first 1100 days. Astrophys. J. **797**, 118 (2014). arXiv:1312.6617. https://doi.org/10.1088/0004-637X/797/2/118

M. Fraser, C. Inserra, A. Jerkstrand, R. Kotak, G. Pignata, S. Benetti, M.T. Botticella, F. Bufano, M. Childress, S. Mattila, A. Pastorello, S.J. Smartt, M. Turatto, F. Yuan, J.P. Anderson, D.D.R. Bayliss, F.E. Bauer, T.W. Chen, F. Förster Burón, A. Gal-Yam, J.B. Haislip, C. Knapic, L. Le Guillou, S. Marchi, P. Mazzali, M. Molinaro, J.P. Moore, D. Reichart, R. Smareglia, K.W. Smith, A. Sternberg, M. Sullivan, K. Takáts, B.E. Tucker, S. Valenti, O. Yaron, D.R. Young, G. Zhou, SN 2009ip à la PESSTO: no evidence for core collapse yet. Mon. Not. R. Astron. Soc. **433**, 1312–1337 (2013). arXiv:1303.3453. https://doi.org/10.1093/mnras/stt813

A.W. Fullerton, D.L. Massa, R.K. Prinja, The discordance of mass-loss estimates for galactic O-type stars. Astrophys. J. **637**, 1025–1039 (2006). arXiv:astro-ph/0510252. https://doi.org/10.1086/498560

A. Gal-Yam, D.C. Leonard, D.B. Fox, S.B. Cenko, A.M. Soderberg, D.S. Moon, D.J. Sand, Caltech Core Collapse Program, W. Li, A.V. Filippenko, G. Aldering, Y. Copin, On the progenitor of SN 2005gl and the nature of type IIn supernovae. Astrophys. J. **656**, 372–381 (2007). arXiv:astro-ph/0608029. https://doi.org/10.1086/510523

J.L. Hoffman, D.C. Leonard, R. Chornock, A.V. Filippenko, A.J. Barth, T. Matheson, The dual-axis circumstellar environment of the type IIn supernova 1997eg. Astrophys. J. **688**, 1186–1209 (2008). arXiv:0709.3258. https://doi.org/10.1086/592261

R.M. Humphreys, K. Davidson, The luminous blue variables: astrophysical geysers. Publ. Astron. Soc. Pac. **106**, 1025–1051 (1994). https://doi.org/10.1086/133478

R.M. Humphreys, K. Davidson, N. Smith, η Carinae's second eruption and the light curves of the η Carinae variables. Publ. Astron. Soc. Pac. **111**, 1124–1131 (1999). https://doi.org/10.1086/316420

S. Immler, X-ray emission from supernovae, in *Supernova 1987A: 20 Years After: Supernovae and Gamma-Ray Bursters*, ed. by S. Immler, K. Weiler, R. McCray. American Institute of Physics Conference Series, vol. 937 (2007), pp. 246–255. https://doi.org/10.1063/1.3682910

S. Katsuda, K. Maeda, T. Nozawa, D. Pooley, S. Immler, SN 2005ip: a luminous type IIn supernova emerging from a dense circumstellar medium as revealed by X-ray observations. Astrophys. J. **780**, 184 (2014). arXiv:1311.7180. https://doi.org/10.1088/0004-637X/780/2/184

S. Katsuda, K. Maeda, A. Bamba, Y. Terada, Y. Fukazawa, K. Kawabata, M. Ohno, Y. Sugawara, Y. Tsuboi, S. Immler, Two distinct-absorption X-ray components from type IIn supernovae: evidence for asphericity in the circumstellar medium. Astrophys. J. **832**, 194 (2016). arXiv:1609.09093. https://doi.org/10.3847/0004-637X/832/2/194

D.C. Leonard, A.V. Filippenko, A.J. Barth, T. Matheson, Evidence for asphericity in the type IIN supernova SN 1998S. Astrophys. J. **536**, 239–254 (2000). arXiv:astro-ph/9908040. https://doi.org/10.1086/308910

W.D. Li, C. Li, A.V. Filippenko, E.C. Moran, Supernova 1998S in NGC 3877. IAU Circ 6829 (1998)

R. Margutti, D. Milisavljevic, A.M. Soderberg, C. Guidorzi, B.J. Morsony, N. Sanders, S. Chakraborti, A. Ray, A. Kamble, M. Drout, J. Parrent, A. Zauderer, L. Chomiuk, Relativistic supernovae have shorter-lived central engines or more extended progenitors: the case of SN 2012ap. Astrophys. J. **797**, 107 (2014). arXiv:1402.6344. https://doi.org/10.1088/0004-637X/797/2/107

R. Margutti, A. Kamble, D. Milisavljevic, E. de Zapartas, S.E. Mink, M. Drout, R. Chornock, G. Risaliti, B.A. Zauderer, M. Bietenholz, M. Cantiello, S. Chakraborti, L. Chomiuk, W. Fong, B. Grefenstette, C. Guidorzi, R. Kirshner, J.T. Parrent, D. Patnaude, A.M. Soderberg, N.C. Gehrels, F. Harrison, Ejection of the massive hydrogen-rich envelope timed with the collapse of the stripped SN 2014C. Astrophys. J. **835**, 140 (2017). arXiv:1601.06806. https://doi.org/10.3847/1538-4357/835/2/140

C.D. Matzner, C.F. McKee, The expulsion of stellar envelopes in core-collapse supernovae. Astrophys. J. **510**, 379–403 (1999). arXiv:astro-ph/9807046. https://doi.org/10.1086/306571

J.C. Mauerhan, N. Smith, A.V. Filippenko, K.B. Blanchard, P.K. Blanchard, C.F.E. Casper, S.B. Cenko, K.I. Clubb, D.P. Cohen, K.L. Fuller, G.Z. Li, J.M. Silverman, The unprecedented 2012 outburst of SN 2009ip: a luminous blue variable star becomes a true supernova. Mon. Not. R. Astron. Soc. **430**, 1801–1810 (2013). arXiv:1209.6320. https://doi.org/10.1093/mnras/stt009

R. McCray, C. Fransson, The remnant of supernova 1987A. Annu. Rev. Astron. Astrophys. **54**, 19–52 (2016). https://doi.org/10.1146/annurev-astro-082615-105405

D. Milisavljevic, R. Margutti, A. Kamble, D.J. Patnaude, J.C. Raymond, J.J. Eldridge, W. Fong, M. Bietenholz, P. Challis, R. Chornock, M.R. Drout, C. Fransson, R.A. Fesen, J.E. Grindlay, R.P. Kirshner, R. Lunnan, J. Mackey, G.F. Miller, J.T. Parrent, N.E. Sanders, A.M. Soderberg, B.A. Zauderer, Metamorphosis of SN 2014C: delayed interaction between a hydrogen poor core-collapse supernova and a nearby circumstellar shell. Astrophys. J. **815**, 120 (2015). arXiv:1511.01907. https://doi.org/10.1088/0004-637X/815/2/120

T.K. Nymark, C. Fransson, C. Kozma, X-ray emission from radiative shocks in type II supernovae. Astron. Astrophys. **449**, 171–192 (2006). arXiv:astro-ph/0510792. https://doi.org/10.1051/0004-6361:20054169

E.O. Ofek, D. Fox, S.B. Cenko, M. Sullivan, O. Gnat, D.A. Frail, A. Horesh, A. Corsi, R.M. Quimby, N. Gehrels, S.R. Kulkarni, A. Gal-Yam, P.E. Nugent, O. Yaron, A.V. Filippenko, M.M. Kasliwal, L. Bildsten, J.S. Bloom, D. Poznanski, I. Arcavi, R.R. Laher, D. Levitan, B. Sesar, J. Surace, X-ray emission from supernovae in dense circumstellar matter environments: a search for collisionless shocks. Astrophys. J. **763**, 42 (2013). arXiv:1206.0748. https://doi.org/10.1088/0004-637X/763/1/42

E.O. Ofek, M. Sullivan, N.J. Shaviv, A. Steinbok, I. Arcavi, A. Gal-Yam, D. Tal, S.R. Kulkarni, P.E. Nugent, S. Ben-Ami, M.M. Kasliwal, S.B. Cenko, R. Laher, J. Surace, J.S. Bloom, A.V. Filippenko, J.M. Silverman, O. Yaron, Precursors prior to type IIn supernova explosions are common: precursor rates, properties, and correlations. Astrophys. J. **789**, 104 (2014a). arXiv:1401.5468. https://doi.org/10.1088/0004-637X/789/2/104

E.O. Ofek, A. Zoglauer, S.E. Boggs, N.M. Barriére, S.P. Reynolds, C.L. Fryer, F.A. Harrison, S.B. Cenko, S.R. Kulkarni, A. Gal-Yam, I. Arcavi, E. Bellm, J.S. Bloom, F. Christensen, W.W. Craig, W. Even, A.V. Filippenko, B. Grefenstette, C.J. Hailey, R. Laher, K. Madsen, E. Nakar, P.E. Nugent, D. Stern, M. Sullivan, J. Surace, W.W. Zhang, SN 2010jl: optical to hard X-ray observations reveal an explosion embedded in a ten solar mass cocoon. Astrophys. J. **781**, 42 (2014b). arXiv:1307.2247. https://doi.org/10.1088/0004-637X/781/1/42

A.G. Pacholczyk, *Radio Astrophysics. Nonthermal Processes in Galactic and Extragalactic Sources*. Series of Books in Astronomy and Astrophysics (Freeman, San Francisco, 1970)

N. Panagia, M. Felli, The spectrum of the free-free radiation from extended envelopes. Astron. Astrophys. **39**, 1–5 (1975)

A. Pastorello, S.J. Smartt, S. Mattila, J.J. Eldridge, D. Young, K. Itagaki, H. Yamaoka, H. Navasardyan, S. Valenti, F. Patat, I. Agnoletto, T. Augusteijn, S. Benetti, E. Cappellaro, T. Boles, J.M. Bonnet-Bidaud, M.T. Botticella, F. Bufano, C. Cao, J. Deng, M. Dennefeld, N. Elias-Rosa, A. Harutyunyan, F.P. Keenan, T. Iijima, V. Lorenzi, P.A. Mazzali, X. Meng, S. Nakano, T.B. Nielsen, J.V. Smoker, V. Stanishev, M. Turatto, D. Xu, L. Zampieri, A giant outburst two years before the core-collapse of a massive star. Nature **447**, 829–832 (2007). arXiv:astro-ph/0703663. https://doi.org/10.1038/nature05825

F. Patat, S. Taubenberger, S. Benetti, A. Pastorello, A. Harutyunyan, Asymmetries in the type IIn SN 2010jl. Astron. Astrophys. **527**, L6 (2011). arXiv:1011.5926. https://doi.org/10.1051/0004-6361/201016217

M. Perez-Torres, A. Alberdi, R.J. Beswick, P. Lundqvist, R. Herrero-Illana, C. Romero-Cañizales, S. Ryder, M. della Valle, J. Conway, J.M. Marcaide, S. Mattila, T. Murphy, E. Ros, Core-collapse and Type Ia supernovae with the SKA, in *Advancing Astrophysics with the Square Kilometre Array*, AASKA14 (2015), 60. arXiv:1409.1827

J. Puls, J.S. Vink, F. Najarro, Mass loss from hot massive stars. Astron. Astrophys. Rev. **16**, 209–325 (2008). arXiv:0811.0487. https://doi.org/10.1007/s00159-008-0015-8

E. Quataert, J. Shiode, Wave-driven mass loss in the last year of stellar evolution: setting the stage for the most luminous core-collapse supernovae. Mon. Not. R. Astron. Soc. **423**, L92–L96 (2012). arXiv:1202.5036. https://doi.org/10.1111/j.1745-3933.2012.01264.x

R.M. Quimby, F. Yuan, C. Akerlof, J.C. Wheeler, Rates of superluminous supernovae at $z \sim 0.2$. Mon. Not. R. Astron. Soc. **431**, 912–922 (2013). arXiv:1302.0911. https://doi.org/10.1093/mnras/stt213

M. Ross, Dwarkadas VV: SNaX—a database of supernova X-ray lightcurves. arXiv:1704.05866 (2017)

I. Salamanca, R. Cid-Fernandes, G. Tenorio-Tagle, E. Telles, R.J. Terlevich, C. Munoz-Tunon, The circumstellar medium of the peculiar supernova SN1997ab. Mon. Not. R. Astron. Soc. **300**, L17–L21 (1998). arXiv:astro-ph/9809208. https://doi.org/10.1046/j.1365-8711.1998.02093.x

N.E. Sanders, A.M. Soderberg, R.J. Foley, R. Chornock, D. Milisavljevic, R. Margutti, M.R. Drout, M. Moe, E. Berger, W.R. Brown, R. Lunnan, S.J. Smartt, M. Fraser, R. Kotak, L. Magill, K.W. Smith, D. Wright, K. Huang, Y. Urata, J.S. Mulchaey, A. Rest, D.J. Sand, L. Chomiuk, A.S. Friedman, R.P. Kirshner, G.H. Marion, J.L. Tonry, W.S. Burgett, K.C. Chambers, K.W. Hodapp, R.P. Kudritzki, P.A. Price, PS1-12sk is a peculiar supernova from a He-rich progenitor system in a brightest cluster galaxy environment. Astrophys. J. **769**, 39 (2013). arXiv:1303.1818. https://doi.org/10.1088/0004-637X/769/1/39

E.M. Schlegel, R. Petre, A Chandra ACIS observation of the X-ray-luminous SN 1988Z. Astrophys. J. **646**, 378–384 (2006). arXiv:astro-ph/0604106. https://doi.org/10.1086/504890

E.M. Schlegel, A. Kong, P. Kaaret, R. DiStefano, S. Murray, Chandra ACIS and XMM-Newton EPIC observations of the X-ray-luminous SN 1978K in NGC 1313. Astrophys. J. **603**, 644–651 (2004). arXiv:astro-ph/0311434. https://doi.org/10.1086/381571

N. Smith, Episodic post-shock dust formation in the colliding winds of Eta Carinae. Mon. Not. R. Astron. Soc. **402**, 145–151 (2010). arXiv:0910.4395. https://doi.org/10.1111/j.1365-2966.2009.15901.x

N. Smith, Mass loss: its effect on the evolution and fate of high-mass stars. Annu. Rev. Astron. Astrophys. **52**, 487–528 (2014). arXiv:1402.1237. https://doi.org/10.1146/annurev-astro-081913-040025

N. Smith, Interacting supernovae: types IIn and Ibn. arXiv:1612.02006 (2016)

N. Smith, R. Tombleson, Luminous blue variables are antisocial: their isolation implies that they are kicked mass gainers in binary evolution. Mon. Not. R. Astron. Soc. **447**, 598–617 (2015). arXiv:1406.7431. https://doi.org/10.1093/mnras/stu2430

N. Smith, K.H. Hinkle, N. Ryde, Red supergiants as potential type IIn supernova progenitors: spatially resolved 4.6 μm CO emission around VY CMa and Betelgeuse. Astron. J. **137**, 3558–3573 (2009). arXiv:0811.3037. https://doi.org/10.1088/0004-6256/137/3/3558

N. Smith, R. Chornock, J.M. Silverman, A.V. Filippenko, R.J. Foley, Spectral evolution of the extraordinary type IIn supernova 2006gy. Astrophys. J. **709**, 856–883 (2010). arXiv:0906.2200. https://doi.org/10.1088/0004-637X/709/2/856

N. Smith, W. Li, A.A. Miller, J.M. Silverman, A.V. Filippenko, J.C. Cuillandre, M.C. Cooper, T. Matheson, S.D. Van Dyk, A massive progenitor of the luminous type IIn supernova 2010jl. Astrophys. J. **732**, 63 (2011). arXiv:1011.4150. https://doi.org/10.1088/0004-637X/732/2/63

N. Smith, J.M. Silverman, A.V. Filippenko, M.C. Cooper, T. Matheson, F. Bian, B.J. Weiner, J.M. Comerford, Systematic blueshift of line profiles in the type IIn supernova 2010jl: evidence for post-shock dust formation? Astron. J. **143**, 17 (2012). arXiv:1108.2869. https://doi.org/10.1088/0004-6256/143/1/17

N. Smith, J.E. Andrews, J.C. Mauerhan, Massive stars dying alone: the extremely remote environment of SN 2009ip. Mon. Not. R. Astron. Soc. **463**, 2904–2911 (2016). arXiv:1607.01056. https://doi.org/10.1093/mnras/stw2190

J. Sollerman, R.J. Cumming, P. Lundqvist, A very low mass of ^{56}Ni in the ejecta of SN 1994W. Astrophys. J. **493**, 933–939 (1998). arXiv:astro-ph/9709061. https://doi.org/10.1086/305163

G. Svirski, E. Nakar, R. Sari, Optical to X-ray supernova light curves following shock breakout through a thick wind. Astrophys. J. **759**, 108 (2012). arXiv:1202.3437. https://doi.org/10.1088/0004-637X/759/2/108

R.F. Temple, S. Raychaudhury, I.R. Stevens, X-ray observations of the edge-on star-forming galaxy NGC 891 and its supernova SN1986J. Mon. Not. R. Astron. Soc. **362**, 581–591 (2005). arXiv:astro-ph/0506657. https://doi.org/10.1111/j.1365-2966.2005.09336.x

M. Turatto, E. Cappellaro, I.J. Danziger, S. Benetti, C. Gouiffes, M. della Valle, The type II supernova 1988Z in MCG + 03-28-022—increasing evidence of interaction of supernova ejecta with a circumstellar wind. Mon. Not. R. Astron. Soc. **262**, 128–140 (1993). https://doi.org/10.1093/mnras/262.1.128

S.D. van Dyk, K.W. Weiler, R.A. Sramek, N. Panagia, SN 1988Z: the most distant radio supernova. Astrophys. J. Lett. **419**, L69 (1993). https://doi.org/10.1086/187139

S.D. van Dyk, K.W. Weiler, R.A. Sramek, E.M. Schlegel, A.V. Filippenko, N. Panagia, B. Leibundgut, Type "IIn" supernovae: a search for radio emission. Astron. J. **111**, 1271 (1996). https://doi.org/10.1086/117872

J.S. Vink, A. de Koter, H.J.G.L.M. Lamers, Mass-loss predictions for O and B stars as a function of metallicity. Astron. Astrophys. **369**, 574–588 (2001). arXiv:astro-ph/0101509. https://doi.org/10.1051/0004-6361:20010127

L. Wang, X. Cui, H. Zhu, W. Tian, Investigations of supernovae and supernova remnants in the era of SKA, in *Advancing Astrophysics with the Square Kilometre Array*, AASKA14, 64 (2015). arXiv:1501.04645

K.W. Weiler, N. Panagia, M.J. Montes, R.A. Sramek, Radio emission from supernovae and gamma-ray bursters. Annu. Rev. Astron. Astrophys. **40**, 387–438 (2002). https://doi.org/10.1146/annurev.astro.40.060401.093744

K.W. Weiler, N. Panagia, R.A. Sramek, Radio emission from supernovae. II—SN 1986J: a different kind of type II. Astrophys. J. **364**, 611–625 (1990). https://doi.org/10.1086/169444

C.L. Williams, N. Panagia, S.D. Van Dyk, C.K. Lacey, K.W. Weiler, R.A. Sramek, Radio emission from SN 1988Z and very massive star evolution. Astrophys. J. **581**, 396–403 (2002). arXiv:astro-ph/0208190. https://doi.org/10.1086/344087

S.E. Woosley, S. Blinnikov, A. Heger, Pulsational pair instability as an explanation for the most luminous supernovae. Nature **450**, 390–392 (2007). arXiv:0710.3314. https://doi.org/10.1038/nature06333

Space Sci Rev (2018) 214:44
https://doi.org/10.1007/s11214-018-0481-x

The Morphologies and Kinematics of Supernova Remnants

Laura A. Lopez[1] · **Robert A. Fesen**[2]

Received: 15 January 2018 / Accepted: 26 January 2018 / Published online: 6 February 2018

Abstract We review the major advances in understanding the morphologies and kinematics of supernova remnants (SNRs). Simulations of SN explosions have improved dramatically over the last few years, and SNRs can be used to test models through comparison of predictions with SNRs' observed large-scale compositional and morphological properties as well as the three-dimensional kinematics of ejecta material. In particular, Cassiopeia A—the youngest known core-collapse SNR in the Milky Way—offers an up-close view of the complexity of these explosive events that cannot be resolved in distant, extragalactic sources. We summarize the progress in tying SNRs to their progenitors' explosions through imaging and spectroscopic observations, and we discuss exciting future prospects for SNR studies, such as X-ray microcalorimeters.

Keywords Supernova remnants · Supernovae

1 Introduction

The predictive power of supernova explosion (SN) simulations has improved dramatically in the last few years. Three-dimensional simulations that properly handle energy-dependent neutrino transport obtain successful explosions via neutrino heating (Lentz et al. 2015; Janka et al. 2016; Roberts et al. 2016; Müller et al. 2017). These models are beginning to make testable predictions, such as the expected light curves for different mass progenitors, neutron

Supernovae
Edited by Andrei Bykov, Roger Chevalier, John Raymond, Friedrich-Karl Thielemann, Maurizio Falanga and Rudolf von Steiger

✉ L.A. Lopez
lopez.513@osu.edu

R.A. Fesen
robert.a.fesen@dartmouth.edu

[1] The Ohio State University, 140 W. 18th Ave., Columbus, OH 43210, USA

[2] Dartmouth College, 6127 Wilder Lab, Hanover, NH 03755, USA

star (NS) kick velocities, and large-scale compositional asymmetries (e.g., Wongwathanarat et al. 2013a, Utrobin et al. 2017, Wongwathanarat et al. 2017).

Although hundreds of SNe are discovered each year by dedicated surveys (e.g., Law et al. 2009, Leaman et al. 2011, Holoien et al. 2017), they are often too distant to resolve the SN ejecta and the immediate surroundings of the exploded stars (e.g., $1'' \approx 50$ pc for a distance of 10 Mpc). While study of the closest SNe, such as SN 1987A (McCray and Fransson 2016), has advanced the field tremendously, our understanding of SN progenitors and the details of the explosion dynamics is severely hampered by the infrequency of nearby events.

Young, nearby supernova remnants (SNRs) offer the means to study SN explosions and dynamics at sub-pc scales, and thus they are crucial to test the predictions of SN models. Evolved SNRs are observable for up to 10^5 years after the explosions across the electromagnetic spectrum, and multiwavelength campaigns have identified over 500 SNRs in the Milky Way and nearby galaxies (Badenes et al. 2010; Sasaki et al. 2012; Green 2014; Maggi et al. 2016; Garofali et al. 2017). Detailed investigations of SNRs that are a few thousand years old or less have yielded valuable insights into the progenitors, environments, and explosion dynamics for all SN subtypes (see e.g., Vink 2012 for a review).

Cassiopeia A (Cas A; $d \simeq 3.4$ kpc) is perhaps the most well-studied SNR in the Milky Way as it is the youngest known (~ 350 years old: Thorstensen and RA van den Bergh S 2001, Fesen et al. 2006a) core-collapse SNR in the galaxy. Cas A resulted from an asymmetric Type IIb explosion (Krause et al. 2008; Rest et al. 2011) associated with a red supergiant progenitor with an initial mass of 15–25 $M_\odot$ (Young et al. 2006). The *Chandra* X-ray Observatory first-light image of Cas A revealed the neutron star produced in the SN (Pavlov et al. 2000), and further X-ray investigations showed metal-rich ejecta primarily distributed in a $\approx 4'$ shell (e.g., Hughes et al. 2000, Hwang et al. 2000, Willingale et al. 2002). Kinematic studies of Cas A have mapped the three-dimensional structure of the ejecta (e.g., DeLaney et al. 2010, Grefenstette et al. 2017), including the primary ring of SN debris as well as the high-velocity knots (up to 15,000 $\mathrm{km\,s^{-1}}$) that protrude beyond the shell in the northeast and southwest (Fesen and Gunderson 1996; Fesen 2001; Fesen et al. 2006a; Fesen and Milisavljevic 2016).

The morphologies and kinematics of young SNRs like Cas A are important observables to test and constrain the SN explosion models being developed now. In this Chapter, we summarize the recent advances in studies of SNR morphologies (Sect. 2) and SNR kinematics (Sect. 3), particularly related to Cas A. We discuss how these results can be utilized to investigate the explosive origins of SNRs, the environments of SNRs, and the instabilities that mix ejecta with the surrounding medium.

2 Morphologies of SNRs

The morphologies of SNRs are complex and varied (see Fig. 1). Consequently, significant effort has been made in terms of observations and interpretation of individual sources. However, through the decades, as more SNRs were imaged at radio, optical, and X-ray wavelengths, astronomers began to classify the sources based on their morphological characteristics (Mathewson et al. 1983; Seward 1990; Williams et al. 1999; Green 2014).

Remnant classifications used most commonly in the field (e.g., as in the Green Catalog of Galactic SNRs: Green 2014) are shell-type, center-filled, and composite. Shell-type SNRs are those whose emission is dominated by a limb-brightened shell (e.g., SN 1006: Winkler et al. 2014). Center-filled SNRs have pulsar wind nebulae (PWNe; see review by Gaensler and Slane 2006) that shine brighter than their shells (e.g., the Crab: Seward et al. 2006).

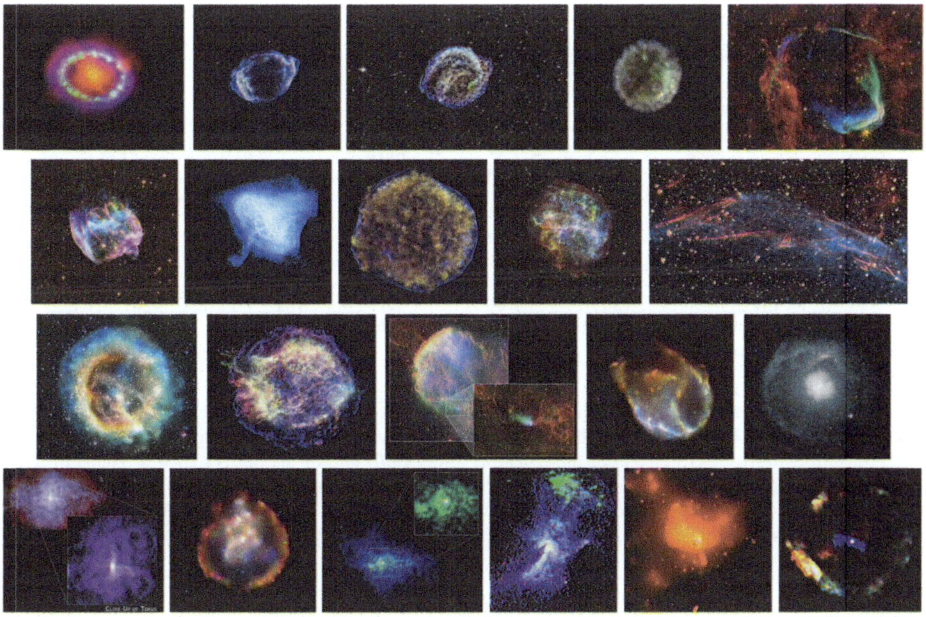

Fig. 1 *Chandra* X-ray Observatory images of a wide array of SNRs. Their morphologies are complex and heterogeneous, yet they are a rich sample from which to learn about explosions and their environments. These images were adapted from the *Chandra* press webpage (http://www.chandra.harvard.edu/press/)

Composite SNRs are sources that have evidence of both a PWN and a shell in their radio and X-ray morphologies (e.g., MSH 11–56: Temim et al. 2013). From X-ray imaging with *Einstein* and *ROSAT*, another class of SNRs was identified with radio shells and center-filled X-rays; these sources are called mixed-morphology or thermal-composite SNRs (Rho and Petre 1998; Shelton et al. 1999).

2.1 Effects of the Explosions on SNR Morphologies

2.1.1 Distinguishing Type Ia and Core-Collapse SNRs

While SNRs were classified based typically on their radio and X-ray morphologies as described above, the relationship of those morphologies to the originating explosions (Type Ia or core-collapse) was unclear. The reason stemmed in part from the challenge of constraining the SN type of SNRs. Generally, SNe are classified phenomenologically based on their optical spectra and light curves near maximum brightness (Minkowski 1941; Filippenko 1997). Given that SNe are typed based on their properties days after an explosion, an alternative means is necessary to discern the explosions responsible for SNRs, which are hundreds to tens of thousands of years old.

The most reliable ways to classify SNR explosion types are the detection of a central neutron star (although chance alignments are possible: Kaspi 1998) or of spectra from surrounding light echoes (e.g., Rest et al. 2005, Krause et al. 2008, Rest et al. 2008). Another means to constrain the progenitor star is to measure the number of OB stars (e.g., Maggi et al. 2016) or to investigate the stellar populations (e.g., Badenes et al. 2009, Jennings et al. 2012) in the vicinities of SNRs.

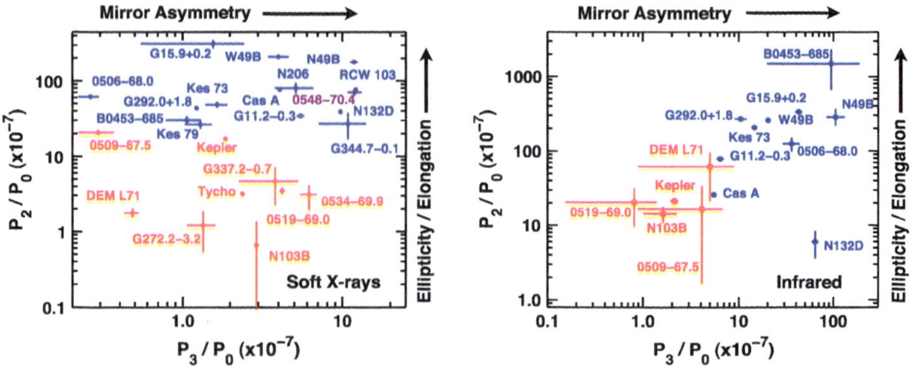

Fig. 2 Results from calculation of the multipole moments (a power-ratio method) to 24 Galactic and LMC SNRs: quadrupole power ratio P_2/P_0 (which measures ellipticity/elongation) vs. octupole power ratio P_3/P_0 (which quantifies mirror asymmetry) of the *Chandra* X-ray (*left*) and of the *Spitzer* 24 μm (*right*) images. Type Ia SNRs are plotted in *red points*, and the CC SNRs are in *blue* (as classified by their abundance ratios, light echo spectra, neutron star detections, and/or their environments). The type Ia SNRs separate naturally from the CC SNRs in both diagrams. Figures are adapted from Lopez et al. (2011) and Peters et al. (2013)

However, the most widely used method to type SNR explosions is to measure metal abundances using X-ray spectroscopy and compare the values to those predicted by SN models. In particular, the oxygen-to-iron ratio (O/Fe) is often employed for this purpose, since Type Ia SNe produce an order of magnitude more Fe than CC SNe, whereas CC SNe yield more intermediate-mass elements (e.g., O, Ne, Mg) than Type Ia SNe (Woosley et al. 2002). Using all of the approaches described above, many dozens of Milky Way and Magellanic Cloud SNRs have been identified as from Type Ia or from CC explosions.

Using a sample of young SNRs with well-constrained explosion types from the literature, Lopez et al. (2009a) measured and compared the morphological asymmetries in the Si XIII *Chandra* X-ray images of Type Ia and CC SNRs using a multipole expansion technique called the power-ratio method (PRM; see Jeltema et al. 2005 and Lopez et al. 2009b for the mathematical formalism). These authors found that the Type Ia SNRs are more circular (have smaller quadrupole power ratios) and symmetric (have smaller octupole power ratios) than CC SNRs. In subsequent works, they extended the PRM to soft (0.5–2.1 keV) *Chandra* X-ray images and to *Spitzer* 24 μm images (Lopez et al. 2011; Peters et al. 2013), and the results were consistent in those bandpasses as well (see Fig. 2). These works suggest that the X-ray and infrared morphologies of SNRs may be used to constrain progenitors' explosions.

The SNR results described above are consistent with mounting evidence that Type Ia and CC SNe have intrinsically different symmetries. For example, spectropolarimetry studies—the measure of the polarization of light as a function of wavelength as it is scattered through the debris layers of expanding SNe—demonstrate that CC SNe have larger deviations from spherical symmetry than Type Ia SNe (e.g., Leonard et al. 2006, Wang and Wheeler 2008. In some cases, the asymmetries of CC SNe can be substantial, altering the nucleosynthetic products of the explosions (e.g., Maeda and Nomoto 2003), ejecting metals at higher velocities (e.g., Mazzali et al. 2007), and kicking newly-formed neutron stars to speeds $\lesssim 1000\ \mathrm{km\,s^{-1}}$ (Lyne and Lorimer 1994; Arzoumanian et al. 2002; Hobbs et al. 2005; Faucher-Giguère and Kaspi 2006).

It is worth noting that a systematic analysis of the Fe K-shell emission in Type Ia and CC SNRs recently showed that the Fe-Kα centroid can also be used to type SNRs (Yamaguchi et al. 2014). Specifically, these authors found that Type Ia SNRs have lower energy Fe-Kα

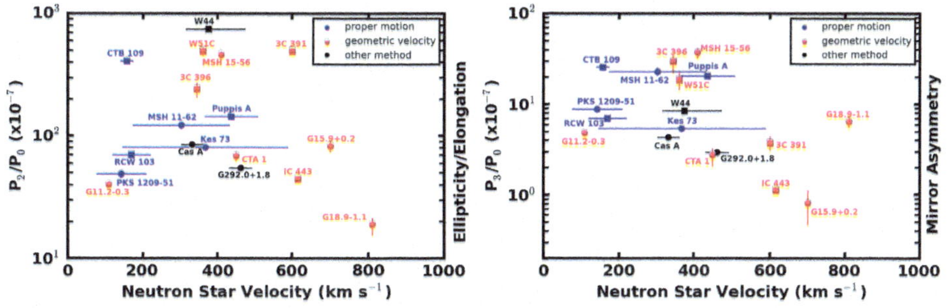

Fig. 3 The quadrupole (*left*) and octupole (*right*) power-ratios vs. neutron star velocities for a sample of 18 Galactic SNRs. *Blue points* are velocities measured via proper motions, and *red points* are velocities estimated from the spatial offset of the NS from the SNRs' geometric centers. Cas A and G292.0+1.8's NS velocities are determined by back-evolved filament motion to find the explosion site; W44's NS velocity is determined by multiple methods. *Circles* are used for SNRs without evidence of CSM/ISM interaction, and *squares* are those with clear evidence of CSM/ISM interaction. Figure is adapted from Holland-Ashford et al. (2017)

centroids than those of CC SNRs, indication that the Type Ia SNRs have less ionized plasmas than CC SNRs. This result indicates that Type Ia progenitors do not modify significantly their environments, and they reside in lower-density media than CC SNRs. Thus, although typing SNRs was a major challenge to the community twenty years ago, researchers now have a variety of tools to diagnose the explosive origins of SNRs.

2.1.2 Connecting Neutron Star Kicks and SNR Morphologies

Neutron stars are thought to possess a substantial "kick" velocity due to an asymmetric, CC explosion. Evidence for this is that NSs in the Milky Way are observed to have velocities of hundreds of km s^{-1} (Lyne and Lorimer 1994; Arzoumanian et al. 2002; Hobbs et al. 2005; Faucher-Giguère and Kaspi 2006), up to $\sim$ 1000 km s^{-1} (e.g., Chatterjee et al. 2005, Winkler and Petre 2007). Such large velocities are more than can be accounted for by the disruption of a close binary system (which produce velocities of $\sim$ 100 km s^{-1}: Lai 2001), suggesting an asymmetry in the dynamics of the explosion is the likely origin of NS kicks.

Two theories have been proposed to explain NS kicks. One scenario is that hydrodynamical instabilities lead to asymmetric mass ejection, accelerating the NS in a direction opposite to the bulk of ejecta (Scheck et al. 2006; Wongwathanarat et al. 2013b; Janka 2017). Such models predict that the heavier elements (e.g., Fe, Ti) are expelled directly opposite to the NS kick, whereas the intermediate-mass elements (e.g., C, O, Ne, Mg) are only marginally affected. In another set of models, a NS kick arises from anisotropic neutrino emission. In this case, the neutrinos carry the bulk of the explosion energy, and both the NS and the ejecta are expelled in the opposite direction (Fryer and Kusenko 2006).

Recently, Holland-Ashford et al. (2017) investigated the relationship between NS kick velocities and their associated SNRs' X-ray morphologies of 18 sources observed by *Chandra* and *ROSAT*. These authors found no correlation between the magnitude of asymmetries (i.e., the quadrupole and octupole power ratios) and the NS speeds (see Fig. 3). However, they also showed that the SNRs' X-ray emission (as measured via the dipole angle of their 0.5–2.1 keV images) was in the opposite direction as the NSs' motions in five out of six SNRs with well-constrained explosion sites (see Fig. 4).

In addition, Holland-Ashford et al. (2017) also noted that the X-ray emission of two of the SNRs (Cas A and G292.0+1.8) is dominated by the ejecta, and the other sources'

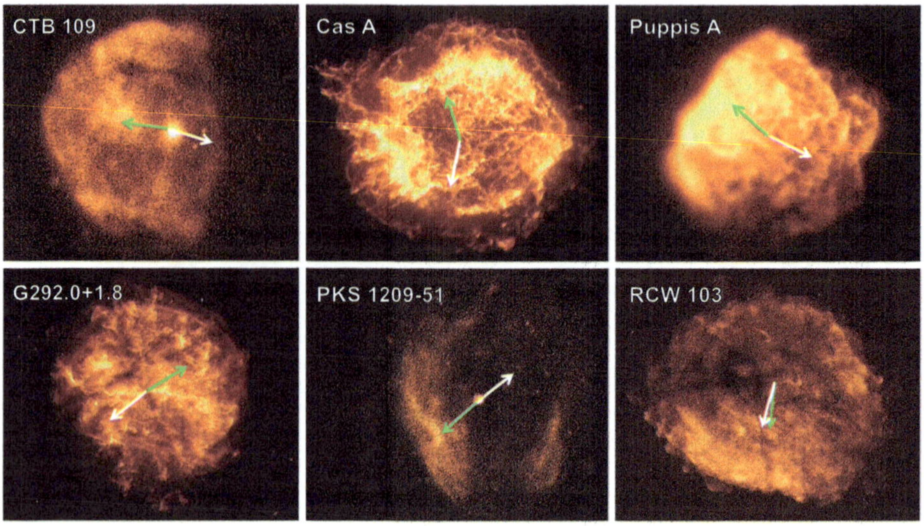

Fig. 4 0.5–2.1 keV *Chandra* and *ROSAT* images of the six Galactic SNRs with robust measures of their explosion sites that are analyzed by Holland-Ashford et al. (2017). The *green arrow* points from the explosion site to the direction of the dipole moment; the *white arrow* represents the direction of the NS motion. In Cas A and G292.0+1.8, the dipole moment direction reflects the bulk of ejecta emission. In CTB 109, Puppis A, and RCW 103, the dipole moment points towards enhanced emission due to interactions with CSM/ISM or a molecular cloud. It is unclear if the emission comes from ejecta or interactions in PKS 1209−51. Figure is adapted from Holland-Ashford et al. (2017)

morphologies are influenced at least in part by their complex environments. In a similar analysis to Holland-Ashford et al. (2017), Katsuda et al. (2017) showed that silicon, sulfur, argon, and calcium are distributed opposite to NS directions of motion in six young SNRs. The results of both works are consistent with predictions from models where NSs are kicked by ejecta asymmetries (Scheck et al. 2006; Wongwathanarat et al. 2013b; Janka 2017).

2.1.3 The Persistent, Elongated Morphologies of Jet-Driven SNRs

The comparative, systematic analysis of SNR samples described in Sects. 2.1.1 and 2.1.2 is also an important means to identify and examine outlier sources. For example, two SNRs from the X-ray morphological studies have anomalously large elongations compared to other young remnants: W49B in the Milky Way and SNR 0104−72.3 in the SMC. Both are extremely elliptical (see the X-ray images in Fig. 5), and the spatial distribution of iron in W49B is remarkable distinct compared to the other elements detected in X-rays (Lopez et al. 2009b). Using spatially-resolved *Chandra* X-ray spectra, Lopez et al. (2013b) and Lopez et al. (2014) showed the mean abundance ratios in these SNRs are consistent with the predicted yields of jet-driven CC SNe. The identification of this kind of explosion in the SMC, where only 23 SNRs are known, may be evidence that jet-driven SNe occur more frequently in low-metallicity environments (since the SMC has a metallicity of $\sim 20\%$ solar). In fact, recent work on long-duration gamma-ray bursts (which are thought to arise from jet-driven explosions) prefer low-metallicity environments (Modjaz et al. 2008). Thus, it is possible that these two SNRs are local analogues to long GRBs observed as cosmological distances.

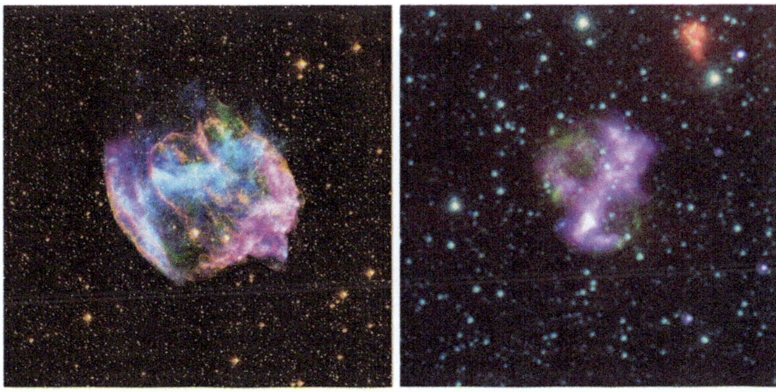

Fig. 5 Multiwavelength images of two extremely elliptical SNRs thought to be from jet-driven explosions (Lopez et al. 2013a,b). *Left*: SNR W49B, with X-rays in *blue*, IR in *yellow*, and radio in *purple*. *Right*: SNR 0104−72.3, with X-rays in *purple* and IR in *green*. Images are from NASA/*Chandra* press releases

To test whether jet-driven explosions would retain their elliptical morphologies a few thousand of years after SN explosion, González-Casanova et al. (2014) performed two-dimensional simulations of a jet-driven SN, following its subsequent evolution as a SNR. Figure 6 shows the synthetic thermal emission maps derived from this work. These authors performed two separate hydrodynamical calculations, each equipped to describe the behavior of the bipolar outflow at two different epochs. They used a simulation with a general equation of state and a reaction network to determine the nucleosynthesis that accompanies a jet propagating through a massive star (left panel). As demonstrated in the abundance distributions in Fig. 6, the jetted outflow that is responsible for synthesizing ^{56}Ni carries more energy and inertia than the products of incomplete silicon and oxygen burning. Thus, the distribution asymmetry in the abundance is preserved as the SN material sweeps up the external medium.

The resulting outputs (i.e., the density, velocity, and compositional structure of the ejecta) were employed as the initial conditions in the calculation of the expansion and associated thermal X-ray emission produced by the hot shocked gas. The Fe XXV and Si XIII maps of the jet-driven SNR inside a wind bubble of a 25 $M_{\rm sun}$ star are shown in Fig. 6, right panel. Even in the presence of an uniform external medium, the X-ray morphology is dominated by the bright iron jet, with explosive oxygen-burning products (such as silicon and sulfur) enclosing the Fe, as observed in W49B.

2.2 Influences of the Environment and the Surrounding Medium on SNR Morphologies

Since the progenitors of CC SNe have short main-sequence lives (∼ 3–50 Myr), their explosions are expected to occur within or near the dense media from which the massive stars formed. Consequently, it is common for CC SNRs to show traits of interaction with an inhomogeneous or dense CSM. For example, roughly one quarter of all Galactic SNRs show evidence of interaction with molecular clouds, such as the coincidence of OH masers (indicating the presence of shocked molecular gas: e.g., Wardle and Yusef-Zadeh 2002).

A SNR-molecular cloud interaction will have a profound influence on the X-ray morphologies and spectra of SNRs. Large-scale density gradients can result in substantial deviations from spherical symmetry, and many SNRs interacting with molecular clouds have

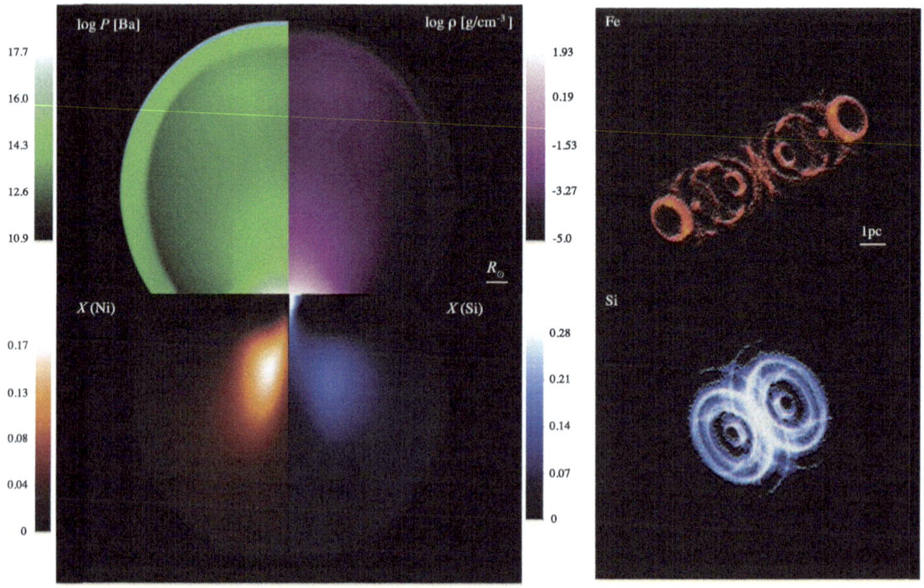

Fig. 6 The X-ray properties of a jet-driven SN remnant as seen in simulations. *Left panel*: A jet-driven supernova remnant. *Left panel*: Jet-driven SN explosion at the onset of the homologous phase, which is achieved 100 seconds after the bipolar ejecta has broken free from the $R_* = 0.43 R_\odot$ stellar progenitor. The pressure P, mass density ρ and the silicon $X(\mathrm{Si})$ and nickel $X(\mathrm{Ni})$ mass fractions are indicated in each frame with corresponding size scales in units of $R_\odot$. This model is then employed to calculate the evolution of a jet-driven SN remnant inside the wind bubble structure expected around a 25 $M_\odot$ massive star for the case of an ISM pressure typical of the high-density molecular region surrounding the bipolar SNR. *Right panel*: Fe XXV and Si XIII X-ray emission maps. The maps are obtained integrating the respective emission coefficients along the line of sight. The images are computed at an evolutionary age of 700 yr. To match the characteristics of W49B, the results of the simulation have been rotated 45° with respect to the plane of the sky (in the direction of the observer), and 120° with respect to the z-axis

highly elliptical/elongated shapes compared to non-interacting CC SNRs (Lopez 2014; Holland-Ashford et al. 2017). A large number of interacting SNRs are centrally bright with thermal X-ray emission, whereas their radio morphologies are shell-like (as described at the beginning of Sect. 2). Known as mixed-morphology (MM) SNRs, ~ 40 of these sources have been identified in the Milky Way (Vink 2012). Based on X-ray observations, MM SNRs can have enhanced metal abundances (Lazendic et al. 2006) and/or isothermal plasmas across their interiors.

In young SNRs, shocks generate ionizing plasmas that slowly reach collisional ionization equilibrium (CIE). However, X-ray observations with *ASCA* first revealed evidence that MM SNRs can have "overionized" plasmas (Kawasaki et al. 2002, 2005). In these cases, the electron temperature kT_e derived from the bremsstrahlung continuum is systematically lower than the effective ionization temperature kT_z given by the line ratios. Subsequent *Suzaku* observations of MM SNRs detected radiative recombination continuum (RRC) features in their spectra (e.g., Yamaguchi et al. 2009, Ozawa et al. 2009, Uchida et al. 2012), conclusive evidence of overionization.

In the collisional plasmas of SNRs, overionization is a signature of rapid electron cooling, and the physical origin of this cooling remains debated. One proposed scenario is thermal condution, where hot ejecta in a SNR interior cools by efficiently exchanging heat with exterior material (e.g., Cox et al. 1999, Shelton et al. 1999). Alternatively, the cooling may

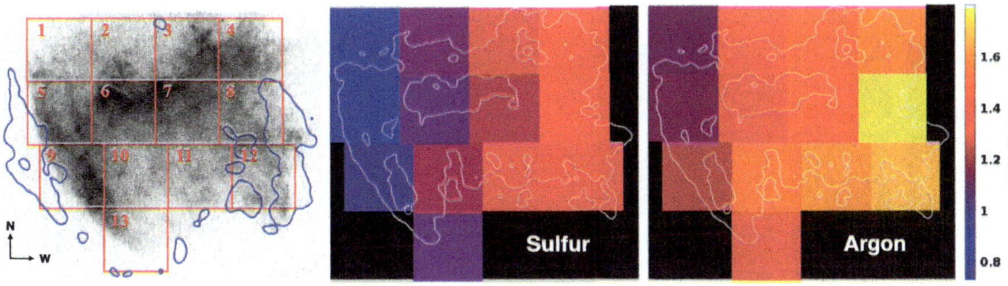

Fig. 7 *Left*: 13 regions analyzed by Lopez et al. (2013a) to map overionization in the SNR W49B. Regions are overplotted on the full-band (0.5–8.0 keV) *Chandra* X-ray ACIS X-ray image, and the warm H_2 gas distribution is marked in blue The ejecta are colliding with the molecular gas in the east, whereas the ejecta are expanding freely in the west where the molecular material is thought to be in the foreground (Lacey et al. 2001). *Other panels*: maps of overionization across W49B. *Middle*: ratio of ionization temperature kT_Z (from the ratio of S XV to S XVI) to the ejecta electron temperature kT_{ej} (from the bremsstrahlung continuum modeling). *Right*: maps of kT_Z/kT_{ej} using the ratio of Ar XVII to Ar XVIII. Regions with $kT_Z/kT_{ej} > 1$ are overionized. Figures are adapted from Lopez et al. (2013a)

arise from adiabatic expansion, where the SN blast wave expands through dense CSM into a rarefied ISM (e.g., Itoh and Masai 1989, Moriya 2012, Shimizu et al. 2012).

To ascertain which cooling scenario is responsible for overionization, the localization of the overionized plasma is crucial. Toward this end, Lopez et al. (2013b) performed a spatially-resolved spectroscopic analysis to compare kT_e (derived by modeling the continuum) to kT_z (measured from the flux ratio of He-like to H-like lines). They found that the overionized plasma is concentrated in the west and has a gradient of increasing ionization from east to west (see Fig. 7). Given that the ejecta has is colliding with molecular material in the east and is expanding unimpeded in the west, the results suggest that adiabatic expansion is the dominant cooling mechanism of the hot plasma.

2.3 Instabilities & Mixing

On small scales, the distribution and scale of individual ejecta clumps can reveal the efficiency of the mixing between metals synthesized in the explosion and the surrounding CSM/ISM. Shock-heated ejecta expanding into lower-density material is subject to hydrodynamical instabilities (e.g., Rayleigh-Taylor instabilities) which are thought to be responsible for efficiently mixing ejecta metals (e.g., Blondin et al. 2001, Wang and Chevalier 2001). X-ray and optical observations of individual SNRs show that these instabilities are capable of mixing ejecta out to and beyond the forward shock, in both Type Ia SNRs (e.g., in Tycho: Hwang et al. 1998) and in CC SNRs (e.g., in Cassiopeia A: Fesen and Gunderson 1996, Hughes et al. 2000, Fesen et al. 2011).

Lopez et al. (2011) investigated the mixing of ejecta in nine SNRs using narrow-band *Chandra* images corresponding to emission lines from metals, including O, Ne, Mg, Si, and Fe. Specifically, they used wavelet-transform analysis to identify substructures and to measure their sizes. For example, Fig. 8 displays the wavelet-transformed images of three emission line complexes (Mg XI, Si XIII, and Fe XXV) in Cas A. They found that all nine sources have well-mixed ejecta: 90% of bright substructures of one element have corresponding substructures of the same size in another element, even if the metals arose from different burning processes. These findings reinforce observationally that hydrodynamical instabilities can efficiently mix ejecta in only a few hundred years.

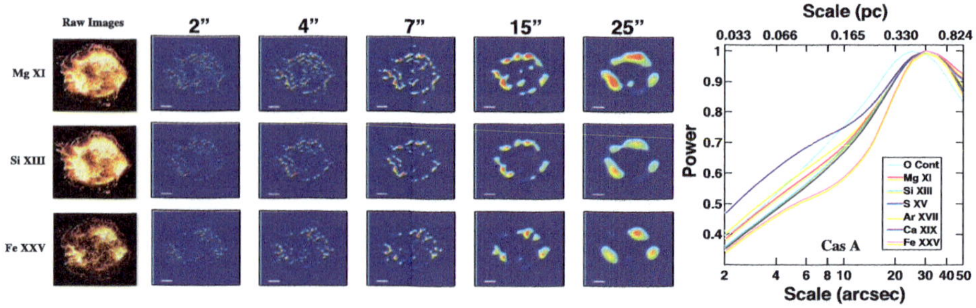

Fig. 8 *Left panels*: Wavelet-transformed images of Cas A showing the power at each scale for the Mg XI (*top row*), Si XIII (*middle row*), and Fe XXV (*bottom row*) as observed by *Chandra*. *Right*: Plot of the relative power as a function of scale for narrow-band images corresponding to emission lines detected in the *Chandra* observations of Cas A. The elements show similar profiles because they are well-mixed throughout the SNR, although some differences are evident (such as the Si XIII emission in the northeast jet and the Fe XXV located exterior to the Mg XI and Si XIII in the southeast). Figures are adapted from Lopez et al. (2011)

2.4 Bilateral SNRs and the Galactic Magnetic Field

In the previous subsections, we have described how the thermal emission morphologies depend on the explosion and the environment. In addition, the morphology of the synchrotron radiation from SNR shocks gives clues about the surrounding medium, particularly the ambient density and the magnetic-field orientation. Specifically, Gaensler (1998) examined a sample of bilateral SNRs at radio frequencies and showed that the bilateral axes tended to be aligned with the Galactic plane. However, some noteworthy exceptions existed, such as SN 1006, where the axis of symmetry is rotated almost 90° from the Galactic plane.

Orlando et al. (2007) performed three-dimensional magnetohydrodynamical (MHD) simulations of a spherically-symmetric shock propagating through a magnetized ISM. They found that a gradient in ambient density or in magnetic-field strength can produce asymmetric bilateral SNRs, as in e.g., objects with two radio limbs of different brightnesses.

Recently, West et al. (2016) investigated the radio morphologies of all Galactic axisymmetric SNRs as well, including those with only one bright limb. They found that using a simple model of SNRs expanding into an ambient Galactic magnetic field, they were able to reproduce the observed radio morphologies of the sample. One implication of this result is that if the large-scale B-field is known near a source, then the distance to that SNR can be inferred. In subsequent work, West et al. (2017) demonstrated that the synchrotron radio morphologies of their bilateral sample are more consistent with quasi-perpendicular cosmic-ray (CR) acceleration than with quasi-parallel case.

3 Kinematics of Young SNRs

As described in the previous section, the morphologies of SNRs are shaped in large part by their explosions but also by their environments. The other principal means to investigate the dynamics of the explosion mechanism and post-explosion expansion of SNRs is through kinematic studies of both radioactive and non-radioactive debris. In this regard, Cas A is a fortuitous, nearby target. As the youngest Galactic core-collapse SNR known (Fesen et al. 2006b), it gives us a relatively high-resolution look at the expansion properties of the metals synthesized in the explosion.

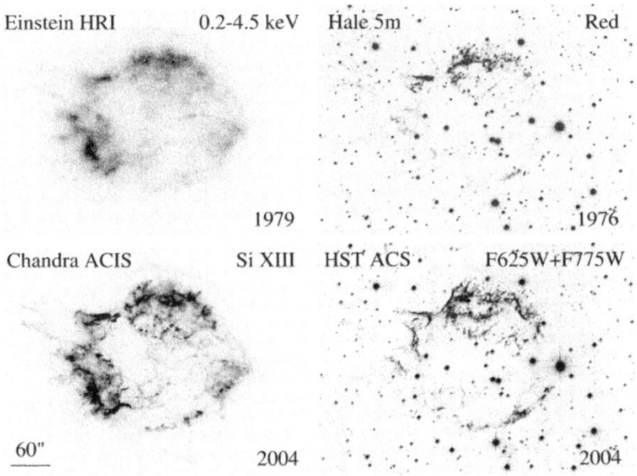

Fig. 9 Comparison of emission differences of Cas A as seen in X-rays and optical. *Left hand panels* show the 1979 *Einstein* X-ray image and a 2004 *Chandra* continuum-subtracted Si XIII X-ray image. *Right hand panels* show broadband red optical images taken at similar epochs as the X-ray images. The optical images are sensitive to line emission from [S II] λλ6716,6731, [O I] λλ6300,6364, and [O II] λλ7320,7330. The Palomar Hale 5-m image is a 1976 July plate while the *HST* image is a co-add of 2004 ACS F625W and F775W filter images (Patnaude and Fesen 2014)

Unless one views the Cas A remnant as peculiar and not representative of typical CC SN explosions, it can serve as a powerful and unique case to investigate the kinetic properties of at least a subset and possibly a significant fraction of CC events. In this Section, we highlight the extensive work that has been done to characterize the kinematics of Cas A and what can be gleaned about the physics of explosions from this case study.

In terms of kinematics, Cas A is composed of three distinct sets of ejecta. The brightest at nearly all wavelengths is its thick main emission shell of ejecta. This material is shock-heated by the remnant's $\simeq 5000\ \mathrm{km\,s^{-1}}$ reverse shock and contains the bulk of Cas A's emitting debris. The shell's expansion is strongly asymmetrical with the rear hemisphere moving at roughly $6000\ \mathrm{km\,s^{-1}}$ while the front hemisphere is expanding around $2000\ \mathrm{km\,s^{-1}}$ slower. Surrounding this shell of shocked ejecta lie hundreds of small ($\leq 0.5''$) ejecta knots with velocities between 7000 and 12,000 $\mathrm{km\,s^{-1}}$ with chemical compositions reflective of the outermost layers of the progenitor star at the time of explosion.

Cas A also exhibits two broad bipolar "jets" of especially high-velocity knots of ejecta ($v = 8000$–$15{,}000\ \mathrm{km\,s^{-1}}$) along its NE and SW limbs. These broad plumes of debris consist of ejecta streams containing hundreds of small, individual ejecta knots. Interestingly, the fastest ejecta knots are richest in S, Ar, Ca—presumably from deep inside the progenitor—whereas ejecta knots near the jets' base are O-rich, suggesting an explosive origin pushing interior material up and out through the outer layers. It seems likely that the energy for the creation of these ejecta plumes/jets is somehow associated with the central explosion engine.

Chandra and *Hubble Space Telescope* (*HST*) images of Cas A taken over the last few decades have revealed a wealth of detail about its expanding main shell of S-, O-, Ar-, Ca-rich ejecta. However, it was obvious from the time of the first high-resolution X-ray image of Cas A taken with the *Einstein* satellite in 1979 that Cas A's X-ray and optical morphologies exhibit significant large-scale differences.

This is shown in Fig. 9 which contrasts the remnant's X-ray and optical emissions in the late 1970s and in 2004, 25 years later. While there is some correlation between certain emission regions seen in the 1976 optical and 1979 X-ray images, especially along the remnant's

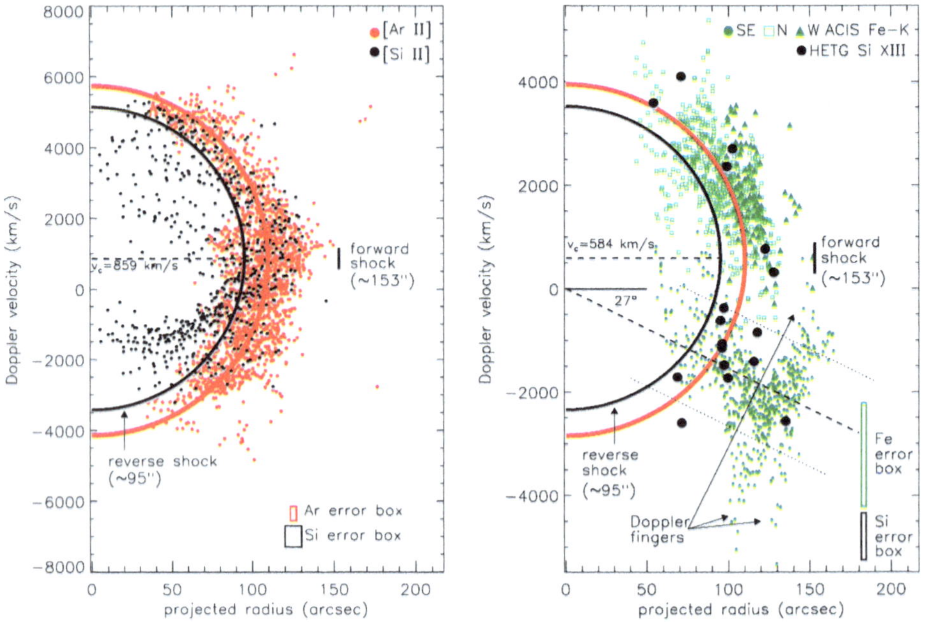

Fig. 10 *Left panel*: Doppler velocity vs. projected radius for infrared [Ar II] emissions in *red* and [Si II] in *gray*. The mean velocity errors and spatial resolution are indicated by *error boxes* at the bottom of the figure. The *red semicircle* represents the best-fit spherical expansion model. The *black semicircle* represents the reverse shock and the projected radius of the forward shock is also indicated. *Right panel*: Doppler velocity vs. projected radius for X-ray Fe-K emissions in *green* and Si XIII emissions in *black*. Different symbols are used to represent the north (*open square*), west (*filled triangle*), and southeast (*filled circle*) Fe-K distributions (DeLaney et al. 2010)

northern limb, there is a poor large-scale optical—X-ray emission correlation over much of the remnant, especially in the south, east, and west quadrants. A better correlation is seen between the X-ray and optical emission in the 2004 *Chandra* and *HST* images.

Such comparisons make it clear that while Cas A's overall X-ray emission has undergone only minor large-scale changes over the last few decades, its optical appearance has gone from one of a faint and sparse emission structure to a bright and extensive emission morphology (Patnaude and Fesen 2014). Differences between X-ray and optical images appear likely to be due to differences in the density of the reverse-shocked material; that is, X-ray emission regions highlight low density, interclump regions, whereas optical emission features are associated with much smaller, denser debris knots.

Although the remnant's X-ray and optical emissions in Cas A's main shell have grossly similar morphologies in certain regions, they possess different expansion velocity ranges and very different detailed structure. Kinematic differences between relatively cool and dense optical/infrared emitting knots and much hotter and more diffuse X-ray emitting debris can be seen in Fig. 10. Such expansion velocity differences are discussed in more detail below in Sect. 3.1.

Outside of the main shell of reverse-shocked material, *HST* images have uncovered nearly 4,000 outlyin fragments around the remnant and largely out ahead of the 4500–5000 $km\,s^{-1}$ blast wave. These knots exhibit proper motions indicating transverse velocities of up to 12,000 $km\,s^{-1}$ (Fesen 2001; Fesen et al. 2006b). They surround nearly all regions about the main shell and are distinctly different in both chemistry and expansion velocity than

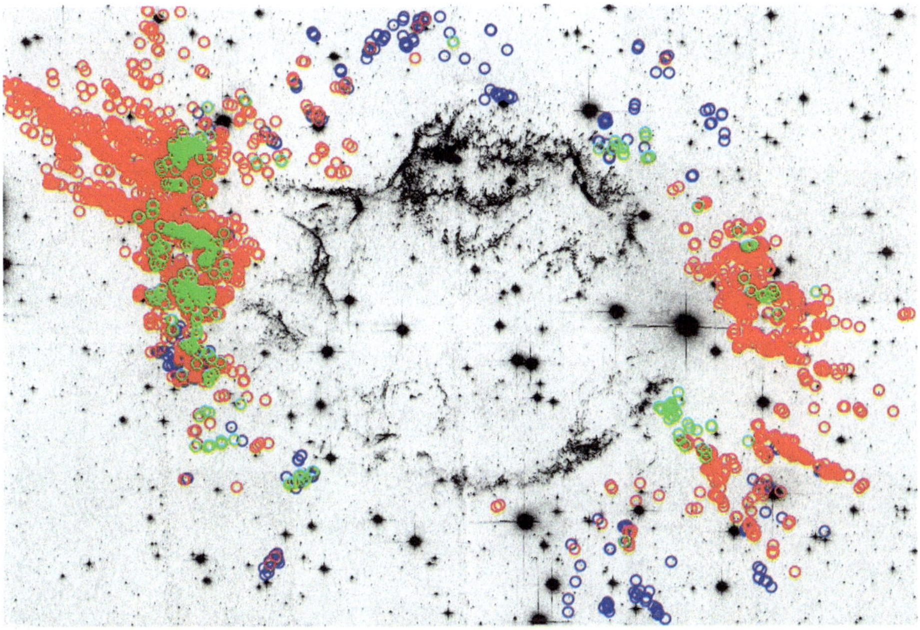

Fig. 11 Locations of high-velocity ejecta outside of the remnant's main shell of ejecta shown on a 2004 HST image. *Blue and green circles* indicate N and O emission knots while red circles mark S emission knots (Fesen and Milisavljevic 2016)

the remnant's NE and SW jets. Most of these knots are too small and faint to detect from the ground. The expansion and chemical properties of these outer ejecta knots is shown in Fig. 11.

As mentioned above, the presence of weak hydrogen emission lines in the spectra of these outer knots (which indicate that the Cas A progenitor had some photospheric hydrogen at the time of outburst) is consistent with the SN IIb classification based on light echo studies. Faint streams of trailing emission indicate ablated material off these outer SN fragments as they move through the local ISM/CSM environment have been detected in deep *HST* images, showcasing the earliest stage of ISM enrichment by SN debris.

3.1 Ejecta Rings and Interior Bubbles

Unlike the more or less random arrangement of debris expelled by a supernova explosion one might have expected, the kinematic and chemical properties of Cas A's main shell of ejecta has been found to consist of a series of large optical and infrared rings of reverse shocked O-, S-, Ar-, Ca-rich ejecta. These rings surround much of the remnant's X-ray emitting Fe-rich ejecta. This unexpected chemical arrangement of ejecta can be seen in Figs. 12 and 13 where the remnant's main shell ejecta can be seen to lie almost exclusively in ring-like structures.

Reed et al. (1995) and Lawrence et al. (1995) were the first to establish the existence of conspicuous rings of reverse-shocked ejecta in Cas A. More recently, DeLaney et al. (2010) and Milisavljevic and Fesen (2013) identified about half a dozen well-defined ring-like structures with angular diameters between 30″ and 120″ (0.5–2 pc). Most of these rings are actually short cylinders giving them a crown-like appearance in 3D projections.

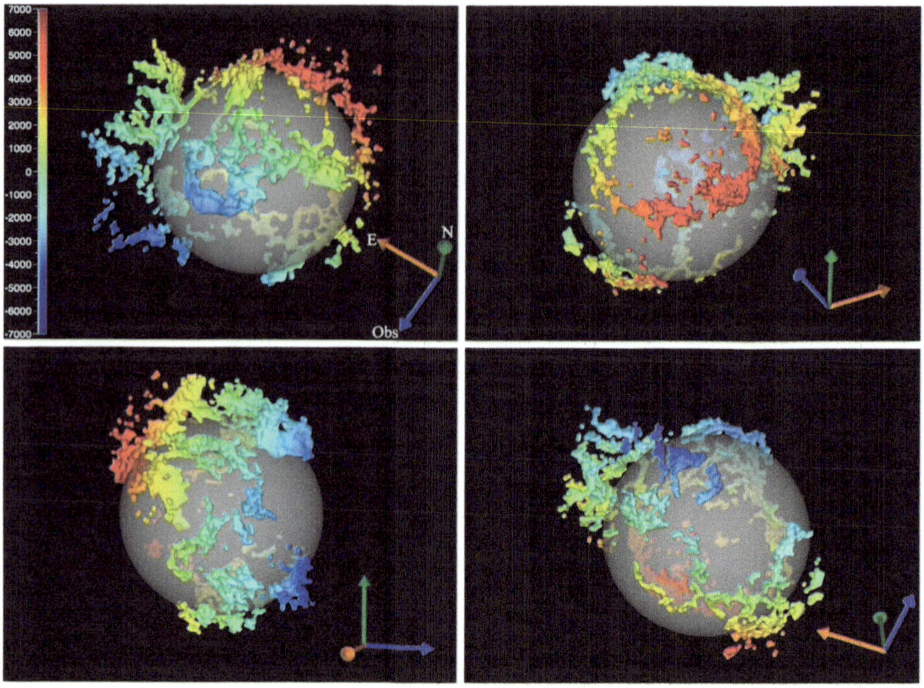

Fig. 12 Three-dimensional projections of the optical emission from Cas A's main shell illustrating the ubiquity of large ejecta rings. Noticeable rings of ejecta include the blueshifted ring in the north and the much larger neighboring redshifted ring (Milisavljevic and Fesen 2013)

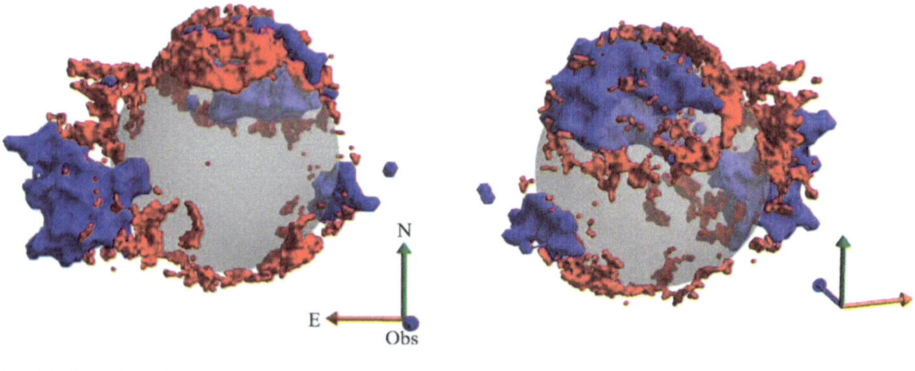

Fig. 13 Location of iron-rich X-ray emitting ejecta (*blue*) with respect to the main shell sulfur- and oxygen-rich optically emitting ejecta (*red*). The X-ray data shown are from DeLaney et al. (2010). Note in the *right-hand panel* the location of the iron-rich (*blue*) ejecta well confined inside the large red ring of S,O-rich ejecta. Adapted from Milisavljevic and Fesen (2013)

The height in velocity space of these crowns radially away from the remnant's center is up to 1000 $\mathrm{km\,s^{-1}}$. This radial extent may be related to the $\simeq$ 20–30 year radiative timescale for cooling of these reverse-shocked ejecta knots.

The presence of these ejecta rings naturally led to questions about their nature. DeLaney et al. (2010) noted the coincidence of large ejecta rings with the three regions of Fe-K X-ray emission and suggested that these and other, less prominent features are regions where the ejecta have emerged from the explosion as "pistons" of faster-than-average ejecta. In this view, the remnant's main shell rings represent the intersection points of these pistons with the reverse shock, similar to the bow-shock structures.

An alternative explanation first suggested by Blondin et al. (2001) is that the observed ejecta rings represent cross-sections of large cavities in the expanding ejecta created by a post-explosion input of energy from plumes of radioactive ^{56}Ni-rich ejecta. From *Spitzer* infrared and *Chandra* images and spectra, DeLaney et al. (2010) were able to show that Fe-rich X-ray emitting material was surprisingly confined inside the remant's optical/IR emitting ejecta rings. This structure can be seen in Fig. 13 where the Fe-rich ejecta (blue) is found inside rings of S-, O-rich ejecta (red).

Turbulent motions that would initiate this Ni bubble structure in Cas A are not unlike recent 3D simulations of the large-scale mixing that takes place in the shock-heated stellar layers ejected in the explosion of a 15.5 $M_{\odot}$ blue supergiant star presented in Hammer et al. (2010). As shown in their models, a progenitor's metal-rich core is partially turned over with nickel-dominated fingers overtaking oxygen-rich bullets. Although the evolution of these simulations is strongly dependent on the internal structure of the progenitor star (Ugliano et al. 2012), it is still tempting to draw an association between the Ni-rich outflows seen in the Hammer et al. (2010) models and the rings of Cas A.

Discovery of Cas A's ejecta rings was followed-up by the detection of unshocked material inside the remnant (Isensee et al. 2010; Milisavljevic and Fesen 2015). Cas A's interior material consists of a handful of large cavities or bubbles, likely the result of plumes of radioactive ^{56}Ni which expand, pushing non-radioactive material like O, S, Ar, and Ca into bubble-like structures. Such an arrangement explains the observed Fe-rich ejecta inside rings of O-, S-rich debris. When the expanding bubbles encounter the remnant's reverse shock, the intersection is in the form of rings filled with Fe-rich ejecta, the product of ^{56}Ni decay.

Consequently, it now appears that the rings of ejecta seen nearly everywhere in Cas A are a product of being the cross-section of large, internal cavities when they encounter the inward moving reverse shock front. Similar rings of ejecta to those seen in Cas A, thought to be due to an interior bubble-like structure, are present in other CC SNRs.

Figure 14 shows a comparison of Mercator maps of emitting ejecta for Cas A (upper planel) and for the much lower expansion velocity and Crab-like SNR 3C 58 (lower panel). In both cases, only a handful of rings—not dozens—are present in these two CCs SNRs despite the difference in their estimated progenitor masses: 15–20 $M_{\odot}$ for Cas A (Young et al. 2006) and 8–10 $M_{\odot}$ for 3C 58.

The presence of large-scale ejecta rings in CC SNe may help explain the clumpy emission-line profiles seen in late-type spectra of SNe II/Ib events. Milisavljevic et al. (2012) showed how Cas A's main shell spectra would appear as an unresolved extragalactic source. Similarities were seen between the integrated Cas A spectrum and several late-time optical spectra of decades-old extragalactic SNe. Particularly well matched with Cas A were SN 1979C, SN 1993J, SN 1980K, and the ultraluminous SNR in NGC 4449.

Since the emission-line substructure observed in the oxygen forbidden line profiles of Cas A is associated with the large-scale rings of ejecta, Milisavljevic et al. (2012) suggested that similar features in intermediate-aged SNe (which have often been interpreted as ejecta "clumps" or "blobs") are, in fact, probable signs that large-scale rings of ejecta are common in these sources.

In summary, kinematic structures in CC SNe are highly dependent on the elemental abundances of the ejecta. This is apparent in Cas A's main shell where Fe-rich ejecta is largely confined to certain regions surrounded by nonradioactive SN debris. It is also true regarding

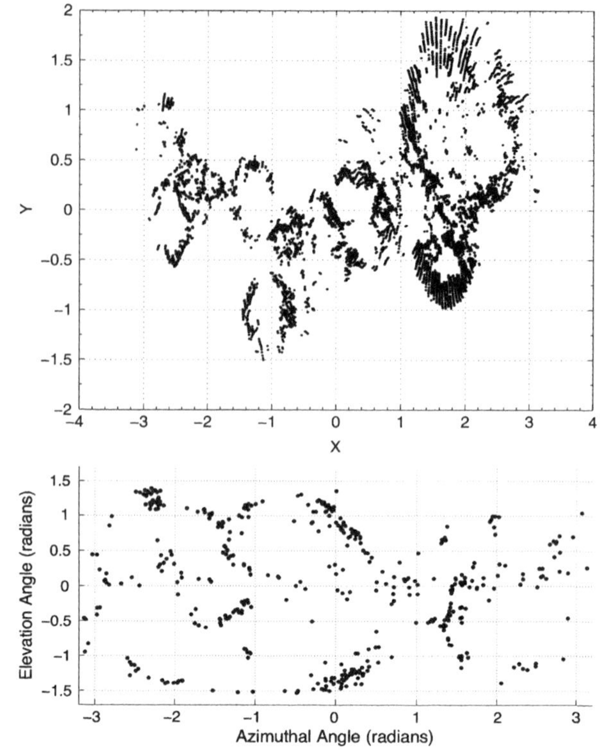

Fig. 14 Mercator projections of ejecta in Cas A and 3C 58. *Upper panel*: The main shell of Cas A's optically emitting ejecta (Milisavljevic and Fesen 2013). *Lower panel*: High-velocity emission knots in the Crab-like SNR 3C 58 (Fesen et al. in preparation). In both projections, the linear scale is equal in all directions around any point and conformal, but the cylindrical map projection distorts the size and shape of large objects, especially toward the poles

the higher-velocity ejecta knots around the remnant and out ahead of the forward shock. These ejecta knots are the only ones which contain any hydrogen, a signature that the progenitor SN had a thin layer of hydrogen, leading to a SN II spectrum at early epochs. Lastly, Cas A's very high-velocity ejecta jets display a strong radial 'upside-down' chemistry unlike anywhere else in the remnant.

3.2 High-Velocity Ejecta "Jets"

It is of interest to consider the presence of jet-like ejecta streams seen in Cas A in the context of recent work on supernovae that has established a continuum of explosion energies extending from broad-lined SNe Ic associated with gamma-ray bursts to more ordinary SNe Ibc (Soderberg et al. 2010; Chakraborti et al. 2015). The existence of jets across such explosive events suggests that a wide variety of jet activity may potentially be occurring at energies that are hidden observationally.

The nature of the anomalously high velocities of ejecta in Cas A's NE and SW regions has long been a puzzle. Asymmetric debris structures produced by circumstellar interaction have been theoretically modeled by Blondin et al. (1996), who showed that a jet-like feature of SN ejecta can be generated in the progenitor's equatorial plane from pole/equator density gradients in the local CSM. Thus, these high-velocity regions could be secondary features caused by instability-powered flows from an equatorial torus, where the explosion axis is loosely defined by the X-ray iron-rich regions found in the SE and NW.

However, the very prominent rupture-like features in Cas A's main shell near the base of the NE jet is visible in X-rays, optical, IR, and radio images and are certainly suggestive of

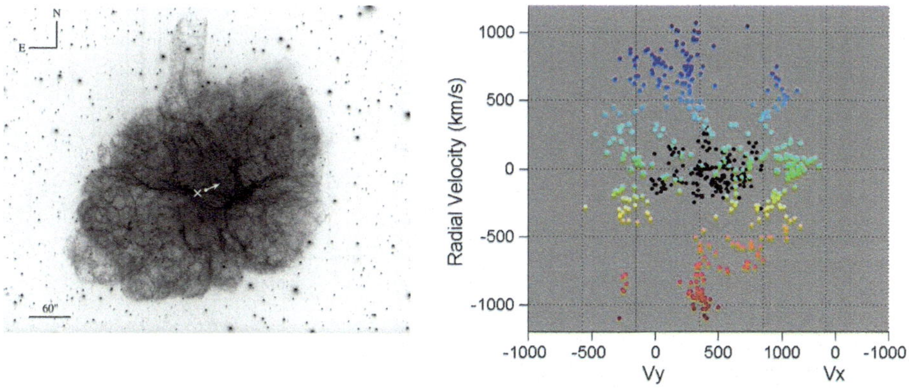

Fig. 15 *Left panel*: An [O III] λλ4959,5007 image of the Crab Nebula showing its northern jet with respect to the entire remnant. The 'X' marks the Nugent (1998) centre of expansion, and the *white dot* and *arrow* marks the pulsar and its proper motion direction as determined by Kaplan et al. (2008). *Right panel*: 2D projection of observed radial velocities (color-coded) vs. transverse RA and Dec velocities for optical emission knots in the young Crab-like SNR 3C 58 assuming an explosion age of 2200 years. Slower moving CSM knots are shown in *black*. Note that the wide north and south openings (jets) exhibit a similar broad opening morphology to that seen in the Crab's northern filamentary jet feature (Fesen et. al. in preparation)

an explosive formation process. Although perhaps not indicative of a jet-induced explosion, there is nonetheless substantial evidence that the NE and SW jets are associated with core-collapse explosion dynamics in some way.

The location of chemically distinct, optically-emitting knots in both jets is also consistent with an unusual, high-velocity ejection of underlying mantle material (Fesen 2001). Knots exhibiting a mix of Hα, [N II], [O II], [S II], and [Ar III] line emissions are only observed in the two jet regions, consistent with an eruptive and turbulent mixing of the underlying S-, O-, and Ar-rich material with photospheric H- and N-rich layers (Fesen and Gunderson 1996; Fesen 2001). Furthermore, spectra of optical knots in the NE and SW jets lying farthest out and possessing the highest ejection velocities show no detectable emission lines other than those of [S II] and [Ca II], suggesting an origin from the Si-, S-, Ar-, Ca-rich layer deep inside the progenitor star.

Kinetic energy estimates for the Cas A jets are around 1×10^{50} erg (Fesen and Milisavljevic 2016). An explosive origin of the jets would mean this value may be an underestimate since it does not include the energy needed to propel the material in both jets up through Cas A's progenitor's outer layers. This added energy, together with the possibility of additional mass from undetected diffuse jet material, means the jets likely contain 10% or more of the expected 10^{51} erg SN explosion energy.

Cas A is not alone in being a young CC SNR with jet-like structures (see e.g., Sect. 2.1.3). As shown in Fig. 15, low-mass progenitor CC SNe, such as the Crab Nebula and 3C 58, exhibit broad, plume-like ejecta streams. The fact that jet-like streams of unusually high-velocity material are present in both low-mass progenitors like the Crab and high-mass progenitors as in GRBs may be an indication that the central engine dynamics are not dissimiliar.

One additional clue as to the nature of Cas A's NE and SW ejecta jets and the explosion dynamics comes from examining the kinematic properties of the inferred motion of the remnant's neutron star. Its location some 7′ south of Cas A's expansion center (see Fig. 4), suggests a transverse velocity "kick" of 350 $\mathrm{km\,s^{-1}}$ imparted during the explosion (Fesen et al. 2006a). Intriguingly, the projected line connecting the NE and SW jets lies nearly

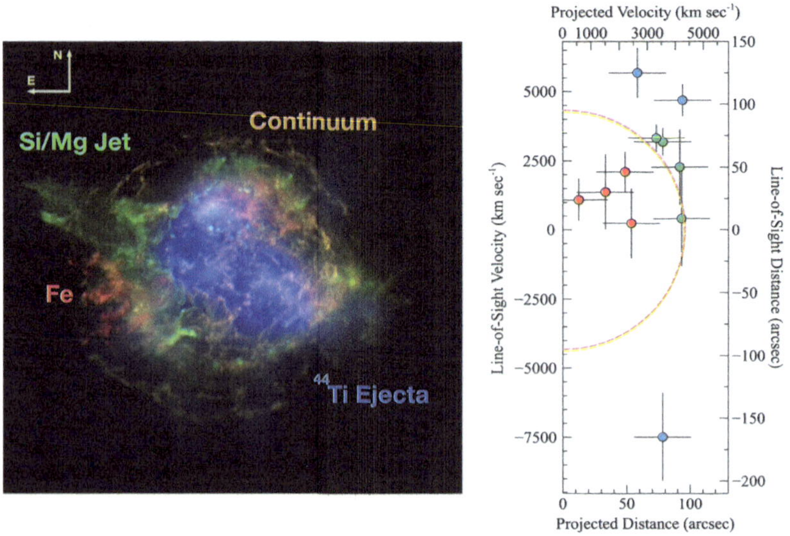

Fig. 16 *Left*: Three-color image of Cas A, with ^{44}Ti (from *NuSTAR* observations) in *blue*, 4–6 keV continuum in *gold*, Fe in *red*, and Si/Mg in *green* (the latter from *Chandra* observations). *Right*: The light-of-sight velocity of the detected ^{44}Ti versus the projected distance from the center of expansion of Cas A, where positive velocities indicate redshift. The *gold dashed circle* represents the location of the reverse shock; thus, the red points are interior to the RS, *green points* are at/near the RS, and blue points are exterior to the RS. Figures are from Grefenstette et al. (2017)

perpendicular to the inferred direction of the NS (see Figs. 4 and 11). Assuming the NE/SW axis to be the most significant in the original core-collapse, this runs counter to most jet-induced explosion models that predict that the neutron star will undergo a kick roughly aligned with the jet axis.

3.3 The Kinematics of Radioactive Titanium in Young SNRs

With the launch of *NuSTAR*, the first focusing, hard X-ray (3–79 keV) observatory (Harrison et al. 2013), it is possible to measure the kinematics of ^{44}Ti from the radioactive decay lines at 68 and 78 keV. To date, these lines have been detected in two young CC SNRs: SN 1987A (Boggs et al. 2015) and Cas A (Grefenstette et al. 2014, 2017). In the case of SN 1987A, the lines were narrow and redshifted with a velocity of $\sim$700 km s^{-1}, and the lack of a blueshifted component suggests large-scale asymmetry of the ejecta.

In Cas A, the ^{44}Ti was localized using the *NuSTAR* observations, and one major surprise was that the ^{44}Ti did not trace the distribution of Fe K (see Fig. 16, left panel), nor does it appear associated with the NE/SW jets (Grefenstette et al. 2014). Using deeper *NuSTAR* data, Grefenstette et al. (2017) performed spatially-resolved spectroscopic analysis of the ^{44}Ti ejecta and found that $\sim$ 40% of the material is interior to the reverse shock, 40% is at/near the reverse shock, and 20% is beyond the reverse shock (see Fig. 16, right panel). Generally, the ^{44}Ti interior to the RS is not correlated with observed features in the optical or infrared. The ^{44}Ti exterior to the RS is co-located with the shock-heated Fe, whereas regions with Fe do not necessarily have associated ^{44}Ti. The authors interpreted this result as evidence that local conditions during the explosive nucleosynthesis varied enough to suppress the production of ^{44}Ti in some regions. Based on the ^{44}Ti line centroids, the majority of the material is receding along the line of sight, with velocities ranging from 1,100–5,700 km s^{-1} (though there is some blueshifted emission as well). The flux-weighted average of the ^{44}Ti velocities gives the direction of this ejecta component that is almost exactly opposite to the direction of the Cas A NS, similar to the results presented in Fig. 4.

4 Conclusions and Future Prospects

The morphology and kinematics of SNRs are an especially powerful means to test and constrain SN explosion models. The predictive power of simulations has improved dramatically over the last few years, and the results can be compared to observational characteristics of SNRs, including large-scale compositional and morphological asymmetries as well as the three-dimensional kinematics of ejecta material. In particular, as the youngest known CC SNR in the Milky Way, Cas A offers an up-close view of the complexity of these explosive events that are unresolved in extragalactic sources.

In the future, an exciting prospect in the study of SNRs is X-ray micro-calorimeters. CCD energy resolution is insufficient to resolve He-like and H-like line complexes of ions at X-ray wavelengths, and grating spectrometers are only useful when objects have minimal angular extension (e.g., SN 1987A: Dewey et al. 2008; SNR 1E 0102.2−7219: Flanagan et al. 2004) or if the ejecta knots are sufficiently bright and isolated (e.g., Cas A: Lazendic et al. 2006; G292.0+1.8: Bhalerao et al. 2015). X-ray microcalorimeters are non-dispersive spectrometers that can take spatially resolved, high-resolution (with a few eV resolution) spectra across extended objects, like SNRs and galaxy clusters.

Before its untimely demise, *Hitomi* observed the LMC SNR N132D for only 3.7 ks (Hitomi Collaboration 2017), and it readily detected line complexes of S, Ar, and Fe. The Fe emission was highly redshifted at $\sim$800 km s^{-1}, but no blueshifted component was detected, suggesting that the Fe-rich ejecta was ejected asymmetrically. In the shorter term (with an anticipated launch date of 2021), the replacement for *Hitomi*, the X-ray Astronomy Recovery Mission (XARM[1]), offers the chance of 5–7 eV energy resolution over a field-of-view of $\sim 3'$.

Looking forward to the 2030s, the proposed *Lynx* X-ray Observatory is intended to have a microcalorimeter with high spatial resolution (1.0″ pixels) over a 5′ field with a few eV energy resolution. These capabilities will enable three-dimensional mapping of the hot metals synthesized in SN explosions, similar to the results presented in Sect. 3 but at X-ray wavelengths. Additionally, with $\sim$50 times the effective area of *Chandra*, *Lynx* will detect fainter and more distant sources, such as SNRs in Local Group galaxies. In this way *Lynx* will dramatically increase the sample size of young SNRs with morphological and kinematic measurements. These observations will be crucial to inform SN models and probe SN feedback and enrichment in different Galactic environments.

Acknowledgements L.A.L. gratefully acknowledges that this work was supported through NSF Astronomy & Astrophysics Grant AST−1517021. The authors thank the organizers of the workshop held in ISSI Bern on 3–7 October 2016 where many helpful discussions took place.

References

Z. Arzoumanian, D.F. Chernoff, J.M. Cordes, The velocity distribution of isolated radio pulsars. Astrophys. J. **568**, 289–301 (2002). astro-ph/0106159. https://doi.org/10.1086/338805

[1] See https://heasarc.gsfc.nasa.gov/docs/xarm/.

C. Badenes, J. Harris, D. Zaritsky, J.L. Prieto, The stellar ancestry of supernovae in the Magellanic Clouds, I: the most recent supernovae in the Large Magellanic Cloud. Astrophys. J. **700**, 727–740 (2009). 0902.2787. https://doi.org/10.1088/0004-637X/700/1/727

C. Badenes, D. Maoz, B.T. Draine, On the size distribution of supernova remnants in the Magellanic Clouds. Mon. Not. R. Astron. Soc. **407**, 1301–1313 (2010). 1003.3030. https://doi.org/10.1111/j.1365-2966.2010.17023.x

J. Bhalerao, S. Park, D. Dewey, J.P. Hughes, K. Mori, J.J. Lee, X-ray ejecta kinematics of the galactic core-collapse supernova remnant G292.0+1.8. Astrophys. J. **800**, 65 (2015). 1408.1949. https://doi.org/10.1088/0004-637X/800/1/65

J.M. Blondin, P. Lundqvist, R.A. Chevalier, Axisymmetric circumstellar interaction in supernovae. Astrophys. J. **472**, 257 (1996). astro-ph/9601137. https://doi.org/10.1086/178060

J.M. Blondin, K.J. Borkowski, S.P. Reynolds, Dynamics of Fe bubbles in young supernova remnants. Astrophys. J. **557**, 782–791 (2001). astro-ph/0010285. https://doi.org/10.1086/321674

S.E. Boggs, F.A. Harrison, H. Miyasaka, B.W. Grefenstette, A. Zoglauer, C.L. Fryer, S.P. Reynolds, D.M. Alexander, H. An, D. Barret, F.E. Christensen, W.W. Craig, K. Forster, P. Giommi, C.J. Hailey, A. Hornstrup, T. Kitaguchi, J.E. Koglin, K.K. Madsen, P.H. Mao, K. Mori, M. Perri, M.J. Pivovaroff, S. Puccetti, V. Rana, D. Stern, N.J. Westergaard, W.W. Zhang, ^{44}Ti gamma-ray emission lines from SN1987A reveal an asymmetric explosion. Science **348**, 670–671 (2015). https://doi.org/10.1126/science.aaa2259

S. Chakraborti, A. Soderberg, L. Chomiuk, A. Kamble, N. Yadav, A. Ray, K. Hurley, R. Margutti, D. Milisavljevic, M. Bietenholz, A. Brunthaler, G. Pignata, E. Pian, P. Mazzali, C. Fransson, N. Bartel, M. Hamuy, E. Levesque, A. MacFadyen, J. Dittmann, M. Krauss, M.S. Briggs, V. Connaughton, K. Yamaoka, T. Takahashi, M. Ohno, Y. Fukazawa, M. Tashiro, Y. Terada, T. Murakami, J. Goldsten, S. Barthelmy, N. Gehrels, J. Cummings, H. Krimm, D. Palmer, S. Golenetskii, R. Aptekar, D. Frederiks, D. Svinkin, T. Cline, I.G. Mitrofanov, D. Golovin, M.L. Litvak, A.B. Sanin, W. Boynton, C. Fellows, K. Harshman, H. von Enos, A. Kienlin, A. Rau, X. Zhang, V. Savchenko, A missing-link in the supernova-GRB connection: the case of SN 2012ap. Astrophys. J. **805**, 187 (2015). 1402.6336. https://doi.org/10.1088/0004-637X/805/2/187

S. Chatterjee, W.H.T. Vlemmings, W.F. Brisken, T.J.W. Lazio, J.M. Cordes, W.M. Goss, S.E. Thorsett, E.B. Fomalont, A.G. Lyne, M. Kramer, Getting its kicks: a VLBA parallax for the hyperfast pulsar B1508+55. Astrophys. J. Lett. **630**, L61–L64 (2005). astro-ph/0509031. https://doi.org/10.1086/491701

D.P. Cox, R.L. Shelton, W. Maciejewski, R.K. Smith, T. Plewa, A. Pawl, M. Rozyczka, Modeling W44 as a supernova remnant in a density gradient with a partially formed dense shell and thermal conduction in the hot interior, I: the analytical model. Astrophys. J. **524**, 179–191 (1999). https://doi.org/10.1086/307781

T. DeLaney, L. Rudnick, M.D. Stage, J.D. Smith, K. Isensee, J. Rho, G.E. Allen, H. Gomez, T. Kozasa, W.T. Reach, J.E. Davis, J.C. Houck, The three-dimensional structure of Cassiopeia A. Astrophys. J. **725**, 2038–2058 (2010). 1011.3858. https://doi.org/10.1088/0004-637X/725/2/2038

D. Dewey, S.A. Zhekov, R. McCray, C.R. Canizares, Chandra HETG spectra of SN 1987A at 20 years. Astrophys. J. Lett. **676**, L131 (2008). 0802.2340. https://doi.org/10.1086/587549

C.A. Faucher-Giguère, V.M. Kaspi, Birth and evolution of isolated radio pulsars. Astrophys. J. **643**, 332–355 (2006). astro-ph/0512585. https://doi.org/10.1086/501516

R.A. Fesen, An optical survey of outlying ejecta in Cassiopeia A: evidence for a turbulent, asymmetric explosion. Astrophys. J. Suppl. Ser. **133**, 161–186 (2001). https://doi.org/10.1086/319181

R.A. Fesen, K.S. Gunderson, Optical spectroscopy and imaging of the northeast jet in the Cassiopeia A supernova remnant. Astrophys. J. **470**, 967 (1996). https://doi.org/10.1086/177923

R.A. Fesen, D. Milisavljevic, An HST survey of the highest-velocity ejecta in Cassiopeia A. Astrophys. J. **818**, 17 (2016). 1512.05049. https://doi.org/10.3847/0004-637X/818/1/17

R.A. Fesen, M.C. Hammell, J. Morse, R.A. Chevalier, K.J. Borkowski, M.A. Dopita, C.L. Gerardy, S.S. Lawrence, J.C. Raymond, S. van den Bergh, Discovery of outlying high-velocity oxygen-rich ejecta in Cassiopeia A. Astrophys. J. **636**, 859–872 (2006a). astro-ph/0509067. https://doi.org/10.1086/498092

R.A. Fesen, M.C. Hammell, J. Morse, R.A. Chevalier, K.J. Borkowski, M.A. Dopita, C.L. Gerardy, S.S. Lawrence, J.C. Raymond, S. van den Bergh, The expansion asymmetry and age of the Cassiopeia A supernova remnant. Astrophys. J. **645**, 283–292 (2006b). astro-ph/0603371. https://doi.org/10.1086/504254

R.A. Fesen, J.A. Zastrow, M.C. Hammell, J.M. Shull, D.W. Silvia, Ejecta knot flickering, mass ablation, and fragmentation in Cassiopeia A. Astrophys. J. **736**, 109 (2011). 1105.3970. https://doi.org/10.1088/0004-637X/736/2/109

A.V. Filippenko, Optical spectra of supernovae. Annu. Rev. Astron. Astrophys. **35**, 309–355 (1997). https://doi.org/10.1146/annurev.astro.35.1.309

K.A. Flanagan, C.R. Canizares, D. Dewey, J.C. Houck, A.C. Fredericks, M.L. Schattenburg, T.H. Markert, D.S. Davis, Chandra high-resolution X-ray spectrum of supernova remnant 1E 0102.2-7219. Astrophys. J. **605**, 230–246 (2004). astro-ph/0312509. https://doi.org/10.1086/382145

C.L. Fryer, A. Kusenko, Effects of neutrino-driven kicks on the supernova explosion mechanism. Astrophys. J. Suppl. Ser. **163**, 335–343 (2006). astro-ph/0512033. https://doi.org/10.1086/500933

B.M. Gaensler, The nature of bilateral supernova remnants. Astrophys. J. **493**, 781–792 (1998). astro-ph/9707149. https://doi.org/10.1086/305146

B.M. Gaensler, P.O. Slane, The evolution and structure of pulsar wind nebulae. Annu. Rev. Astron. Astrophys. **44**, 17–47 (2006). astro-ph/0601081. https://doi.org/10.1146/annurev.astro.44.051905.092528

K. Garofali, B.F. Williams, P.P. Plucinsky, T.J. Gaetz, B. Wold, F. Haberl, K.S. Long, W.P. Blair, T.G. Pannuti, P.F. Winkler, J. Gross, Supernova remnants in M33: X-ray properties as observed by XMM-Newton. Mon. Not. R. Astron. Soc. **472**, 308–333 (2017). 1708.01239. https://doi.org/10.1093/mnras/stx1905

D.F. González-Casanova, F. De Colle, E. Ramirez-Ruiz, L.A. Lopez, The morphology and dynamics of jet-driven supernova remnants: the case of W49B. Astrophys. J. Lett. **781**, L26 (2014). 1309.7049. https://doi.org/10.1088/2041-8205/781/2/L26

D.A. Green, A catalogue of 294 Galactic supernova remnants. Bull. Astron. Soc. India **42**, 47–58 (2014). 1409.0637

B.W. Grefenstette, F.A. Harrison, S.E. Boggs, S.P. Reynolds, C.L. Fryer, K.K. Madsen, D.R. Wik, A. Zoglauer, C.I. Ellinger, D.M. Alexander, H. An, D. Barret, F.E. Christensen, W.W. Craig, K. Forster, P. Giommi, C.J. Hailey, A. Hornstrup, V.M. Kaspi, T. Kitaguchi, J.E. Koglin, P.H. Mao, H. Miyasaka, K. Mori, M. Perri, M.J. Pivovaroff, S. Puccetti, V. Rana, D. Stern, N.J. Westergaard, W.W. Zhang, Asymmetries in core-collapse supernovae from maps of radioactive ^{44}Ti in Cassiopeia A. Nature **506**, 339–342 (2014). 1403.4978. https://doi.org/10.1038/nature12997

B.W. Grefenstette, C.L. Fryer, F.A. Harrison, S.E. Boggs, T. DeLaney, J.M. Laming, S.P. Reynolds, D.M. Alexander, D. Barret, F.E. Christensen, W.W. Craig, K. Forster, P. Giommi, C.J. Hailey, A. Hornstrup, T. Kitaguchi, J.E. Koglin, L. Lopez, P.H. Mao, K.K. Madsen, H. Miyasaka, K. Mori, M. Perri, M.J. Pivovaroff, S. Puccetti, V. Rana, D. Stern, N.J. Westergaard, D.R. Wik, W.W. Zhang, A. Zoglauer, The distribution of radioactive ^{44}Ti in Cassiopeia A. Astrophys. J. **834**, 19 (2017). 1612.02774. https://doi.org/10.3847/1538-4357/834/1/19

N.J. Hammer, H.T. Janka, E. Müller, Three-dimensional simulations of mixing instabilities in supernova explosions. Astrophys. J. **714**, 1371–1385 (2010). 0908.3474. https://doi.org/10.1088/0004-637X/714/2/1371

F.A. Harrison, W.W. Craig, F.E. Christensen, C.J. Hailey, W.W. Zhang, S.E. Boggs, D. Stern, W.R. Cook, K. Forster, P. Giommi, B.W. Grefenstette, Y. Kim, T. Kitaguchi, J.E. Koglin, K.K. Madsen, P.H. Mao, H. Miyasaka, K. Mori, M. Perri, M.J. Pivovaroff, S. Puccetti, V.R. Rana, N.J. Westergaard, J. Willis, A. Zoglauer, H. An, M. Bachetti, N.M. Barrière, E.C. Bellm, V. Bhalerao, N.F. Brejnholt, F. Fuerst, C.C. Liebe, C.B. Markwardt, M. Nynka, J.K. Vogel, D.J. Walton, D.R. Wik, D.M. Alexander, L.R. Cominsky, A.E. Hornschemeier, A. Hornstrup, V.M. Kaspi, G.M. Madejski, G. Matt, S. Molendi, D.M. Smith, J.A. Tomsick, M. Ajello, D.R. Ballantyne, M. Baloković, D. Barret, F.E. Bauer, R.D. Blandford, W.N. Brandt, L.W. Brenneman, J. Chiang, D. Chakrabarty, J. Chenevez, A. Comastri, F. Dufour, M. Elvis, A.C. Fabian, D. Farrah, C.L. Fryer, E.V. Gotthelf, J.E. Grindlay, D.J. Helfand, R. Krivonos, D.L. Meier, J.M. Miller, L. Natalucci, P. Ogle, E.O. Ofek, A. Ptak, S.P. Reynolds, J.R. Rigby, G. Tagliaferri, S.E. Thorsett, E. Treister, C.M. Urry, The Nuclear Spectroscopic Telescope Array (NuSTAR) high-energy X-ray mission. Astrophys. J. **770**, 103 (2013). 1301.7307. https://doi.org/10.1088/0004-637X/770/2/103

Hitomi Collaboration, F. Aharonian, H. Akamatsu, F. Akimoto, S.W. Allen, L. Angelini, M. Audard, H. Awaki, M. Axelsson, A. Bamba, M.W. Bautz, R. Blandford, L.W. Brenneman, G.V. Brown, E. Bulbul, E.M. Cackett, M. Chernyakova, M.P. Chiao, P.S. Coppi, E. Costantini, J. de Plaa, C.P. de Vries, J.W. den Herder, C. Done, T. Dotani, K. Ebisawa, M.E. Eckart, T. Enoto, Y. Ezoe, A.C. Fabian, C. Ferrigno, A.R. Foster, R. Fujimoto, Y. Fukazawa, A. Furuzawa, M. Galeazzi, L.C. Gallo, P. Gandhi, M. Giustini, A. Goldwurm, L. Gu, M. Guainazzi, Y. Haba, K. Hagino, K. Hamaguchi, I.M. Harrus, I. Hatsukade, K. Hayashi, T. Hayashi, K. Hayashida, J.S. Hiraga, A. Hornschemeier, A. Hoshino, J.P. Hughes, Y. Ichinohe, R. Iizuka, H. Inoue, Y. Inoue, M. Ishida, K. Ishikawa, Y. Ishisaki, M. Iwai, J. Kaastra, T. Kallman, T. Kamae, J. Kataoka, S. Katsuda, N. Kawai, R.L. Kelley, C.A. Kilbourne, T. Kitaguchi, S. Kitamoto, T. Kitayama, T. Kohmura, M. Kokubun, K. Koyama, S. Koyama, P. Kretschmar, H.A. Krimm, A. Kubota, H. Kunieda, P. Laurent, S.H. Lee, M.A. Leutenegger, O. Limousin, M. Loewenstein, K.S. Long, D. Lumb, G. Madejski, Y. Maeda, D. Maier, K. Makishima, M. Markevitch, H. Matsumoto, K. Matsushita, D. McCammon, B.R. McNamara, M. Mehdipour, E.D. Miller, J.M. Miller, S. Mineshige, K. Mitsuda, I. Mitsuishi, T. Miyazawa, T. Mizuno, H. Mori, K. Mori, K. Mukai, H. Murakami, R.F. Mushotzky, T. Nakagawa, H. Nakajima, T. Nakamori, S. Nakashima, K. Nakazawa, K.K. Nobukawa, M. Nobukawa, H. Noda, H. Odaka, T. Ohashi, M. Ohno, T. Okajima, N. Ota, M. Ozaki, F. Paerels, S. Paltani, R. Petre, C. Pinto, F.S. Porter, K. Pottschmidt, C.S. Reynolds, S. Safi-Harb, S. Saito, K. Sakai, T. Sasaki, G. Sato, K. Sato, R. Sato, T. Sato, M. Sawada, N. Schartel, P.J. Serlemtsos, H. Seta, M. Shidatsu, A. Simionescu, R.K. Smith, Y. Soong, Ł. Stawarz, Y. Sugawara, S. Sugita, A. Szymkowiak, H. Tajima, H. Takahashi, T. Takahashi, S. Takeda, Y. Takei, T. Tamagawa, T. Tamura, T. Tanaka, Y. Tanaka, Y.T.

Tanaka, M.S. Tashiro, Y. Tawara, Y. Terada, Y. Terashima, F. Tombesi, H. Tomida, Y. Tsuboi, M. Tsujimoto, H. Tsunemi, T.G. Tsuru, H. Uchida, H. Uchiyama, Y. Uchiyama, S. Ueda, Y. Ueda, S. Uno, C.M. Urry, E. Ursino, S. Watanabe, N. Werner, D.R. Wilkins, B.J. Williams, S. Yamada, H. Yamaguchi, K. Yamaoka, N.Y. Yamasaki, M. Yamauchi, S. Yamauchi, T. Yaqoob, Y. Yatsu, D. Yonetoku, I. Zhuravleva, Hitomi observations of the LMC SNR N132D: highly redshifted X-ray emission from iron ejecta. 1712.02365 (2017)

G. Hobbs, D.R. Lorimer, A.G. Lyne, M. Kramer, A statistical study of 233 pulsar proper motions. Mon. Not. R. Astron. Soc. **360**, 974–992 (2005). astro-ph/0504584. https://doi.org/10.1111/j.1365-2966.2005.09087.x

T. Holland-Ashford, L.A. Lopez, K. Auchettl, T. Temim, E. Ramirez-Ruiz, Comparing neutron star kicks to supernova remnant asymmetries. Astrophys. J. **844**, 84 (2017). 1705.08454. https://doi.org/10.3847/1538-4357/aa7a5c

T.W.S. Holoien, K.Z. Stanek, C.S. Kochanek, B.J. Shappee, J.L. Prieto, J. Brimacombe, D. Bersier, D.W. Bishop, S. Dong, J.S. Brown, A.B. Danilet, G.V. Simonian, U. Basu, J.F. Beacom, E. Falco, G. Pojmanski, D.M. Skowron, P.R. Woźniak, C.G. Ávila, E. Conseil, C. Contreras, I. Cruz, J.M. Fernández, R.A. Koff, Z. Guo, G.J. Herczeg, J. Hissong, E.Y. Hsiao, J. Jose, S. Kiyota, F. Long, L.A.G. Monard, B. Nicholls, J. Nicolas, W.S. Wiethoff, The ASAS-SN bright supernova catalogue, I: 2013–2014. Mon. Not. R. Astron. Soc. **464**, 2672–2686 (2017). 1604.00396. https://doi.org/10.1093/mnras/stw2273

J.P. Hughes, C.E. Rakowski, D.N. Burrows, P.O. Slane, Nucleosynthesis and mixing in Cassiopeia A. Astrophys. J. Lett. **528**, L109–L113 (2000). astro-ph/9910474. https://doi.org/10.1086/312438

U. Hwang, J.P. Hughes, R. Petre, The X-ray iron emission from Tycho's Supernova remnant. Astrophys. J. **497**, 833 (1998). astro-ph/9712241. https://doi.org/10.1086/305501

U. Hwang, S.S. Holt, R. Petre, Mapping the X-ray-emitting ejecta in Cassiopeia A with Chandra. Astrophys. J. Lett. **537**, L119–L122 (2000). astro-ph/0005560. https://doi.org/10.1086/312776

K. Isensee, L. Rudnick, T. DeLaney, J.D. Smith, J. Rho, W.T. Reach, T. Kozasa, H. Gomez, The three-dimensional structure of interior ejecta in Cassiopeia A at high spectral resolution. Astrophys. J. **725**, 2059–2070 (2010). 1006.0918. https://doi.org/10.1088/0004-637X/725/2/2059

H. Itoh, K. Masai, The effect of a circumstellar medium on the X-ray emission of young remnants of type II supernovae. Mon. Not. R. Astron. Soc. **236**, 885–899 (1989). https://doi.org/10.1093/mnras/236.4.885

H.T. Janka, Neutron star kicks by the gravitational tug-boat mechanism in asymmetric supernova explosions: progenitor and explosion dependence. Astrophys. J. **837**, 84 (2017). 1611.07562. https://doi.org/10.3847/1538-4357/aa618e

H.T. Janka, T. Melson, A. Summa, Physics of core-collapse supernovae in three dimensions: a sneak preview. Annu. Rev. Nucl. Part. Sci. **66**, 341–375 (2016). 1602.05576. https://doi.org/10.1146/annurev-nucl-102115-044747

T.E. Jeltema, C.R. Canizares, M.W. Bautz, D.A. Buote, The evolution of structure in X-ray clusters of galaxies. Astrophys. J. **624**, 606–629 (2005). astro-ph/0501360. https://doi.org/10.1086/428940

Z.G. Jennings, B.F. Williams, J.W. Murphy, J.J. Dalcanton, K.M. Gilbert, A.E. Dolphin, M. Fouesneau, D.R. Weisz, Supernova remnant progenitor masses in M31. Astrophys. J. **761**, 26 (2012). 1210.6353. https://doi.org/10.1088/0004-637X/761/1/26

D.L. Kaplan, S. Chatterjee, B.M. Gaensler, J. Anderson, A precise proper motion for the Crab pulsar, and the difficulty of testing spin-kick alignment for young neutron stars. Astrophys. J. **677**, 1201–1215 (2008). 0801.1142. https://doi.org/10.1086/529026

V.M. Kaspi, Radio pulsar/supernova remnant associations. Adv. Space Res. **21**, 167–176 (1998). https://doi.org/10.1016/S0273-1177(97)00801-6

S. Katsuda, M. Morii, H.T. Janka, A. Wongwathanarat, K. Nakamura, K. Kotake, K. Mori, E. Müller, T. Takiwaki, M. Tanaka, N. Tominaga, H. Tsunemi, Intermediate-mass-elements in young supernova remnants reveal neutron star kicks by asymmetric explosions. 1710.10372 (2017)

M.T. Kawasaki, M. Ozaki, F. Nagase, K. Masai, M. Ishida, R. Petre, ASCA observations of the supernova remnant IC 443: thermal structure and detection of overionized plasma. Astrophys. J. **572**, 897–905 (2002). astro-ph/0202484. https://doi.org/10.1086/340383

M. Kawasaki, M. Ozaki, F. Nagase, H. Inoue, R. Petre, Ionization states and plasma structures of mixed-morphology supernova remnants observed with ASCA. Astrophys. J. **631**, 935–946 (2005). astro-ph/0507348. https://doi.org/10.1086/432591

O. Krause, S.M. Birkmann, T. Usuda, T. Hattori, M. Goto, G.H. Rieke, K.A. Misselt, The Cassiopeia A supernova was of type IIb. Science **320**, 1195 (2008). 0805.4557. https://doi.org/10.1126/science.1155788

C.K. Lacey, T.J.W. Lazio, N.E. Kassim, N. Duric, D.S. Briggs, K.K. Dyer, Spatially resolved thermal continuum absorption against supernova remnant W49B. Astrophys. J. **559**, 954–962 (2001). astro-ph/0105451. https://doi.org/10.1086/322372

D. Lai, Neutron star kicks and asymmetric supernovae, in *Physics of Neutron Star Interiors*, ed. by D. Blaschke, N.K. Glendenning, A. Sedrakian. Lecture Notes in Physics, vol. 578 (Springer, Berlin, 2001), p. 424. astro-ph/0012049

N.M. Law, S.R. Kulkarni, R.G. Dekany, E.O. Ofek, R.M. Quimby, P.E. Nugent, J. Surace, C.C. Grillmair, J.S. Bloom, M.M. Kasliwal, L. Bildsten, T. Brown, S.B. Cenko, D. Ciardi, E. Croner, S.G. Djorgovski, J. van Eyken, A.V. Filippenko, D.B. Fox, A. Gal-Yam, D. Hale, N. Hamam, G. Helou, J. Henning, D.A. Howell, J. Jacobsen, R. Laher, S. Mattingly, D. McKenna, A. Pickles, D. Poznanski, G. Rahmer, A. Rau, W. Rosing, M. Shara, R. Smith, D. Starr, M. Sullivan, V. Velur, R. Walters, J. Zolkower, The palomar transient factory: system overview, performance, and first results. Publ. Astron. Soc. Pac. **121**, 1395 (2009). 0906.5350. https://doi.org/10.1086/648598

S.S. Lawrence, G.M. MacAlpine, A. Uomoto, B.E. Woodgate, L.W. Brown, R.J. Oliversen, J.D. Lowenthal, C. Liu, Three-dimensional Fabry-Perot imaging spectroscopy of the Crab Nebula, Cassiopeia A, and Nova GK Persei. Astron. J. **109**, 2635 (1995). https://doi.org/10.1086/117477

J.S. Lazendic, D. Dewey, N.S. Schulz, C.R. Canizares, The kinematic and plasma properties of X-ray knots in Cassiopeia A from the Chandra HETGS. Astrophys. J. **651**, 250–267 (2006). astro-ph/0605078. https://doi.org/10.1086/507481

J. Leaman, W. Li, R. Chornock, A.V. Filippenko, Nearby supernova rates from the Lick Observatory Supernova Search, I: the methods and data base. Mon. Not. R. Astron. Soc. **412**, 1419–1440 (2011). 1006.4611. https://doi.org/10.1111/j.1365-2966.2011.18158.x

E.J. Lentz, S.W. Bruenn, W.R. Hix, A. Mezzacappa, O.E.B. Messer, E. Endeve, J.M. Blondin, J.A. Harris, P. Marronetti, K.N. Yakunin, Three-dimensional core-collapse supernova simulated using a 15 M progenitor. Astrophys. J. Lett. **807**, L31 (2015). 1505.05110. https://doi.org/10.1088/2041-8205/807/2/L31

D.C. Leonard, A.V. Filippenko, M. Ganeshalingam, F.J.D. Serduke, W. Li, B.J. Swift, A. Gal-Yam, R.J. Foley, D.B. Fox, S. Park, J.L. Hoffman, D.S. Wong, A non-spherical core in the explosion of supernova SN 2004dj. Nature **440**, 505–507 (2006). arXiv:astro-ph/0603297. https://doi.org/10.1038/nature04558

L.A. Lopez, What shapes supernova remnants? in *Supernova Environmental Impacts*, ed. by A. Ray, R.A. McCray. IAU Symposium, vol. 296, (2014), pp. 239–244. 1306.2638. https://doi.org/10.1017/S1743921313009538

L.A. Lopez, E. Ramirez-Ruiz, C. Badenes, D. Huppenkothen, T.E. Jeltema, D.A. Pooley, Typing supernova remnants using X-ray line emission morphologies. Astrophys. J. Lett. **706**, L106–L109 (2009a). 0910.3208. https://doi.org/10.1088/0004-637X/706/1/L106

L.A. Lopez, E. Ramirez-Ruiz, D.A. Pooley, T.E. Jeltema, Tools for dissecting supernova remnants observed with Chandra: methods and application to the galactic remnant W49B. Astrophys. J. **691**, 875–893 (2009b). 0810.0009. https://doi.org/10.1088/0004-637X/691/1/875

L.A. Lopez, E. Ramirez-Ruiz, D. Huppenkothen, C. Badenes, D.A. Pooley, Using the X-ray morphology of young supernova remnants to constrain explosion type, ejecta distribution, and chemical mixing. Astrophys. J. **732**, 114 (2011). 1011.0731. https://doi.org/10.1088/0004-637X/732/2/114

L.A. Lopez, S. Pearson, E. Ramirez-Ruiz, D. Castro, H. Yamaguchi, P.O. Slane, R.K. Smith, Unraveling the origin of overionized plasma in the galactic supernova remnant W49B. Astrophys. J. **777**, 145 (2013a). 1309.1464. https://doi.org/10.1088/0004-637X/777/2/145

L.A. Lopez, E. Ramirez-Ruiz, D. Castro, S. Pearson, The galactic supernova remnant W49B likely originates from a jet-driven, core-collapse explosion. Astrophys. J. **764**, 50 (2013b). 1301.0618. https://doi.org/10.1088/0004-637X/764/1/50

L.A. Lopez, D. Castro, P.O. Slane, E. Ramirez-Ruiz, C. Badenes, Identification of a jet-driven supernova remnant in the Small Magellanic Cloud: possible evidence for the enhancement of bipolar explosions at low metallicity. Astrophys. J. **788**, 5 (2014). 1310.4498. https://doi.org/10.1088/0004-637X/788/1/5

A.G. Lyne, D.R. Lorimer, High birth velocities of radio pulsars. Nature **369**, 127–129 (1994). https://doi.org/10.1038/369127a0

K. Maeda, K. Nomoto, Bipolar supernova explosions: nucleosynthesis and implications for abundances in extremely metal-poor stars. Astrophys. J. **598**, 1163–1200 (2003). arXiv:astro-ph/0304172. https://doi.org/10.1086/378948

P. Maggi, F. Haberl, P.J. Kavanagh, M. Sasaki, L.M. Bozzetto, M.D. Filipović, G. Vasilopoulos, W. Pietsch, S.D. Points, Y.H. Chu, J. Dickel, M. Ehle, R. Williams, J. Greiner, The population of X-ray supernova remnants in the Large Magellanic Cloud. Astron. Astrophys. **585**, A162 (2016). 1509.09223. https://doi.org/10.1051/0004-6361/201526932

D.S. Mathewson, V.L. Ford, M.A. Dopita, I.R. Tuohy, K.S. Long, D.J. Helfand, Supernova remnants in the Magellanic Clouds. Astrophys. J. Suppl. Ser. **51**, 345–355 (1983). https://doi.org/10.1086/190854

P.A. Mazzali, K.S. Kawabata, K. Maeda, R.J. Foley, K. Nomoto, J. Deng, T. Suzuki, M. Iye, N. Kashikawa, Y. Ohyama, A.V. Filippenko, Y. Qiu, J. Wei, The aspherical properties of the energetic type Ic SN 2002ap as inferred from its nebular spectra. Astrophys. J. **670**, 592–599 (2007). 0708.0966. https://doi.org/10.1086/521873

R. McCray, C. Fransson, The remnant of supernova 1987A. Annu. Rev. Astron. Astrophys. **54**, 19–52 (2016). https://doi.org/10.1146/annurev-astro-082615-105405

D. Milisavljevic, R.A. Fesen, A detailed kinematic map of Cassiopeia A's optical main shell and outer high-velocity ejecta. Astrophys. J. **772**, 134 (2013). 1306.2310. https://doi.org/10.1088/0004-637X/772/2/134

D. Milisavljevic, R.A. Fesen, The bubble-like interior of the core-collapse supernova remnant Cassiopeia A. Science **347**, 526–530 (2015). 1501.07283. https://doi.org/10.1126/science.1261949

D. Milisavljevic, R.A. Fesen, R.A. Chevalier, R.P. Kirshner, P. Challis, M. Turatto, Late-time optical emission from core-collapse supernovae. Astrophys. J. **751**, 25 (2012). 1203.0006. https://doi.org/10.1088/0004-637X/751/1/25

R. Minkowski, Spectra of supernovae. Publ. Astron. Soc. Pac. **53**, 224 (1941). https://doi.org/10.1086/125315

M. Modjaz, L. Kewley, R.P. Kirshner, K.Z. Stanek, P. Challis, P.M. Garnavich, J.E. Greene, P.L. Kelly, J.L. Prieto, Measured metallicities at the sites of nearby broad-lined type Ic supernovae and implications for the supernovae gamma-ray burst connection. Astron. J. **135**, 1136–1150 (2008). astro-ph/0701246. https://doi.org/10.1088/0004-6256/135/4/1136

T.J. Moriya, Progenitors of recombining supernova remnants. Astrophys. J. Lett. **750**, L13 (2012). 1203.5799. https://doi.org/10.1088/2041-8205/750/1/L13

B. Müller, T. Melson, A. Heger, H.T. Janka, Supernova simulations from a 3D progenitor model—impact of perturbations and evolution of explosion properties. Mon. Not. R. Astron. Soc. **472**, 491–513 (2017). 1705.00620. https://doi.org/10.1093/mnras/stx1962

R.L. Nugent, New measurements of the expansion of the Crab Nebula. Publ. Astron. Soc. Pac. **110**, 831–836 (1998). https://doi.org/10.1086/316199

S. Orlando, F. Bocchino, F. Reale, G. Peres, O. Petruk, On the origin of asymmetries in bilateral supernova remnants. Astron. Astrophys. **470**, 927–939 (2007). 0704.0890. https://doi.org/10.1051/0004-6361:20066045

M. Ozawa, K. Koyama, H. Yamaguchi, K. Masai, T. Tamagawa, Suzaku discovery of the strong radiative recombination continuum of iron from the supernova remnant W49B. Astrophys. J. Lett. **706**, L71–L75 (2009). 0910.3302. https://doi.org/10.1088/0004-637X/706/1/L71

D.J. Patnaude, R.A. Fesen, A comparison of X-ray and optical emission in Cassiopeia A. Astrophys. J. **789**, 138 (2014). 1403.5812. https://doi.org/10.1088/0004-637X/789/2/138

G.G. Pavlov, V.E. Zavlin, B. Aschenbach, J. Trümper, D. Sanwal, The compact central object in Cassiopeia A: a neutron star with hot polar caps or a black hole? Astrophys. J. Lett. **531**, L53–L56 (2000). astro-ph/9912024. https://doi.org/10.1086/312521

C.L. Peters, L.A. Lopez, E. Ramirez-Ruiz, K.G. Stassun, E. Figueroa-Feliciano, Constraining explosion type of young supernova remnants using 24 μm emission morphology. Astrophys. J. Lett. **771**, L38 (2013). 1306.3503. https://doi.org/10.1088/2041-8205/771/2/L38

J.E. Reed, J.J. Hester, A.C. Fabian, P.F. Winkler, The three-dimensional structure of the Cassiopeia A supernova remnant, I: the spherical shell. Astrophys. J. **440**, 706 (1995). https://doi.org/10.1086/175308

A. Rest, N.B. Suntzeff, K. Olsen, J.L. Prieto, R.C. Smith, D.L. Welch, A. Becker, M. Bergmann, A. Clocchiatti, K. Cook, A. Garg, M. Huber, G. Miknaitis, D. Minniti, S. Nikolaev, C. Stubbs, Light echoes from ancient supernovae in the Large Magellanic Cloud. Nature **438**, 1132–1134 (2005). astro-ph/0510738. https://doi.org/10.1038/nature04365

A. Rest, T. Matheson, S. Blondin, M. Bergmann, D.L. Welch, N.B. Suntzeff, R.C. Smith, K. Olsen, J.L. Prieto, A. Garg, P. Challis, C. Stubbs, M. Hicken, M. Modjaz, W.M. Wood-Vasey, A. Zenteno, G. Damke, A. Newman, M. Huber, K.H. Cook, S. Nikolaev, A.C. Becker, A. Miceli, R. Covarrubias, L. Morelli, G. Pignata, A. Clocchiatti, D. Minniti, R.J. Foley, Spectral identification of an ancient supernova using light echoes in the Large Magellanic Cloud. Astrophys. J. **680**, 1137–1148 (2008). 0801.4762. https://doi.org/10.1086/587158

A. Rest, R.J. Foley, B. Sinnott, D.L. Welch, C. Badenes, A.V. Filippenko, M. Bergmann, W.A. Bhatti, S. Blondin, P. Challis, G. Damke, H. Finley, M.E. Huber, D. Kasen, R.P. Kirshner, T. Matheson, P. Mazzali, D. Minniti, R. Nakajima, G. Narayan, K. Olsen, D. Sauer, R.C. Smith, N.B. Suntzeff, Direct confirmation of the asymmetry of the Cas A supernova with light echoes. Astrophys. J. **732**, 3 (2011). 1003.5660. https://doi.org/10.1088/0004-637X/732/1/3

J. Rho, R. Petre, Mixed-morphology supernova remnants. Astrophys. J. Lett. **503**, L167–L170 (1998). https://doi.org/10.1086/311538

L.F. Roberts, C.D. Ott, R. Haas, E.P. O'Connor, P. Diener, E. Schnetter, General-relativistic three-dimensional multi-group neutrino radiation-hydrodynamics simulations of core-collapse supernovae. Astrophys. J. **831**, 98 (2016). 1604.07848. https://doi.org/10.3847/0004-637X/831/1/98

M. Sasaki, W. Pietsch, F. Haberl, D. Hatzidimitriou, H. Stiele, B. Williams, A. Kong, U. Kolb, Supernova remnants and candidates detected in the XMM-Newton M 31 large survey. Astron. Astrophys. **544**, A144 (2012). 1206.4789. https://doi.org/10.1051/0004-6361/201219025

L. Scheck, K. Kifonidis, H.T. Janka, E. Müller, Multidimensional supernova simulations with approximative neutrino transport, I: neutron star kicks and the anisotropy of neutrino-driven explosions in two spatial dimensions. Astron. Astrophys. **457**, 963–986 (2006). astro-ph/0601302. https://doi.org/10.1051/0004-6361:20064855

F.D. Seward, Einstein observations of galactic supernova remnants. Astrophys. J. Suppl. Ser. **73**, 781–819 (1990). https://doi.org/10.1086/191489

F.D. Seward, W.H. Tucker, R.A. Fesen, Faint X-ray structure in the Crab Pulsar Wind Nebula. Astrophys. J. **652**, 1277–1287 (2006). astro-ph/0608485. https://doi.org/10.1086/508532,

R.L. Shelton, D.P. Cox, W. Maciejewski, R.K. Smith, T. Plewa, A. Pawl, M. Rozyczka, Modeling W44 as a Supernova remnant in a density gradient with a partially formed dense shell and thermal conduction in the hot interior, II: the hydrodynamic models. Astrophys. J. **524**, 192–212 (1999). astro-ph/9806090. https://doi.org/10.1086/307799

T. Shimizu, K. Masai, K. Koyama, Evolution of supernova remnants expanding out of the dense circumstellar matter into the rarefied interstellar medium. Publ. Astron. Soc. Jpn. **64**, 24 (2012). 1110.4948. https://doi.org/10.1093/pasj/64.2.24

A.M. Soderberg, S. Chakraborti, G. Pignata, R.A. Chevalier, P. Chandra, A. Ray, M.H. Wieringa, A. Copete, V. Chaplin, V. Connaughton, S.D. Barthelmy, M.F. Bietenholz, N. Chugai, M.D. Stritzinger, M. Hamuy, C. Fransson, O. Fox, E.M. Levesque, J.E. Grindlay, P. Challis, R.J. Foley, R.P. Kirshner, P.A. Milne, M.A.P. Torres, A relativistic type Ibc supernova without a detected γ-ray burst. Nature **463**, 513–515 (2010). 0908.2817. https://doi.org/10.1038/nature08714

T. Temim, P. Slane, D. Castro, P.P. Plucinsky, J. Gelfand, J.R. Dickel, High-energy emission from the composite supernova remnant MSH 15-56. Astrophys. J. **768**, 61 (2013). 1303.2425. https://doi.org/10.1088/0004-637X/768/1/61

J.R. Thorstensen, F. RA van den Bergh S, The expansion center and dynamical age of the galactic supernova remnant Cassiopeia A. Astron. J. **122**, 297–307 (2001). astro-ph/0104188. https://doi.org/10.1086/321138

H. Uchida, K. Koyama, H. Yamaguchi, M. Sawada, T. Ohnishi, T.G. Tsuru, T. Tanaka, S. Yoshiike, Y. Fukui, Recombining plasma and hard X-ray filament in the mixed-morphology supernova remnant W 44. Publ. Astron. Soc. Jpn. **64**, 141 (2012). 1208.0113. https://doi.org/10.1093/pasj/64.6.141

M. Ugliano, H.T. Janka, A. Marek, A. Arcones, Progenitor-explosion connection and remnant birth masses for neutrino-driven supernovae of iron-core progenitors. Astrophys. J. **757**, 69 (2012). 1205.3657. https://doi.org/10.1088/0004-637X/757/1/69

V.P. Utrobin, A. Wongwathanarat, H.T. Janka, E. Müller, Light-curve analysis of ordinary type IIP supernovae based on neutrino-driven explosion simulations in three dimensions. Astrophys. J. **846**, 37 (2017). 1704.03800. https://doi.org/10.3847/1538-4357/aa8594

J. Vink, Supernova remnants: the X-ray perspective. Astron. Astrophys. Rev. **20**, 49 (2012). 1112.0576. https://doi.org/10.1007/s00159-011-0049-1

C.Y. Wang, R.A. Chevalier, Instabilities and clumping in type IA supernova remnants. Astrophys. J. **549**, 1119–1134 (2001). astro-ph/0005105. https://doi.org/10.1086/319439

L. Wang, J.C. Wheeler, Spectropolarimetry of supernovae. Annu. Rev. Astron. Astrophys. **46**, 433–474 (2008). 0811.1054. https://doi.org/10.1146/annurev.astro.46.060407.145139

M. Wardle, F. Yusef-Zadeh, Supernova remnant OH masers: signposts of cosmic collision. Science **296**, 2350–2354 (2002). https://doi.org/10.1126/science.1068168

J.L. West, S. Safi-Harb, T. Jaffe, R. Kothes, T.L. Landecker, T. Foster, The connection between supernova remnants and the Galactic magnetic field: a global radio study of the axisymmetric sample. Astron. Astrophys. **587**, A148 (2016). 1510.08536. https://doi.org/10.1051/0004-6361/201527001

J.L. West, S. Safi-Harb, G. Ferrand, The connection between supernova remnants and the Galactic magnetic field: an analysis of quasi-parallel and quasi-perpendicular cosmic-ray acceleration for the axisymmetric sample. Astron. Astrophys. **597**, A121 (2017). 1609.02887. https://doi.org/10.1051/0004-6361/201628079

R.M. Williams, Y.H. Chu, J.R. Dickel, R. Petre, R.C. Smith, M. Tavarez, Supernova remnants in the Magellanic Clouds, III: an X-ray atlas of Large Magellanic Cloud supernova remnants. Astrophys. J. Suppl. Ser. **123**, 467–485 (1999). astro-ph/9905073. https://doi.org/10.1086/313246

R. Willingale, B. JAM van der Heyden KJ, J.S. Kaastra, J. Vink, X-ray spectral imaging and Doppler mapping of Cassiopeia A. Astron. Astrophys. **381**, 1039–1048 (2002). astro-ph/0107270. https://doi.org/10.1051/0004-6361:20011614

P.F. Winkler, R. Petre, Direct measurement of neutron star recoil in the oxygen-rich supernova remnant Puppis A. Astrophys. J. **670**, 635–642 (2007). astro-ph/0608205. https://doi.org/10.1086/522101

P.F. Winkler, B.J. Williams, S.P. Reynolds, R. Petre, K.S. Long, S. Katsuda, U. Hwang, A high-resolution X-ray and optical study of SN 1006: asymmetric expansion and small-scale structure in a type Ia supernova remnant. Astrophys. J. **781**, 65 (2014). 1312.3367. https://doi.org/10.1088/0004-637X/781/2/65

A. Wongwathanarat, H.T. Janka, E. Müller, Three-dimensional neutrino-driven supernovae: neutron star kicks, spins, and asymmetric ejection of nucleosynthesis products. Astron. Astrophys. **552**, A126 (2013a). 1210.8148. https://doi.org/10.1051/0004-6361/201220636

A. Wongwathanarat, H.T. Janka, E. Müller, Three-dimensional neutrino-driven supernovae: neutron star kicks, spins, and asymmetric ejection of nucleosynthesis products. Astron. Astrophys. **552**, A126 (2013b). 1210.8148. https://doi.org/10.1051/0004-6361/201220636

A. Wongwathanarat, H.T. Janka, E. Müller, E. Pllumbi, S. Wanajo, Production and distribution of ^{44}Ti and ^{56}Ni in a three-dimensional supernova model resembling Cassiopeia A. Astrophys. J. **842**, 13 (2017). 1610.05643. https://doi.org/10.3847/1538-4357/aa72de

S.E. Woosley, A. Heger, T.A. Weaver, The evolution and explosion of massive stars. Rev. Mod. Phys. **74**, 1015–1071 (2002). https://doi.org/10.1103/RevModPhys.74.1015

H. Yamaguchi, M. Ozawa, K. Koyama, K. Masai, J.S. Hiraga, M. Ozaki, D. Yonetoku, Discovery of strong radiative recombination continua from the supernova remnant IC 443 with Suzaku. Astrophys. J. Lett. **705**, L6–L9 (2009). 0909.3848. https://doi.org/10.1088/0004-637X/705/1/L6

H. Yamaguchi, C. Badenes, R. Petre, T. Nakano, D. Castro, T. Enoto, J.S. Hiraga, J.P. Hughes, Y. Maeda, M. Nobukawa, S. Safi-Harb, P.O. Slane, R.K. Smith, H. Uchida, Discriminating the progenitor type of supernova remnants with iron K-shell emission. Astrophys. J. Lett. **785**, L27 (2014). 1403.5154. https://doi.org/10.1088/2041-8205/785/2/L27

P.A. Young, C.L. Fryer, A. Hungerford, D. Arnett, G. Rockefeller, F.X. Timmes, B. Voit, C. Meakin, K.A. Eriksen, Constraints on the progenitor of Cassiopeia A. Astrophys. J. **640**, 891–900 (2006). astro-ph/0511806. https://doi.org/10.1086/500108

Space Sci Rev (2018) 214:28
https://doi.org/10.1007/s11214-017-0453-6

Shock Waves in Supernova Ejecta

J.C. Raymond[1]

Received: 8 June 2017 / Accepted: 22 November 2017 / Published online: 9 January 2018

Abstract Astrophysical shock waves are a major mechanism for dissipating energy, and by heating and ionizing the gas they produce emission spectra that provide valuable diagnostics for the shock parameters, for the physics of collisionless shocks, and for the composition of the shocked material. Shocks in SN ejecta in which H and He have been burned to heavier elements behave differently than shocks in ordinary astrophysical gas because of their very large radiative cooling rates. In particular, extreme departures from thermal equilibrium among ions and electrons and from ionization equilibrium may arise. This paper discusses the consequences of the enhanced metal abundances for the structure and emission spectra of those shocks.

Keywords Supernova · Shock waves · Supernova remnants

1 Introduction

Most of the shock waves familiar to astrophysicists sweep up material of normal astrophysical abundances, which is to say mostly H and He with 0.1% by number of heavier elements. The material in supernova ejecta can be much different, however. CNO processing can eliminate the hydrogen, but if most of the gas is helium the basic shock physics will not be greatly affected. On the other hand, if both H and He are burned to heavier elements, the behavior and appearance of the shock will be drastically altered. This paper reviews shock waves in highly enriched material in supernova remnants (SNRs) and the nebular phase of some supernovae. It does not deal with the shocks in the interior of the exploding supernova where nuclear processing occurs.

Supernovae
Edited by Andrei Bykov, Roger Chevalier, John Raymond, Friedrich-Karl Thielemann, Maurizio Falanga and Rudolf von Steiger

✉ J.C. Raymond
jraymond@cfa.harvard.edu

1 Center for Astrophysics, 60 Garden St., Cambridge, MA 02138, USA

The first difference between normal shocks and ejecta shocks is the larger mean particle mass, μ. For fully ionized gas of astrophysical abundances, $\mu = (n_H + 4n_{\rm He})/(2n_{\rm H} + 3n_{\rm He})$ times the proton mass, where the denominator is the number of electrons plus the number of nuclei, or 0.6. For a fully ionized oxygen plasma, however, μ is $16/9 = 1.78$, and for fully ionized iron it is $56/27 = 2.07$. The difference becomes much more severe if the gas is partially ionized, for instance at a temperature of 10^5 K. In that case there are far fewer electrons per atom, perhaps 2 for the oxygen plasma and 4 for Fe, so that μ is 5.3 or 11.2, respectively.

The value of μ enters the shock jump conditions directly. For a strong shock,

$$T = \frac{3}{16}\frac{\mu m_p}{k} V_s^2 \tag{1}$$

for a ratio of specific heats $\gamma = 5/3$, where k is the Boltzmann constant, or

$$T = 1.36 \times 10^5 \mu_{0.6} V_{100}^2 \tag{2}$$

where V_{100} is the shock speed in units of 100 km/s. Thus for normal abundances, a 100 km/s shock heats the gas to 1.36×10^5 K. For fully ionized oxygen it would be 3.8×10^5 K, but if the oxygen is only doubly ionized it would be 1.1×10^6 K. A 100 km/s shock in pure iron would produce $T = 4.5 \times 10^5$ K if the plasma is fully ionized, or 2.4×10^6 K if it is 4 times ionized. These estimates assume equal electron and ion temperatures, but as discussed below that is unlikely to be the case. Moreover, the ionization state will change rapidly once the gas is heated, and the immediate post-shock temperature will have little relation to the ionization state seen by an observer.

The second difference between shocks in gas of normal abundances and those in ejecta is the far larger cooling rate in the latter. At temperatures above about 30,000 K, hydrogen is fully ionized, and it emits radiation mainly by the inefficient processes of Bremsstrahlung and radiative recombination. The heavy elements that retain bound electrons, on the other hand, produce strong emission lines, and the radiative cooling rate in the shocked ejecta is 1 to 3 orders of magnitude larger than in normal abundance gas at the same temperature.

Figure 1 shows the cooling rate coefficients Λ as functions of electron temperature for normal astrophysical abundances and for pure oxygen, silicon and iron. The power emitted is

$$P = \Lambda(T_e) n_i n_e \ \mathrm{erg}/\left(\mathrm{cm}^3\ \mathrm{s}\right) \tag{3}$$

computed with CHIANTI version 8 (Del Zanna et al. 2015). n_i is the density of H nuclei for the normal abundance case, and the density of O, Si or Fe nuclei in the others. Note, however, that these curves assume that the plasma is in ionization equilibrium, and as discussed below this is often a poor approximation in shocked ejecta gas.

The character of the emission is apparent from Fig. 1. At the highest temperatures, all the ions are stripped bare, and the emission is dominated by the bremsstrahlung continuum, which is approximately proportional to Z^2 and increases as $T^{1/2}$. Thus Λ is roughly proportional to Z^2. At lower temperatures where emission lines dominate, the ratio of Λ to the solar abundance cooling rate is closer to ratio of the abundance to the solar value, or around 2000 where oxygen dominates at Log $T \sim 5.2$ to5.5, even at normal astrophysical abundances. The line-to-continuum ratio, or equivalent width (EW) is the basis for determining elemental abundances from X-ray spectra. When the continuum is dominated by bremsstrahlung, it is proportional to the abundance divided by Z^2.

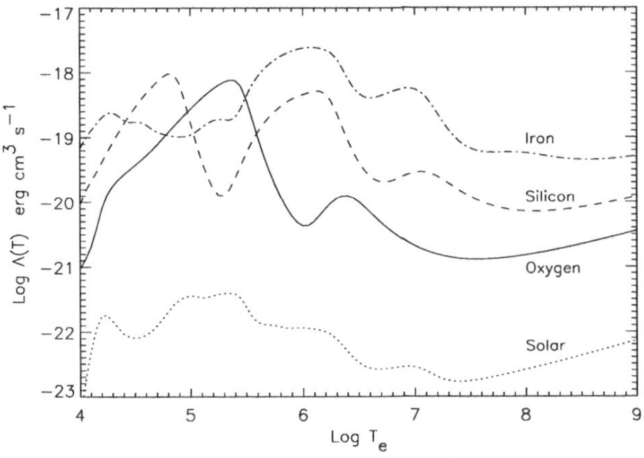

Fig. 1 Cooling rate coefficients for solar abundances (Grevesse et al. 2007) and for pure O, Si and Fe as functions of electron temperature. These rates assume collisional ionization equilibrium, which is generally a poor assumption for shocks, but they give a general idea of the relative magnitudes and rough temperature dependence. The smoothly rising rates at the highest temperatures are dominated by bremsstrahlung emission with a contribution from radiative recombination, while the cooling at lower temperatures is dominated by line emission. The bumps and dips in the curves correspond to the shell structure of the dominant ions. The drop in $\Lambda_{\rm O}$ between 2×10^5 K and 10^6 K occurs because oxygen is ionized to the stable He-like and H-like states that have no electrons that are easy to excite. The dip in $\Lambda_{\rm Si}$ at about 1.5×10^5 K is where Si is Ne-like, and the drop above 2×10^6 K is where it becomes He-like. The $\Lambda_{\rm Fe}$ curve dips where Fe is Ne-like at 3×10^6 K and above 1.5×10^7 K where it becomes He-like. These cooling rates were computed with CHIANTI (Del Zanna et al. 2015)

The consequence of the large cooling rate is a very short cooling time scale. Under normal conditions, the radiative cooling time is generally comparable to or longer than the Coulomb collisional time scales for energy transfer between electrons and ions and among different ion species, but in ejecta shocks that is not true (Itoh 1981a). Radiative cooling time scales are typically longer than ionization and recombination times at temperatures above 10^6 K in normal abundance plasma, while they are much shorter in the ejecta, meaning that time-dependent ionization (non-equilibrium ionization or NEI) is much more important. While the rapid cooling might be expected to make radiative cooling dominate over energy transfer processes such as thermal conduction, the rapid cooling makes the cooling length very short, and the strong temperature gradient makes the divergence of thermal conduction quite important in the absence of a magnetic field (Borkowski and Shull 1990).

Another implication of the rapid cooling is a change in the character of fluid instabilities. The cooling zone behind a shock is subject to thermal instability over a wide range of shock speeds in normal gas (Chevalier and Imamura 1982; Innes et al. 1992; Gaetz et al. 1988) This instability leads to violent collapse, secondary shocks and large swings in the shock speed itself, though magnetic fields can cushion the effect (Innes 1992). A higher cooling rate will enhance this instability. Another instability, called the thin shell instability (Vishniac 1983), occurs when a thin, dense shell created by a radiatively cooling shock is driven by hot, high pressure gas behind it. Ripples in the shell grow because the ram pressure of gas entering the shell lies along a different direction than the pressure force from the inside. A high cooling rate means that a shell prone to the Vishniac instability is established very quickly.

This paper discusses the above points as they relate to observations of the reverse shocks in SNRs and supernovae, paying particular attention to their implications for the accuracy with which one can infer elemental abundances and other physical parameters from spectra. Sections 2 and 3 distinguish between collisional vs. collisionless shocks and radiative vs. non-radiative shocks. Sections 4 and 5 discuss electron-ion and ion-ion temperature equilibration. Sections 6, 7 and 8 discuss the roles of the photoionization precursor, fluid instabilities and cosmic ray acceleration in the reverse shock, and Sect. 9 covers the reliability of shock parameters inferred from spectra and models. Those sections address shocks in supernova remnants, while shocks in the ejecta of supernova themselves (taken here to be less than about 100 years old) are discussed in Sect. 10. Section 11 is a summary discussion of the reliability of abundances and other parameters inferred from spectra of shocks in highly enriched ejecta. A brief Appendix lists O-rich SNRs and some recent observations. The important question of dust destruction by sputtering and grain-grain collisions is covered in separate papers in this volume (Micelotta et al. 2017; Sarangi et al. 2017).

2 Collisional and Collisionless Shocks

A shock wave converts a supersonic flow to a subsonic one both by decreasing the flow speed in the rest frame of the shock and by increasing the sound speed by increasing the temperature. The decrease in flow speed must be accompanied by compression in order to conserve mass flux. The increase in sound speed results from the conversion of kinetic energy of the bulk flow into thermal energy. The transition from pre-shock to post-shock conditions is sharp, because the supersonic speed does not allow information about the approaching shock to be carried upstream by sound waves.

In a collisional shock, such as a weak shock in air, the transition is mediated by collisions between particles, so the thickness is approximately a collisional mean free path. Collisions lead naturally to a Maxwellian distribution in order to maximize entropy, and for the same reason they lead to thermal equilibrium among different particle species; all the species reach the same temperature.

In a plasma, the transition can occur on a scale much shorter than the collision scale. One natural scale is the proton Larmor radius, R_L. For typical ISM conditions, the collision scale is on the order of 10^{15} cm, while R_L is on the order of 10^8 cm. Another scale is the proton skin depth, given by the speed of light divided by the proton plasma frequency. It characterizes electrostatic waves that interact with the protons, and $L_s = c/\omega_p = 1.4 \times 10^8/n^{1/2}$ cm, which is comparable to the Larmor radius for typical conditions in the interstellar medium (ISM).

In collisionless shocks, a particle interacts not with another particle, but with the collective effects of a large group of particles. There is no obvious reason to expect Maxwellian distributions or equal temperatures of different species, and in fact particle acceleration and differences among proton, electron and heavier ion temperatures are often observed in shocks in the solar wind and in the ISM. A large range of complex plasma processes comes into play. Some protons are reflected from the shock, and the resulting two-stream instabilities generate plasma waves. The more energetic particles can scatter between the shock and magnetic turbulence in a shock precursor, gaining energy each time and reaching relativistic speeds. Magnetic fields can be amplified beyond the values given by compression of the upstream field.

Shocks in plasmas are divided into quasi-perpendicular and quasi-parallel based on the angle between the shock normal and the local magnetic field. The former tend to be more laminar, while the latter are more turbulent. The quasi-parallel shocks are generally thought

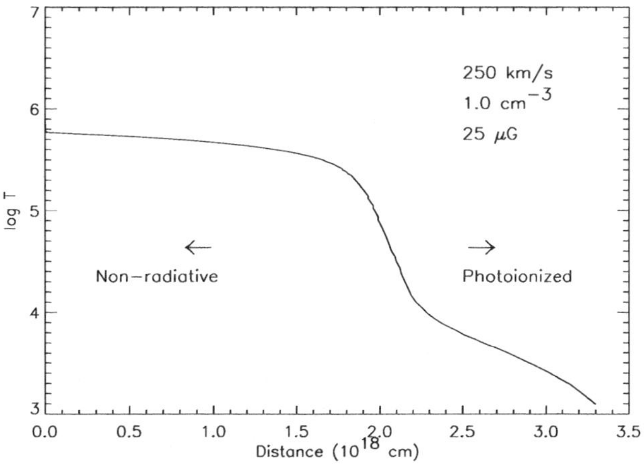

Fig. 2 Temperature structure of the post-shock flow in a radiative shock. This steady flow model from an updated version of the Raymond (1979) model does not include thermal conduction or the thermal instability that affects shocks above about 150 km/s

to be able to accelerate more particles, while the quasi-perpendicular ones accelerate particles more quickly, though turbulence, magnetic field irregularities and ripples in the shock can blur the distinction (Giacalone 2013).

Observations of the consequences of collisionless shock physics in ejecta shocks are sparse. In subsequent sections we describe what is known about shocks in normal abundance plasmas from observation and simulation, then what is known about the shocks in ejecta.

3 Radiative and Non-Radiative Shocks

It is convenient to divide shocks into radiative and non-radiative classes based on the global situation rather than the physics in the shock itself. A non-radiative shock is one in which the radiative cooling time of the shocked gas is long compared to other dynamical time scales. For instance, the blast wave in a young SNR produces temperatures of order 10^8 K, and the radiative cooling time is much longer than the age of the SNR. Therefore, radiative cooling does not have time to convert much of the post-shock thermal energy into radiation or to affect the dynamics of the flow. The X-ray emission in SNRs is mostly produced by non-radiative shocks. The gas that has gone through a radiative shock, on the other hand, has time to cool to something like the pre-shock temperature.

The radiative cooling time for normal abundance gas (at constant pressure) is given by

$$t_{cool} = \frac{5}{2}(n_e + n_i)kT/(n_e n_i \Lambda) \sim 5\, kT/(n_e \Lambda), \tag{4}$$

where $n_e \simeq n_i$ in a plasma dominated by hydrogen. Over much of the interesting temperature range above $T \sim 3 \times 10^5$ K, Λ decreases with increasing T, so that t_{cool} increases rapidly with V_S. As a result, there is a strong tendency for SNR shocks in the ISM faster than about 300 km s^{-1} to be non-radiative and slower shocks to be radiative. Because of the fact that by definition radiative shocks convert much of their thermal energy to photons, radiative shocks tend to be much brighter and easier to observe in the optical, infrared (IR) and ultraviolet (UV).

From the point of view of plasma kinetics, emission that is produced in the neighborhood of a non-radiative collisionless shock will reflect such things as departures from Maxwellian distributions or temperature differences among particle species. Radiative shocks, on the other hand, produce radiation by collisions at somewhat comparable rates to the Coulomb collisions that bring the gas toward thermal equilibrium. Thus most of the signatures of plasma processes in the shock will have been erased by the time the gas produces the bright emission that dominates the spectrum.

Figure 2 illustrates the structure of a radiative shock in the normal ISM, in this case a 250 km/s shock with a pre-shock density of 1 cm^{-3}. If the shock has progressed a short distance into the medium so that a post-shock layer thinner than about 10^{18} cm has accumulated, it is effectively non-radiative in that cooling has not greatly changed the temperature or dynamics of the flow. Beyond that, cooling abruptly drops the temperature to about 10^4 K. Still farther behind the shock, photoionization maintains an H II region-like zone, often supported by the pressure of the compressed magnetic field, where the temperature gradually declines as the ionizing flux is absorbed away. Note that this is an idealized steady-flow model. Shocks in this velocity range are subject to drastic thermal instability after they become radiative (e.g. Innes 1992), as discussed below.

Ejecta shocks can be divided in the same way into radiative and non-radiative classes, but of course t_{cool} is orders of magnitude shorter for a given density and temperature. This is offset to a limited extent by the higher temperature produced by a given shock speed. In general the shock in the ejecta is the reverse shock in an SNR or type IIn SN, produced when the unshocked ejecta encounters the high pressure shell formed by the blast wave. To the extent that the pressure in that region is uniform, the ram pressures ρV_S^2 of different parts of the reverse shock are the same. The ejecta in Cas A and other O-rich SNRs consist of dense clumps embedded in a lower density gas. The reverse shock in the low density gas is non-radiative, and it produces the strong X-ray emission. The slower reverse shocks in the small, dense clumps are radiative, and they as seen as the optical knots. The optical knots make up a small fraction of the volume and mass. In type I SNRs like Tycho and SN1006, the ejecta are more uniform, and the ejecta shocks are entirely non-radiative, though diffuse clumps are seen, probably as a result of the Rayleigh-Taylor instability of the contact discontinuity between the shocked ejecta and shocked ISM (e.g., Blondin and Ellison 2001).

Because of the enormous cooling rate, even radiative shocks in the ejecta can maintain signatures of the plasma processes in the shock. Most importantly, the cooling time can be shorter than the Coulomb equilibration time even at modest temperatures, so electron and ion temperatures can be drastically different.

4 Electron-ion Thermal Equilibration

As mentioned above, there is no a priori reason to expect thermal equilibrium among different particle species behind a collisionless shock, and there is strong observational evidence for large differences between electron and ion temperatures in shock waves in the solar wind and in fast shocks in the ISM. In principle, one might expect that since the protons carry 1840 times as much kinetic energy as the electrons when they enter the shock, they would be up to 1840 times hotter when the kinetic energy is transformed into thermal energy. Itoh (1978) computed models of SNR blastwaves in which the electrons and ions were treated as separate fluids coupled by Coulomb collisions. In those models the electrons were cool at the shock, and the Coulomb collisions heated them fairly rapidly to $kT = 1$–2 keV in young SNRs, at shock speeds that gave ion temperatures 10 times higher or more.

Several theories have been advanced that provide varying degrees of collisionless electron heating at the shock. Cargill and Papadopoulos (1988) proposed a sequence of instabilities driven by ions reflected from a fast shock (Mach number $M > 20$). The Buneman instability heats the electrons quickly, but it saturates at a fairly low electron temperature. It can, however, permit the ion acoustic instability to heat the electrons to perhaps 30% of the ion temperature. Bykov and Uvarov (1999) examined electron heating by diffusion in the turbulent magnetic precursor of a quasi-parallel shock predicted by a hybrid code. Shimada and Hoshino (2000) performed particle-in-cell (PIC) simulations and found that electrostatic waves driven by streaming instabilities could heat the electrons to about 20% of the ion temperature. Ghavamian et al. (2007a) and Rakowski et al. (2008) showed that streaming cosmic rays in the shock precursor can drive lower hybrid waves that can effectively couple electrons to ions. Their model predicts $T_e/T_i \propto 1/V_S^2$ for shocks faster than about 300 km/s, though Laming et al. (2014) find that there can be additional electron heating in very fast shocks.

Other theories rely on the electric field on the shock. In general, the electrons must interact with electromagnetic fluctuations or turbulence of some sort in order to be confined in the region where the electric field exists if they to be significantly heated (Hull et al. 2000; Ohira and Takahara 2007; Mozer and Sundkvist 2013; Schwartz 2014). The electron heating is closely tied to the injection of electrons into the non-thermal acceleration mechanism. Numerical simulations by Riquelme and Spitkovsky (2011) show an electron distribution with a Maxwellian core and a power law tail. In their simulations there is a high degree of electron-ion equilibration close to the shock for some shock parameters.

Detailed reviews of electron-ion thermal equilibration are given by Ghavamian et al. (2013) for shocks in normal astrophysical gas and Bykov et al. (2008) for galaxy clusters. A thermodynamic approach to thermal equilibration in shocks is given by Vink et al. (2015). Here we summarize the major points and consider the case of shocks in SN ejecta.

4.1 Non-Radiative Shocks

Electron-ion temperature ratios can be measured in several ways. Instruments on spacecraft in the solar wind measure electron and ion temperatures directly from the velocity distribution functions. Bow shocks form around the magnetospheres of the Earth and other planets, and both coronal mass ejections (CMEs) and interactions between solar wind flows of different speeds drive shock waves in the solar wind. At Earth, the solar wind temperature is on the order of 10^5 K, and the shocks range from very weak to those of Mach numbers (shock speed divided by sound speed) of order 10. Schwartz et al. (1988) measured the heating of electrons and ions that passed through these shock waves and found on average $T_e/T_p \sim 0.2$ with considerable scatter. The scatter could be due to differences in the shock normal angle, since quasi-parallel and quasi-perpendicular shocks are expected to behave very differently.

T_e/T_p has also been determined by remote sensing observations of a few CME-driven shocks in the solar corona. Bemporad et al. (2014) combined UV spectra and White Light coronagraph images and found a range of about 0.16 to 0.8 for shocks with Mach numbers below or about 2. On the other hand, Ma et al. (2011) determined the electron heating in another low Mach number CME-driven shock in the corona based on narrow band EUV images from the AIA instrument aboard SDO. They found an electron temperature consistent with the shock speed and full electron-ion temperature equilibration. In this case the electron heating is determined from the evolution of the ionization state in the downstream gas.

Another way to measure electron and ion temperatures uses the Hα profiles of collisionless shocks in partly neutral gas. The profiles consist of a narrow component produced by atoms that pass through the shock unaffected by the fields and turbulence in the narrow shock layer, and a broad component produced by atoms that experience charge transfer collisions with post-shock protons (Chevalier and Raymond 1978). The broad component gives the proton temperature fairly directly, since its width is given by the proton temperature convolved with the charge transfer cross section times the speed. The intensity ratio of broad and narrow components is sensitive to the electron temperature because it depends on the ratio of ionization rate (by electrons) to charge transfer rate. The velocity width of the narrow component is generally higher than would be consistent with the ambient ISM temperature for neutral gas, indicating the presence of a shock precursor (Hester et al. 1994; Smith et al. 1994; Sollerman et al. 2003; Ohira 2014; Lee et al. 2007, 2010).

An advantage of this method is that it applies to the region close to the shock where H has not yet been ionized, and that means that there has not been time for Coulomb collisions to affect T_e/T_i. In terms of column density, the Hα emission from the Balmer line shocks comes from a region within about 10^{15} cm^{-2} from the shock. The possible contribution of the precursor to the narrow line intensity is the main limitation to the accuracy of the T_e determination, but radiative transfer of the Lyβ photons (Chevalier et al. 1980) also enters. This diagnostic has been applied to supernova remnant shocks between about 350 and 2500 km/s, and T_e/T_p is near one in the slower shocks and less than 0.1 in the fast ones (e.g. Ghavamian et al. 2001; Medina et al. 2014; Ghavamian et al. 2013).

A third means of measuring electron heating is by measuring the shape of the X-ray continuum, since bremsstrahlung produces an $e^{-h\nu/kT_e}$ spectral shape. The proton temperature can be obtained from the Hα profile method described above or from the shock speed obtained from proper motions. Again, shocks near 300 km/s show $T_e/T_p \sim 1$ (Salvesen et al. 2009), while shocks above 1000 km/s show $T_e/T_p \sim 0.1$ or less (Rakowski et al. 2003; Long et al. 2003; Kosenko et al. 2008; Helder et al. 2011; Hovey et al. 2015). Hughes et al. (2000) estimated a shock speed of 6000 km/s from the expansion proper motion of 1E102.2-7219 in the SMC, and that resulted in $T_e/T_p < 0.04$, but a recent measurement of the expansion rate from Chandra images indicates a slower shock and an electron-ion temperature ratio several times higher (Plucinsky and Gaetz, private communication). Thus these results agree with those from Balmer line profiles, and they can be applied to shocks in fully ionized plasma as well as partly neutral gas. A limitation of this method is the finite resolution of the X-ray telescope, generally implying a region at least 10 to 100 times thicker than the region where Hα emission is produced. Therefore, it is subject to the effects of Coulomb collisions.

Figure 3 shows some results from the above methods for shocks in the solar wind and the normal ISM (Ghavamian et al. 2013). The X-axes in the three panels show how T_e/T_i scales with shock speed and Mach number (V_S divided by characteristic signal speed) for the Alfven speed or fast mode speed. All three of the scalings show some disconnect between the solar wind shocks and the generally faster ISM shocks, but the general trend for fast shocks to be less equilibrated than slow ones is seen in both sets of data. A few points show T_e/T_i ratios above 1, which is not generally predicted by shock models, but those points may result from non-local effects such as thermal conduction or non-Maxwellian electron velocity distributions which occur in the solar wind.

Yet another way to infer T_e/T_p is presented by Laming et al. (1996), who used the UV spectrum of SN1006 obtained by the Hopkins Ultraviolet Telescope to measure the intensities of lines of He II, C IV, N V and O VI. Because of its high excitation potential, the He II λ1640 Å line is dominated by electron collisions, while the other lines are excited mainly by collisions with protons and α particles. In agreement with the results of other studies of the northwestern (NW) region of SN1006 based on Hα profiles and X-ray spectra, Laming et al. (1996) found $T_e/T_p \leq 0.2$.

It is more difficult to measure T_e/T_i in shocks in SN ejecta. The electron temperature can be measured from the X-ray continuum shape, but it is difficult to determine the ion

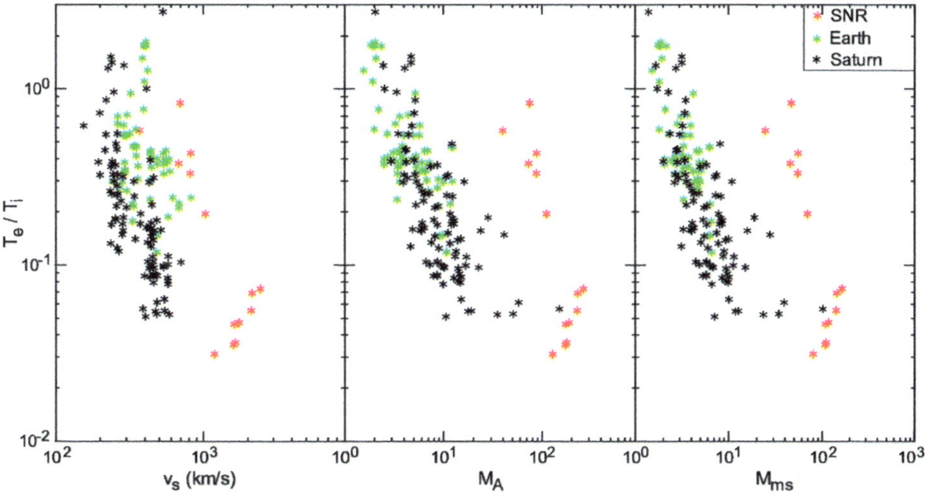

Fig. 3 Comparison of electron-to-ion temperature ratios for solar wind shocks (*black* and *green*) and supernova remnant (*red*) shocks. The X-axes show how the values depend on shock speed, Alfven Mach number and fast mode (magnetosonic) Mach number. From Ghavamian et al. (2013)

temperature. Measurements of the O VII emission line profile in an ejecta knot in SN1006 give a value of the kinetic temperature T_O of 275 keV, compared with an electron temperature of 1.35 keV (Vink et al. 2003; Broersen and Vink 2014), so that T_e/T_O is about 0.005 in this 3000 km/s shock. Another method was introduced by Yamaguchi et al. (2014), who measured the Fe Kα and Kβ centroids and the intensity ratio of the two lines. These depend on the ionization state of the Fe because of different energies and fluorescence yields of ions less ionized than He-like, as shown in Fig. 4. This allowed Yamaguchi et al. (2014) to determine the electron temperature in the reverse shock in Tycho's SNR. They find that T_e/T_{Fe} is about 0.01 in the $\sim$4000 km s^{-1} reverse shock. Kosenko et al. (2008) observed the type Ia SNR 0509-76.5 in the LMC with XMM-Newton. They used a global model for the continuum and Fe emission and found T_e/T_i in the range 0.003 to 0.03, but this includes both the blast wave and the reverse shock.

A very direct upper limit to the electron heating in a reverse shock of SN1006 is given by Hamilton et al. (2007). They measured the absorption in the Si II λ1260 line from both the shocked and unshocked ejecta of SN1006 against a background star. A fit to the line profile yields the pre- and post-shock speeds and the post-shock temperature. The 2700 km/s shock speed derived from the difference between pre-and post-shock speeds and the post-shock temperature are consistent with the shock jump conditions with no electron heating, and Hamilton et al. (2007) find a 3σ upper limit on the fraction of energy that goes into electrons of 0.26.

Figure 5 shows models in which plasma turbulence heats the electrons to 1% of the ion temperature, and then Coulomb collisions gradually bring the electron and ion temperatures together. It is sensible that only Coulumb collisions should act in the region away from the shock transition because the turbulence associated with the shock should die out rapidly, but this assumption has not been tested. The shock speeds were chosen to give similar T_e and T_i downstream, and the pre-shock ion densities are the same. An unsubtle difference between the normal abundance shock on the left and the pure oxygen shock on the right is the order of magnitude difference in the scales of the X-axes. This is mostly due to the Z^2 factor in

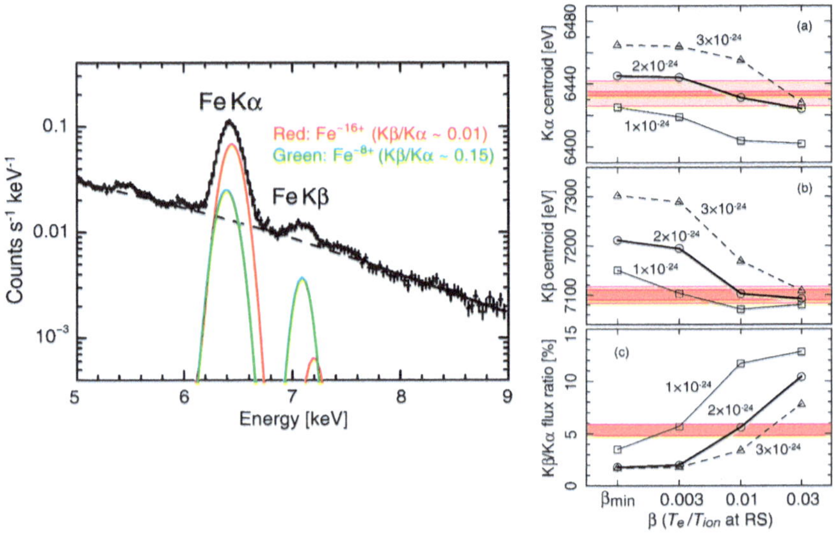

Fig. 4 Dependence of the centroids and the intensity ratio of the Fe Kα and Kβ lines on the ionization state of Fe in the reverse shock of Tycho's SNR (Yamaguchi et al. 2014). The *lines in the right hand plots* correspond to ambient densities of 1, 2 and 3 × 10^{-24} g/cm^3

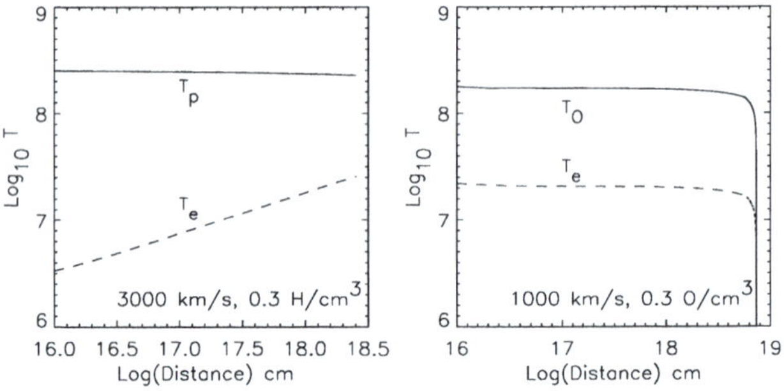

Fig. 5 Non-radiative shocks in normal abundance plasma (*left*) and pure oxygen (*right*). Note the different scales on the X-axes. The models assume that $T_e = 0.01 T_i$ at the shock and that Coulomb collisions subsequently bring the plasma toward equilibrium

the Coulomb collision rate. A more subtle difference is the drop in T_O shortly behind the pure oxygen shock. It occurs as the O ions share their energy with 8 electrons per ion and as radiative cooling by the L-shell oxygen ions O II–VI converts some of the thermal energy into photons. In the constant T_e region in the ejecta shock model, radiative cooling of the electrons nearly balances the transfer of heat from ions to electrons by Coulomb collisions.

Itoh (1977) constructed hydrodynamic models of the non-radiative shocks in Cas A assuming either $T_e = T_i$ or no electron-ion equilibration other than by Coulomb collisions. The latter gave good agreement with the X-ray spectra available at the time, while the complete equilibration model predicted far too much emission at high energies. It is now known that a substantial fraction of the higher energy X-ray spectrum arises from synchrotron emission from non-thermal electrons (Helder and Vink 2008), so the constraints on T_e are even more severe.

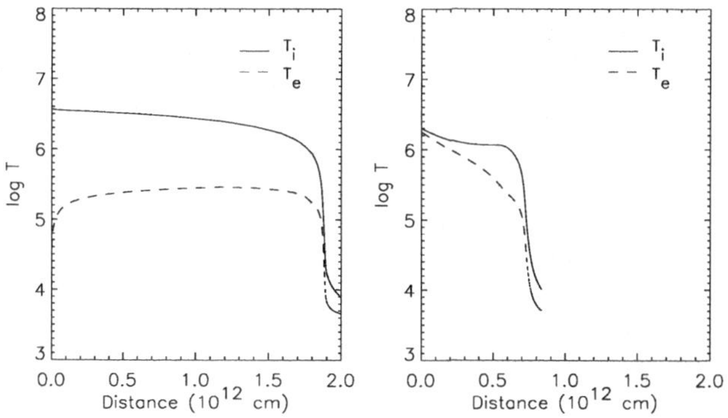

Fig. 6 Radiative 100 km/s shocks in oxygen at a density of 100 cm^{-3}. The *left panel* shows a case with $T_e = 0.01 T_O$ at the shock, and *the right* shows $T_e = 0.90 T_O$. In the *left panel*, the initial low T_e gives very rapid Coulomb heating of the electrons until they reach about 2×10^5 K. There is a long plateau in which heating by Coulomb collisions nearly balances the radiative losses, but it ends when the oxygen ions run out of energy, and the temperature collapses. In the *right panel*, the electron and ion temperatures start out nearly equal, but the rapid radiative cooling causes T_e to drop to a fraction of T_O. The initial drop in T_O results from the increase in the number of electrons sharing the energy, and the subsequent plateau in T_O is analogous to that in the *left panel*. The cooling length is shorter because of faster cooling at the higher electron temperatures. Again, it collapses, and both temperatures drop rapidly

4.2 Radiative Shocks

Itoh (1981a,b) performed the first calculations of a radiative shock in pure oxygen and compared them with observations of Cas A. He found that whether or not the electron and ion temperatures were equilibrated at the shock, the intense cooling of the electrons caused T_e to be well below T_O through most of the flow. Figure 6 shows the electron and ion temperatures for 100 km/s shocks in pure oxygen. The temperature structures and even the cooling lengths depend upon the degree of electron-ion thermal equilibration at the shock front, but in either case the electrons are much cooler than the ions in most of the flow. Itoh (1986) found some difficulties in matching the spectra of observed O-rich SNRs. In particular, the flow had to be artificially truncated in order to avoid overpredicting the permitted O I recombination lines at λλ7774, 8446 Å, and the models had difficulty predicting strong enough emission in the [O I] λλ6303, 6363 Å forbidden lines. This led Itoh (1988) to suggest that simple planar models were not adequate, and that unsteady flows or turbulent stripping in the shearing flows at the edge of the knots were important.

Laboratory experiments and their supporting numerical simulations provide confirmation of the large differences between ion and electron temperatures when shocks travel through material with large cooling rates. Measurements of laser-produced shocks in xenon by Grun et al. (1991) were successfully modeled by Laming and Grun (2002) with simulations in which only Coulomb collisions heat the electrons and T_e/T_{Xe} ranges from about 0.01 to 0.1 in the post-shock flow. Suzuki-Vidal et al. (2015) studied bow shocks driven by electrical current pulses in aluminum foils, and their models showed T_e/T_{Al} around 0.1 in the post-shock flow. Both of these experiments focused on the fluid instabilities that result from the intense cooling, and those instabilities are discussed below.

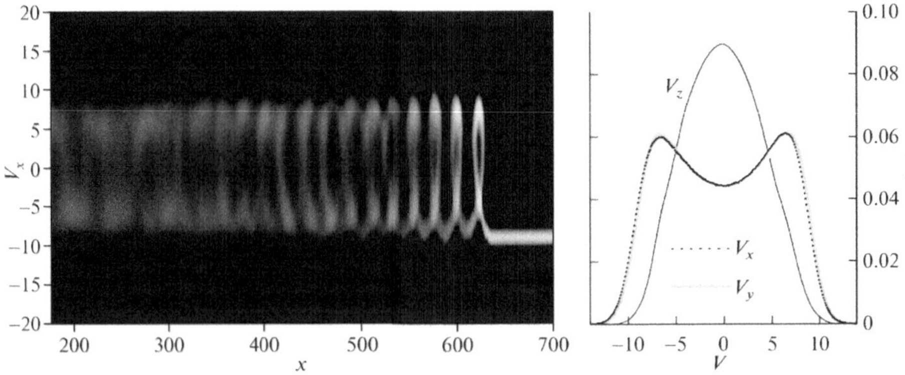

Fig. 7 Phase space and integrated velocity distributions for Fe II impurity ions in Si III background gas from a hybrid simulation by Kropotina et al. (2015). In the *left panel*, x is in units of the Si III inertial length. The V normalization is the inertial length times the cyclotron frequency, i.e., the Alfvén speed. This figure shows a quasi-perpendicular shock with Alfvén Mach number $M_A = 9$ and plasma beta parameter $\beta = 1.0$

5 Ion-Ion Thermal Equilibration

Ion-ion thermal equilibration could be different from electron-ion equilibration. For instance, the charge-to-mass ratio Z/A is typically 2 to 5 times smaller for heavy ions than for protons, and this is a very small difference compared to the value of 1840 for electrons vs. protons. Thus energy transfer that depends on plasma waves that interact resonantly with both species is much more difficult for electrons than for ions (Kropotina et al. 2015, 2016). Measurements of ion heating in shocks in the solar wind generally show $T_i > T_p$ (Berdichevsky et al. 1977; Korreck et al. 2007), and some models predict strong preferential heating, with $T_i > (m_i/m_p)T_p$ and asymmetry such that $T_\perp > T_\parallel$ (Zimbardo 2011).

Thus ion temperatures present interesting diagnostic capabilities, and the degree of equilibration between He and H ions influences the interpretation of Hα line widths in Balmer-dominated shocks in terms of shock speed (Chevalier et al. 1980). Figure 7 shows the distribution of Fe II impurity ions behind a quasi-perpendicular shock (Kropotina et al. 2015).

The Hopkins Ultraviolet Telescope (Kruk et al. 1995) was used to observe lines of He II, C IV, N V and O VI in a 2300 km/s shock in the NW part of SN1006 blast wave (Raymond et al. 1995). Line widths were measured for the He II λ1640 and C IV λ1550 lines and compared with the Hα line width from Smith (1991). The result was that the thermal speeds were all the same to within measurement uncertainties of 10 to 24%, indicating that the temperatures are mass-proportional; $T_a = (m_a/m_p)T_p$. Subsequently Korreck et al. (2004) used FUSE to observe the same position in O VI and obtain a kinetic temperature that was mass-proportional to the He and C temperatures within the uncertainties, but formally only half the 2.9×10^9 K value that would be mass-proportional to the proton temperature. That paper assumed a 3000 km/s shock speed based on Ghavamian et al. (2002), and the formal uncertainty of the O VI line width may be optimistic due to severe blending with Lyβ.

While the very fast blast wave in SN1006 showed little or no ion-ion thermal equilibration, the 350 km/s blast wave in the Cygnus Loop shows very strong equilibration, with $T_{\rm He}$, $T_{\rm C}$ and $T_{\rm O}$ equal to T_p to within the uncertainties (Raymond et al. 2015). This is in accord with the effective electron-ion equilibration in this section of the Cygnus Loop (Ghavamian et al. 2002; Salvesen et al. 2009; Medina et al. 2014). Estimates of $T_{\rm O}/T_p$ in DEML71, which has intermediate shock speeds of about 650 and 900 km/s, are 6.5 and 16, suggesting partial equilibration at 650 km/s and no equilibration at 900 km/s (Ghavamian et al. 2007b).

Broersen et al. (2013) measured the width of the O VII X-ray lines in an ejecta knot in SN1006 using the RGS on XMM-Newton. As mentioned above, it indicated very little electron-ion temperature equilibration. The line width corresponds to a shock speed of 3000 km/s if there is no equilibration. They could not measure the widths of other lines, but a low resolution X-ray spectrum showed strong Si emission, implying that the Si abundance was enhanced 6 or 7 times more that the O abundance. That would mean that if there is significant ion-ion equilibration, O would be heated by the Si and a slower shock would give the observed line width. Unfortunately, while the speed of the shock driven into the ISM by the ejecta knot can be determined from its Hα profile or its proper motion, we do not know the speed of the reverse shock that heats the ejecta.

An interesting possible modification of ion-ion equilibration can occur if some of the elements passing through the shock are neutral while others are completely ionized. For instance, some of the hydrogen going into Balmer line shocks is neutral, while elements whose ionization potentials are less than the energy of Lyα photons (10.2 eV) or of the Lyman edge at 13.6 eV are likely to be singly ionized in the pre-shock gas. The ions interact with the fields and turbulence at the shock, while the neutrals are not affected. When they become ionized well downstream, the turbulence is likely to have decayed away, and the newly formed ions behave as pickup ions (Williams and Zank 1994). When they are ionized, the velocity components parallel and perpendicular to the field are conserved, and the perpendicular component quickly evolves into a shell in velocity space. In the process the ions generate plasma waves at their Larmor frequency, but those waves do not interact strongly with other ion species. This pickup ion behavior has been suggested as a possible explanation for the non-Gaussian line profile seen at one position in Tycho's SNR (Raymond et al. 2008, 2010), but that is only one of four plausible explanations. Something similar could occur in an ejecta shock propagating through a mixture of O (I.P. = 13.6 eV), Ne (I.P. = 21.6 eV), Si (I.P. = 8.2 eV) and Fe (I.P = 7.9 eV), but as yet there are no observations of ejecta shocks that could test this idea. However, even the higher IP neutrals may be photoionized before they reach the reverse shock, as discussed in the next section.

6 Photoionized Shocks and Precursors

Photoionization enters the picture in three ways. First, radiative shocks by definition effectively convert the thermal energy of the shocked gas into radiation, and if the electron temperature is above about 10^5 K, much of that is ionizing radiation. That radiation is always included in computing the spectrum emitted by the cooling gas, and it can also produce a photoionized precursor whose strength depends on the evolution of the shock as it passes through the cloud. Second, non-radiative shocks produce some emission, both in the X-rays and in UV lines from lower ionization as the shocked gas becomes ionized (Hamilton and Fesen 1988). That radiation can be important in the unshocked ejecta. And, third, a pulsar wind nebula (PWN) can ionize the shocked gas. Though all three can occur simultaneously, we consider them separately for simplicity.

6.1 Photoionized Precursors to Radiative Shocks

Photoionization precursors can be seen ahead of radiative shocks in normal abundance gas. Figure 8 shows a deep Hα image of a portion of the Cygnus Loop, and the faint, diffuse emission at about $20^h\ 57^m\ 30^s$ is the pre-shock cloud photoionized by the EUV radiation from the bright radiative shocks at lower RA. In this case, the precursor is easily spatially

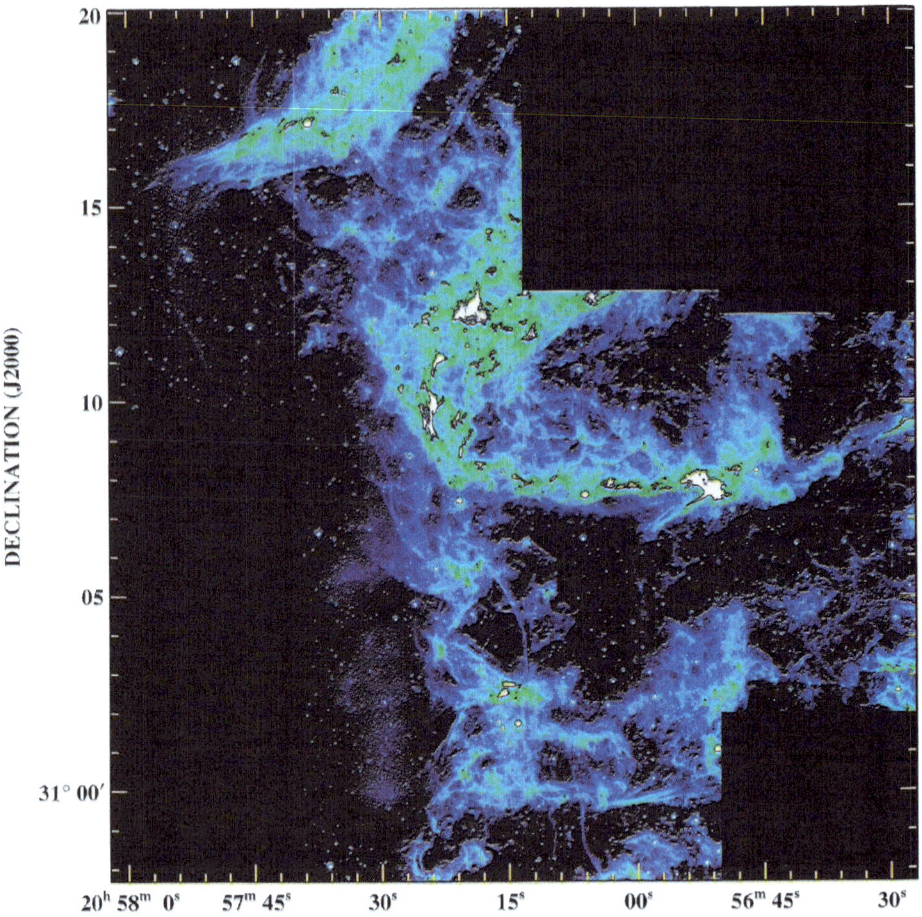

Fig. 8 Hα image of a portion of the Cygnus Loop showing vertical stripe of faint diffuse emission ahead of the shock (at higher RA) near $20^h\ 57^m\ 30^s$ between about $+31°0'$ and $+31°3'$ in declination (Szentgyorgyi et al. 2000). At the distance to the Cygnus Loop, $5'$ corresponds to roughly 1 pc

resolved from the shocked gas, but in more distant SNRs in denser gas that may not be the case.

The early models of shocks in pure oxygen by Itoh (1981a) were not able to match all the line ratios. Itoh (1981b) included a photoionization precursor in his models, but they still had difficulties with some line ratios. Subsequent calculations by Dopita et al. (1984) included other elements besides oxygen. They also found that the models failed to match observed spectra. In particular, the temperature-sensitive ratios [O III] I(4363)/I(5007) and [O II] I(7320,30)/I(3726, 29) were predicted to be much too high in the models. Blair et al. (1989) and Blair et al. (1994) obtained UV spectra of 1E102-7219 and N132D with IUE, and they found basically the same difficulty that Dopita et al. (1984) pointed out. If T_e is high enough to produce the observed ions by collisional ionization, then the models predict UV to optical line ratios larger than observed. This led Blair et al. to suggest that photoionization of the unshocked ejecta plays an important role. Sutherland and Dopita (1995) included more elements and produced a grid of models for which they tabulated the emission from the shocked gas and the precursor separately. They found that the precursor could make a very substantial contribution to the optical emission lines if the structure reaches a steady state between photoionizing flux and the flow of gas into the shock.

On the other hand, Blair et al. (2000) observed the O-rich remnants in the Magellanic Clouds N132D and 1E102-7219 with the FOS on HST, and they covered the 1200–8000 Årange. They found that a power law distribution of shock speeds within the FOS aperture could reproduce the UV to optical line ratios and the temperature-sensitive optical line ratios without a photoionization precursor, though a low temperature cutoff was still necessary to avoid overpredicting the lowest ionization lines. Because the knots are very small and subject to shearing and other instabilities when they encounter the reverse shock (Eriksen 2009), it is plausible and even probable that a range of shock speeds is present even within a small knot, and it is unlikely that a single steady-flow shock model is appropriate.

Part of the difficulty in assessing the importance of the precursor stems from the small size of the knots. The 10 year typical lifetime of knots in Cas A corresponds to a size smaller than 10^{16} cm, so the amount of photoionized, unshocked ejecta drops rapidly as the shock moves through the knot. For normal abundance SNR shocks, it is relatively easy to separate the emission from the pre- and post-shock gas. The pre-shock emission shows up as low surface brightness, diffuse emission ahead of the shock filaments in relatively nearby SNRs such as the Cygnus Loop (Szentgyorgyi et al. 2000; Medina et al. 2014), while in LMC SNRs it can show up as a narrow spectral feature at the rest velocity, distinct from the broad, Doppler-shifted emission from the shocked gas (Vancura et al. 1992). In the case of ejecta shocks, the spatial separation is tiny, and both pre- and post-shock gas are moving at thousands of $\mathrm{km\,s^{-1}}$, while the pre- and post-shock velocities only differ by something like 100 $\mathrm{km\,s^{-1}}$. A possible way to resolve the question is by comparing proper motions with the age of the remnant, since the shock decelerates the post-shock gas but not the precursor. In Cas A, explosion dates of 1662 ± 27 AD and 1672 ± 18 AD have been inferred from the positions and proper motions of all the outer ejecta knots and of a subsample of bright, compact knots, respectively (Fesen et al. 2006), while the SN might have been observed by the Reverend John Flamsteed in 1680. Taken at face value, the two proper motion dates would indicate decelerations, and therefore shock speeds, of 0.055 or 0.025 times the speeds of the knots, or about 150 to 550 km/s in the Cas A knots. In practice, the uncertainties in the explosion dates from proper motions are large enough to include the possible observation date, and the Flamsteed sighting is controversial (Stephenson and Green 2002).

6.2 Photoionized Precursors to Non-Radiative Shocks in Ejecta

Hamilton and Fesen (1988) studied the UV absorption line spectrum of the Schweizer-Middleditch star located behind the type Ia SNR SN1006. While it shows very strong Fe II absorption, the corresponding mass is much smaller than expected for a type Ia explosion. That led Hamilton and Fesen (1988) to investigate whether the unshocked ejecta could be photoionized to Fe III and above, which are not detectable in their wavelength range. While non-radiative shocks are inefficient in producing radiation, if they produced none at all they would be invisible. The photoabsorption cross sections at X-ray wavelengths are relatively small, but non-radiative shocks also produce UV emission as the elements are progressively ionized. Each ion below the ones found in equilibrium at T_e takes a time $1/(n_e q_i)$ to be ionized, where q_i is the ionization rate coefficient, typically a weak function of temperature for ions well below the ones that would be present at T_e. During that time, each spectral line of the ion is excited at a rate $n_e q_{ex}$, so each atom passing though the shock produces q_{ex}/q_i photons in the line. For strong permitted lines, that number is a few tens, while forbidden lines have much smaller q_{ex} at high temperatures and the number is below one. Shocks in normal gas produce enough ionizing He I and He II photons to partially ionize the pre-shock gas (Morse et al. 1996; Ghavamian et al. 2000; Medina et al. 2014). Ejecta shocks produce quite a few ionizing photons per O atom, for instance, so they can produce many ionizing photons per atom passing through the shock.

Hamilton and Fesen (1988) computed the number of ionizing photons for the ions of O and Fe, and they found of order 1 to 10 ionizing UV photons for most ions. Putting those numbers into hydrodynamic simulations, they found that the unshocked Fe ejecta is mostly Fe IV and Fe V. Blair et al. (1996) observed the Schweizer-Middleditch star with the Hopkins Ultraviolet Telescope to search for the Fe III line at 1123 Å. Their 3σ upper limit corresponds to 0.054 $M_\odot$of Fe III, which is consistent with the Hamilton and Fesen (1988) model. An additional process could enhance the photoionization rate beyond that predicted by Hamilton and Fesen (1988). Laming et al. (1996) found that in non-radiative shocks faster than about 1000 km s^{-1}, protons and alpha particles could excite UV emission lines as effectively as do electrons. This process, along with the related ionization rates, is sensitive to the threshold energy of the transition, and in the 1000 to 3000 km s^{-1} range, the protons could excite UV transitions around 10 eV, but they could not ionize the corresponding ions. Thus they increased the number of UV photons per atom passing through the shock. The excitation rates scale roughly as Z^2, while the number of electrons per ion is Z, so the ratio of ion to electron excitation increases as Z.

Another interesting case of photoionization of the unshocked ejecta is the recovered supernova S And, also known as SN 1885A in M31 (Fesen et al. 1999, 2015). It is seen in absorption against the bulge of M31 in lines of Ca I, Ca II, Fe I and Fe II, along with probable lines of Mg and other elements. The absorption lines show expansion velocities of 13,000 km/s and relative abundances consistent with a type Ia explosion. The ejecta are less ionized than those in SN1006, perhaps because of the small age of S And or because a low ambient density means that relatively little material has gone through the reverse shock thus far.

6.3 Photoionization by PWNe

The obvious case of photoionization by the synchrotron emission from a PWN is the Crab Nebula. The ejecta show high He, C and O abundances due to CNO processing, but not the more completely burned composition of the O-rich remnants discussed above (Sibley et al. 2016). While both shocked and unshocked ejecta are ionized, the gas that has been compressed by the shock driven by the expanding PWN has a much higher emissivity, and it produces the prominent filaments of the Crab (Sankrit and Hester 1997; Sankrit et al. 1998; Loll et al. 2013).

The Crab's optical emission lines are basically explained by photoionization of the compressed gas without any major contribution from the thermal energy of the shock, though the high temperature region behind the shock can contribute significantly to the UV lines because of their sensitivity to temperature. The radiative shock creates a dense shell that is accelerated by the PWN, leading to development of the Rayleigh-Taylor instability and the radial structures and strong clumping seen in some parts of the Crab (Hester et al. 1996; Loll et al. 2013). One characteristic of photoionization by the PWN power law emission is the simultaneous presence of high and low ionization emission from the outer and inner parts of the filaments, respectively. The cool cores of the filaments produce strong emission in low temperature lines, but not enough to account for the strong emission near 24 μm (Temim et al. 2006). The remnant 3C 58 may be similar, but has been less extensively studied (Fesen et al. 2008).

A case of a PWN inside O-rich SN ejecta is B0540-69.3 (Serafimovich et al. 2005; Morse et al. 2006; Williams et al. 2008). There, it appears that the pre-shock gas is photoionized to such an extent that a very slow shock shows relatively high ionization species and appears much like a faster shock (Williams et al. 2008), though faster shocks which might produce the observed emission without the assistance of photoionization by the PWN might also be present (Sandin et al. 2013). The young SNR G54.1+0.3 appears to be another case of shocks driven into the ejecta by the PWN, but the role of photoionization by the PWN is poorly constrained (Temim et al. 2010).

7 Instabilities

Radiative shock waves in SN ejecta are subject to the same instabilities that affect shocks in normal astrophysical plasma. Unless the cooling coefficient Λ increases rapidly with temperature, the shocks are thermally unstable (McCray et al. 1975; Chevalier and Imamura 1982; Innes et al. 1992; Innes 1992). Figure 1 indicates that shocks with electron temperatures above 10^6 K will be thermally unstable, and lower temperature shocks may be unstable depending upon the composition. The instability can be violent even in normal composition shocks, because the radiative cooling time can be shorter than the sound-crossing time in the cooling zone. That leads to extreme departures from pressure balance that cause the shock speed to oscillate. They also produce secondary shock waves that repressurize the gas to restore equilibrium. The enormous cooling rate of the shocked ejecta should exacerbate the departures from pressure equilibrium, though the time lag associated with the Coulomb collisional time scale for energy transfer from ions to electrons will limit the effect. Thermal conduction and non-thermal pressure from magnetic fields or cosmic rays may also limit the strengths of secondary shocks (Innes 1992). In any case it is likely that thermal instabilities play a major role in the discrepancies between observed spectra and planar, steady-flow models of shocks in O-rich SNRs.

Another important instability is the thin shell instability first studied by Vishniac (1983). Radiative cooling produces a thin, dense shell that is subject to the ram pressure of the upstream gas and thermal pressure of hot, low density gas in the SNR interior. When the shell is rippled, those forces are not aligned, and the ripples grow. The rapid cooling of shocks in the ejecta means that such shells are produced very quickly. However, the instability takes some time to develop, and if the ejecta are clumpy, as seems to be the case in the O-rich SNRs, the pre-shock density inhomogeneity may be a more important factor in the appearance of the radiative shocks.

Two other fluid instabilities are also important. Rayleigh-Taylor instabilities, or their shock analogue Richtmeyer-Meshkov instabilities, pertain to a dense fluid accelerated by a less dense one. The Rayleigh-Taylor instability at the contact discontinuity has been investigated for many years (Gull 1975; Blondin and Ellison 2001), and it can be even more severe when the ejecta are strongly clumped before they encounter the reverse shock. Either Rayleigh-Taylor fingers or pre-existing clumps form strong shear layers as they pass trough the surrounding gas, generating Kelvin-Helmholtz instabilities that shred the dense clumps and mix the plasmas of different density, temperature and composition. The process has been explored numerically for ordinary astrophysical plasmas (Klein et al. 1994; MacLow et al. 1994; Hansen et al. 2017), but the enormous cooling rates make it challenging to simulate for strongly enriched SN ejecta. Itoh (1988) appealed to turbulent stripping of dense knots to explain the lack of O I recombination lines in the optical spectrum, but the only numerical simulation is that of Eriksen (2009). Figure 9 shows a 2D model that assumes a spherical

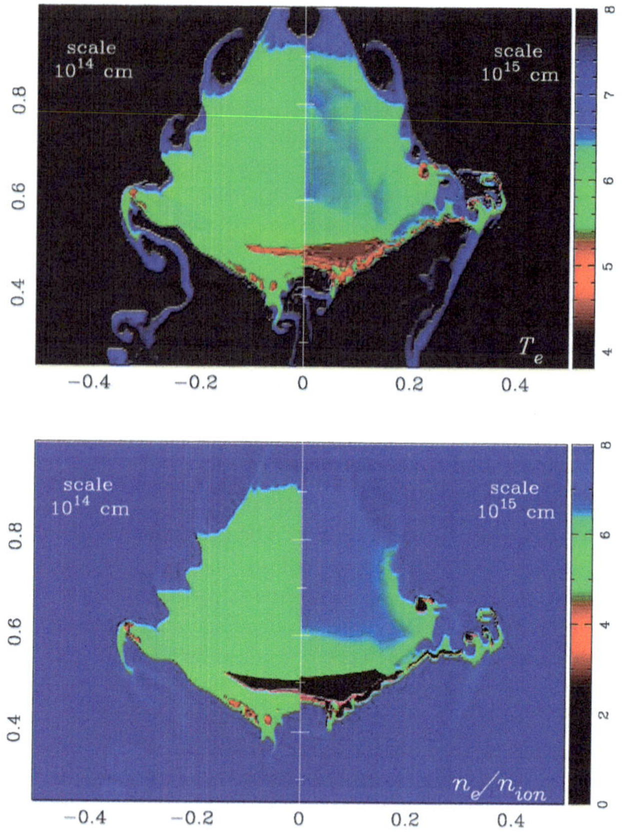

Fig. 9 Simulations of two similar dense oxygen knots of 10^{14} and 10^{15} cm size passing through the reverse shock (Eriksen 2009). In this 2D model, the spherical cloud is 200 times denser than the surrounding medium. The *upper panel* gives $\log(T)$, and the *lower panel* gives the ionization state; for instance, $n_e/n_i = 6$ corresponds to O VII

cloud of pure oxygen gas with a density contrast of 200 and an incident shock of 2500 km/s. As in the normal abundance cases, the cloud rapidly develops a complex structure due to the fluid instabilities, and a wide range of ionization states is present, but the computer resources were insufficient to cover the large dynamic range of the problem.

8 Particle Acceleration

It is often assumed that particle acceleration at SNR reverse shocks is inefficient. First, the elemental abundances of cosmic rays (Cummings et al. 2016) are enhanced relative to protons, but not nearly to the extent observed in SN ejecta. Rather, the cosmic ray abundances look like ISM abundances modified by the charge-to-mass ratio effects expected from Diffusive Shock Acceleration with some further enhancement of species sputtered from dust grains (Drury 2001). Second, no X-ray synchrotron emission is generally seen from the reverse shock region. However, an inner ring of X-ray synchrotron radio emission is observed in Cas A (Gotthelf et al. 2001). Helder and Vink (2008) used Lucy-Richardson deconvolution to show that the inner X-ray ring arises from the region of the reverse shock, and it is not a projection of emission from a distorted blast wave. Also, the theoretical analysis by Zirakashvili et al. (2014) somewhat favors acceleration at the reverse shock in Cas A. One other SNR has been suggested as an example of X-ray synchrotron emission from the reverse shock, RX J1713.7-3946 (Zirakashvili and Aharonian 2010), but the position of the reverse shock is not independently known and the interpretation is ambiguous.

Young type Ia SNRs such as Tycho and SN1006 show no evidence of X-ray synchrotron emission from the reverse shock in radio or X-rays, though Dickel et al. (1991) show that the radio emission from Tycho consists of a thin shell at the forward shock and a thicker shell in the region of the turbulent contact discontinuity between shocked ejecta and shocked ISM. Kosenko et al. (2011) found that acceleration at the reverse shock in Tycho must be less efficient than at the forward shock in order to match observations, while Schure et al. (2010) pointed out that particles accelerated at the forward shock could be re-accelerated at the reverse shock if the magnetic field there is sufficiently strong. Telezhinsky et al. (2012) construct models assuming a significant magnetic field at the reverse shock, and they find that acceleration at the reverse shock can affect the morphology, the spectrum and the evolution of synchrotron emission in type Ia SNRs. Telezhinsky et al. (2013) construct similar models for core-collapse SNRs.

To the extent that it is true that Cas A is fairly unique in showing strong synchrotron X-ray emission from the reverse shock, it is likely a result of the unusually strong perturbations of the reverse shock front due to the strong density inhomogeneities in the ejecta, both the large scale rings (Milisavljevic and Fesen 2015) and the small scale knots (Fesen et al. 2011). Either of these might amplify the magnetic field dramatically, but the small scale knots especially can generate strong small-scale vorticity that can amplify the field very rapidly. It is also possible that vorticity and magnetic fields were generated when the dense knots were first formed during the early stages of the explosion.

9 Abundance, Shock Speeds and Pressures in Shocks in Ejecta

The main applications of shock models for SN ejecta are to derive the elemental abundances for comparison with SN explosion models and to derive the shock parameters for comparison with dynamical models of the SNR evolution. Thus it is important to know how accurately the abundances, density and reverse shock speed can be estimated from observed spectra. We consider first non-radiative, then radiative shocks, which are equivalent to X-ray or to UV/optical/IR observations, respectively. It should also be remembered that the mass of unshocked ejecta can only be determined in a few cases; Cas A has about 0.4 $M_\odot$ (Isensee et al. 2010; DeLaney et al. 2014), SN1006 has about 0.04 $M_\odot$ of Fe and 0.25 $M_\odot$ of Si, with about half as much S and Si (Hamilton et al. 1997; Blair et al. 1996), and S And shows 0.1 to 1 $M_\odot$ of Fe (Fesen et al. 2015). There may also be a significant amount of mass locked up in dust, for instance 0.1 $M_\odot$ in Cas A (De Looze et al. 2017).

9.1 Parameters from X-Ray Spectra

As discussed in Sect. 4.1, a fast reverse shock provides some heat to the electrons (Yamaguchi et al. 2014), but Coulomb collisions soon dominate and bring kT_e to 1–3 keV. X-ray spectra are typically compared with models in which T_e is suddenly increased. The model parameters are T_e, the elemental abundances, and the ionization age, $n_e t$, which determines how far the plasma progresses toward ionization equilibrium. Smith and Hughes (2010) show the characteristic ionization age scales for different elements as functions of temperature. In effect, T_e is determined from the Boltzmann factor $e^{-h\nu/T}$ which enters most of the atomic processes that produce the X-ray spectrum, and $n_e t$ is determined by the ratios of different stages of ionization. To the extent that those numbers are accurate, the elemental abundances should also be accurate. There are some caveats, however. An X-ray synchrotron contribution could affect the T_e estimate, and both T_e and $n_e t$ are averages over the observed volume. Problems might arise if $n_e t$ is different for different elements, as would be expected in a layered explosion.

The dynamical parameters also carry some uncertainty. It is easy to measure T_e from the X-ray spectrum, but very difficult to measure the ion temperature, and the relation between T_e and shock speed is model-dependent. The density estimate contains a factor of $f^{1/2}$, where f is the filling factor. The filling factor also affects estimates of the total ejecta mass.

A partial solution to both problems is increasing the spatial resolution of the observations, which often means increasing the exposure time in order to obtain enough counts in smaller resolution elements. Higher spectral resolution with instruments such as LYNX (http://adsabs.harvard.edu/abs/2016SPIE.9904E..0NG) and ATHENA (http://adsabs.harvard.edu/abs/2012SPIE.8443E..28R) will help by adding velocity information and helping to separate different contributions along the line of sight.

9.2 Parameters from UV/Optical/IR Spectra

As discussed above, there are substantial uncertainties in models of radiative shocks in highly enriched gas regarding electron-ion temperature equilibration and the effects of instabilities and density inhomogeneities on the emission spectra. One approach, if one has a spectrum covering a large wavelength range, is to add the spectra of shocks over a range of velocities, with the weighting chosen to match the relative intensities of lines from different ionization states. This especially makes sense if one is observing an extragalactic object, because the field of view is bound to include a range of shock parameters. Vancura et al. (1992) introduced this technique for the normal abundance SNR N49 in the LMC, and Blair et al. (2000) and France et al. (2009) applied it to the O-rich Magellanic Cloud remnants N132D and 1E102.2-7219.

Blair et al. (2000) were able to match the relative intensities of UV and optical lines of O I, O II, O III and O IV to establish the relative ionization states, along with the temperature-sensitive ratios of the O III lines λλ1664, 4363, 5007. They then assumed that the model that matched the ionization state and temperature for O gave the correct set of temperature and ionization for the other elements to determine the abundances of C, Ne and Mg. This technique obviously works best if lines of several ionization states of the other elements are available. In this case Ne III, Ne IV and Ne V and C II, C III and C IV were observed, but the only observable ion of Mg was Mg II. Blair et al. (2000) concluded that the O:C ratios were 16 and 32 for 1E102 and N132D and the O:Ne ratios were 3.2 and 5.6, while the O:Mg ratios were 10–20 and 20–40, respectively. The match between models and observations was not perfect, of course, especially at the low (O I) and high (Ne V) ends, but the general agreement of the C II/C III and Ne III/Ne IV ratios with the models indicates that procedure works well for the dominant ion states and the abundance determinations are reliable.

The same cannot be said for the distribution of shock speeds used in the model itself. It is probable that instabilities and the shredding of dense knots lead to a much different distribution of temperatures than is given by a 1D shock model, and the assumed distribution of shock speeds is a way of mocking up that distribution. All one can say with confidence is that speeds of order 100 km/s are needed to produce the observed ionization states and temperatures.

If only a modest portion of the spectrum is available, it is difficult to observe more than 1 or 2 ionization stages for a given element. In that case if there are ions of 2 elements that have very similar ionization fractions and which have lines of similar excitation potential, the abundance can be estimated from the atomic rates alone, without recourse of hydrodynamic models. The [P II] and [Fe II] lines in the near infrared present one example. The relative abundances of P and Fe were derived for dozens of knots in Cas A by Koo et al. (2013).

10 Shocks in Supernovae

The emission from most supernovae is powered by radioactive decay, but in some core-collapse SNe the interaction between the ejecta and a dense circumstellar medium (CSM) dominates. In type II-L and type IIn SNe, the ejecta come from the red supergiant envelope, and the abundances are fairly normal. In type Ib, the abundances resemble those of Wolf-Rayet stars.

As time goes on, the reverse shock caused by the interaction penetrates more and more deeply into the ejecta, and it eventually reaches the layers devoid of H. Late-time spectra sometimes show a progression, in which Hα fades and [O III] brightens, presumably due to the changing abundances (Milisavljevic et al. 2012). An example is SN 2014C, in which the spectrum evolved from type Ib to type IIn, with weak Hα absorption near maximum light. At later times, the SN developed strong Hα emission. The narrow line widths showed that the H and the highly ionized Fe emission originated in a photoionized circumstellar shell, while a component with an intermediate line width ($\sim$1200 km/s) may arise from shocked CSM material (Milisavljevic et al. 2015).

SN 1993J, like SN 2014C, showed weak Hα at early times, but it quickly evolved to type IIb. Over the course of a few years the [O III]/Hα ratio increased by an order of magnitude (Milisavljevic et al. 2012). X-ray studies by Nymark and Fransson (2009) and Chandra et al. (2009) found that most of the X-rays come from the reverse shock, that the lower temperature components are from radiative shocks, and that the X-ray spectrum indicates that the reverse shock was encountering gas that is N-rich and O-poor. The broad feature near Hα is likely to have a strong contribution from [N II] (Jerkstrand et al. 2015). Given that at least parts of the reverse shock are radiative, a partial dense shell will form. In SN 1993J, the density of the shell is expected to be above 10^{10} cm^{-3}, a density so high that forbidden lines are suppressed, and the [O III] I(4363)/I(5007) ratio would greatly exceed the observed upper limit if the [O III] were produced in the shell. Therefore, it seems that the [O III] comes from a photoionization precursor to the reverse shock of the sort discussed in Sect. 6.2. The [O III] luminosity is on the order of 10% of the X-ray luminosity, which suggests very efficient reprocessing of the EUV and X-ray radiation from the reverse shock.

Eventually, the reverse shock will encounter the layers devoid of H and He. If the density is still high and the shock is radiative, it might be observable in the UV if the reddening is not too severe. If oxygen dominates, the intercombination transitions of [O III] and [O IV] should be bright.

11 Summary

Shock waves in highly enriched SN ejecta differ from those in astrophysical plasma of normal abundances because of the enormous cooling rates. The rapid cooling can lead to extreme departures from electron-ion temperature equilibrium as well as collisional ionization equilibrium. It can also enhance fluid instabilities which disrupt simple 1D flows.

For the purposes of inferring elemental abundances, shock wave models that match the observed ratios of ionization states and the observed temperature-sensitive line ratios should give reliable results. However, the uncertainty in electron-ion temperature ratios makes it difficult to infer shock speeds in either radiative or non-radiative shocks from optical/UV or X-ray spectra.

Acknowledgements This work was supported by GO grant HST-GO-13436.01-A from the Space Telescope Science Institute and by the International Space Sciences Institute. It grew out of the ISSI meeting on Supernovae in October, 2016.

Appendix

For the reader's convenience, we list the SNRs showing shocks in O-rich material, along with some recent references that cite earlier works. The list of references is not intended to be complete or to include studies not directly related to shocks in enriched ejecta.

Cas A is the brightest and best known O-rich SNR. A comprehensive study of the X-ray spectra is given by Hwang and Laming (2012), and optical studies include Patnaude and Fesen (2014) and Milisavljevic and Fesen (2015). A near IR study is described by Koo et al. (2013) and mid-IR spectra by Isensee et al. (2010).

The bright O-rich SNR N132D is in the LMC. Schenck et al. (2016) analyze the X-ray spectrum, and France et al. (2009) analyze a UV spectrum from COS. Blair et al. (2000) discuss optical and UV spectra obtained with the FOS on HST. Tappe et al. (2012) obtained Spitzer spectra of N132D. They concentrate on PAH emission from the blast wave, but they also show a spectrum of the ejecta knots dominated by [Ne II] and [Ne III].

The SMC supernova remnant 1E102.2-7219 was observed in X-rays by Plucinsky et al. (2017) and Flanagan et al. (2004). The far UV spectrum was obtained by FUSE (Sasaki et al. 2006), and the rest of the UV spectrum is given by Blair et al. (2000). Blair et al. (2000) also give the optical spectrum, and Vogt et al. (2017) present an IFU study of the optical structure. Spitzer IR data are presented by Rho et al. (2009).

The LMC remnant B0540-69.3 is an O-rich remnant photoionized by a PWN. Brantseg et al. (2014) present the X-ray spectra, along with radio and optical images. Morse et al. (2006) give optical images and spectra from HST, and Williams et al. (2008) analyze the IR spectrum.

The X-ray spectrum of the very young O-rich SNR in NGC 4449 was analyzed by Patnaude and Fesen (2003), and its optical spectrum is given by Milisavljevic and Fesen (2010).

The Galactic SNR G292.0+1.8 has been observed in X-rays by Bhalerao et al. (2015) and Kamitsukasa et al. (2014), and in the optical by Winkler et al. (2009). Ghavamian et al. (2012) analyze Spitzer spectra and images.

The Puppis A supernova remnant is brightest in normal abundance material, but it contains a small region of shocked ejecta. Katsuda et al. (2010) give the X-ray spectrum and Winkler and Kirshner (1985) give the optical.

G54.1+0.3 is a PWN-dominated supernova remnant similar to B0540-69.3, except that no thermal X-ray emission is seen, and the IR spectrum is dominated by intermediate mass elements (Temim et al. 2010).

References

A. Bemporad, R. Susino, G. Lapenta, Astrophys. J. **784**, 102 (2014)
D. Berdichevsky, J. Giess, G. Gloeckler, U. Marshall, J. Geophys. Res. **102**, 2623 (1977)
J. Bhalerao, S. park, D. Dewey, J.P. Hughes, K. Mori, J.-J. Lee, Astrophys. J. **800**, 65 (2015)

W.P. Blair, J.C. Raymond, J. Danziger, F. Mateucci, Astrophys. J. **338**, 812 (1989)
W.P. Blair, J.C. Raymond, K.S. Long, Astrophys. J. **423**, 334 (1994)
W.P. Blair, K.S. Long, J.C. Raymond, Astrophys. J. **468**, 871 (1996)
W.P. Blair et al., Astrophys. J. **537**, 667 (2000)
J.M. Blondin, D.C. Ellison, Astrophys. J. **560**, 244 (2001)
K.J. Borkowski, J.M. Shull, Astrophys. J. **348**, 169 (1990)
T. Brantseg, R.L. McEntaffer, L.M. Bozzetto, M. Filipovic, N. Grieves, Astrophys. J. **780**, 50 (2014)
S. Broersen, J. Vink, Mon. Not. R. Astron. Soc. **446**, 3885 (2014)
S. Broersen, J. Vink, M. Miceli, F. Bocchino, G. Maurin, A. Decourchelle, Astron. Astrophys. **552**, 9 (2013)
A.M. Bykov, Y.A. Uvarov, J. Exp. Theor. Phys. **88**, 465 (1999)
A.M. Bykov, F.B.S. Paerels, V. Petrosian, Space Sci. Rev. **134**, 141 (2008)
P.J. Cargill, K. Papadopoulos, Astrophys. J. Lett. **329**, 29 (1988)
P. Chandra, V.V. Dwarkadas, A. Ray, S. Immler, D. Pooley, Astrophys. J. **699**, 388 (2009)
R.A. Chevalier, J.N. Imamura, Astrophys. J. **261**, 543 (1982)
R.A. Chevalier, J.C. Raymond, Astrophys. J. Lett. **225**, L27 (1978)
R.A. Chevalier, R.P. Kirshner, J.C. Raymond, Astrophys. J. **235**, 186 (1980)
A.C. Cummings et al., Astrophys. J. **831**, 18 (2016)
I. De Looze, M.J. Barlow, B.M. Swinyard, J. Rho, H.L. Gomez, M. Matsuura, R. Wesson, Mon. Not. R. Astron. Soc. **465**, 3309 (2017)
g. Del Zanna, K.P. Dere, P. Young, E. Landi, H.E. Mason, Astron. Astrophys. **582**, 56 (2015)
T. DeLaney, N.E. Kassim, L. Rudnick, R.A. Perley, Astrophys. J. **785**, 7 (2014)
J.R. Dickel, W.J.M. van Breugel, R.G. Strom, Astron. J. **101**, 2151 (1991)
M.A. Dopita, L. Binette, I.R. Tuohy, Astrophys. J. **282**, 142 (1984)
L.OC. Drury, Nature **99**, 329 (2001)
K.A. Eriksen Ph.D. thesis, University Arizona, 2009
R.A. Fesen, C.L. Gerardy, K.M. McLin, A.J.S. Hamilton, Astrophys. J. **514**, 195 (1999)
R.A. Fesen et al., Astrophys. J. **645**, 283 (2006)
R. Fesen, G. Rudie, A. Hurford, A. Soto, Astrophys. J. **174**, 379 (2008)
R.A. Fesen, J.A. Zastrow, M.C. Hammell, J.M. Shull, D.W. Silvia, Astrophys. J. **736**, 109 (2011)
R.A. Fesen, P.A. Höflich, A.J.S. Hamilton, Astrophys. J. **804**, 140 (2015)
K.A. Flanagan et al., Astrophys. J. **605**, 230 (2004)
K. France et al., Astrophys. J. Lett. **707**, L27 (2009)
T.J. Gaetz, R.J. Edgar, R.A. Chevalier, Astrophys. J. **329**, 927 (1988)
P. Ghavamian, J. Raymond, P. Hartigan, W.P. Blair, Astrophys. J. **535**, 266 (2000)
P. Ghavamian, J. Raymond, R.C. Smith, P. Hartigan, Astrophys. J. **547**, 995 (2001)
P. Ghavamian, P.F. Winkler, J.C. Raymond, K.S. Long, Astrophys. J. **572**, 888 (2002)
P. Ghavamian, J.M. Laming, C.E. Rakowski, Astrophys. J. Lett. **654**, 69 (2007a)
P. Ghavamian, W.P. Blair, R. Sankrit, J.C. Raymond, J.P. Hughes, Astrophys. J. **664**, 304 (2007b)
P. Ghavamian et al., Astrophys. J. **750**, 39 (2012)
P. Ghavamian, S.J. Schwartz, J. Mitchell, A. Masters, J.M. Laming, Space Sci. Rev. **47**, 557 (2013)
J. Giacalone, Space Sci. Rev. **176**, 73 (2013)
E.V. Gotthelf, B. Koralesky, L. Rudnick, T.W. Jones, U. Hwang, R. Petre, Astrophys. J. Lett. **522**, L39 (2001)
N. Grevesse, M. Asplund, A.J. Sauval, Space Sci. Rev. **130**, 105 (2007)
J. Grun, J. Stamper, C. Manka, R. Resnick, J. Crawford, B.H. Ripin, Phys. Rev. Lett. **66**, 2738 (1991)
S.F. Gull, Mon. Not. R. Astron. Soc. **171**, 263 (1975)
A.J.S. Hamilton, R.A. Fesen, Astrophys. J. **327**, 178 (1988)
A.J.S. Hamilton, R.A. Fesen, C.-C. Wu, D.M. Crenshaw, C.L. Sarazin, Astrophys. J. **481**, 838 (1997)
A.J.S. Hamilton, R.A. Fesen, W.P. Blair, Mon. Not. R. Astron. Soc. **381**, 771 (2007)
E.C. Hansen, A. Frank, P. Hartigan, S.V. lebedev, Astrophys. J. **837**, 143 (2017)
E.A. Helder, J. Vink, Astrophys. J. **686**, 1094 (2008)
E.A. Helder, J. Vink, C.G. Bassa, Astrophys. J. **737**, 85 (2011)
J.J. Hester, J.C. Raymond, W.P. Blair, Astrophys. J. **420**, 721 (1994)
J.J. Hester et al., Astrophys. J. **456**, 225 (1996)
L. Hovey, J.P. Hughes, K. Eriksen, Astrophys. J. **809**, 119 (2015)
J.P. Hughes, C.E. Rakowski, A. Decourchelle, Astrophys. J. Lett. **543**, 61 (2000)
A.J. Hull, J.D. Scudder, R.J. Fitzenreiter, K.W. Ogilvie, J.A. Newbury, C.T. Russell, J. Geophys. Res., Atmos. **105**, 20957 (2000)
U. Hwang, J.M. Laming, Astrophys. J. **746**, 130 (2012)
D.E. Innes, Astron. Astrophys. **256**, 660 (1992)
D.E. Innes, J.R. Giddings, S.A.E.G. Falle, Mon. Not. R. Astron. Soc. **224**, 179 (1992)
K. Isensee et al., Astrophys. J. **725**, 2059 (2010)

H. Itoh, Publ. Astron. Soc. Jpn. **29**, 813 (1977)
H. Itoh, Publ. Astron. Soc. Jpn. **30**, 489 (1978)
H. Itoh, Publ. Astron. Soc. Jpn. **33**, 1 (1981a)
H. Itoh, Publ. Astron. Soc. Jpn. **33**, 521 (1981b)
H. Itoh, Publ. Astron. Soc. Jpn. **38**, 717 (1986)
H. Itoh, Publ. Astron. Soc. Jpn. **40**, 673 (1988)
A. Jerkstrand, M. Ergon, S.J. Smartt et al., Astron. Astrophys. **573**, A12 (2015)
F. Kamitsukasa et al., Nature **66**, 64 (2014)
S. Katsuda, U. Hwang, R. Petre, S. Park, K. Mori, H. Tsunemi, Astrophys. J. **714**, 1725 (2010)
R.I. Klein, C.F. McKee, P. Colella, Astrophys. J. **420**, 213 (1994)
B.-C. Koo, Y.-H. Lee, D.-S. Moon, S.-C. Yoon, J.C. Raymond, Science **342**, 1346 (2013)
K.E. Korreck, J.C. Raymond, T.H. Zurbuchen, P. Ghavamian, Astrophys. J. **615**, 280 (2004)
K.E. Korreck, T.H. Zurbuchen, S.T. Lepri, J.M. Raines, Astrophys. J. **659**, 773 (2007)
D. Kosenko, J. Vink, S. Blinnikov, A. Rasmussen, Astron. Astrophys. **490**, 223 (2008)
D. Kosenko, S.I. Blinnikov, J. Vink, Astron. Astrophys. **532**, A114 (2011)
Yu.A. Kropotina, A.M. Bykov, M.Yu. Gustov, A.M. Krassilchtchikov, K.P. Levenfish, J. Tech. Phys. **60**, 231 (2015)
Yu.A. Kropotina, A.M. Bykov, A.M. Krassilchtchikov, K.P. Levenfish, J. Tech. Phys. **61**, 517 (2016)
J.W. Kruk, S.T. Durrance, G.A. Kriss, A.F. Davidson, W.P. Blair, B.R. Espey, D.S. Finley, Astrophys. J. Lett. **454**, L1 (1995)
J.M. Laming, J. Grun, Phys. Rev. Lett. **89**, 125002 (2002)
J.M. Laming, J.C. Raymond, B.M. McLaughlin, w.P. Blair, Astrophys. J. **472**, 267 (1996)
J.M. Laming, U. Hwang, P. Ghavamian, C. Rakowski, Astrophys. J. **790**, 11 (2014)
J.-J. Lee, B.-C. Koo, J.C. Raymond, P. Ghavamian, T.-S. Pyo, A. Tajitsu, M. Hayashi, Astrophys. J. Lett. **659**, L133 (2007)
J.-J. Lee, J.C. Raymond, S. Park, W.P. Blair, P. Ghavamian, P.F. Winkler, K. Korreck, Astrophys. J. Lett. **715**, 146 (2010)
A.M. Loll, S.J. Desch, P.A. Scowen, J.P. Foy, Astrophys. J. **765**, 152 (2013)
K.S. Long et al., Astrophys. J. **586**, 1162 (2003)
S. Ma et al., Astrophys. J. **738**, 160 (2011)
M.-M. MacLow, C.F. McKee, R.I. Klein, J.M. Stone, M. Norman, Astrophys. J. **433**, 757 (1994)
R. McCray, R.F. Stein, M. Kafatos, Astrophys. J. **196**, 565 (1975)
A. Medina, J.C. Raymond, R.J. Edgar, N. Caldwell, R.A. Fesen, D. Milisavljevic, Astrophys. J. **791**, 30 (2014)
E.R. Micelotta et al., Space Sci. Rev. (2017, submitted)
D. Milisavljevic, R.A. Fesen, Astrophys. J. **677**, 306 (2010)
D. Milisavljevic, R.A. Fesen, Science **347**, 526 (2015)
D. Milisavljevic, R.A. Fesen, R.A. Chevalier, R.P. Kirshner, P. Challis, M. Turatto, Astrophys. J. **751**, 25 (2012)
D. Milisavljevic et al., Astrophys. J. **815**, 120 (2015)
J.A. Morse et al., Astron. J. **112**, 509 (1996)
J.A. Morse, N. Smith, W.P. Blair, R.P. Kirshner, P.F. Winkler, J.P. Hughes, Astrophys. J. **644**, 188 (2006)
F.S. Mozer, D. Sundkvist, J. Geophys. Res. **118**, 5413 (2013)
T.K. Nymark, C. Fransson, Astron. Astrophys. **494**, 179 (2009)
Y. Ohira, Mon. Not. R. Astron. Soc. **440**, 514 (2014)
Y. Ohira, F. Takahara, Astrophys. J. Lett. **661**, L171 (2007)
D.J. Patnaude, R.A. Fesen, Astrophys. J. **587**, 221 (2003)
D.J. Patnaude, R.A. Fesen, Astrophys. J. **789**, 138 (2014)
P. Plucinsky et al., Astron. Astrophys. **597**, 83 (2017)
C.E. Rakowski, P. Ghavamian, J.P. Hughes, Astrophys. J. **590**, 846 (2003)
C.E. Rakowski, J.M. Laming, P. Ghavamian, Astrophys. J. **684**, 348 (2008)
J.C. Raymond, Astrophys. J. Suppl. Ser. **39**, 1 (1979)
J.C. Raymond, W.P. Blair, K.S. Long, Astrophys. J. Lett. **454**, L31 (1995)
J.C. Raymond, P.A. Isenberg, J.M. Laming, Astrophys. J. **682**, 408 (2008)
J.C. Raymond, P.F. Winkler, W.P. Blair, J.-J. Lee, S. Park, Astrophys. J. **712**, 901 (2010)
J.C. Raymond, R.J. Edgar, P. Ghavamian, W.P. Blair, Astrophys. J. **805**, 152 (2015)
J. Rho, W.T. reach, A. Tappe, U. Hwang, J.D. Slavin, T. Kozasa, L. Dunne, Astrophys. J. **700**, 579 (2009)
M.A. Riquelme, A. Spitkovsky, Astrophys. J. **733**(1), 63 (2011). https://doi.org/10.1088/0004-637X/733/1/63
G. Salvesen, J.C. Raymond, R.J. Edgar, Astrophys. J. **702**, 327 (2009)

C. Sandin, P. Lundqvist, N. Lundqvist, C.-I. Björnsson, G. Olofsson, Yu.A. shibanov, Mon. Not. R. Astron. Soc. **432**, 2854 (2013)
R. Sankrit, J.J. Hester, Astrophys. J. **491**, 796 (1997)
R. Sankrit et al., Astrophys. J. **504**, 344 (1998)
A. Sarangi et al., Space Sci. Rev. (2017, submitted)
M. Sasaki, T.J. Gaetz, W.P. Blair, R.J. Edgar, J.A. Morse, P.P. Plucinsky, R.K. Smith, Astrophys. J. **642**, 260 (2006)
A. Schenck, S. Park, S. Post, Astrophys. J. **151**, 161 (2016)
K.M. Schure, A. Achterberg, R. Keppens, J. Vink, Mon. Not. R. Astron. Soc. **406**, 2633 (2010)
S.J. Schwartz, J. Geophys. Res. **119**, 1507 (2014)
S.J. Schwartz, M.F. Thomsen, S.J. Bame, J. Stansberry, J. Geophys. Res. **93**, 12923 (1988)
N.I. Serafimovich, P. Lundqvist, A. Shibanov Yu, J. Sollerman, Adv. Space Res. **35**, 1106 (2005)
N. Shimada, M. Hoshino, Astrophys. J. Lett. **543**, L71 (2000)
A.R. Sibley, A.M. Katz, T.J. Satterfield, S.J. Vanderveer, G.M. MacAlpine, Astron. J. **152**, 93 (2016)
R.C. Smith Ph.D. thesis, Harvard University, 1991
R.K. Smith, J.P. Hughes, Astrophys. J. **718**, 583 (2010)
R.C. Smith, J.C. Raymond, J.M. Laming, Astrophys. J. **420**, 286 (1994)
J. Sollerman, P. Ghavamian, P. Lundqvist, R.C. Smith, Astron. Astrophys. **407**, 249 (2003)
F.R. Stephenson, D.A. Green, in *Historical Supernovae and Their Remnants* (Clarendon, Oxford, 2002)
R.S. Sutherland, M.A. Dopita, Astrophys. J. **439**, 381 (1995)
F. Suzuki-Vidal et al., Astrophys. J. **815**, 96 (2015)
A.H. Szentgyorgyi, J.C. Raymond, J.J. Hester, S. Curiel, Astrophys. J. **529**, 279 (2000)
A. Tappe, J. Rho, C. Boersma, E.R. Micelotta, Astrophys. J. **754**, 132 (2012)
I. Telezhinsky, V.V. Dwarkadas, M. Pohl, Astropart. Phys. **35**, 300 (2012)
I. Telezhinsky, V.V. Dwarkadas, M. Pohl, Astron. Astrophys. **552**, 102 (2013)
T. Temim et al., Astron. J. **132**, 1610 (2006)
T. Temim, P. Slane, S.P. Reynolds, J.C. Raymond, K.J. Borkowski, Astrophys. J. **710**, 309 (2010)
O. Vancura, W.P. Blair, K.S. Long, J.C. Raymond, Astrophys. J. **394**, 158 (1992)
J. Vink, J.M. Laming, F.M. Gu, A. Rasmussen, J. Kaastra, Astrophys. J. Lett. **587**, L31 (2003)
J. Vink, S. Broersen, A. Bykov, S. Gabici, Astron. Astrophys. **579**, 13 (2015)
E.T. Vishniac, Astrophys. J. **274**, 152 (1983)
F.P.A. Vogt, I. Seitenzahl, M.A. Dopita, A.J. Ruiter, Publ. Astron. Soc. Pac. **129**, 058012 (2017)
L.L. Williams, G.P. Zank, J. Geophys. Res. **99**, 19229 (1994)
B.J. Williams et al., Astrophys. J. **687**, 1054 (2008)
P.F. Winkler, R.P. Kirshner, Astrophys. J. **299**, 981 (1985)
P.F. Winkler, K. Twelker, C.N. Reith, K.S. Long, Astrophys. J. **692**, 1489 (2009)
H. Yamaguchi, K. Eriksen, C. Badenes, J.O. Hughes, N. Brickhouse, A. Foster, D.J. Patnaude, R. Petre, P.O. Slane, R.K. Smith, Astrophys. J. **780**, 136 (2014)
G. Zimbardo, Planet. Space Sci. **59**, 468 (2011)
V.N. Zirakashvili, F.A. Aharonian, Astrophys. J. **708**, 965 (2010)
V.N. Zirakashvili, F.A. Aharonian, R. Yang, E. Oña-Wilhelmi, R.J. Tuffs, Astrophys. J. **785**, 130 (2014)

Space Sci Rev (2018) 214:62
https://doi.org/10.1007/s11214-018-0494-5

Nucleosynthesis in Supernovae

Friedrich-Karl Thielemann[1,2] · Jordi Isern[3] · Albino Perego[4] · Peter von Ballmoos[5]

Received: 29 December 2017 / Accepted: 12 March 2018 / Published online: 26 March 2018

Abstract We present the status and open problems of nucleosynthesis in supernova explosions of both types, responsible for the production of the intermediate mass, Fe-group and heavier elements (with the exception of the main s-process). Constraints from observations can be provided through individual supernovae (SNe) or their remnants (e.g. via spectra and gamma-rays of decaying unstable isotopes) and through surface abundances of stars which witness the composition of the interstellar gas at their formation. With a changing fraction of elements heavier than He in these stars (known as metallicity) the evolution of the nucleosynthesis in galaxies over time can be determined. A complementary way, related to gamma-rays from radioactive decays, is the observation of positrons released in β^+-decays, as e.g. from ^{26}Al, ^{44}Ti, ^{56,57}Ni and possibly further isotopes of their decay chains (in competition with the production of e^+e^- pairs in acceleration shocks from SN remnants, pulsars, magnetars or even of particle physics origin). We discuss (a) the role of the core-collapse

Supernovae
Edited by Andrei Bykov, Roger Chevalier, John Raymond, Friedrich-Karl Thielemann, Maurizio Falanga and Rudolf von Steiger

✉ F.-K. Thielemann
f-k.thielemann@unibas.ch

J. Isern
isern@ice.cat

A. Perego
albino.perego@mib.infn.it

P. von Ballmoos
pvb@irap.omp.eu

1 Dept of Physics, Univ. of Basel, 4056 Basel, Switzerland

2 GSI Darmstadt, 64291 Darmstadt, Germany

3 Institut de Ciències de l'Espai (ICE, CSIC) & Institut d'Estudis Espacials de Catalunya (IEEC), Campus UAB Bellaterra, 08193 Barcelona, Spain

4 INFN, Sezione di Milano-Bicocca, gruppo collegato di Parma, Parco Area delle Scienze 7/A, 43124 Parma, Italy

5 Institut de Recherche en Astrophysique et Planétologie (UPS/CNRS), 31028 Toulouse, France

supernova explosion mechanism for the composition of intermediate mass, Fe-group (and heavier?) ejecta, (b) the transition from neutron stars to black holes as the final result of the collapse of massive stars, and the relation of the latter to supernovae, faint supernovae, and gamma-ray bursts/hypernovae, (c) Type Ia supernovae and their nucleosynthesis (e.g. addressing the ^{55}Mn puzzle), plus (d) further constraints from galactic evolution, γ-ray and positron observations. This is complemented by the role of rare magneto-rotational supernovae (related to magnetars) in comparison with the nucleosynthesis of compact binary mergers, especially with respect to forming the heaviest r-process elements in galactic evolution.

Keywords Massive stars · Core-collapse supernovae · Type Ia supernovae · Nucleosynthesis · γ-ray observations · Positron sources · Galactic chemical evolution

1 Introduction

The observation of low metallicity stars tells us that all elements, except for H, He, and Li (produced in the Big Bang), have a stellar origin. The production of C, N, and s-process (slow neutron capture) elements is related to **low and intermediate mass stars**, their winds and their wind ejecta in planetary nebulae. Especially the s-process elements (and also some lighter odd-Z elements) are linked to metallicity-dependent neutron sources and/or seed distributions on which neutron captures can act. ^{22}Ne (acting via ^{22}Ne$(\alpha, n)^{25}$Mg) goes back to the initial CNO content, i.e. metallicity of the star, turning all CNO-isotopes via the CNO-cycle to ^{14}N during H-burning, which is processed in subsequent He-burning into ^{22}Ne via ^{14}N$(\alpha,\gamma)^{18}$F$(\beta^+)^{18}$O(α,γ). The alternative neutron source ^{13}C, produced via mixing of H into He-burning zones, is not metallicity-dependent, but the metallicity-dependent seed composition (including possible neutron poisons) and the amount of seeds is decisive for the resulting s-abundance distribution. The observed ratio of alpha-elements, like O, Ne, Mg, Si, S, Ar, Ca, and Ti, to Fe ($[\alpha/\text{Fe}] = \log_{10}((\alpha/\text{Fe})/(\alpha/\text{Fe})_\odot)$) as a function of galactic time or metallicity can be explained by the **early dominance of massive stars**, reaching their evolutionary endpoints faster than low and intermediate mass stars (ending as white dwarfs), which produce in core-collapse supernovae (CCSNe) α/Fe ratios on average a factor of 2–3 higher than in the solar composition, i.e. $[\alpha/\text{Fe}] \approx 0.5$. This turns over when delayed binary evolution—involving white dwarfs (the endpoint of stars with initial masses below 8 $M_\odot$)—leads to Type Ia supernova (SNIa) events, which produce large amounts of ^{56}Ni (decaying to ^{56}Fe) and small amounts of alpha-elements (and among them only elements from Si onwards).

While this general picture has been understood for quite a while, major open questions are related to the precise amounts of alpha-elements produced in CCSNe, the detailed composition of the Fe-group and the respective role of SNeIa vs. the core-collapse of massive stars. Such aspects can be constrained by observations of γ-ray lines from radioactive decays and positron emission from β^+-decays related to individual explosions, complemented by the element abundances of (old) stars with varying metallicities. The latter can clarify the origin of the change from fast evolving progenitors (i.e. massive stars) to later emerging progenitors (related to a binary origin), being accompanied by the changing ratio of their α/Fe ejecta.

Especially the origin of r-process nuclei has been riddled with uncertainties. They have (when averaged) a similar behavior as alpha-elements with respect to Fe. However, the large scatter of up to two orders of magnitude, observed e.g. for [Eu/Fe] at low metallicities, points to rare events being responsible for their production. This is also supported by the fact that we see recent additions of ^{60}Fe (of massive star origin) in deep sea sediments, while this is not the case for ^{244}Pu, a heavy r-process element.

The present article addresses these puzzles and how they enter the element abundances in the "chemical" evolution of galaxies. This includes the role of Type Ia supernovae in producing heavy alpha- and Fe-group elements, the transition from supernova explosions to failed/faint supernovae and hypernovae (forming black holes), the role of core-collapse supernovae in the understanding of some intermediate mass and the Fe-group elements, and explores also the origin of heavier nuclei, which are related to the νp-process and the r-process, passing through all possibly responsible sites. This also includes the finding from low metallicity stars that (i) Zn and Ge are correlated with the Fe-group (i.e. must be co-produced with Ni/Fe), (ii) Sr, Y, Zr show a weak correlation (i.e. they are co-produced with the Fe-group to some extend), while (iii) heavy r-process elements show no correlation with the Fe-group (i.e. their major production seems not related to regular supernovae).

2 End-Stages of Massive Stars and Core-Collapse Supernovae

Stars beyond 8 $M_\odot$ undergo core collapse, either up to about 10 $M_\odot$ due to electron capture on C-burning products in the O-Ne-Mg core, resulting in fast contraction and the formation of an Fe-core during collapse (dubbed electron capture or EC supernovae) or via core collapse after central Si-burning. EC supernovae produce small amounts of alpha-elements and Fe-peak elements (see e.g. Wanajo et al. 2009, 2011; Hüdepohl et al. 2010; Moriya et al. 2014; Jones et al. 2016; Müller et al. 2017). 10–90 $M_\odot$ stars undergo Fe-core collapse. A major question is how the transition occurs from a "regular" core-collapse supernova (CCSNe), producing a neutron star, to the formation of a central black hole. This depends strongly on the properties of the stellar progenitor, among other properties the compactness, i.e. the central mass concentration is important. In case of fast rotation and strong magnetic fields a so-called hypernovae/long duration gamma-ray bursts (GRBs) possibly occurs after black hole formation, otherwise this leads to a failed or faint supernova and "quiet" black hole formation. For the nucleosynthesis in GRBs/hypernovae, aspherical explosions are important. This is also the case if fast rotation and strong magnetic fields are present in core collapse, leading, however, finally to a central neutron star with magnetic fields as high as 10^{15} Gauss, known as magnetar (Winteler et al. 2012; Mösta et al. 2014, 2015, 2017; Nishimura et al. 2015, 2017; Halevi and Mösta 2018). 90–140 $M_\odot$ stars undergo pulsational nuclear instabilities at various nuclear burning stages, including O and Si-burning. 140–300 $M_\odot$ stars become pair-instability supernovae, if the mass loss is small enough to permit this final end-stage. Very massive stars greater than 300 $M_\odot$ undergo core-collapse to form intermediate mass black holes. Detailed reviews and recent findings on the present understanding of the end stages of massive stars are given e.g. in Heger and Woosley (2010), Chieffi and Limongi (2013), Nomoto et al. (2006, 2013), Karakas and Lattanzio (2014), Sukhbold et al. (2016), Ebinger et al. (2017).

CCSNe contribute to galactic evolution (a) via their wind ejecta during stellar evolution, and in case of a final explosion via (b) ejecta of essentially unburned matter from the outer stellar zones and (c) explosively processed matter from the innermost ejected mass zones. ^{60}Fe (half-life 2.6×10^6 y) is an example for (b) and goes back to hydrostatic burning stages (Limongi and Chieffi 2006, 2012; Woosley and Heger 2007; Maeder and Meynet 2012). It is produced in He-shell burning of massive stars in late phases after core C-burning and ejected afterwards in CCSNe. Recent findings show that it can witness the last CCSNe near the solar system about 2 to 3 million years ago (Knie et al. 2004; Firestone 2014; Wallner et al. 2015, 2016; Fimiani et al. 2016; Ludwig et al. 2016; Sørensen et al. 2017).

The nucleosynthesis yields of CCSNe are characterized by an average ejected ^{56}Ni mass of about 0.1 $M_\odot$ and strong contributions to the so-called alpha elements O, Ne, Mg, Si, S, Ar, Ca, and Ti. They result either from hydrostatic burning in stellar evolution (O, Ne, Mg, and some Si) or from explosive burning (Si, S, Ar, Ca, and Ti) (Chieffi and Limongi 2013, 2017). There exist observational indications that also Ge and Zn are co-produced with the Fe-group (Cowan et al. 2005). In the νp-process CCSNe can contribute some nuclei beyond the Fe-group like Sr, Y, Zr, (e.g. Fröhlich et al. 2006a,b; Wanajo 2006; Pruet et al. 2006; Eichler et al. 2018). In a weak s-process massive stars can produce nuclei up to $A = 100$, and even beyond (Frischknecht et al. 2016). One of the major open questions is whether they can also be a dominant site for r-process nucleosynthesis.

While most of this information comes from observations (e.g. Branch and Wheeler 2017; Cayrel et al. 2004; Kretschmer et al. 2013; Diehl et al. 2015; Isern et al. 2016), the goal is to predict and explain the composition of nucleosynthetic ejecta from self-consistent modeling. Thus, the computation of comprehensive nucleosynthesis yields from core-collapse supernovae relies on detailed numerical models of stellar explosion and matter ejection. Tracer particles (in case of multi-dimensional simulations) or spherically symmetric, radial mass zones (in case of one-dimensional simulations), modelled as Lagrangian mass elements within the fluid, are usually employed inside these models to record the thermodynamic history of ejected fluid elements. In most cases this step is done in the form of "postprocessing", after (or in parallel to) the successful radiation-hydrodynamic modelling with a limited nuclear network or within nuclear statistical equilibrium (if applicable), being sufficient to provide the correct energy generation for the hydrodynamic evolution. The precise changes of the nuclear composition in these tracer fluid elements is obtained during the postprocessing via large nuclear networks that comprise the relevant nuclear species and reactions. Such an analysis permits to address fundamental questions, like the following: What are the conditions for explosive nucleosynthesis in stellar layers as a function of progenitor properties? How do the nucleosynthesis yields vary as a function of progenitor mass and metallicity? What is the connection between the progenitor model and the compact remnant (neutron star or black hole)?

Neutrino-matter interactions and a microphysical description of the hot and dense matter are key ingredients in these calculations, since they directly affect the composition of the innermost ejecta and, in particular, the relative abundances of neutrons and protons at the onset of the expansion. The strength of the shock wave (i.e., the explosion energy) sets the conditions at which explosive nucleosynthesis takes place during its passage through the different stellar layers, for several seconds after the explosion has been launched and hundreds up to thousands of kilometers from the center. Adequate numerical models are ideally required to catch all the relevant aspects influencing the properties of the ejecta and of the explosion, as well as their dependence on the progenitor stars.

Soon after the discovery that the photodissociation of nuclei and the intense neutrino emission, associated with the shock expansion inside the iron core, prevent a prompt explosion and lead to shock stalling, it was recognized that neutrino energy deposition just behind the shock can help reviving the explosion. However, self-consistent spherically symmetric models, including detailed microphysics and neutrino transport, fail in general to explode, with the exception of a few light progenitors. Multi-dimensional effects, including convection, turbulence and shock hydrodynamic instabilities (e.g., SASI), increase the time that a fluid particle spends inside the so-called gain region, in which electron neutrinos (ν_e's) and antineutrinos ($\bar{\nu}_e$) are absorbed and their energy is deposited, heating up matter in this

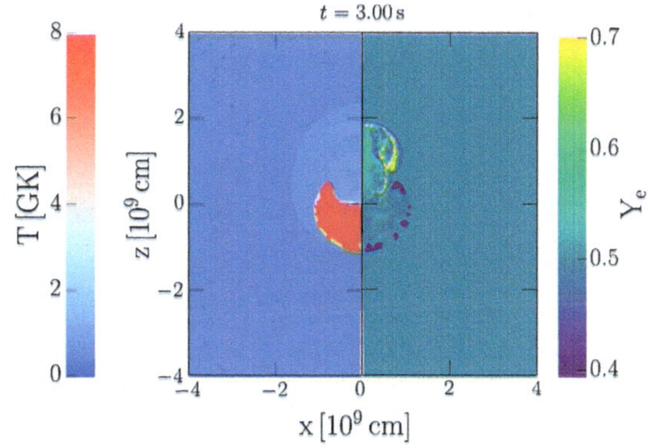

Fig. 1 Distribution of temperature and Y_e during the SN explosion of a 12.7 $M_\odot$ star at $t = 3$ s, based on a 2D simulation (Nakamura et al. 2015). It can be seen that Y_e values > 0.5 and < 0.5 exists, dependent on the initial core-collapse phase with high electron capture effects and electron neutrino and antineutrino absorption on neutrons and protons during the explosion, see Eichler et al. (2018)

region. This causes a pressure increase and shock revival. Recent multi-dimensional simulations have confirmed that these effects provide more favorable conditions for successful explosions than in spherically symmetric simulations (see next subsection).

2.1 Present Multi-Dimensional Approaches for Neutrino-Driven Explosions

2.1.1 SN-Explosions

The full solution to the CCSN problem in a self-consistent way has not converged, yet. There exists a growing set of 2D and 3D CCSN explosions, see e.g. Janka (2012), Burrows (2013), Nakamura et al. (2015), Janka et al. (2016), Bruenn et al. (2016) and the progress of active groups in Garching/Belfast/Monash, Princeton/Caltech/MSU, Oak Ridge/U. of Tennessee/Berkeley, Tokyo/Kyushu, Paris, and Basel, but it is still too early to provide complete nucleosynthesis predictions from self-consistent multi-D simulations. A few preliminary nucleosynthesis results exist, e.g. Pruet et al. (2006), Wanajo et al. (2011), Harris et al. (2017), Wongwathanarat et al. (2017), Yoshida et al. (2017), Eichler et al. (2018). We show here the results of a detailed nucleosynthesis study by Eichler et al. (2018), based on long-term, two-dimensional core-collapse supernova simulations of an 11.2 $M_\odot$ star (Nakamura et al. 2015). This CCSN model shows an axis-symmetric neutrino-driven explosions of a non-rotating, solar metallicity model. The numerical treatment was described in Nakamura et al. (2015), including spectral transport of electron and anti-electron neutrinos, using the isotropic diffusion source approximation (IDSA Liebendörfer et al. 2009) and a Lattimer & Swesty equation of state (Lattimer and Douglas Swesty 1991) with compressibility $K = 220$ MeV. After a successful shock revival, the axis-symmetric nature of the simulation causes a preference of the outflow towards polar directions (see Fig. 1).

This behavior is usually not observed in 3D simulations of regular CCSNe, since they do not have an imposed symmetry. Explosion times and energies are also different between 2D and 3D simulations, with 2D simulations being usually "more optimistic" and on the high side, as is shown e.g., in Takiwaki et al. (2014), Lentz et al. (2015), Melson et al. (2015), Janka et al. (2016), Hix et al. (2016). In the discussed simulations it is found that nuclei well beyond the iron group (up to $Z \approx 44$) can be produced, including the p-nuclei ^{92,94}Mo and ^{96,98}Ru (see Fig. 2). While ^{92}Mo and ^{94}Mo can be produced in slightly neutron-rich conditions, ^{96,98}Ru can only be produced efficiently via the νp-process, which depends heavily on the presence of very proton-rich material in the ejecta, obtained in polar outflows. Fig. 1 shows the Y_e distribution of ejecta 3 s after bounce. One realizes low Y_e material

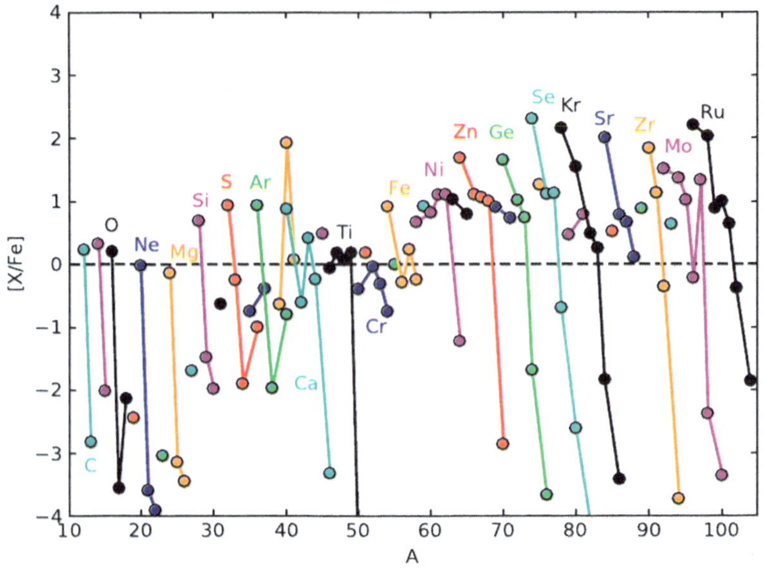

Fig. 2 Nucleosynthesis features of a 12.7 $M_\odot$ SN explosion from a 2D simulation (Nakamura et al. 2015), featuring a strong νp-process with the production of nuclei up to $A = 100$ (see also Eichler et al. 2018)

(< 0.5) in early outflows (typically at the outer edges of ejecta) which still have values close to the ones inherited from core collapse. High electron densities, caused by high Fermi energies of (degenerate) electrons, lead to neutron-rich conditions due to electron captures on free protons. High Y_e material is produced in subsequent ejections, following behind the earlier ejecta, which are strongly affected by neutrino and anti-neutrino captures on protons and neutrons, according to the following reaction sequences

$$\nu_e + n \to p + e^- \tag{1}$$

$$\bar{\nu}_e + p \to n + e^+. \tag{2}$$

These reactions turn matter only neutron-rich if the average anti-neutrino energy $\langle \epsilon_{\bar{\nu}_e} \rangle$ is higher than the average neutrino energy $\langle \epsilon_{\nu_e} \rangle$ by 4 times the neutron-proton mass difference Δ for similar (electron) neutrino L_{ν_e} and anti-neutrino $L_{\bar{\nu}_e}$ luminosities. This was pointed out initially in Qian and Woosley (1996), leading—when approaching equilibrium conditions for neutrino and antineutrino captures—to

$$Y_e = \left[1 + \frac{L_{\bar{\nu}_e}(\langle \epsilon_{\bar{\nu}_e} \rangle - 2\Delta + 1.2\Delta^2 / \langle \epsilon_{\bar{\nu}_e} \rangle)}{L_{\nu_e}(\langle \epsilon_{\nu_e} \rangle + 2\Delta + 1.2\Delta^2 / \langle \epsilon_{\nu_e} \rangle)} \right]^{-1}. \tag{3}$$

Further details and in-medium corrections for neutrons and protons in comparison to their treatment as free particles are given in Martínez-Pinedo et al. (2012), Roberts et al. (2012). Thus, in most cases the energetically favorable first reaction (1) wins, changing Y_e from the initial (neutron-rich) conditions towards values beyond $Y_e = 0.5$. The example shown here features a nice/strong νp-process caused by $Y_e > 0.5$ conditions along polar outflows (see Fig. 2). In such a case ^{64}Ge is produced in slightly proton-rich conditions of explosive Si-burning. The long beta-decay half-life of ^{64}Ge (which would prevent further processing via proton captures on ^{64}Ga to higher nuclear mass numbers) can be circumvented by an

(n, p)-reaction, made possible by neutrons created in antineutrino captures on existing protons in this proton-rich matter. Further proton-captures and fast beta-decays can lead to the production of nuclear masses up to $A = 100$ (Fröhlich et al. 2006a,b).

2.1.2 *Black Hole Formation*

While the above result is an example for the possible outcome of core-collapse supernova explosions, one of the major questions is at which progenitor mass the turnover takes place from successful explosions with neutron star formation to black holes as the final outcome (if there exits such a clear limit!). Here we show another case of a 2D simulation which leads to black hole formation after core-collapse of a non-rotating 40 $M_\odot$ progenitor star (Pan et al. 2017). This study also makes use of the isotropic diffusion source approximation (IDSA) for the transport of electron flavor neutrinos and a modified gravitational potential for general relativistic effects. It is performed for four different neutron star equations of state (EoS), including LS220, SFHo, BHB$\Lambda\Phi$ and DD2, examining the impact of the equation of state on black hole formation dynamics and gravitational wave emissions. The simulations utilize the FLASH code (Fryxell et al. 2000).

It is found that the black hole formation time is sensitive to the equation of state and is delayed in multiple dimensions due to finite entropy effects that enlarge the maximum proto-neutron star (PNS) mass and via proto-neutron star convection. Depending on the equation of state, these simulations also show the possibility that the shock is revived together with the formation of a black hole. Figure 3 displays the behavior of the average shock radius as a function of time. In Fig. 4 the multi-D behavior is indicated, utilizing the EoS SFHo (Oertel et al. 2017). Convective regions, where neutrino heating takes place, are characterized by negative Brunt-Väisälä frequencies. Here not only the average but also the maximum shock radius is given as a function of time. We see that the radius of the PNS recedes, finally forming a black hole.

Based on the simulations for four different nuclear EoSs, also the gravitational waveforms are provided for four different EoSs in Fig. 5. The gravitational waveforms contain four major features: (1) a low frequency signal from the core-bounce and prompt convection, (2) a strong signal from the PNS g-mode oscillation, (3) a high frequency signal from the PNS inner core convection, and (4) signals from the standing accretion shock instabilities and convections. The peak frequency at the onset of black hole formation reaches 2.3 kHz. We see that the signal extends over different time intervals up to black hole formation, dependent on the EoS utilized.

The characteristic amplitude of such a signal is detectable at peak frequency up to 10 kpc, but close to the noise threshold of the Advanced LIGO and KAGRA gravitational wave detectors, suggesting that the next generation gravitational wave detectors will need to improve the sensitivity in the kHz domain to better observe stellar mass black hole formation from failed supernovae.

2.2 1D Spherically Symmetric Simulations

As discussed in the previous subsection and some recent literature, some individual predictions for multi-D supernovae nucleosynthesis are presently available. But large-scale nucleosynthesis predictions for huge sets of stellar progenitors are still too time-consuming and somewhat risky, as the methods seem not converged, yet. In addition, multi-dimensional simulations are significantly more computationally intensive than spherically symmetric

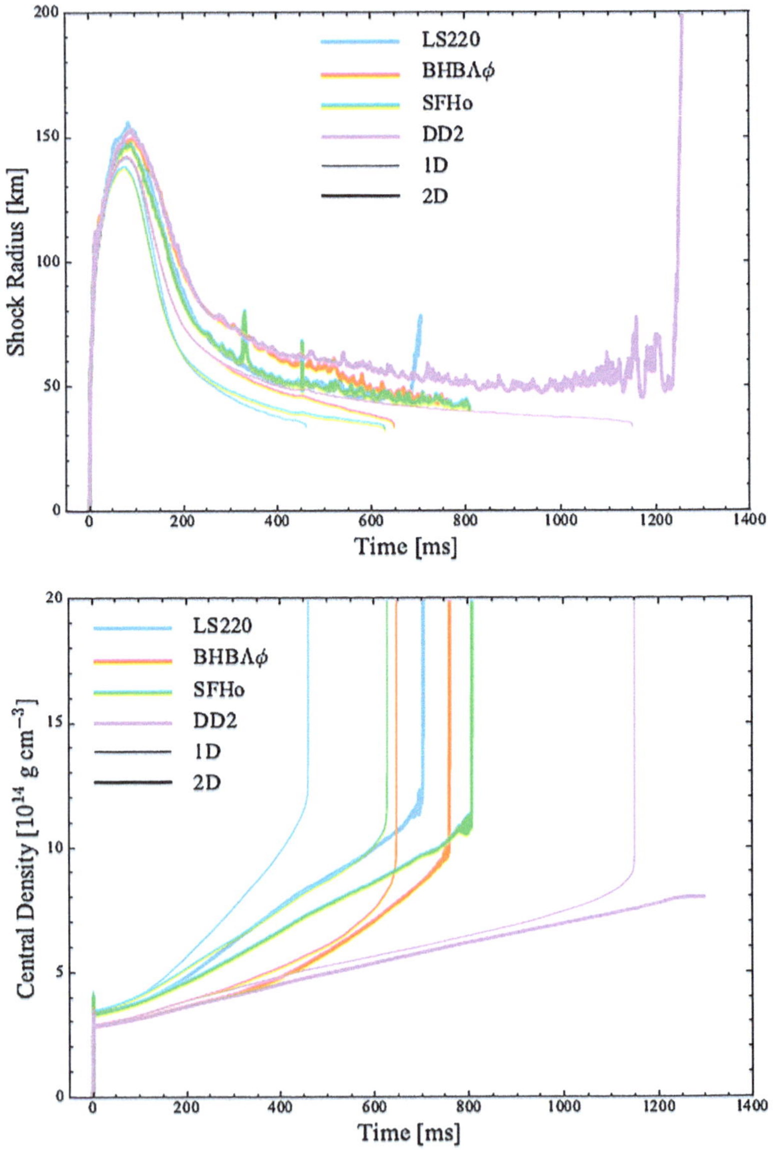

Fig. 3 Time evolution of averaged shock radius (top) and central density (bottom). Different colors represent simulations with different nuclear EoS. Thick and thin lines indicate simulations in 2D and 1D respectively. Except for EoS DD2, all utilized EoSs lead to black hole formation (Pan et al. 2017)

ones, in particular if large spatial domains and simulations are required to follow the evolution for long timescales until the explosion sets in. Therefore, the need of exploring large sets of progenitors makes systematic studies in multi-dimension still prohibitive. Parametrized explosions induced in spherically symmetric models are still a pragmatic and valuable approach to study large numbers of stellar progenitors, from the onset of the explosion up to several seconds after core bounce. But these methods have difficulties in reproducing

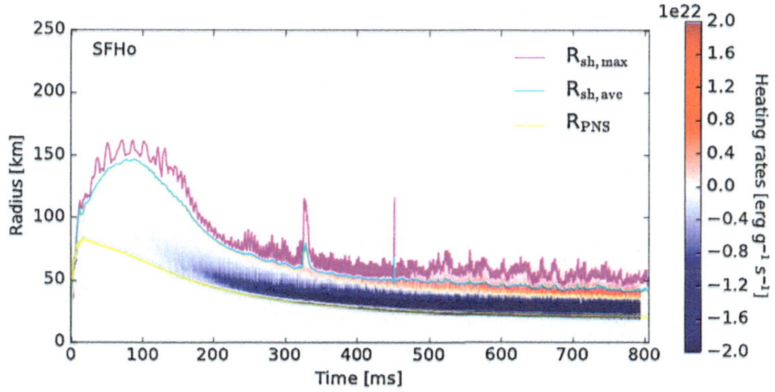

Fig. 4 Color maps of the Brunt-Väisälä frequency as a function of time and radius. Negative values, indicated by the red color represent unstable regions based on the Ledoux criterion for convection. These are the regions where convective overturn takes place and neutrino heating is active. The magenta and light blue lines show the maximum and averaged shock radius. The yellow line indicates the radius of the PNS (Pan et al. 2017)

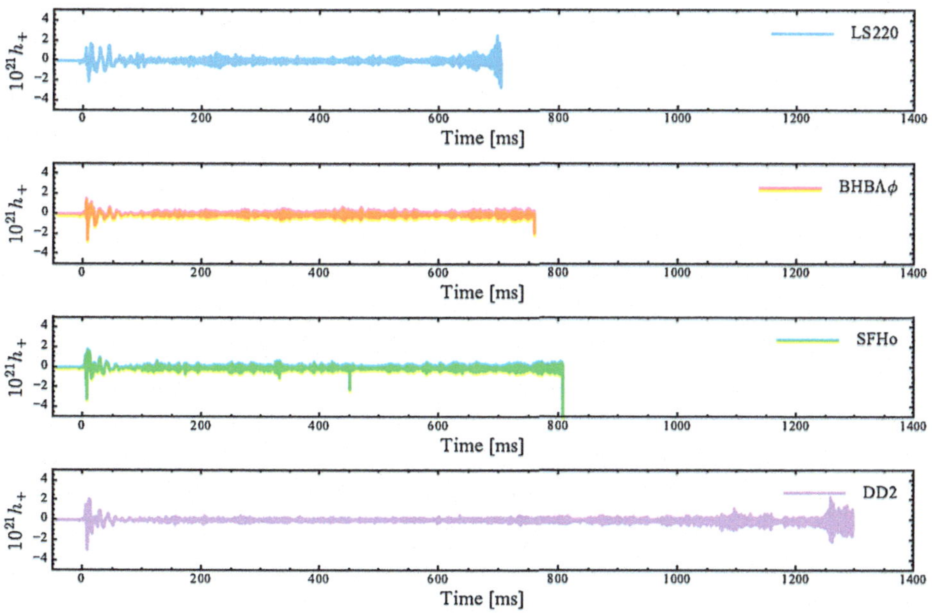

Fig. 5 GW amplitudes as functions of time after core bounce with an assumption of 10 kpc distance from the source. From top to bottom, each panel shows 2D simulations with LS220, BHB$\Lambda\Phi$, SFHo, and DD2 EoS, respectively. In the top three panels, a BH is formed at the end of the simulation (Pan et al. 2017)

some characteristic features of multi-dimensional explosions like, for example, the coexistence of accretion flows and expanding plumes for a few seconds after the onset of the explosion, and the related injection of the explosion energy. However, despite their marked fluid instabilities, the most realistic three dimensional simulations show a more pronounced sphericity than axis-symmetric simulations. Hence their average conditions resemble more closely the shock expansion that would be obtained by an exploding spherically symmet-

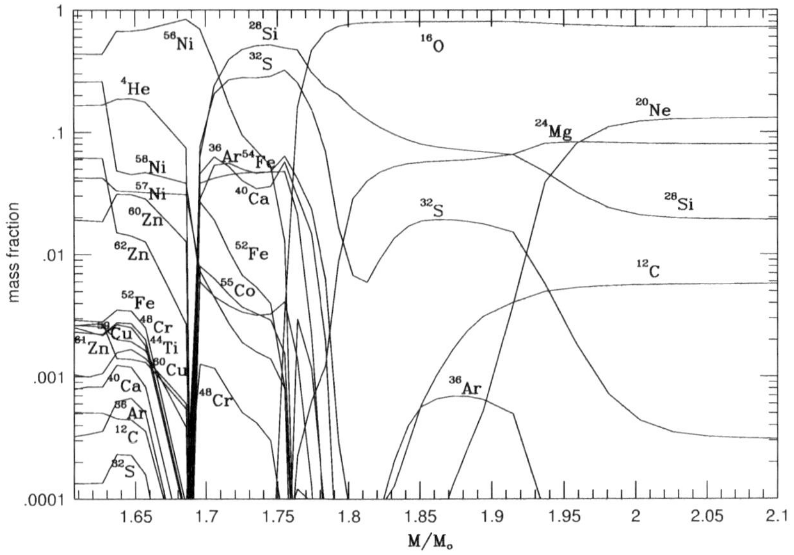

Fig. 6 Mass fractions of a few major nuclei after passage of the supernova shock front through a star with an initial mass of 20 $M_\odot$. Matter outside 2 $M_\odot$ is essentially unaltered. Mass zones further in experience explosive Si, O, Ne, and C-burning. For ejecting 0.07 $M_\odot$ of ^{56}Ni the mass cut between neutron star and ejecta is required to be located at 1.6 $M_\odot$ (from Thielemann et al. 1996)

ric model. Thus, this supports spherically symmetric studies, but the enhanced heating via neutrino and antineutrino absorption has to be incorporated.

2.2.1 Early Spherically Symmetric Approaches

In the past some simplified approaches have been used, which artificially induce explosions with estimated typical explosion energies. The first attempts to predict supernova nucleosynthesis from artificially triggered explosions were performed in the context of induced pistons (e.g. Woosley and Weaver 1995), thermal energy bomb models (e.g. Thielemann et al. 1996; Umeda and Nomoto 2008, and Fig. 6), or kinetic energy bomb models (e.g. Limongi and Chieffi 2006, 2012; Chieffi and Limongi 2013, 2017). In the first case, the motion of a mass shell is imposed along a ballistic trajectory with a typical explosion energy of $E = 1.2\mathrm{B} = 1.2 \times 10^{51}$ erg at a radial position related to a value of the entropy that was expected to result in the most realistic mass cut (i.e., the bifurcation between the PNS and the ejecta). In the latter cases, explosions are triggered by adding typical kinetic or thermal energies to a specific mass zone (usually in deeper zones). The mass cut was then determined by integrating the nucleosynthesis yields from the outside inwards, up to the point where the observationally indicated amount of ^{56}Ni was reached. In all cases the explosion energy is not determined self-consistently, and the physics of the explosion is not included. The mass cut and the explosion energy are free parameters of the model and have to be constrained from the yields of the innermost ejecta, separately for each progenitor. These approaches are suitable for the outer layers, where the nucleosynthesis mostly depends on the strength of the shock wave.

These outer layers are characterized by explosive C and Ne-burning (producing O, Mg, and some Si), further in by explosive O-burning (producing Si, S, and also Ca and Ar), while

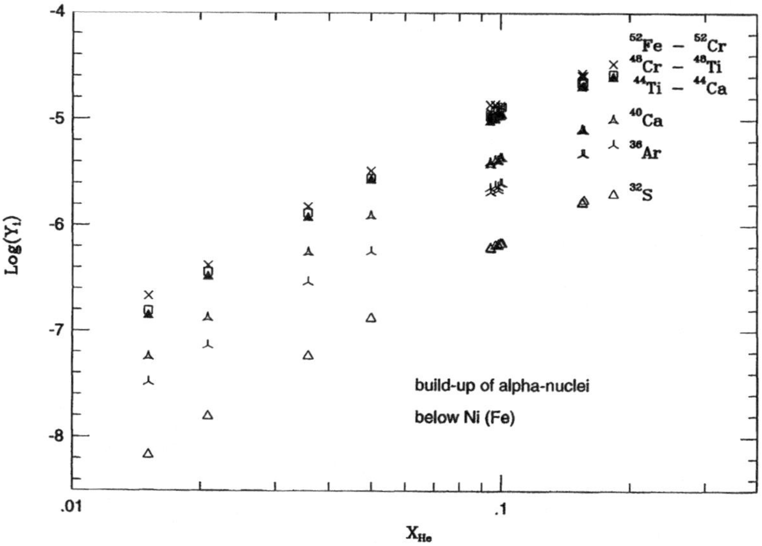

Fig. 7 Final abundances of alpha-nuclei from ^{32}S to ^{52}Fe as a function of the final He mass fraction, which is a measure of the alpha-rich freeze-out from explosive Si-burning and thus the experienced entropy (from Thielemann et al. 1996)

explosive Si-burning has ^{56}Ni as the major burning product. However, it can be realized, that in the inner ejecta large amounts of He are retained. This is due to explosive Si-burning with a so-called alpha-rich freeze-out, related to the fact that in the shock most matter is first disintegrated in neutrons, protons, and alpha-particles, and the build-up occurs only during the subsequent cooling and expansion. If the densities are too low, the triple-alpha reaction and competing reactions like e.g. $\alpha\alpha n$, which eventually pass the bottle neck to carbon production, are not effective enough to permit all matter to reach nuclei in the Fe-group with the highest binding energy, like ^{56}Ni. Such an alpha-rich freeze-out of matter which initially attained nuclear statistical equilibrium will retain also alpha-nuclei along the chain to ^{56}Ni, like ^{32}S, ^{36}Ar, ^{40}Ca, ^{44}Ti. ^{48}Cr, and ^{52}Fe, as can be seen in the inner ejecta displayed in Fig. 6 and in more detail in Fig. 7, where the final abundances of these nuclei are plotted as a function of the final He abundance, which again is related to T^3/ρ, a quantity proportional to the entropy of a radiation-dominated plasma. Thus, an increasing entropy of matter during explosive Si-burning stands for an increasing strength of the alpha-rich freeze-out, and thus the abundances of these nuclei, in tandem with the He-abundance.

However, the conditions in the innermost layers depend also significantly on the physics of the collapse, on the details of the explosion, and, in particular, on the interaction between matter and neutrinos, which were ignored in these approaches. As neutrinos play an important role, the neutrino and anti-neutrino interaction with matter, which determines Y_e, is missing. When comparing piston results (e.g. Heger and Woosley 2010), integrated over a standard IMF from 10 to 100 $M_\odot$, with a representative sample of low metallicity stars (Cayrel et al. 2004), i.e. representing CCSN yields, this leads to an underproduction of Sc, Ti, Co and Zn when utilizing zero metallicity stellar models. For higher metallicities, nuclei beyond Ni are produced (in the s-process), but the dominant fraction of solar Zn (^{64}Zn) cannot by made this way. All such prior results made use of the initial stellar structure (and Y_e!), when inducing artificial explosions. Neglecting the effect of the explosion mechanism on the innermost zones, causes strange overproductions of some neutron-rich Ni isotopes

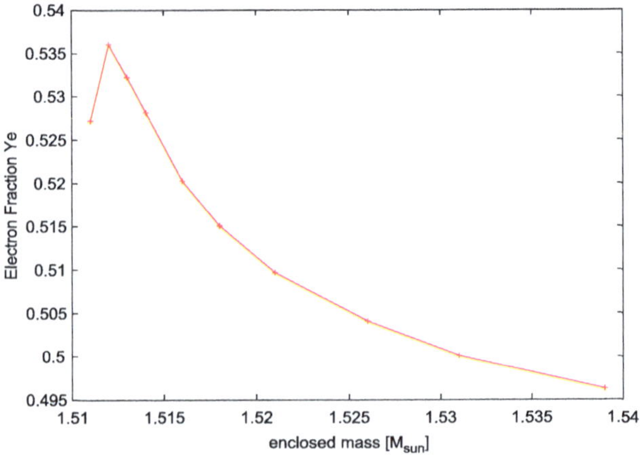

Fig. 8 Y_e of the innermost ejecta due to neutrino interactions with matter. At high temperatures electrons are not degenerate, thus the reduction of Y_e due to electron captures is ineffective. For similar neutrino and antineutrino spectra the neutron-proton mass difference favors $\nu_e + n \leftrightarrow p + e^-$ over $\bar{\nu}_e + p \leftrightarrow n + e^+$ (from Fröhlich et al. 2006a)

and does not permit go much beyond Ni. Three aspects are important and can in principle only be solved by self-consistent explosion models in 3D: (a) predicting a consistent explosion energy, (b) considering the neutrino interactions with nucleons and nuclei, which strongly effect Y_e, even in spherical symmetry, and (c) on top of this, including the role of multi-D effects.

Interim approaches beyond piston or thermal bomb models of Woosley and Heger (2007), Heger and Woosley (2010), Limongi and Chieffi (2006, 2012), Nomoto et al. (2006, 2013), Nomoto (2017), Chieffi and Limongi (2017) exist by now, trying to mimic multi-D neutrino heating in a spherical approach, in order to obtain more appropriate predictions of the explosion energy, mass cut between neutron star and ejecta, as well as nucleosynthesis (including the effects of neutrinos on Y_e, the proton/nucleon ratio). This includes the "neutrino light-bulb" method, where the PNS is excised and replaced with an inner boundary condition which contains an analytical prescription for the neutrino luminosities. Suitable choices of the neutrino luminosities and energies can trigger neutrino-driven explosions (e.g. Yamasaki and Yamada 2005; Iwakami et al. 2008; Yamamoto et al. 2013). In "absorption methods" (Fröhlich et al. 2006a,b; Fischer et al. 2010) the increase in the neutrino energy deposition is obtained by modifying the neutrino opacities in spherically symmetric models with detailed Boltzmann neutrino transport, i.e. this means that neutrino-capture rates are multiplied by a factor, causing additional ν-heating, in order to trigger explosions. This led to the prediction of Y_e-values affected by the central neutrino flux (see Fig. 8), improving strongly the Fe-group composition due to the discovery that proton-rich conditions are attained in the innermost ejecta (see the discussion on the effect of neutrino and antineutrino capture in Sect. 2.2.2).

Another outcome of these investigations was the discovery of the νp-process, also discussed in Sect. 2.2.2. Both aspects led to an improvement of nucleosynthesis predictions for Fe-group nuclei and opened an explanation for understanding light p-nuclei, which cannot be produced in the typical p/γ-process in explosive burning of outer stellar shells, i.e. the Ne-shell. The major drawback of these methods is, however, either the explicit modification of the electron neutrino and antineutrino luminosities or the modification of neutrino opacities, which (while apparently improving nucleosynthesis results) could directly impact the composition of the innermost ejecta in a non-consistent way.

Thus, one is either constrained by artificially initiated explosions of the piston or bomb type, with estimated typical explosion energies of 10^{51} erg, which do not treat at all the innermost ejecta correctly (utilizing inconsistent Y_e-values from the pre-collapse stellar models), or one suffers from treatments which include neutrino interactions with the innermost ejecta, but the estimated neutrino luminosities or opacities do not yet treat these effects correctly. Realistic explosion scenarios, including a full treatment of weak interactions, are affected by (electron) neutrino and anti-neutrino captures on neutrons and protons. If the neutrino flux is sufficient to have an effect (as it scales with $1/r^2$), and the total luminosities are comparable for neutrinos and anti-neutrinos, only conditions with $\langle \epsilon_{\bar{\nu}_e} \rangle - \langle \epsilon_{\nu_e} \rangle > 4(m_n - m_p)c^2 = 4\Delta$ lead to $Y_e < 0.5$ (see Eq. (3) when expanding the ratio in a series). Present calculations (Martínez-Pinedo et al. 2012; Roberts et al. 2012; Arcones and Thielemann 2013) seem therefore to exclude (at least strong) r-process conditions in CCSNe. Otherwise the interaction with neutrinos leads to proton-rich conditions. The latter favors improvements in the Fe-group composition of elements like Sc, Ti, Co, as well as for ^{64}Ge (decaying to ^{64}Zn!), and the νp-process, which can produce nuclei up to Sr, Y, Zr and Mo (see e.g. Fröhlich et al. 2006a), being a treatment within the type of the above-mentioned absorption methods. Figure 2 has shown this effect from a more realistic multi-D approach. In general, explosive Si-burning with a strong alpha-rich freeze-out (dependent on the explosion energy/entropy attained) favors—for identical Y_e conditions—those Fe-group nuclei which would result from additional alpha-captures on the nuclei produced in a normal freeze-out from explosive Si-burning. Thus, ^{56}Ni can be moved up to ^{64}Ge (decaying to ^{64}Zn) or ^{54}Fe up to ^{58}Ni. The first effect takes place for very high explosion energies, only attained in hypernovae, the second one in all regular CCSNe. Thus, ^{64}Ge (resp. ^{64}Zn) can result from Y_e values > 0.5 in regular CCSNe (due to neutrino interactions with the innermost ejecta, see above), as well as in highly energetic hypernovae.

2.2.2 *Recent Improved Approaches*

Recent (spherically symmetric) approaches try to mimic the effect of multi-D neutrino transport in a way adapted more consistently to core-collapse and proto-neutron star accretion. They need, however, calibrations which can be provided by comparison with a variety of observations of explosion energies, deduced ejected ^{56}Ni-masses, and progenitor properties. Ugliano et al. (2012) presented a more sophisticated light-bulb method to explode spherically symmetric models, using neutrino energy deposition in post-shock layers. They used an approximate, grey neutrino transport and replaced the innermost 1.1 $M_\odot$ of the PNS by an inner boundary. The evolution of the neutrino boundary luminosity was based on an analytic cooling model of the PNS, which depends on a set of free parameters. These parameters (within the so-called PHOTB approach) are set by fitting observational properties of SN 1987A for progenitor masses around 20 $M_\odot$ (see also Ertl et al. 2016; Sukhbold et al. 2016). Perego et al. (2015) utilize the energy in muon and tau neutrinos as an additional energy source that approximately captures the essential effects of (3D) neutrino transport (the PUSH method which is discussed in more detail below). Both approaches make it possible to predict the variation of explosion energies (and other parameters like neutron star mass cuts) as a function of stellar mass and thus can provide improved nucleosynthesis yields for chemical evolution modeling. There exist detailed results by now from PHOTB, but only PUSH includes the Y_e-effects due to neutrino interactions with nuclei, which is highly important for the Fe-group composition.

A major open question is whether core collapse leads finally to a supernova explosion with a neutron star remnant or whether the final outcome is a central black hole. Expectations from observations and their interpretation (e.g. Nomoto et al. 2006, 2013) would argue for a gradual transition between these two regimes as a function of initial stellar mass. PHOTB (Sukhbold et al. 2016) shows that both possible outcomes can occur within the same mass interval, mainly dependent on the pre-collapse stellar model and its compactness parameter. This needs further investigations and might pose questions about the stellar models. Could such scatter of the pre-core-collapse models in a narrow mass range, which is due to the properties of the last Si-shell burning stages, be avoided in general by the inclusion of rotation, leading possibly to a smearing out of radial gradients? The transition from central neutron star to black hole remnants can involve the effect of fallback of material from the innermost regions as a result of the outgoing shock wave being reflected by density gradients, causing a reverse shock. A pragmatic solution, utilizing a combination of mixing and fallback, can on the one hand minimize the effect of inconsistent Y_e-values in the innermost regions for approximated spherical models which do not include neutrino interactions (i.e. of the piston or bomb type). This mixes some Fe-group nuclei to larger radii, while the fallback reduces in total the ejected amount of Fe-group elements and can this way create overall C-rich ejecta, as observed in some extremely low metallicity stars. Such behavior is suggested to stem from some very massive stars, possible being more frequent in the early Galaxy (Nomoto et al. 2006, 2013). However, in the following we want to concentrate on the recent advances within the PUSH method.

The PUSH method provides a computationally efficient and physically motivated framework to explode massive stars in spherical symmetry. It allows to study multiple aspects related to core-collapse supernovae that require modeling of the explosion for several seconds after its onset and for extended sets of progenitors. The method was primarily designed to study explosive nucleosynthesis, but it is also well suited to explore other relevant aspects, including the effects of the shock passage through the star, the neutron star mass distribution, and the distribution of the explosion energies. PUSH relies on the so-called delayed neutrino-driven mechanism as a central engine of core-collapse supernovae. In particular, it provides an artificially enhanced neutrino energy deposition inside the gain region, inspired by the increase of the net neutrino heating that a fluid element experiences due to the presence of multi-dimensional effects. Unlike other methods that employ external energy sources or that use modified electron flavor neutrino luminosities, it deposits a fraction of the energy carried away by heavy flavor neutrinos ($\nu_x = \nu_\mu, \bar{\nu}_\mu, \nu_\tau, \bar{\nu}_\tau$) behind the shock to ultimately provide successful explosion conditions. In self-consistent core-collapse models, the ν_x's present a marginal dependence on the temporal evolution of the accretion rate (e.g. Liebendörfer et al. 2004), and their contribution to the energy deposition inside the gain region is negligible. However, their usage in PUSH presents a number of advantages. On the one hand, the properties of the ν_x emission, which includes dynamical feedback from accretion history, as well as cooling properties of the forming compact object, correlate significantly with the main features of the ν_e and $\bar{\nu}_e$ emission (O'Connor and Ott 2013). Moreover, the accretion luminosity depends not only on the accretion rate but also on the evolution of the mass and radius of the PNS, which is treated accurately and self-consistently in this method. This allows to trigger explosions in 1D simulations without modifying ν_e and $\bar{\nu}_e$ luminosities nor changing charged current reactions. It increases the accuracy of treating the electron fraction for the innermost ejecta, which is a crucial ingredient for nucleosynthesis calculations. In addition, unlike the electron (anti-)neutrino luminosities, which decrease suddenly once the shock has been revived in spherically symmetric models, ν_x luminosities are only marginally affected by the development of an explosion. This allows PUSH to continue injecting energy inside the expanding shock for a few hundreds of milliseconds after the explosion has set in. However, also the PUSH method is still parametrized, in the sense that the additional energy deposition is calibrated by comparing the obtained explosion energies and nucleosynthesis yields with observations of nearby supernovae (in particular, SN 1987A). This ensures that the artificially increased heating efficiency has an empirical foundation and the model has a predictive power in the sense of an effective model.

A first implementation of the PUSH method and its calibration strategy was extensively documented in Perego et al. (2015). The hydrodynamic evolution in spherical symmetry was performed with the general relativistic hydrodynamics code AGILE (Liebendörfer et al. 2001). For the stellar collapse, the deleptonisation scheme of Liebendörfer (2005) was applied. The Isotropic Diffusion Source Approximation (IDSA) was employed for the electron neutrino and anti-neutrino transport (Liebendörfer et al. 2009), while the heavy-lepton flavor neutrinos were modelled by an Advanced Spectral Leakage scheme (ASL, Perego et al. 2016). An extended Equation of State (EOS), comprising both nuclear statistical equilibrium (NSE) and non-NSE conditions, was included in the model. For the former, the tabulated microphysical EOS HS(DD2) was employed (Hempel and Schaffner-Bielich 2010), while for the latter an ideal gas of electrons, positrons, photons and 25 representative nuclei from neutrons and protons to iron-group nuclei was used. For conditions not permitting to use NSE, the changes in nuclear composition were followed by an approximate α-network.

The two free parameters of the method and require a calibration procedure. In the first study, 16 pre-explosion models with M_{ZAMS} between 18.0 $M_\odot$ and 21.0 $M_\odot$, at solar-metallicity, were used (Woosley et al. 2002). These models were selected to have ZAMS mass similar to the progenitor of SN 1987A. Each progenitor model was labeled by its compactness parameter ξ_M (O'Connor and Ott 2011), which is defined as $\xi_M \equiv (M/\mathrm{M}_\odot)/(R(M)/1000\,\mathrm{km})$, where $R(M)$ is the radius of the enclosed mass M, computed for $M = 1.75\ \mathrm{M}_\odot$ at the onset of the collapse. The considered progenitor models fall into two distinct families of compactness: low compactness ($\xi_{1.75} < 0.45$; LC models) and high compactness ($\xi_{1.75} > 0.45$; HC models). All the simulations included the progenitor star up to the helium shell, corresponding to a radius of $R \approx (1.3\text{–}1.5) \times 10^{10}$ cm, and they were run for a total time of 5 s, corresponding to $\gtrsim 4.6$ s after core bounce. For this time duration the shock was always contained inside the computational domain. Tracer mass elements (shells) were extracted from the simulation and post-processed with the `WINNET` (Winteler et al. 2012) nuclear network.

The analysis of explosion models with a broad parameter exploration revealed a distinction between HC and LC progenitor models, i.e. a clear dependence on the compactness of the progenitors. Explosions tend to set in earlier, with lower explosion energies and lower remnant masses in the case of LC progenitors. For HC progenitors, explosions are more difficult to archive and require more time to develop. However, when they happen, they are more energetic and produce more massive PNSs. These differences can be explained in terms of different accretion histories. The HC models produce larger neutrino luminosities and harder neutrino spectra due to the larger accretion rates. In order to overcome the stronger ram pressure at the shock front, a more intense neutrino energy deposition is required. In Fig. 9, the temporal evolution of several key quantities of exploding models from one HC and one LC progenitors are presented. A close investigation of the evolution of the explosion energy revealed that the energy deposition by neutrinos is the main source of the explosion energy. The net nuclear binding energy, released by the ejecta during the collapse and the subsequent explosion (including both the initial endothermic photodissociation and the final exothermic explosive burning), is positive, but significantly smaller than the energy provided by neutrinos.

As mentioned above, PUSH is not a fully self-consistent method, it requires a calibration of the free parameters k_{push} and t_{rise}, which has been performed with explosion and nucleosynthetic properties of SN 1987A. The requirement of an explosion energy around 1 Bethe was achieved only by HC progenitors, with a moderately high compactness parameter. Initial calibration attempts found a systematic overproduction of ^{56}Ni for runs with an explosion energy around and above 1 Bethe. This discrepancy had to be cured by invoking a relatively large amount of fallback (0.1 $M_\odot$). However, more recent developments

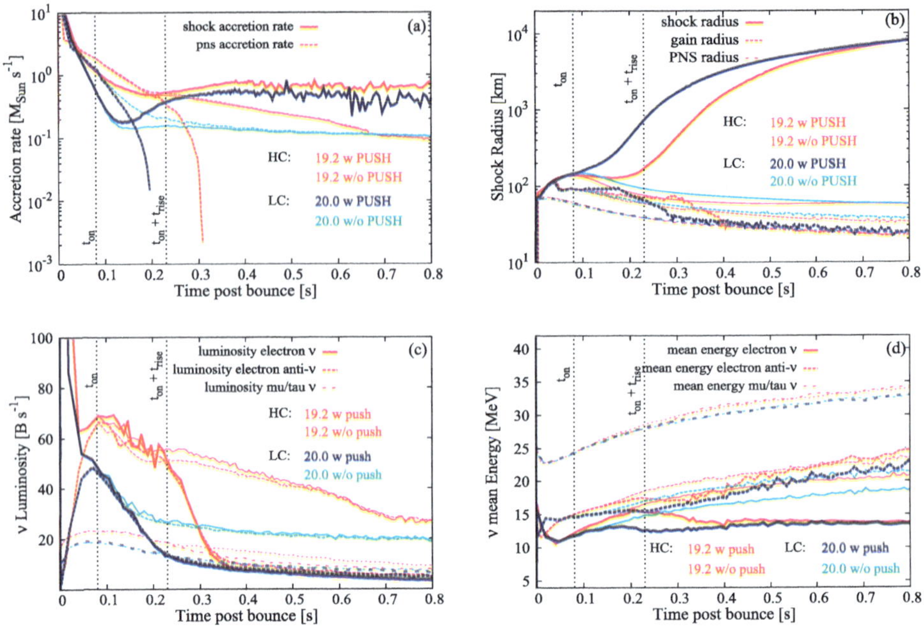

Fig. 9 Temporal evolution of (**a**) the accretion rate at the PNS and at the shock, (**b**) the shock, the gain, and the PNS radii, (**c**) the neutrino luminosities, and (**d**) the neutrino mean energies, for all modeled neutrino flavors. In all panels, we present exploding runs for the 19.2 M$_\odot$ (red lines) and then 20.0 M$_\odot$ (blue lines) ZAMS mass models obtained with PUSH Perego et al. (2015). We also plot the corresponding non-exploding runs obtained by setting $k_{\rm push} = 0$ for the 19.2 M$_\odot$ (light red) and 20.0 M$_\odot$ (light blue) ZAMS mass progenitor models

of the PUSH method found that a satisfactory agreement with SN 1987A observables can be achieved, using dedicated blue supergiant progenitors and slightly different parameters, compared with the values reported in Perego et al. (2015), without invoking any fallback (Ebinger et al. 2017). According to the calibrated model, the SN 1987A remnant should host a neutron star with a baryonic mass of ~ 1.66 M$_\odot$, corresponding to a gravitational mass of ~ 1.50 M$_\odot$ for a cold neutron star with the HS(DD2) EOS. The formation of a BH seems to be unlikely, since it would require a very large fallback ($\gtrsim 0.5$ M$_\odot$) to reach the observed neutron star maximum mass limit, and an even larger fallback (~ 1.3 M$_\odot$) to reach the maximum neutron star baryonic mass predicted by the HS(DD2) EOS.

This calibration includes the reproduction of the observed explosion energy and some key abundances which could be observed in the explosion. ^{56}Ni plays a key role, but also ^{57,58}Ni and ^{44}Ti should match the observations. A detailed analysis of the neutron-rich nickel isotopes showed that the production of $^{57-58}$Ni is highly sensitive to the electron fraction of the innermost ejecta. A final mass cut initially located inside the silicon shell can provide slightly neutron rich ejecta ($Y_e < 0.5$), corresponding to the conditions required to fit the $^{57-58}$Ni yields of SN 1987A. Under these conditions, the alpha-rich freeze-out leads to nuclei only one or two neutron units away from the $N = Z$ line. In many HC models, characterized by larger $\xi_{1.75}$, the location of the mass cut inside the oxygen shell does not provide favorable conditions for the synthesis of these isotopes. Figure 10 shows a comparison of the Y_e (top panels) and composition profiles (bottom panels) for the 18.0 M$_\odot$ (left panels) and the 20.6 M$_\odot$ (right panels) progenitors. In the case of the 18.0 M$_\odot$ model, the cutoff

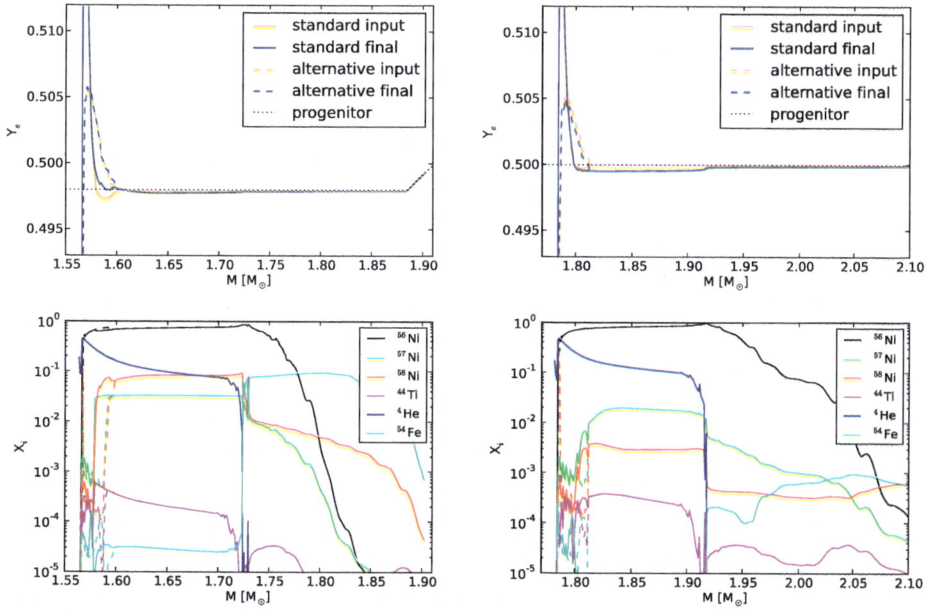

Fig. 10 Electron fraction profiles (top) and nuclear compositions at 100 s (bottom) above the mass cut for the 18.0 M$_\odot$ (left) and the 20.6 M$_\odot$ (right) progenitors exploded with PUSH Perego et al. (2015). The solid lines correspond to the case, where the initial ("input") electron fraction is taken from the hydrodynamic simulation when the temperature drops below 10 GK. The dashed lines correspond to the alternative case, where the electron fraction at the end of the hydrodynamic simulation is considered as the initial electron fraction in the network. The electron fraction after post-processing ("final") is also shown for the two cases

mass is 1.56 M$_\odot$ and a large part of the silicon shell is ejected. In this shell, the initial matter composition is slightly neutron-rich ($Y_e \simeq 0.498$) and the conditions for the production of ^{57}Ni and ^{58}Ni are matched. The increase in Y_e around 1.9 M$_\odot$ marks the transition to the oxygen shell. The same transition for the 20.6 M$_\odot$ model happens inside the mass cut, around 1.74 M$_\odot$. Therefore, this model ejected significantly less ^{57}Ni and ^{58}Ni. In all our models, ^{44}Ti is produced within the innermost 0.15 M$_\odot$ of the ejecta (see Fig. 10). A comparison with the expected ^{44}Ti abundances revealed that an underproduction of this isotope was predicted in all models. Uncertainties in the relevant nuclear reaction rates, and in the mixing occurring in the innermost ejecta, can help reducing this discrepancy and explaining the observed abundances, but to fully explain ^{44}Ti, the effects of real 3D modelling are probably needed.

More recently, further advances were attained, utilizing the PUSH approach (Sinha et al. 2017; Ebinger et al. 2017), with Sinha et al. (2017) focusing on the nucleosynthesis yields of Fe-group elements from CCSNe, based on models from two progenitor series (Woosley et al. 2002; Woosley and Heger 2007). It turns out that explosions with detailed Y_e-evolution, based on the explosion mechanism, give a better match with observational abundances of metal-poor stars, witnessing the explosive yields of single stars (CCSNe) in the early Galaxy before the delayed SNeIa from binary systems enter their ejecta. It was also realized that the structure and the evolution of the pre-supernova star appear to influence significantly the Fe-group nucleosynthesis yields. Figure 11 presents the observed abundance ratios of Fe-group elements (Sneden et al. 2016) for the very metal-poor main sequence turnoff star HD 84937, along with the PUSH nucleosynthesis yields for Fe-group elements. PUSH yields are shown

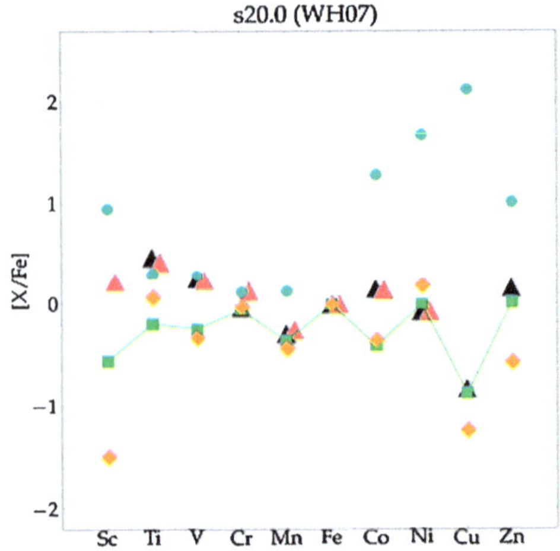

Fig. 11 Observed abundances of HD 84937 Sneden et al. (2016) (black and magenta) compared with the yields obtained with progenitor models of a 20 $M_\odot$ star from the Woosley and Heger (2007) progenitor series and reported in Sinha et al. (2017). Shown are piston results Woosley and Weaver (1995) (blue), thermal bomb results Thielemann et al. (1996) (yellow), and the present PUSH results (green). Due to consistent explosion energies and Y_e-values based on neutrino interactions during the explosion, it can be recognized that especially Sc and Zn show highly improved values in comparison to these older approaches (a first improvement, when including neutrino interactions during the explosion was already obtained in Fröhlich et al. 2006a). Cu is mostly due to s-processing during stellar evolution, which is not part of the explosion calculation

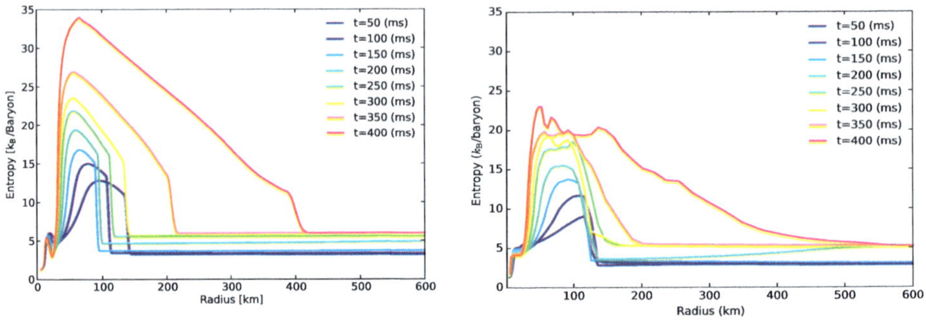

Fig. 12 Radial profiles of the spherically averaged entropy per baryon obtained from a spherically symmetric simulation with PUSH (left panel) and from an axis-symmetric Flash simulation (right panel) of a 20 $M_\odot$ progenitor, see Ebinger et al. (2017)

along with piston yields (Woosley and Weaver 1995) and thermal bomb yields (Thielemann et al. 1996) for a 20 $M_\odot$ model from Woosley and Heger (2007) progenitor set.

Ebinger et al. (2017) compared profiles of the innermost stellar layers obtained in spherically symmetric models with PUSH to profiles obtained in exploding multi-dimensional models. They showed that the PUSH method presents a behavior more consistent with multi-dimensional models than older methods (e.g. pistons and thermal bombs). In Fig. 12 the spherically averaged entropy per baryon as a function of radius is shown, obtained (a) from a 2D Flash simulation (Pan et al. 2016), see right panel, and (b) from a 1D simulation performed with PUSH for the same progenitor, see left panel.

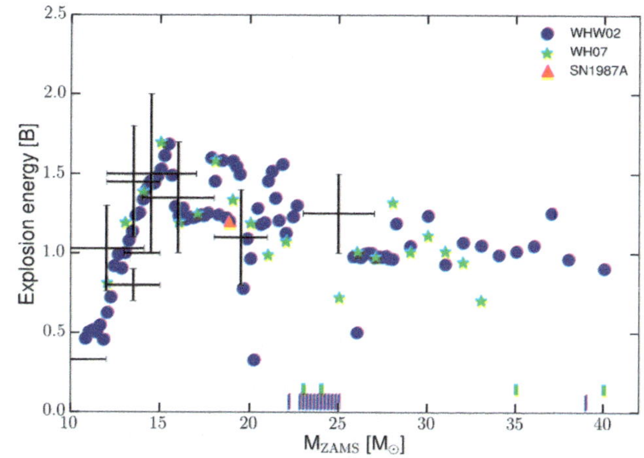

Fig. 13 Explosion energies as a function of progenitor mass deduced from observations (Nomoto et al. 2013) (black crosses) and predicted from PUSH simulations (Ebinger et al. 2017) for two sets of progenitor models (blue dots Woosley et al. 2002) and (green stars Woosley and Heger 2007). Dashed vertical lines close to the abscissa indicate no explosions (in blue and green for the two sets, mentioned above), i.e. black hole formation in PUSH simulations in the range 22–25 $M_\odot$ and beyond 33 $M_\odot$

Moreover, the capability of PUSH to explore additional stellar collapse features, like the explodability of a large set of progenitors or the formation of a stellar black hole from a failed supernova, was investigated. Figure 13 shows the explosion energy obtained with PUSH for a set of progenitor models, covering a large portion of the IMF. We see there is a rise in explosion energies up to about 18–20 $M_\odot$ (accompanied by an increasing compactness of the stellar models). Similar to Sukhbold et al. (2016) one can see that there is a transition to BH formation beyond 20 $M_\odot$, but that still individual models above that mass range can experience explosions, while BHs can also form for some lower mass progenitors. The window above 20 $M_\odot$, where black holes are formed is followed by a range where explosions are possible again, and finally a general tendency to black hole formation above 35 $M_\odot$. This procedure has been carried through for a number of stellar samples, also from Heger and Woosley (and for different metallicities). The results differ in detail, especially—dependent on the compactness of stellar models—black hole formation can also occur in some cases for lower initial masses, but the general tendency remains (Ebinger et al. 2017). This shows the strong dependence of the core-collapse outcome on the detailed structure of progenitor models and thus also the dependence on accurate stellar modeling.

Concluding, we can say that the improved spherically symmetric approaches like PHOTB and PUSH seem to provide for the first time quasi self-consistent values for supernova explosion energies and resulting neutron star masses, as well as the transition to black hole formation. Thus, this modeling also permits to predict the transition from regular CCSNe to faint or failed supernovae. As the present models do not include rotation and magnetic fields, predictions of the hypernova branch, leading to long-duration gamma-ray bursts, cannot be made within this approach. It should be mentioned, however, that the additional advantage of PUSH (over other methods) is that the correct inclusion of electron neutrino and antineutrino interactions with matter permit also to predict reliable values of the electron fraction Y_e and therefore also a reliable prediction for the detailed isotopic composition of the Fe-group.

2.3 The Effect of Strong Magnetic Fields

Recent observations (and their interpretation) (Greiner et al. 2015; Bernardini 2015) underline that there exist core-collapse supernova explosions whose light curves are not determined by (large) amounts of ^{56}Ni ejecta, but rather by the energy release of a fast rotating neutron star (pulsar) with extremely strong magnetic fields of the order 10^{15} Gauss (magnetars). The question is how can neutron stars of such extremely high magnetic fields (in

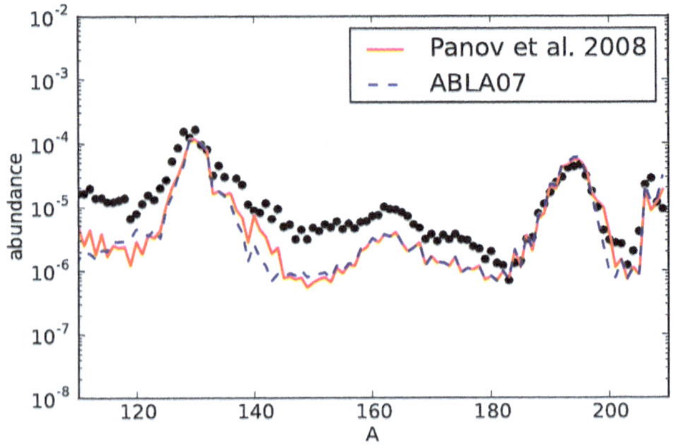

Fig. 14 Nucleosynthesis features of an MHD-jet supernova, originating from a 15 $M_\odot$ progenitor with a 5×10^{12} Gauss magnetic field of the collapsing Fe-core. The original simulation (Winteler et al. 2012) showed that the 2nd and 3rd t-process peak could be reproduced very well with a $Y_e = 0.1$–0.15 of the ejecta, but big troughs occurred below and above the 2nd r-process peak at $A = 130$. Utilizing more modern fission fragment distributions, e.g. from the statistical de-excitation code ABLA07 (Kelic et al. 2008), vs. prior treatments (Panov et al. 2008) improved these features strongly for not extremely neutron-rich environments with only weak fission-cycling

comparison to the typical 10^{12} Gauss) emerge from a supernova explosion? A reasonable assumption is that such objects originate from massive stars which are fast rotators with initially also rather strong magnetic fields. Such objects, with assumptions made for the initial rotation rate and magnetic fields, have been modeled extensively (Fujimoto et al. 2007, 2008; Winteler et al. 2012; Mösta et al. 2014, 2015, 2017; Nishimura et al. 2015, 2017; Halevi and Mösta 2018) and will be called here magneto-rotational or MHD-jet supernovae. The result is typically (when starting with initial fields of the order 10^{12} Gauss) that the winding up of magnetic fields results in strong magnetic pressure along the polar rotation axis and jet-like ejection of matter. This matter has experienced high densities (and thus degenerate electrons with high Fermi energies), leading via electron captures on protons and nuclei to strongly neutron-rich matter with a Y_e of the order 0.1–0.15. The fast ejection along the poles avoids that the interaction with neutrinos and anti-neutrinos causes a major rise of Y_e (see Eqs. (2) to (3)). Such environments permit a strong r-process, dependent on the initial conditions (Winteler et al. 2012; Nishimura et al. 2015, 2017; Mösta et al. 2015, 2017; Halevi and Mösta 2018). In this process, due to a high fraction of neutrons per nucleus after the freeze-out of charged-particle reactions during the expansion at $\approx 3 \times 10^9$ K, rapid neutron capture can occur, lading to nuclei far from stability with extremely short beta-decay half-lives and the production of heavy nuclei up tp the actinides. In Fig. 14 we show the nucleosynthesis results of the 3D collapse of a fast rotator with a strong initial magnetic field of 5×10^{12} Gauss in z-direction before core collapse. A 15 $M_\odot$ progenitor with an initial shellular rotation with period of 2 s at a 1000 km radius results in a rare class of supernovae with a central magnetar and negligible amounts of Fe-group ejecta. Initial results (Winteler et al. 2012) made use of older fission fragment distributions. Figure 14 shows the effect of more modern fission yields which avoid the abundance troughs below and above the $A = 130$ peak. This result should be taken into account with respect to other investigations (Shibagaki et al. 2016), which rely on quite different abundance features for MHD-jet supernovae and neutron star mergers, based on fission barriers which introduce fission in the r-process only above $A = 300$.

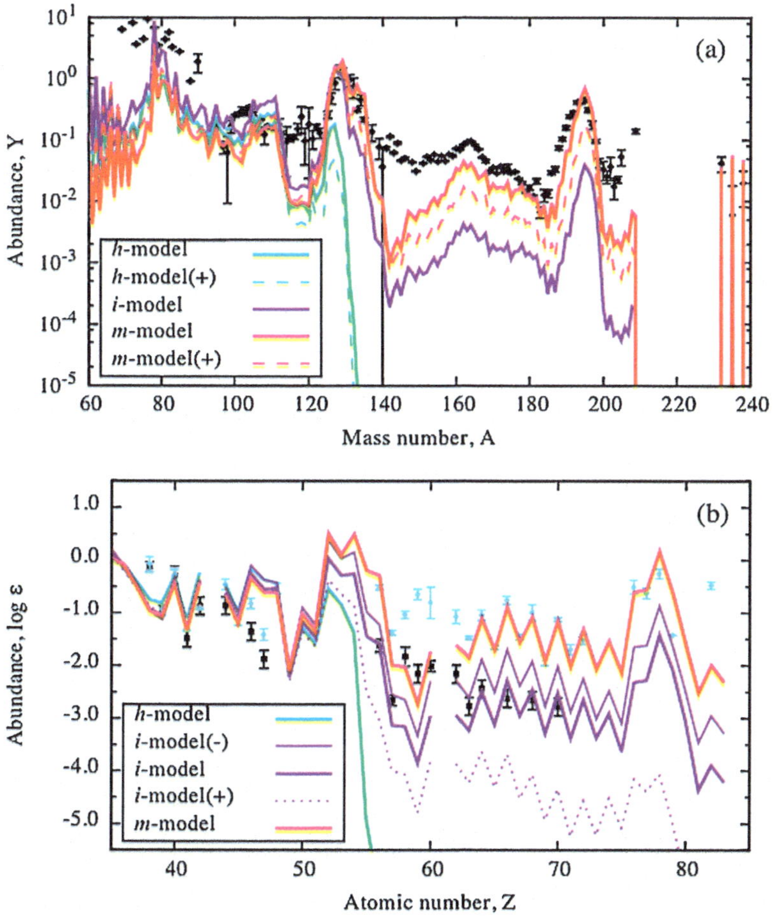

Fig. 15 Abundances from nucleosynthesis calculations with varying ratios of magnetic field strength with respect to the neutrino heating mechanisms of regular CCSNe, increasing for the models h, i−, i, i+, and m (for details see Nishimura et al. 2017). For comparison also (**a**) solar r-process abundances are shown (black dots Arlandini et al. 1999), (**b**) abundances from metal-poor stars with a weak r-process, i.e. HD122563 (black dots Honda et al. 2006), and solar-type r-process observations from CS22892-052 (blue dots Sneden et al. 1996). Abundances are normalized for $Z = 40$ of HD122563. Observations of low metallicity stars with strong r-process contributions vary for abundances below $Z = 50$ (Sneden et al. 2008)

A fully self-consistent treatment would require high resolution simulations which can resolve magneto-rotational instabilities (MRI) and would predict reliably the possible amplification of magnetic fields during the explosion. While the latter is actually possible by now (Nishimura et al. 2017), present calculations depend on the uncertain and therefore assumed initial conditions, that either cause strong jet ejection or can develop kink instabilities of the jets (Mösta et al. 2015). Based on initial conditions, somewhat parametrized in terms of the impact either neutrino heating or magnetic pressure has in causing the supernova explosion, the production for heavy neutron capture elements varies strongly. This is shown in Fig. 15, which underlines that results can range from a weak r-process, barely proceeding up to the second peak at $A = 130$, to a full r-process with a strong contribution to the actinides.

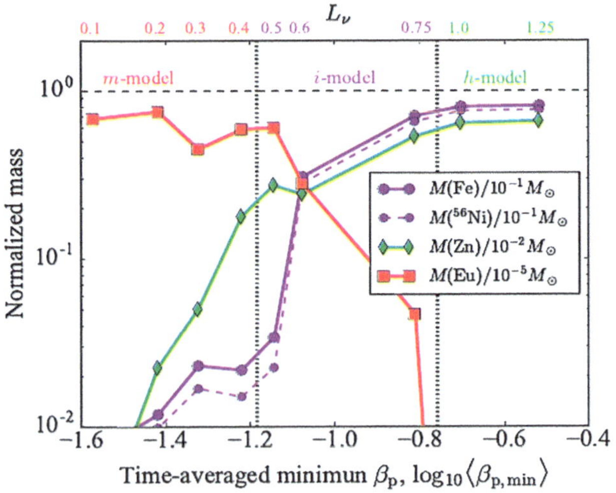

Fig. 16 Nucleosynthesis features of rotating CCSN models (h, i−, i, i+, m) with varying ratios of neutrino luminosity and magnetic field strengths as in Fig. 15. Model m represents a strong MHD-jet supernova. One can see the transition from a regular CCSN pattern, dominated by ^{56}Ni, total Fe (after decay), and Zn to a strong r-process pattern with a high Eu abundance (for details see Nishimura et al. 2017)

In terms of applications to galactic chemical evolution it should be noticed that the MHD-jet supernovae discussed here are expected to occur as a fraction of 0.1–1 percent of all CC-SNe, probably being somewhat metallicity dependent. Higher metallicities lead to stronger stellar wind loss which will be accompanied by a loss of angular momentum, thus reducing the fast rotation necessary for this type of SN explosions. Another feature is that these events can lead to small amounts of Fe-group ejecta for the cases of strong r-processing (Nishimura et al. 2015, 2017). Plotting the results of Fig. 15 with respect to the relative influence of neutrino heating vs. magnetic field effects, one can see that the Ni/Eu-ratio (and similarly the Fe/Eu-ratio) varies strongly (see Fig. 16). Thus, if these types of supernovae would contribute already at low metallicities, they alone would be able to provide a large spread in Eu/Fe and might even explain the variations in actinides vs. Eu, seen in a number of cases at low metallicities (see e.g. Wehmeyer et al. 2015; Thielemann et al. 2017a).

2.4 Long Duration Gamma-Ray Bursts and Hypernovae

Section 2.2 discussed the final outcome of the evolution of massive stars after core-collapse. For non-rotating cases only the supernova branch (with neutron stars as final outcome) and the faint or failed supernova branch (leading eventually to black holes, but without gamma-ray bursts and high ejecta masses) of Nomoto et al. (2006, 2013) can be attained. Thus, massive stars, which fail to explode as CCSNe via neutrino-powered explosions, will eventually experience the formation of central black hole (BH) remnants. However, rotating BHs and the formation of accretion disks with accretion rates of about $\approx 0.1\ \mathrm{M}_\odot/\mathrm{s}$ can lead—for certain conditions (strong magnetic fields)—to long duration gamma-ray bursts (lGRBs) or hypernovae (Nomoto's hypernova branch). Many authors have contributed to the discovery and laying out first ideas for theoretical explanations (e.g. Klebesadel, Bloom, Paczynski, Mezaros, Rees, Piran, see the review by Piran 2004). The collapsar model was proposed by Woosley, MacFadyen and others (see also MacFadyen and Woosley 1999; MacFadyen et al. 2001; Nagataki et al. 2007; Sekiguchi and Shibata 2011; Nagataki 2011), based on neutrino heating from the accretion disk and/or the winding of strong magnetic fields and MHD jets (McKinney et al. 2013; Ono et al. 2012). Hydrodynamic simulations (injecting explosion energies artificially) were performed by Umeda, Nomoto, Maeda, Iwamoto, Nagataki, either by introducing high explosion energies (up to 10^{52} erg) in a spherically symmetric way or aspherically in order to understand jet-like explosions (Nakamura et al. 2001; Nomoto et al. 2006, 2013; Nomoto 2017). The role of weak interactions and resulting nucleosynthesis was the focus of contributions by Pruet, Surman, McLaughlin, Hoffman, Hix, Janka, Ruffert, Fuller, Lemoine, Inoue, Fujimoto, Beloborodov, Janiuk (for specific nucleosynthesis results see e.g. Surman et al. 2006; Janiuk 2014). The basic (consensus) picture is the following: explosion energies can be found up to 5×10^{52} erg, ^{56}Ni ejecta up to 0.5 $M_{\odot}$, and the ejecta are beamed with relativistic jets. Many attempts have been undertaken to model such events. There exists uncertainty in predicting Y_e, following weak interactions and especially neutrino transport in disks and jets, but there exists also the constraint of high ^{56}Ni ejecta. Therefore, the dominant Y_e in matter has to be of the order of 0.5. High explosion energies lead to high entropies and a strong alpha-rich freeze-out, including interesting amounts of ^{45}Sc, ^{64}Zn (from ^{64}Ge-decay) and other Fe-group elements. For general and more detailed considerations see Nomoto (2017), concluding that larger abundance ratios for (Zn, Co, V, Ti)/Fe and smaller (Mn, Cr)/Fe ratios are expected than for normal SNe, which seems to be consistent with observations in extremely metal-poor (EMP) stars, as will be discussed in Sect. 4.

3 Binary Systems with Compact Objects

This very general topic would include systems with white dwarfs (WDs), and thus novae as well as Type Ia supernovae, and also systems involving neutron stars, where accretion of matter in binary systems can lead to X-ray bursts and possibly also superbursts. Mergers of neutron stars can lead to short-duration gamma-ray bursts sGRBs. With the exception of Type Ia supernovae and neutron star mergers, these topics are not strongly relevant with respect to nucleosynthesis contributions. Therefore, we want to focus here just on these two types of explosive events.

3.1 Type Ia Supernovae and Their Nucleosynthesis

After initial suggestions by Hoyle and Fowler (1960) and (layed out in more detail by Iben & Tutokov and Webbink, Iben and Tutukov 1984; Webbink 1984, and followed by many more publications of these authors), two explanations existed for Type Ia supernovae, both involving white dwarfs in binary stellar systems: the single-degenerate scenario (involving only one accreting WD), and the double-degenerate scenario (involving spiraling in WD mergers). For a while the Chandrasekhar mass models (single degenerates), causing growth towards the Chandrasekhar mass and then central ignition, leading initially to deflagration, possibly changing into a detonation in the outer layers, was the favorite explanation (see e.g. the W7 model and its precursors by Nomoto and collaborators, Nomoto 1982; Nomoto et al. 1984). Latest updates on this scenario have been given by Hoeflich et al. (2017) and Leung and Nomoto (2017). But in the meantime the spectrum of scenarios has widened a lot, involving also single-degenerates with off-center ignition, leading to gravitationally defined detonations or pulsating reverse detonations for near Chandrasekhar mass models (Plewa et al. 2004; Bravo et al. 2009), to double detonations via He-accretion for sub-Chandrasekhar models (Livne 1990; Woosley and Weaver 1994; Shen and Bildsten 2014; Shen et al. 2017), detonating ONe white dwarfs (Marquardt et al. 2015), as well as double degenerate mergers (Pakmor et al. 2010, 2012, 2013; Sim et al. 2010; Dan et al. 2012, 2015), and even WD collisions (Benz et al. 1989; Rosswog et al. 2009; Raskin et al. 2009; García-Senz et al.

2013; Papish and Perets 2016), which might be all responsible for the observational diversity with multiple subclasses from recent extended surveys (Maoz et al. 2014; Goldstein and Kasen 2018). Especially, there is growing evidence for sub-Chandra double detonations due to He-shell ignition (Jiang et al. 2017), which was ruled out in the early days, permitting (explosive) He-ignition only after a critical layer was accumulated. Shen et al. (2010) showed that He-detonation can set in earlier via turbulent motions, thus preventing the above mentioned delay until the accumulation of larger He-shells. Such models tend to display enhanced abundances of the heavy alpha elements beyond ^{40}Ca and up to ^{56}Ni. Nucleosynthesis calculations for this diversity of models have been performed (Thielemann et al. 1986, 2004; Iwamoto et al. 1999; Brachwitz et al. 2000; Travaglio et al. 2004; Meakin et al. 2009; Kromer et al. 2010; Maeda et al. 2010; Seitenzahl et al. 2011, 2013, 2016; Waldman et al. 2011; Moore et al. 2013; Shen and Bildsten 2014; Papish and Perets 2016). Light curves and spectra show faint SNeIa with weak intermediate-mass elements (IME), bright cases with early emergence of Fe-group lines (IGE), faint and fast evolving cases, as well the normal ones utilized for cosmology. Although there has been further progress since then, the reviews by Hillebrandt et al. (2013), Maeda and Terada (2016), Roepke (2017), Seitenzahl and Townsley (2017) give an enlightening overview over all these options.

Nucleosynthesis-wise there are essentially two different features, which characterize the burning products: (a) is burning the result of a detonation or deflagration and (b) what is the density at which matter ignites? A detonation of a near-Chandrasekhar mass WD with ignition densities clearly in excess of 10^7 g/cm^3 will produce essentially only iron group elements (as a detonation with supersonic speed does not permit to expand matter of outer layers to lower densities before the burning front arrives). A sub-Chandrasekhar mass WDs will contain extended regions with densities below 10^7 g/cm^3, and will (also in a detonation) produce substantial amounts of intermediate mass elements from Si to Ar and Ti. Objects starting with initial high central densities, but an early deflagration (subsonic) burning front, have the chance to expand during the early phase, and therefore, if it turns—with a delay—into a detonation at larger radii (delayed detonation), also for near-Chandrasekhar mass models substantial amounts of intermediate mass elements are produced. High central densities relate to high electron Fermi energies and permit electron capture (the reverse to Eq. (2)). This is especially the case during deflagrations, when such conditions are experienced for extended time scales due to the slow propagation of the burning front. We want to depict this specific behavior, occurring only in the single-degenerate near Chandrasekhar mass scenario, here in more detail. Figure 17 gives an indication of the type of burning processes taking place in such initial central deflagrations (from Thielemann et al. 1986) and displays also the resulting Y_e (from Thielemann et al. 2004), due to electron captures in the central high density regions (for two sets of electron capture rate predictions FFN, Fuller et al. 1982 and the advanced shell model results LMP, Langanke and Martínez-Pinedo 2001).

Corresponding to Fig. 17 we display in Fig. 18 the resulting abundances of the classic near-Chandrasekhar mass model W7 (Nomoto et al. 1984; Thielemann et al. 1986). As indicated, we have in the center two important features: (a) the normal freeze-out causes only Fe-group elements in the final abundances, (b) the reduced Y_e due to electron captures leads to a composition not dominated by ^{56}Ni (later decaying to ^{56}Fe and powering the light curve), instead the more neutron-rich ^{56}Fe is directly produced. ^{54}Fe is taking on that role in intermediate regions. In the alpha-rich freeze-out zones ^{54}Fe is turned to ^{58}Ni, distorting the normal freeze-out composition by a final alpha-capture. The other burning products are easy to understand. The only effect to realize is that the Y_e there is determined by metallicity. Pre-existing CNO isotopes (metallicity) in the initial stellar model have been turned into ^{14}N during H-burning via the CNO-cycle, which was transferred to slightly neutron-rich ^{22}Ne in

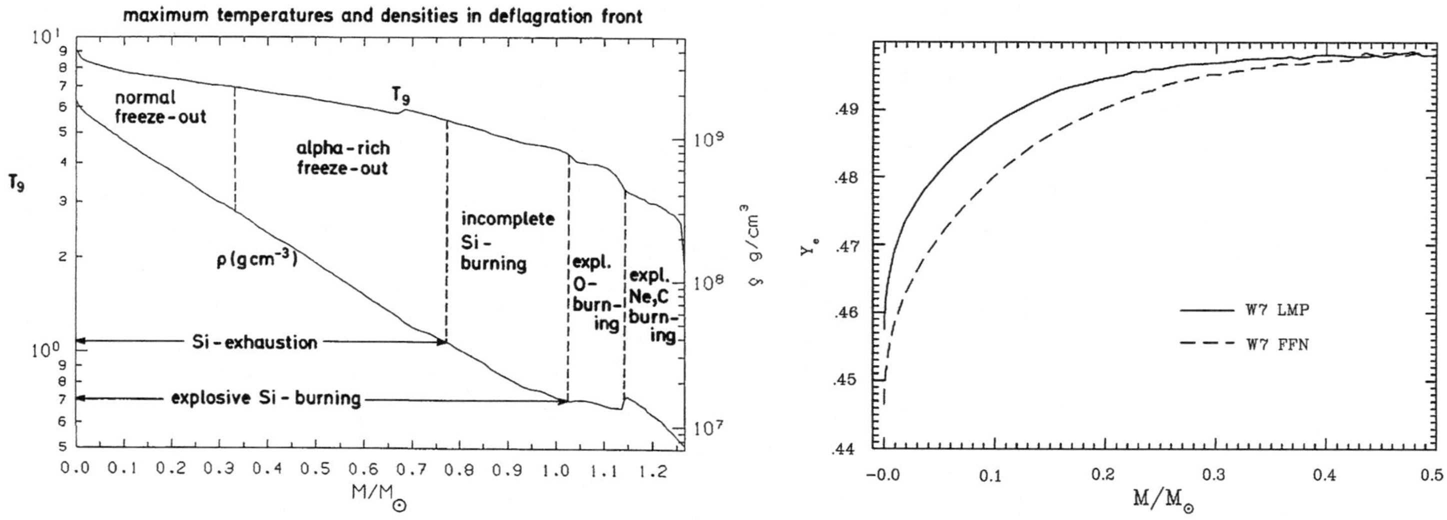

Fig. 17 left: maximum temperatures and densities attained during a deflagration in a near-Chandrasekhar mass model, combined with the explosive burning phases taking place under such conditions. While in the center only Fe-group nuclei remain after the thermal freeze-out of charge particle reactions (normal freeze-out), this changes at lower densities (higher entropies) to an alpha-rich freeze-out; right: for a near-Chandrasekhar mass model with initially subsonic burning front (deflagration), substantial electron captures can take place, reducing Y_e to values below 0.5. With the old FFN (Fuller et al. 1982) electron capture rates also captures on heavy nuclei contributed, the more recent large-scale shell model rates LMP (Langanke and Martínez-Pinedo 2001) result essentially only in electron captures on free protons during explosive burning (from Thielemann et al. 1986; Brachwitz et al. 2000)

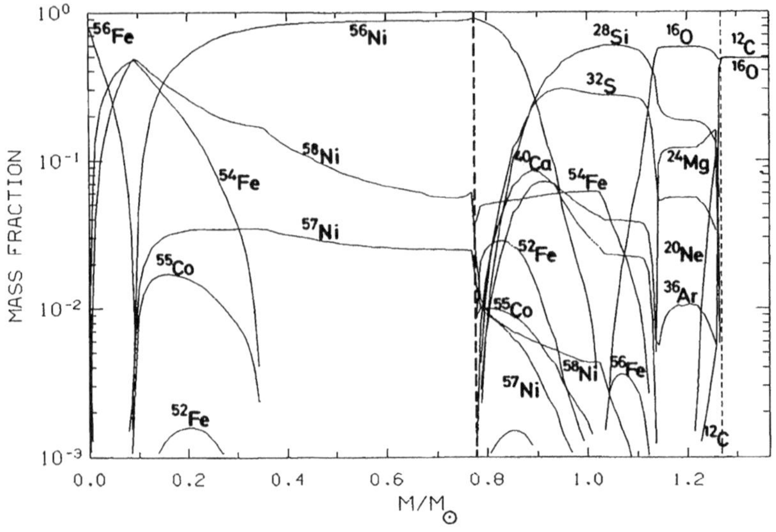

Fig. 18 Abundance composition after explosion of the classic near-Chandrasekhar deflagration model W7 (from Thielemann et al. 1986)

He-burning via $^{14}N(\alpha,\gamma)^{18}F(\beta^+)^{18}O(\alpha,\gamma)^{22}Ne$. Thus, the Y_e slightly below 0.5 for non-zero metallicities permits the existence of other isotopes with $Z/A < 0.5$.

The Mn/Fe ratio of the ejecta seems to be a sensitive measure whether near Chandrasekhar mass single-degenerate models, either in form of deflagrations like W7 (Nomoto et al. 1984) or delayed detonations (Iwamoto et al. 1999; Seitenzahl et al. 2013), occur at all, because only they show low Y_e-regions close to the center of the WD. Mn comes in form of its only stable isotope ^{55}Mn, and is the decay product of ^{55}Co, produced in incomplete and complete Si-burning under optimal conditions with $Y_e = Z/A = 0.491$. In alpha-rich freeze-out of explosive Si-burning, determined by the entropy $S \propto T^3/\rho$ of explosive burning layers with values of T_9 and ρ_8 exceeding $T_9^3/\rho_8 > 180$, ^{55}Co is moved to ^{59}Cu ($\rightarrow$ ^{59}Co) in the radial mass range of 0.3 to 0.8 $M_\odot$ (see Fig. 17), not shown here. The inner zones of M_{Ch}-models experience only moderate entropies and such lower Y_e's are attained via electron capture, due to degenerate electrons with high Fermi energy. In the outer zones it can be approached via metallicity, as CNO nuclei are burned in He-burning to (slightly neutron-rich) ^{22}Ne. For initial metallicities of the WD progenitor with values of [Fe/H] $= -\infty$, 0, 0.25, and 0.5 this leads to $Y_e = 0.5$, 0.499, 0.498, and 0.496, attained in the outer SNIa ejecta. For $Y_e = 0.5$ the ^{55}Co appearing in the outer part of Fig. 18 would disappear completely. This leads to an increase of [Mn/Fe] as a function of metallicity, and also the appearance of ^{54}Fe (moved to ^{58}Ni in alpha-rich freeze-out). Changing metallicities can also have an effect on the total ^{56}Ni mass ejected and thus light curve features, with the effect that higher metallicities decrease Y_e slightly in layers unaffected by electron captures (Calder et al. 2013).

While we have emphasized in this subsection the detailed features of near-Chandrasekhar mass models, and especially the role electron captures play in the central high density regions, sub-Chandrasekhar models behave similar to the conditions displayed in Fig. 17, excluding essentially the central lower Y_e regions. Detonations cause higher entropies and lead to alpha-rich freeze-out conditions with enhanced abundances of the heavier alpha-nuclei (similar to the effect discussed in Fig. 7 for core-collapse supernovae).

In addition to the dominant Fe-group composition and a smaller fraction of heavy alpha-nuclei (beyond Si), there have also been suggestions that SNeIa contribute to p-process nuclei. This process is normally associated with explosive Ne-burning zones of core-collapse SNe, where "photo-spallation" of pre-existing heavy nuclei can produce proton-rich isotopes via (γ, n) and (γ, α) reactions, typically on a level of the order 1% of the total element abundances. This seems to fail for the lighter p-isotopes of Mo and Ru, where they amount to about or more than 10% of the element abundances (Rauscher et al. 2013). In Sect. 2.1 a possible alternative via the νp-process has been presented. However, another option was proposed that the regular features of the p-process (but including the light p-isotopes) work in explosive Ne-burning in the outer zones of SNeIa, if these zones contain a strongly enhanced prior s-process contribution (Travaglio et al. 2011, 2015). This depends on the working of an s-process during the pre-SN accretion phase of the later exploding WD (see however Denissenkov et al. 2017).

3.2 Neutron Star Mergers

Short-duration GRBs (sGRBs, with a light curve decline of less than about two seconds) are due to relativistic jets created by the merger of two compact stellar objects (specifically two neutron stars or a neutron star and a black hole). Mergers of this kind are also expected to produce significant quantities of neutron-rich radioactive species, whose decay should result in a faint transient, known as kilonova, in the days following the burst. Recent calculations suggest that much of the kilonova energy should appear in the near-infrared, because of the high optical opacity created by these heavy r-process elements. Optical and near-infrared observations of such an event, accompanying the short-duration GRB130603B have been reported in recent years by Tanvir et al. (2013), (see also Barnes et al. 2016). The first gravitational wave detection of such an event (GW170817), combined with a short GRB (GRB170817A) and the optical and infrared afterglow (see e.g. Abbott et al. 2017) has clearly underlined this r-process production site (Metzger 2017). After the first detailed nucleosynthesis predictions (following ideas of Lattimer and Schramm 1974 and Eichler et al. 1989 of such an event) by Freiburghaus et al. (1999), many more and more sophisticated investigations have been undertaken (for a review see Thielemann et al. 2017b). Because this book chapter is dedicated to nucleosynthesis in supernovae, we do not want to treat such neutron star merger events in great detail, just to mention them here as a valid (and possibly dominant) site of the astrophysical r-process.

Figure 19 displays the early dynamical ejecta and their dependence on nuclear properties. Non-relativistic simulations (e.g. Korobkin et al. 2012; Rosswog et al. 2014) lead to large amounts of ejecta of the order 10^{-2} $M_\odot$ with very small Y_e-values of 0.04 and less. This causes a very strong r-process with fission cycling. The utilized mass model, beta-decay half-lives, fission barriers, and fission yield prescriptions have a strong effect on the final abundance distribution. While during the r-process (when reactions are still in $n, \gamma - \gamma, n$ equilibrium) the 2nd and 3rd r-process peak are exactly at the right position, the neutron capture of large amounts of fission neutrons (after freeze-out from this equilibrium) has some effect on abundances below $A = 165$, and the third peak seems always shifted to heavier nuclei. Deviations (troughs) in the mass range $A = 130$–165 can be improved with modern fission fragment distribution (Kelic et al. 2008, ABLA07,), but not the shift of the 3rd peak. One option to remedy this effect are variations in beta-decay rates. Shorter half-lives (Marketin et al. 2016) of heavies release neutrons from fission earlier, when $n, \gamma - \gamma, n$ equilibrium is still in place and can avoid or strongly reduce the late shift of the 3rd peak (Eichler et al. 2015). This effect is also seen with the HFB mass model (see also Goriely 2015;

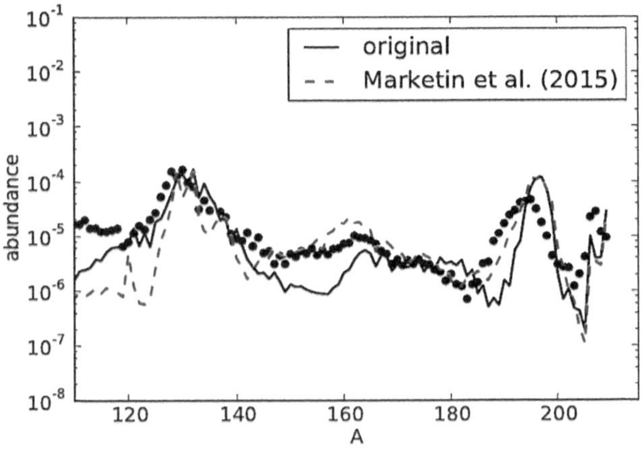

Fig. 19 r-process abundance distribution in dynamical ejecta of neutron star mergers, utilizing different mass models, fission fragment distributions, and half-lives (Eichler et al. 2015). Here the effect of improved half-life calculations (Marketin et al. 2016) in comparison to the original FRDM half-lives (Möller et al. 2003) is shown (from Eichler et al. 2015)

Goriely et al. 2015), while Mendoza-Temis et al. (2015) analyzed additional nuclear uncertainties and showed especially an improvement with the Duflo-Zuker mass model (Duflo and Zuker 1995). Alternative ways to cure the problems of the 3rd r-process peak discussed above, come from full general relativistic modeling of the merger event. This leads to deeper gravitational potentials, higher temperatures (including neutrino energies), electron-positron pairs, which—via positron captures and neutrino interaction with nuclei/nucleons—increase Y_e to values of the order 0.15 in equatorial dynamic ejecta, comparable to those mentioned above in MHD-supernova (magnetar) jets (Wanajo et al. 2014; Goriely et al. 2015; Martin et al. 2018), where the 3rd peak shift did not occur.

After ballistic/hydrodynamic ejection of matter (i.e. the dynamic ejecta), a hot and massive combined neutron star, forming (dependent on the equation of state) before collapsing to a black hole, evaporates a neutrino wind (Rosswog et al. 2014; Perego et al. 2014; Martin et al. 2015) with neutrino wind contributions from matter in more polar directions with Y_e's up to 0.4. These can contribute also the lighter r-process nuclei. In a similar way outflow from the black hole accretion disk, also powered by neutrinos (and viscous disk heating), can provide, in addition to heavy r-process nuclei, the abundance component of light r-process nuclei (Wanajo et al. 2014; Just et al. 2015; Goriely et al. 2015; Wu et al. 2016).

4 Observational Constraints

4.1 Individual Supernovae and Gamma-Ray Detection

The rapid rise of the optical light curves of Type Ia supernovae (SNeIa), ~ 3 weeks, suggested immediately that a compact object was responsible for the explosion, possibly a carbon-oxygen white dwarf in a close binary system as it was advanced initially by Hoyle and Fowler (1960), see also Sect. 3.1. In such a scenario, the major part of the energy released by the explosion is invested in the adiabatic expansion of the debris and, consequently, an additional source of energy has to be invoked to account for the observed properties of the light curve. This fact, together with the presence of a long lasting optical tail, $\tau \simeq 70$ days, let Truran et al. (1967), Clayton et al. (1969), Colgate and McKee (1969) to propose the radioactive decay of $^{56}Ni \rightarrow {}^{56}Co \rightarrow {}^{56}Fe$ ($q_{Ni} \sim 7 \times 10^{49}$ erg/M$_\odot$, $q_{Co} \sim 1.5 \times 10^{50}$ erg/M$_\odot$, and $\tau_{1/2}(Ni) = 6.1$ days and $\tau_{1/2}(Co) = 77.1$ days) as the source of energy necessary to power the supernova light curve.

The decay of ^{56}Ni and ^{56}Co produces gamma photons in the range of a few keVs to few MeVs, as well as positrons. As the debris expand, more and more photons avoid thermalization and escape in such a way that they can be used as a diagnostic tool of the explosion. The main advantages are the penetrating power of gammas, the relative simplicity of their transport and the possibility to associate them to specific isotopes synthesized during the explosion. The problem of using these medium energy gamma-rays as a diagnostic tool is their detection. CGRO only obtained an upper limit in the case of SN1998bu (Georgii et al. 2002) and performed a controversial detection in the case of SN1991T (Lichti et al. 1994; Morris et al. 1997), while INTEGRAL obtained un upper limit in the case of SN2011fe (Isern et al. 2013) and finally, twelve years after its launch, a true measurement in the case of SN2014J (Churazov et al. 2014, 2015; Diehl et al. 2014, 2015; Isern et al. 2016) that definitively confirmed the hypothesis that ^{56}Ni was powering the light curve of Type Ia supernovae. The origin of this difficulty is that the cross section of ~ 1 MeV photons lies between the photoelectric effect and pair creation, and is dominated by Compton scattering, which deposits small amounts of energy during the interaction and creates secondary photons.

The gamma-ray light curves and line shapes are sensitive to the distribution of the radioactive elements in velocity and position. These distributions are determined by the mode in which the flame propagates on its way through the entire star and by the extension of electron capture processes in the ashes left by the flame. The properties of the burning front are extremely sensitive to the way it begins and the initial conditions of temperature, density and chemical composition. These, in turn, depend on the way the white dwarf approaches the Chandrasekhar mass, which depends on the initial mass and structure of the white dwarf, the accretion rate and the chemical composition of the accreted material.

Core collapse supernovae (CCSNe) also synthesize ^{56}Ni. Typical values found in nearby supernovae—SN1987A (Arnett et al. 1989), SN1993J (Woosley et al. 1994a), SN1994I (Iwamoto et al. 1994)—are of the order of 0.07 $M_{\odot}$ as obtained from the theoretical analysis of the exponential behavior of the optical light curve tail (see e.g. Nomoto et al. 2013, for an overview). In this case, the importance of the detection of ^{56}Ni comes from the fact that it provides a good diagnostic of the energy involved in the explosion (Suwa et al. 2017) as well as on the physical conditions leading to the formation of a neutron star or a black hole.

Another isotope that can provide interesting information is ^{44}Ti. This isotope is the result of an alpha-rich freeze-out of explosive Si-burning (see Sect. 2.2 and Fig. 7) and is synthesized near the mass-cut between the proto-neutron star and the ejecta in CCSNe, and by SNeIa in the sub-Chandrasekhar scenarios (Shen and Bildsten 2014; Shen et al. 2017) or in the collision of a CO white dwarf with a He white dwarf (see Sect. 3.1).

A further radioactive burning product emitting gamma-rays, ^{26}Al, is due to hydrostatic and explosive burning in massive stars, leading to CCSNe. It is produced in core and shell H-burning via the NeNaMgAl-cycle and ejected in stellar wind phases of massive Wolf-Rayet stars. While He-burning is destructive due to n-production and neutron capture of existing ^{26}Al (but produces ^{25}Mg in the neutron-producing reaction $^{22}\mathrm{Ne}(\alpha, n)^{25}\mathrm{Mg}$), C-burning (with protons released from C-fusion) acts as a ^{26}Al-source via Mg-production and proton captures on ^{25}Mg. Explosive Ne/C-burning during a supernova explosion produces ^{26}Al as well (Tur et al. 2010; Thielemann et al. 2011), being actually of higher importance than the hydrostatic burning stages and leading to a typical production of about 10^{-4} $M_{\odot}$ per supernova. Due to its long half-life of 720000 y it is not only seen in supernova explosions or their remnants, but gives rather an integrated picture of the interstellar medium around massive star sources, as observed e.g. with the INTEGRAL satellite, indicating its distribution along the galactic disk and bulge (see e.g. Kretschmer et al. 2013). For this reason we will not discuss it here along with individual supernovae and their remnants, but refer to Sect. 4.2 and especially Sects. 4.2.2 and 4.2.4, where it is discussed as a positron source.

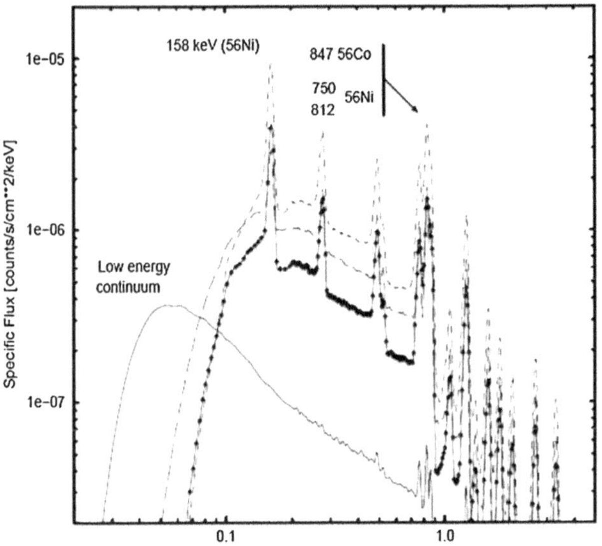

Fig. 20 Gamma-ray spectrum for four models of a SNIa explosion at 5 Mpc 20 days after the explosion. Pure deflagration model (solid line), delayed detonation model (long-dashed line), detonation model (dashed line) and sub-Chandrasekhar model (starred line) (from Gómez-Gomar et al. 1998)

4.1.1 Properties of the Gamma-Signal: Type Ia Case

At present there is not a self-consistent theory of the propagation of nuclear flames through the degenerate interior of a white dwarf and, in one way or another, all calculations rely on a parameterization of the burning front. In 1D geometry, the predicted gamma-emission can be roughly described as follows (Gómez-Gomar et al. 1998):

- The maximum intensity of these lines strongly depends on the expansion velocity and distribution of ^{56}Ni in the models. Those models involving prompt and delayed detonations display strong lines twenty days after the explosion, when the optical maximum occurs, because their high expansion velocity induces a rapid decrease of the optical depth. These lines are particularly strong in pure detonation and sub-Chandrasekhar models since they contain ^{56}Ni and ^{56}Co in the outer layers. Pure deflagration models only display a continuum since they efficiently comptonize the high-energy gamma-rays. The shape of the continuum at low energies is limited in all the models by the competing photoelectric absorption, which imposes a cut-off below 40–100 keV. The energy of the cut-off is determined by the chemical composition of the external layers where most of the emergent continuum is formed at this epoch (Fig. 20). Because of the Doppler broadening of the lines, the ^{56}Ni 812 keV and the ^{56}Co 847 keV lines overlap and it is necessary to obtain the amount and distribution of ^{56}Co at late epochs to disentangle them. The 750 keV line is weaker, but could be useful in the case of very bright sources. Consequently, the detection and measurement of the ^{56}Ni 158 keV line is the most reliable way to interpret the emission of this early epoch.
- Two months after the explosion (see Fig. 21), the ^{56}Ni nuclei have disappeared and the emission is dominated by the ^{56}Co lines, which reach their maximum of intensity roughly two months after the explosion in all models except for the pure deflagration ones. At maximum, the intensity of the lines in pure detonations, delayed detonations and sub-Chandrasekhar models is determined by the total mass of radioactive isotopes, while the differences caused by the expansion velocity are secondary. The 122 and 136 keV lines of ^{57}Co are already visible although faint.

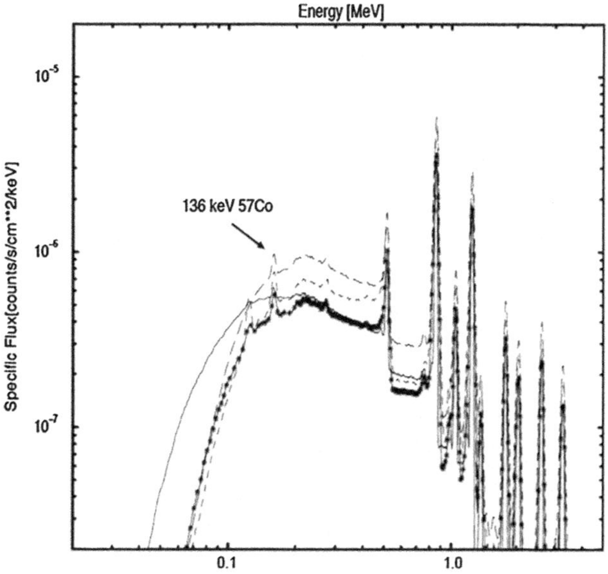

Fig. 21 Gamma-ray spectrum for four models of a SNIa explosion at 5 Mpc 60 days after the explosion. Pure deflagration model (solid line), delayed detonation model (long-dashed line), detonation model (dashed line) and sub-Chandrasekhar model (starred line) (from Gómez-Gomar et al. 1998)

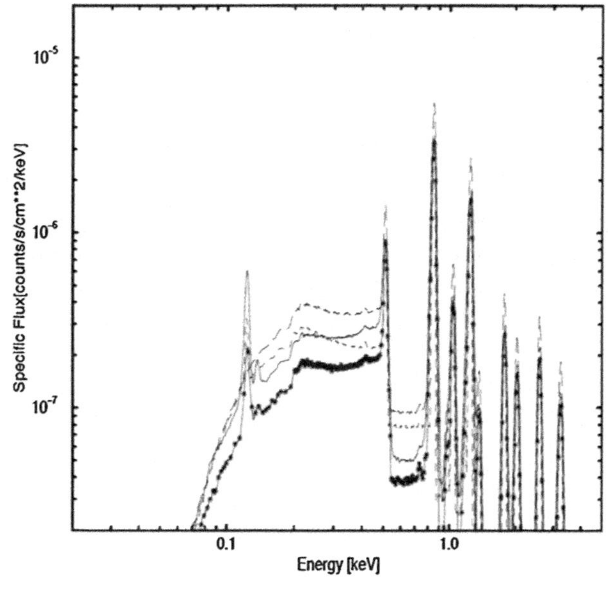

Fig. 22 Gamma-ray spectrum for four models of a SNIa explosion at 5 Mpc 120 days after the explosion. Pure deflagration model (solid line), delayed detonation model (long-dashed line), detonation model (dashed line) and sub-Chandrasekhar model (starred line) (from Gómez-Gomar et al. 1998)

– Four months after the explosion, the ejecta are optically thin in all cases and the intensity of the lines is proportional to the total mass of the parent isotopes (Fig. 22). The continuum is faint and dominated by the positronium annihilation component, which shows a step below 170 keV, the energy of the backscattered 511 photons, plus a contribution of photons scattered once.

Two months after the explosion, line profiles reveal the distribution in velocity of their parent isotopes in all the layers of the ejecta. The profiles corresponding to the pure detonation case are the broadest ones and display a peculiar truncated core produced by the absence of ^{56}Co in the central layers, while those corresponding to the sub-Chandrasekhar models display characteristic wings.

The decay chain ^{44}Ti $\rightarrow$ ^{44}Sc $\rightarrow$ ^{44}Ca ($\tau_{1/2} = 60.0$ years and $\tau_{1/2} = 3.97$ hours, respectively) offers an additional opportunity to obtain information about the explosion although, as a consequence of the relatively long lifetime of ^{44}Ti, the observation has to be performed in young supernova remnants. As it has been already said, this isotope is synthesized during the α-rich freeze out in a low density/high entropy environment, similar to those that occur in delayed detonation and sub-Chandrasekhar models.

4.1.2 Properties of the Gamma-Signal: The CCSN Case

The size of the C+O cores of massive stars strongly depends on the initial mass of the star and it can reach ~ 14 M$_{\odot}$ in stars with initial masses larger than ~ 40 M$_{\odot}$. The total amount of ^{56}Ni injected into the envelope is still unknown and its value is critical to understand the explosion mechanism. Conventional observations suggest that this mass is in the range of 0.4 to 0.07 M$_{\odot}$, or even less. Stars more massive than 20 M$_{\odot}$ seem to be divided into two families (Maeda and Nomoto 2003; Nomoto et al. 2013). One corresponds to the so called hypernovae, with kinetic energies $K \sim 10^{52}$ ergs and ^{56}Ni masses in the range of up to 0.4 M$_{\odot}$, and the other to the so called faint supernovae, with $K \sim 10^{51}$ ergs and ^{56}Ni mass in he range of 0.01 M$_{\odot}$ or less.

In a similar way as in SNeIa, the more promising lines to be observed are those of ^{56}Co during the exponential tail phase of the light curve. In principle they can only be detected in nearby supernovae, since the total amount of ^{56}Ni synthesized is small and, according to the conventional models, it is buried in the innermost layers. However, the high velocity of neutron stars indicates that they are born with a kick. Further observations support that spherical symmetry is not necessarily valid. Therefore, if some events contain radioactive clumps or 'holes' in the debris, the number of gamma photons able to escape could increase. The regular CCSN Cas A shows such a behavior (Wongwathanarat et al. 2017). Another example is the Type Ib supernova SN2005bf, where the optical tail declined much more quickly than the one corresponding to the completely trapped gamma photons, suggesting that ^{56}Ni was either placed in the outer layers or that there were holes in the mass distribution (Tominaga et al. 2005) or (similarly) that ^{56}Ni was initially ejected in a jet-like feature, followed by a magnetar-powered light curve in the later phase (Nomoto et al. 2011, see also Sect. 2.3).

^{44}Ti is produced in the innermost ejecta (Si-burning layer) just on the top of the collapsing iron core and provides a direct probe of the explosion mechanism. The must useful lines are the 67.87 and 78.32 keV lines for the ^{44}Ti and the 1157 keV line for the ^{44}Sc decays.

4.1.3 Positrons

One of the more prominent features is the 511 keV annihilation line. Positrons emitted during the decay of ^{56}Co thermalize as a consequence of ionization and excitation energy losses as well as other mechanisms and eventually annihilate either directly or through the formation of positronium. After 200 days, almost all the high energy photons can escape and the energy deposited by positrons is the only available source of energy to power the light curve. Therefore, a careful determination of the 511 keV is fundamental to understand the evolution of the supernova debris. Furthermore, they could noticeably contribute to the galactic content of positrons if an important fraction of them escape (see also Sect. 4.2).

4.1.4 Individual Supernovae: SN2014J

SN2014J was discovered by Fossey et al. (2014) on January 21st 2014 in M82 ($d = 3.5 \pm 0.3$ Mpc) The moment of the explosion was estimated to be on January 14.72 UT 2014 or JD 2456672.22 (Zheng et al. 2014) and was observed three times by the *INTEGRAL* instruments SPI and IBIS/ISGRI. During the first observation run, 16.5–35.2 days after the explosion, *INTEGRAL* obtained a robust detection of the gamma emission despite the fact that its interpretation is controversial due to the weakness of the signal and the presence of instrumental lines[1] very near of the 158 and 812 keV lines in the SPI spectrograph (Churazov et al. 2014, 2015; Diehl et al. 2014, 2015; Isern et al. 2016).

The analysis of the data obtained by SPI during this epoch, at the position of the supernova, reveals the existence of an emission excess in the energy bands 70–190 and 650–1380 keV that was not present in the observations performed before the explosion. The significance of this excess was 5σ, and in the low energy region it was found a broad and completely unexpected redshifted feature associated to the ^{56}Ni 158 keV gamma ray line (Isern et al. 2016). When the temporal evolution of the spectrum is taken into account and the secondary photons are removed, a line of intensity $(1.59 \pm 0.57) \times 10^{-4}$ photons cm^{-2} s^{-1} centered at 154 ± 0.64 keV and a width of 3.7 ± 1.5 keV appears. This line almost disappears during the period 22.6–28.2 days after the explosion and reappears in the period 28.6–35.2 after the explosion, although it is too weak to obtain any conclusion (e.g. Isern et al. 2016, Fig. 7). Interestingly enough, the IBIS/ISGRI displays a similar behavior: 0.149 ± 0.039, 0.078 ± 0.041, 0.143 ± 0.037 count/s in the energy band 67.5–189 keV, during the same time intervals, but with a better signal to noise ratio, 3.8, 1.9, 3.8, respectively (e.g. Isern et al. 2016, Fig. 11). Diehl et al. (2014) also found an excess at the same epoch in the SPI data although their analysis was different. The ^{56}Ni 158 and 812 keV lines were very near to the laboratory value, the Doppler shift was below 2100 km s^{-1} and the broadening modest, less than 6000 km s^{-1}.

The observations during the late period, 50–162 days after the explosion allowed the detection of the ^{56}Co 847 and 1238 keV lines at 4.7σ and 4.3σ of confidence level, thus confirming the hypothesis that the light curves of SNeIa are powered by the ^{56}Ni decay chain. The spectra and the light curve obtained in this way, were broadly consistent with the standard spherical deflagration or delayed detonation models near the Chandrasekhar's mass (Churazov et al. 2014, 2015).

The existence of a gamma emission excess, as compared with the predictions of classical 1D models, before the optical maximum seems robust and strongly suggests the presence of radioactive material in the outer layers of the supernova debris. A conservative point of view suggests that the data are consistent with the standard delayed detonation model (Churazov et al. 2015) without excluding the presence of small amount of ^{56}Ni at the surface if the lines are broad. However, the concordance between the SPI and IBIS/ISGRI data opens the possibility to speculate about the presence in the outer layers of rapidly moving matter containing ^{56}Ni and no longer being able to thermalize gammas at the moment of the observation, followed by an upturn that could correspond to the exposition of the innermost radioactive layers (Isern et al. 2016).

[1]The isotopes ^{47}Sc and ^{58}Co produce two lines at 159 and 811 keV, respectively, that can affect the measurement of the ^{56}Ni lines depending on the energy shift with respect to their value at rest.

4.1.5 Individual Supernovae: SN1987A and Cas A

SN 1987A is a core collapse supernova that exploded in the Large Magellanic cloud on February 24th, 1987. Because of the distance, this supernova has been extensively followed with almost all the existing instruments. The first detection with hard X-rays was obtained from the Mir-Kvant observatory, when they found a very hard spectrum in the 20–300 keV band and a cut-off between 10 and 25 keV (Sunyaev et al. 1987). The first detection of the ^{56}Co was obtained with the gamma-ray spectrometer of the Solar Maximum emission (Matz et al. 1988). The early appearance together with the shape of the profiles indicated the presence of asymmetries and the rapid mixing of ^{56}Ni to the outer ejecta. The GRIS balloon mission allowed to obtain a Gaussian fit of the two main ^{56}Co lines, suggesting velocity dispersions of the order of 3500 km s^{-1} but line intensities smaller than the ones predicted from the bolometric light curve. These facts opened the possibility of considering fragmented and no spherical geometries (Teegarden et al. 1989; Tueller et al. 1990). The data obtained by *INTEGRAL* (Grebenev et al. 2012), and NuSTAR (Boggs et al. 2015) from the ^{44}Ti 68 and 78 keV lines also suggest the existence of a large scale asymmetry in the explosion.

Cas A is the remnant of a CCSN that exploded ~ 350 years ago. The ^{44}Ti emission has been detected with COMPTEL/CGRO (Iyudin et al. 1994), Beppo-SAX (Vink et al. 2001), IBIS/ISGRI of *INTEGRAL* (Renaud et al. 2006) and NuSTAR (Grefenstette et al. 2014). All these measurements are consistent with a yield of $(1.37 \pm 0.19) \times 10^{-4}$ $M_\odot$ (Siegert et al. 2015). Recently it has been shown (Grefenstette et al. 2017) that NuSTAR has an energy resolution able to discriminate the Doppler broadening of the 67.87 and 78.32 keV lines with enough precision to provide a detailed map of this isotope. The results provide evidences that the explosion asymmetry was caused by a low-mode convective instability.

4.1.6 Three Dimensional Effects

The low level of polarization and the rather homogeneous profile of the absorption line of Si II observed in normal SNeIa suggest that the departures from spherical symmetry are small, while the infrared observations indicate that the total amount of ejected unburnt matter is also small. However, three dimensional pure deflagration models show interesting deviations from spherical symmetry since ^{56}Ni is localized in pockets distributed along the radius of the star while large amounts of not burned carbon and oxygen are left in the outer and inner layers (García-Senz and Bravo 2005). Delayed detonation models convert the unburnt mixture of carbon and oxygen into ^{56}Ni, erasing the differences between burnt and not burnt regions. In principle, the profiles of γ-lines could provide some evidences of the burning regime. In a deflagration blobs moving at different velocities should introduce irregularities in the profile of the line, while in the delayed detonation case the profile should be smoother. However, the influence of such effect in the ^{56}Co lines seems to be small (Isern et al. 2006) and is beyond the capabilities of the present instruments, except perhaps in the case of a supernova that would explode in our Galaxy.

Since low energy lines of ^{56}Ni, like the 158 keV one, can only fully develop if significant amounts of this isotope are placed at small optical depths (Gómez-Gomar et al. 1998), one would expect important differences from the different models, scenarios and, in case of asymmetries, lines of sight. The early emission excess found in S2014J could be a good example. Models like the sub-Chandrasekhar ones (Woosley and Weaver 1994), Pulsating Reverse Detonations (Bravo and García-Senz 2009), Gravitational Confined Detonations (Plewa et al. 2004) and collisions of white dwarfs (Kushnir et al. 2013; Aznar-Siguán et al.

2013) (for further reference see Sect. 3.1) can provide scenarios with ^{56}Ni present in the outer layers and, consequently, produce prominent lines at low energies that can be used as a diagnostic tool.

Recently a prominent anomaly in he growing epoch of the light curve of SN2016jhr it has been detected, just ~ 0.5 days after the explosion. Interestingly enough, this SNIa displayed features of different types, a light curve similar to that of the normal ones and absorption lines of titanium similar to those found in the subluminous ones. These properties do not seem a consequence of the interaction with the circumstellar material but the result of the detonation of a thin shell of helium around a massive white dwarf or the merging of two low mass white dwarfs (Jiang et al. 2017). It is evident that the detection in this case of nuclear lines could provide a definitive argument in favor of a helium detonation.

In any case, it is necessary to realize that in order to use the gamma emission as a diagnostic tool we not only need to have spectra with good resolution and high signal-to-noise ratio, but also a statistically significant sample of supernovae in order to disentangle, for each SNIa subtype, which properties are due to the flame itself and which to the asymmetries of the scenario. To achieve such a goal, high sensitivity detectors as those of DUAL (von Ballmoos et al. 2012), e-ASTROGAM (De Angelis et al. 2017) or LOX (R.S. Miller, private communication) are necessary.

4.2 Supernova Remnants as Positron Sources

For over five decades, balloon and satellite experiments have detected positrons both directly, as a cosmic ray component, and indirectly, through the observation of Galactic 511 keV annihilation radiation. As positron production is ubiquitous in a large number of high-energy processes, it was soon supposed that they should be produced in a multitude of astrophysical objects—besides being abundantly generated by cosmic ray interactions in the interstellar medium, and last but not least in the atmosphere and our instruments... Today, the question on the origin of the "astrophysical" positrons is still far from being settled—neither for the high-energy cosmic ray positrons, nor for the positrons annihilating in the Galaxy. However, the most plausible production scenarios all involve processes occurring at the endpoints of stellar evolution, and particularly in supernovae remnants. After passing through a wealth of observational methods, we will discuss the latter in more detail in Sects. 4.2.2 and 4.2.4.

4.2.1 Direct Positron Detection

Balloon-borne cosmic ray detectors have been observing GeV positrons above the Earth's atmosphere since the early 1960's. Initially, the measured ratio of positrons to the combination of positrons and electrons seemed consistent with what was expected from cosmic rays colliding with the interstellar medium, i.e. $e^+/(e^- + e^+) \simeq 0.1$. While some of those early measurements already hinted an enhancement of the positron fraction with energy, the modest statistics of the balloon data made this "anomaly" controversial at best. The picture changed radically in 2006, when the satellite-borne experiment PAMELA started operation, measuring positrons and electrons in the energy range 1.5–100 GeV with largely improved statistics (Adriani et al. 2009). PAMELA firmly established that the positron fraction increases sharply at energies above 10 GeV (see Fig. 23). The anomaly was confirmed up to 200 GeV by the FERMI gamma-ray telescope, in which the absence of a magnet was compensated by the use of the Earth's magnetic field (Ackermann et al. 2012). Finally, precision measurements by the Alpha Magnetic Spectrometer (AMS-02) extended the range from

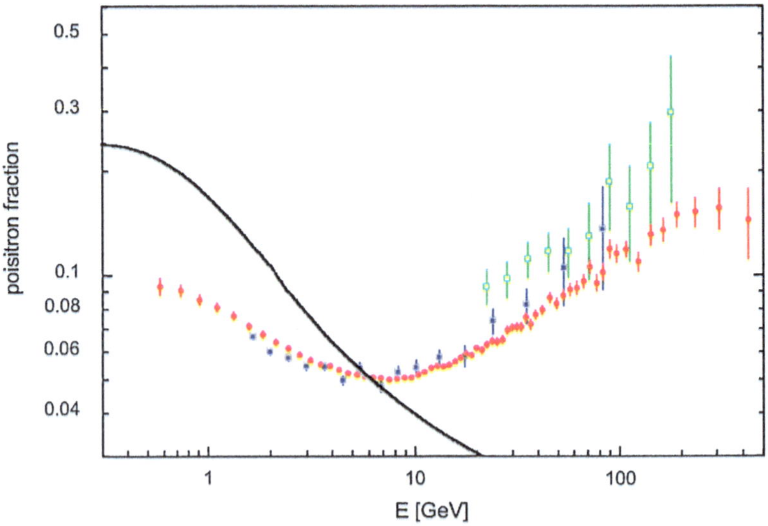

Fig. 23 The most reliable measurements of the positron fraction: full squares (blue) are form PAMELA (Adriani et al. 2009), empty squares (green) are from FERMI-LAT (Ackermann et al. 2012) and full circles (red) are from AMS-02 Krafczyk (2016). The data are compared with the model of Moskalenko and Strong (1998) for positron production through primary CRs interacting with the nuclei of the interstellar medium (solid black line)

0.5–500 GeV. AMS-02 has been taking data since 2011 on the International Space Station (ISS). The latest AMS-02 data (see e.g Krafczyk 2016; Caroff 2016) confirm the monotonic increase of the positron fraction beyond 8 GeV with high statistics and hence unprecedented accuracy. The $e^+/(e^- + e^+)$ ratio grows up to an energy of $\simeq 350$ GeV, after which it shows a tendency to decrease (Fig. 23).

4.2.2 The "Anomalous" Positron Fraction

Before the discovery of the "PAMELA anomaly" it was assumed that the positrons we detect above the atmosphere must have been produced as cosmic ray secondaries, i.e. through spallation reactions of primary CRs with the nuclei of the interstellar medium (ISM) such as the in the typical decay chain: $\pi^+ \rightarrow \mu^+ \rightarrow e^+ + \nu_e + \bar{\nu}_\mu$. The secondary pions created in the collisions of CR protons with the ISM decay in fully polarized muons, resulting in an electron/positron decay asymmetry. This causes a difference in the production spectra of electrons and positrons.

An increase of the positron fraction beyond 8 GeV, however, is not explained by secondary sources. The latter being predicted to generate a decreasing positron fraction—as for example in the GALPROP model of Moskalenko and Strong (1998). Their prediction is compared to the data in Fig. 23, it represents a pure secondary e^+ production due to the propagation and interaction of cosmic rays in the Galaxy without reacceleration. While essentially all the studies agree that the measured "anomaly" requires the presence of a primary source of positrons, a small number of authors pursue modeling of secondary cosmic-ray production. Cowsik et al. (2014) claim that the rise in the positron fraction can be explained entirely as cosmic-ray secondaries, if positrons have a 1–2 Myr residence time in the ISM.

Today, there is a general agreement that the measured "anomaly" requires the presence of a primary source of positrons. Della Torre et al. (2015) estimate the amount of excess positrons (electrons) above ~ 10 GeV as follows: They subtract a model of CR induced secondaries[2] from the positron (electron) spectra observed by AMS-02. The authors find a ratio of excess positrons to excess electrons which is about unity above 50 GeV. The fact that *equal amounts of excess electrons and positrons* are observed above the atmosphere, pinpoints the dominant physical process(es) at work in generating—and transporting—the required leptons. e^-e^+ pair production is hence a natural mechanism for producing the "PAMELA anomaly". Possible positron sources are nearby pulsars and Pulsar Wind Nebulae (PWNe) as well as Supernova Remnants (SNRs); it also has soon been suggested that the excess positrons were associated with the annihilation—or decay—of dark matter.

Dark Matter: Since the discovery of the positron anomaly by PAMELA, hundreds of dark matter related studies have been released (a considerable number of such papers are listed in the works of e.g. Profumo 2011; Ting 2016). Annihilation of Dark Matter could produce positrons ($\chi + \chi \rightarrow e^+ + \cdots$). The telltale signature in the positron spectrum is a cut-off at the mass of dark matter, i.e. $E \sim (\frac{1}{2} - \frac{1}{4})M_\chi$. M_χ is the mass of the dark matter particle. A fit to the AMS-02 data indicates a tentative $M_\chi \simeq 0.5$–1 TeV (Boudaud 2016; Ting 2016). Although receiving enormous attention from the scientific community, a discussion of the positron anomaly produced by a hypothetical dark matter origin is not within the scope of this short review. All the more so, when considering the many effective but unspectacular astrophysical positron sources that can "save the phenomena" without having to invent new physics (Entia non sunt multiplicanda praeter necessitatem or, "When you hear hoofbeats, think of horses not zebras").

Pulsars: Nearby pulsars and pulsar wind nebulae are know to be potentially powerful sources of primary electrons and positrons; they form the largest family of source models not relying on physics from the "dark side of the Universe". Typical pulsar scenario consist in positron production through electromagnetic cascades in the magnetosphere—both in the polar gap and outer gap model for relativistic particle acceleration. In either case, electrons are accelerated in the strong electric fields of the magnetosphere, resulting in the emission of curvature radiation with energies above the threshold for pair production. The gamma rays create e^-e^+ pairs through interactions with the intense magnetic fields, and/or through scattering off the dense thermal X-rays. Before propagating through the ISM, the electrons and positrons produced are trapped in the PWN or in the SNR that envelopes it. The accumulated time-lag is of the order of 10^4–10^5 years (Profumo 2011), implying that the relevant pulsars are "mature", i.e. having typical ages of $> 10^5$ years. Galactic magnetic fields regulate the diffusion of the relativistic leptons ($E_{e^\pm} \gtrsim 10$ GeV) which loose their energy mainly through synchrotron radiation and inverse Compton interactions with background radiation photons. With a lifetime of $t \sim 10^5$ yr (i.e. for cooling from injection energy to rest energy m_e), and a canonical diffusion coefficient D, the electrons and positrons must have been produced within a radius $R \sim \sqrt{D \times t} \sim 100$–500 pc.

A rich literature describes various pulsar models (single pulsars, superposition of nearby pulsars, a galactic population) being able to quantitatively explain the observed positron fraction. Hooper et al. (2009), Linden and Profumo (2013), Profumo (2011), Boudaud et al. (2015) argue that a small number of nearby mature pulsars—or even a single pulsar—can satisfactorily reproduce the spectral features and intensities reported by AMS-02 (Geminga, Monogem, Vela and B0656+14 being among the dominant source candidates). The same argument has actually already been presented by Aharonian et al. (1995) for the Geminga (which had just revealed itself as pulsar) on the basis of balloon data from 1990.

[2] I.e. the "local interstellar spectrum" computed with the GALPROP model of Moskalenko and Strong (1998).

Much more recently, the discussion of the pulsar scenario has been vitalized by observational evidence from very high-energy gamma-ray observations of the HAWC (and Milagro) telescopes. Thanks to their linear propagation, the spectral and spatial information conveyed by gamma-rays can constrain the contribution of nearby pulsars to the positron flux at Earth. The interpretation of the data, however, seems to be far less straightforward: While Hooper et al. (2017) entitle their paper "HAWC Observations Strongly Favor Pulsar Interpretations of the Cosmic-Ray Positron Excess" (based on HAWC measurements of the Geminga and B0656+14 pulsars) the HAWC collaboration (Abeysekara et al. 2017) concludes that the detection of extended gamma-ray emission from the same Geminga and B0656+14 pulsars "disfavors them as origin of the positron excess above GeV energies detected at the Earth".

Supernovae: Here we want to discuss in more detail a source which is more closely related to the overall topic of this review. Various classes of processes within nearby supernova remnants can produce positrons relevant in our context: (a) β^+ decay of the SN ejecta nuclei, with the positrons accelerated in the SNR shock; (b) the acceleration of secondary electrons and positrons injected by way of hadronic or electromagnetic interactions inside the remnant; (c) charge-sign dependent particle injection and shock acceleration

(a) Diffusive shock acceleration (DSA) of decay positrons in supernova remnants has first been proposed by Ellison et al. (1990). While positrons being abundantly created in supernovae through β^+-decay (such as ^{56}Co, ^{44}Ti, ^{26}Al etc.), the escape probability from the dense initial SN environment will be low for positrons produced by isotopes with short half lives—only 0.5%–2% for ^{56}Co with its $\simeq 70$ days half lives (Chan and Lingenfelter 1993). Knowledge of the configuration of the magnetic field lines (open/closed) would be crucial to better constrain the amount of injected positrons (see Sect. 4.2.4). A DSA model for decay positrons in a single SNR has been used by Erlykin and Wolfendale (2002). The SNR responsible for the observed positron excess was located between 250 and 320 pc from the Sun and is 170–380 kyear old. The most probable candidate for such a source is the SNR which contained the Geminga pulsar.
(b) Berezhko and Ksenofontov (2013) study the acceleration of CR, including electrons and positrons, based on a nonlinear time-dependent kinetic model, both inside the SNR and in the diffuse interstellar medium. For the injection into the DSA, two particle populations are studied: (I) postshock thermal particles and (II) pre-existing Galactic Cosmic Rays. The self-consistent CR spectra obtained demonstrate that reacceleration of positrons created in SNR and ISM produces a satisfactory agreement with the existing measurements.
(c) Malkov et al. (2016) have developed a SNR scenario featuring a charge-sign dependent particle injection and shock acceleration. The model is based on a subclass of SNRs with shock accelerated protons, illuminating molecular clumps of weakly ionized dense molecular gas. As the (high energy) protons penetrate into the clumps situated upstream of the shock, they charge them positively. The positrons inside these molecular clumps, produced during pp collisions, are pre-accelerated by the electric potential and driven out of the clumps. While the electrons (together with other negatively charged secondaries) remain locked inside the molecular clumps, positrons are accelerated in the upstream plasma by diffuse shock acceleration. Positrons accelerated in the subclass of SNRs, propagating into clumpy gas media therefore develop harder spectra than those accelerated in other SNRs. An important aspect in the scenario of Malkov et al. (2016) concerns the non-monotonic behavior of the positron fraction, i.e. its v-shaped spectrum with a minimum at roughly 8 GeV. This feature is naturally reproduced through nonlinear shock modification by CR protons. In the subclass of SNRs expanding into molecular clumps, the spectrum is modified with respect to the p^{-4} spectrum produced

by unmodified shocks: it is steeper for low-energies and flatter for high-energy particles, and necessarily contains a point where $p = p^{-4}$. The minimum in the $e^+/(e^- + e^-)$ ratio as well as the upturn at higher energies is obtained by mixing different subsets of SNR shocks (with/without) molecular clumps.

4.2.3 Indirect Positron Detection

Line emission at 511 keV from the Galactic Center region has been observed since the early seventies with balloon experiments. The eighties were marked by ups and downs in the measured 511 keV flux by Ge-detectors flown on stratospheric balloons and on the HEAO-3 satellite. The variable results were initially interpreted as the signature of a compact source of annihilation radiation at the Galactic Center. Yet in 1990, neither the eight years of SMM data (Share et al. 1990), nor the revisited data of the HEAO-3 Ge detectors (Mahoney and Ling 1993), showed evidence for variability in the 511 keV flux. In the early nineties, the imaging SIGMA telescope detected a broad spectral feature (300–700 keV) in the spectrum of 1E 1740.7-2942, a microquasar located close to the galactic center (Bouchet et al. 1991). As the emission vanished within months, the source was supposedly compact, and therefore earned the surname "great annihilator". Throughout the nineties, however, CGRO-OSSE measured steady fluxes from the Galactic center region, and rough skymaps of the Galactic Center region (Purcell et al. 1997) became available based on data from OSSE, SMM and TGRS.

With the INTEGRAL observatory (Vedrenne et al. 2003) that was launched in October 2002—and still is operational as of today—our knowledge in positron astrophysics has significantly improved: using data from its spectro-imager SPI, a first *all*-sky map of electron-positron radiation has been drawn and the physical conditions of the medium in which the positrons annihilate have been constrained by spectroscopy.

The main spectral features of the Galactic e^-e^+ emission measured by INTEGRAL/SPI are shown in Fig. 24 (Jean et al. 2006). The narrow and broad 511 keV line plus the orthopositronium continuum indicate that most positrons annihilate after positronium formation (positronium fraction of $f_{Ps} = 97 \pm 2\%$). The sky map displayed in Fig. 25 represents a best-fit model based on ten years of SPI data (Skinner 2013). The 511 keV emission is strongly concentrated in a bulge region of $\sim 6°$ FWHM diameter in the direction of the Galactic Center. There is evidence for a slight offset (0.5°) of the bulge to negative Galactic longitudes. The Galactic disk is detected with a lower surface brightness, implying a high bulge-to-disk ratio. The total Galactic e^+ annihilation rate is of the order of $2 \times 10^{43}\ e^+\ \mathrm{s}^{-1}$. No evidence has been found for point sources, a "Galactic fountain", or from source regions other than our Galactic disk and bulge.

Recently, Siegert et al. (2016) have analyzed the INTEGRAL/SPI data using a slightly (1 year) better statistics. Their best fit model favors a much thicker and brighter disk than previous models. The disk having an extent of 160^{+10}_{-5} degrees in longitude, and $10.5^{+2.5}_{-1.5}$ degrees in latitude, corresponding to a scale height of ~ 1 kpc. The bulge-to-disk flux ratio is found to be 0.58 ± 0.13, i.e. considerably smaller than in earlier fits.

4.2.4 The Inaccessible Sources of Galactic Positrons

Galactic positron annihilation has been studied with balloon- and satellite-experiments for over five decades. The sky map of electron-positron radiation has been drawn and redrawn by INTEGRAL/SPI since 2002. Thanks to SPI's high resolution spectroscopy (Fig. 24), the

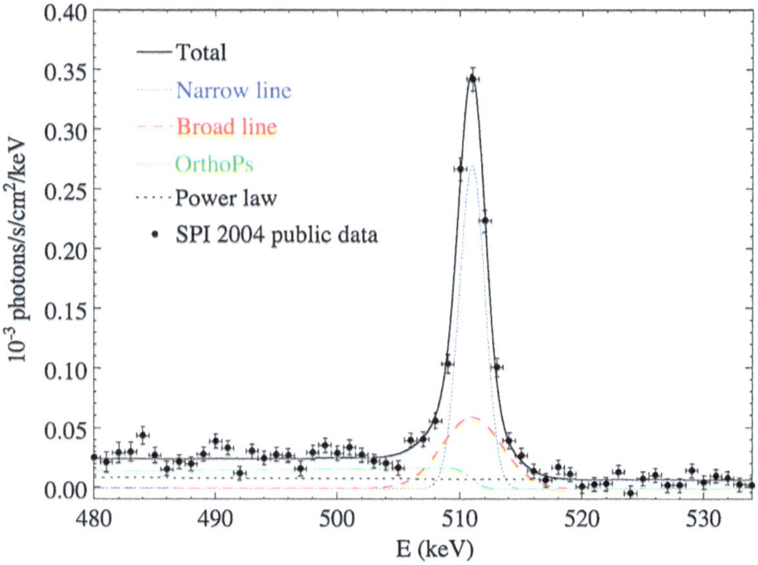

Fig. 24 The spectrum of the Galactic center e^-e^+ annihilation observed by INTEGRAL/SPI (Jean et al. 2006). The fit of the spectrum—i.e. the narrow and a broad Gaussian line—is constraining the physical conditions in the sites where annihilation occurs, i.e. the neutral and ionized warm phases of the interstellar medium

Fig. 25 Best-fit model of the 511-keV e^-e^+ annihilation line in the Galactic Center region (Skinner 2013). The bright central bulge is by far the most prominent feature. The map is based on observations with the imaging spectrometer SPI on board the INTEGRAL satellite, and uses 10 years of mission data

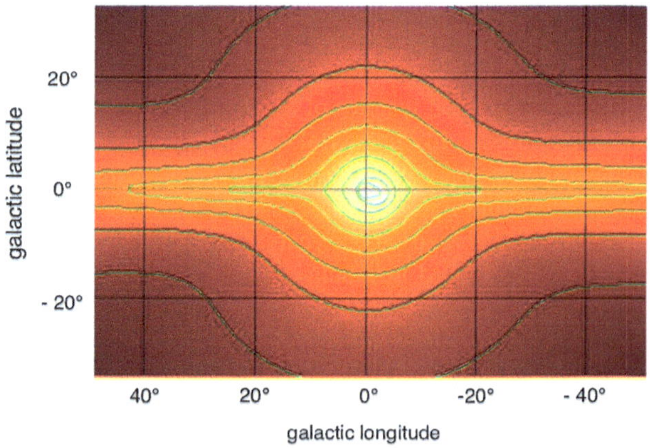

physical conditions in the sites where annihilation occurs are now reasonably well understood. Still, the very origin of the positrons has remained as enigmatic as ever. The spectrum can be interpreted as the signature of positrons annihilating in the warm neutral phase and the warm ionized phase of the interstellar medium ($T \sim 8000$ K). But while spectroscopy characterizes the physical conditions in which the positrons *perish*, it can't help us with finding their birthplace. Likewise, the 511 keV skymap (Fig. 25), although inviting for being compared and fitted with familiar astrophysical source distributions, is most likely quite different from the "true" map of positron sources. Is it the fast variability of the 1990's "great annihilator" that made scholars believe that *compact* 511 keV could exist—i.e. that positrons would annihilate close to their sources? Only if the positrons *do not travel* too far before an-

Table 1 Principal astrophysical sources proposed for Galactic positrons

Galactic Supernovae	(Lingenfelter and Ramaty 1979)
Galactic Supernovae, escape from disk	(Higdon and Lingenfelter 2006)
SN 1991bg-like sub-luminous Supernovae	(Crocker et al. 2017)
Hypernovae/GRB in the Galaxy	(Hueter and Lingenfelter 1983)
Novae	(Clayton and Hoyle 1974)
LMXBs	(Prantzos 2004)
HMXBs/Micro-Quasars	(Guessoum et al. 2006)
Wolf-Rayet stars	(Dearborn and Blake 1985)
Red Giants	(Norgaard 1980)
Pulsars	(Sturrock 1971)
Millisecond pulsars	(Wang et al. 2006)
Cosmic Ray interactions with matter	(Ramaty et al. 1970)
Light (MeV) dark matter	(Boehm et al. 2004)
Heavy 500 GeV DM + de-excitation	(Finkbeiner and Weiner 2007)
Sgr A*	(Titarchuk and Chardonnet 2006)

nihilating, their origin can be constrained by the morphology of the 511 keV radiation. Here is one of the principal challenges for all studies invoking the origin of the positrons: depending on the still poorly understood properties the ISM and the Galactic magnetic fields, positrons can propagate for several 10^5 years away from their sources before annihilating (Prantzos et al. 2011).

If the source distribution of positrons is difficult to come by, what about the source mechanism? Positrons are the most common and easily produced form of antimatter, hence a multitude of processes is available for their generation. Amongst the principal mechanisms that have been proposed for their production in the Galaxy are (a) β^+-decaying radioactive isotopes, (b) the decay of π^+'s produced in cosmic-ray proton interaction with interstellar nuclei, (c) pair creation by high-energy photons, and (d) annihilation or decay of "light dark matter". Accordingly, the conditions for generating positrons are realized in an impressive number of objects—Table 1 summarizes the principal astrophysical sources together with the first (or most relevant) mention in the literature. Many other exotic positrons sources are missing, such as superconducting cosmic strings, decaying axinos, evaporating primordial black holes, color superconducting dark matter droplets (strangelets), not to speak of Q balls, dark energy stars, moduli, sterile neutrinos, mirror matter, millicharged fermions, unstable branons...

In the following paragraph the origin of the positrons will be discussed briefly, with the emphasis on positrons from supernovae. It is assumed that the 511 keV emission of the large scale features in the 511 keV skymap (Fig. 25), namely the Galactic disk and bulge, represent genuine positron source regions (i.e. little or no propagation of the positrons).

The Galactic Bulge positrons do not fit any 'classic' source distribution. Besides non-stationary explanations such as a recent starburst episode close the Galactic Center black hole Sag A* (Alexis et al. 2014), cusp shaped distributions of dark matter distributions have been proposed for producing the bulge positrons. This latter hypothesis recently has gotten renewed interest as certain dark matter halo models (Kuhlen et al. 2013) invoke an offset in the central density peak towards negative Galactic longitudes... The origin of the positrons annihilating in the Galactic disk could readily be explained by several astrophysical objects from Table 1, however the most plausible source remain radioactive isotopes (^{56}Ni, ^{44}Ti, ^{26}Al) produced by supernovae.

The observed ^{26}Al alone (see the last paragraph in the opening of Sect. 4.1) can account for $\sim$ 4–5 $\times$ 10^{-4} photons cm^{-2} s^{-1} since its disintegration is accompanied by the emission of a positron in 85% of the decays. This represents already 1/5 to 1/4 of the positrons required by the observations. The decay of ^{56}Ni and ^{56}Co in Type Ia supernovae creates a wealth of positrons. Together with the nuclear gamma-rays produced in the decay, they power the light curves of SNeIa by injecting their energy into the expanding SNR. Whether or not the positrons are able to escape from the ejecta depends on the strength and geometry of the magnetic field. Positron transport through SNIa models was simulated by Chan and Lingenfelter (1993) who estimate that typically only 0.5%–2% of ^{56}Co-decay positrons escaping the ejecta. However, in the case of radially combed magnetic fields and fully mixed ejecta a much more substantial fraction of ^{56}Co positrons (10%) may escape (see also Sect. 4.1.3). Milne et al. (2001) followed up that study by comparing a sample of light curves of SNeIa with simulations, the underlying idea being that positron escape leads to kinetic energy loss from the system and a drop in the late light curve. According to this study, the number of positrons ejected by a typical SNIa $n_{e^+} \sim 8 \times 10^{52}$ (i.e. an escape fraction of 3%). The positron production rate from SNeIa can hence be estimated $\dot{N}_{e^+} = n_{e^+} \times R_{SNIa} \simeq 1.3\text{–}1.5 \times 10^{43}$ s^{-1} with the SNIa frequency $R_{SNIa} \simeq 0.5\text{–}0.6$ SN/century (Prantzos et al. 2011). This high a production rate would explain virtually all the observed annihilation radiation. Aside from failing to explain the high bulge/disk ratio (Galactic supernova predominantly occur in the disk), such a production rate is clearly a very optimistic scenario, at best an upper limit. A more recent study of the energy production and retention in the ejecta of SN 2011fe by Kerzendorf et al. (2014) reveals that the optical light curve is explained by the *full* thermalization of the ^{56}Co decay positrons. In other words: all positrons created remain trapped and annihilate within the thick supernova remnant...

The latest proposed positron source is a rare type of thermonuclear supernovae known as SN 1991bg-like (Crocker et al. 2017), resulting from the merger of a CO white dwarf and a He white dwarf. 1991bg-like supernovae are several times less luminous than average SNIa; they represent $\sim$ 15% of all SNIa. The essential difference to other SNe lies in the fact that 1991bg-like SNe are supposed to produce considerable amounts of ^{44}Ti. Because of its long lifetime of $\sim$ 60 days, the ^{44}Ti positrons will escape easily from the expanding supernova shell which will have become tenuous by the time the ^{44}Ti decays. If 1991bg-like SNeIa produce as a matter of fact $\sim$ 0.03 $M_\odot$ of ^{44}Ti as claimed by the authors of the study, this scenario would become a very interesting alternative.

4.3 Chemical Evolution

The general tendency of the dominance of CCSN ejecta during low metallicities (showing increased [α/Fe] ratios, α standing for the so-called alpha-elements O, Ne, Mg, Si, S, Ar, Ca, Ti) until the emergence of SNeIa, contributing large amounts of ^{56}Ni (decaying to Fe), has been discussed already in the introduction. Figure 26 shows this tendency for the alpha-element Mg and the r-process element Eu. Let us first focus on the behavior of the alpha-elements. Massive stars show increasing shell sizes for H, He, C, and Ne-burning. As can bee seen from Fig. 6, the products of these hydrostatic burning stages are not "strongly" affected by explosive burning, i.e. only smaller fractions of these elements are destroyed in the shock wave. This means that e.g. O, Ne, and Mg ejecta are dominated by matter from the initial stellar model, where the involved mass zones increase also with the overall stellar mass. This is different for Si, S, Ar, Ca, Ti, and also Ni(Fe), which are created in explosive burning. If,

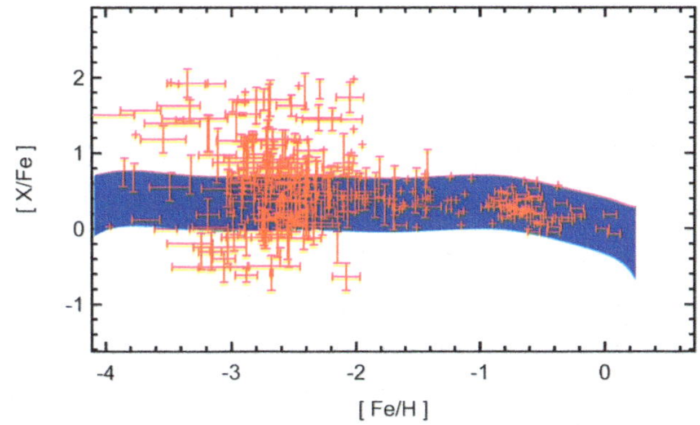

Fig. 26 Ratios of [Mg/Fe] (blue uncertainty range, indicating 95% of observations) and [Eu/Fe] (individual stellar observations shown as red error bars) as a function of "metallicity" [Fe/H] for stars in our Galaxy, taken from the database (Suda et al. 2008, 2011). [X/Y] stands for $\log_{10}$ [(X/Y)/(X/Y)$_\odot$], i.e. [Mg/Fe] $= 0$ or [Fe/H] $= 0$ for solar ratios, -1 for 1/10 of solar etc. Mg shows a relatively flat behavior up to [Fe/H] ≤ -1, similar to other alpha elements like O, Mg, Si, S, Ar, Ca, Ti, changing the slope at [Fe/H] $= -1$ and attaining solar values at [Fe/H] $= 0$. This is explained by the early contributions of core-collapse supernovae before Type Ia supernovae set in. The real scatter is probably smaller than indicated by the blue region, as this is a collection of many observations from different telescopes, different observers and different analysis techniques. To the contrary, the scatter of Eu/Fe is larger than two orders of magnitude at low metallicities, indicating rare production sites with a low event rate, which take long to arrive at average values only in the interval $-2 \leq$ [Fe/H] ≤ -1. Such average values are seen for alpha-elements (with core-collapse supernova origin) already in the range -4 $\leq$ [Fe/H] ≤ -3. This effect can be due to an early averaging or also due to very similar element ratios obtained for a large fraction of all CCSNe

as discussed in Sect. 2.2, the explosion energy increases with compactness of the stellar model, i.e. also with stellar mass up to about 20 $M_\odot$, this would also lead to an increase of the heavier alpha element production as a function of initial stellar mass, including also Ni and Fe. Thus, although we do not expect the same yields from core-collapse supernovae with different progenitor masses, the ratio of these elements will tend to be similar. This might be an explanation why (even in inhomogeneous galactic evolution models and observed reality) no large scatter is seen in [alpha/Fe], even at lowest metallicities, where not much mixing or averaging could have occurred, yet, and the varying contribution of individual SN-explosions could still be detected. This could be an explanation for the observations, here shown for the alpha-element Mg.

In Sect. 3.1 we have already discussed that the behavior of [Mn/Fe] can be based on a similar behavior, reflecting the roles which CCSNe and SNeIa play in galactic evolution, i.e. a [Mn/Fe] < 0 ratio in CCSNe and an increase as a function of [Fe/H] due to the production of ^{55}Mn in SNeIa. In addition, the latter would also increase as a function of metallicity. The observed evolution of [Mn/Fe] as function of [Fe/H] from recent studies is displayed in Fig. 27 (from Mishenina et al. 2015). [Mn/Fe] from CCSNe results in about ≈ -0.4. The old W7-model predicts for SNeIa ejecta [Mn/Fe] $= 0.067, 0.227, 0.30, 0.38$ at [Fe/H] $= -\infty$, 0, 0.25, 0.5. Seitenzahl et al. (2013) find [Mn/Fe] $= 0.4$ already for solar [Fe/H] values and conclude that M_{Ch} models have to contribute in order to explain the observed trend (see also Kobayashi et al. 2015 who conclude that 50% of Type Ia supernovae require deflagration models with low central Y_e-values and 50% can be explained by He-detonations, possibly consistent with Shen et al. 2017). Thus, these trends in galactic chemical evolution give an additional indication for the diversity of SNeIa progenitor models.

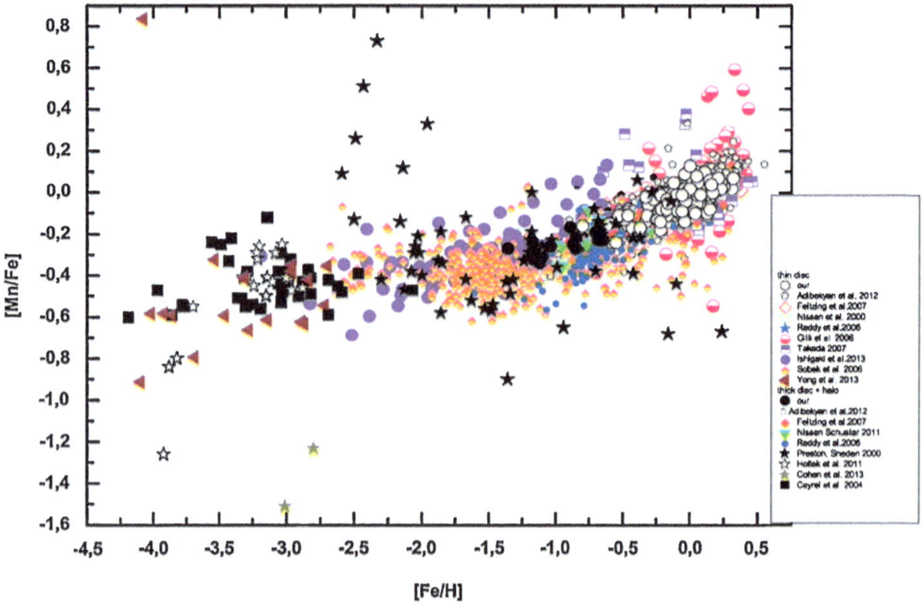

Fig. 27 The evolution of [Mn/Fe] as function of [Fe/H] from recent studies/observations (Mishenina et al. 2015). Notice the rise of [Mn/Fe] from subsolar values at metallicities close to -1 to ratios exceeding solar beyond solar metallicities. This behavior has a similar origin (but acting in the opposite way) as the starting decline of [α/Fe] from supersolar values at the same metallicity. In both cases the change is due to the appearance of SNe Ia after the early dominance of CCSNe. The rise of [Mn/Fe] seems to require some contribution from single degenerate central deflagration models of SNeIa

There exist additional tendencies beyond the major ones for alpha elements and Mn, shown here. Lighter odd-Z elements are affected by s-processing (not only the heavy s-process nuclei), which is indirectly also a measure of metallicity. (The Fe-group element Cu displays this effect as well.) Thus, opposite to the (even-Z) alpha elements, one can observe a metallicity dependence of lighter odd-Z elements (and Cu). Most of the Fe-group nuclei are essentially determined by explosive Si-burning. They are thus the products of this type of processing in CCSNe as well as SNeIa, and could thus feature the different burning conditions for these two types of objects. The case of [Mn/Fe] discussed above is an example for this. Lower metallicities should be dominated by core-collapse events. The early problems to explain near-solar ratios for [Sc/Fe], [Cr/Fe], [Co/Fe], and [Zn/Fe] by CCSNe via piston or thermal bomb models seems to have been solved with improved explosion models which include the effect of neutrino interactions during the supernova explosion Fröhlich et al. (2006b), Ebinger et al. (2017) as discussed in Sect. 2.2. There have been reports of non-solar ratios for Cr, Co, and Zn at low metallicities which could be explained by different fractions of ejected matter from explosive Si-burning with alpha-rich freeze-out or incomplete Si-burning (Nakamura et al. 1999; Nomoto et al. 2013, see e.g.). However, there exist also more recent observations which question the results for Cr and Co. This is not the case for Zn, which shows a clear rise above solar for low metallicities with [Fe/H] < -3. While solar Zn/Fe ratios can be explained with improved core-collapse supernova models (see Sect. 2.2), the high Zn/Fe ratios at lowest metallicities need another origin. Very high explosion energies of hypernovae, resulting in high entropies and a very strong alpha-rich freeze-out could explain this behavior (Nomoto et al. 2013; Nomoto 2017).

In the remaining part we want to focus here on the origin of the r-process. Abundances in low metallicity stars (Sneden et al. 2008; Honda et al. 2006; Roederer et al. 2014) seem to show two types of (extreme) patterns: (a) so-called Sneden stars come with a solar r-process pattern (i.e. the responsible site dominates the solar r-process and apparently produces it in each single event), Thus, solar r-abundances, which are in principle an average over many possible different signatures, are dominated by one identical signature. (b) So-called Honda stars, with abundance distributions which look like a weak r-process and include some Eu but no 3rd r-process peak, can be seen at low metallicities. (c) Apparently some combinations of (a) and (b) can be found before the solar r-process pattern of (a) is also the general pattern beyond metallicities of [Fe/H] ≈ -2. Eu, as the best observed (almost pure) r-process element, shows a big scatter in [Eu/Fe] at low metallicities, indicating that the r-process is a rare event and it takes longer to arrive at average values than for element which are produced in frequent events like supernovae. A second indication that the r-process is produced in sites of rare occurrence is the recent addition of ^{60}Fe found in deep-sea sediments on earth by probably the last nearby CCSN about 2 million years ago, while ^{244}Pu has not been added at this point in time and its low value must come from rarer and earlier events which permitted for some subsequent decay (Knie et al. 2004; Wallner et al. 2015, 2016; Ludwig et al. 2016). A final indication that supernovae (and at low metallicities these are regular CCSNe) are not responsible for the heavy r-process elements is the non-correlation of r-process elements with the Fe-group, which is produced by CCSNe (Cowan et al. 2005).

Historically a number of possible astrophysical sites for the r-process have been discussed: 1. Neutrino-driven winds in regular CCSNe (e.g. Woosley et al. 1994b; Takahashi et al. 1994; Farouqi et al. 2010), but this site is contradicted by observational constraints and present simulations result at most in a weak r-process (Martínez-Pinedo et al. 2012; Roberts et al. 2012). 2. Electron capture supernovae, do not produce much Fe/Ni, thus, the non-correlation with r-elements would hold, but this mass range in the IMF leads to frequent events, and simulations (e.g. Wanajo et al. 2011) also indicate that at most a weak r-process can be obtained. 3. Neutron star mergers produce large amounts of r-elements (as discussed in detail in Sect. 3.2), negligible amounts of Fe/Ni and they are rare events, of the order of 1% of CCSNe. 4. Black hole accretion disks are the final stage of neutron star mergers and can also provide the lower mass (or even all) r-process elements (Just et al. 2015; Wu et al. 2016). They will also occur for the collapse of massive stars resulting in lGRBs/Hypernovae, but are not expected to produce heavy r-elements (unless special neutrino properties permit so (Surman et al. 2006)). But the latter clearly produce large amounts of Ni/Fe which would be in contradiction with the non-correlation of r-process elements with Fe/Ni. 5. Explosive He-burning in outer shells of massive stars has been suggested in the past, but realistic simulations did not attain the required conditions, unless this occurs at very low metallicities (Banerjee et al. 2011). In that case charge-current neutrino interactions would permit sufficient amounts of neutrons, but with low number densities, producing an abundance patter with peaks shifted to lower mass numbers, thus tending more towards an s-process pattern. In addition, these would not be rare events (at these low metallicities). 6. Finally, polar jets from fast rotators with strong magnetic fields, leading to a rare class of CCSNe producing magnetars. These objects have been discussed in detail in Sect. 2.3 (Winteler et al. 2012; Mösta et al. 2015, 2017; Nishimura et al. 2015, 2017; Halevi and Mösta 2018), they are rare (1% to 1 permille of regular CCSNe), do not produce much Ni/Fe, but large/comparable amounts of heavy r-elements as neutron star mergers.

As a result of this discussion only 3/4 and 6 remain as possible sites to produce heavy r-elements and we will discuss them in the following with respect to their appearance in the chemical evolution of galaxies. It should be made clear that 3 and 4 are part of the same scenario, neutron star mergers (Thielemann et al. 2017b), and this scenario has obviously been proven as a valid one by the recent observations of GW170817 (see e.g. Metzger 2017, for a clear interpretation of the observed nucleosynthesis results). The question remains whether 6 is needed as well.

Chemical evolution modelling of rare events requires inhomogeneous models, which do not automatically assume instantaneous mixing of explosive ejecta with the (whole) interstellar medium. Such approaches have recently been undertaken by a number of authors (Argast et al. 2004; Cescutti et al. 2015; van de Voort et al. 2015; Shen et al. 2015; Wehmeyer et al. 2015; Hirai et al. 2015). While Argast et al. (2004), Cescutti et al. (2015), Wehmeyer et al. (2015) came to the conclusion that the contribution of neutron star mergers with respect to heavy r-process elements shows up too late, i.e. only at metallicities around [Fe/H] $= -2$, different from observations which see them already for [Fe/H] < -3, van de Voort et al. (2015), Shen et al. (2015), Hirai et al. (2015) argue otherwise. The question is how the delayed binary evolution and explosion enters into the interstellar medium, after the related supernovae producing the two neutron stars already ejected ample amounts of Fe. The explanation is essentially due to the assumptions of ejecta mixing with the interstellar medium. If e.g. one assumes that only the typical mass swept up by a Taylor-Sedov blast wave is polluted (of the order a few 10^{-5} $M_{\odot}$), then the results of Argast et al. (2004), Cescutti et al. (2015), Wehmeyer et al. (2015) are reproduced (but seen also in the high resolution run of van de Voort et al. 2015). There are three ways to avoid such conclusions (and there is of course an uncertainty related to them): (1) permitting large scale mixing (like turbulent mixing as argued in van de Voort et al. 2015; Shen et al. 2015), (2) proposing that our Galaxy results from a mixture of initially different galactic substructures with different star formation rates which can shift [Eu/Fe] as a function of [Fe/H] to smaller and larger values Hirai et al. (2015), and (3) neutron star kicks during the last supernova explosion can move the binary neutron star system to locations which did not experience, yet, an Fe-pollution by supernovae, i.e. this could lead to r-process injection in pristine matter with very low [Fe/H] ratios, avoiding the impact of neutron star mergers only a metallicities above -2. (1) is either an artifact of too large star particles in SPH codes, which automatically assume a well mixed abundance distribution in each one of such star particles, while smaller masses of star particles in high resolution runs of van de Voort et al. (2015) actually reproduce the results of Argast et al. (2004), Cescutti et al. (2015), Wehmeyer et al. (2015) or Nature actually provided such large-scale turbulent mixing, a feature which we still need to understand. There is no doubt that compact binary mergers are an essential or even the dominant r-process site, if (3) solves the above mentioned problems. However, other (also rare) additional events like magnetars recently observed by Greiner et al. (2015) could help avoiding problems or add to the explanation of r-process observations at low metallicities.

In fact, the combination of neutron star mergers and magneto-rotational supernovae with MHD-jets can reproduce the observations nicely in (stochastic) inhomogeneous galactic chemical evolution models (Wehmeyer et al. 2015, and see Fig. 28). Both events which produce the strong r-process, including the actinides, are rare events with a frequency of about 1% to 1 permille of regular CCSNe, consistent with constraints from recent additions to the solar neighborhood as well as the early Galaxy, and they fulfill the non-correlation with Ni/Fe production.

Summarizing this extended chapter on supernovae and their nucleosynthesis, we have passed through the evolution of massive stars and their catastrophic endpoints as core-collapse supernovae, their transition to hypernovae, magneto-rotationally driven supernovae, as well as events in binary systems like Type Ia supernovae (requiring white dwarfs), and neutron star mergers (the final endpoint of two core-collapse supernovae in a binary system

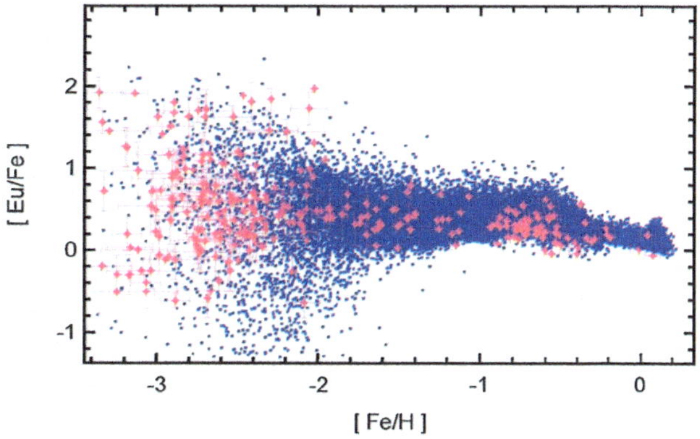

Fig. 28 The evolution of [Eu/Fe] as function of [Fe/H] from observations (SAGA data base Suda et al. 2008, 2011, red crosses) and inhomogeneous chemical evolution predictions (Wehmeyer et al. 2015, blue dots). The combination of MHD-jet supernovae (leading to magnetars) at low metallicities, and neutron star mergers, contributing in later, reproduces nicely the large observational scatter (due to the rare nature of r-process events), and the overall trend

which survives the earlier explosions). With the exception of light elements originating from the big bang, as well as elements up to C and N or beyond Fe, due to s-processing in low and intermediate mass stars, the sites discussed here are the dominant ones permitting the understanding of the evolution of element abundances in galaxies.

Acknowledgements This research was supported by an ERC Advanced Grant (FISH), the Swiss SNF, the MINECO-FEDER grant ESP2015-66134-R (JI), the grant 2009SGR315 and the CERCA program of the Generalitat de Catalunya (JI), the Italian INFN, and the French CNRS. Thanks go to all our present and past collaborators from Barcelona (Eduardo Bravo), Basel (M. Liebendörfer, M. Hempel, T. Rauscher), Darmstadt (A. Arcones, K. Ebinger, M. Eichler, G. Martinez-Pinedo), Kyoto (N. Nishimura). Kyushu (K. Kotake, K. Nakamura), Mainz (K.-L. Kratz), Michigan (K.-C. Pan), Oklahoma (J. Cowan), Moscow (I. Panov, L. Mashonkina), North Carolina (C. Fröhlich, S. Curtis, B. Wehmeyer), Odessa (T. Mishenina), Tokyo (K. Nomoto, T. Takiwaki), as well as Toulouse (Pierre Jean, Jürgen Knoedlseder), and their respective groups (with joint results displayed in a number of figures).

References

B.P. Abbott, R. Abbott, T.D. Abbott, F. Acernese, K. Ackley, C. Adams, T. Adams, P. Addesso, R.X. Adhikari, V.B. Adya et al., Multi-messenger observations of a binary neutron star merger. Astrophys. J. Lett. **848**, L12 (2017). arXiv:1710.05833. https://doi.org/10.3847/2041-8213/aa91c9

A.U. Abeysekara, A. Albert, R. Alfaro, C. Alvarez, J.D. Álvarez, R. Arceo, J.C. Arteaga-Velázquez, D. Avila Rojas, H.A. Ayala Solares, A.S. Barber, N. Bautista-Elivar, A. Becerril, E. Belmont-Moreno, S.Y. BenZvi, D. Berley, A. Bernal, J. Braun, C. Brisbois, K.S. Caballero-Mora, T. Capistrán, A. Carramiñana, S. Casanova, M. Castillo, U. Cotti, J. Cotzomi, S. Coutiño de León, C. De León, E. De la Fuente, B.L. Dingus, M.A. DuVernois, J.C. Díaz-Vélez, R.W. Ellsworth, K. Engel, O. Enríquez-Rivera, D.W. Fiorino, N. Fraija, J.A. García-González, F. Garfias, M. Gerhardt, A. González Muñoz, M.M. González, J.A. Goodman, Z. Hampel-Arias, J.P. Harding, S. Hernández, A. Hernández-Almada, J. Hinton, B. Hona, C.M. Hui, P. Hüntemeyer, A. Iriarte, A. Jardin-Blicq, V. Joshi, S. Kaufmann, D. Kieda, A. Lara, R.J. Lauer, W.H. Lee, D. Lennarz, H.L. Vargas, J.T. Linnemann, A.L. Longinotti, G. Luis Raya, R. Luna-García, R. López-Coto, K. Malone, S.S. Marinelli, O. Martinez, I. Martinez-Castellanos, J. Martínez-Castro, H. Martínez-Huerta, J.A. Matthews, P. Miranda-Romagnoli, E. Moreno, M. Mostafá, L. Nellen, M. Newbold, M.U. Nisa, R. Noriega-Papaqui, R. Pelayo, J. Pretz, E.G. Pérez-Pérez, Z. Ren, C.D. Rho, C. Rivière, D. Rosa-González, M. Rosenberg, E. Ruiz-Velasco, H. Salazar, F. Salesa Greus, A. Sandoval, M. Schneider, H. Schoorlemmer, G. Sinnis, A.J. Smith, R.W. Springer, P. Surajbali, I. Taboada, O. Tibolla, K. Tollefson, I. Torres, T.N. Ukwatta, G. Vianello, T. Weisgarber, S. Westerhoff, I.G. Wisher, J. Wood, T. Yapici, G. Yodh, P.W. Younk, A. Zepeda, H. Zhou, F. Guo, J. Hahn, H. Li, H. Zhang, Extended gamma-ray sources around pulsars constrain the origin of the positron flux at Earth. Science **358**, 911–914 (2017). arXiv:1711.06223. https://doi.org/10.1126/science.aan4880

M. Ackermann, M. Ajello, A. Allafort, W.B. Atwood, L. Baldini, G. Barbiellini, D. Bastieri, K. Bechtol, R. Bellazzini, B. Berenji, R.D. Blandford, E.D. Bloom, E. Bonamente, A.W. Borgland, A. Bouvier, J.

Bregeon, M. Brigida, P. Bruel, R. Buehler, S. Buson, G.A. Caliandro, R.A. Cameron, P.A. Caraveo, J.M. Casandjian, C. Cecchi, E. Charles, A. Chekhtman, C.C. Cheung, J. Chiang, S. Ciprini, R. Claus, J. Cohen-Tanugi, J. Conrad, S. Cutini, A. de Angelis, F. de Palma, C.D. Dermer, S.W. Digel, E. Do Couto E Silva, P.S. Drell, A. Drlica-Wagner, C. Favuzzi, S.J. Fegan, E.C. Ferrara, W.B. Focke, P. Fortin, Y. Fukazawa, S. Funk, P. Fusco, F. Gargano, D. Gasparrini, S. Germani, N. Giglietto, P. Giommi, F. Giordano, M. Giroletti, T. Glanzman, G. Godfrey, I.A. Grenier, J.E. Grove, S. Guiriec, M. Gustafsson, D. Hadasch, A.K. Harding, M. Hayashida, R.E. Hughes, G. Jóhannesson, A.S. Johnson, T. Kamae, H. Katagiri, J. Kataoka, J. Knödlseder, M. Kuss, J. Lande, L. Latronico, M. Lemoine-Goumard, M. Llena Garde, F. Longo, F. Loparco, M.N. Lovellette, P. Lubrano, G.M. Madejski, M.N. Mazziotta, J.E. McEnery, P.F. Michelson, W. Mitthumsiri, T. Mizuno, A.A. Moiseev, C. Monte, M.E. Monzani, A. Morselli, I.V. Moskalenko, S. Murgia, T. Nakamori, P.L. Nolan, J.P. Norris, E. Nuss, M. Ohno, T. Ohsugi, A. Okumura, N. Omodei, E. Orlando, J.F. Ormes, M. Ozaki, D. Paneque, D. Parent, M. Pesce-Rollins, M. Pierbattista, F. Piron, G. Pivato, T.A. Porter, S. Rainò, R. Rando, M. Razzano, S. Razzaque, A. Reimer, O. Reimer, T. Reposeur, S. Ritz, R.W. Romani, M. Roth, H.F.W. Sadrozinski, C. Sbarra, T.L. Schalk, C. Sgrò, E.J. Siskind, G. Spandre, P. Spinelli, A.W. Strong, H. Takahashi, T. Takahashi, T. Tanaka, J.G. Thayer, J.B. Thayer, L. Tibaldo, M. Tinivella, D.F. Torres, G. Tosti, E. Troja, Y. Uchiyama, T.L. Usher, J. Vandenbroucke, V. Vasileiou, G. Vianello, V. Vitale, A.P. Waite, B.L. Winer, K.S. Wood, M. Wood, Z. Yang, Measurement of separate cosmic-ray electron and positron spectra with the Fermi Large Area Telescope. Phys. Rev. Lett. **108**(1), 011103 (2012). arXiv:1109.0521. https://doi.org/10.1103/PhysRevLett.108.011103

O. Adriani, G.C. Barbarino, G.A. Bazilevskaya, R. Bellotti, M. Boezio, E.A. Bogomolov, L. Bonechi, M. Bongi, V. Bonvicini, S. Bottai, A. Bruno, F. Cafagna, D. Campana, P. Carlson, M. Casolino, G. Castellini, M.P. de Pascale, G. de Rosa, N. de Simone, V. di Felice, A.M. Galper, L. Grishantseva, P. Hofverberg, S.V. Koldashov, S.Y. Krutkov, A.N. Kvashnin, A. Leonov, V. Malvezzi, L. Marcelli, W. Menn, V.V. Mikhailov, E. Mocchiutti, S. Orsi, G. Osteria, P. Papini, M. Pearce, P. Picozza, M. Ricci, S.B. Ricciarini, M. Simon, R. Sparvoli, P. Spillantini, Y.I. Stozhkov, A. Vacchi, E. Vannuccini, G. Vasilyev, S.A. Voronov, Y.T. Yurkin, G. Zampa, N. Zampa, An anomalous positron abundance in cosmic rays with energies 1.5–100 GeV. Nature **458**, 607–609 (2009). arXiv:0810.4995. https://doi.org/10.1038/nature07942

F.A. Aharonian, A.M. Atoyan, H.J. Voelk, High energy electrons and positrons in cosmic rays as an indicator of the existence of a nearby cosmic tevatron. Astron. Astrophys. **294**, L41–L44 (1995)

A. Alexis, P. Jean, P. Martin, K. Ferrière, Monte Carlo modelling of the propagation and annihilation of nucleosynthesis positrons in the Galaxy. Astron. Astrophys. **564**, A108 (2014). arXiv:1402.6110. https://doi.org/10.1051/0004-6361/201322393

A. Arcones, F.K. Thielemann, Neutrino-driven wind simulations and nucleosynthesis of heavy elements. J. Phys. G, Nucl. Part. Phys. **40**(1), 013201 (2013). arXiv:1207.2527. https://doi.org/10.1088/0954-3899/40/1/013201

D. Argast, M. Samland, F.K. Thielemann, Y.Z. Qian, Neutron star mergers versus core-collapse supernovae as dominant r-process sites in the early Galaxy. Astron. Astrophys. **416**, 997–1011 (2004). arXiv:astro-ph/0309237. https://doi.org/10.1051/0004-6361:20034265

C. Arlandini, F. Käppeler, K. Wisshak, R. Gallino, M. Lugaro, M. Busso, O. Straniero, Neutron capture in low-mass asymptotic giant branch stars: cross sections and abundance signatures. Astrophys. J. **525**, 886–900 (1999). arXiv:astro-ph/9906266. https://doi.org/10.1086/307938

W.D. Arnett, J.N. Bahcall, R.P. Kirshner, S.E. Woosley, Supernova 1987A. Annu. Rev. Astron. Astrophys. **27**, 629–700 (1989). https://doi.org/10.1146/annurev.aa.27.090189.003213

G. Aznar-Siguán, E. García-Berro, P. Lorén-Aguilar, J. José, J. Isern, Detonations in white dwarf dynamical interactions. Mon. Not. R. Astron. Soc. **434**, 2539–2555 (2013). arXiv:1306.6559. https://doi.org/10.1093/mnras/stt1198

P. Banerjee, W.C. Haxton, Y.Z. Qian, Long, cold, early r process? Neutrino-induced nucleosynthesis in He shells revisited. Phys. Rev. Lett. **106**(20), 201104 (2011). arXiv:1103.1193. https://doi.org/10.1103/PhysRevLett.106.201104

J. Barnes, D. Kasen, M.R. Wu, G. Martínez-Pinedo, Radioactivity and thermalization in the ejecta of compact object mergers and their impact on kilonova light curves. Astrophys. J. **829**, 110 (2016). arXiv:1605.07218. https://doi.org/10.3847/0004-637X/829/2/110

W. Benz, F.K. Thielemann, J.G. Hills, Three-dimensional hydrodynamical simulations of stellar collisions. II—white dwarfs. Astrophys. J. **342**, 986–998 (1989). https://doi.org/10.1086/167656

E.G. Berezhko, L.T. Ksenofontov, Energy spectra of electrons and positrons produced in supernova remnants. J. Phys. Conf. Ser. **012**, 025 (2013). http://stacks.iop.org/1742-6596/409/i=1/a=012025

M.G. Bernardini, Gamma-ray bursts and magnetars: observational signatures and predictions. J. High Energy Astrophys. **7**, 64–72 (2015). https://doi.org/10.1016/j.jheap.2015.05.003

C. Boehm, D. Hooper, J. Silk, M. Casse, J. Paul, MeV dark matter: has it been detected? Phys. Rev. Lett. **92**(10), 101301 (2004). arXiv:astro-ph/0309686. https://doi.org/10.1103/PhysRevLett.92.101301

S.E. Boggs, F.A. Harrison, H. Miyasaka, B.W. Grefenstette, A. Zoglauer, C.L. Fryer, S.P. Reynolds, D.M. Alexander, H. An, D. Barret, F.E. Christensen, W.W. Craig, K. Forster, P. Giommi, C.J. Hailey, A. Hornstrup, T. Kitaguchi, J.E. Koglin, K.K. Madsen, P.H. Mao, K. Mori, M. Perri, M.J. Pivovaroff, S. Puccetti, V. Rana, D. Stern, N.J. Westergaard, W.W. Zhang, ^{44}Ti gamma-ray emission lines from SN1987A reveal an asymmetric explosion. Science **348**, 670 (2015)

L. Bouchet, P. Mandrou, J.P. Roques, G. Vedrenne, B. Cordier, A. Goldwurm, F. Lebrun, J. Paul, R. Sunyaev, E. Churazov, M. Gilfanov, M. Pavlinsky, S. Grebenev, G. Babalyan, I. Dekhanov, N. Khavenson, Sigma discovery of variable e(+)-e(-) annihilation radiation from the near Galactic center variable compact source 1E 1740.7-2942. Astrophys. J. Lett. **383**, L45–L48 (1991). https://doi.org/10.1086/186237

M. Boudaud, A new look at the cosmic ray positron fraction, in *PoS ICRC2015* (2016), p. 1183. arXiv:1510.07492

M. Boudaud et al., A new look at the cosmic ray positron fraction. Astron. Astrophys. **575**, A67 (2015). arXiv:1410.3799. https://doi.org/10.1051/0004-6361/201425197

F. Brachwitz, D.J. Dean, W.R. Hix, K. Iwamoto, K. Langanke, G. Martínez-Pinedo, K. Nomoto, M.R. Strayer, F.K. Thielemann, H. Umeda, The role of electron captures in Chandrasekhar-mass models for Type IA supernovae. Astrophys. J. **536**, 934–947 (2000). arXiv:astro-ph/0001464. https://doi.org/10.1086/308968

D. Branch, J.C. Wheeler, *Supernova Explosions* (2017). https://doi.org/10.1007/978-3-662-55054-0

E. Bravo, D. García-Senz, Pulsating reverse detonation models of Type Ia supernovae. I. Detonation ignition. Astrophys. J. **695**, 1244–1256 (2009). arXiv:0901.3008. https://doi.org/10.1088/0004-637X/695/2/1244

E. Bravo, D. García-Senz, R.M. Cabezón, I. Domínguez, Pulsating reverse detonation models of Type Ia supernovae. II. Explosion. Astrophys. J. **695**, 1257–1272 (2009). arXiv:0901.3013. https://doi.org/10.1088/0004-637X/695/2/1257

S.W. Bruenn, E.J. Lentz, W.R. Hix, A. Mezzacappa, J.A. Harris, O.E.B. Messer, E. Endeve, J.M. Blondin, M.A. Chertkow, E.J. Lingerfelt, P. Marronetti, K.N. Yakunin, The development of explosions in axisymmetric ab initio core-collapse supernova simulations of 12–25 M stars. Astrophys. J. **818**, 123 (2016). arXiv:1409.5779. https://doi.org/10.3847/0004-637X/818/2/123

A. Burrows, Colloquium: perspectives on core-collapse supernova theory. Rev. Mod. Phys. **85**, 245 (2013). arXiv:1210.4921

A.C. Calder, B.K. Krueger, A.P. Jackson, D.M. Townsley, The influence of chemical composition on models of Type Ia supernovae. Front. Phys. **8**, 168–188 (2013). arXiv:1303.2207. https://doi.org/10.1007/s11467-013-0301-4

S. Caroff, High statistics measurement of the positron fraction in primary cosmic rays with the Alpha Magnetic Spectrometer on the International Space Station, in *25th European Cosmic Ray Symposium (ECRS 2016)*, Turin, Italy, September 04–09, 2016 (2016). arXiv:1612.09579

R. Cayrel, E. Depagne, M. Spite, V. Hill, F. Spite, P. François, B. Plez, T. Beers, F. Primas, J. Andersen, B. Barbuy, P. Bonifacio, P. Molaro, B. Nordström, First stars V—abundance patterns from C to Zn and supernova yields in the early Galaxy. Astron. Astrophys. **416**, 1117–1138 (2004). arXiv:astro-ph/0311082. https://doi.org/10.1051/0004-6361:20034074

G. Cescutti, D. Romano, F. Matteucci, C. Chiappini, R. Hirschi, The role of neutron star mergers in the chemical evolution of the Galactic halo. Astron. Astrophys. **577**, A139 (2015). arXiv:1503.02954. https://doi.org/10.1051/0004-6361/201525698

K.W. Chan, R.E. Lingenfelter, Positrons from supernovae. Astrophys. J. **405**, 614–636 (1993). https://doi.org/10.1086/172393

A. Chieffi, M. Limongi, Pre-supernova evolution of rotating solar metallicity stars in the mass range 13–120 $M_\odot$ and their explosive yields. Astrophys. J. **764**, 21 (2013). https://doi.org/10.1088/0004-637X/764/1/21

A. Chieffi, M. Limongi, The synthesis of ^{44}Ti and ^{56}Ni in massive stars. Astrophys. J. **836**, 79 (2017). arXiv:1701.02914. https://doi.org/10.3847/1538-4357/836/1/79

E. Churazov, R. Sunyaev, J. Isern, J. Knödlseder, P. Jean, F. Lebrun, N. Chugai, S. Grebenev, E. Bravo, S. Sazonov, M. Renaud, Cobalt-56 γ-ray emission lines from the Type Ia supernova 2014J. Nature **512**, 406–408 (2014). arXiv:1405.3332. https://doi.org/10.1038/nature13672

E. Churazov, R. Sunyaev, J. Isern, I. Bikmaev, E. Bravo, N. Chugai, S. Grebenev, P. Jean, J. Knödlseder, F. Lebrun, E. Kuulkers, Gamma-rays from Type Ia supernova SN2014J. Astrophys. J. **812**, 62 (2015). arXiv:1502.00255. https://doi.org/10.1088/0004-637X/812/1/62

D.D. Clayton, F. Hoyle, Gamma-ray lines from novae. Astrophys. J. Lett. **187**, L101 (1974). https://doi.org/10.1086/181406

D.D. Clayton, S.A. Colgate, G.J. Fishman, Gamma-ray lines from young supernova remnants. Astrophys. J. **155**, 75 (1969). https://doi.org/10.1086/149849

S.A. Colgate, C. McKee, Early supernova luminosity. Astrophys. J. **157**, 623 (1969). https://doi.org/10.1086/150102

J.J. Cowan, C. Sneden, T.C. Beers, J.E. Lawler, J. Simmerer, J.W. Truran, F. Primas, J. Collier, S. Burles, Hubble Space Telescope observations of heavy elements in metal-poor Galactic halo stars. Astrophys. J. **627**, 238–250 (2005). arXiv:astro-ph/0502591. https://doi.org/10.1086/429952

R. Cowsik, B. Burch, T. Madziwa-Nussinov, The origin of the spectral intensities of cosmic-ray positrons. Astrophys. J. **786**, 124 (2014). arXiv:1305.1242. https://doi.org/10.1088/0004-637X/786/2/124

R.M. Crocker, A.J. Ruiter, I.R. Seitenzahl, F.H. Panther, S. Sim, H. Baumgardt, A. Möller, D.M. Nataf, L. Ferrario, J.J. Eldridge, M. White, B.E. Tucker, F. Aharonian, Diffuse Galactic antimatter from faint thermonuclear supernovae in old stellar populations. Nat. Astron. **1**, 0135 (2017). arXiv:1607.03495. https://doi.org/10.1038/s41550-017-0135

M. Dan, S. Rosswog, J. Guillochon, E. Ramirez-Ruiz, How the merger of two white dwarfs depends on their mass ratio: orbital stability and detonations at contact. Mon. Not. R. Astron. Soc. **422**, 2417–2428 (2012). arXiv:1201.2406. https://doi.org/10.1111/j.1365-2966.2012.20794.x

M. Dan, J. Guillochon, M. Brüggen, E. Ramirez-Ruiz, S. Rosswog, Thermonuclear detonations ensuing white dwarf mergers. Mon. Not. R. Astron. Soc. **454**, 4411–4428 (2015). arXiv:1508.02402. https://doi.org/10.1093/mnras/stv2289

A. De Angelis, V. Tatischeff, M. Tavani, U. Oberlack, I. Grenier, L. Hanlon, R. Walter, A. Argan, P. von Ballmoos, A. Bulgarelli, I. Donnarumma, M. Hernanz, I. Kuvvetli, M. Pearce, A. Zdziarski, A. Aboudan, M. Ajello, G. Ambrosi, D. Bernard, E. Bernardini, V. Bonvicini, A. Brogna, M. Branchesi, C. Budtz-Jorgensen, A. Bykov, R. Campana, M. Cardillo, P. Coppi, D. De Martino, R. Diehl, M. Doro, V. Fioretti, S. Funk, G. Ghisellini, E. Grove, C. Hamadache, D.H. Hartmann, M. Hayashida, J. Isern, G. Kanbach, J. Kiener, J. Knödlseder, C. Labanti, P. Laurent, O. Limousin, F. Longo, K. Mannheim, M. Marisaldi, M. Martinez, M.N. Mazziotta, J. McEnery, S. Mereghetti, G. Minervini, A. Moiseev, A. Morselli, K. Nakazawa, P. Orleanski, J.M. Paredes, B. Patricelli, J. Peyré, G. Piano, M. Pohl, H. Ramarijaona, R. Rando, I. Reichardt, M. Roncadelli, R. Silva, F. Tavecchio, D.J. Thompson, R. Turolla, A. Ulyanov, A. Vacchi, X. Wu, The e-ASTROGAM mission. Exploring the extreme Universe with gamma rays in the MeV-GeV range. Exp. Astron. **44**, 25–82 (2017). arXiv:1611.02232. https://doi.org/10.1007/s10686-017-9533-6

D.S.P. Dearborn, J.B. Blake, On the source of the Al-26 observed in the interstellar medium. Astrophys. J. Lett. **288**, L21–L24 (1985). https://doi.org/10.1086/184413

S. Della Torre, M. Gervasi, P.G. Rancoita, D. Rozza, A. Treves, Pulsar Wind Nebulae as a source of the observed electron and positron excess at high energy: the case of Vela-X. J. High Energy Astrophys. **8**, 27–34 (2015). arXiv:1508.01457. https://doi.org/10.1016/j.jheap.2015.08.001

P.A. Denissenkov, F. Herwig, U. Battino, C. Ritter, M. Pignatari, S. Jones, B. Paxton, i-Process nucleosynthesis and mass retention efficiency in He-shell flash evolution of rapidly accreting white dwarfs. Astrophys. J. Lett. **834**, L10 (2017). arXiv:1610.08541. https://doi.org/10.3847/2041-8213/834/2/L10

R. Diehl, T. Siegert, W. Hillebrandt, S.A. Grebenev, J. Greiner, M. Krause, M. Kromer, K. Maeda, F. Röpke, S. Taubenberger, Early ^{56}Ni decay gamma rays from SN2014J suggest an unusual explosion. Science **345**, 1162–1165 (2014). arXiv:1407.3061. https://doi.org/10.1126/science.1254738

R. Diehl, T. Siegert, W. Hillebrandt, M. Krause, J. Greiner, K. Maeda, F.K. Röpke, S.A. Sim, W. Wang, X. Zhang, SN2014J gamma rays from the ^{56}Ni decay chain. Astron. Astrophys. **574**, A72 (2015). arXiv:1409.5477. https://doi.org/10.1051/0004-6361/201424991

J. Duflo, A.P. Zuker, Microscopic mass formulas. Phys. Rev. C **52**, R23–R27 (1995). arXiv:nucl-th/9505011. https://doi.org/10.1103/PhysRevC.52.R23

K. Ebinger, S. Sinha, C. Fröhlich, A. Perego, M. Hempel, M. Eichler, J. Casanova, M. Liebendörfer, F.K. Thielemann, Explosion dynamics of parametrized spherically symmetric core-collapse supernova simulations, in *14th International Symposium on Nuclei in the Cosmos (NIC2016)*, ed. by S. Kubono, T. Kajino, S. Nishimura, T. Isobe, S. Nagataki, T. Shima, Y. Takeda (2017), 020611. arXiv:1610.05629. https://doi.org/10.7566/JPSCP.14.020611

D. Eichler, M. Livio, T. Piran, D.N. Schramm, Nucleosynthesis, neutrino bursts and gamma-rays from coalescing neutron stars. Nature **340**, 126–128 (1989). https://doi.org/10.1038/340126a0

M. Eichler, A. Arcones, A. Kelic, O. Korobkin, K. Langanke, T. Marketin, G. Martinez-Pinedo, I. Panov, T. Rauscher, S. Rosswog, C. Winteler, N.T. Zinner, F.K. Thielemann, The role of fission in neutron star mergers and its impact on the r-process peaks. Astrophys. J. **808**, 30 (2015). arXiv:1411.0974. https://doi.org/10.1088/0004-637X/808/1/30

M. Eichler, K. Nakamura, T. Takiwaki, T. Kuroda, K. Kotake, M. Hempel, R. Cabezón, M. Liebendörfer, F.K. Thielemann, Nucleosynthesis in 2D core-collapse supernovae of 11.2 and 17.0 $M_\odot$ progenitors: implications for Mo and Ru production. J. Phys. G, Nucl. Part. Phys. **014**, 001 (2018). arXiv:1708.08393. https://doi.org/10.1088/1361-6471/aa8891

C.D. Ellison, C.F. Jones, R. Ramaty, Diffusive shock acceleration of decay positrons in supernovae, in *International Cosmic Ray Conference*, vol. 4 (1990), p. 68

A.D. Erlykin, A.W. Wolfendale, Supernova remnants and the origin of the cosmic radiation: the electron component. J. Phys. G, Nucl. Part. Phys. **28**, 359–378 (2002). https://doi.org/10.1088/0954-3899/28/3/301

T. Ertl, H.T. Janka, S.E. Woosley, T. Sukhbold, M. Ugliano, A two-parameter criterion for classifying the explodability of massive stars by the neutrino-driven mechanism. Astrophys. J. **818**, 124 (2016). arXiv:1503.07522. https://doi.org/10.3847/0004-637X/818/2/124

K. Farouqi, K.L. Kratz, B. Pfeiffer, T. Rauscher, F.K. Thielemann, J.W. Truran, Charged-particle and neutron-capture processes in the high-entropy wind of core-collapse supernovae. Astrophys. J. **712**, 1359–1377 (2010). arXiv:1002.2346. https://doi.org/10.1088/0004-637X/712/2/1359

L. Fimiani, D.L. Cook, T. Faestermann, J.M. Gómez-Guzmán, K. Hain, G. Herzog, K. Knie, G. Korschinek, P. Ludwig, J. Park, R.C. Reedy, G. Rugel, Interstellar Fe 60 on the surface of the Moon. Phys. Rev. Lett. **116**(15), 151104 (2016). https://doi.org/10.1103/PhysRevLett.116.151104

D.P. Finkbeiner, N. Weiner, Exciting dark matter and the INTEGRAL/SPI 511 keV signal. Phys. Rev. D **76**(8), 083519 (2007). arXiv:astro-ph/0702587. https://doi.org/10.1103/PhysRevD.76.083519

R.B. Firestone, Observation of 23 supernovae that exploded $<$ 300 pc from Earth during the past 300 kyr. Astrophys. J. **789**, 29 (2014). https://doi.org/10.1088/0004-637X/789/1/29

T. Fischer, S.C. Whitehouse, A. Mezzacappa, F.K. Thielemann, M. Liebendörfer, Protoneutron star evolution and the neutrino-driven wind in general relativistic neutrino radiation hydrodynamics simulations. Astron. Astrophys. **517**, A80 (2010). arXiv:0908.1871. https://doi.org/10.1051/0004-6361/200913106

S.J. Fossey, B. Cooke, G. Pollack, M. Wilde, T. Wright, Supernova 2014J in M82 = Psn J09554214+6940260. Central Bureau Electronic Telegrams 3792 (2014)

C. Freiburghaus, S. Rosswog, F.K. Thielemann, R-process in neutron star mergers. Astrophys. J. Lett. **525**, L121–L124 (1999). https://doi.org/10.1086/312343

U. Frischknecht, R. Hirschi, M. Pignatari, A. Maeder, G. Meynet, C. Chiappini, F.K. Thielemann, T. Rauscher, C. Georgy, S. Ekström, s-Process production in rotating massive stars at solar and low metallicities. Mon. Not. R. Astron. Soc. **456**, 1803–1825 (2016). arXiv:1511.05730. https://doi.org/10.1093/mnras/stv2723

C. Fröhlich, P. Hauser, M. Liebendörfer, G. Martínez-Pinedo, F.K. Thielemann, E. Bravo, N.T. Zinner, W.R. Hix, K. Langanke, A. Mezzacappa, K. Nomoto, Composition of the innermost core-collapse supernova ejecta. Astrophys. J. **637**, 415–426 (2006a). arXiv:astro-ph/0410208. https://doi.org/10.1086/498224

C. Fröhlich, G. Martínez-Pinedo, M. Liebendörfer, F.K. Thielemann, E. Bravo, W.R. Hix, K. Langanke, N.T. Zinner, Neutrino-induced nucleosynthesis of $A > 64$ nuclei: the νp process. Phys. Rev. Lett. **96**(14), 142502 (2006b). arXiv:astro-ph/0511376. https://doi.org/10.1103/PhysRevLett.96.142502

B. Fryxell, K. Olson, P. Ricker, F.X. Timmes, M. Zingale, D.Q. Lamb, P. MacNeice, R. Rosner, J.W. Truran, H. Tufo, FLASH: an adaptive mesh hydrodynamics code for modeling astrophysical thermonuclear flashes. Astrophys. J. Suppl. Ser. **131**, 273–334 (2000). https://doi.org/10.1086/317361

S-i. Fujimoto, M.-a. Hashimoto, K. Kotake, S. Yamada, Heavy-element nucleosynthesis in a collapsar. Astrophys. J. **656**, 382–392 (2007). arXiv:astro-ph/0602460. https://doi.org/10.1086/509908

S-i. Fujimoto, N. Nishimura, M.-a. Hashimoto, Nucleosynthesis in magnetically driven jets from collapsars. Astrophys. J. **680**, 1350–1358 (2008). arXiv:0804.0969. https://doi.org/10.1086/529416

G.M. Fuller, W.A. Fowler, M.J. Newman, Stellar weak interaction rates for intermediate mass nuclei. III—rate tables for the free nucleons and nuclei with $A = 21$ to $A = 60$. Astrophys. J. **Suppl**(48), 279–319 (1982). https://doi.org/10.1086/190779

D. García-Senz, E. Bravo, Type Ia supernova models arising from different distributions of igniting points. Astron. Astrophys. **430**, 585–602 (2005). arXiv:astro-ph/0409480. https://doi.org/10.1051/0004-6361:20041628

D. García-Senz, R.M. Cabezón, A. Arcones, A. Relaño, F.K. Thielemann, High-resolution simulations of the head-on collision of white dwarfs. Mon. Not. R. Astron. Soc. **436**, 3413–3429 (2013). arXiv:1309.6884. https://doi.org/10.1093/mnras/stt1821

R. Georgii, S. Plüschke, R. Diehl, G.G. Lichti, V. Schönfelder, H. Bloemen, W. Hermsen, J. Ryan, K. Bennett, COMPTEL upper limits for the ^{56}Co gamma-ray emission from SN1998bu. Astron. Astrophys. **394**, 517–523 (2002). arXiv:astro-ph/0208152. https://doi.org/10.1051/0004-6361:20021133

D.A. Goldstein, D. Kasen, Evidence for sub-Chandrasekhar mass Type Ia supernovae from an extensive survey of radiative transfer models. Astrophys. J. Lett. **852**, L33 (2018). arXiv:1801.00789. https://doi.org/10.3847/2041-8213/aaa409

J. Gómez-Gomar, J. Isern, P. Jean, Prospects for Type IA supernova explosion mechanism identification with gamma rays. Mon. Not. R. Astron. Soc. **295**, 1 (1998). arXiv:astro-ph/9709048. https://doi.org/10.1046/j.1365-8711.1998.29511115.x

S. Goriely, Towards more accurate and reliable predictions for nuclear applications. Eur. Phys. J. A **51**, 172 (2015). https://doi.org/10.1140/epja/i2015-15172-2

S. Goriely, A. Bauswein, O. Just, E. Pllumbi, H.T. Janka, Impact of weak interactions of free nucleons on the r-process in dynamical ejecta from neutron star mergers. Mon. Not. R. Astron. Soc. **452**, 3894–3904 (2015). arXiv:1504.04377. https://doi.org/10.1093/mnras/stv1526

S.A. Grebenev, A.A. Lutovinov, S.S. Tsygankov, C. Winkler, Hard-X-ray emission lines from the decay of ^{44}Ti in the remnant of supernova 1987A. Nature **490**, 373–375 (2012). arXiv:1211.2656. https://doi.org/10.1038/nature11473

B.W. Grefenstette, F.A. Harrison, S.E. Boggs, S.P. Reynolds, C.L. Fryer, K.K. Madsen, D.R. Wik, A. Zoglauer, C.I. Ellinger, D.M. Alexander, H. An, D. Barret, F.E. Christensen, W.W. Craig, K. Forster, P. Giommi, C.J. Hailey, A. Hornstrup, V.M. Kaspi, T. Kitaguchi, J.E. Koglin, P.H. Mao, H. Miyasaka, K. Mori, M. Perri, M.J. Pivovaroff, S. Puccetti, V. Rana, D. Stern, N.J. Westergaard, W.W. Zhang, Asymmetries in core-collapse supernovae from maps of radioactive ^{44}Ti in Cassiopeia A. Nature **506**, 339–342 (2014). arXiv:1403.4978. https://doi.org/10.1038/nature12997

B.W. Grefenstette, C.L. Fryer, F.A. Harrison, S.E. Boggs, T. DeLaney, J.M. Laming, S.P. Reynolds, D.M. Alexander, D. Barret, F.E. Christensen, W.W. Craig, K. Forster, P. Giommi, C.J. Hailey, A. Hornstrup, T. Kitaguchi, J.E. Koglin, L. Lopez, P.H. Mao, K.K. Madsen, H. Miyasaka, K. Mori, M. Perri, M.J. Pivovaroff, S. Puccetti, V. Rana, D. Stern, N.J. Westergaard, D.R. Wik, W.W. Zhang, A. Zoglauer, The distribution of radioactive ^{44}Ti in Cassiopeia A. Astrophys. J. **834**, 19 (2017). arXiv:1612.02774. https://doi.org/10.3847/1538-4357/834/1/19

J. Greiner, P.A. Mazzali, D.A. Kann, T. Krühler, E. Pian, S. Prentice, E.F. Olivares, A. Rossi, S. Klose, S. Taubenberger, F. Knust, P.M.J. Afonso, C. Ashall, J. Bolmer, C. Delvaux, R. Diehl, J. Elliott, R. Filgas, J.P.U. Fynbo, J.F. Graham, A.N. Guelbenzu, S. Kobayashi, G. Leloudas, S. Savaglio, P. Schady, S. Schmidl, T. Schweyer, V. Sudilovsky, M. Tanga, A.C. Updike, H. van Eerten, K. Varela, A very luminous magnetar-powered supernova associated with an ultra-long γ-ray burst. Nature **523**, 189–192 (2015). arXiv:1509.03279. https://doi.org/10.1038/nature14579

N. Guessoum, P. Jean, N. Prantzos, Microquasars as sources of positron annihilation radiation. Astron. Astrophys. **457**, 753–762 (2006). arXiv:astro-ph/0607296. https://doi.org/10.1051/0004-6361:20065240

G. Halevi, P. Mösta, r-Process nucleosynthesis from three-dimensional jet-driven core-collapse supernovae with magnetic misalignments (2018). ArXiv e-prints arXiv:1801.08943

J.A. Harris, W.R. Hix, M.A. Chertkow, C.T. Lee, E.J. Lentz, O.E.B. Messer, Implications for post-processing nucleosynthesis of core-collapse supernova models with Lagrangian particles. Astrophys. J. **843**, 2 (2017). arXiv:1701.08876. https://doi.org/10.3847/1538-4357/aa76de

A. Heger, S.E. Woosley, Nucleosynthesis and evolution of massive metal-free stars. Astrophys. J. **724**, 341–373 (2010). arXiv:0803.3161. https://doi.org/10.1088/0004-637X/724/1/341

M. Hempel, J. Schaffner-Bielich, A statistical model for a complete supernova equation of state. Nucl. Phys. A **837**, 210 (2010). arXiv:0911.4073

J.C. Higdon, R.E. Lingenfelter, The superbubble origin for galactic cosmic rays. Adv. Space Res. **37**, 1913–1917 (2006). https://doi.org/10.1016/j.asr.2005.07.071

W. Hillebrandt, M. Kromer, F.K. Röpke, A.J. Ruiter, Towards an understanding of Type Ia supernovae from a synthesis of theory and observations. Front. Phys. **8**, 116–143 (2013). arXiv:1302.6420. https://doi.org/10.1007/s11467-013-0303-2

Y. Hirai, Y. Ishimaru, T.R. Saitoh, M.S. Fujii, J. Hidaka, T. Kajino, Enrichment of r-process elements in dwarf spheroidal galaxies in chemo-dynamical evolution model. Astrophys. J. **814**, 41 (2015). arXiv:1509.08934. https://doi.org/10.1088/0004-637X/814/1/41

W.R. Hix, E.J. Lentz, S.W. Bruenn, A. Mezzacappa, O.E.B. Messer, E. Endeve, J.M. Blondin, J.A. Harris, P. Marronetti, K.N. Yakunin, The multi-dimensional character of core-collapse supernovae. Acta Phys. Pol. B **47**, 645 (2016). arxiv:1602.05553. https://doi.org/10.5506/APhysPolB.47.645

P. Hoeflich, E.Y. Hsiao, C. Ashall, C.R. Burns, T.R. Diamond, M.M. Phillips, D. Sand, M.D. Stritzinger, N. Suntzeff, C. Contreras, K. Krisciunas, N. Morrell, L. Wang, Light and color curve properties of Type Ia supernovae: theory versus observations. Astrophys. J. **846**, 58 (2017). arXiv:1707.05350. https://doi.org/10.3847/1538-4357/aa84b2

S. Honda, W. Aoki, Y. Ishimaru, S. Wanajo, S.G. Ryan, Neutron-capture elements in the very metal poor star HD 122563. Astrophys. J. **643**, 1180–1189 (2006). arXiv:astro-ph/0602107. https://doi.org/10.1086/503195

D. Hooper, P. Blasi, P.D. Serpico, Pulsars as the sources of high energy cosmic ray positrons. J. Cosmol. Astropart. Phys. **0901**, 025 (2009). arXiv:0810.1527. https://doi.org/10.1088/1475-7516/2009/01/025

D. Hooper, I. Cholis, T. Linden, K. Fang, HAWC observations strongly favor pulsar interpretations of the cosmic-ray positron excess. Phys. Rev. D **96**(10), 103013 (2017). arXiv:1702.08436. https://doi.org/10.1103/PhysRevD.96.103013

F. Hoyle, W.A. Fowler, Nucleosynthesis in supernovae. Astrophys. J. **132**, 565–590 (1960). https://doi.org/10.1086/146963

L. Hüdepohl, B. Müller, H.T. Janka, A. Marek, G.G. Raffelt, Neutrino signal of electron-capture supernovae from core collapse to cooling. Phys. Rev. Lett. **104**(25), 251101 (2010). arXiv:0912.0260. https://doi.org/10.1103/PhysRevLett.104.251101

G.J. Hueter, R.E. Lingenfelter, A fireball model for the March 25, 1978 gamma ray burst, in *Positron-Electron Pairs in Astrophysics*, ed. by M.L. Burns, A.K. Harding, R. Ramaty. American Institute of Physics Conference Series, vol. 101 (1983), pp. 89–93. https://doi.org/10.1063/1.34134

I. Iben Jr., A.V. Tutukov, Supernovae of Type I as end products of the evolution of binaries with components of moderate initial mass (M not greater than about 9 solar masses). Astrophys. J. Suppl. Ser. **54**, 335–372 (1984). https://doi.org/10.1086/190932

J. Isern, E. Bravo, A. Hirschmann, The science of γ-ray spectroscopy. Adv. Space Res. **38**, 1434–1438 (2006). https://doi.org/10.1016/j.asr.2005.07.037

J. Isern, P. Jean, E. Bravo, R. Diehl, J. Knödlseder, A. Domingo, A. Hirschmann, P. Hoeflich, F. Lebrun, M. Renaud, S. Soldi, N. Elias-Rosa, M. Hernanz, B. Kulebi, X. Zhang, C. Badenes, I. Domínguez, D. Garcia-Senz, C. Jordi, G. Lichti, G. Vedrenne, P. Von Ballmoos, Observation of SN2011fe with INTEGRAL. I. Pre-maximum phase. Astron. Astrophys. **552**, A97 (2013). arXiv:1302.3381. https://doi.org/10.1051/0004-6361/201220303

J. Isern, P. Jean, E. Bravo, J. Knödlseder, F. Lebrun, E. Churazov, R. Sunyaev, A. Domingo, C. Badenes, D.H. Hartmann, P. Hoeflich, M. Renaud, S. Soldi, N. Elias-Rosa, M. Hernanz, I. Domínguez, D. García-Senz, G.G. Lichti, G. Vedrenne, P. Von Ballmoos, Gamma-ray emission from SN2014J near maximum optical light. Astron. Astrophys. **588**, A67 (2016). arXiv:1602.02918. https://doi.org/10.1051/0004-6361/201526941

W. Iwakami, K. Kotake, N. Ohnishi, S. Yamada, K. Sawada, Three-dimensional simulations of standing accretion shock instability in core-collapse supernovae. Astrophys. J. **678**, 1207 (2008). arXiv:0710.2191

K. Iwamoto, K. Nomoto, P. Höflich, H. Yamaoka, S. Kumagai, T. Shigeyama, Theoretical light curves for the Type IC supernova SN 1994I. Astrophys. J. Lett. **437**, L115–L118 (1994). https://doi.org/10.1086/187696

K. Iwamoto, F. Brachwitz, K. Nomoto, N. Kishimoto, H. Umeda, W.R. Hix, F.K. Thielemann, Nucleosynthesis in Chandrasekhar mass models for Type IA supernovae and constraints on progenitor systems and burning-front propagation. Astrophys. J. Suppl. Ser. **125**, 439–462 (1999). arXiv:astro-ph/0002337. https://doi.org/10.1086/313278

A.F. Iyudin, R. Diehl, H. Bloemen, W. Hermsen, G.G. Lichti, D. Morris, J. Ryan, V. Schoenfelder, H. Steinle, M. Varendorff, C. de Vries, C. Winkler, COMPTEL observations of Ti-44 gamma-ray line emission from CAS A. Astron. Astrophys. **284**, L1–L4 (1994)

A. Janiuk, Nucleosynthesis of elements in gamma-ray burst engines. Astron. Astrophys. **568**, A105 (2014). arXiv:1406.4440. https://doi.org/10.1051/0004-6361/201423822

H.T. Janka, Explosion mechanisms of core-collapse supernovae. Annu. Rev. Nucl. Part. Sci. **62**, 407–451 (2012). arXiv:1206.2503. https://doi.org/10.1146/annurev-nucl-102711-094901

H.T. Janka, T. Melson, A. Summa, Physics of core-collapse supernovae in three dimensions: a sneak preview. Annu. Rev. Nucl. Part. Sci. **66**, 341–375 (2016). arXiv:1602.05576. https://doi.org/10.1146/annurev-nucl-102115-044747

P. Jean, J. Knödlseder, W. Gillard, N. Guessoum, K. Ferrière, A. Marcowith, V. Lonjou, J.P. Roques, Spectral analysis of the Galactic e^+e^- annihilation emission. Astron. Astrophys. **445**(2), 579–589 (2006)

J.A. Jiang, M. Doi, K. Maeda, T. Shigeyama, K. Nomoto, N. Yasuda, S.W. Jha, M. Tanaka, T. Morokuma, N. Tominaga, Ž. Ivezić, P. Ruiz-Lapuente, M.D. Stritzinger, P.A. Mazzali, C. Ashall, J. Mould, D. Baade, N. Suzuki, A.J. Connolly, F. Patat, L. Wang, P. Yoachim, D. Jones, H. Furusawa, S. Miyazaki, A hybrid Type Ia supernova with an early flash triggered by helium-shell detonation. Nature **550**, 80–83 (2017). arXiv:1710.01824. https://doi.org/10.1038/nature23908

S. Jones, F.K. Röpke, R. Pakmor, I.R. Seitenzahl, S.T. Ohlmann, P.V.F. Edelmann, Do electron-capture supernovae make neutron stars? First multidimensional hydrodynamic simulations of the oxygen deflagration. Astron. Astrophys. **593**, A72 (2016). arXiv:1602.05771. https://doi.org/10.1051/0004-6361/201628321

O. Just, A. Bauswein, R.A. Pulpillo, S. Goriely, H.T. Janka, Comprehensive nucleosynthesis analysis for ejecta of compact binary mergers. Mon. Not. R. Astron. Soc. **448**, 541–567 (2015). arXiv:1406.2687. https://doi.org/10.1093/mnras/stv009

A.I. Karakas, J.C. Lattanzio, The Dawes review 2: nucleosynthesis and stellar yields of low- and intermediate-mass single stars. Publ. Astron. Soc. Aust. **31**, e030 (2014). arXiv:1405.0062. https://doi.org/10.1017/pasa.2014.21

A. Kelic, M.V. Ricciardi, K.H. Schmidt, New insight into the fission process from experiments with relativistic heavy-ion beams, in *Dynamical Aspects of Nuclear Fission*, ed. by J. Kliman, M.G. Itkis, Š. Gmuca (2008), pp. 203–215. https://doi.org/10.1142/9789812837530_0016

W.E. Kerzendorf, S. Taubenberger, I.R. Seitenzahl, A.J. Ruiter, Very late photometry of SN 2011fe. Astrophys. J. Lett. **796**, L26 (2014). arXiv:1406.6050. https://doi.org/10.1088/2041-8205/796/2/L26

K. Knie, G. Korschinek, T. Faestermann, E.A. Dorfi, G. Rugel, A. Wallner, ^{60}Fe anomaly in a deep-sea manganese crust and implications for a nearby supernova source. Phys. Rev. Lett. **93**(17), 171103 (2004). https://doi.org/10.1103/PhysRevLett.93.171103

C. Kobayashi, K. Nomoto, I. Hachisu, Subclasses of Type Ia supernovae as the origin of $[\alpha/\mathrm{Fe}]$ ratios in dwarf spheroidal galaxies. Astrophys. J. Lett. **804**, L24 (2015). arXiv:1503.06739. https://doi.org/10.1088/2041-8205/804/1/L24

O. Korobkin, S. Rosswog, A. Arcones, C. Winteler, On the astrophysical robustness of the neutron star merger r-process. Mon. Not. R. Astron. Soc. **426**, 1940–1949 (2012). arXiv:1206.2379. https://doi.org/10.1111/j.1365-2966.2012.21859.x

M. Krafczyk, A precision measurement of the cosmic ray positron fraction on the international space station. PhD thesis, Massachusetts Institute of Technology (2016)

K. Kretschmer, R. Diehl, M. Krause, A. Burkert, K. Fierlinger, O. Gerhard, J. Greiner, W. Wang, Kinematics of massive star ejecta in the Milky Way as traced by ^{26}Al. Astron. Astrophys. **559**, A99 (2013). arXiv:1309.4980. https://doi.org/10.1051/0004-6361/201322563

M. Kromer, S.A. Sim, M. Fink, F.K. Röpke, I.R. Seitenzahl, W. Hillebrandt, Double-detonation sub-Chandrasekhar supernovae: synthetic observables for minimum helium shell mass models. Astrophys. J. **719**, 1067–1082 (2010). arXiv:1006.4489. https://doi.org/10.1088/0004-637X/719/2/1067

M. Kuhlen, J. Guedes, A. Pillepich, P. Madau, L. Mayer, An off-center density peak in the Milky Way's dark matter halo? Astrophys. J. **765**, 10 (2013). arXiv:1208.4844. https://doi.org/10.1088/0004-637X/765/1/10

D. Kushnir, B. Katz, S. Dong, E. Livne, R. Fernández, Head-on collisions of white dwarfs in triple systems could explain Type Ia supernovae. Astrophys. J. Lett. **778**, L37 (2013). arXiv:1303.1180. https://doi.org/10.1088/2041-8205/778/2/L37

K. Langanke, G. Martínez-Pinedo, Rate tables for the weak processes of pf-SHELL nuclei in stellar environments. At. Data Nucl. Data Tables **79**, 1 (2001)

J.M. Lattimer, F. Douglas Swesty, A generalized equation of state for hot, dense matter. Nucl. Phys. A **535**, 331–376 (1991). https://doi.org/10.1016/0375-9474(91)90452-C

J.M. Lattimer, D.N. Schramm, Black-hole-neutron-star collisions. Astrophys. J. Lett. **192**, L145–L147 (1974). https://doi.org/10.1086/181612

E.J. Lentz, S.W. Bruenn, W.R. Hix, A. Mezzacappa, O.E.B. Messer, E. Endeve, J.M. Blondin, J.A. Harris, P. Marronetti, K.N. Yakunin, Three-dimensional core-collapse supernova simulated using a 15 $M_\odot$ progenitor. Astrophys. J. Lett. **807**, L31 (2015). arXiv:1505.05110. https://doi.org/10.1088/2041-8205/807/2/L31

S.C. Leung, K. Nomoto, Explosive nucleosynthesis in near-Chandrasekhar mass white dwarf models for Type Ia supernovae: dependence on model parameters (2017). ArXiv e-prints arXiv:1710.04254

G.G. Lichti, K. den Bennett, J.W. Herder, R. Diehl, D. Morris, J. Ryan, V. Schoenfelder, H. Steinle, A.W. Strong, C. Winkler, COMPTEL upper limits on gamma-ray line emission from Supernova 1991T. Astron. Astrophys. **292**, 569–579 (1994)

M. Liebendörfer, A simple parameterization of the consequences of deleptonization for simulations of stellar core collapse. Astrophys. J. **633**, 1042–1051 (2005). arXiv:astro-ph/0504072. https://doi.org/10.1086/466517

M. Liebendörfer, A. Mezzacappa, F.K. Thielemann, Conservative general relativistic radiation hydrodynamics in spherical symmetry and comoving coordinates. Phys. Rev. D **104**, 003 (2001). arXiv:astro-ph/0012201. https://doi.org/10.1103/PhysRevD.63.104003

M. Liebendörfer, O.E.B. Messer, A. Mezzacappa, S.W. Bruenn, C.Y. Cardall, F.K. Thielemann, A finite difference representation of neutrino radiation hydrodynamics in spherically symmetric general relativistic spacetime. Astrophys. J. Suppl. Ser. **150**, 263–316 (2004). arXiv:astro-ph/0207036. https://doi.org/10.1086/380191

M. Liebendörfer, S.C. Whitehouse, T. Fischer, The isotropic diffusion source approximation for supernova neutrino transport. Astrophys. J. **698**, 1174–1190 (2009). arXiv:0711.2929. https://doi.org/10.1088/0004-637X/698/2/1174

M. Limongi, A. Chieffi, The nucleosynthesis of ^{26}Al and ^{60}Fe in solar metallicity stars extending in mass from 11 to 120 $M_\odot$: the hydrostatic and explosive contributions. Astrophys. J. **647**, 483–500 (2006). arXiv:astro-ph/0604297. https://doi.org/10.1086/505164

M. Limongi, A. Chieffi, Presupernova evolution and explosive nucleosynthesis of zero metal massive stars. Astrophys. J. Suppl. Ser. **199**, 38 (2012). arXiv:1202.4581. https://doi.org/10.1088/0067-0049/199/2/38

T. Linden, S. Profumo, Probing the pulsar origin of the anomalous positron fraction with AMS-02 and atmospheric Cherenkov telescopes. Astrophys. J. **772**, 18 (2013). arXiv:1304.1791. https://doi.org/10.1088/0004-637X/772/1/18

R.L. Lingenfelter, R. Ramaty, Primary cosmic ray positrons and galactic annihilation radiation, in *International Cosmic Ray Conference*, vol. 1 (1979), p. 501

E. Livne, Successive detonations in accreting white dwarfs as an alternative mechanism for Type I supernovae. Astrophys. J. Lett. **354**, L53–L55 (1990). https://doi.org/10.1086/185721

P. Ludwig, S. Bishop, R. Egli, V. Chernenko, B. Deneva, T. Faestermann, N. Famulok, L. Fimiani, J.M. Gómez-Guzmán, K. Hain, G. Korschinek, M. Hanzlik, S. Merchel, G. Rugel, Time-resolved 2-million-year-old supernova activity discovered in Earth's microfossil record. Proc. Natl. Acad. Sci. USA **113**, 9232–9237 (2016). https://doi.org/10.1073/pnas.1601040113

A.I. MacFadyen, S.E. Woosley, Collapsars: gamma-ray bursts and explosions in "failed supernovae". Astrophys. J. **524**, 262–289 (1999). arXiv:astro-ph/9810274. https://doi.org/10.1086/307790

A.I. MacFadyen, S.E. Woosley, A. Heger, Supernovae, jets, and collapsars. Astrophys. J. **550**, 410–425 (2001). arXiv:astro-ph/9910034. https://doi.org/10.1086/319698

K. Maeda, K. Nomoto, Bipolar supernova explosions: nucleosynthesis and implications for abundances in extremely metal-poor stars. Astrophys. J. **598**, 1163–1200 (2003). arXiv:astro-ph/0304172. https://doi.org/10.1086/378948

K. Maeda, Y. Terada, Progenitors of Type Ia supernovae. Int. J. Mod. Phys. D **25**, 1630024 (2016). arXiv:1609.03639

K. Maeda, F.K. Röpke, M. Fink, W. Hillebrandt, C. Travaglio, F.K. Thielemann, Nucleosynthesis in two-dimensional delayed detonation models of Type Ia supernova explosions. Astrophys. J. **712**, 624–638 (2010). arXiv:1002.2153. https://doi.org/10.1088/0004-637X/712/1/624

A. Maeder, G. Meynet, Rotating massive stars: from first stars to gamma ray bursts. Rev. Mod. Phys. **84**, 25–63 (2012). https://doi.org/10.1103/RevModPhys.84.25

W.A. Mahoney, J.C. Ling, High-resolution spectrum of the Galactic center. Astron. Astrophys. Suppl. Ser. **97**, 159–163 (1993)

M.A. Malkov, P.H. Diamond, R.Z. Sagdeev, Positive charge prevalence in cosmic rays: room for dark matter in the positron spectrum. Phys. Rev. D **94**(6), 063006 (2016). arXiv:1607.01820. https://doi.org/10.1103/PhysRevD.94.063006

D. Maoz, F. Mannucci, G. Nelemans, Observational clues to the progenitors of Type Ia supernovae. Annu. Rev. Astron. Astrophys. **52**, 107–170 (2014). arXiv:1312.0628. https://doi.org/10.1146/annurev-astro-082812-141031

T. Marketin, L. Huther, G. Martínez-Pinedo, Large-scale evaluation of β-decay rates of r-process nuclei with the inclusion of first-forbidden transitions. Phys. Rev. C **93**(2), 025805 (2016). arXiv:1507.07442. https://doi.org/10.1103/PhysRevC.93.025805

K.S. Marquardt, S.A. Sim, A.J. Ruiter, I.R. Seitenzahl, S.T. Ohlmann, M. Kromer, R. Pakmor, F.K. Röpke, Type Ia supernovae from exploding oxygen-neon white dwarfs. Astron. Astrophys. **580**, A118 (2015). arXiv:1506.05809. https://doi.org/10.1051/0004-6361/201525761

D. Martin, A. Perego, A. Arcones, F.K. Thielemann, O. Korobkin, S. Rosswog, Neutrino-driven winds in the aftermath of a neutron star merger: nucleosynthesis and electromagnetic transients. Astrophys. J. **813**, 2 (2015). arXiv:1506.05048. https://doi.org/10.1088/0004-637X/813/1/2

D. Martin, A. Perego, W. Kastaun, A. Arcones, The role of weak interactions in dynamic ejecta from binary neutron star mergers. Class. Quantum Gravity **35**(3), 034001 (2018). arXiv:1710.04900. https://doi.org/10.1088/1361-6382/aa9f5a

G. Martínez-Pinedo, T. Fischer, A. Lohs, L. Huther, Charged-current weak interaction processes in hot and dense matter and its impact on the spectra of neutrinos emitted from protoneutron star cooling. Phys. Rev. Lett. **109**(25), 251104 (2012). arXiv:1205.2793. https://doi.org/10.1103/PhysRevLett.109.251104

S.M. Matz, G.H. Share, M.D. Leising, E.L. Chupp, W.T. Vestrand, Gamma-ray line emission from SN1987A. Nature **331**, 416–418 (1988). https://doi.org/10.1038/331416a0

J.C. McKinney, A. Tchekhovskoy, R.D. Blandford, Alignment of magnetized accretion disks and relativistic jets with spinning black holes. Science **339**, 49 (2013). arXiv:1211.3651

C.A. Meakin, I. Seitenzahl, D. Townsley, G.C. Jordan IV., J. Truran, D. Lamb, Study of the detonation phase in the gravitationally confined detonation model of Type Ia supernovae. Astrophys. J. **693**, 1188–1208 (2009). arXiv:0806.4972. https://doi.org/10.1088/0004-637X/693/2/1188

T. Melson, H.T. Janka, R. Bollig, F. Hanke, A. Marek, B. Müller, Neutrino-driven explosion of a 20 solar-mass star in three dimensions enabled by strange-quark contributions to neutrino-nucleon scattering. Astrophys. J. Lett. **808**, L42 (2015). arXiv:1504.07631. https://doi.org/10.1088/2041-8205/808/2/L42

J.d.J. Mendoza-Temis, M.R. Wu, K. Langanke, G. Martínez-Pinedo, A. Bauswein, H.T. Janka, Nuclear robustness of the r process in neutron-star mergers. Phys. Rev. C **92**(5), 055805 (2015). https://doi.org/10.1103/PhysRevC.92.055805

B.D. Metzger, Kilonovae. Living Rev. Relativ. **20**, 3 (2017). arXiv:1610.09381. https://doi.org/10.1007/s41114-017-0006-z

P.A. Milne, L.S. The, M.D. Leising, Late light curves of Type Ia supernovae. Astrophys. J. **559**, 1019–1031 (2001). arXiv:astro-ph/0104185. https://doi.org/10.1086/322352

T. Mishenina, T. Gorbaneva, M. Pignatari, F.K. Thielemann, S.A. Korotin, Mn abundances in the stars of the Galactic disc with metallicities $-1.0 < [Fe/H] < 0.3$. Mon. Not. R. Astron. Soc. **454**, 1585–1594 (2015). arXiv:1509.05341. https://doi.org/10.1093/mnras/stv2038

P. Möller, B. Pfeiffer, K.L. Kratz, New calculations of gross β-decay properties for astrophysical applications: speeding-up the classical r process. Phys. Rev. C **67**(5), 055802 (2003). https://doi.org/10.1103/PhysRevC.67.055802

K. Moore, D.M. Townsley, L. Bildsten, The effects of curvature and expansion on helium detonations on white dwarf surfaces. Astrophys. J. **776**, 97 (2013). arXiv:1308.4193. https://doi.org/10.1088/0004-637X/776/2/97

T.J. Moriya, N. Tominaga, N. Langer, K. Nomoto, S.I. Blinnikov, E.I. Sorokina, Electron-capture supernovae exploding within their progenitor wind. Astron. Astrophys. **569**, A57 (2014). arXiv:1407.4563. https://doi.org/10.1051/0004-6361/201424264

D.J. Morris, K. Bennett, H. Bloemen, R. Diehl, W. Hermsen, G.G. Lichti, M.L. McConnell, J.M. Ryan, V. Schönfelder, Reassessment of the ^{56}Co emission from SN 1991T, in *Proceedings of the Fourth Compton Symposium*, ed. by C.D. Dermer, M.S. Strickman, J.D. Kurfess. American Institute of Physics Conference Series, vol. 410 (1997), pp. 1084–1088. https://doi.org/10.1063/1.54174

I.V. Moskalenko, A.W. Strong, Production and propagation of cosmic ray positrons and electrons. Astrophys. J. **493**, 694–707 (1998). arXiv:astro-ph/9710124. https://doi.org/10.1086/305152

P. Mösta, S. Richers, C.D. Ott, R. Haas, A.L. Piro, K. Boydstun, E. Abdikamalov, C. Reisswig, E. Schnetter, Magnetorotational core-collapse supernovae in three dimensions. Astrophys. J. Lett. **785**, L29 (2014). arXiv:1403.1230. https://doi.org/10.1088/2041-8205/785/2/L29

P. Mösta, C.D. Ott, D. Radice, L.F. Roberts, E. Schnetter, R. Haas, A large-scale dynamo and magnetoturbulence in rapidly rotating core-collapse supernovae. Nature **528**, 376–379 (2015). arXiv:1512.00838. https://doi.org/10.1038/nature15755

P. Mösta, L.F. Roberts, G. Halevi, C.D. Ott, J. Lippuner, R. Haas, E. Schnetter, R-process nucleosynthesis from three-dimensional magnetorotational core-collapse supernovae (2017). ArXiv e-prints arXiv:1712.09370

B. Müller, S. Wanajo, H.T. Janka, A. Heger, D. Gay, S.A. Sim, Simulations of electron capture and low-mass iron core supernovae. Mem. Soc. Astron. Ital. **88**, 288 (2017). arXiv:1710.02641

S. Nagataki, Grb-Sn connection: central engine of long GRBs and explosive nucleosynthesis. Int. J. Mod. Phys. D **20**, 1975–1978 (2011). https://doi.org/10.1142/S0218271811020032

S. Nagataki, R. Takahashi, A. Mizuta, T. Takiwaki, Numerical study of gamma-ray burst jet formation in collapsars. Astrophys. J. **659**, 512–529 (2007). arXiv:astro-ph/0608233. https://doi.org/10.1086/512057

K. Nakamura, T. Takiwaki, T. Kuroda, K. Kotake, Systematic features of axisymmetric neutrino-driven core-collapse supernova models in multiple progenitors. Publ. Astron. Soc. Jpn. **67**, 107 (2015). arXiv:1406.2415

T. Nakamura, H. Umeda, K. Nomoto, F.K. Thielemann, A. Burrows, Nucleosynthesis in Type II supernovae and the abundances in metal-poor stars. Astrophys. J. **517**, 193–208 (1999). arXiv:astro-ph/9809307. https://doi.org/10.1086/307167

T. Nakamura, H. Umeda, K. Iwamoto, K. Nomoto, H. Ma, W.R. Hix, F.K. Thielemann, Explosive nucleosynthesis in hypernovae. Astrophys. J. **555**, 880–899 (2001). arXiv:astro-ph/0011184. https://doi.org/10.1086/321495

N. Nishimura, T. Takiwaki, F.K. Thielemann, The r-process nucleosynthesis in the various jet-like explosions of magnetorotational core-collapse supernovae. Astrophys. J. **810**, 109 (2015). arXiv:1501.06567. https://doi.org/10.1088/0004-637X/810/2/109

N. Nishimura, H. Sawai, T. Takiwaki, S. Yamada, F.K. Thielemann, The intermediate r-process in core-collapse supernovae driven by the magneto-rotational instability. Astrophys. J. Lett. **836**, L21 (2017). arXiv:1611.02280. https://doi.org/10.3847/2041-8213/aa5dee

K. Nomoto, Accreting white dwarf models for Type 1 supernovae. II—off-center detonation supernovae. Astrophys. J. **257**, 780–792 (1982). https://doi.org/10.1086/160031

K. Nomoto, Nucleosynthesis in hypernovae associated with gamma ray bursts, in *Handbook of Supernovae*, ed. by A.W. Alsabti, P. Murdin (Springer, Berlin, 2017). https://doi.org/10.1007/978-3-319-20794-0_86-1

K. Nomoto, F.K. Thielemann, K. Yokoi, Accreting white dwarf models of Type I supernovae. III—carbon deflagration supernovae. Astrophys. J. **286**, 644–658 (1984). https://doi.org/10.1086/162639

K. Nomoto, N. Tominaga, H. Umeda, C. Kobayashi, K. Maeda, Nucleosynthesis yields of core-collapse supernovae and hypernovae, and galactic chemical evolution. Nucl. Phys. A **777**, 424–458 (2006). arXiv:astro-ph/0605725. https://doi.org/10.1016/j.nuclphysa.2006.05.008

K. Nomoto, K. Maeda, M. Tanaka, T. Suzuki, Gamma-ray bursts and magnetar-forming supernovae. Astrophys. Space Sci. **336**, 129–137 (2011). https://doi.org/10.1007/s10509-011-0658-1

K. Nomoto, C. Kobayashi, N. Tominaga, Nucleosynthesis in stars and the chemical enrichment of galaxies. Annu. Rev. Astron. Astrophys. **51**, 457–509 (2013). https://doi.org/10.1146/annurev-astro-082812-140956

H. Norgaard, Al-26 from red giants. Astrophys. J. **236**, 895–898 (1980). https://doi.org/10.1086/157815

E. O'Connor, C.D. Ott, Black hole formation in failing core-collapse supernovae. Astrophys. J. **730**, 70 (2011). arXiv:1010.5550. https://doi.org/10.1088/0004-637X/730/2/70

E. O'Connor, C.D. Ott, The progenitor dependence of the pre-explosion neutrino emission in core-collapse supernovae. Astrophys. J. **762**, 126 (2013). arXiv:1207.1100. https://doi.org/10.1088/0004-637X/762/2/126

M. Oertel, M. Hempel, T. Klähn, S. Typel, Equations of state for supernovae and compact stars. Rev. Mod. Phys. **89**(1), 015007 (2017). arXiv:1610.03361. https://doi.org/10.1103/RevModPhys.89.015007

M. Ono, M. Hashimoto, S. Fujimoto, K. Kotake, S. Yamada, Explosive nucleosynthesis in magnetohydrodynamical jets from collapsars. II—heavy-element nucleosynthesis of s, p, r-processes. Prog. Theor. Phys. **128**, 741–765 (2012). arXiv:1203.6488

R. Pakmor, M. Kromer, F.K. Röpke, S.A. Sim, A.J. Ruiter, W. Hillebrandt, Sub-luminous Type Ia supernovae from the mergers of equal-mass white dwarfs with mass $\sim 0.9\ M_{\odot}$. Nature **463**, 61–64 (2010). arXiv:0911.0926. https://doi.org/10.1038/nature08642

R. Pakmor, M. Kromer, S. Taubenberger, S.A. Sim, F.K. Röpke, W. Hillebrandt, Normal Type Ia supernovae from violent mergers of white dwarf binaries. Astrophys. J. Lett. **747**, L10 (2012). arXiv:1201.5123. https://doi.org/10.1088/2041-8205/747/1/L10

R. Pakmor, M. Kromer, S. Taubenberger, V. Springel, Helium-ignited violent mergers as a unified model for normal and rapidly declining Type Ia supernovae. Astrophys. J. Lett. **770**, L8 (2013). arXiv:1302.2913. https://doi.org/10.1088/2041-8205/770/1/L8

K.C. Pan, M. Liebendörfer, M. Hempel, F.K. Thielemann, Two-dimensional core-collapse supernova simulations with the isotropic diffusion source approximation for neutrino transport. Astrophys. J. **817**, 72 (2016). arXiv:1505.02513. https://doi.org/10.3847/0004-637X/817/1/72

K.C. Pan, M. Liebendörfer, S.M. Couch, F.K. Thielemann, Equation of state dependent dynamics and multi-messenger signals from stellar-mass black hole formation (2017). ArXiv e-prints arXiv:1710.01690

I.V. Panov, I.Y. Korneev, F.K. Thielemann, The r-process in the region of transuranium elements and the contribution of fission products to the nucleosynthesis of nuclei with $A \leq 130$. Astron. Lett. **34**, 189–197 (2008). https://doi.org/10.1007/s11443-008-3006-1

O. Papish, H.B. Perets, Supernovae from direct collisions of white dwarfs and the role of helium shell ignition. Astrophys. J. **822**, 19 (2016). arXiv:1502.03453. https://doi.org/10.3847/0004-637X/822/1/19

A. Perego, S. Rosswog, R.M. Cabezón, O. Korobkin, R. Käppeli, A. Arcones, M. Liebendörfer, Neutrino-driven winds from neutron star merger remnants. Mon. Not. R. Astron. Soc. **443**, 3134–3156 (2014). arXiv:1405.6730. https://doi.org/10.1093/mnras/stu1352

A. Perego, M. Hempel, C. Fröhlich, K. Ebinger, M. Eichler, J. Casanova, M. Liebendörfer, F.K. Thielemann, PUSHing core-collapse supernovae to explosions in spherical symmetry I: the model and the case of SN 1987A. Astrophys. J. **806**, 275 (2015). arXiv:1501.02845. https://doi.org/10.1088/0004-637X/806/2/275

A. Perego, R.M. Cabezón, R. Käppeli, An advanced leakage scheme for neutrino treatment in astrophysical simulations. Astrophys. J. Suppl. Ser. **223**, 22 (2016). arXiv:1511.08519. https://doi.org/10.3847/0067-0049/223/2/22

T. Piran, The physics of gamma-ray bursts. Rev. Mod. Phys. **76**, 1143–1210 (2004). arXiv:astro-ph/0405503. https://doi.org/10.1103/RevModPhys.76.1143

T. Plewa, A.C. Calder, D.Q. Lamb, Type Ia supernova explosion: gravitationally confined detonation. Astrophys. J. Lett. **612**, L37–L40 (2004). arXiv:astro-ph/0405163. https://doi.org/10.1086/424036

N. Prantzos, Astrophysical gamma-ray lines: a probe of stellar nucleosynthesis and star formation. ESA SP **552**, 15 (2004). arXiv:astro-ph/0404501

N. Prantzos, C. Boehm, A.M. Bykov, R. Diehl, K. Ferrière, N. Guessoum, P. Jean, J. Knoedlseder, A. Marcowith, I.V. Moskalenko, A. Strong, G. Weidenspointner, The 511 keV emission from positron annihilation in the Galaxy. Rev. Mod. Phys. **83**, 1001–1056 (2011). arXiv:1009.4620. https://doi.org/10.1103/RevModPhys.83.1001

S. Profumo, Dissecting cosmic-ray electron-positron data with Occam's Razor: the role of known pulsars. Cent. Eur. J. Phys. **10**, 1–31 (2011). arXiv:0812.4457. https://doi.org/10.2478/s11534-011-0099-z

J. Pruet, R.D. Hoffman, S.E. Woosley, H.T. Janka, R. Buras, Nucleosynthesis in early supernova winds. II. The role of neutrinos. Astrophys. J. **644**, 1028–1039 (2006). arXiv:astro-ph/0511194. https://doi.org/10.1086/503891

W.R. Purcell, L.X. Cheng, D.D. Dixon, R.L. Kinzer, J.D. Kurfess, M. Leventhal, M.A. Saunders, J.G. Skibo, D.M. Smith, J. Tueller, OSSE mapping of galactic 511 keV positron annihilation line emission. Astrophys. J. **491**, 725–748 (1997). https://doi.org/10.1086/304994

Y.Z. Qian, S.E. Woosley, Nucleosynthesis in neutrino-driven winds. I. The physical conditions. Astrophys. J. **471**, 331 (1996). arXiv:astro-ph/9611094. https://doi.org/10.1086/177973

R. Ramaty, F.W. Stecker, D. Misra, Low-energy cosmic ray positrons and 0.51-Mev gamma rays from the galaxy. J. Geophys. Res. **75**, 1141 (1970). https://doi.org/10.1029/JA075i007p01141

C. Raskin, F.X. Timmes, E. Scannapieco, S. Diehl, C. Fryer, On Type Ia supernovae from the collisions of two white dwarfs. Mon. Not. R. Astron. Soc. **399**, L156–L159 (2009). arXiv:0907.3915. https://doi.org/10.1111/j.1745-3933.2009.00743.x

T. Rauscher, N. Dauphas, I. Dillmann, C. Fröhlich, Z. Fülöp, G. Gyürky, Constraining the astrophysical origin of the p-nuclei through nuclear physics and meteoritic data. Rep. Prog. Phys. **76**(6), 066201 (2013). arXiv:1303.2666. https://doi.org/10.1088/0034-4885/76/6/066201

M. Renaud, J. Vink, A. Decourchelle, F. Lebrun, P.R. den Hartog, R. Terrier, C. Couvreur, J. Knödlseder, P. Martin, N. Prantzos, A.M. Bykov, The signature of ^{44}Ti in Cassiopeia A revealed by IBIS/ISGRI on INTEGRAL. Astrophys. J. Lett. **647**, L41–L44 (2006). arXiv:astro-ph/0606736. https://doi.org/10.1086/507300

L.F. Roberts, S. Reddy, G. Shen, Medium modification of the charged-current neutrino opacity and its implications. Phys. Rev. C **86**(6), 065803 (2012). arXiv:1205.4066. https://doi.org/10.1103/PhysRevC.86.065803

I.U. Roederer, H. Schatz, J.E. Lawler, T.C. Beers, J.J. Cowan, A. Frebel, I.I. Ivans, C. Sneden, J.S. Sobeck, New detections of arsenic, selenium, and other heavy elements in two metal-poor stars. Astrophys. J. **791**, 32 (2014). arXiv:1406.4554. https://doi.org/10.1088/0004-637X/791/1/32

F.K. Roepke, Combustion in thermonuclear supernova explosions, in *Handbook of Supernovae*, ed. by A.W. Alsabti, P. Murdin (Springer, Berlin, 2017). arXiv:1703.09274. https://doi.org/10.1007/978-3-319-20794-0_58-1

S. Rosswog, D. Kasen, J. Guillochon, E. Ramirez-Ruiz, Collisions of white dwarfs as a new progenitor channel for Type Ia supernovae. Astrophys. J. Lett. **705**, L128–L132 (2009). arXiv:0907.3196. https://doi.org/10.1088/0004-637X/705/2/L128

S. Rosswog, O. Korobkin, A. Arcones, F.K. Thielemann, T. Piran, The long-term evolution of neutron star merger remnants—I. The impact of r-process nucleosynthesis. Mon. Not. R. Astron. Soc. **439**, 744–756 (2014). arXiv:1307.2939. https://doi.org/10.1093/mnras/stt2502

I. Seitenzahl, D. Townsley, Nucleosynthesis in thermonuclear supernovae, in *Handbook of Supernovae*, ed. by A.W. Alsabti, P. Murdin (Springer, Berlin, 2017). https://doi.org/10.1007/978-3-319-20794-0_87-1

I.R. Seitenzahl, F. Ciaraldi-Schoolmann, F.K. Röpke, Type Ia supernova diversity: white dwarf central density as a secondary parameter in three-dimensional delayed detonation models. Mon. Not. R. Astron. Soc. **414**, 2709–2715 (2011). arXiv:1012.4929. https://doi.org/10.1111/j.1365-2966.2011.18588.x

I.R. Seitenzahl, F. Ciaraldi-Schoolmann, F.K. Röpke, M. Fink, W. Hillebrandt, M. Kromer, R. Pakmor, A.J. Ruiter, S.A. Sim, S. Taubenberger, Three-dimensional delayed-detonation models with nucleosynthesis for Type Ia supernovae. Mon. Not. R. Astron. Soc. **429**, 1156–1172 (2013). arXiv:1211.3015. https://doi.org/10.1093/mnras/sts402

I.R. Seitenzahl, M. Kromer, S.T. Ohlmann, F. Ciaraldi-Schoolmann, K. Marquardt, M. Fink, W. Hillebrandt, R. Pakmor, F.K. Röpke, A.J. Ruiter, S.A. Sim, S. Taubenberger, Three-dimensional simulations of gravitationally confined detonations compared to observations of SN 1991T. Astron. Astrophys. **592**, A57 (2016). arXiv:1606.00089. https://doi.org/10.1051/0004-6361/201527251

Y. Sekiguchi, M. Shibata, Formation of black hole and accretion disk in a massive high-entropy stellar core collapse. Astrophys. J. **737**, 6 (2011). arXiv:1009.5303. https://doi.org/10.1088/0004-637X/737/1/6

G.H. Share, M.D. Leising, D.C. Messina, W.R. Purcell, Limits on a variable source of 511 keV annihilation radiation near the Galactic center. Astrophys. J. Lett. **358**, L45–L48 (1990). https://doi.org/10.1086/185776

K.J. Shen, L. Bildsten, The ignition of carbon detonations via converging shock waves in white dwarfs. Astrophys. J. **785**, 61 (2014). arXiv:1305.6925. https://doi.org/10.1088/0004-637X/785/1/61

K.J. Shen, D. Kasen, N.N. Weinberg, L. Bildsten, E. Scannapieco, Thermonuclear Ia supernovae from helium shell detonations: explosion models and observables. Astrophys. J. **715**, 767–774 (2010). arXiv:1002.2258. https://doi.org/10.1088/0004-637X/715/2/767

K.J. Shen, S. Toonen, O. Graur, The evolution of the Type Ia supernova luminosity function. Astrophys. J. Lett. **851**, L50 (2017). arXiv:1710.09384. https://doi.org/10.3847/2041-8213/aaa015

S. Shen, R.J. Cooke, E. Ramirez-Ruiz, P. Madau, L. Mayer, J. Guedes, The history of R-process enrichment in the Milky Way. Astrophys. J. **807**, 115 (2015). arXiv:1407.3796. https://doi.org/10.1088/0004-637X/807/2/115

S. Shibagaki, T. Kajino, G.J. Mathews, S. Chiba, S. Nishimura, G. Lorusso, Relative contributions of the weak, main, and fission-recycling r-process. Astrophys. J. **816**, 79 (2016). arXiv:1505.02257. https://doi.org/10.3847/0004-637X/816/2/79

T. Siegert, R. Diehl, M.G.H. Krause, J. Greiner, Revisiting INTEGRAL/SPI observations of ^{44}Ti from Cassiopeia A. Astron. Astrophys. **579**, A124 (2015). arXiv:1505.05999. https://doi.org/10.1051/0004-6361/201525877

T. Siegert, R. Diehl, G. Khachatryan, M.G.H. Krause, F. Guglielmetti, J. Greiner, A.W. Strong, X. Zhang, Gamma-ray spectroscopy of positron annihilation in the Milky Way. Astron. Astrophys. **586**, A84 (2016). arXiv:1512.00325. https://doi.org/10.1051/0004-6361/201527510

S.A. Sim, F.K. Röpke, W. Hillebrandt, M. Kromer, R. Pakmor, M. Fink, A.J. Ruiter, I.R. Seitenzahl, Detonations in sub-Chandrasekhar-mass C+O white dwarfs. Astrophys. J. Lett. **714**, L52–L57 (2010). arXiv:1003.2917. https://doi.org/10.1088/2041-8205/714/1/L52

S. Sinha, C. Fröhlich, K. Ebinger, A. Perego, M. Hempel, M. Eichler, M. Liebendörfer, F.K. Thielemann, PUSHing core-collapse supernovae to explosions in spherical symmetry: nucleosynthesis yields, in *14th International Symposium on Nuclei in the Cosmos (NIC2016)*, ed. by S. Kubono, T. Kajino, S. Nishimura, T. Isobe, S. Nagataki, T. Shima, Y. Takeda (2017), 020608, arXiv:1701.05203. https://doi.org/10.7566/JPSCP.14.020608

G. Skinner, Positron annihilation in the galaxy and integral observations (2013). http://www.iaps.inaf.it/sz/integral2013/talks-posters/39-gks_rome_final.pdf

C. Sneden, A. McWilliam, G.W. Preston, J.J. Cowan, D.L. Burris, B.J. Armosky, The ultra–metal-poor, neutron-capture–rich giant star CS 22892-052. Astrophys. J. **467**, 819 (1996). https://doi.org/10.1086/177656

C. Sneden, J.J. Cowan, R. Gallino, Neutron-capture elements in the early galaxy. Annu. Rev. Astron. Astrophys. **46**, 241–288 (2008). https://doi.org/10.1146/annurev.astro.46.060407.145207

C. Sneden, J.J. Cowan, C. Kobayashi, M. Pignatari, J.E. Lawler, E.A. Den Hartog, M.P. Wood, Iron-group abundances in the metal-poor main-sequence turnoff star HD 84937. Astrophys. J. **817**, 53 (2016). arXiv:1511.05985. https://doi.org/10.3847/0004-637X/817/1/53

M. Sørensen, H. Svensmark, U. Gråe Jørgensen, Near-Earth supernova activity during the past 35 Myr (2017). ArXiv e-prints arXiv:1708.08248

P.A. Sturrock, A model of pulsars. Astrophys. J. **164**, 529 (1971). https://doi.org/10.1086/150865

T. Suda, Y. Katsuta, S. Yamada, T. Suwa, C. Ishizuka, Y. Komiya, K. Sorai, M. Aikawa, M.Y. Fujimoto, Stellar abundances for the galactic archeology (SAGA) database—compilation of the characteristics of known extremely metal-poor stars. Publ. Astron. Soc. Jpn. **60**, 1159–1171 (2008). arXiv:0806.3697. https://doi.org/10.1093/pasj/60.5.1159

T. Suda, S. Yamada, Y. Katsuta, Y. Komiya, C. Ishizuka, W. Aoki, M.Y. Fujimoto, The Stellar Abundances for Galactic Archaeology (SAGA) data base—II. implications for mixing and nucleosynthesis in extremely metal-poor stars and chemical enrichment of the Galaxy. Mon. Not. R. Astron. Soc. **412**, 843–874 (2011). arXiv:1010.6272. https://doi.org/10.1111/j.1365-2966.2011.17943.x

T. Sukhbold, T. Ertl, S.E. Woosley, J.M. Brown, H.T. Janka, Core-collapse supernovae from 9 to 120 solar masses based on neutrino-powered explosions. Astrophys. J. **821**, 38 (2016). arXiv:1510.04643. https://doi.org/10.3847/0004-637X/821/1/38

R. Sunyaev, A. Kaniovsky, V. Efremov, M. Gilfanov, E. Churazov, S. Grebenev, A. Kuznetsov, A. Melioranskiy, N. Yamburenko, S. Yunin, D. Stepanov, I. Chulkov, N. Pappe, M. Boyarskiy, E. Gavrilova, V. Loznikov, A. Prudkoglyad, V. Rodin, C. Reppin, W. Pietsch, J. Engelhauser, J. Truemper, W. Voges, E. Kendziorra, M. Bezler, R. Staubert, A.C. Brinkman, J. Heise, W.A. Mels, R. Jager, G.K. Skinner, O. Al-Emam, T.G. Patterson, A.P. Willmore, M. Gilfanov, E. Churazov, Discovery of hard X-ray emission from supernova 1987A. Nature **330**, 227–229 (1987). https://doi.org/10.1038/330227a0

R. Surman, G.C. McLaughlin, W.R. Hix, Nucleosynthesis in the outflow from gamma-ray burst accretion disks. Astrophys. J. **643**, 1057–1064 (2006). arXiv:astro-ph/0509365. https://doi.org/10.1086/501116

Y. Suwa, N. Tominaga, K. Maeda, Importance of ^{56}Ni production on diagnosing explosion mechanism of core-collapse supernova (2017). ArXiv e-prints arXiv:1704.04780

K. Takahashi, J. Witti, H.T. Janka, Nucleosynthesis in neutrino-driven winds from protoneutron stars II. The r-process. Astron. Astrophys. **286** (1994)

T. Takiwaki, K. Kotake, Y. Suwa, A comparison of two- and three-dimensional neutrino-hydrodynamics simulations of core-collapse supernovae. Astrophys. J. **786**, 83 (2014). arXiv:1308.5755. https://doi.org/10.1088/0004-637X/786/2/83

N.R. Tanvir, A.J. Levan, A.S. Fruchter, J. Hjorth, R.A. Hounsell, K. Wiersema, R.L. Tunnicliffe, A 'kilonova' associated with the short-duration γ-ray burst GRB 130603B. Nature **500**, 547–549 (2013). arXiv:1306.4971. https://doi.org/10.1038/nature12505

B.J. Teegarden, S.D. Barthelmy, N. Gehrels, J. Tueller, M. Leventhal, Resolution of the 1,238-keV gamma-ray line from supernova 1987A. Nature **339**, 122 (1989). https://doi.org/10.1038/339122a0

F.K. Thielemann, K. Nomoto, K. Yokoi, Explosive nucleosynthesis in carbon deflagration models of Type I supernovae. Astron. Astrophys. **158**, 17–33 (1986)

F.K. Thielemann, K. Nomoto, M.A. Hashimoto, Core-collapse supernovae and their ejecta. Astrophys. J. **460**, 408 (1996). https://doi.org/10.1086/176980

F.K. Thielemann, F. Brachwitz, P. Höflich, G. Martinez-Pinedo, K. Nomoto, The physics of Type Ia supernovae. New Astron. Rev. **48**, 605–610 (2004). https://doi.org/10.1016/j.newar.2003.12.038

F.K. Thielemann, R. Hirschi, M. Liebendörfer, R. Diehl, Massive stars and their supernovae, in *Lecture Notes in Physics*, vol. 812, ed. by R. Diehl, D.H. Hartmann, N. Prantzos (Springer, Berlin, 2011), pp. 153–232. arXiv:1008.2144

F.K. Thielemann, M. Eichler, I.V. Panov, M. Pignatrari, B. Wehmeyer, Making the heaviest elements in a rare class of supernovae, in *Handbook of Supernovae*, ed. by A.W. Alsabti, P. Murdin (Springer, Berlin, 2017a). https://doi.org/10.1007/978-3-319-20794-0_81-1

F.K. Thielemann, M. Eichler, I.V. Panov, B. Wehmeyer, Neutron star mergers and nucleosynthesis of heavy elements. Annu. Rev. Nucl. Part. Sci. **67**, 253–274 (2017b). arXiv:1710.02142. https://doi.org/10.1146/annurev-nucl-101916-123246

S. Ting, The first five years of the alpha magnetic spectrometer on the international space station. CERN Colloquium of 08/12/2016 (2016). https://indico.cern.ch/event/592392/

L. Titarchuk, P. Chardonnet, The observed galactic annihilation line: possible signature of accreting small-mass black holes in the galactic center. Astrophys. J. **641**, 293–301 (2006). arXiv:astro-ph/0511333. https://doi.org/10.1086/499394

N. Tominaga, M. Tanaka, K. Nomoto, P.A. Mazzali, J. Deng, K. Maeda, H. Umeda, M. Modjaz, M. Hicken, P. Challis, R.P. Kirshner, W.M. Wood-Vasey, C.H. Blake, J.S. Bloom, M.F. Skrutskie, A. Szentgyorgyi, E.E. Falco, N. Inada, T. Minezaki, Y. Yoshii, K. Kawabata, M. Iye, G.C. Anupama, D.K. Sahu, T.P. Prabhu, The unique Type Ib supernova 2005bf: a WN star explosion model for peculiar light curves and spectra. Astrophys. J. Lett. **633**, L97–L100 (2005). arXiv:astro-ph/0509557. https://doi.org/10.1086/498570

C. Travaglio, W. Hillebrandt, M. Reinecke, F.K. Thielemann, Nucleosynthesis in multi-dimensional SN Ia explosions. Astron. Astrophys. **425**, 1029–1040 (2004). arXiv:astro-ph/0406281. https://doi.org/10.1051/0004-6361:20041108

C. Travaglio, F.K. Röpke, R. Gallino, W. Hillebrandt, Type Ia supernovae as sites of the p-process: two-dimensional models coupled to nucleosynthesis. Astrophys. J. **739**, 93 (2011). arXiv:1106.0582. https://doi.org/10.1088/0004-637X/739/2/93

C. Travaglio, R. Gallino, T. Rauscher, F.K. Röpke, W. Hillebrandt, Testing the role of SNe Ia for galactic chemical evolution of p-nuclei with two-dimensional models and with s-process seeds at different metallicities. Astrophys. J. **799**, 54 (2015). arXiv:1411.2399. https://doi.org/10.1088/0004-637X/799/1/54

J.W. Truran, W.D. Arnett, A.G.W. Cameron, Nucleosynthesis in supernova shock waves. Can. J. Phys. **45**, 2315 (1967)

J. Tueller, S. Barthelmy, N. Gehrels, B.J. Teegarden, M. Leventhal, C.J. MacCallum, Observations of gamma-ray line profiles from SN 1987A. Astrophys. J. Lett. **351**, L41–L44 (1990). https://doi.org/10.1086/185675

C. Tur, A. Heger, S.M. Austin, Production of ^{26}Al, ^{44}Ti, and ^{60}Fe in core-collapse supernovae: sensitivity to the rates of the triple alpha and $^{12}\mathrm{C}(\alpha,\gamma)^{16}\mathrm{O}$ reactions. Astrophys. J. **718**, 357–367 (2010). arXiv:0908.4283. https://doi.org/10.1088/0004-637X/718/1/357

M. Ugliano, H.T. Janka, A. Marek, A. Arcones, Progenitor-explosion connection and remnant birth masses for neutrino-driven supernovae of iron-core progenitors. Astrophys. J. **757**, 69 (2012). arXiv:1205.3657. https://doi.org/10.1088/0004-637X/757/1/69

H. Umeda, K. Nomoto, How much ^{56}Ni can be produced in core-collapse supernovae? Evolution and explosions of 30–100 $M_\odot$ stars. Astrophys. J. **673**, 1014 (2008). arXiv:0707.2598

F. van de Voort, E. Quataert, P.F. Hopkins, D. Kereš, C.A. Faucher-Giguère, Galactic r-process enrichment by neutron star mergers in cosmological simulations of a Milky Way-mass galaxy. Mon. Not. R. Astron. Soc. **447**, 140–148 (2015). arXiv:1407.7039. https://doi.org/10.1093/mnras/stu2404

G. Vedrenne, J.P. Roques, V. Schönfelder, P. Mandrou, G.G. von Lichti, A. Kienlin, B. Cordier, S. Schanne, J. Knödlseder, G. Skinner, P. Jean, F. Sanchez, P. Caraveo, B. Teegarden, P. von Ballmoos, L. Bouchet, P. Paul, J. Matteson, S. Boggs, C. Wunderer, P. Leleux, G. Weidenspointner, P. Durouchoux, R. Diehl, A. Strong, M. Cassé, M.A. Clair, SPI: the spectrometer aboard INTEGRAL. Astron. Astrophys. **411**, L63–L70 (2003). https://doi.org/10.1051/0004-6361:20031482

J. Vink, J.M. Laming, J.S. Kaastra, J.A.M. Bleeker, H. Bloemen, U. Oberlack, Detection of the 67.9 and 78.4 keV lines associated with the radioactive decay of ^{44}Ti in Cassiopeia A. Astrophys. J. Lett. **560**, L79–L82 (2001). arXiv:astro-ph/0107468. https://doi.org/10.1086/324172

P. von Ballmoos, J. Alvarez, N. Barrière, S. Boggs, A. Bykov, J.M. Del Cura Velayos, F. Frontera, L. Hanlon, M. Hernanz, E. Hinglais, J. Isern, P. Jean, J. Knödlseder, L. Kuiper, M. Leising, B. Pirard, J.P. Prost, R.M.C. da Silva, T Takahashi, J. Tomsick, R. Walter, A. Zoglauer, A DUAL mission for nuclear astrophysics. Exp. Astron. **34**, 583–622 (2012). https://doi.org/10.1007/s10686-011-9286-6

R. Waldman, D. Sauer, E. Livne, H. Perets, A. Glasner, P. Mazzali, J.W. Truran, A. Gal-Yam, Helium shell detonations on low-mass white dwarfs as a possible explanation for SN 2005E. Astrophys. J. **738**, 21 (2011). arXiv:1009.3829. https://doi.org/10.1088/0004-637X/738/1/21

A. Wallner, T. Faestermann, J. Feige, C. Feldstein, K. Knie, G. Korschinek, W. Kutschera, A. Ofan, M. Paul, F. Quinto, G. Rugel, P. Steier, Abundance of live ^{244}Pu in deep-sea reservoirs on Earth points to rarity of actinide nucleosynthesis. Nat. Commun. **6**, 5956 (2015). arXiv:1509.08054

A. Wallner, J. Feige, N. Kinoshita, M. Paul, L.K. Fifield, R. Golser, M. Honda, U. Linnemann, H. Matsuzaki, S. Merchel, G. Rugel, S.G. Tims, P. Steier, T. Yamagata, S.R. Winkler, Recent near-Earth supernovae probed by global deposition of interstellar radioactive ^{60}Fe. Nature **532**, 69–72 (2016). https://doi.org/10.1038/nature17196

S. Wanajo, The rp-process in neutrino-driven winds. Astrophys. J. **647**, 1323–1340 (2006). arXiv:astro-ph/0602488. https://doi.org/10.1086/505483

S. Wanajo, K. Nomoto, H.T. Janka, F.S. Kitaura, B. Müller, Nucleosynthesis in electron capture supernovae of asymptotic giant branch stars. Astrophys. J. **695**, 208–220 (2009). arXiv:0810.3999. https://doi.org/10.1088/0004-637X/695/1/208

S. Wanajo, H.T. Janka, B. Müller, Electron-capture supernovae as the origin of elements beyond iron. Astrophys. J. Lett. **726**, L15 (2011). arXiv:1009.1000. https://doi.org/10.1088/2041-8205/726/2/L15

S. Wanajo, Y. Sekiguchi, N. Nishimura, K. Kiuchi, K. Kyutoku, M. Shibata, Production of all the r-process nuclides in the dynamical ejecta of neutron star mergers. Astrophys. J. Lett. **789**, L39 (2014). arXiv:1402.7317. https://doi.org/10.1088/2041-8205/789/2/L39

W. Wang, C.S.J. Pun, K.S. Cheng, Could electron-positron annihilation lines in the Galactic center result from pulsar winds? Astron. Astrophys. **446**, 943–948 (2006). arXiv:astro-ph/0509760. https://doi.org/10.1051/0004-6361:20053559

R.F. Webbink, Double white dwarfs as progenitors of R Coronae Borealis stars and Type I supernovae. Astrophys. J. **277**, 355–360 (1984). https://doi.org/10.1086/161701

B. Wehmeyer, M. Pignatari, F.K. Thielemann, Galactic evolution of rapid neutron capture process abundances: the inhomogeneous approach. Mon. Not. R. Astron. Soc. **452**, 1970–1981 (2015). arXiv:1501.07749. https://doi.org/10.1093/mnras/stv1352

C. Winteler, R. Käppeli, A. Perego, A. Arcones, N. Vasset, N. Nishimura, M. Liebendörfer, F.K. Thielemann, Magnetorotationally driven supernovae as the origin of early galaxy r-process elements? Astrophys. J. Lett. **750**, L22 (2012). arXiv:1203.0616. https://doi.org/10.1088/2041-8205/750/1/L22

A. Wongwathanarat, H.T. Janka, E. Müller, E. Pllumbi, S. Wanajo, Production and distribution of ^{44}Ti and ^{56}Ni in a three-dimensional supernova model resembling Cassiopeia A. Astrophys. J. **842**, 13 (2017). arXiv:1610.05643. https://doi.org/10.3847/1538-4357/aa72de

S.E. Woosley, A. Heger, Nucleosynthesis and remnants in massive stars of solar metallicity. Phys. Rep. **442**, 269–283 (2007). arXiv:astro-ph/0702176. https://doi.org/10.1016/j.physrep.2007.02.009

S.E. Woosley, T.A. Weaver, Sub-Chandrasekhar mass models for Type IA supernovae. Astrophys. J. **423**, 371–379 (1994). https://doi.org/10.1086/173813

S.E. Woosley, T.A. Weaver, The evolution and explosion of massive stars. II. Explosive hydrodynamics and nucleosynthesis. Astrophys. J. Suppl. Ser. **101**, 181 (1995). https://doi.org/10.1086/192237

S.E. Woosley, R.G. Eastman, T.A. Weaver, P.A. Pinto, SN 1993J: a Type IIb supernova. Astrophys. J. **429**, 300–318 (1994a). https://doi.org/10.1086/174319

S.E. Woosley, J.R. Wilson, G.J. Mathews, R.D. Hoffman, B.S. Meyer, The r-process and neutrino-heated supernova ejecta. Astrophys. J. **433**, 229–246 (1994b). https://doi.org/10.1086/174638

S.E. Woosley, A. Heger, T.A. Weaver, The evolution and explosion of massive stars. Rev. Mod. Phys. **74**, 1015–1071 (2002). https://doi.org/10.1103/RevModPhys.74.1015

M.R. Wu, R. Fernández, G. Martínez-Pinedo, B.D. Metzger, Production of the entire range of r-process nuclides by black hole accretion disc outflows from neutron star mergers. Mon. Not. R. Astron. Soc. **463**, 2323–2334 (2016). arxiv:1607.05290. https://doi.org/10.1093/mnras/stw2156

Y. Yamamoto, F. Si, H. Nagakura, S. Yamada, Post-shock-revival evolution in the neutrino-heating mechanism of core-collapse supernovae. Astrophys. J. **771**, 27 (2013). arXiv:1209.4824. https://doi.org/10.1088/0004-637X/771/1/27

T. Yamasaki, S. Yamada, Effects of rotation on the revival of a stalled shock in supernova explosions. Astrophys. J. **623**, 1000–1010 (2005). arXiv:astro-ph/0412625. https://doi.org/10.1086/428496

T. Yoshida, Y. Suwa, H. Umeda, M. Shibata, K. Takahashi, Explosive nucleosynthesis of ultra-stripped Type Ic supernovae: application to light trans-iron elements. Mon. Not. R. Astron. Soc. **471**, 4275–4285 (2017). arXiv:1707.02685. https://doi.org/10.1093/mnras/stx1738

W. Zheng, I. Shivvers, A.V. Filippenko, K. Itagaki, K.I. Clubb, O.D. Fox, M.L. Graham, P.L. Kelly, J.C. Mauerhan, Estimating the first-light time of the Type Ia supernova 2014J in M82. Astrophys. J. Lett. **783**, L24 (2014). arXiv:1401.7968. https://doi.org/10.1088/2041-8205/783/1/L24

Space Sci Rev (2018) 214:63
https://doi.org/10.1007/s11214-018-0492-7

Dust in Supernovae and Supernova Remnants I: Formation Scenarios

A. Sarangi[1,2] · M. Matsuura[3] · E.R. Micelotta[4]

Received: 17 June 2017 / Accepted: 17 February 2018 / Published online: 27 March 2018

Abstract Supernovae are considered as prime sources of dust in space. Observations of local supernovae over the past couple of decades have detected the presence of dust in supernova ejecta. The reddening of the high redshift quasars also indicate the presence of large masses of dust in early galaxies. Considering the top heavy IMF in the early galaxies, supernovae are assumed to be the major contributor to these large amounts of dust. However, the composition and morphology of dust grains formed in a supernova ejecta is yet to be understood with clarity. Moreover, the dust masses inferred from observations in mid-infrared and submillimeter wavelength regimes differ by two orders of magnitude or more. Therefore, the mechanism responsible for the synthesis of molecules and dust in such environments plays a crucial role in studying the evolution of cosmic dust in galaxies. This review summarises our current knowledge of dust formation in supernova ejecta and tries to quantify the role of supernovae as dust producers in a galaxy.

Keywords Dust · Supernovae · Chemical kinetics · Infrared · Submillimeter · Presolar grains · SN1987A

Supernovae
Edited by Andrei Bykov, Roger Chevalier, John Raymond, Friedrich-Karl Thielemann, Maurizio Falanga and Rudolf von Steiger

✉ A. Sarangi
arkaprabha.sarangi@nasa.gov

✉ M. Matsuura
matsuuram@cardiff.ac.uk

E.R. Micelotta
elisabetta.micelotta@helsinki.fi

[1] NASA Goddard Space Flight Center, Greenbelt, MD, USA

[2] CRESST II, The Catholic University of America, Washington DC, USA

[3] School of Physics and Astronomy, Cardiff University, Cardiff, UK

[4] Department of Physics, University of Helsinki, Helsinki, Finland

1 Introduction

The solid grains in space which may vary in size from the dimension of a few molecules to the order of a few microns, are broadly classified as cosmic dust. Cosmic dust can be of diverse chemical types, which includes carbon (amorphous and graphite), silicates (olivines, pyroxenes, both amorphous and crystalline), metal oxides (alumina, fayalite, magnesia, spinel), silica, pure metals (iron, magnesium, silicon), metal sulphides, carbides and some other species yet to be determined with certainty (Molster et al. 2010; Cherchneff 2013). Dust accounts for about 1–2% of the total baryonic mass in the universe. On the other hand, it is responsible for up to 50% of the total radiation from some galaxies, which rightfully justifies its importance. These dust grains are efficient in absorbing and scattering UV, optical and near-infrared (IR) radiations, and re-emitting the absorbed energy in the mid-IR and submillimeter (submm) wavelengths (Martin 1978; Li 2005). Owing to their continuum opacities, the dust grains absorb and emit over a large swathe of the electromagnetic spectrum (Draine 2003). Therefore, dust in space profoundly influences the spectra and colours of stars and galaxies. Moreover, cosmic dust acts as the basic building block for the planets in protoplanetary disks around main-sequence (MS) stellar systems (Brownlee 1985; Dwek 2006). The branches of science related to the study of cosmic dust has flourished over last four decades (Grün et al. 2001). It has brought together scientists (astronomers, meteoriticists, physicists, chemists, mineralogists and many more) from various disciplines and expertise to focus on the development of the field.

The earliest discoveries of molecule and dust clouds, leading to interstellar reddening were inferred in the first part of the 20th century (Barnard 1919; Trumpler 1930). However, only after the advent of IR astronomy in the 1960s, we have made a fast progress towards the current state of the art. The infrared region of the electromagnetic spectrum ranges from 700 nm (the red edge of the visible spectrum) to approximately 1 mm in wavelengths and is further subdivided into near-IR, mid-IR and far-IR regions. The biggest challenge to the development of IR astronomy, is that only a small part of the IR spectrum can actually be observed from ground-based facilities, owing to the absorption by earth's atmosphere, mainly by water vapour. In order to overcome the impediment of atmospheric shielding, the following schemes have been adopted: (a) to set up telescopes on the top of high altitude mountain peaks, (b) to raise the telescope above the atmosphere by helium balloons, (c) to attach the telescopes in jet transport aeroplanes, and (d) to launch IR satellite telescopes (since 1983) (Walker 2000). Later on, the ground-based telescope facilities with IR capabilities have also been developed. Some of the most important telescopes, operating in the near-IR, mid-IR, far-IR and submm, launched in the 21st century are the Hubble Space Telescope (optical and near-IR), Spitzer space telescope (3–180 μm), AKARI (2–200 μm), Herschel space observatory (55-672 μm) (until 2013) and WISE (3–25 μm). The Atacama Large Millimeter/submillimeter Array (ALMA), situated high on the Chajnantor plateau in Chile, is a high resolution millimeter-submillimeter telescope, which can study radiations from the coldest objects in the universe. The James Webb Space Telescope, under preparation for launch in 2018, is designed to boost IR astronomy with its unprecedented resolution and sensitivity in 0.6–27 μm range. The energy radiated in the IR is largely thermal in nature. Therefore, IR photometry and spectroscopy have become the crucial tools to study the dust continuum from various sources, vibrational transitions in molecules, signatures from exoplanets and the high redshift universe.

Using the newly developed techniques, confirmed evidence for the presence of cosmic dust has been established through various studies over the past couple of decades. Having said that, it is equally necessary at this point to investigate the potential sources of origin

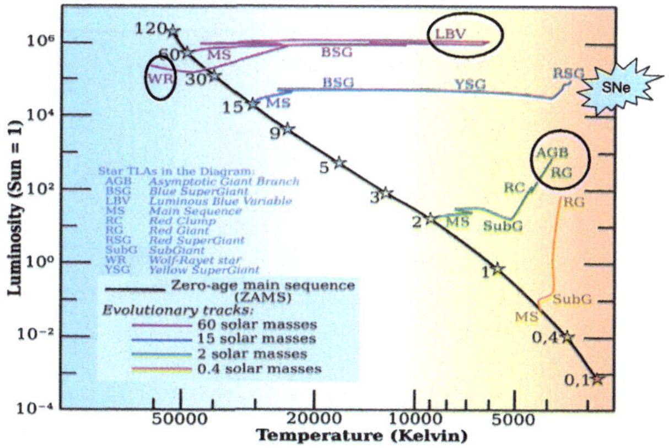

Fig. 1 Hertzsprung-Russell diagram showing luminosity-temperature relation of stars at their evolutionary phases as function of main sequence initial mass. The potential stellar sources of cosmic dust are marked with circles. The core-collapse supernova (SN), the final fate of the RSG (red super giant) phase as shown in the diagram, is a major source of dust formation

for dust present in space. Cosmic dust is made up of large clusters of chemical compounds. Hence its synthesis requires chemical processes, which are sensitive to the local gas densities and temperatures. The chemical processes that are prevalent at moderately high temperatures ∼1000–2000 K and high gas densities govern the dust formation scenario. Such physical conditions are mainly encountered in evolved stellar outflows or winds and ejecta of novae and supernovae (Cherchneff 2010, 2013; Kozasa et al. 2009). However, the required temperature and densities in these evolving circumstellar environments persist only for a short timescale making the entire mechanism more complex.

The evolved stellar environments responsible for dust formation in galaxies are presented in Fig. 1 with their respective position on the Hertzsprung-Russell diagram. The quiescent mass outflows of the low mass ($<8\ M_\odot$) stars in their Asymptotic Giant Branch (AGB) phase are thought to be major dust producers in galaxies (Whittet 2003; Cherchneff 2012; Gobrecht et al. 2016). Core-collapse SNe, the ultimate fate of massive (8–30 $M_\odot$ MS mass) stars, marked in the figure (not in the scale of the H-R diagram), are a prime source for dust formation (Wooden et al. 1993; Sugerman et al. 2006; Dwek 2006; Kozasa et al. 2009; Kotak et al. 2009; Cherchneff 2013; Todini and Ferrara 2001; Szalai et al. 2011). Additionally, periodic mass ejections in massive Luminous Blue Variable (LBV) stars (Hillier et al. 2001; Ferland et al. 2005) and mass-loss events in the Wolf-Rayet stars (Harries et al. 2004) are also associated to dust formation in the galaxies. However, AGB stars and CCSNe are the known to be the main stellar sources. In case of supernova ejecta, where the stellar core has already collapsed, the physical conditions evolve very fast. Therefore, the entire process of dust synthesis occurs in less than a decade. Synthesis of dust in a suitable environment proceeds in two steps: the nucleation and the condensation (Cherchneff and Dwek 2009; Sarangi and Cherchneff 2013). Out of all the potential sources, this chapter aims to discuss and quantify the contribution of supernovae as dust producer in the galaxies.

Dust in dense clouds with a visual extinction exceeding ∼3 mag is characterized by solid cores surrounded by ice mantles. The presence of ice mantles in the dense interstellar clouds compared to the small dust present in the diffuse ISM confirms the growth of grains in the clouds. The grains grow in mass over the lifetime of the molecular clouds. For significant growth of dust grains in the ISM, the necessary conditions are: (a) the presence of preexisting grain cores on which accretion takes place, (b) the accretion timescales must be shorter than the lifetime of the clouds (Dwek and Cherchneff 2011). Interstellar dust grains are subject to various turbulent forces over the span of its lifetime, which either reduce the sizes of the grains or completely destroy them (Dwek 1998; Jones 2004). According to theoretical models, for dust grains to survive in the ISM, a continuous source of newly formed

Fig. 2 The current classification of supernovae done on the basis of the spectral types and light curves. On the basis of explosion mechanism, supernova explosion categorised as thermonuclear or core-collapse (Turatto 2003)

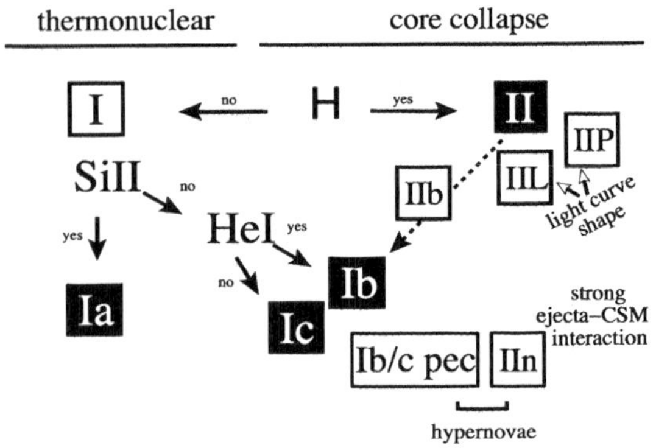

grains must be present in the galaxy which can compensate for the loss due to destruction processes in the clouds (Dwek 2006). Interstellar silicates are predominantly amorphous, whereas stardust silicate samples from meteorites are about 20% crystalline (Li and Draine 2001; Kemper et al. 2005; Li et al. 2007) in nature. This is consistent with the notion that dust in the ISM might have originated at a different source. Therefore, synthesis of new seeds from the gas phase is highly unlikely in the ISM due to low densities and temperatures.

Even though there remains no ambivalence on supernovae being a prime dust factory in the universe, the mechanism of dust formation still remains unclear. Further, depending on the subclasses of SNe, they are subject to diverse physical conditions directly impacting the chemical processes. SNe are classified broadly on the basis of their optical spectra and some subclasses are defined by the nature of their light curves. The taxonomy is progressively developing since 1941 (Minkowski 1941) and is extensively reviewed by Filippenko (1997) & Wheeler and Benetti (2000). A schematic presentation of the SNe types is shown in Fig. 2. Here we discuss a few important subclasses of SNe, in light of their role in dust formation.

Pair-instability SNe (PISNe): Based on the explosion mechanism, a large section of massive stars ($>100\ M_\odot$) are categorised as pair-instability supernova, a case common among Population III and some Population II stars. There is no observational constraint on such objects and neither there is any confirmed evidence for dust formation in the ejecta. However, to account for the dust content in some high-z galaxies, PISNe are considered as prime dust forming agents in those environments. Theoretical studies dealing with nucleosynthesis and hydrodynamics in the pre- and post-explosion massive stars provide the backbone to understand the physical conditions relevant to these systems (Umeda and Nomoto 2002; Heger and Woosley 2002; Fryer et al. 2001). A dust mass of $0.1\ M_\odot$ per supernova event has been estimated by Cherchneff and Dwek (2009) adopting a chemical kinetic approach.

Type Ia: Supernovae type Ia is characterised by the absence of hydrogen lines in the spectra. The mechanism of explosion is thermonuclear, which occurs in low mass stars ($<8\ M_\odot$) in binary systems (Branch et al. 1995). Several studies have advocated in support of type Ia's as a probable dust producer (Clayton et al. 1997; Travaglio et al. 1999). Gomez et al. (2012) have investigated two Type Ia remnants, Tycho and Kepler, using *Herschel* PACS and SPIRE photometry at 70–500 μm. Their findings suggest a presence of a warm dust component in the spectra at ∼82 K, resulting from a dust mass of $\sim 3.0 \times 10^{-3}\ M_\odot$. The IR emission features are found to be spatially consistent with the X-ray emitting region and the circumstellar region swept up by the SN blast-wave. There has been no observation till date of new dust formation in the ejecta of Type Ia's (Gomez et al. 2012). On the other hand, theoretical study by Nozawa et al. (2011) has predicted a possibility of new dust formation in the ejecta, with a bottom heavy size distribution. However, that being the case, the newly formed dust grains are almost likely to get completely destroyed by the reverse shock in course of their injection into the ISM (Nozawa et al. 2011).

Type Ib/c: These SNe types are characterised by the absence of H and Si II lines and the presence of He I in case of Type Ib, where as in Type Ic He I lines are also absent (Barbon et al. 1999; Matheson et al. 2000b). These objects are identified as core collapse SNe of massive stars ($>25\ M_\odot$) which have been stripped of their hydrogen envelope. They show evidence of shock interactions with dense circumstellar environments (Chevalier 1982) at early times after explosion. Presence of dust continuum in the spectra of Type Ib SN 1990I was first reported by Elmhamdi et al. (2004). Further, peculiar Type Ib SN2006jc has shown dust signatures as early as day 50 post explosion (Nozawa et al. 2008; Smith et al. 2008; Di Carlo et al. 2008), which originate from the dense shell formed in the post-shock CSM. Mid-IR continuum recorded by *Spitzer* from Type Ic SN2005at and SN2007gr confirm the presence of dust in the environment (Kankare et al. 2014). The source of the emission can be either an echo or the expanding ejecta, it is yet to be understood with clarity.

Type IIb: Type IIb's are core collapse supernovae which have early time signatures similar to Type II's (presence of prominent H lines) and a late time spectra similar to Type Ib's (Matheson et al. 2000a). The decline of the lightcurves in Type IIb SNe are much faster compared to other Type II, which implies a smaller mass of the ejecta. Due to the absence of the outer H-envelope after explosion, the ejecta expands faster and the typical gas densities are lower (Nozawa et al. 2010) when compared to a Type IIP. Therefore, Type IIb supernovae do not form dust as efficiently as Type IIP's (Biscaro and Cherchneff 2014, 2016). IR and submm observations of galactic SN remnant Cassiopeia A of Type IIb have revealed the presence of dust ranging between $10^{-4}\ M_\odot$ (Douvion et al. 2001; Dunne et al. 2003) to $0.08\ M_\odot$ (Barlow et al. 2010; Sibthorpe et al. 2010).

Type IIP: The SNe Type II-P (Plateau) is the most abundant type of core-collapse supernovae in the local universe (Barbon et al. 1979). It differs from Type I's by the presence of prominent H lines in the spectra. The light curve is characterised by a flat plateau type nature during the first few months after explosion. Massive stars in the 8–25 $M_\odot$ range can result to Type II-P supernovae at the end of their lives. The presence of a large H-envelope as the outer shell of the ejecta makes the outflowing ejecta massive and slow moving. Moreover, the densities of the expanding He core remain $\sim 10^2/10^3$ higher than Type IIb's. Based on the degree of stripping of the envelope, SNe can be arranged in sequence as Type IIP-IIL-IIb-Ib-Ic (Nomoto et al. 1995) in the ascending order. Also, this is the order of importance in terms of dust synthesis. Therefore, the ejecta of Type IIP SNe is the most efficient and copious source of dust among all SNe types (Szalai and Vinkó 2013; Sarangi and Cherchneff 2015; Cherchneff 2013; Kotak et al. 2009) and our discussion in this chapter will be largely based on Type IIP's.

Type IIn: SNe Type IIn, introduced by Schlegel (1990), are characterised by the presence of narrow ($\sim 100\ \mathrm{km\,s^{-1}}$), intermediate ($1\text{–}4 \times 10^3\ \mathrm{km\,s^{-1}}$) and broad ($10\text{–}15 \times 10^3\ \mathrm{km\,s^{-1}}$) velocity width components. The narrow component originates from the slow moving CSM before being traversed by the supernova shock. The source of intermediate width component is the post-shock region of the CSM, where as the broad component arises from the fast expanding SN ejecta (Smith et al. 2008). Observations indicate early interactions between the ejecta and the circumstellar medium (Smith et al. 2012), which lead to a velocity shear followed by shocks. Many recent studies have reported the presence of warm dust in such environments (Fassia et al. 2001; Mauerhan et al. 2013; Andrews et al. 2011; Smith et al. 2012; Gall et al. 2014).

2 Observations of Dust Formation in Core-Collapse Supernovae

Using modern high-resolution telescopes, a lot of important information can be derived from a single observation and a general trend for dust formation in supernovae can be studied. The formation of warm dust in the ejecta is characterised by the following: (a) an excess in the mid-IR region of the spectral energy distribution due to thermal emission (Roche et al. 1991; Bouchet and Danziger 1993), (b) a sharp decline in the optical light curve (Danziger et al. 1991), (c) a progressive and systematic blue-shift of emission line profiles when the receding part of the ejecta is increasingly blocked by newly formed dust (Lucy et al. 1989), (d) diminishing line emission of elements and molecules compared to the adjacent continuum, owing to depletion in dust (Bouchet and Danziger 1993; Kotak et al. 2009). The first two cases might also arise either due to an IR echo from pre-existing dust in the circumstellar material (Smith et al. 2012) or from the escape of radioactive luminosity due to decrease in optical depth in the ejecta. But the occurrence of three or all of the four phenomena concurrently, clearly indicates the formation of new dust in the ejecta.

There are several local SNe that observers have closely monitored over the past few decades. We shall now discuss the findings from some important SNe with regard to their history of dust formation and its evolution in the ejecta.

2.1 Dust Formation in the Ejecta of SN 1987A

The explosion of a core-collapse SN was detected in the Large Magellanic Cloud, only 50 kpc away on the 23rd of February, 1987 (Kunkel et al. 1987). Since its detection, studies of SN 1987A have led to many discoveries of SNe and SNRs.

2.1.1 Historical Observations of Dust in SN 1987A

One of the major discoveries associated with the SN 1987A is the detection of newly formed dust (Danziger et al. 1989). From about 615 days, thermal emission of dust has been detected at near and mid-infrared (IR) with an increase in contrast and a shift towards longer wavelength in time (Fig. 3). The reported dust mass was 10^{-4} $M_\odot$ at day 775 with the temperature of 300–400 K. However, the possibility of larger dust mass was suggested because dust could be optically thick (Wooden et al. 1993). At almost the same time ($\sim$530 days), the optical light curve declined at a much steeper rate than the predicted light curve accounting for the decay of ^{56}Co and ^{57}Co (Fig. 4: Whitelock et al. 1989). It has been interpreted that, the dust grains that are formed, are causing the extinction in optical wavelengths (Danziger et al. 1989). Furthermore, asymmetric line-profile has also been detected simultaneously. Lucy et al. (1989) modelled the blue-shifted line profile absorbed by the dust that has condensed in the ejecta.

It had been suggested that dust could be present in clumps instead of being smoothly distributed across the ejecta (Lucy et al. 1989; Wooden et al. 1993). If a large mass had been evenly distributed in the ejecta, UV and optical energy would have been totally absorbed resulting in a black out (Hoyle and Wickramasinghe 1970; Dwek 1988). In order to avoid such black out, dust must be present in clumps so that UV and optical light can escape between the clumps. If dust is located in clumps, and if individual clumps are assumed to be optically thick, it is difficult to estimate the actual dust mass. It possibly suggests that the dust masses as large as 0.5 $M_\odot$ (Lucy et al. 1989) or 1 $M_\odot$ (Dwek 1988). Whether it is feasible to accommodate such a large amount of dust in early days still remains a matter of

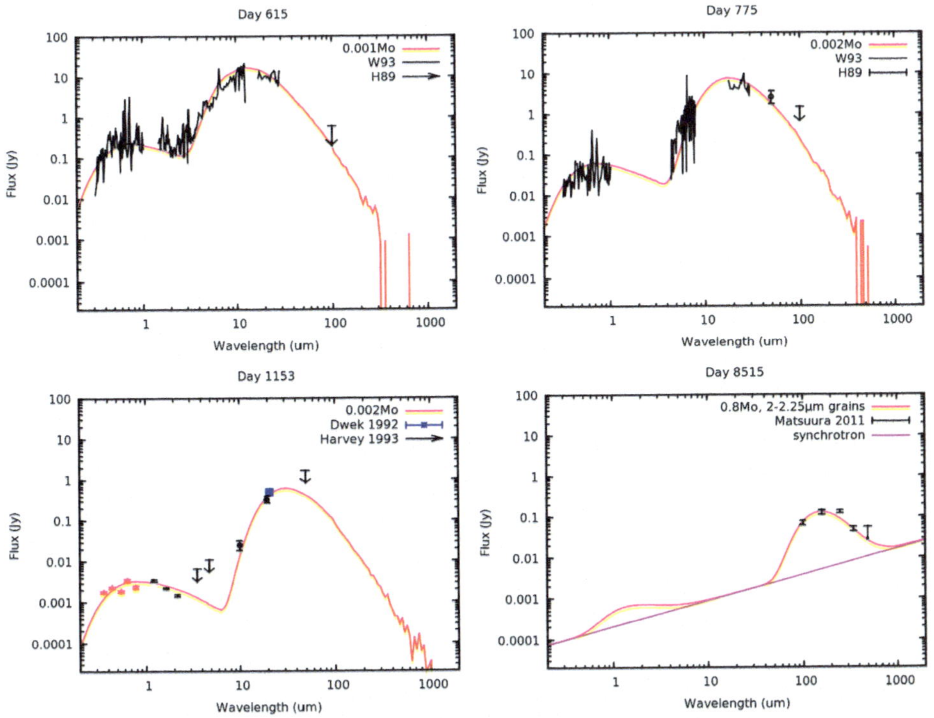

Fig. 3 Time evolution of thermal dust emission of SN 1987A. A broad thermal emission appears at mid-infrared, indicating formation of dust. The peak is shifting towards the longer wavelength in time. The figure is reproduced from Wesson et al. (2014), based on observational data from Wooden et al. (1993), Harvey et al. (1989), Bouchet and Danziger (1993), Dwek et al. (1992), Matsuura et al. (2011)

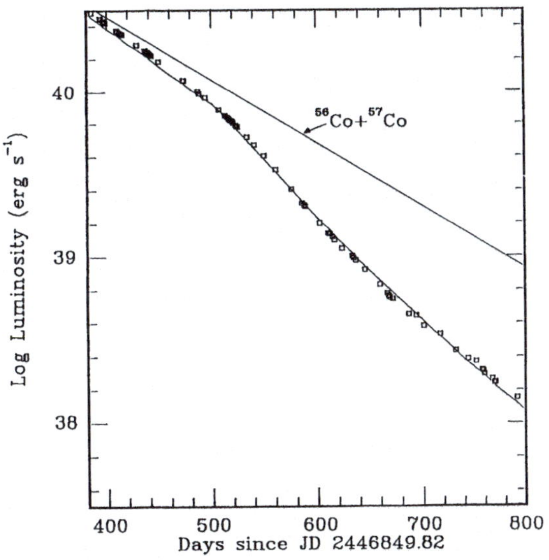

Fig. 4 The light curve of SN 1987A. The measured luminosity (squares) more sharply declined at about day 530, than the predicted light-curve from ^{56}Co and ^{57}Co decay. This further decline was attributed to the newly formed dust in the ejecta (Whitelock et al. 1989)

debate (Ercolano et al. 2007; Wesson et al. 2014; Bevan and Barlow 2016; Dwek and Arendt 2015).

Since its detection, thermal emission of dust had been monitored at mid-infrared up to day 1,316 (Suntzeff and Bouchet 1990; Bouchet et al. 1991; Dwek et al. 1992). The following study was at day 6000 with nearly a 13 year gap in between the observations (Bouchet et al. 2004). By day $\sim$1150, the temperature had dropped to 150 K with an inferred dust mass of 5×10^{-4} $M_\odot$ (Dwek et al. 1992).

Ten years later, mid-infrared observations found dust thermal emission from the circumstellar ring (Bouchet et al. 2006). The ring composed of the material that were expelled from the progenitor star when it was in the red supergiant phase about 20,000 years ago (McCray 1993). The fast ($\sim$2000 $km\,s^{-1}$) blast wave from the explosion has collided into the slowly expanding ($\sim$100 km^{-1}) circumstellar ring. The collisions and subsequent shocks yield X-radiation heating dust grains with 300–400 K radiation at mid-infrared. The infrared luminosity of the ring dust continued to increase concurrently with a constant increase of the X-ray brightness, suggesting destruction of dust grains in the ring (Dwek et al. 2008). However, since day 8,500, infrared brightness started declining, while X-ray luminosity still increased. That is almost the time when the knots around the known ring structure started to brighten up, so that SN waves appeared to have overtaken the main part of the existing ring (Fransson et al. 2015).

2.1.2 Detection of a Large Mass of Dust in the Ejecta

In 2010, the HERSCHEL SPACE OBSERVATORY discovered far-infrared emission towards SN 1987A (Fig. 5: Matsuura et al. 2011). The inferred dust mass was 0.4–0.7 $M_\odot$ with the temperature of $\sim$20–30 K. The only place that can form such a large mass of dust would be the ejecta, where a large mass of refractory elements are present. The confirmation of the presence of dust in the ejecta was made by ALMA, soon after the HERSCHEL observations. ALMA high angular resolution image (angular resolution of 0.3″) enabled to resolve SN 1987A system at submillimeter wavelengths, detecting thermal emission from cold dust towards the ejecta (Indebetouw et al. 2014), with a mass of $\sim$0.5 $M_\odot$ (Matsuura et al. 2015). Detailed analysis of ALMA data by Zanardo et al. (2014) suggested that there could be an extra component beyond dust at millimetre wavelength, which can be associated with the pulsar winds.

Importantly, the ejecta's expansion velocity of SN 1987A might appear to be too fast for dust formation. Typically dust grains were predicted to be destroyed by SN winds with velocities exceeding 100 $km\,s^{-1}$ (Jones et al. 1994). On the other hand, the velocities of SN 1987A ejecta are typically 2000 $km\,s^{-1}$ (full width of half maximum (FWHM); McCray 1993) much larger than the dust destruction velocities. Dust grains can form and survive in such a high velocity environment, because in the rest frame of dust grains, the velocity is much smaller. The expansion velocities reach 2000 $km\,s^{-1}$ a few months after the explosion, and following that, the velocities were measured to be constant (McCray 1993). Dust grains have formed only after these expansion velocities settled. Therefore currently in SN1987A, the dust grains are expanding at much faster speed without being impacted by shocks.

2.1.3 Look Back at the Historical Measurements of Dust in SN 1987A

The detection of a large mass ($\sim$0.5 $M_\odot$) of ejecta dust at $\sim$ day 8500 reignites the debate if such a large mass of dust is present from early days, starting from day 650. The inferred

Fig. 5 Image of SN 1987A and surrounding ISM (Matsuura et al. 2011). Spitzer 8 μm in blue, Spitzer 24 μm in red and Herschel 250 μm in green. LMC field stars are seen in blue, while SN 1987A, which contain significant mass of cold dust, is seen as a green point source in this image

dust mass in early days was 10^{-4} $M_{\odot}$ at day 775, while measurements at late day (~ day 10,000) suggests ~0.5 $M_{\odot}$. There are two possibilities that can resolve the disparity. The first possibility is that inferred dust masses in early days were underestimated because the fitting assumed optically thin dust, but the emission was actually optically thick and a large mass of dust was present from the early days itself (Dwek and Arendt 2015). Alternative possibility is that the dust mass has increased over time by growing dust grains (Wesson et al. 2014; Bevan and Barlow 2016). Using currently available observational data, it is not possible to disentangle these two cases.

2.1.4 Molecules in SN 1987A

SN 1987A is also the first SN where molecule formation was reported. The emission of CO $\nu = 2$ bands at 2.3 μm started appearing possibly 112–136 days after the explosion, with much more recognisable features from day 192 (Spyromilio et al. 1988; Catchpole et al. 1988). CO $\nu = 2$ bands at 2.3 μm and $\nu = 1$ bands at 4.6 μm continued to be detected until about day 600 and then disappeared from infrared spectra (Wooden et al. 1993).

More than 20 years after, CO was detected at millimetre wavelength after high sensitivity millimetre telescope, the ALMA, started operation in 2011 (Kamenetzky et al. 2013; Matsuura et al. 2017). The analysis of these rotational bands suggests an excitation temperature of 20–50 K, which is much lower than the estimated temperature of 1200–4000 K in early days (up to day ~600; Liu et al. 1992; Liu and Dalgarno 1995). It has been perceived that initially the gas temperature is high enough to excite vibrational-rotational bands. When the gas temperature drops, CO population is mainly in rotational transitions.

Not only CO but also detections of a few other molecules have been reported in SN 1987A. SiO fundamental bands at 4 and 8 μm appeared at about day 200 (Rank et al. 1988; Aitken et al. 1988; Roche et al. 1991; Meikle et al. 1989) and SiO rotational lines at millimetre wavelength were detected at day 9,173 onwards (Kamenetzky et al. 2013; Matsuura et al. 2017). Potential detections of CO^+, CS and H_3^+ were also reported (Meikle et al. 1989; Miller et al. 1992) though not yet well established (Cherchneff and Dwek 2009). In later epoch, vibrationally excited H_2 at 2 μm was found since day 6,832 (Fransson et al. 2016), and HCO^+ and SO at millimetre wavelength at day 10,037 (Matsuura et al. 2017).

The ALMA offers angular resolution of 50 miliarcsec images of CO and SiO in the ejecta (Abellán et al. 2017) as shown in Fig. 6. It is found that the CO and SiO clumps in the

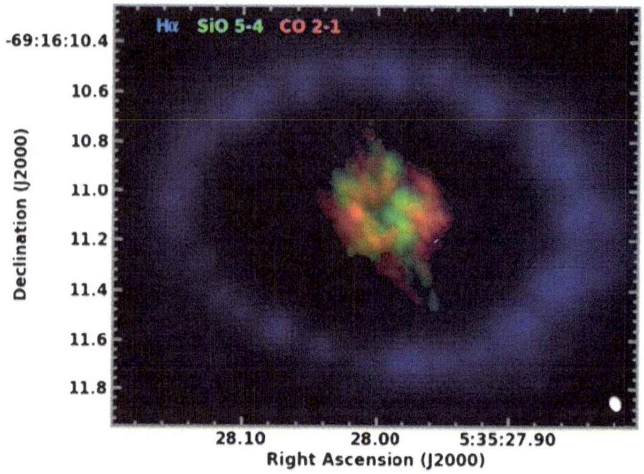

Fig. 6 ALMA CO $J = 2$–1 (red) and SiO $J = 5$–4 (green) images overlaid on the HUBBLE SPACE TELESCOPE Hα image of the circumstellar ring of SN1987A. The figure is published by Abellán et al. (2017) and the HST image taken from Larsson et al. (2016)

ejecta forms torus shape. These molecules fill in clumps. Three-dimensional models of SN explosions predict that Rayleigh-Taylor instabilities, triggered by the explosion, break the ejecta material into clumps (Wongwathanarat et al. 2015). After over 20 years of evolution in the post-explosion era, high resolution of the ALMA enabled us to resolve these clumps.

2.2 Cassiopeia A—Forward Shocks, Reverse Shocks and Unshocked Dust

Cassiopeia (Cas) A is the remnant of Type IIb SN, with an approximate age of 335 years (Fesen et al. 2006; Krause et al. 2008). IRAS (Infrared Astronomical Satellite), an all sky survey mission launched in 1983, detected dust in the SNR, Cas A (Dwek et al. 1987). Together with SN 1987A, detection of dust in this young SNR was one of the milestone discoveries which confirmed that dust can be formed in core collapse SNe.

2.2.1 SN Dust or ISM Dust?

The total dust mass in Cas A is largely debated. This is because there is heavy extinction of ISM dust towards Cas A, and evaluation of foreground and background ISM dust could affect Cas A dust mass estimates. The reported dust mass from IRAS was 7×10^{-3} $M_\odot$ (Dwek et al. 1987). Using SCUBA/JCMT, a dust mass of 0.2–4 $M_\odot$ was reported in submillimeter wavelength (Dunne et al. 2003). Fitting of the 8–850 μm observations including SPITZER MIPS, indicates a presence of 3×10^{-3} $M_\odot$ of dust (Hines et al. 2004). AKARI and BLAST 65–500 μm observations deduced a 0.003 and 10 $M_\odot$ of hot (99 K) and cold (17 K) dust (Sibthorpe et al. 2010). The analysis of HERSCHEL observations resulted in 0.075 $M_\odot$ of dust with a temperature of ∼35 K (Barlow et al. 2010). Developing SN template analysis on SPITZER and HERSCHEL found 0.04 $M_\odot$ of warm SN dust and <0.1 $M_\odot$ of cold dust, whose origin was unclear. It can either be SN ejecta or the ISM (Arendt et al. 2014). After subtracting ISM dust templates from the SED of SPITZER and HERSCHEL, 0.4–0.7 $M_\odot$ of cold SN dust was still present in Cas A in the unshocked region (De Looze et al. 2017).

2.2.2 Dust Compositions

Another important finding in Cas A was the spatially varying dust compositions distributed within the SNR and the type of dust compositions correlated to specific atomic lines (Lagage

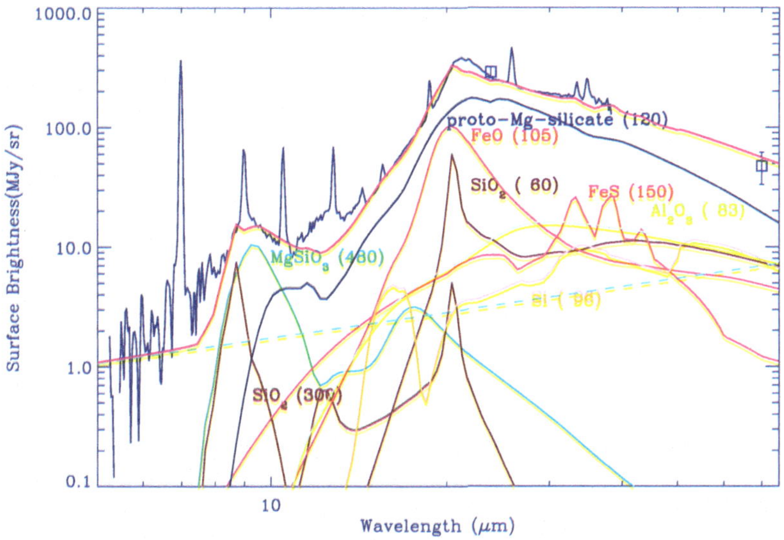

Fig. 7 One of the SPITZER spectra obtained in Cas A. Strong 21 μm feature was detected, which was attributed to Mg protosilicate, amorphous SiO_2 and FeO (Rho et al. 2008)

et al. 1996; Rho et al. 2008; Arendt et al. 2014). The spatially resolved spectra of Cas A showed that the spectra with the '21 μm feature' are accompanied with a strong [Ar II] line, while the spectra with featureless dust emission is correlated to the [Si II] line (Fig. 7; Arendt et al. 2014; Rho et al. 2008). The '21 μm feature' could be attributed to a combination of Mg protosilicate (Arendt et al. 1999), amorphous SiO_2 and FeO (Rho et al. 2008). This is an unique case where a specific dust feature has been identified in SNRs, with recent finding in another Galactic SNR, G54.1+0.3 (Temim et al. 2017).

2.2.3 CO Molecules in Cas A—Survived Reverse Shocks or Reformed in Shocked Region

CO molecules have also been detected in Cas A. Filter measurements at 2 and 4 μm suggested the presence of CO emission at the reverse shocked region of Cas A (Rho et al. 2009) and was later spectroscopically confirmed by Rho et al. (2012). One possibility is that the CO formed in early days might have survived reverse shocks. Alternative possibility is that CO is re-formed. Wallström et al. (2013) detected high J lines in Cas A. The wide range of temperatures (400–200 K) and high column density (5×10^{17} cm^{-2}) suggests that these CO molecules are from the dense regions of the post-shocked gas, where CO molecules were reformed.

Chemical models suggest that molecules can re-form in the post-shock region, but not dust (Biscaro and Cherchneff 2014).

2.3 Crab Nebula

2.3.1 Dust in the Crab Nebula

The SN explosion of the Crab Nebula was recorded in 1054. Unlike Cas A, whose bright optical and X-ray nebula is collisionally heated by reverse shocks (Gotthelf et al. 2001), the major heating source of the Crab Nebula is synchrotron radiation (Hester 2008). Dust found in the Crab Nebula was also heated by the synchrotron radiation (Owen and Barlow 2015).

The Crab Nebula is composed of many filament like structures. The wind from the pulsar compresses the expanding ejecta to form these dense filaments. Not only gas but also dust is associated with these filaments (Fig. 8: Fesen and Blair 1990; Gomez et al. 2012; Grenman et al. 2017).

There has been a range of dust masses, that is inferred from observation studies of the Crab Nebula. Marsden et al. (1984) reported thermal emission of dust from IRAS flux measurements at 25, 60 and 100 μm. The dust mass was inferred to be between 5×10^{-3}–0.03 $M_\odot$. Green et al. (2004) analysed the INFRARED SPACE OBSERVATORY (ISO)'s 60, 100 and 170 μm images, together with the 850 μm synchrotron image from James Clerk Maxwell Telescope (JCMT) and estimated a dust mass of 3×10^{-3}–0.07 $M_\odot$ with temperatures of 45–50 K. Temim et al. (2006) used SPITZER 24 and 70 μm flux, together with ISO's photometric data, resulting in the estimated dust mass of 4×10^{-3}–1.5×10^{-2} $M_\odot$ with a temperature of 74 K. Adding the photometric measurements at longer wavelength of 70–500 μm from HERSCHEL SPACE OBSERVATORY, Gomez et al. (2012) found that the dust emission from the Crab Nebula is much broader than a single temperature black body/grey body, which previous authors have used. Using a two component black body, Gomez et al. (2012) found that while their hot component has similar dust temperature and mass as those derived by Temim et al. (2006). Gomez et al. (2012) found that the additional cold component, emitting at longer wavelengths, is at a mass of 0.11–0.24 $M_\odot$ with temperatures of 28–34 K. However, the question is if mid-infrared emission from warm dust can hide emission from cold dust, which dominates the total dust mass.

In order to fit the broad emission, Temim and Dwek (2013) invoked temperature gradient of dust grains because synchrotron radiations can make small grains hotter than larger grains due to difference in heat capacities and absorption coefficients. The resultant dust mass is comparable to the estimates from two black bodies (Gomez et al. 2012), as far as the same dust emissivities are chosen. Temim and Dwek (2013)'s model assumed that all dust grains are located at a fixed distance. Owen and Barlow (2015) advanced the distance issue, by applying full radiative transfer modelling to shell structure. Their fitted results on both ionised lines and dust emissions show that the dust mass ranges between 0.17–1.5 $M_\odot$ depending on the morphologies of the dust species and their emissivities.

2.3.2 Molecules in the Filaments of the Crab Nebula

Vibrationally excited molecular hydrogen lines have been detected in the Crab Nebula (Graham et al. 1990), and the H_2 emission originates in the filaments (Loh et al. 2011). Loh et al. (2012) estimated the excitation temperature of H_2 to be 2000–3000 K, which is most likely collisionally excited.

The first detection of noble molecular gas in space was made in the Crab Nebula, by the detection of 618 and 1235 GHz lines of $^{36}ArH^+$ found in the HERSCHEL Space Observatory's spectra (Barlow et al. 2013). This is an unusual case because majority of molecules had been found first in interstellar medium (ISM) or circumstellar medium of evolved stars and later identified in SNRs. On the other hand, this molecule was reported in a SNR first and then identified later in the ISM (Schilke et al. 2014). Argon is a noble gas, i.e. non-reactive, but its ionised state Ar^+ is extremely reactive and its reaction with H_2 leads to the formation of ArH^+.

2.4 Galactic and Magellanic SNRs

Infrared surveys enabled to detect and analyse Galactic and Magellanic clouds SNRs with possibilities to analyse the dust within. Launched in 1983, the IRAS is the first whole sky survey at 12, 25, 60 and 100 μm and detections of 51 and 11 SNRs were reported from the survey (Arendt 1989; Saken et al. 1992). These achieved systematic studies of infrared emission from SNRs.

Launched in 2003, the SPITZER SPACE TELESCOPE used Galactic plane surveys ($|b| < 1^\circ$ and $10^\circ < |l| < 65^\circ$; e.g. Benjamin et al. 2003) with two different instruments, IRAC and MIPS. These survey data were used to detect infrared emission from SNRs. Compared to the pixel scale of 2' of the IRAS map used by Arendt (1989), ~2.5 arcsec resolution at 3.5–8.0 μm, 6 arcsec at 24 μm and 20 arcsec at 70 μm are great improvements, which helped in resolving ISM cirrus confusion towards the SNRs. Reach et al. (2006) found 17 detections at 3.5–8.0 μm and colour analysis suggested that four of the SNRs appeared to have infrared emission associated with polycyclic aromatic hydrocarbons (PAHs). The emission of the remaining thirteen SNe were found to be due to molecular lines (H_2 and CO) or atomic lines ([Fe II], [Ar II] and Brα) caused by shocks. Pinheiro Gonçalves et al. (2011) searched for counterparts of Green's SNR catalogue (Green 2004) and reported the detection of 39 SNRs at 24 and 70 μm. The mid-infrared emissions of these SNRs have various origins that include dust and lines.

SPITZER's sensitivities and angular resolutions allowed us to study SNRs beyond the Galaxy such as those in the Magellanic Clouds. That includes dedicated studies of individual SNRs, 1E 0102.2−7219 (Stanimirović et al. 2005; Sandstrom et al. 2009) and the studies (Seok et al. 2013; Lakicevic et al. 2015) using the Magellanic Clouds survey (Meixner et al. 2006). Seok et al. (2008) also studied Magellanic Clouds SNRs using the AKARI infrared whole sky survey. Correlation between infrared emission and X-ray emission suggests that 24 and 70 μm flux of the SNRs are collisionally heated (Seok et al. 2015) with the possibility of dust destruction by sputtering (Williams et al. 2006).

3 Extra-Galactic SNe

The detections of ejecta dust in SNe beyond the magellanic clouds have been reported by finding asymmetric structures in the optical emission lines (e.g. Elmhamdi et al. 2003b). After the explosion, the ejecta is characterised by broad emission lines and at some stage the blue wing of the broad ejecta lines might be affected by extinction. This extinction indicates the formation of dust in the ejecta. Additionally, sharp decrease in the optical light curve can be found (Elmhamdi et al. 2004). These two methods have been effective to detect ejecta dust in extragalactic SNe.

Presence of high sensitivity instruments in Spitzer Space Telescope and the development of mid-infrared instruments on eight meter telescopes enabled to detect mid-infrared thermal emission of dust from extragalactic SNe (e.g. Sugerman et al. 2006). Detections of mid-infrared dust emission help to quantify the dust masses and temperatures. Typical ejecta dust masses reported in SNe younger than day 10,000 are between 10^{-6}–10^{-3} $M_\odot$ (e.g. Sugerman et al. 2006; Meikle et al. 2007; Kotak et al. 2006; Fabbri et al. 2011; Szalai et al. 2016) and it has been suggested that the dust masses increase in time (Gall et al. 2011, 2014). Because the measurements of late phase dust masses are limited to a few SNe (e.g. SN 1987A and 1995N; Matsuura et al. 2015; van Dyk 2013) with a large gap in epochs, it

Fig. 8 The HUBBLE SPACE TELESCOPE optical image (blue-white) with Herschel Space Observatory image (red) of the Crab Nebula. Copy right: Oli Usher (UCL); Herschel Space Observatory, Hubble Space Telescope: ESA, NASA. Published in Owen and Barlow (2015). Publication of HST image?

is not yet clear whether the dust masses in the early phase is actually such a small quantity or a large mass of dust was missed due to being optically thick (Sect. 2.1.3).

The dust thermal emission can be associated with ejecta dust, light echo, circumstellar dust or dust from circumstellar material. Often the timing of the appearance together with optical line extinction or optical light curves are used in order to identify the origin of dust emission.

3.1 Molecules as Precursor to Dust Formation?

By exploiting extra-galactic SNe, the number of SNe where molecules and dust have been detected has increased. Chemical models predict, the time sequence of formation of molecules in SNe is followed by dust formation. The binding energy of CO is one of the highest among molecules found in astronomical environment, and formation of CO starts about day $\sim$200 with a sharp increase in mass up to day 500, followed by a gradual increase up to day 800 (Sect. 6.2.2: Sarangi and Cherchneff 2013). After that, the CO mass does not increase so much. Whereas the dust condensation starts at about day 200, followed by silica formation at day 400. We test this scenario by looking into the past observations of molecules, mainly CO, and dust.

First, we test link between molecule and dust formation. So far, molecular detections were reported in the spectra of twelve SNe (Summarised in Table 1; extracted from Matsuura 2017). CO detections suggested in photometry measurement only are excluded from the analysis. SN 1987A is included among the twelve, but not SNRs such as Cas A and the Crab Nebula. Figure 9 shows the statistics of dust detections out of twelve SNe. We divide the sample into two parts—six SNe before the Spitzer mission and six SNe after the Spitzer mission. The latter includes SN 2002hh, whose explosion was recorded before Spitzer launch in 2003, but has been observed with Spitzer, and SN 2013eg, which was type Ic and detected after Spitzer cryogenic phase.

Table 1 SNe with spectroscopic detections of CO and SiO molecules in early days, followed by their status of subsequent detections in the ejecta

SNe	Type	Molecules			Dust		Ref.
			Days	$M_\odot$	Days	$M_\odot$	
Pre *Spitzer*							
SN 1987A	IIpec	CO	112?, 130–615[a]	10^{-4}–0.45	~600–775[b]	3×10^{-4}–0.5	[1–7]
		SiO	192–517	3×10^{-6}–2×10^{-5}			[8–11]
SN 1995ad	II	CO	105	5×10^{-2}–10^{-4}			[12]
SN 1998S	IIn	CO	95–225	10^{-3}	260–355		[13, 14]
SN 1998dl	II	CO	~150	510			[15, 16]
SN 1999em	II	CO	~170	10^{-4}			[15]
SN 2000ew	Ic	CO	97				[16]
During *Spitzer*							
SN 2002hh	IIP	CO	192–349		c		[17]
SN 2004dj	IIP	CO	89–119		281?, 461–996	0.2×10^{-4}–10^{-4}	[18, 19]
SN 2004et	IIP	CO	294–446	10^{-4}	300–795	0.5×10^{-4}–1.5×10^{-3}	[20, 21]
		SiO	294–446	0.5×10^{-4}–5.7×10^{-4}			[20]
SN 2005af	IIP	CO	194–399		194–940	1.1×10^{-5}–1.8×10^{-5}	[22, 23]
		SiO	214	2×10^{-4}			[22]
SN2007gr	Ic	CO	70–175				[24]
Post cold *Spitzer*							
SN2013eg	Ib/c	CO	48–123				[25]

[a] CO and SiO detections at late epoch were reported at millimetre wavelength at day 9,173–10,053 (Kamenetzky et al. 2013; Matsuura et al. 2017)

[b] Dust detections at late epoch were reported at day 6067 (Bouchet et al. 2004) and at day 8500 and 9090 (Matsuura et al. 2011, 2015)

[c] Detection of dust thermal emission was reported (Barlow et al. 2005). Most likely it is associated with progenitor dust, but potentially ejecta dust

References: [1] Spyromilio et al. (1988), [2] Catchpole et al. (1988), [3] Wooden et al. (1993), [4] Liu et al. (1992), [5] Liu and Dalgarno (1995), [6] Danziger et al. (1989), [7] Lucy et al. (1989), [8] Aitken et al. (1988), [9] Roche et al. (1991), [10] Roche et al. (1989), [11] Meikle et al. (1989), [12] Spyromilio and Leibundgut (1996), [13] Gerardy et al. (2000), [14] Fassia et al. (2001), [15] Spyromilio et al. (2001), [16] Gerardy et al. (2002), [17] Pozzo et al. (2006), [18] Kotak et al. (2005), [19] Meikle et al. (2011), [20] Kotak et al. (2009), [21] Fabbri et al. (2011), [22] Kotak et al. (2006), [23] Szalai and Vinkó (2013), [24] Hunter et al. (2009), [25] Drout et al. (2016)

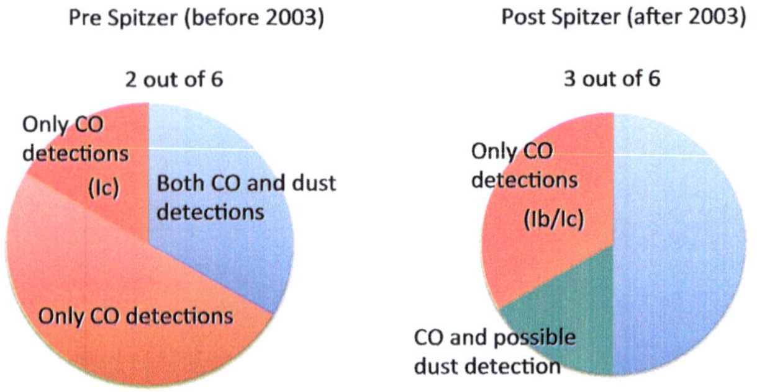

Fig. 9 Statistics of CO detected SNe, highlighting whether they were followed by dust detections or not

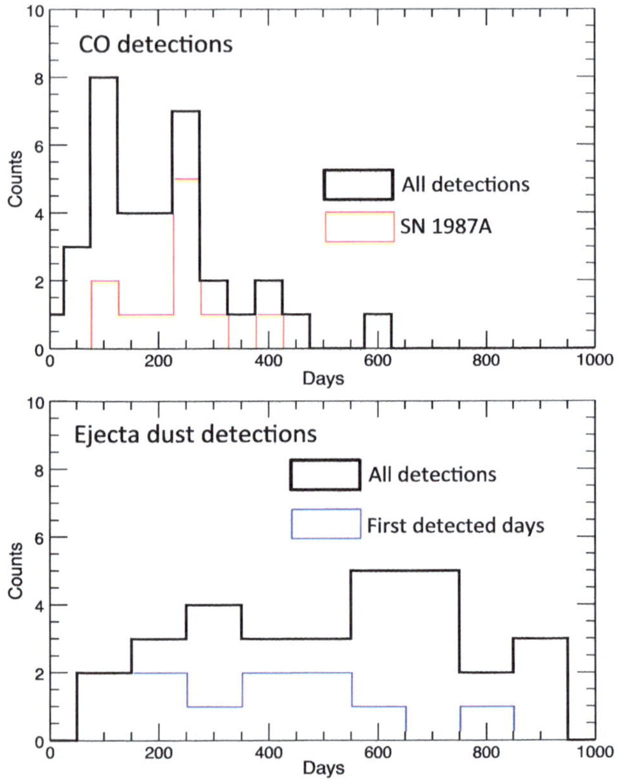

Fig. 10 Histograms indicating the reported days of molecular and dust detections in SNe

Out of the twelve SNe with molecular detections, dust detection was reported in six SNe (Fig. 9). Three of the no-dust detected SNe are of Type Ic and Ib/c SNe, which have fewer dust detections reported so far (Matsuura 2017). Other SNe are all Type II, mainly Type IIP, with one IIn and unclassified subclass (Table 1). From Type II, dust detection was reported in six SNe out of nine, making the succession of molecule formation followed by dust formation very likely. In particular, after the launch of Spitzer, all Type II SNe with molecular detections ended with dust detections, so this scenario seems very likely.

Next, we test the timing of molecule and dust formation/detections. We assumed the reported days of CO detections and ejecta dust detections from Matsuura (2017) and the histogram of these days are plotted in Fig. 10. CO detections from one SN at different epochs are counted individually. In these panels, CO and ejecta dust detections in late epoch (>1000 days) of SN 1987A are not plotted.

The histogram (Fig. 10) shows that CO detections have been reported about 100–300 days. These are mainly detections at 2.3 μm and 4.6 μm CO fundamental and the first-overtone band. After that period, CO emission fades, making it difficult to detect at near-infrared wavelengths. The second highest peak at 225–275 days is due to five separate CO observations of SN 1987A, distorting the frequency of CO detection rate during this period. While the ejecta gas cools down, the gas can not excite vibrational level of CO, so that near-infrared vibrational CO bands disappear by ∼600 days. Instead, CO rotational lines appear at millimetre wavelength ranges and is at least reported by day ∼9000 (Kamenetzky et al. 2013; Matsuura et al. 2017).

While CO detections have a gentle peak between 100–300 days, any such peak is not found in the timings of the ejecta dust detections. In Fig. 10, records of first ejecta dust detection per SN are plotted in blue and these reported first detections spread across 100–800 days. Dust can be associated with ejecta, light echo, circumstellar and ISM dust, and it is not always easy to disentangle these cases. Nevertheless, typically the highest chance of the ejecta dust detection appears to be between 600–700 days.

4 Indirect Evidence of Supernova Dust

Apart from direct observations of dust signatures in the mid-IR and submillimeter wavelengths in different local SNe as discussed above, there are also some indirect yet strong evidences of dust in SNe, which helps us understand and constrain the mechanism of dust formation in such environments. Among them, the dust case of high redshift galaxies (Maiolino et al. 2004; Dwek and Cherchneff 2011) and the isotopic anomalies in presolar grains from meteorites (Clayton and Nittler 2004; Ott 2001) provide us with a lot of important information on SNe dust, as explained below.

4.1 The High Red-Shift Galaxies

The reddening of the background quasars and Lyα systems at high redshift indicates the presence of dust in those galaxies (Pei et al. 1991; Pettini et al. 1994; Hines et al. 2006). Large masses of dust of the order 1–5 × 10^8 $M_\odot$ is estimated to be present in Sloan Digital Sky Survey (SDSS) J1148+5251, a hyperluminous quasar at $z = 6.4$ (Bertoldi et al. 2003; Dwek et al. 2007). Figure 11 points to the high redshift galaxies on the cosmic time scale where presence of large dust masses is reported. The presence of large masses of dust plays pivotal role in the cooling of the primeval large gas clouds, eventually leading to the formation of next generation of stars (Cherchneff 2013). The Population III stars present in these galaxies are predominantly massive and a large section is assumed to be in binary form. The sources of such large masses of dust and the formation mechanisms are yet to be understood with confidence. Generally supernovae and AGB stars are taken as the prime dust producers in the galaxy. In order to explain the large masses of dust at high redshift galaxies, Dwek and Cherchneff (2011) assumed the contribution from all possible dust producers in the galaxy and its balance with the galactic star formation rate. Massive stars evolve much faster compared to the low mass stars. Therefore, in the timescales of the lifetime of the early galaxies, supernovae are estimated to be the main source of dust. If 1 $M_\odot$ of dust is produced by core-collapse supernovae before getting reprocessed in the remnant phase and a top-heavy initial

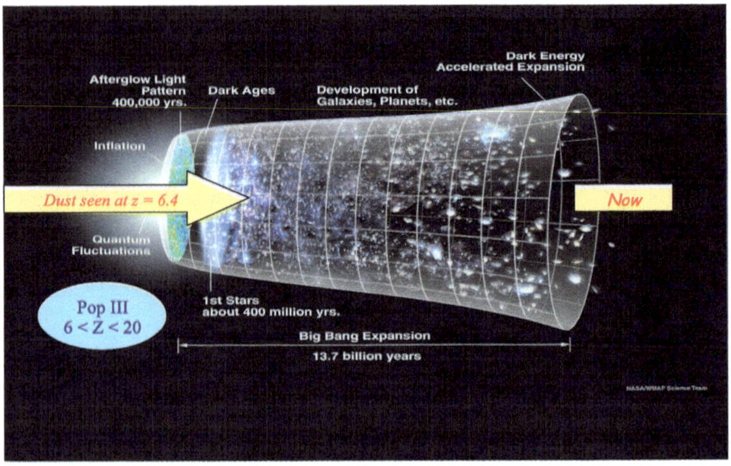

Fig. 11 Evidence of large masses of dust reported from the far-IR spectrum of hyperluminous galaxy SDSS J1148+5251 at high redshift $z = 6.4$ (Bertoldi et al. 2003; Dwek et al. 2007) (Photo courtesy: NASA WMAP)

mass function is assumed for the primeval stars, then the dust masses observed at high redshift galaxies can be explained. However, theoretical models infer a mass of 0.1–0.15 $M_{\odot}$ of silicate dust formed per massive pair-instability supernova as an upper limit (Cherchneff and Dwek 2009; Cherchneff 2010). Moreover, the destruction of dust by the reverse shock in the ejecta or by UV photolysis in the ISM are somewhat underestimated. AGB stars can only contribute as a major dust producer if the galactic age is assumed to be much larger (400 Myrs). Therefore, in case of high z galaxies, SNe are expected to have obvious advantages (Maiolino et al. 2004; Calura et al. 2014; Gall et al. 2011) over AGB stars in terms of relative contributions owing to the large abundances of massive stars and their shorter evolutionary timescales (Dwek 1998; Dwek and Cherchneff 2011).

4.2 Presolar Stardust from Meteorites

Laboratory studies have revealed the presence of surviving dust grains of circumstellar origin within primitive meteorites (Bernatowicz and Zinner 1997). The deviation from solar isotopic composition confirms the presolar origin of these dust particles. Further, depending on the isotopic abundances, the stellar source of the grains as well as some limitations on the physical conditions of the environment can also be inferred (Hoppe and Ott 1997). Dust formed in the ejecta of a supernova carries signatures of the nuclear processes from the pre-explosion and post-explosion phases of the progenitor (Ott 2001).

Out of several types of chemical species found in the meteorites, silicon carbide grains have been profoundly studied and grouped into categories (A, B, Y, Z, X etc.) depending on their possible sources of origin (Hoppe and Ott 1997). The presolar SiC grains of type X ($\sim$1% of all presolar SiC samples collected from meteorites) has been attributed to have SN origin (Amari et al. 1997), as shown in Fig. 12, based on the enrichment of ^{12}C and ^{15}N, high abundance of ^{28}Si, and importantly the presence of ^{44}Ca, a remnant of short-lived radioisotope ^{44}Ti (Ott 2001; Hoppe 2010). Similar to silicon carbide, a few of the silicon nitride (Si_3N_4) samples found also have isotopic signatures of that of SNe (Amari et al. 1997). Apart from these, a small subsection of presolar aluminum oxide grains (Nittler et al. 1998;

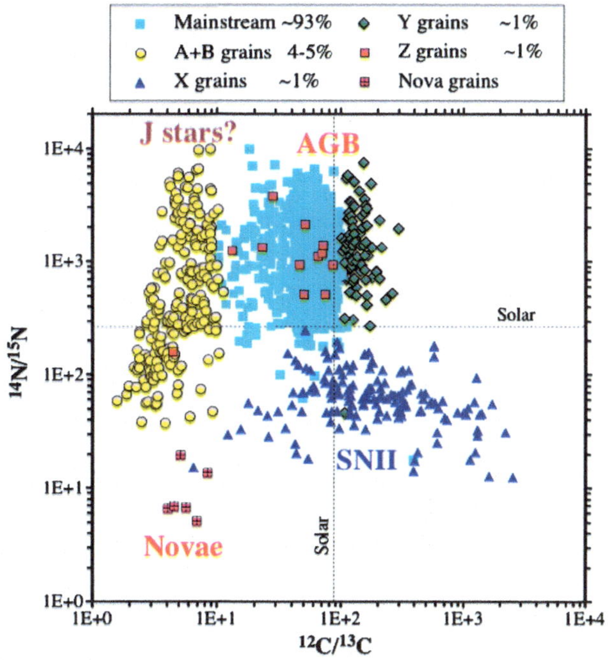

Fig. 12 The categorisation of presolar SiC grains based on $^{14}N/^{15}N$ and $^{12}C/^{13}C$ ratios (Figure adopted from: Zinner et al. 2010, Hynes and Gyngard 2009)

Choi et al. 1998) and some of the low-density graphite grains from Murchison carbonaceous meteorite (Travaglio et al. 1999) are estimated to be of SN origin. Recent detection of $^{18}O/^{16}O$ enhancement in a few of the silicate samples from meteorites additionally claims the presence of Mg-O-silicates in SNe (Leitner et al. 2015).

Even though, SN has been established as a source of cosmic dust in many previous analyses, the isotopic compositions of the grains are not exactly coherent with the existing theoretical models (Herant and Woosley 1994) and also observational studies of the SN ejecta (Hughes et al. 2000). In order to match the isotopic signatures, selective yet microscopic mixing between the ejected materials from different zones of the ejecta is necessary (Nittler 2003; Zinner et al. 2010). In order to understand the mixing scales compatible with the isotopic compositions in the grains, observations of the SN ejecta at high spatial resolutions are required. New theoretical studies of the post-explosion ejecta using 3D hydrodynamic models (Hammer et al. 2010b; Wongwathanarat et al. 2015) also hint at more intense mixing between the stratified zones.

5 The Chemical Nature of Dust Grains

Compared to dust in earth's atmosphere, dust grains in space have relatively simpler chemical forms as they exist in more hostile and pristine surroundings. The chemical types of cosmic dust and their precursor molecules are not well defined through observations due to lack of high-resolution mid-IR data (Cherchneff 2013). Composition of dust in space can be analysed by (a) the laboratory study of presolar grains from meteorites, (b) fitting of mid-IR spectral energy distribution using known opacities of probable dust species, (c) modelling of dust condensation from its gas phase precursors. Cosmic dust from circumstellar sources can be classified chemically as (a) oxides (silicates of quartz or metal oxides), (b) carbides,

Table 2 Circumstellar dust types derived from mid-IR (5–40 μm) spectra through laboratory experiments (Adopted from Cherchneff 2013)

Family	Name	Formula	Bands (μm)	Reference
Oxides				
Silicates	Pyroxene	$Mg_xFe_{1-x}SiO_3$	10; 20	Dorschner et al. (1995)
	Enstatite	$MgSiO_3$	9–12; 15.4; 19.5; 36.2	Chihara et al. (2002)
	Ferrosilite	$FeSiO_3$	11.3; 20.4; 31.7	Chihara et al. (2002)
	Olivine	$Mg_2xFe_{2-2x}SiO_4$	10; 20	Dorschner et al. (1995)
	Fosterite	Mg_2SiO_4	10–12; 16.3; 19.5–24	Koike et al. (2003)
	Fayalite	Fe_2SiO_4	10; 18–22; 27; 32	Suto et al. (2002)
Quartz	Silica	SiO_2	9.1; 12.6; 20.4–21.2; 26.1	Fabian et al. (2000)
Metal oxides	Alumina	Al_2O_3	13	Koike et al. (1995)
	Spinel	$MgAl_2O_4$	13; 16.8; 32	Fabian et al. (2001)
	Magnesia	MgO	19	Henning et al. (1995)
	Wüstite	FeO	23.4	Henning et al. (1995)
	Hematite	Fe_2O_3	9.2; 18; 21; 20	Koike et al. (1981)
	Magnetite	Fe_3O_4	17; 25	Koike et al. (1981)
	Calcium oxide	CaO	31.4	Hofmeister et al. (2003)
Carbon				
	Amorphous carbon	C	6.2; 8	Colangeli et al. (1995)
	Graphite	C	6.3; 11.52	Draine (1984)
Carbides				
	Silicon carbide	SiC	11.3	Mutschke et al. (1999)
Sulphides				
	Magnesium sulphide	MgS	25–39	Hofmeister et al. (2003)
	Iron sulphide	FeS	23; 34; 39	Hofmeister et al. (2003)

(c) sulphides, (d) homogeneous elemental clusters. Further they can be either crystalline or amorphous in nature depending on the physical conditions of the environment at the time of their synthesis. Stretching and bending modes of vibration within the grains of a given type give rise to their unique emission or absorption features. The dust species can be identified by their mid-IR spectroscopic signatures through laboratory measurements, which are heavily dependent on the proper estimation of the optical properties of the grains. The HJPDOC database (developed by the Astrophysical Institute of the Friedrich Schiller University, Jena, Germany and the Astronomical Institute of the St. Petersburg University, Russia) provides information about a large set of optical constants necessary to analyse cosmic dust and its properties. Table 2, adopted from Cherchneff (2013), gives a summary of known dust species and their mid-IR bands with the references.

In case of SN ejecta, hydrogen mixing is understood to occur in macroscopic scales only (Joggerst et al. 2010; McCray 1993). Therefore, dust formation pathways are controlled by simplified H-free chemistry leading to a small set of possible dust species. From bottom-up perspectives, the ejecta chemistry is modelled to predict the types and abundances of molecules and dust grains in the metal-rich ejecta (Bianchi and Schneider 2007; Cherchneff and Dwek 2009; Sarangi and Cherchneff 2013; Sluder et al. 2016). However, in order to conclusively comment on the type(s) of dust present in a certain environment, radiative transfer

(RT) modelling of the ejecta and fitting of the observed spectra are required (Sugerman et al. 2006; Ercolano et al. 2007). For optically thin media, the spectral features of the dust grains are prominent whereas in case of large optical depths, the dust continuum behaves more like a black body spectrum making the detection of a particular dust type more ambiguous. In the following sections of this chapter, the theoretical aspects of the study of SNe have been addressed in more detail.

6 Theoretical Models

Observational studies have confirmed the presence of dust signatures in SN ejecta, which can either be contribution of echo from pre-existing dust or newly formed dust in the ejecta. In order to understand the origin of dust in such environments, we shall now discuss the mechanisms which aid the synthesis of dust grains in SN ejecta.

Attempts to theoretically model an astrophysical environment or event focus on the temporal and spatial variance of the following: (a) the morphology/geometry, (b) relevant physical processes, (c) chemical pathways. To study these aspects with respect to a typical SN ejecta, several simplifications/ approximations are necessarily imposed on the models. The following subsections describe the existing models of SN ejecta and their pertinence compared to the actual environments.

6.1 Models Based on CNT and LTE

The earliest approach to model dust synthesis in SN ejecta involved a Classical Nucleation Theory (CNT) based formalism. The CNT based approach was developed using the 'liquid-drop model' introduced to study water droplets in earth's atmosphere (Feder et al. 1966). In this model, the nucleation of gas-phase metals is addressed by a free-energy controlled nucleation current that leads to the formation of critical clusters. The initial growth of these clusters by the attachment of new molecules is an energetically unfavorable process as the surface energy of these species (the missing binding energy of surface molecules) exceeds the volume binding energy. The work required to form a single cluster of k-monomers is a function of (a) the change in the energy of molecules due to its transition from gas to solid phase, (b) the binding energy of the involved species, and (c) the surface energy of the cluster. In macroscopic scale, the energy to form a cluster of k-monomers is equal to the change in the Gibbs free energy of a gas containing k-number of monomers that is undergoing a phase transition at a given pressure and temperature, which is given as,

$$W(k) = \Delta G_k = -k_B T \ln \mathcal{S} k + u_s k^{2/3}, \tag{1}$$

where $W(k)$ is the work required, k_B is the Boltzmann constant, $\mathcal{S} \equiv P/P_v$ is the supersaturation ratio and u_s is the surface energy term.

These supersaturated compounds (the critical clusters), synthesized in this process, are assumed to be in thermodynamic equilibrium (TE) with the solid phase of dust. The accretion and detachment rates of gas monomers are therefore calculated considering a steady state environment.

Dust formation in SNe was discussed in an empirical approach by Clayton (1979). Thereafter, a pioneering attempt to model dust formation was made by Kozasa et al. (1989) based on the early results of SN1987A. This study considered a fully microscopically mixed ejecta where the gas and dust phase was connected by CNT and TE assumptions. Following that,

CNT based models were applied to study dust formation in SNe (CCSN & PISN) by Todini and Ferrara (2001), Schneider et al. (2004), Bianchi and Schneider (2007), and more recently by Marassi et al. (2015) for various metallicities and mass of progenitors as well as for various levels of mixing.

Models based on kinetic nucleation theory (KNT) provide an alternative formalism to study dust formation in such environments. In this approach, the number densities of clusters are traced explicitly. The condensation of grains is computed using accretion, where as the destruction of clusters occur through evaporation. The rate coefficients are calculated using kinetic balance. Therefore, it does not necessarily consider a steady state environment for the nucleation processes. Several studies over the last decade have applied this approach to model dust formation in SN ejecta (Nozawa et al. 2003, 2008; Nozawa and Kozasa 2013).

However, the use of nucleation theory to study dust formation in environments such as SN ejecta was questioned first by Donn and Nuth (1985) and later elaborated by Cherchneff and Lilly (2008). CNT is applicable to steady-state conditions, which are usually not found in dynamic environments out of equilibrium (Cherchneff and Dwek 2009). All CNT-based models ignore the non-equilibrium chemistry related to the formation of molecules and dust clusters and the specific physics of ejecta where radioactivity greatly impacts the gas-phase chemistry through cascades of Compton electrons. Moreover, the CNT approach includes the quantities like specific surface energy, sticking coefficient, supersaturation ratio, partial vapour pressure, critical clusters, etc. in analysing the dust formation scenario. These all are essentially bulk properties to matter. However, the critical clusters sizes are of molecular dimensions and therefore, application of the bulk properties like surface energy or sticking coefficient on such small particles is not appropriate (Cherchneff 2013). Additionally, in CNT-models, the condensation sequence is based on assuming condensation temperatures at equilibrium as references. Estimating the condensation temperatures at equilibrium for the seed nuclei is also not an appropriate approach in most cases, as the temperatures of the newly formed grains heavily depend on the ambient gas temperatures and densities, controlled by the amount of radioactive ^{56}Ni produced at the explosion. Even though the KNT based models can take the non-steady state environment into account, they still ignore the actual chemical pathways that lead to the synthesis of the dust-seed nuclei. Further, important condensation mechanisms such as coagulation are also not included in such analyses.

6.2 Chemical Kinetics Model

A formalism based on chemical kinetics can overcome most of the limitations imposed by previous models. In this case, dynamics and chemistry control the synthesis of condensates (Cherchneff and Lilly 2008; Cherchneff and Dwek 2009, 2010; Sarangi and Cherchneff 2013, 2015; Biscaro and Cherchneff 2014, 2016; Sluder et al. 2016). A chemical kinetic approach deals with the dynamics of chemical reactions and the conditions affecting the reaction rates. It is controlled by the concentration of the reactants, temperature of the medium and the enthalpy of the individual reactions. The SN ejecta has a very dynamic and unstable physical environment. The passage of the supernovae blast wave through the helium core and the deposition of radioactive energy in the ejecta creates a complex physical environment whose chemistry is neither at steady state nor can be assumed to be at thermodynamic equilibrium (TE). Following the explosion, the ejecta moves at a velocity ∼2000 km/s and the density and temperature of the ejecta drop rapidly as power-laws. The rich chemical composition of the ejecta is controlled by a large set of chemical reactions with diverse reaction rates. All these chemical reactions occur simultaneously and are coupled to each other. Each individual species is linked to the chemical network through a series of formation and destruction pathways, which comprise of both thermal and non-thermal processes.

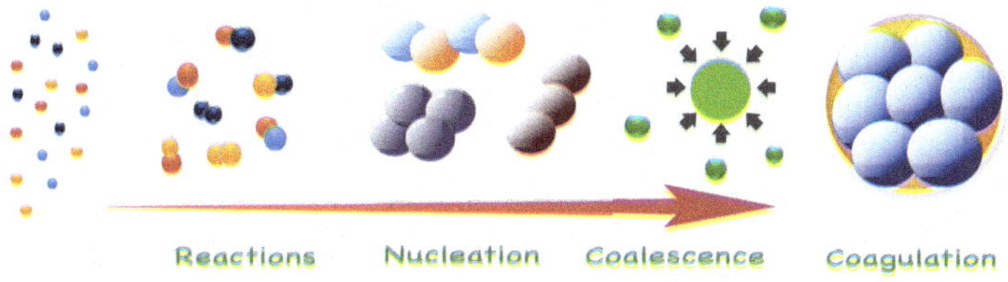

Fig. 13 A cartoon shows the chemical paths leading to the synthesis of molecules, clusters and dust grains in the ejecta through nucleation and condensation phases

Therefore, with the alteration of temperature and density in the ejecta, the relative importance of specific reactions in the network may also change. These factors not only ensure the complexity of the expected chemical yields but also create a case which is far from the steady state condition.

The chemical pathway proceeds through simultaneous phases of nucleation and condensation (Fig. 13). The nucleation phase comprises of chemical reactions which lead to the synthesis of molecules and small clusters. The condensation phase couples the gas phase chemistry to the solid phase of dust. The small clusters formed in the gas phase condense through coagulation and coalescence to form large grains of dust, provided suitable conditions are prevalent. Therefore, we first discuss the physical environment of a typical SN ejecta before going through the detail chemical scheme.

6.2.1 Physical Conditions in the Ejecta

We choose a typical 15 $M_\odot$ Type IIP SN as a standard case study considering Type IIP SN ejecta to be the most conducive to dust formation. The ejecta of a typical PISN has some morphological differences with a CCSNe owing to the metallicity of the Population III stars. The contrasts in the physical conditions of such objects will be addressed separately. For the sake of simplicity, a 1D stratified model of the ejecta is assumed and we shall ignore the details of the explosion mechanism in this discussion. The abundances of post-explosion elements are provided by nucleosynthesis studies and hydrodynamic simulation models of the ejecta (Woosley and Weaver 1995; Thielemann et al. 1996, 2011; Thielemann 2017; Limongi et al. 2000; Rauscher et al. 2002). The final yields from these models correspond to a timescale of 6–7 hours from the explosion. These elemental yields are then used as initial inputs to the chemical models of the ejecta (Rauscher et al. 2002).

Following the expulsion of the helium core, the ejected materials are crossed by the blast wave, which deposits energy in the gas. The neutrino-driven convection during the collapse leads to the Rayleigh-Taylor instabilities and mixing in the helium core. The mixing scenario continues over a few hours after the explosion and leads to partial fragmentation of the core (Joggerst et al. 2010). Thereafter, the chemical species produced by nucleosynthesis flows ballistically as carried by the blast wave. Hence the abundances of the elements presented as functions of mass zones (Rauscher et al. 2002) in the ejecta (Fig. 14) can be safely assumed to remain constant over the next few days until chemical reactions becomes relevant. The explosion energy, E_{exp} ($\sim 10^{51}$ ergs) is assessed using SN1987A as reference (Woosley 1988), which gives an estimate of the ejecta velocity. Based on the distribution of elements, the ejecta can be stratified into several regions or zones, assuming no further microscopic mixing takes place after a few hours from explosion. The ejecta of the 15 $M_\odot$ progenitor model is divided into 8 zones according to the abundances of important elements: 1A (Si/S), 1B(Si/O), 2(O/Si/Mg), 3(O/Ne/Mg), 4A(C/O), 4B(C/O/He), 5(He/C) and 6(He/N) (Sarangi

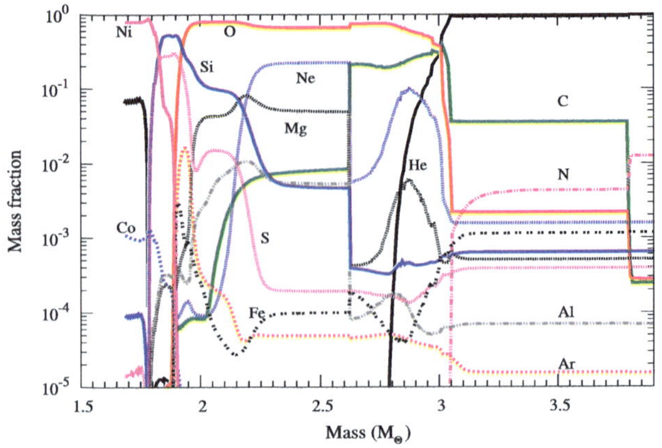

Fig. 14 The post-explosion elemental abundances inside the He-core as a function of mass zones for a 15 M☉ progenitor star which has exploded as a Type IIP supernova (Rauscher et al. 2002)

and Cherchneff 2013). In this study, the species in individual zones are assumed to be microscopically mixed. The dominant chemical processes in each zone vary depending on the temperature, density and the abundances.

The SN light curve is powered by radioactive decays (Woosley et al. 1989). The energy deposited by the SN explosion-flash is radiated away in a couple of weeks (Woosley 1988; Blinnikov et al. 2000; Ensman and Burrows 1992; Dwek and Arendt 2008). After that, the bolometric luminosity, which includes the optical, IR and UV emissions, is controlled by interactions between high energy radiations with the supernova ejecta. There are several radioactive nuclei produced in the explosion. On the basis of the abundances and the half-lives, the most important radioactive nuclei in the ejecta are ^{56}Ni, ^{57}Ni, ^{55}Co, ^{44}Ti and ^{22}Na (Kozasa et al. 2009). ^{56}Ni has a very short lifetime of 6 days and quickly decays to ^{56}Co through electron capture. ^{56}Co in turn decays to ^{56}Fe with a half-life is 77.12 days. The light curve of the first few years of SN1987A is dominated by ^{56}Co and a mass of 0.075 $M_\odot$ reproduces the bolometric luminosity (Woosley et al. 1989). Hence an equal amount of ^{56}Ni is estimated to be produced in the explosion. The ^{56}Co decays through electron capture ($^{56}\mathrm{Co} \longrightarrow {}^{56}\mathrm{Fe} + \gamma + \nu_e$) and produces energetic γ-rays and electron neutrinos. The energetic γ-rays interact with the ejecta and get degraded to hard X-rays through Compton scattering. The fast Compton electrons thermalize the ejecta through heating, excitation and ionisation (Liu and Dalgarno 1995). The hard X-ray gets further degraded to soft X-rays and UV photons through cascades of inverse Compton scattering and ionization-recombination processes. The entire energy of the γ-rays is not deposited in the ejecta. Study of the SN1987A light curve indicates the presence of macroscopic mixing of ^{56}Ni in the ejecta, which is also confirmed by the early emergence of γ-rays and X-rays within a few hundred years from explosion. The ejecta is presumably clumpy and hence a small fraction of the high energy photons can possibly pass through the less dense interclumps unabsorbed (McCray 1993). However, the opacity of the ejecta is found to increase rapidly at early times. Hence, to a first order approximation, the fraction of escaping energy becomes negligible. For the sake of generality, it can be assumed that the entire radioactive energy is uniformly deposited in the ejecta (Cherchneff and Dwek 2009; Sarangi and Cherchneff 2013). The rate of energy deposition (L_γ) in the ejecta by ^{56}Co is calculated following the study by Woosley et al. (1989) in terms of the γ-ray optical depth (τ_0), γ-ray mass-absorption coefficient (κ_γ), the deposition function ($f(\tau)$) and the energy emitted per ^{56}Co nucleus (E_γ) (Liu and Dalgarno 1994; Cherchneff and Dwek 2009; Sugerman et al. 2006). The rate varies linearly with the

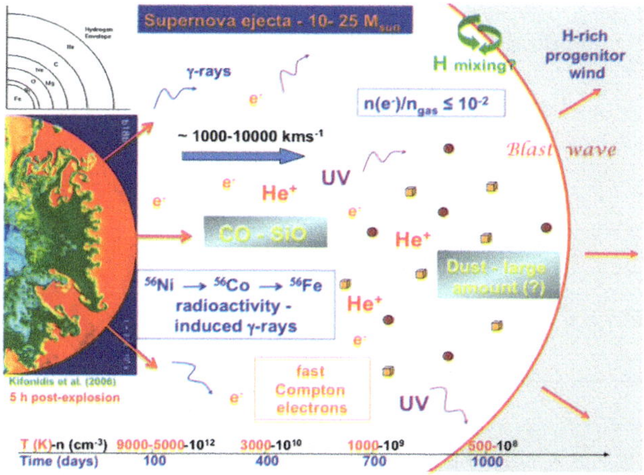

Fig. 15 The schematic diagram, adopted from Cherchneff and Sarangi (2011), represents the time evolution of a typical CCSN ejecta. The scale in the bottom of the figure illustrates the temperature/density variation with post-explosion time. On the top left corner, the stratification of the stellar core and envelope is shown which retains the signatures from the burning phases of the massive star. The mixing and the Rayleigh-Taylor instabilities in the ejecta that occur until a few hours after the explosion, is also presented in the picture, adopted from Kifonidis et al. (2006)

total mass of radioactive ^{56}Ni produced by the explosion (Sarangi and Cherchneff 2013, 2015).

The ejecta is assumed to undergo a quasi-adiabatic expansion (Cherchneff and Dwek 2009) over a span of few years. The temperature profile is obtained with reference to observations and models based on SN1987A (Fransson and Chevalier 1989; Kozasa et al. 1989). The temporal variation of gas temperature in the ejecta is represented by an 'adiabatic' index γ ($\sim$1.42) for SN Type IIP, as derived by Umeda and Nomoto (2002) & Nozawa et al. (2010). The propagation of the shock causes a velocity shear in the post-explosion ejecta and within one day of explosion, the ejecta shell expands homologously (Shigeyama and Nomoto 1990; Nozawa et al. 2003). The initial gas densities and temperatures at $\sim$ day 100 (Sarangi and Cherchneff 2013; Biscaro and Cherchneff 2014) used for the model are derived from Nozawa et al. (2010). Figure 15, adopted from Cherchneff and Sarangi (2011), presents a cartoon illustrating the post-explosion ejecta expanding outwards from the stellar remnant.

6.2.2 Nucleation Phase

Molecules and clusters act as the building block of dust in supernova ejecta. The nucleation scheme, which deals with the formation and destruction of such species in the gas phase, therefore acts the bottleneck to the entire mechanism (Cherchneff and Dwek 2010; Sarangi and Cherchneff 2013). The He-core, rich in metal concentration and poor in H abundance, is the prime site of dust synthesis in SN ejecta. The environment in concern being H-poor, simplifies the relevant chemical pathways to a great extent. Moreover it diminishes the variety of chemical compounds synthesized as intermediate or end products. Table 3 provides the list of all possible chemical species ($\sim$120) in the gas phase that are important for this study (Cherchneff and Dwek 2009; Sarangi and Cherchneff 2013; Biscaro and Cherchneff 2014). A chemical network is developed for the study comprising of all feasible chemical pathways

Table 3 The set of atoms, ions, molecules, and molecular clusters important for the study (Sarangi and Cherchneff 2013, 2015; Biscaro and Cherchneff 2014, 2016)

Elements						
He	C	N	O	Ne	Mg	Al
Si	S	Ar	Ti	Fe	Co	Ni
Molecules						
CO	CO_2	NO	O_2	MgO	AlO	SiO
SO	FeO	S_2	CN	SiC	CS	SiS
FeS	MgS	N_2				
Ions						
He^+	C^+	N^+	O^+	Ne^+	Mg^+	Al^+
Si^+	S^+	Ar^+	Fe^+	CO^+	SiO^+	SO^+
$O_2{}^+$	$N_2{}^+$	$C_2{}^+$				
Molecular clusters						
C_2	C_3	C_4	C_5	C_6	C_7	C_8
C_9	C_{10}	C_{11}	C_{12}	C_{13}	C_{14}	C_{15}
C_{16}	C_{17}	C_{18}	C_{19}	C_{20}	C_{21}	C_{22}
C_{23}	C_{24}	C_{25}	C_{26}	C_{27}	$\mathbf{C_{28}}$	
$(SiO)_2$	Si_2O_3	$(SiO)_3$	Si_3O_4	$(SiO)_4$	Si_4O_5	$(SiO)_5$
$MgSi_2O_3$	$MgSi_2O_4$	$Mg_2Si_2O_4$	$Mg_2Si_2O_5$	$(MgSiO_3)_2$		
$Mg_3Si_2O_6$	$Mg_3Si_2O_7$	$Mg_4Si_2O_7$	$\mathbf{(Mg_2SiO_4)_2}$			
Al_2	$(AlO)_2$	$(AlO)_3$	AlO_2	Al_2O	Al_2O_3	
$(Al_2O_3)_2$	$(Al_2O_3)_3$	$\mathbf{(Al_2O_3)_4}$				
Si_2	Si_3	$\mathbf{Si_4}$	Mg_2	Mg_3	$\mathbf{Mg_4}$	
Fe_2	Fe_3	$\mathbf{Fe_4}$	$(FeS)_2$	$(FeS)_3$	$\mathbf{(FeS)_4}$	
$(MgS)_2$	$(MgS)_3$	$(MgS)_4$	$(SiC)_2$	$(SiC)_3$	$\mathbf{(SiC)_4}$	
$(MgO)_2$	$(MgO)_3$	$(MgO)_4$	$(FeO)_2$	$(FeO)_3$	$(FeO)_4$	

relevant to the hot and dense environments. The network consists of about 600 reactions of different kinds in the gas phase. They can be further categorized into thermal (termolecular or 3-body processes, thermal fragmentation, bimolecular reactions, ion-molecule reactions, radiative association etc.) and non-thermal processes (photon induced, Compton-electron induced etc.) (Cherchneff and Dwek 2009). The reactions are characterized by the respective reactions rates. Hence the temporal variation of number density of a certain species i located at mass zone M_r in the ejecta at a certain time t, is given by,

$$\frac{\partial n_i(M_r,t)}{\partial t} = P_i - L_i = \sum_j k_{ji} n_j n_i - \sum_k k_{ik} n_i n_k \tag{2}$$

$$k_{ij}(T) = A_{ij} \times \left(\frac{T}{T_0}\right)^{\nu} \times \exp(-E_{ij}/T), \tag{3}$$

where P_i represents the sum over the all the production (formation) processes whereas L_i represents the sum over all the destruction processes. The term k_{ij} represents the tempera-

Fig. 16 The nucleation scheme for silicates leading to the formation of the enstatite and forsterite dimers, which act as the seed in the condensation formalism. The reactant species are given on the side of the directional arrows in the schematic diagram (Goumans and Bromley 2012; Zachariah and Tsang 1993)

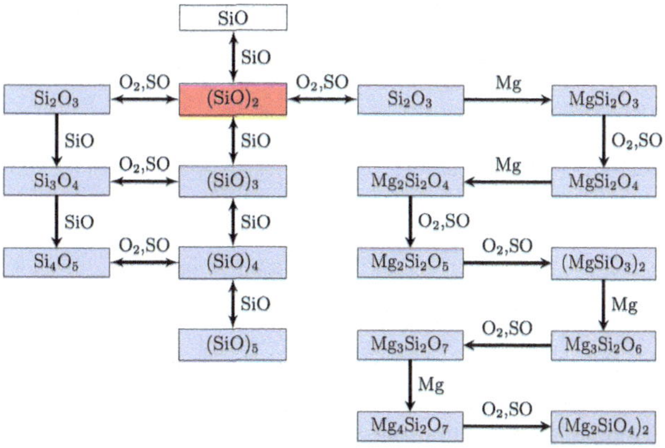

ture dependent reaction rates between the ith and jth species (Cherchneff and Dwek 2009). T_0 is a reference temperature chosen as 300 K for the study. A_{ij} is the Arrhenius rate coefficient expressed in s^{-1}, $cm^3\,s^{-1}$ or $cm^6\,s^{-1}$ depending on the order of reaction and E_{ij} is the activation energy in the unit of K^{-1}. The factor ν represents the time dependence of the pre-exponential factor of the general Arrhenius equation. Typical values of the ν-factor range between -1 to 1. The gas phase chemistry is overall controlled by N number of stiff coupled nonlinear ordinary differential equations similar to (2), where N is the number of species that has been included in the study. This entire set of differential equations is solved simultaneously and the time evolution of the abundances for each species is estimated. The reaction rates are either theoretically calculated using translation theory or estimated in the laboratory. Standard chemical databases such as NIST, KIDA, UMIST, etc., are used as sources for the compilation of the network. The rate coefficients for the non-thermal processes are estimated using the rate of deposition of radioactive energy by the thermalized γ-rays in the ejecta (Cherchneff and Dwek 2009; Liu and Dalgarno 1994, 1995). The branching ratio between the energy deposited by electron through excitation, ionisation ($A \rightarrow A^+ + e^-$) and fragmentation ($AB \rightarrow A + B^+ + e^-$) depends on the mean energy per ion pair ($W_d = E_p X_i / N_d$, where N_d and X_i are the number of ion pairs produced by the species in this process and the abundance of that species in the ejecta). The detailed account of all the non-thermal processes related to SN ejecta can be found in Cherchneff and Dwek (2009), Liu and Dalgarno (1995) & Clayton et al. (1999).

A stochastic, kinetically driven approach has been adopted to study the nucleation of gas molecules to small clusters. The vaporisation experiments in the laboratory and the study of ceramic synthesis in the flames provide insights into similar nucleation processes under non-steady state environment. The important species that demand detailed analysis are oxygen rich clusters of magnesium silicates, alumina and Si_mO_n molecules, carbon chains, rings and fullerenes, silicon carbide, metallic sulphides and oxides, and pure metallic clusters. The upper limit of the cluster size in the nucleation network is not defined very specifically. The geometry of the molecules is the backbone to determine the final product of nucleation and the dust precursor in the network (Table 3 in bold font). The clusters that attain a three dimensional structure like a cage or twisted rings, has a larger collision cross section for coagulation (Cherchneff and Dwek 2010; Sarangi and Cherchneff 2015). Further, they are energetically more stable, and hence likely to be resilient under the destruction processes in the gas phase. In some cases, the favoured chemical routes for nucleation network are derived from quantum chemical calculations. Figure 17 presents the structures of a few stable

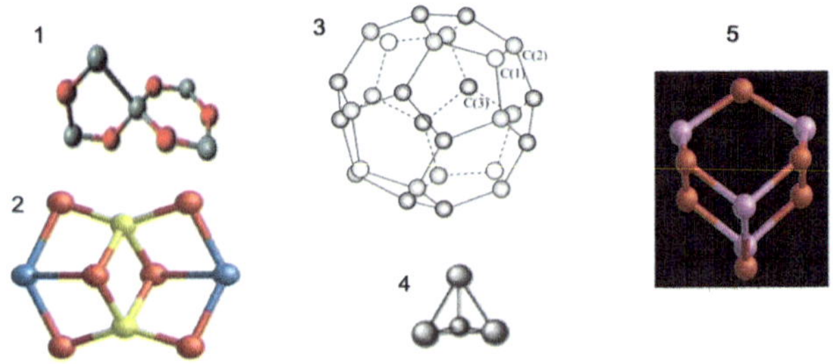

Fig. 17 (**1**) $(SiO)_5$ (Reber et al. 2008), (**2**) $(MgSiO_3)_2$ (Goumans and Bromley 2012), (**3**) C_{28} (Makurin et al. 2001), (**4**) Mg_4 (Cherchneff and Dwek 2010), (**5**) $(Al_2O_3)_2$ (Gobrecht et al. 2016, private communications)

gas phase molecular clusters, which acts as precursor seeds to the condensation phase. The nucleation schemes vary vastly depending on the type of species. Figure 16 is a schematic presentation of the nucleation pathways that lead to stable $(SiO)_n$ compounds and magnesium silicates $(Mg_2SiO_4)_2$. In a hydrogen free environment, pure carbon chains are formed of sp-hybridization up to C_9 followed by monocyclic and fullerenes. Cluster beam experiments suggest, the smallest stable structure of fullerene present in carbon vapour is C_{28} (Dunk et al. 2012). All the possible nucleation networks relevant for such environments are described in Cherchneff and Dwek (2009) & Sarangi and Cherchneff (2013).

The model predicts the formation of several diatomic molecules in the ejecta, such as CO, SiO, SO, O_2, CS, SiS, N_2 etc., over a span of a few (between 1 and 4) years after the explosion. The epoch of synthesis and the abundances vary with the zones in the ejecta. The findings are congruent with the observations of CO (Danziger et al. 1988; Catchpole et al. 1988) and SiO (Roche et al. 1991; Kotak et al. 2005, 2009) molecules in local supernovae, mainly SN1987A. Figure 18 shows the time evolution of CO and SiO masses in a 15 $M_\odot$ progenitor SN ejecta compared to the observations. CO molecules are essentially synthesized in the C/O rich zone 4, followed by the central region on the O-core (zone 2). CO formation follows very efficient chemical routes. Hence, the abundance increases as early as 150 days and becomes constant in mass by 700 days post-explosion. Due to the strong chemical bonds, once they are formed in the ejecta, the CO molecules are found to withstand the harsh environment conditions and possible destruction mechanisms. The chemical pathways important for CO formation are mainly: $C + O_2 \rightarrow CO + O$, $C + O \rightarrow CO + h\nu$ etc. The final masses estimated by the model are in accord with the findings of Kamenetzky et al. (2013) and also it matches well at $\sim$300 days (4×10^{-3} $M_\odot$) with the IR observations of SN1987A. There is no evidence of depletion of CO molecules with time. Hence it can be safely claimed that CO does not act as a precursor to the carbon dust (Sarangi and Cherchneff 2013; Biscaro and Cherchneff 2014).

SiO molecules are formed mainly in the central zones of the O-core, named as zones 1B, 2 & 3. The SiO increases rapidly to 8×10^{-3} $M_\odot$ at day 200 from explosion in these regions. Reactions with CO, O_2 and radiative association of Si and O are the dominant chemical processes for the formation of SiO in the ejecta (Sarangi and Cherchneff 2013). Unlike CO, the rapid formation SiO molecules in the ejecta is subsequently also followed by decline of the abundances after day $\sim$300. This is attributed to the rapid conversion into small clusters, which can be correlated to the IR observations of SN2004et and SN1987A (Kotak et al. 2009).

Table 4 summarises the final masses of all the important molecules, small clusters and residual dust precursors in the gas phase, at day 2000 post-explosion. Even though O_2 is the most abundant molecule in the ejecta, due to symmetry and zero dipole moment, the vibrational transitions in the IR are absent. The model further predicts the presence of SO and SiS in the gas phase, where the presence of SO is most likely (Matsuura et al. 2017), while

Table 4 Masses ($M_{\odot}$) of molecules, intermediate clusters and uncondensed dust-clusters in the gas phase at 2000 days post-explosion present in the SN ejecta of a 15 $M_{\odot}$progenitor (Sarangi and Cherchneff 2013, 2015) are tabulated

Zones	Z1A	Z1B	Z2	Z3	Z4A	Z4B	Z5	Z6	Total
	Molecules								
Mass	9.6(−2)	9.5(−2)	0.292	0.347	0.195	0.225	0.75	0.347	2.35
SiO	2.0(−8)	1.3(−8)	1.3(−8)	2.1(−9)	...	...	...	...	4.8(−8)
O_2	...	2.8(−5)	0.15	0.16	6.2(−2)	4.9(−3)	...	...	0.38
CO	7.5(−7)	1.7(−5)	2.2(−3)	6.6(−3)	9.5(−2)	0.14	2.9(−3)	...	0.25
SO	...	1.5(−2)	3.8(−3)	1.0(−4)	1.1(−4)	...	...	...	1.9(−2)
SiS	4.3(−2)	2.1(−7)	...	...	...	...	...	...	4.3(−2)
CO_2	...	...	1.7(−5)	2.7(−5)	...	...	...	...	4.4(−5)
CS	...	...	...	...	...	...	1.5(−4)	...	1.5(−4)
N_2	...	...	...	3.2(−7)	3.5(−7)	...	3.1(−3)	...	3.1(−3)
Total	4.3(−2)	1.5(−2)	0.156	0.167	0.157	0.15	6.2(−3)	0	0.695
Efficiency (%)	44.8	15.8	53.4	48.1	80.5	66.7	0.8	0	29.6
	Intermediate molecular clusters								
Si_mO_n	...	5.5(−2)	3.4(−2)	2.9(−3)	...	...	...	...	9.0(−2)
C_{2n}	...	...	...	...	...	...	5.5(−3)	...	5.5(−3)
Fe_n	8.2(−5)	...	...	...	...	...	...	...	8.2(−5)
Total	8.2(−5)	5.5(−2)	3.4(−2)	2.9(−3)	...	...	5.5(−3)	...	9.6(−2)
	Uncondensed dust clusters								
$(Mg_2SiO_4)_2$	...	9.9(−6)	4.1(−5)	1.7(−5)	2.2(−6)	2.0(−6)	1.9(−5)	...	9.1(−5)
$(Al_2O_3)_4$	...	3.4(−6)	4.8(−5)	4.7(−5)	...	...	...	...	9.9(−5)
C_{28}	...	...	...	...	...	...	1.0(−8)	...	1.0(−8)
Mg_4	...	...	1.4(−5)	1.8(−5)	...	...	...	...	3.2(−5)
Si_4	9.7(−6)	...	...	...	...	...	...	...	9.7(−6)
Fe_4	6.5(−6)	...	...	...	...	...	...	...	6.5(−6)
SiC_4	...	...	...	...	...	...	1.1(−5)	...	1.1(−5)
FeS_4	2.2(−7)	...	...	...	...	...	...	...	2.2(−7)
Total	1.6(−4)	1.3(−5)	1.0(−4)	8.2(−5)	2.2(−6)	2.0(−6)	3.0(−5)	...	2.5(−4)

SiS yet to be confirmed from observations. Small clusters such as $(Mg_2SiO_4)_2$, $(Al_2O_3)_4$, or C_{28} fullerenes are stable gas-phase monomers of dust; therefore they undergo quick depletion immediately after synthesis. Other than that, there are a few intermediate clusters of the form Si_nO_m which remain in the ejecta as uncondensed residue. About 30% of all the gas in the ejecta is in molecular form after a few years of the explosion.

6.2.3 Condensation Phase

The dust precursors act as seeds in the condensation scheme, where they grow by coagulation and coalescence to form large grains. The gas phase reactions and grain condensation occur in tandem. The model is designed to maintain the balance between gas phase

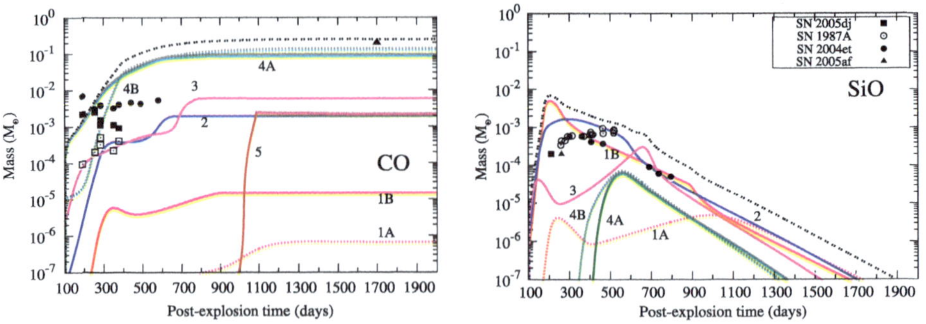

Fig. 18 *Left-panel:* Mass of CO in a 15 M$_\odot$ model presented in M$_\odot$ as a function of various ejecta zones as labelled by the side of the lines. The black dotted line corresponds to the total over all the zones. CO masses derived from the observations of SN1987A are shown as, LTE: open squares, non-LTE: squares (Liu et al. 1992), thermal assumption: circles (Liu and Dalgarno 1995), ALMA data: triangle (Kamenetzky et al. 2013). The ALMA data is not true to the time axis, but rather indicates the final mass at late times as compared to the model results. *Right panel:* Evolution of SiO mass (M$_\odot$) with post-explosion time in days as a function of different zones in the ejecta. The black dotted black line corresponds to the sum over the SiO mass from all the zones. The solid square, circle, and triangle indicates the mass of SiO reported from observations in IR from several supernovae as specified in the legend box (Sarangi and Cherchneff 2013)

processes, the solid phase interactions and the smooth transition between the two interactive phases at all times. The formalism has been developed by Jacobson (2005), where the variation of the grain number density of volume v is described by the integro-differential coagulation equation,

$$\frac{dn_v(t)}{dt} = \frac{1}{2}\int_0^v \beta_{v-v',v} n_{v-v'} n_{v'} dv' - n_v \int_0^\infty \beta_{v,v'} n_{v'} dv' \tag{4}$$

where β is the rate coefficient of coagulation between particles. The rate coefficient is controlled by physical processes such as Brownian diffusion, convective Brownian motion enhancement, gravitational collection, turbulent inertial motion and Van der Waals forces. In free molecular regimes (where mean free-path of a particle in the gas is much larger compared to the mean radius of a grain), Brownian diffusion prevails as the most important phenomena along with Van der Waals interactions. Brownian diffusion accounts for the scattering, collision and coalescence of the grains through Brownian motion. On the other hand, the Van der Waals forces develop weak, local charge fluctuations that enhance the rate of coagulation for particles with sizes in the molecular range. Solving the appropriate equations, the concentrations and the grain size distribution function for individual dust types are estimated. This formalism has been used to model dust in SN ejecta and winds of AGB stars by Sarangi and Cherchneff (2015), Biscaro and Cherchneff (2016) & Gobrecht et al. (2016).

The dust components comprise of silicates, alumina, carbon, iron sulphide, silicon carbide and pure metallic clusters of magnesium, iron and silicon. The analysis traces the evolution mechanism of all these dust components individually. The formation of dust grains in the ejecta is found to be highly sensitive to the gas densities and the concentration of the specific clusters in consideration. Hence, the dust formation scenario varies to a great extent from one zone to the other even for a single dust type. The estimated mass of each dust component and the size distributions of the grains were calculated as a function of post-explosion time. There is no specific threshold in size to differentiate between the large

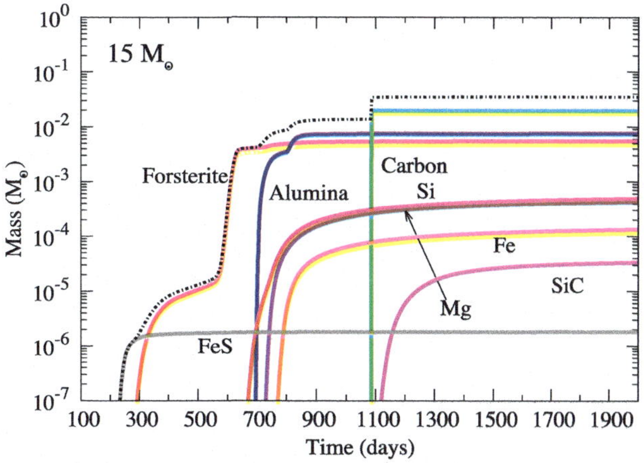

Fig. 19 Dust masses for grains with radius larger than 10 Å summed over all ejecta zones as a function of post-explosion time and dust type. The dotted black line corresponds to the total dust mass formed in the homogeneous ejecta with progenitor mass 15 $M_\odot$ (Sarangi and Cherchneff 2015)

Table 5 Dust masses ($M_\odot$) at different post-explosion epochs for the 15 $M_\odot$ progenitor model. The final peak grain size of each dust type is also indicated in the table (Sarangi and Cherchneff 2015)

Dust type	Post-explosion time (days)							a_{peak} (Å)
	500	700	900	1100	1200	1500	2000	
Forsterite	1.1(−5)	4.2(−3)	5.3(−3)	5.5(−3)	5.5(−3)	5.6(−3)	5.6(−3)	64; 168
Alumina	–	6.1(−6)	7.4(−3)	7.6(−3)	7.6(−3)	7.7(−3)	7.7(−3)	96
Carbon	–	–	–	2.0(−2)	2.0(−2)	2.0(−2)	2.1(−2)	134
Pure Magnesium	–	2.5(−6)	1.9(−4)	3.2(−4)	3.6(−4)	4.4(−4)	5.0(−4)	29
Pure Silicon	–	–	1.5(−4)	2.7(−4)	3.1(−4)	3.8(−4)	4.4(−4)	69
Pure Iron	–	–	4.1(−5)	9.1(−5)	9.4(−5)	1.2(−4)	1.4(−4)	36
Silicon Carbide	–	–	–	–	6.1(−6)	2.6(−5)	3.4(−5)	9
Iron Sulphide	1.7(−6)	1.8(−6)	1.8(−6)	1.8(−6)	1.8(−6)	1.8(−6)	1.8(−6)	21
Total	1.3(−5)	4.3(−3)	0.013	0.034	0.034	0.035	0.035	

molecular clusters and the dust particles. From the observational point of view, the dust formation in the ejecta is characterised by an excess in IR spectral energy distribution and blue shift of optical emission lines. From the modeling perspective, a fiducial size of 10 Å has been chosen as the minimum size of grains in the ejecta, which can be categorised as dust. The grains smaller in size are considered as uncondensed molecular clusters, which remain in gas phase. Figure 19 illustrates the evolution of dust masses of all the dust components from day 100 to 2000 after explosion. The masses of dust components in the figure are presented as the sum over all the ejecta zones.

Formation of dust in the ejecta dawns after 200 days post-explosion time with the formation of iron sulphide grains. Later on in the period of 300–400 days silicates (forsterite) starts to form mainly in the inner zones of the O-core. The mass of dust surges up at ∼550 days due to synthesis of silicates in the central zone 2. The process of gradual increase of dust mass continue with the advent of alumina at day 800 and lately by amorphous carbon at 1100 days. This way the dust mass increases from 10^{-5} $M_\odot$ at 400 days to 0.035 $M_\odot$ at the end of four years. Table 5 presents all the evolution pattern of each dust type over a span of a 2000 days after explosion. Amorphous carbon, alumina and forsterite are the most abun-

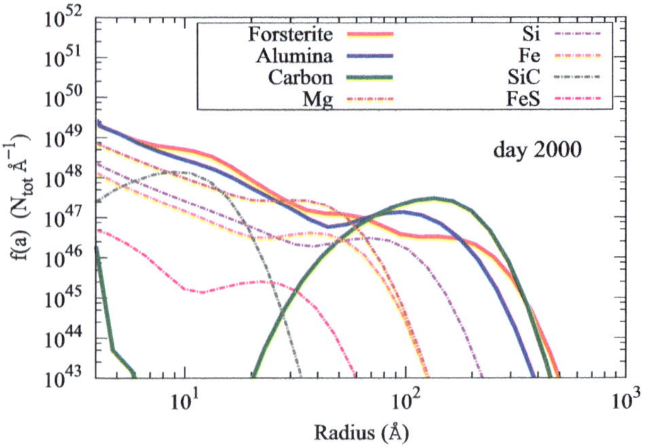

Fig. 20 The final size distribution $f(a)$ of dust grains at 2000 days post-explosion is presented for all the dust components (Sarangi and Cherchneff 2015). The distribution function is expressed in terms of total number of particles per unit radius ($N_{tot}(a)$. Å^{-1})

dant dust types in the ejecta. The range of dust masses closely agrees with the estimated masses of dust derived from observational data in the IR. At times between 200–600 days, the mid-IR observations predict the presence of dust in the ejecta with masses typically in the order of 10^{-5} to 10^{-3} $M_\odot$ (Kotak et al. 2009, 2005; Elmhamdi et al. 2003b; Wooden et al. 1993; Inserra et al. 2011; Sugerman et al. 2006; Gallagher et al. 2012). These findings are consistent with the model results of the total dust mass over time (Fig. 19). Submm observations however report the presence of much larger masses of dust in the remnants. In Cas A, ~0.08 $M_\odot$ of dust is inferred from Herschel data (Barlow et al. 2010). In case of SN1987A, Matsuura et al. 2011 has derived masses of dust in the range of 0.4–0.7 $M_\odot$ to be present in the ejecta after 28 years from explosion. Recent ALMA observation concluded the final dust mass to be about 0.5 $M_\odot$ (Matsuura et al. 2015). Also for the filaments of Crab nebula an estimated mass of 0.1–0.2 $M_\odot$ is required to fit the Herschel data as explained by Gomez et al. 2012. The final mass of dust estimated in this study ends up in 0.035 $M_\odot$, and we can suppose the dust mass will remain consistent till the advent of reverse shock in the remnant. These trends derived from the model match well with the results of submm observations (Sarangi and Cherchneff 2015; Biscaro and Cherchneff 2014). However the final masses reported from these late time observational data are 2–10 times larger than our assessment. Importantly though, all the masses reported from observations are derived by fitting the spectral energy distributions at IR and submm wavelengths where some composition of dust has been assumed. The dust masses from observations hence can be sensitive to the choice of dust components, their compositions and size at $a \ll \lambda$. Finally, there are several options for the choice of optical constants related to each dust type, and the results vary significantly from one to the other. In most of the cases, the size distributions were assumed to either follow a power law profile like the standard Mathis-Rumpl-Nordsieck (MRN) distribution (Mathis et al. 1977) or the grains were chosen to be of identical sizes. All these factors add to some uncertainty in the estimation of dust masses fitting the observed spectra.

The size distribution function defines the number of grains per unit of particle radius, assuming a spherical geometry. The size distribution function $f(a)$, for all the important dust species at day 2000 is presented in Fig. 20. The main three dust components carbon, alumina and forsterite are presented in solid colours in the figure. The distribution function at a given time depends on (a) the number of monomers present in the gas phase, (b) the rate of coagulation at the given temperature and (c) the concentration of grains from the previous time step. The prime dust components being more abundant, also grow to form larger grains in the order of 200–500 Å. The size distribution function of one dust species

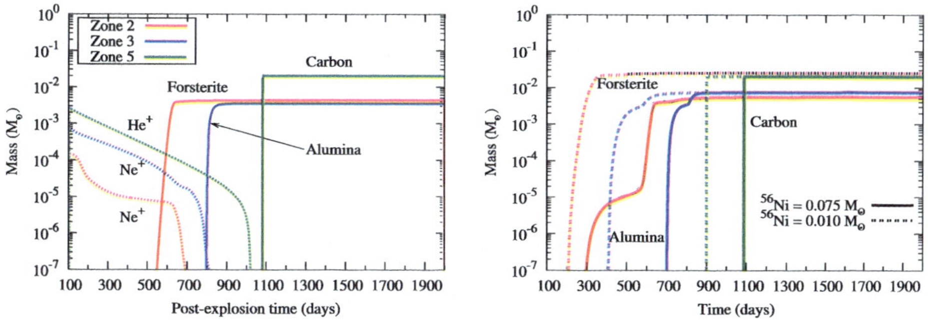

Fig. 21 *Left-panel:* The figure presents the correlation between the ions of the nobel gases and the epochs of dust formation. The presence of He^+, Ne^+ or Ar^+ ions in the gas tend to hinder the dust formation processes in the respective ejecta zones, as illustrated in the figure and the text. *Right panel:* The time evolution of forsterite, alumina and carbon dust masses for standard ($^{56}Ni = 0.075$ $M_\odot$) model and a lower ^{56}Ni mass ($^{56}Ni = 0.01$ $M_\odot$) model are compared in this figure (Sarangi and Cherchneff 2013)

differs in behaviour from the other and hence cannot be defined by a single mathematical expression. Depending on the formation zones in the ejecta, the distribution functions can have a single peak or sometimes are found to have multiple peaks. The findings suggest that the dust masses reside mainly in the larger grains even though the population of large grains may be less than that of relatively smaller grains in the ejecta. Moreover, the condensation phase is an efficient process in the ejecta and once the large grains are formed, they are unlikely to segregate to smaller size fragments.

The size distributions of dust produced by SNe do not follow a MRN size variation both for single dust type and for the total dust distribution. The size distributions derived from the study have less small grains with radius $a < 30$ Å and are skewed towards large grains in the size range 40–1000 Å, compared with the MRN distribution. This result clearly indicates that the use of a MRN size distribution for modeling the fluxes from a SN ejecta is inappropriate. Further, using MRN for dust sputtering calculations in the SNe remnants may lead to erroneous results on the derived dust masses and the dust mass that survives shock processing in the remnants. Ideally if there at all exists a distribution function for supernovae dust grains, that should be defined by the dust size patterns of a single dust component in a given zone, and not over a sum of all the species. The analysis clearly indicates none of the dust components follow a power law pattern of dust sizes at any epoch.

The ions of the nobel gases, namely helium, neon and argon present in the ejecta, play an important role in moulding the chemistry (Lepp et al. 1990). The γ-rays in the ejecta produced by radioactivity undergo Compton scattering and degrade to X-rays. The energetic Compton electrons further lose energy by excitation and ionisation of the gas. By day 100, most of the atoms in the ejecta gets ionised to their corresponding cation through collision with the fast Compton electrons. The ground state energy level of their inert parents being very low, the cations of He, Ne and Ar actively participate in charge exchange processes with newly synthesised molecules, leading to their destruction (e.g. $AB + He^+ \rightarrow A + B^+ + He$). The impact is clearly evident in the outermost zones of the He-core, where the detrimental effects of He^+ delay the appearance of stable molecules and clusters in the ejecta till $\sim$ day 1000. Figure 21 (left-panel) presents the correlation between the epoch of dust formation to the abundances of He^+, Ne^+ or Ar^+ in the ejecta zones. With the passage of time, effect of radioactivity decreases as most of the ^{56}Co already gets converted to stable ^{56}Fe. The energy from the γ-rays and X-rays get deposited in the ejecta and a small fraction escapes outside the core. Thereafter, the abundance of Compton electrons also falls rapidly and all the cations eventually recombine back to their stable inert parents. The molecule and dust synthesis gain in efficiency when the ions of He, Ne and Ar are no longer present in the ejecta.

The amount of radioactive elements produced by the explosion also directly influences the chemical yields. The radioactive ^{56}Ni mass produced by a SN can be derived from the variation of the optical light curves and H_α luminosities in the nebula phase (Elmhamdi et al. 2003a). Sarangi and Cherchneff (2013) provides a list of local SNe with their estimated progenitor masses as well as the masses of ^{56}Ni produced by the explosion. Observational findings support the trend derived by Hamuy (2003) that more massive supernovae produce more energetic explosions, and supernovae with greater energies produce larger ^{56}Ni masses. The energy deposited by the radioactive ^{56}Ni in the ejecta is linearly proportional to the its mass. Therefore, the rate coefficients for the destructions by energetic Compton electrons are directly connected to the mass of ^{56}Ni. Physically, a smaller ^{56}Ni mass in the ejecta reduces the number of Compton electrons resulting from the degradation of a lower amount of γ-rays. Therefore, fewer ions such as He^+, Ne^+ or Ar^+ are produced, thereby enhancing the rate of recombination. The destruction rates for molecules and clusters (e.g., SiO, C_2, AlO) are much less severe for a low ^{56}Ni mass case, because of the lower Ni^+ and He^+ concentration. So the chances of survival for the molecules and clusters increase, thereby also leading to early condensation of dust grains. Figure 21 (right-panel) shows the evolution of silicates, alumina and carbon dust masses corresponding to $^{56}\mathrm{Ni} = 0.075\ M_\odot$ and $^{56}\mathrm{Ni} = 0.01\ M_\odot$ cases and compares the respective yields.

7 Mixing and Clumpiness

A homogeneous, 1D model of the ejecta is capable to indicate trends, which are closely similar to the actual event. Nevertheless, a real supernova ejecta is neither homogeneous nor spherically symmetric. Observational findings suggest the possibility of macroscopic mixing of ^{56}Ni in the outer layers of the H-He core (McCray 1993; Hanuschik et al. 1988). Detection of hard X-rays and γ-ray lines in SN1987A, immediately after the explosion, indicates a presence of a high-velocity component of radioactive ^{56}Ni (Dotani et al. 1987; Syunyaev et al. 1987; Matz et al. 1988). The mass of the high velocity (>3000 km/s) ^{56}Ni clump is estimated to be about $10^{-3}\ M_\odot$, which is between 4-17 % of the total ^{56}Ni produced at the explosion (Haas et al. 1990; Utrobin et al. 1995), as explained in Fig. 22 (Wongwathanarat et al. 2015). Therefore, the mixing of ^{56}Ni in the earliest days post-explosion is intense yet not complete. As a result, a two-phase SN-ejecta structure is formed. Low-density blobs consisting of Fe-bubbles are created by the energy input from radioactive ^{56}Co and ^{56}Ni, which is surrounded by compressed high density metal-rich ejecta (Blondin et al. 2001).

The multi-D explosion models study the presence of Rayleigh-Taylor unstable filaments in the non-homogeneous ejecta (Kifonidis et al. 2003, 2006; Joggerst et al. 2010; Hammer et al. 2010a; Wongwathanarat et al. 2015) of Type II supernovae. Figure 22 presents the isosurfaces of various metal-rich blobs after a few minutes post-explosion (Wongwathanarat et al. 2015). Moreover, the deviation from spherical symmetry is also evident from the structure of the ejecta.

The explosion energy is imparted in the ejecta in form of kinetic energy, which triggers an outward flow with velocities of around 2000–2500 km/s. If all the materials in the He-core travel at a constant velocity, the density of the outflowing matter should decrease as

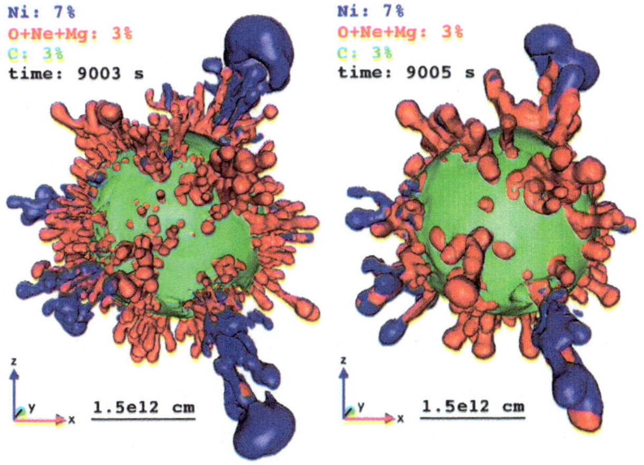

Fig. 22 The isosurfaces of constant mass-fraction of Ni plumes and O/Mg zone at 9000 s post outburst is shown in the Figure adopted from Wongwathanarat et al. (2015). The ^{56}Ni blob is found to travel at velocities between 3800 and 4000 km/s

t^{-2} law (Woosley 1988). However, the active radioactive processes (e.g. $^{56}\mathrm{Ni} \rightarrow {}^{56}\mathrm{Co} \rightarrow {}^{56}\mathrm{Fe}$) deposit the extra energy in the innermost Ni-Co core. Further, the neutrino-driven convection in the ejecta creates inhomogeneity between the layers (Joggerst et al. 2010). This creates a velocity shear across the zones of the ejecta, and the inner zone with extra energy tends to expand faster (Woosley 1988). The velocity shear induces the homologous expansion in the ejecta. Therefore, the fast moving low-density Ni-layers penetrate in the high density overlying zones of the ejecta. This density inversion occurs within a day of the explosion, thereby creating the Rayleigh-Taylor instabilities. The mixed Ni-bubbles further deposit energy in the O-core, which is in turn accelerated through the outer He/N/C zones. These instabilities lead to the fragmentation of the He-core, but the chemical stratification persists over time. Importantly though, this velocity shear results in clumping of the ejecta very early after the explosion.

Radiative transfer models trying to fit the IR contribution from the ejecta dust have repeatedly emphasized on the clumpy nature of the medium (Ercolano et al. 2007; Gallagher et al. 2012). Further, in order to account for the isotopic anomalies in the presolar grains in meteorites of SN origin, a small percentage of element mixing in microscopic scales is necessary. The scenario of mixing and inhomogeneity in the ejecta can only be modelled using hydrodynamic simulations, which is a very computationally rigorous process when all the important metals in the ejecta are being traced. Macroscopic mixing between the ejecta zones following the explosion is evident from observations and simulations. However, the possibility or the mechanism of microscopic diffusion of metals between the zone boundaries is not yet well understood.

A quasi-clumpy model based on SN1987A was formulated by Sarangi and Cherchneff (2015) using the volume filling factors for individual zones derived by Jerkstrand et al. (2011) through the study of the ultraviolet, optical, and near-IR emission lines. Dust formation is found to be faster and more efficient inside a high-density clump. However, depending on the optical depth of the clumpy region and rapid cooling of the newly formed grains, the dust can remain hidden and escape mid-IR detection. Microscopic mixing between inner (Si/S) and outer (H/He/C) zones would lead to formation of SiC grains (important constituent in presolar grains) in larger abundance and H-bearing species such as recently detected HCO^+ (Matsuura et al. 2017).

8 Summary of Current State-of-the-Art

Dust formation scenario in SNe is a complex interplay between several physical and chemical processes, which may vary depending on minimal alteration of conditions. Therefore, formulation of an all-inclusive single model of dust in core-collapse SNe most of the time over-simplifies the actual scenario. On the contrary, a more acceptable approach is to understand the general trends and then apply the knowledge to study any specific SN with its given constraints. Here we shall summarise the aspects related to the parameter space, which characterises the physics and chemistry of dust formation in a typical SN-ejecta.

(a) Core-collapse supernovae, mainly Type II-P, are the most efficient dust producers among all SN-types. On the other hand, the amount of dust formed in Type Ia SN is negligible.
(b) Dust formation in the ejecta is characterised by the chemistry prevailing in a non-steady state non-thermodynamic equilibrium environment.
(c) In a typical SN ejecta, dust formation commences within a year from the explosion and the dust mass continues to increase for a couple of years. The span depends essentially on the density and clumpiness of the medium.
(d) Dust synthesis is controlled by simultaneous processes of nucleation and condensation. Nucleation of stable dust clusters act as the bottleneck to the entire chain of chemical network. These molecular clusters resulting from nucleation determines the chemical composition of the dust as well as the earliest epochs of dust formation.
(e) Theory predicts of about 30–40% of the total mass in the ejecta. CO, SiO, O_2, CS, AlO, SO etc. are important molecular species. Dust, on the other hand, is about 3–10% of the total ejecta mass.
(f) The dust budget in the SN ejecta is comprises of silicates, amorphous carbon, alumina and pure metallic clusters of silicon, iron and magnesium. There is also the possibility of mixed species as the coexisting grains in a certain zone might coagulate heterogeneously.
(g) Dust masses and grain sizes are inversely related to the amount of radioactive ^{56}Ni produced at the time of explosion. On the other hand, they are directly proportional to the degree of clumpiness.
(h) Relative abundances of C-rich and O-rich dust depend on the mass of the progenitor. The larger the progenitor, the larger is the O-core, and therefore O-rich dust dominates over other components.
(i) The size distributions of dust grains in SN ejecta are functions of dust type and epochs of formation. They cannot be generalised as power laws of any exponent, contrary to the general assumption.
(j) A clumpy ejecta of supernova forms larger grains at high densities. These grains are most likely to survive the passage of the reverse shock in the later phase of the evolution.
(k) Theory predicts that a typical type II-P supernova can produce dust masses between 0.02 to 0.2 $M_{\odot}$, while measured dust masses range from 10^{-6} to 1 $M_{\odot}$.
(l) A 1-D model can precisely determine the trends and upper or lower limits for the scenario. However, in order to incorporate the impact of mixing and inhomogeneous deposition of radioactive energy in the ejecta, multi-D dynamical modeling of the ejecta is necessary.

Acknowledgements The authors deeply acknowledge the support of the International Space Science Institute in Bern, Switzerland and their successful endeavor to organise the workshop on Supernovae. We also sincerely thank our collaborator, Dr. Isabelle Cherchneff, for the discussions and feedbacks at the workshop and also while writing. A.S. acknowledges the support of NASA Astrophysics Theory grant NNX15AD27G and the CUA grant 361209. Also, E.R.M. wishes to acknowledge the support of Academy of Finland grant 1285769 and M.M. acknowledges support from an STFC Ernest Rutherford fellowship (ST/L003597/1).

References

F. Abellán, R. Indebetouw, J.M. Marcaide et al., Very deep inside the sn 1987a core ejecta: molecular structures seen in 3d. Astrophys. J. Lett. (2017)

D.K. Aitken, C.H. Smith, S.D. James, P.F. Roche, A.R. Hyland, P.J. McGregor, 10 micron spectral observations of SN1987A—the first year. Mon. Not. R. Astron. Soc. **235**, 19P–31P (1988)

S. Amari, L.R. Nittler, E. Zinner, R.S. Lewis, Continued search for rare types of presolar SiC—grains X and Y, in *Lunar and Planetary Science Conference*, vol. 28 (1997), p. 33

J.E. Andrews, B.E.K. Sugerman, G.C. Clayton, J.S. Gallagher, M.J. Barlow, J. Clem, B. Ercolano, J. Fabbri, M. Meixner, M. Otsuka, D.L. Welch, R. Wesson, Photometric and spectroscopic evolution of the IIP SN 2007it to day 944. Astrophys. J. **731**, 47 (2011). 1102.2431. https://doi.org/10.1088/0004-637X/731/1/47

R.G. Arendt, An infrared survey of Galactic supernova remnants. Astrophys. J. Suppl. Ser. **70**, 181–212 (1989)

R.G. Arendt, E. Dwek, S.H. Moseley, Newly synthesized elements and pristine dust in the Cassiopeia A supernova remnant. Astrophys. J. **521**(1), 234–245 (1999)

R.G. Arendt, E. Dwek, G. Kober, J. Rho, U. Hwang, Interstellar and ejecta dust in the Cas A supernova remnant. Astrophys. J. **786**(1), 55 (2014)

R. Barbon, F. Ciatti, L. Rosino, Photometric properties of type II supernovae. Astron. Astrophys. **72**, 287–292 (1979)

R. Barbon, V. Buondí, E. Cappellaro, M. Turatto, The Asiago Supernova Catalogue—10 years after. Astron. Astrophys. Suppl. **139**, 531–536 (1999). astro-ph/9908046. https://doi.org/10.1051/aas:1999404

M.J. Barlow, B.E.K. Sugerman, J. Fabbri, M. Meixner, R.S. Fisher, J.E. Bowey, N. Panagia, B. Ercolano, G.C. Clayton, M. Cohen, T.M. Gledhill, K.D. Gordon, A.G.G.M. Tielens, A.A. Zijlstra, Detection of a massive dust shell around the type II supernova SN 2002hh. Astrophys. J. **627**(2), L113–L116 (2005)

M.J. Barlow, O. Krause, B.M. Swinyard, B. Sibthorpe, M.A. Besel, R. Wesson, R.J. Ivison, L. Dunne, W.K. Gear, H.L. Gomez, P.C. Hargrave, T. Henning, S.J. Leeks, T.L. Lim, G. Olofsson, E.T. Polehampton, A Herschel PACS and SPIRE study of the dust content of the Cassiopeia A supernova remnant. Astron. Astrophys. **518**, L138 (2010). 1005.2688. https://doi.org/10.1051/0004-6361/201014585

M.J. Barlow, B.M. Swinyard, P.J. Owen, J. Cernicharo, H.L. Gomez, R.J. Ivison, O. Krause, T.L. Lim, M. Matsuura, S. Miller, G. Olofsson, E.T. Polehampton, Detection of a noble gas molecular ion, 36ArH+, in the Crab Nebula. Science **342**(6164), 1343–1345 (2013)

E.E. Barnard, On the dark markings of the sky, with a catalogue of 182 such objects. Astrophys. J. **49**, 1–24 (1919). https://doi.org/10.1086/142439

R.A. Benjamin, E. Churchwell, B.L. Babler, T.M. Bania, D.P. Clemens, M. Cohen, J.M. Dickey, R. Indebetouw, J.M. Jackson, H.A. Kobulnicky, A. Lazarian, A.P. Marston, J.S. Mathis, M.R. Meade, S. Seager, S.R. Stolovy, C. Watson, B.A. Whitney, M.J. Wolff, M.G. Wolfire, GLIMPSE. I. An SIRTF legacy project to map the inner galaxy. Publ. Astron. Soc. Pac. **115**, 953 (2003)

T.J. Bernatowicz, E. Zinner (eds.), *Astrophysical Implications of the Laboratory Study of Presolar Materials*. American Institute of Physics Conference Series, vol. 402 (1997)

F. Bertoldi, C.L. Carilli, P. Cox, X. Fan, M.A. Strauss, A. Beelen, A. Omont, R. Zylka, Dust emission from the most distant quasars. Astron. Astrophys. **406**, L55–L58 (2003). astro-ph/0305116. https://doi.org/10.1051/0004-6361:20030710

A. Bevan, M.J. Barlow, Modelling supernova line profile asymmetries to determine ejecta dust masses: SN 1987A from days 714 to 3604. Mon. Not. R. Astron. Soc. **456**(2), 1269–1293 (2016)

S. Bianchi, R. Schneider, Dust formation and survival in supernova ejecta. Mon. Not. R. Astron. Soc. **378**, 973–982 (2007). 0704.0586. https://doi.org/10.1111/j.1365-2966.2007.11829.x

C. Biscaro, I. Cherchneff, Molecules and dust in Cassiopeia A. I. Synthesis in the supernova phase and processing by the reverse shock in the clumpy remnant. Astron. Astrophys. **564**, A25 (2014). 1401.5594. https://doi.org/10.1051/0004-6361/201322932

C. Biscaro, I. Cherchneff, Molecules and dust in Cassiopeia A. II. Dust sputtering and diagnosis of supernova dust survival in remnants. Astron. Astrophys. **589**, A132 (2016). 1511.05487. https://doi.org/10.1051/0004-6361/201527769

S. Blinnikov, P. Lundqvist, O. Bartunov, K. Nomoto, K. Iwamoto, Radiation hydrodynamics of SN 1987A. I. Global analysis of the light curve for the first 4 months. Astrophys. J. **532**, 1132–1149 (2000). astro-ph/9911205. https://doi.org/10.1086/308588

J.M. Blondin, K.J. Borkowski, S.P. Reynolds, Dynamics of Fe bubbles in young supernova remnants. Astrophys. J. **557**, 782–791 (2001). astro-ph/0010285. https://doi.org/10.1086/321674

P. Bouchet, I.J. Danziger, Infrared photometry and spectrophotometry of supernova 1987A—part two—1987NOV to 1991MAR observations. Astron. Astrophys. **273**, 451 (1993)

P. Bouchet, I.J. Danziger, L.B. Lucy, Bolometric light curve of SN 1987A—results from day 616 to 1316 after outburst. Astron. J. **102**, 1135–1146 (1991)

P. Bouchet, J.M. De Buizer, N.B. Suntzeff, I.J. Danziger, T.L. Hayward, C.M. Telesco, C. Packham, High-resolution mid-infrared imaging of SN 1987A. Astrophys. J. **611**(1), 394–398 (2004)

P. Bouchet, E. Dwek, J. Danziger, R.G. Arendt, I.J.M. De Buizer, S. Park, N.B. Suntzeff, R.P. Kirshner, P. Challis, SN 1987A after 18 years: mid-infrared Gemini and Spitzer observations of the remnant. Astrophys. J. **650**, 212 (2006)

D. Branch, M. Livio, L.R. Yungelson, F.R. Boffi, E. Baron, In search of the progenitors of type IA supernovae. Publ. Astron. Soc. Pac. **107**, 1019 (1995). https://doi.org/10.1086/133657

D.E. Brownlee, Cosmic dust—collection and research. Annu. Rev. Earth Planet. Sci. **13**, 147–173 (1985). https://doi.org/10.1146/annurev.ea.13.050185.001051

F. Calura, R. Gilli, C. Vignali, F. Pozzi, A. Pipino, F. Matteucci, The dust content of QSO hosts at high redshift. Mon. Not. R. Astron. Soc. **438**, 2765–2783 (2014). 1312.1087. https://doi.org/10.1093/mnras/stt2329

R.M. Catchpole, P.A. Whitelock, M.W. Feast, J.M. Menzies, I.S. Glass, F. Marang, J.D. Laing, J.H. Spencer Jones, G. Roberts, L.A. Balona, B.S. Carter, C.D. Laney, L.T. Evans, K. Sekiguchi, G.G. Hutchinson, R. Maddison, J. Albinson, A. Evans, F.A. Allen, H. Winkler, A. Fairall, C. Corbally, J.K. Davies, Q.A. Parker, Spectroscopic and photometric observations of SN 1987A. III—days 135 to 260. Mon. Not. R. Astron. Soc. **231**, 75P (1988)

I. Cherchneff, Dust formation in massive stars and their explosive ends, in *Hot and Cool: Bridging Gaps in Massive Star Evolution*, ed. by C. Leitherer, P.D. Bennett, P.W. Morris, J.T. Van Loon. Astronomical Society of the Pacific Conference Series, vol. 425 (2010), p. 237. 0909.0164

I. Cherchneff, The inner wind of IRC+10216 revisited: new exotic chemistry and diagnostic for dust condensation in carbon stars. Astron. Astrophys. **545**, A12 (2012). 1111.6809. https://doi.org/10.1051/0004-6361/201118542

I. Cherchneff, The chemistry of dust formation in red supergiants, in *Betelgeuse Workshop 2012*, ed. by P. Kervella, T. Le Bertre, G. Perrin. EAS Publications Series, vol. 60 (2013), pp. 175–184. 1303.2805. https://doi.org/10.1051/eas/1360020

I. Cherchneff, E. Dwek, The chemistry of population III supernova ejecta. I. Formation of molecules in the early universe. Astrophys. J. **703**, 642–661 (2009). 0907.3621. https://doi.org/10.1088/0004-637X/703/1/642

I. Cherchneff, E. Dwek, The chemistry of population III supernova ejecta. II. The nucleation of molecular clusters as a diagnostic for dust in the early universe. Astrophys. J. **713**, 1–24 (2010). 1002.3060. https://doi.org/10.1088/0004-637X/713/1/1

I. Cherchneff, S. Lilly, Primordial massive supernovae as the first molecular factories in the early universe. Astrophys. J. Lett. **683**, L123 (2008). 0807.2511. https://doi.org/10.1086/591906

I. Cherchneff, A. Sarangi, Molecules in supernova ejecta, in *IAU Symposium*, vol. 280, ed. by J. Cernicharo, R. Bachiller (2011), pp. 228–236. 1107.5668. https://doi.org/10.1017/S1743921311025002

R.A. Chevalier, The radio and X-ray emission from type II supernovae. Astrophys. J. **259**, 302–310 (1982). https://doi.org/10.1086/160167

H. Chihara, C. Koike, A. Tsuchiyama, S. Tachibana, D. Sakamoto, Compositional dependence of infrared absorption spectra of crystalline silicates. I. Mg-Fe pyroxenes. Astron. Astrophys. **391**, 267–273 (2002). https://doi.org/10.1051/0004-6361:20020791

B.G. Choi, G.R. Huss, G.J. Wasserburg, Oxygen, magnesium, calcium, and titanium isotopes in asymptotic giant branch and supernova oxides. Meteorit. Planet. Sci. Suppl. **33**, A32 (1998)

D.D. Clayton, Sudden grain nucleation and growth in supernova and nova ejecta. Astrophys. Space Sci. **65**, 179–189 (1979). https://doi.org/10.1007/BF00643499

D.D. Clayton, L.R. Nittler, Astrophysics with presolar stardust. Annu. Rev. Astron. Astrophys. **42**, 39–78 (2004). https://doi.org/10.1146/annurev.astro.42.053102.134022

D.D. Clayton, D. Arnett, J. Kane, B.S. Meyer, Type X silicon carbide presolar grains: type Ia supernova condensates? Astrophys. J. **486**, 824–834 (1997). https://doi.org/10.1086/304545

D.D. Clayton, W. Liu, A. Dalgarno, Condensation of carbon in radioactive supernova gas. Science **283**, 1290 (1999). https://doi.org/10.1126/science.283.5406.1290

L. Colangeli, V. Mennella, P. Palumbo, A. Rotundi, E. Bussoletti, Mass extinction coefficients of various submicron amorphous carbon grains: tabulated values from 40 NM to 2 mm. Astron. Astrophys. Suppl. **113**, 561 (1995)

I.J. Danziger, P. Bouchet, R.A.E. Fosbury, C. Gouiffes, L.B. Lucy, SN 1987A—observational results obtained at ESO, in *Supernova 1987A in the Large Magellanic Cloud*, ed. by M. Kafatos, A.G. Michalitsianos (1988), pp. 37–50

I. Danziger, C. Gouiffes, P. Bouchet, L.B. Lucy, Supernova 1987a in the large magellanic cloud, in *IAU Circular*, vol. 4747 (1989)

I.J. Danziger, L.B. Lucy, P. Bouchet, C. Gouiffes, Molecules dust and ionic abundances in supernova 1987A, in *Supernovae*, ed. by S.E. Woosley (1991), p. 69

I. De Looze, M.J. Barlow, B.M. Swinyard, J. Rho, H.L. Gomez, M. Matsuura, R. Wesson, The dust mass in Cassiopeia A from a spatially resolved Herschel analysis. Mon. Not. R. Astron. Soc. **465**(3), 3309–3342 (2017)

E. Di Carlo, C. Corsi, A.A. Arkharov, F. Massi, V.M. Larionov, N.V. Efimova, M. Dolci, N. Napoleone, A. Di Paola, Near-infrared observations of the type Ib supernova SN 2006jc: evidence of interactions with dust. Astrophys. J. **684**, 471–480 (2008). 0712.3855. https://doi.org/10.1086/590051

B. Donn, J.A. Nuth, Does nucleation theory apply to the formation of refractory circumstellar grains? Astrophys. J. **288**, 187–190 (1985). https://doi.org/10.1086/162779

J. Dorschner, B. Begemann, T. Henning, C. Jaeger, H. Mutschke, Steps toward interstellar silicate mineralogy. II. Study of Mg-Fe-silicate glasses of variable composition. Astron. Astrophys. **300**, 503 (1995)

T. Dotani, K. Hayashida, H. Inoue, M. Itoh, K. Koyama, F. Makino, K. Mitsuda, T. Murakami, M. Oda, Y. Ogawara, S. Takano, Y. Tanaka, A. Yoshida, K. Makishima, T. Ohashi, N. Kawai, M. Matsuoka, R. Hoshi, S. Hayakawa, T. Kii, H. Kunieda, F. Nagase, Y. Tawara, I. Hatsukade, S. Kitamoto, S. Miyamoto, H. Tsunemi, K. Yamashita, M. Nakagawa, M. Yamauchi, M.J.L. Turner, K.A. Pounds, H.D. Thomas, G.C. Stewart, A.M. Cruise, B.E. Patchett, D.H. Reading, Discovery of an unusual hard X-ray source in the region of supernova 1987A. Nature **330**, 230 (1987). https://doi.org/10.1038/330230a0

T. Douvion, P.O. Lagage, E. Pantin, Cassiopeia A dust composition and heating. Astron. Astrophys. **369**, 589–593 (2001). https://doi.org/10.1051/0004-6361:20010053

B.T. Draine, The infrared signature of graphite grains. Astrophys. J. Lett. **277**, L71–L74 (1984). https://doi.org/10.1086/184206

B.T. Draine, Interstellar dust grains. Annu. Rev. Astron. Astrophys. **41**, 241–289 (2003). astro-ph/0304489. https://doi.org/10.1146/annurev.astro.41.011802.094840

M.R. Drout, D. Milisavljevic, J. Parrent, R. Margutti, A. Kamble, A.M. Soderberg, P. Challis, R. Chornock, W. Fong, S. Frank, N. Gehrels, M.L. Graham, E. Hsiao, K. Itagaki, M. Kasliwal, R.P. Kirshner, D. Macomb, G.H. Marion, J. Norris, M.M. Phillips, The double-peaked sn 2013ge: a type Ib/c SN with an asymmetric mass ejection or an extended progenitor envelope. Astrophys. J. **821**(1), 57 (2016)

P. Dunk, N. Kaiser, M. Mullet-Gas et al., The smallest stable fullerene, M@C_{28} (M = Ti, Zr, U): stabilization and growth from carbon vapor. J. Am. Chem. Soc. **134**, 9380 (2012). https://doi.org/10.1021/ja302398h

L. Dunne, S. Eales, R. Ivison, H. Morgan, M. Edmunds, Type II supernovae as a significant source of interstellar dust. Nature **424**, 285–287 (2003). astro-ph/0307320. https://doi.org/10.1038/nature01792

E. Dwek, Will dust black out SN 1987A? Astrophys. J. **329**, 814 (1988)

E. Dwek, The evolution of the elemental abundances in the gas and dust phases of the galaxy. Astrophys. J. **501**, 643 (1998). astro-ph/9707024. https://doi.org/10.1086/305829

E. Dwek, The supernova origin of interstellar dust. Science **313**(5784), 178–180 (2006). http://science.sciencemag.org/content/313/5784/178.full.pdf. https://doi.org/10.1126/science.1130423

E. Dwek, R.G. Arendt, Infrared echoes reveal the shock breakout of the Cas A supernova. Astrophys. J. **685**, 976–987 (2008). 0802.0221. https://doi.org/10.1086/589988

E. Dwek, R.G. Arendt, The evolution of dust mass in the ejecta of SN1987A. Astrophys. J. **810**(1), 75 (2015)

E. Dwek, I. Cherchneff, The origin of dust in the early universe: probing the star formation history of galaxies by their dust content. Astrophys. J. **727**, 63 (2011). 1011.1303. https://doi.org/10.1088/0004-637X/727/2/63

E. Dwek, M.G. Hauser, H.L. Dinerstein, F.C. Gillett, W.L. Rice, Physical processes and infrared emission from the Cassiopeia A supernova remnant. Astrophys. J. **315**, 571–579 (1987)

E. Dwek, S.H. Moseley, W. Glaccum, J.R. Graham, R.F. Loewenstein, R.F. Silverberg, R.K. Smith, Dust and gas contributions to the energy output of SN 1987A on day 1153. Astrophys. J. **389**, L21–L24 (1992)

E. Dwek, F. Galliano, A.P. Jones, The evolution of dust in the early universe with applications to the galaxy SDSS J1148+5251. Astrophys. J. **662**, 927–939 (2007). 0705.3799. https://doi.org/10.1086/518430

E. Dwek, R.G. Arendt, P. Bouchet, D.N. Burrows, P. Challis, I.J. Danziger, J.M. De Buizer, R.D. Gehrz, R.P. Kirshner, R.A. McCray, S. Park, E.F. Polomski, C.E. Woodward, Infrared and X-ray evidence for circumstellar grain destruction by the blast wave of supernova 1987A. Astrophys. J. **676**, 1029 (2008)

A. Elmhamdi, N.N. Chugai, I.J. Danziger, Light curves and Hα luminosities as indicators of ^{56}Ni mass in type IIP supernovae. Astron. Astrophys. **404**, 1077–1086 (2003a). astro-ph/0304144. https://doi.org/10.1051/0004-6361:20030522
A. Elmhamdi, I.J. Danziger, N. Chugai, A. Pastorello, M. Turatto, E. Cappellaro, G. Altavilla, S. Benetti, F. Patat, M. Salvo, Photometry and spectroscopy of the type IIP SN 1999em from outburst to dust formation. Mon. Not. R. Astron. Soc. **338**, 939–956 (2003b). astro-ph/0209623. https://doi.org/10.1046/j.1365-8711.2003.06150.x
A. Elmhamdi, I.J. Danziger, E. Cappellaro, M. Della Valle, C. Gouiffes, M.M. Phillips, M. Turatto, SN Ib 1990I: clumping and dust in the ejecta? Astron. Astrophys. **426**, 963–977 (2004). astro-ph/0407145. https://doi.org/10.1051/0004-6361:20041318
L. Ensman, A. Burrows, Shock breakout in SN 1987A. Astrophys. J. **393**, 742–755 (1992). https://doi.org/10.1086/171542
B. Ercolano, M.J. Barlow, B.E.K. Sugerman, Dust yields in clumpy supernova shells: SN 1987A revisited. Mon. Not. R. Astron. Soc. **375**, 753–763 (2007). astro-ph/0611719. https://doi.org/10.1111/j.1365-2966.2006.11336.x
J. Fabbri, M. Otsuka, M.J. Barlow, J.S. Gallagher, R. Wesson, B.E.K. Sugerman, G.C. Clayton, M. Meixner, J.E.E. Andrews, D.L. Welch, B. Ercolano, The effects of dust on the optical and infrared evolution of SN 2004et. Mon. Not. R. Astron. Soc. **418**(2), 1285–1307 (2011)
D. Fabian, C. Jäger, T. Henning, J. Dorschner, H. Mutschke, Steps toward interstellar silicate mineralogy. V. Thermal evolution of amorphous magnesium silicates and silica. Astron. Astrophys. **364**, 282–292 (2000)
D. Fabian, T. Posch, H. Mutschke, F. Kerschbaum, J. Dorschner, Infrared optical properties of spinels. A study of the carrier of the 13, 17 and 32 μm emission features observed in ISO-SWS spectra of oxygen-rich AGB stars. Astron. Astrophys. **373**, 1125–1138 (2001). https://doi.org/10.1051/0004-6361:20010657
A. Fassia, W.P.S. Meikle, N. Chugai, T.R. Geballe, P. Lundqvist, N.A. Walton, D. Pollacco, S. Veilleux, G.S. Wright, M. Pettini, T. Kerr, E. Puchnarewicz, P. Puxley, M. Irwin, C. Packham, S.J. Smartt, D. Harmer, Optical and infrared spectroscopy of the type IIn SN 1998S: days 3–127. Mon. Not. R. Astron. Soc. **325**, 907–930 (2001). astro-ph/0011340. https://doi.org/10.1046/j.1365-8711.2001.04282.x
J. Feder, K. Russell, J. Lothe, G. Pound, Homogeneous nucleation and growth of droplets in vapours. Adv. Phys. **15**(57), 111–178 (1966). https://doi.org/10.1080/00018736600101264
G.J. Ferland, N. Abel, K. Davidson, N. Smith, Physical conditions in the homunculus, in *The Fate of the Most Massive Stars*, ed. by R. Humphreys, K. Stanek. Astronomical Society of the Pacific Conference Series, vol. 332 (2005), p. 294. astro-ph/0503633
R. Fesen, W.P. Blair, Optical identification of dust within the Crab Nebula's filaments. Astrophys. J. **351**, L45 (1990)
R.A. Fesen, M.C. Hammell, J. Morse, R.A. Chevalier, K.J. Borkowski, M.A. Dopita, C.L. Gerardy, S.S. Lawrence, J.C. Raymond, S. van den Bergh, The expansion asymmetry and age of the Cassiopeia A supernova remnant. Astrophys. J. **645**(1), 283–292 (2006)
A.V. Filippenko, Optical spectra of supernovae. Annu. Rev. Astron. Astrophys. **35**, 309–355 (1997). https://doi.org/10.1146/annurev.astro.35.1.309
C. Fransson, R.A. Chevalier, Late emission from supernovae—a window on stellar nucleosynthesis. Astrophys. J. **343**, 323–342 (1989). https://doi.org/10.1086/167707
C. Fransson, J. Larsson, K. Migotto, D. Pesce, P. Challis, R.A. Chevalier, K. France, R.P. Kirshner, B. Leibundgut, P. Lundqvist, R.A. McCray, J. Spyromilio, F. Taddia, A. Jerkstrand, S. Mattila, N. Smith, J. Sollerman, J.C. Wheeler, A. Crotts, P. Garnavich, K. Heng, S.S. Lawrence, N. Panagia, C.S.J. Pun, G. Sonneborn, B.E.K. Sugerman, The destruction of the circumstellar ring of SN 1987A. Astrophys. J. **806**(1), L19 (2015)
C. Fransson, J. Larsson, J. Spyromilio, B. Leibundgut, R.A. McCray, A. Jerkstrand, Discovery of molecular hydrogen in SN 1987A. Astrophys. J. **821**(1), L5 (2016)
C.L. Fryer, S.E. Woosley, A. Heger, Pair-instability supernovae, gravity waves, and gamma-ray transients. Astrophys. J. **550**, 372–382 (2001). astro-ph/0007176. https://doi.org/10.1086/319719
C. Gall, J. Hjorth, A.C. Andersen, Production of dust by massive stars at high redshift. Astron. Astrophys. Rev. **19**, 43 (2011). 1108.0403. https://doi.org/10.1007/s00159-011-0043-7
C. Gall, J. Hjorth, D. Watson, E. Dwek, J.R. Maund, O.D. Fox, G. Leloudas, D. Malesani, A.C. Day-Jones, Rapid formation of large dust grains in the luminous supernova 2010jl. Nature **511**(7509), 326–329 (2014)
J.S. Gallagher, B.E.K. Sugerman, G.C. Clayton, J.E. Andrews, J. Clem, M.J. Barlow, B. Ercolano, J. Fabbri, M. Otsuka, R. Wesson, M. Meixner, Optical and infrared analysis of type II SN 2006bc. Astrophys. J. **753**, 109 (2012). 1205.5517. https://doi.org/10.1088/0004-637X/753/2/109
C.L. Gerardy, R.A. Fesen, P. Höflich, J.C. Wheeler, Detection of CO and dust emission in near-infrared spectra of SN 1998S. Astron. J. **119**(6), 2968–2981 (2000)

C.L. Gerardy, R.A. Fesen, K. Nomoto, K. Maeda, P. Höflich, J.C. Wheeler, Carbon monoxide in the type Ic supernova 2000ew. Publ. Astron. Soc. Jpn. **54**(6), 905–910 (2002)

D. Gobrecht, I. Cherchneff, A. Sarangi, J.M.C. Plane, S.T. Bromley, Dust formation in the oxygen-rich AGB star IK Tauri. Astron. Astrophys. **585**, A6 (2016). 1509.07613. https://doi.org/10.1051/0004-6361/201425363

H.L. Gomez, O. Krause, M.J. Barlow, B.M. Swinyard, P.J. Owen, C.J.R. Clark, M. Matsuura, E.L. Gomez, J. Rho, M.A. Besel, J. Bouwman, W.K. Gear, T. Henning, R.J. Ivison, E.T. Polehampton, B. Sibthorpe, A cool dust factory in the Crab Nebula: a Herschel study of the filaments. Astrophys. J. **760**, 96 (2012). 1209.5677. https://doi.org/10.1088/0004-637X/760/1/96

E.V. Gotthelf, B. Koralesky, L. Rudnick, T.W. Jones, U. Hwang, R. Petre, Chandra detection of the forward and reverse shocks in Cassiopeia A. Astrophys. J. **552**(1), L39–L43 (2001)

T.P.M. Goumans, S.T. Bromley, Efficient nucleation of stardust silicates via heteromolecular homogeneous condensation. Mon. Not. R. Astron. Soc. **420**, 3344–3349 (2012). https://doi.org/10.1111/j.1365-2966.2011.20255.x

J.R. Graham, G.S. Wright, A.J. Longmore, Infrared spectroscopy and imaging of the Crab Nebula. Astrophys. J. **352**, 172–183 (1990)

D.A. Green, Galactic supernova remnants: an updated catalogue and some statistics. Bull. Astron. Soc. India **32**, 335–370 (2004)

D.A. Green, R.J. Tuffs, C.C. Popescu, Far-infrared and submillimetre observations of the Crab Nebula. Mon. Not. R. Astron. Soc. **355**(4), 1315–1326 (2004)

T. Grenman, G.F. Gahm, E. Elfgren, Dusty globules in the Crab Nebula. Astron. Astrophys. **599**, A110 (2017). 1610.08449. https://doi.org/10.1051/0004-6361/201629693

E. Grün, B.A.S. Gustafson, S. Dermott, H. Fechtig, in *Interplanetary Dust*, ed. by E. Grün, B.A.S. Gustafson, S. Dermott, H. Fechtig Astronomy and Astrophysics Library (Springer, Berlin, 2001), 804 p. ISBN 3-540-42067-3

M.R. Haas, E.F. Erickson, S.D. Lord, D.J. Hollenbach, S.W.J. Colgan, M.G. Burton, Velocity-resolved far-infrared spectra of forbidden Fe II—evidence for mixing and clumping in SN 1987A. Astrophys. J. **360**, 257–266 (1990). https://doi.org/10.1086/169115

N.J. Hammer, H.T. Janka, E. Müller, Three-dimensional simulations of mixing instabilities in supernova explosions. Astrophys. J. **714**, 1371–1385 (2010a). 0908.3474. https://doi.org/10.1088/0004-637X/714/2/1371

N.J. Hammer, H.T.T. Janka, E. Muller, Three-dimensional simulations of mixing instabilities in supernova explosions. Astrophys. J. **714**(2), 1371–1385 (2010b)

M. Hamuy, Observed and physical properties of core-collapse supernovae. Astrophys. J. **582**, 905–914 (2003). astro-ph/0209174. https://doi.org/10.1086/344689

R.W. Hanuschik, G. Thimm, J. Dachs, H-alpha fine-structure in SN 1987A within the first 111 days. Mon. Not. R. Astron. Soc. **234**, 41P–49P (1988). https://doi.org/10.1093/mnras/234.1.41P

T.J. Harries, J.D. Monnier, N.H. Symington, R. Kurosawa, Three-dimensional dust radiative-transfer models: the Pinwheel Nebula of WR 104. Mon. Not. R. Astron. Soc. **350**, 565–574 (2004). astro-ph/0401574. https://doi.org/10.1111/j.1365-2966.2004.07668.x

P. Harvey, D. Lester, H. Dinerstein, B. Smith, C. Colome, Far-IR monitoring of SN1987A. Bull. Am. Astron. Soc. **21**, 1215 (1989)

A. Heger, S.E. Woosley, The nucleosynthetic signature of population III. Astrophys. J. **567**, 532–543 (2002). astro-ph/0107037. https://doi.org/10.1086/338487

T. Henning, B. Begemann, H. Mutschke, J. Dorschner, Optical properties of oxide dust grains. Astron. Astrophys. Suppl. **112**, 143 (1995)

M. Herant, S.E. Woosley, Postexplosion hydrodynamics of supernovae in red supergiants. Astrophys. J. **425**, 814–828 (1994). https://doi.org/10.1086/174026

J.J. Hester, The Crab Nebula: an astrophysical chimera. Annu. Rev. Astron. Astrophys. **46**(1), 127–155 (2008)

D.J. Hillier, K. Davidson, K. Ishibashi, T. Gull, On the nature of the central source in η Carinae. Astrophys. J. **553**, 837–860 (2001). https://doi.org/10.1086/320948

D.C. Hines, G.H. Rieke, K.D. Gordon, J. Rho, K.A. Misselt, C.E. Woodward, M.W. Werner, O. Krause, W.B. Latter, C.W. Engelbracht, E. Egami, D.M. Kelly, J. Muzerolle, J.A. Stansberry, K.Y.L. Su, J.E. Morrison, E.T. Young, A. Noriega-Crespo, D.L. Padgett, R.D. Gehrz, E. Polomski, J.W. Beeman, E.E. Haller, Imaging of the supernova remnant Cassiopeia A with the multiband imaging photometer for Spitzer (MIPS). Astrophys. J. Suppl. Ser. **154**(1), 290–295 (2004)

D.C. Hines, O. Krause, G.H. Rieke, X. Fan, M. Blaylock, G. Neugebauer, Spitzer observations of high-redshift QSOs. Astrophys. J. Lett. **641**, L85–L88 (2006). astro-ph/0604347. https://doi.org/10.1086/504109

A.M. Hofmeister, E. Keppel, A.K. Speck, Absorption and reflection infrared spectra of MgO and other diatomic compounds. Mon. Not. R. Astron. Soc. **345**, 16–38 (2003). https://doi.org/10.1046/j.1365-8711.2003.06899.x

P. Hoppe, Measurements of presolar grains, in *Nuclei in the Cosmos* (2010)

P. Hoppe, U. Ott, Mainstream silicon carbide grains from meteorites, in *American Institute of Physics Conference Series*, vol. 402 ed. by T.J. Bernatowicz, E. Zinner. (1997), pp. 27–58. https://doi.org/10.1063/1.53314

F. Hoyle, N.C. Wickramasinghe, Dust in supernova explosions. Nature **226**(5240), 62–63 (1970)

J.P. Hughes, C.E. Rakowski, D.N. Burrows, P.O. Slane, Nucleosynthesis and mixing in Cassiopeia A. Astrophys. J. Lett. **528**, L109–L113 (2000). astro-ph/9910474. https://doi.org/10.1086/312438

D.J. Hunter, S. Valenti, R. Kotak, W.P.S. Meikle, S. Taubenberger, A. Pastorello, S. Benetti, V. Stanishev, S.J. Smartt, C. Trundle, A.A. Arkharov, F. Bufano, E. Cappellaro, E. Di Carlo, M. Dolci, N. Elias-Rosa, S. Frandsen, J.U. Fynbo, U. Hopp, V.M. Larionov, P. Laursen, P. Mazzali, H. Navasardyan, C. Ries, A. Riffeser, L. Rizzi, D.Y. Tsvetkov, M. Turatto, S. Wilke, Extensive optical and near-infrared observations of the nearby, narrow-lined type Ic SN 2007gr: days 5 to 415. Astron. Astrophys. **508**(1), 371–389 (2009)

K.M. Hynes, F. Gyngard, in *Lunar and Planetary Science Conference*, vol. 40 (2009), p. 1198. The presolar grain database: http://presolar.wustl.edu/~pgd,

R. Indebetouw, M. Matsuura, E. Dwek, G. Zanardo, M.J. Barlow, M. Baes, P. Bouchet, D.N. Burrows, R.A. Chevalier, G.C. Clayton, C. Fransson, R.P. Kirshner, M. Lakicevic, K.S. Long, P. Lundqvist, I. Martí-Vidal, J. Marcaide, R.A. McCray, M. Meixner, C.Y. Ng, S. Park, G. Sonneborn, L. Staveley-Smith, C. Vlahakis, Dust production and particle acceleration in supernova 1987a revealed with ALMA. Astrophys. J. **782**(1), L2 (2014)

C. Inserra, M. Turatto, A. Pastorello, S. Benetti, E. Cappellaro, M.L. Pumo, L. Zampieri, I. Agnoletto, F. Bufano, M.T. Botticella, M. Della Valle, N. Elias Rosa, T. Iijima, S. Spiro, S. Valenti, The type IIP SN 2007od in UGC 12846: from a bright maximum to dust formation in the nebular phase. Mon. Not. R. Astron. Soc. **417**, 261–279 (2011). 1102.5468. https://doi.org/10.1111/j.1365-2966.2011.19128.x

M.Z. Jacobson, *Fundamentals of Atmospheric Modeling*, 2nd edn. (Cambridge University Press, Cambridge, 2005)

A. Jerkstrand, C. Fransson, C. Kozma, The ^{44}Ti-powered spectrum of SN 1987A. Astron. Astrophys. **530**, A45 (2011). 1103.3653. https://doi.org/10.1051/0004-6361/201015937

C.C. Joggerst, A. Almgren, S.E. Woosley, Three-dimensional simulations of Rayleigh-Taylor mixing in core-collapse supernovae with Castro. Astrophys. J. **723**, 353–363 (2010). 1009.3336. https://doi.org/10.1088/0004-637X/723/1/353

A.P. Jones, Dust destruction processes, in *Astrophysics of Dust*, ed. by A.N. Witt, G.C. Clayton, B.T. Draine. Astronomical Society of the Pacific Conference Series, vol. 309 (2004), p. 347

A.P. Jones, A.G.G.M. Tielens, D.J. Hollenbach, C.F. McKee, Grain destruction in shocks in the interstellar medium. Astrophys. J. **433**, 797 (1994)

J. Kamenetzky, R. McCray, R. Indebetouw, M.J. Barlow, M. Matsuura, M. Baes, J.A.D.L. Blommaert, A. Bolatto, L. Decin, L. Dunne, C. Fransson, J. Glenn, H.L. Gomez, M.A.T. Groenewegen, R. Hopwood, R.P. Kirshner, M. Lakicevic, J. Marcaide, I. Marti-Vidal, M. Meixner, P. Royer, A. Soderberg, G. Sonneborn, L. Staveley-Smith, B.M. Swinyard, G. Van de Steene, P.A.M. van Hoof, J.T. van Loon, J. Yates, G. Zanardo, Carbon monoxide in the cold debris of supernova 1987A. Astrophys. J. Lett. **773**, L34 (2013). 1307.6561. https://doi.org/10.1088/2041-8205/773/2/L34

E. Kankare, M. Fraser, S. Ryder, C. Romero-Cañizales, S. Mattila, R. Kotak, P. Laursen, L.A.G. Monard, M. Salvo, P. Väisänen, SN 2005at—a neglected type Ic supernova at 10 Mpc. Astron. Astrophys. **572**, A75 (2014). 1409.1758. https://doi.org/10.1051/0004-6361/201424563

F. Kemper, W.J. Vriend, A.G.G.M. Tielens, Erratum: "The absence of crystalline silicates in the diffuse interstellar medium" (ApJ, 609, 826 [2004]). Astrophys. J. **633**, 534 (2005). https://doi.org/10.1086/447764

K. Kifonidis, T. Plewa, H.T. Janka, E. Müller, Non-spherical core collapse supernovae. I. Neutrino-driven convection, Rayleigh-Taylor instabilities, and the formation and propagation of metal clumps. Astron. Astrophys. **408**, 621–649 (2003). astro-ph/0302239. https://doi.org/10.1051/0004-6361:20030863

K. Kifonidis, T. Plewa, L. Scheck, H.T. Janka, E. Müller, Non-spherical core collapse supernovae. II. The late-time evolution of globally anisotropic neutrino-driven explosions and their implications for SN 1987 A. Astron. Astrophys. **453**, 661–678 (2006). astro-ph/0511369. https://doi.org/10.1051/0004-6361:20054512

C. Koike, H. Hasegawa, N. Asada, T. Hattori, The extinction coefficients in mid- and far-infrared of silicate and iron-oxide minerals of interest for astronomical observations. Astrophys. Space Sci. **79**, 77–85 (1981). https://doi.org/10.1007/BF00655906

C. Koike, C. Kaito, T. Yamamoto, H. Shibai, S. Kimura, H. Suto, Extinction spectra of corundum in the wavelengths from UV to FIR. Icarus **114**, 203–214 (1995). https://doi.org/10.1006/icar.1995.1055

C. Koike, H. Chihara, A. Tsuchiyama, H. Suto, H. Sogawa, H. Okuda, Compositional dependence of infrared absorption spectra of crystalline silicate. II. Natural and synthetic olivines. Astron. Astrophys. **399**, 1101–1107 (2003). https://doi.org/10.1051/0004-6361:20021831

R. Kotak, P. Meikle, S.D. van Dyk, P.A. Höflich, S. Mattila, Early-time Spitzer observations of the type II Plateau supernova SN 2004dj. Astrophys. J. Lett. **628**, L123–L126 (2005). astro-ph/0506407. https://doi.org/10.1086/432719

R. Kotak, W.P.S. Meikle, M. Pozzo, S.D. van Dyk, D. Farrah, R. Fesen, A.V. Filippenko, R.J. Foley, C. Fransson, P.A. Höflich, P. Lundqvist, S. Mattila, J. Sollerman, J.C. Wheeler, Spitzer measurements of atomic and molecular abundances in the type IIP SN 2005af. Astrophys. J. **651**(2), L117–L120 (2006)

R. Kotak, W.P.S. Meikle, D. Farrah, C.L. Gerardy, R.J. Foley, S.D. Van Dyk, C. Fransson, P. Lundqvist, J. Sollerman, R. Fesen, A.V. Filippenko, S. Mattila, J.M. Silverman, A.C. Andersen, P.A. Höflich, M. Pozzo, J.C. Wheeler, Dust and the type II-Plateau supernova 2004et. Astrophys. J. **704**, 306–323 (2009). 0904.3737. https://doi.org/10.1088/0004-637X/704/1/306

T. Kozasa, H. Hasegawa, K. Nomoto, Formation of dust grains in the ejecta of SN 1987A. Astrophys. J. **344**, 325–331 (1989). https://doi.org/10.1086/167801

T. Kozasa, T. Nozawa, N. Tominaga, H. Umeda, K. Maeda, K. Nomoto, Dust in supernovae: formation and evolution, in *Cosmic Dust—Near and Far*, ed. by T. Henning, E. Grün, J. Steinacker. Astronomical Society of the Pacific Conference Series, vol. 414 (2009), p. 43. 0903.0217

O. Krause, S.M. Birkmann, T. Usuda, T. Hattori, M. Goto, G.H. Rieke, K.A. Misselt, The Cassiopeia A supernova was of type IIb. Science **320**(5880), 1195–1197 (2008)

W. Kunkel, B. Madore, I. Shelton, O. Duhalde, F.M. Bateson, A. Jones, B. Moreno, S. Walker, G. Garradd, B. Warner, J. Menzies, Supernova 1987A in the large magellanic cloud. IAU Circ **4316**, 1 (1987)

P.O. Lagage, A. Claret, J. Ballet, F. Boulanger, C.J. Cesarsky, D. Cesarsky, C. Fransson, A. Pollock, Dust formation in the Cassiopeia A supernova. Astron. Astrophys. **315**, L273–L276 (1996)

Lakicevic et al., Astrophys. J. **799**, 50 (2015)

J. Larsson, C. Fransson, J. Spyromilio, B. Leibundgut, P. Challis, R.A. Chevalier, K. France, A. Jerkstrand, R.P. Kirshner, P. Lundqvist, M. Matsuura, R.A. McCray, N. Smith, J. Sollerman, P. Garnavich, K. Heng, S. Lawrence, S. Mattila, K. Migotto, G. Sonneborn, F. Taddia, J.C. Wheeler, Three-dimensional distribution of ejecta in supernova 1987A at 10,000 days. Astrophys. J. **833**(2), 147 (2016)

J. Leitner, P. Hoppe, J. Zipfel, Distribution and abundance of presolar silicate and oxide stardust in CR chondrites, in *Lunar and Planetary Science Conference*, vol. 46 (2015), p. 1874

S. Lepp, A. Dalgarno, R. McCray, Molecules in the ejecta of SN 1987A. Astrophys. J. **358**, 262–265 (1990). https://doi.org/10.1086/168981

A. Li, The warm, cold and very cold dusty universe. ArXiv Astrophysics e-prints astro-ph/0503571 (2005)

A. Li, B.T. Draine, On ultrasmall silicate grains in the diffuse interstellar medium. Astrophys. J. Lett. **550**, L213–L217 (2001). astro-ph/0012147. https://doi.org/10.1086/319640

M.P. Li, G. Zhao, A. Li, On the crystallinity of silicate dust in the interstellar medium. Mon. Not. R. Astron. Soc. **382**, L26–L29 (2007). 0808.4129. https://doi.org/10.1111/j.1745-3933.2007.00382.x

M. Limongi, O. Straniero, A. Chieffi, Massive stars in the range 13-25 M_{solar}: evolution and nucleosynthesis. II. The solar metallicity models. Astrophys. J. Suppl. Ser. **129**, 625–664 (2000). astro-ph/0003401. https://doi.org/10.1086/313424

W. Liu, A. Dalgarno, Silicon monoxide in SN 1987A. Astrophys. J. **428**, 769–776 (1994). https://doi.org/10.1086/174285

W. Liu, A. Dalgarno, The oxygen temperature of SN 1987A. Astrophys. J. **454**, 472 (1995). https://doi.org/10.1086/176498

W. Liu, A. Dalgarno, S. Lepp, Carbon monoxide in SN 1987A. Astrophys. J. **396**, 679–685 (1992). https://doi.org/10.1086/171749

E.D. Loh, J.A. Baldwin, Z.K. Curtis, G.J. Ferland, C.R. O'Dell, A.C. Fabian, P. Salomé, A survey of molecular hydrogen in the Crab Nebula. Astrophys. J. Suppl. Ser. **194**(2), 30 (2011)

E.D. Loh, J.A. Baldwin, G.J. Ferland, Z.K. Curtis, C.T. Richardson, A.C. Fabian, P. Salomé, H2 temperatures in the Crab Nebula. Mon. Not. R. Astron. Soc. (2012)

L.B. Lucy, I.J. Danziger, C. Gouiffes, P. Bouchet, Dust condensation in the ejecta of SN 1987 A, in *IAU Colloq. 120: Structure and Dynamics of the Interstellar Medium*, ed. by G. Tenorio-Tagle, M. Moles, J. Melnick. Lecture Notes in Physics, vol. 350 (Springer, Berlin, 1989), p. 164. https://doi.org/10.1007/BFb0114861

R. Maiolino, R. Schneider, E. Oliva, S. Bianchi, A. Ferrara, F. Mannucci, M. Pedani, M. Roca Sogorb, A supernova origin for dust in a high-redshift quasar. Nature **431**, 533–535 (2004). astro-ph/0409577. https://doi.org/10.1038/nature02930

Y.N. Makurin, A.A. Sofronov, A.I. Gusev, A.L. Ivanovsky, Electronic structure and chemical stabilization of C_{28} fullerene. Chem. Phys. **270**, 293–308 (2001). https://doi.org/10.1016/S0301-0104(01)00342-1

S. Marassi, R. Schneider, M. Limongi, A. Chieffi, M. Bocchio, S. Bianchi, The metal and dust yields of the first massive stars. Mon. Not. R. Astron. Soc. **454**, 4250–4266 (2015). 1509.08923. https://doi.org/10.1093/mnras/stv2267

P.L. Marsden, F.C. Gillett, R.E. Jennings, J.P. Emerson, T. de Jong, F.M. Olnon, Far-infrared observations of the Crab Nebula. Astrophys. J. **278**, L29 (1984)

P.G. Martin, *Cosmic Dust. Its Impact on Astronomy*. Oxford Studies in Physics (Clarendon, Oxford, 1978)

T. Matheson, A.V. Filippenko, R. Chornock, D.C. Leonard, W. Li, Helium emission lines in the type IC supernova 1999CQ. Astron. J. **119**, 2303–2310 (2000a). astro-ph/0001270. https://doi.org/10.1086/301352

T. Matheson, A.V. Filippenko, L.C. Ho, A.J. Barth, D.C. Leonard, Detailed analysis of early to late-time spectra of supernova 1993J. Astron. J. **120**, 1499–1515 (2000b). astro-ph/0006264. https://doi.org/10.1086/301519

J.S. Mathis, W. Rumpl, K.H. Nordsieck, The size distribution of interstellar grains. Astrophys. J. **217**, 425–433 (1977). https://doi.org/10.1086/155591

M. Matsuura, Dust and molecular formation in supernovae, in *Handbook of Supernovae*, ed. by W.A. Athem, M. Paul (Springer, Berlin, 2017)

M. Matsuura, E. Dwek, M. Meixner, M. Otsuka, B. Babler, M.J. Barlow, J. Roman-Duval, C. Engelbracht, K. Sandstrom, M. Lakićević, J.T. van Loon, G. Sonneborn, G.C. Clayton, K.S. Long, P. Lundqvist, T. Nozawa, K.D. Gordon, S. Hony, P. Panuzzo, K. Okumura, K.A. Misselt, E. Montiel, M. Sauvage, Herschel detects a massive dust reservoir in supernova 1987A. Science **333**, 1258 (2011). 1107.1477. https://doi.org/10.1126/science.1205983

M. Matsuura, E. Dwek, M.J. Barlow, B. Babler, M. Baes, M. Meixner, J. Cernicharo, G.C. Clayton, L. Dunne, C. Fransson, J. Fritz, W. Gear, H.L. Gomez née Morgan, M.A.T. Groenewegen, R. Indebetouw, R.J. Ivison, A. Jerkstrand, V. Lebouteiller, T.L. Lim, P. Lundqvist, C.P. Pearson, J. Roman-Duval, P. Royer, L. Staveley-Smith, B.M. Swinyard, P.A.M. van Hoof, J.T. van Loon, J. Verstappen, R. Wesson, G. Zanardo, J.A.D.L. Blommaert, L. Decin, W.T. Reach, G. Sonneborn, G.C. Van de Steene, J.A. Yates, A stubbornly large mass of cold dust in the ejecta of supernova 1987A. Astrophys. J. **800**(1), 50 (2015)

M. Matsuura, R. Indebetouw, S. Woosley, V. Bujarrabal, F.J. Abellan, R. McCray, J. Kamenetzky, C. Fransson, M.J. Barlow, H.L. Gomez, P. Cigan, I. De Looze, J. Spyromilio, L. Staveley-Smith, G. Zanardo, P. Roche, J. Larsson, S. Viti, J.T. van Loon, J.C. Wheeler, M. Baes, R. Chevalier, P. Lundqvist, J.M. Marcaide, E. Dwek, M. Meixner, C.Y. Ng, G. Sonneborn, J. Yates, ALMA spectral survey of Supernova 1987A—molecular inventory, chemistry, dynamics and explosive nucleosynthesis. ArXiv e-prints 1704.02324 (2017)

S.M. Matz, G.H. Share, M.D. Leising, E.L. Chupp, W.T. Vestrand, Gamma-ray line emission from SN1987A. Nature **331**, 416–418 (1988). https://doi.org/10.1038/331416a0

J.C. Mauerhan, N. Smith, J.M. Silverman, A.V. Filippenko, A.N. Morgan, S.B. Cenko, M. Ganeshalingam, K.I. Clubb, J.S. Bloom, T. Matheson, P. Milne, SN 2011ht: confirming a class of interacting supernovae with plateau light curves (type IIn-P). Mon. Not. R. Astron. Soc. **431**, 2599–2611 (2013). 1209.0821. https://doi.org/10.1093/mnras/stt360

R. McCray, Supernova 1987A revisited. Annu. Rev. Astron. Astrophys. **31**, 175–216 (1993). https://doi.org/10.1146/annurev.aa.31.090193.001135

W.P.S. Meikle, J. Spyromilio, G.F. Varani, D.A. Allen, Spectroscopy of supernova 1987A at 1–5 microns. I—The first year. Mon. Not. R. Astron. Soc. **238**, 193–223 (1989)

W.P.S. Meikle, S. Mattila, A. Pastorello, C.L. Gerardy, R. Kotak, J. Sollerman, S.D. van Dyk, D. Farrah, A.V. Filippenko, P. Höflich, P. Lundqvist, M. Pozzo, J.C. Wheeler, A Spitzer space TelescopeStudy of SN 2003gd: still no direct evidence that core-collapse supernovae are major dust factories. Astrophys. J. **665**(1), 608–617 (2007)

W.P.S. Meikle, R. Kotak, D. Farrah, S. Mattila, S.D. van Dyk, A.C. Andersen, R. Fesen, A.V. Filippenko, R.J. Foley, C. Fransson, C.L. Gerardy, P.A. Höflich, P. Lundqvist, M. Pozzo, J. Sollerman, J.C. Wheeler, Dust and the type II-Plateau supernova 2004dj. Astrophys. J. **732**(2), 109 (2011)

M. Meixner, K.D. Gordon, R. Indebetouw, J.L. Hora, B.A. Whitney, R.D. Blum, W. Reach, J.P. Bernard, M. Meade, B. Babler, C.W. Engelbracht, B.Q. For, K. Misselt, U. Vijh, C. Leitherer, M. Cohen, E.B. Churchwell, F. Boulanger, J.A. Frogel, Y. Fukui, J.S. Gallagher III, V. Gorjian, J. Harris, D. Kelly, A. Kawamura, S. Kim, W.B. Latter, S.C. Madden, F. Kemper, A. Mizuno, N. Mizuno, J. Mould, A. Nota, M.S. Oey, K. Olsen, T. Onishi, R. Paladini, N. Panagia, P. Perez-Gonzalez, H. Shibai, S. Sato, L.J. Smith, L. Staveley-Smith, A.G.G.M. Tielens, T. Ueta, S.V. Dyk, K.M. Volk, M.W. Werner, D. Zaritsky, Spitzer survey of the large magellanic cloud: surveying the agents of a galaxy's evolution (SAGE). I. Overview and initial results. Astron. J. **132**, 2268 (2006)

S. Miller, J. Tennyson, S. Lepp, A. Dalgarno, Identification of features due to H3 + in the infrared spectrum of supernova 1987A. Nature **355**(6359), 420–422 (1992)

R. Minkowski, Spectra of supernovae. Publ. Astron. Soc. Pac. **53**, 224 (1941). https://doi.org/10.1086/125315

F.J. Molster, L.B.F.M. Waters, F. Kemper, The mineralogy of interstellar and circumstellar dust in galaxies, in *Lecture Notes in Physics*, vol. 815, ed. by T. Henning (Springer, Berlin, 2010), pp. 143–201

H. Mutschke, A.C. Andersen, D. Clément, T. Henning, G. Peiter, Infrared properties of SiC particles. Astron. Astrophys. **345**, 187–202 (1999). astro-ph/9903031

L.R. Nittler, Presolar stardust in meteorites: recent advances and scientific frontiers. Earth Planet. Sci. Lett. **209**, 259–273 (2003). https://doi.org/10.1016/S0012-821X(02)01153-6

L.R. Nittler, C.M.O. Alexander, J. Wang, X. Gao, Meteoritic oxide grain from supernova found. Nature **393**, 222 (1998). https://doi.org/10.1038/30377

K.I. Nomoto, K. Iwamoto, T. Suzuki, The evolution and explosion of massive binary stars and type Ib-Ic-IIb-IIL supernovae. Phys. Rep. **256**, 173–191 (1995). https://doi.org/10.1016/0370-1573(94)00107-E

T. Nozawa, T. Kozasa, Formulation of non-steady-state dust formation process in astrophysical environments. Astrophys. J. **776**, 24 (2013). 1308.1873. https://doi.org/10.1088/0004-637X/776/1/24

T. Nozawa, T. Kozasa, H. Umeda, K. Maeda, K. Nomoto, Dust in the early universe: dust formation in the ejecta of population III supernovae. Astrophys. J. **598**, 785–803 (2003). astro-ph/0307108. https://doi.org/10.1086/379011

T. Nozawa, T. Kozasa, N. Tominaga, I. Sakon, M. Tanaka, T. Suzuki, K. Nomoto, K. Maeda, H. Umeda, M. Limongi, T. Onaka, Early formation of dust in the ejecta of type Ib SN 2006jc and temperature and mass of the dust. Astrophys. J. **684**(2), 1343–1350 (2008)

T. Nozawa, T. Kozasa, N. Tominaga, K. Maeda, H. Umeda, K. Nomoto, O. Krause, Formation and evolution of dust in type IIb supernovae with application to the Cassiopeia A supernova remnant. Astrophys. J. **713**, 356–373 (2010). 0909.4145. https://doi.org/10.1088/0004-637X/713/1/356

T. Nozawa, K. Maeda, T. Kozasa, M. Tanaka, K. Nomoto, H. Umeda, Formation of dust in the ejecta of type Ia supernovae. Astrophys. J. **736**, 45 (2011). 1105.0973. https://doi.org/10.1088/0004-637X/736/1/45

U. Ott, Presolar grains in meteorites: an overview and some implications. Planet. Space Sci. **49**, 763–767 (2001). https://doi.org/10.1016/S0032-0633(01)00025-3

P.J. Owen, M.J. Barlow, The dust and gas content of the Crab Nebula. Astrophys. J. **801**(2), 141 (2015)

Y.C. Pei, S.M. Fall, J. Bechtold, Confirmation of dust in damped Lyman-alpha systems. Astrophys. J. **378**, 6–16 (1991). https://doi.org/10.1086/170401

M. Pettini, L.J. Smith, R.W. Hunstead, D.L. King, Metal enrichment, dust, and star formation in galaxies at high redshifts. 3: Zn and CR abundances for 17 damped Lyman-alpha systems. Astrophys. J. **426**, 79–96 (1994). https://doi.org/10.1086/174041

D. Pinheiro Gonçalves, A. Noriega-Crespo, R. Paladini, P.G. Martin, S.J. Carey, The MIPSGAL view of supernova remnants in the galactic plane. Astron. J. **142**(2), 47 (2011)

M. Pozzo, W.P.S. Meikle, J.T. Rayner, R.D. Joseph, A.V. Filippenko, R.J. Foley, W. Li, S. Mattila, J. Sollerman, Optical and infrared observations of the TypeIIP SN2002hh from days 3 to 397. Mon. Not. R. Astron. Soc. **368**(3), 1169–1195 (2006)

D.M. Rank, P.A. Pinto, S.E. Woosley, J.D. Bregman, F.C. Witteborn, Nickel, argon and cobalt in the infrared spectrum of SN1987A—the core becomes visible. Nature **331**(6156), 505–506 (1988)

T. Rauscher, A. Heger, R.D. Hoffman, S.E. Woosley, Nucleosynthesis in massive stars with improved nuclear and stellar physics. Astrophys. J. **576**, 323–348 (2002). astro-ph/0112478. https://doi.org/10.1086/341728

W.T. Reach, J. Rho, A. Tappe, T.G. Pannuti, C.L. Brogan, E.B. Churchwell, M.R. Meade, B. Babler, R. Indebetouw, B.A. Whitney, A Spitzer space telescope infrared survey of supernova remnants in the inner galaxy. Astron. J. **131**(3), 1479–1500 (2006)

A.C. Reber, S. Paranthaman, P.A. Clayborne et al., From SiO molecules to silicates in circumstellar space: atomic structures, growth patterns, and optical signatures of Si_nO_m clusters. ACS Nano **2**, 1729 (2008). https://doi.org/10.1021/nn7003958

J. Rho, T. Kozasa, W.T. Reach, J.D. Smith, L. Rudnick, T.A. DeLaney, J.A. Ennis, H.L. Gomez, A. Tappe, Freshly formed dust in the Cassiopeia A supernova remnant as revealed by the Spitzer space telescope. Astrophys. J. **673**, 271 (2008)

J. Rho, T.H. Jarrett, W.T. Reach, H. Gomez, M. Andersen, Carbon monoxide in the Cassiopeia A supernova remnant. Astrophys. J. Lett. **693**, L39–L43 (2009). 0901.2308. https://doi.org/10.1088/0004-637X/693/1/L39

J. Rho, T. Onaka, J. Cami, W.T. Reach, Spectroscopic detection of carbon monoxide in the young supernova remnant Cassiopeia A. Astrophys. J. **747**(1), L6 (2012)

P.F. Roche, D.K. Aitken, C.H. Smith, S.D. James, Old cold dust heated by supernova 1987A. Nature **337**(6207), 533–535 (1989)

P.F. Roche, D.K. Aitken, C.H. Smith, Silicon monoxide in supernova 1987A. Mon. Not. R. Astron. Soc. **252**, 39P–42P (1991)

J.M. Saken, R.A. Fesen, J.M. Shull, An IRAS survey of galactic supernova remnants. Astrophys. J. Suppl. Ser. **81**, 715 (1992)

K.M. Sandstrom, A.D. Bolatto, S. Stanimirović, J.T. van Loon, J.D.T. Smith, Measuring dust production in the small magellanic cloud core-collapse supernova remnant 1E 0102.2-7219. Astrophys. J. **696**, 2138 (2009)
A. Sarangi, I. Cherchneff, The chemically controlled synthesis of dust in type II-P supernovae. Astrophys. J. **776**, 107 (2013). 1309.5887. https://doi.org/10.1088/0004-637X/776/2/107
A. Sarangi, I. Cherchneff, Condensation of dust in the ejecta of type II-P supernovae. Astron. Astrophys. **575**, A95 (2015). 1412.5522. https://doi.org/10.1051/0004-6361/201424969
P. Schilke, D.A. Neufeld, H.S.P. Müller, C. Comito, E.A. Bergin, D.C. Lis, M. Gerin, J.H. Black, M.G. Wolfire, N. Indriolo, J.C. Pearson, K.M. Menten, B. Winkel, Á. Sánchez Monge, T. Möller, B. Godard, E. Falgarone, Ubiquitous argonium (ArH +) in the diffuse interstellar medium: a molecular tracer of almost purely atomic gas. Astron. Astrophys. **566**, A29 (2014)
E.M. Schlegel, A new subclass of type II supernovae? Mon. Not. R. Astron. Soc. **244**, 269–271 (1990)
R. Schneider, A. Ferrara, R. Salvaterra, Dust formation in very massive primordial supernovae. Mon. Not. R. Astron. Soc. **351**, 1379–1386 (2004). astro-ph/0307087. https://doi.org/10.1111/j.1365-2966.2004.07876.x
J.Y. Seok, B.C.C. Koo, T. Onaka, Y. Ita, H.G. Lee, J.J. Lee, D.S. Moon, I. Sakon, H. Kaneda, H.M. Lee, M.G. Lee, S.E. Kim, Supernova remnants in the AKARI IRC survey of the large magellanic cloud. Publ. Astron. Soc. Jpn. **60**, 453 (2008)
J.Y. Seok, B.C.C. Koo, T. Onaka, A survey of infrared supernova remnants in the large magellanic cloud. Astrophys. J. **779**(2), 134 (2013)
J.Y. Seok, B.C.C. Koo, H. Hirashita, Dust cooling in supernova remnants in the large magellanic cloud. Astrophys. J. **807**(1), 100 (2015)
T. Shigeyama, K. Nomoto, Theoretical light curve of SN 1987A and mixing of hydrogen and nickel in the ejecta. Astrophys. J. **360**, 242–256 (1990). https://doi.org/10.1086/169114
B. Sibthorpe, P.A.R. Ade, J.J. Bock, E.L. Chapin, M.J. Devlin, S. Dicker, M. Griffin, J.O. Gundersen, M. Halpern, P.C. Hargrave, D.H. Hughes, W.S. Jeong, H. Kaneda, J. Klein, B.C. Koo, H.G. Lee, G. Marsden, P.G. Martin, P. Mauskopf, D.S. Moon, C.B. Netterfield, L. Olmi, E. Pascale, G. Patanchon, M. Rex, A. Roy, D. Scott, C. Semisch, M.D.P. Truch, C. Tucker, G.S. Tucker, M.P. Viero, D.V. Wiebe, AKARI and BLAST observations of the Cassiopeia A supernova remnant and surrounding interstellar medium. Astrophys. J. **719**, 1553–1564 (2010). 0910.1094. https://doi.org/10.1088/0004-637X/719/2/1553
A. Sluder, M. Milosavljevic, M.H. Montgomery, Molecular nucleation theory of dust formation in core-collapse supernovae applied to SN 1987A. ArXiv e-prints. 1612.09013 (2016)
N. Smith, R. Chornock, W. Li, M. Ganeshalingam, J.M. Silverman, R.J. Foley, A.V. Filippenko, A.J. Barth, SN 2006tf: precursor eruptions and the optically thick regime of extremely luminous type IIn supernovae. Astrophys. J. **686**, 467–484 (2008). 0804.0042. https://doi.org/10.1086/591021
N. Smith, J.M. Silverman, A.V. Filippenko, M.C. Cooper, T. Matheson, F. Bian, B.J. Weiner, J.M. Comerford, Systematic blueshift of line profiles in the type IIn supernova 2010jl: evidence for post-shock dust formation? Astron. J. **143**, 17–20 (2012). 1108.2869. https://doi.org/10.1088/0004-6256/143/1/17
J. Spyromilio, B. Leibundgut, Carbon monoxide in supernova 1995ad. Mon. Not. R. Astron. Soc. **283**, L89–L93 (1996)
J. Spyromilio, W.P.S. Meikle, R.C.M. Learner, D.A. Allen, Carbon monoxide in supernova 1987A. Nature **334**(6180), 327–329 (1988)
J. Spyromilio, B. Leibundgut, R. Gilmozzi, Carbon monoxide in type II supernovae. Astron. Astrophys. **376**(1), 188–193 (2001)
S. Stanimirović, A.D. Bolatto, K.M. Sandstrom, A.K. Leroy, J.D. Simon, B.M. Gaensler, R.Y. Shah, J.M. Jackson, Spitzer space telescope detection of the young supernova remnant 1E 0102.2-7219. Astrophys. J. **632**(2), L103–L106 (2005)
B.E.K. Sugerman, B. Ercolano, M.J. Barlow, A.G.G.M. Tielens, G.C. Clayton, A.A. Zijlstra, M. Meixner, A. Speck, T.M. Gledhill, N. Panagia, M. Cohen, K.D. Gordon, M. Meyer, J. Fabbri, J.E. Bowey, D.L. Welch, M.W. Regan, R.C. Kennicutt, Massive-star supernovae as major dust factories. Science **313**, 196–200 (2006). astro-ph/0606132. https://doi.org/10.1126/science.1128131
N.B. Suntzeff, P. Bouchet, The bolometric light curve of SN 1987A. I—results from ESO and CTIO U to Q0 photometry. Astron. J. **99**, 650–663 (1990)
H. Suto, C. Koike, H. Sogawa, A. Tsuchiyama, H. Chihara, K. Mizutani, Infrared spectra of fayalite crystal. Astron. Astrophys. **389**, 568–571 (2002). https://doi.org/10.1051/0004-6361:20020501
R.A. Syunyaev, A. Kaniovskii, V. Efremov, M. Gilfanov, E. Dhurazov, S.A. Grebenev, A.V. Kuznetsov, A.S. Melioranskii, N.S. Yamburenko, S. Yunin, D. Stepanov, I. Chulkov, N. Pappe, M.N. Boyarskii, E.A. Gavrilova, V.M. Loznikov, A. Prudkoglyad, V.G. Rodin, C. Reppin, W. Pietsch, J. Engelhauser, J. Trumper, W. Voges, E. Kendziorra, M. Bezler, R. Staubert, A.C. Brinkman, J. Heise, W.A. Mels, R. Jager, G.K. Skinner, O. Al-Emam, T.G. Patterson, A.P. Willmore, Detection of hard X-rays from supernova 1987A—preliminary Mir-Kvant results. Sov. Astron. Lett. **13**, 431 (1987)

T. Szalai, J. Vinkó, Twelve type II-P supernovae seen with the eyes of Spitzer. Astron. Astrophys. **549**, A79 (2013). 1211.0854. https://doi.org/10.1051/0004-6361/201220015

T. Szalai, J. Vinkó, Z. Balog, A. Gáspár, M. Block, L.L. Kiss, Dust formation in the ejecta of the type II-P supernova 2004dj. Astron. Astrophys. **527**, A61 (2011). 1012.2035. https://doi.org/10.1051/0004-6361/201015624

T. Szalai, J. Vinkó, A.P. Nagy, J.M. Silverman, J.C. Wheeler, G. Dhungana, G.H. Marion, R. Kehoe, O.D. Fox, K. Sárneczky, G. Marschalkó, B.I. Bíró, T. Borkovits, T. Hegedüs, R. Szakáts, F.V. Ferrante, E. Bányai, G. Hodosán, J. Kelemen, A. Pál, The continuing story of SN IIb 2013df: new optical and IR observations and analysis. Mon. Not. R. Astron. Soc. **460**(2), 1500–1518 (2016)

T. Temim, E. Dwek, The importance of physical models for deriving dust masses and grain size distributions in supernova ejecta. I. Radiatively heated dust in the Crab Nebula. Astrophys. J. **774**, 8 (2013). 1302.5452. https://doi.org/10.1088/0004-637X/774/1/8

T. Temim, R.D. Gehrz, C.E. Woodward, T.L. Roellig, N. Smith, L. Rudnick, E.F. Polomski, K. Davidson, L. Yuen, T. Onaka, Spitzer space TelescopeInfrared imaging and spectroscopy of the Crab Nebula. Astron. J. **132**(4), 1610–1623 (2006)

T. Temim, E. Dwek, R.G. Arendt, K.J. Borkowski, S.P. Reynolds, P. Slane, J.D. Gelfand, J.C. Raymond, A massive shell of supernova-formed dust in SNR G54.1+0.3. Astrophys. J. **836**(1), 129 (2017)

F.K. Thielemann, Nucleosynthesis in supernovae, hypernovae/gamma-ray bursts and compact binary mergers, in *14th International Symposium on Nuclei in the Cosmos (NIC2016)*, ed. by S. Kubono, T. Kajino, S. Nishimura, T. Isobe, S. Nagataki, T. Shima, Y. Takeda (2017), p. 010605. https://doi.org/10.7566/JPSCP.14.010605

F.K. Thielemann, K. Nomoto, M.A. Hashimoto, Core-collapse supernovae and their ejecta. Astrophys. J. **460**, 408 (1996). https://doi.org/10.1086/176980

F.K. Thielemann, R. Hirschi, M. Liebendörfer, R. Diehl, Massive stars and their supernovae, in *Lecture Notes in Physics*, vol. 812, ed. by R. Diehl, D.H. Hartmann, N. Prantzos (Springer, Berlin, 2011), pp. 153–232. 1008.2144

P. Todini, A. Ferrara, Dust formation in primordial type II supernovae. Mon. Not. R. Astron. Soc. **325**, 726–736 (2001). astro-ph/0009176. https://doi.org/10.1046/j.1365-8711.2001.04486.x

C. Travaglio, R. Gallino, S. Amari, E. Zinner, S. Woosley, R.S. Lewis, Low-density graphite grains and mixing in type II supernovae. Astrophys. J. **510**, 325–354 (1999). https://doi.org/10.1086/306551

R.J. Trumpler, Absorption of light in the galactic system. Publ. Astron. Soc. Pac. **42**, 214 (1930). https://doi.org/10.1086/124039

M. Turatto, Classification of supernovae, in *Supernovae and Gamma-Ray Bursters*, ed. by K. Weiler. Lecture Notes in Physics, vol. 598 (Springer, Berlin, 2003), pp. 21–36. astro-ph/0301107

H. Umeda, K. Nomoto, Nucleosynthesis of zinc and iron peak elements in population III type II supernovae: comparison with abundances of very metal poor halo stars. Astrophys. J. **565**, 385–404 (2002). astro-ph/0103241. https://doi.org/10.1086/323946

V.P. Utrobin, N.N. Chugai, A.A. Andronova, Asymmetry of SN 1987A: fast Ni-56 clump. Astron. Astrophys. **295**, 129–135 (1995)

S.D. van Dyk, Late-time dust emission from the type IIn supernova 1995N. Astron. J. **145**(5), 118 (2013)

H.J. Walker, A brief history of infrared astronomy. Astron. Geophys. **41**(5), 10 (2000). https://doi.org/10.1046/j.1468-4004.2000.41510.x

S.H.J. Wallström, C. Biscaro, F. Salgado, J.H. Black, I. Cherchneff, S. Muller, O. Berné, J. Rho, A.G.G.M. Tielens, CO rotational line emission from a dense knot in Cassiopeia A. Astron. Astrophys. **558**, L2 (2013)

R. Wesson, M.J. Barlow, M. Matsuura, B. Ercolano, The timing and location of dust formation in the remnant of SN 1987A. Mon. Not. R. Astron. Soc. **446**(2), 2089–2101 (2014)

J.C. Wheeler, S. Benetti Supernovae, Cox, A.N., p. 451 (2000)

P.A. Whitelock, R.M. Catchpole, J.W. Menzies, M.W. Feast, S.E. Woosley, D. Allen, F. van Wyk, F. Marang, C.D. Laney, H. Winkler, K. Sekiguchi, L.A. Balona, B.S. Carter, J.H. Spencer Jones, J.D. Laing, T.L. Evans, A.P. Fairall, D.A.H. Buckley, I.S. Glass, M.V. Penston, L.N. da Costa, S.A. Bell, C. Hellier, M. Shara, A.F.J. Moffat, Spectroscopic and photometric observations of SN1987A. VI—days 617 to 792. R. Astron. Soc. **240**, 7 (1989)

D.C.B. Whittet (ed.), Dust in the galactic environment (2003)

B.J. Williams, K.J. Borkowski, S.P. Reynolds, W.P. Blair, P. Ghavamian, S.P. Hendrick, K.S. Long, S. Points, J.C. Raymond, R. Sankrit, R.C. Smith, P.F. Winkler, Dust destruction in fast shocks of core-collapse supernova remnants in the large magellanic cloud. Astrophys. J. **652**(1), L33–L36 (2006)

A. Wongwathanarat, E. Müller, H.T. Janka, Three-dimensional simulations of core-collapse supernovae: from shock revival to shock breakout. Astron. Astrophys. **577**, A48 (2015). 1409.5431. https://doi.org/10.1051/0004-6361/201425025

D.H. Wooden, D.M. Rank, J.D. Bregman, F.C. Witteborn, A.G.G.M. Tielens, M. Cohen, P.A. Pinto, T.S. Axelrod, Airborne spectrophotometry of SN 1987A from 1.7 to 12.6 microns—time history of the dust continuum and line emission. Astrophys. J. Suppl. Ser. **88**, 477–507 (1993). https://doi.org/10.1086/191830

S.E. Woosley, SN 1987A—after the peak. Astrophys. J. **330**, 218–253 (1988). https://doi.org/10.1086/166468

S.E. Woosley, T.A. Weaver, The evolution and explosion of massive stars. II. Explosive hydrodynamics and nucleosynthesis. Astrophys. J. Suppl. **101**, 181 (1995). https://doi.org/10.1086/192237

S.E. Woosley, D. Hartmann, P.A. Pinto, Hard emission at late times from SN 1987A. Astrophys. J. **346**, 395–404 (1989). https://doi.org/10.1086/168019

M.R. Zachariah, W. Tsang, Application of ab-initio molecular orbital and reaction rate theories to nucleation kinetics. Aerosol Sci. Technol. **19**, 499 (1993)

G. Zanardo, L. Staveley-Smith, R. Indebetouw, R.A. Chevalier, M. Matsuura, B.M. Gaensler, M.J. Barlow, C. Fransson, R.N. Manchester, M. Baes, J.R. Kamenetzky, M. Lakićević, P. Lundqvist, J.M. Marcaide, I. Marti-Vidal, M. Meixner, C.Y. Ng, S. Park, G. Sonneborn, J. Spyromilio, J.T. van Loon, Spectral and morphological analysis of the remnant of supernova 1987A with ALMA and ATCA. Astrophys. J. **796**(2), 82 (2014)

E. Zinner, F. Gyngard, Y. Lin, Silicon carbide grains of type X and supernova nucleosynthesis, in *Nuclei in the Cosmos* (2010), p. 148

Space Sci Rev (2018) 214:53
https://doi.org/10.1007/s11214-018-0484-7

Dust in Supernovae and Supernova Remnants II: Processing and Survival

E.R. Micelotta[1] · M. Matsuura[2] · A. Sarangi[3,4]

Received: 19 June 2017 / Accepted: 2 February 2018 / Published online: 5 March 2018

Abstract Observations have recently shown that supernovae are efficient dust factories, as predicted for a long time by theoretical models. The rapid evolution of their stellar progenitors combined with their efficiency in precipitating refractory elements from the gas phase into dust grains make supernovae the major potential suppliers of dust in the early Universe, where more conventional sources like Asymptotic Giant Branch (AGB) stars did not have time to evolve. However, dust yields inferred from observations of young supernovae or derived from models do not reflect the net amount of supernova-condensed dust able to be expelled from the remnants and reach the interstellar medium. The cavity where the dust is formed and initially resides is crossed by the high velocity reverse shock which is generated by the pressure of the circumstellar material shocked by the expanding supernova blast wave. Depending on grain composition and initial size, processing by the reverse shock may lead to substantial dust erosion and even complete destruction. The goal of this review is to present the state of the art about processing and survival of dust *inside* supernova remnants, in terms of theoretical modelling and comparison to observations.

Keywords Dust · Supernova remnants · Supernovae · Shock waves · Early Universe

Supernovae
Edited by Andrei Bykov, Roger Chevalier, John Raymond, Friedrich-Karl Thielemann, Maurizio Falanga and Rudolf von Steiger

✉ E.R. Micelotta
elisabetta.micelotta@helsinki.fi

M. Matsuura
MatsuuraM@cardiff.ac.uk

A. Sarangi
arkaprabha.sarangi@nasa.gov

1 Department of Physics, PO Box 64, 00014 University of Helsinki, Finland

2 School of Physics & Astronomy, Cardiff University, The Parade, Cardiff, CF24 3AA, UK

3 NASA Goddard Space Flight Center, Greenbelt, MD, 20771, USA

4 Physics Department, The Catholic University of America, Washington, DC, 20064, USA

1 Introduction

The solid state component of the Interstellar Medium (ISM) is identified with the generic term of "interstellar dust". It is composed by large molecules, mainly Polycyclic Aromatic Hydrocarbons (PAHs) and fullerenes C_{60} and C_{70}, nanoparticles (Very Small Grains—VSGs) and larger carbonaceous/silicate grains. Although very scarce (1–2% of the total mass of the ISM), dust is the key agent that drives and regulates the core physical and chemical processes responsible for star formation and galaxy evolution. Dust absorbs visual and ultraviolet radiation from stars, and re-emits it in the infrared (IR) and sub-millimetre (sub-mm) part of the electromagnetic spectrum, dominating the emission from all types of galaxies, from local to distant, from regular to starburst and active. Dust contributes to the thermal balance of the ISM via photo-electric heating while the thermal (far-infrared) emission from dust strips away a relevant fraction of the gravitational energy of a collapsing cloud, providing the necessary conditions for star formation to occur. Many fundamental chemical reactions, including the formation of molecular hydrogen H_2, occur on the surface of dust grains acting as catalyst. At the same time, dust grains shield these fragile molecules against strong stellar radiation fields allowing them to survive. Dust grains lock up the majority of the refractory elements (e.g. C, Si, Mg, Fe, Al, Ti, Ca) removing important ISM coolants from the gas phase. In addition, they represent the seeds for the formation of planetesimals in protoplanetary disks, which will eventually lead to planets.

Star formation shapes the formation history of galaxies and their evolution. The way dust traces star formation is ultimately determined by its interaction with the other components of the ISM: radiation fields (excitation, ionisation, dissociation), gas (through surface chemical reactions and larger scale phenomena like shocks, winds, turbulence), cosmic rays and magnetic fields. These processes modify the optical properties and the dust size distribution (through coagulation/accretion, erosion, fragmentation), which in turn leave distinct signatures in the shape of the spectral energy distribution. The comprehension of all the astronomical phenomena connected/regulated by dust thus requires the precise determination of the origin and physical properties of dust particles, and a deeper understanding of the response of dust to the different processes occurring in space at all redshifts.

It is widely accepted that dust is mainly formed at high densities and temperatures in the ejecta of evolved stars such as those populating the Red Giant Branch and the Asymptotic Giant Branch (RGB and AGB stars). However, the detection of a large amount of dust ($\gtrsim 10^7$ $M_\odot$) at very high redshift ($z > 6$, Bertoldi et al. 2003; Watson et al. 2015) when RGB and AGB stars did not have time to evolve, raises questions about the origin of cosmic dust in the distant Universe. From an evolutionary point of view, young supernovae (SNe) could represent a viable source of dust in high-redshift galaxies, although their role is still controversial.

Dust condensed in supernovae has been detected in meteorites (Clayton and Nittler 2004; Zinner 2008) and there is observational evidence of dust emission in young unmixed supernova remnants (SNRs) such as Cassiopeia A (Cas A; Lagage et al. 1996; Rho et al. 2008; Dunne et al. 2009; Barlow et al. 2010b; Arendt et al. 2014) or the Crab Nebula (Hester 2008; Gomez et al. 2012; Temim and Dwek 2013). Dust emission has been also observed in other remnants: N132D (Rho et al. 2009c), G292.0+1.8 (Ghavamian et al. 2009), 1E0102.2-7219 (E0102—Sandstrom et al. 2009; Rho et al. 2009b) and G54.1+0.3 (Temim et al. 2010, 2017). This latter is a Crab-like SNR with a dust mass comparable to SN 1987A (see below), but no evidence of a reverse shock.

Observations of the young supernova SN 1987A have revealed the presence of an increasingly large amount of dust (Danziger et al. 1989; Moseley et al. 1989; Dwek et al.

1992; Wooden 1997; Bouchet et al. 2004; Matsuura et al. 2011; Indebetouw et al. 2014; Matsuura et al. 2015, and references therein). Observations with Herschel (Matsuura et al. 2011, 2015) and ALMA (Indebetouw et al. 2014) in particular have shown that a large amount of freshly formed dust resides in the inner (cold) ejecta of SN 1987A (0.4–0.7 $M_\odot$ and >0.2 $M_\odot$ respectively). These detections are in agreement with theoretical predictions (0.1–1 $M_\odot$, Todini and Ferrara 2001; Nozawa et al. 2003) and seem to confirm that SNe are indeed efficient dust factories. At the same time, this discovery raises a fundamental question about the fate of dust: how much is injected into the ISM and can therefore be detected in galaxies at all redshifts?

The dust resides in the supernova cavity and is heavily processed by the reverse shock, generated from the impact of the SN blast wave with the circumstellar medium (CSM) and propagating towards the centre of the remnant. The gas particles accelerated and heated by the reverse shock collide with the dust grains, progressively eroding them through a process called sputtering. It might be expected therefore that only a fraction of the dust mass originally synthesised by SNe will be able to reach the ISM, with composition and size distribution different from those expected shortly after formation. The goal of this review, which is the second paper of a series of two, is to provide the state of the art about the processing and survival of dust *inside* supernova remnants, in terms of theoretical modelling and comparison with observations. The theoretical and observational aspects related to the formation and detection of dust in supernovae are discussed in our companion Paper I (Sarangi et al. 2018).

This review is organised as follows. Section 2 introduces some basic concepts about the evolution of SNRs and the formation of the forward and reverse shocks. In Sect. 3 we present the observational evidence of dust processing and destruction in supernova remnants and in Sect. 4 we review some of the methods used to estimate dust destruction from observational data. Section 5 discusses the theoretical modelling of dust processing by the reverse shock in SNRs, examining some of the key ingredients required by this task and comparing how different works include such ingredients in their implementations. The theoretical results are then contrasted with observations. In Sect. 6 we discuss briefly the global role of SNRs as dust source and sink and in Sect. 7 we present our summary and conclusions.

2 The Evolution of SNRs and the Development of Shockwaves

2.1 The Evolutionary Stages of SNRs

The evolution of SNRs originating from core collapse SNe (CCSNe) is characterised by four different phases (Fig. 1; e.g. Weiler and Sramek 1988; Padmanabhan 2001; Gaensler and Slane 2006).

1. *Free expansion* or *ejecta-dominated (ED)* phase, which lasts for a few hundred years after explosion.
2. *Adiabatic* or *Sedov-Taylor (ST)* phase for the next ~20,000 years.
3. *Radiative* or *snow-plow* phase up to 500,000 years.
4. *Merging* phase.

The explosion of the supernova progenitor star accelerates the envelope, causing turbulence (e.g. Wongwathanarat et al. 2015). The turbulence settles after a few months after the explosion (McCray 1993). After that, the ejecta material expands freely with a nearly constant velocity. The mass of the expelled ejecta dominates over the swept-up circumstellar

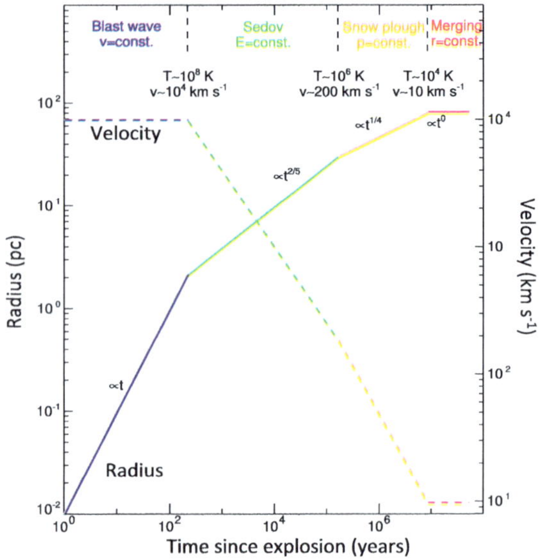

Fig. 1 Phases of supernova remnants resulting from core collapse SNe. While the radius (*left axis*) of a SNR increases with time, the blast-wave velocity (*right axis*) decreases. The evolution of the remnant radius was originally published by Padmanabhan (2001), while that of the velocity has been added by Wilms (2012)

material. The ejecta travel at supersonic speed, which implies the formation of the so-called blast-wave (or forward) shock ahead of them. The ambient medium shocked by the blast-wave reacts pushing back on the ejecta, with the consequent formation of the reverse shock. This latter decelerates, compresses and heats the ejecta. During the free-expansion phase, the expanding blast wave sweeps up the ambient ISM material, and eventually, the total mass of the swept-up material exceeds the original SN ejecta mass. The remnant enters then the adiabatic phase, and its radius, r, is generally described by the expression $r \propto t^{2/5}$, where t is the age of the SNR. This phase is also called Sedov-Taylor because it can be shown that the blast-wave of the remnant will tend to evolve towards the so-called Sedov-Taylor solution for the expanding blast-wave generated by a "point explosion", i.e. the sudden release of a large amount of energy into a background fluid. Such a solution was found independently by Taylor (1950) and Sedov (1959).

When the radiative losses of the hot shocked gas become important, the remnant enters the radiative phase. Emission lines from neutrals and ions cool down the gas, and the temperature drops sharply. The cool gas accumulates into a dense shell enclosing a central volume of hot gas where radiative losses are negligible. This phase is also called snow-plow because the sweeping-up of the ambient medium progressively increases the mass of the dense shell. The velocity of the shock front decreases gradually until it reaches the sound speed of the gas which it is propagating through. The shock wave fades away into a sound wave and the SNR finally merges into the ISM.

Starting from the Sedov-Taylor phase, the velocity of the expanding shell decreases with time from a few thousands $\mathrm{km\,s^{-1}}$ to a few hundred $\mathrm{km\,s^{-1}}$, and finally integrates into the ISM gas which has velocities of a few $\mathrm{km\,s^{-1}}$, as illustrated in Fig. 1.

2.2 Forward and Reverse Shocks in SNRs

In SNRs, the blast-wave or forward shock is generated by the ejecta moving supersonically into the circumstellar and interstellar medium. The reaction of the shocked ambient medium results in the formation of a reverse shock, which propagates inwards in the frame of the

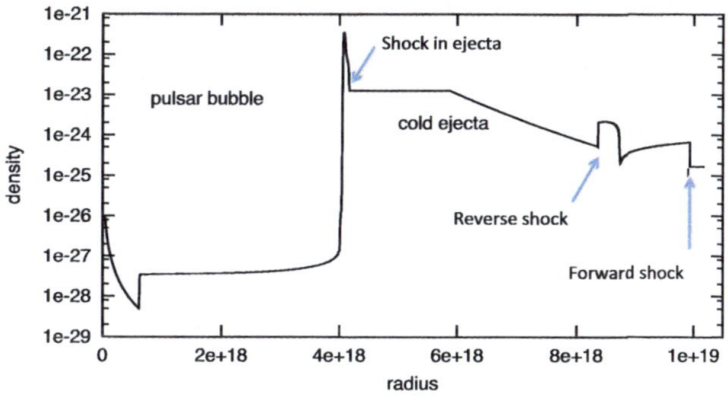

Fig. 2 Density structure of a SNR as a function of radius (Blondin et al. 2001). The forward and reverse shocks are indicated by the *arrows*. The model includes a pulsar wind nebula in the inner region, responsible for the formation of an additional shock in the ejecta. This shock has not always been detected in SNRs originating from core collapse SNe

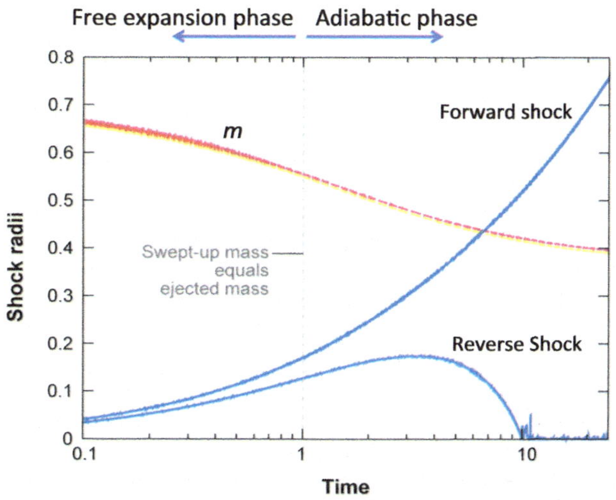

Fig. 3 Radii of the forward and reverse shocks (*blue lines*), together with the parameter $m = vt/R$ (*red line*) as a function of time. Figure from Reynolds (2008)

ejected gas. Figure 2 shows the density structure of a SNR as a function of radius, where the location of these two shocks is highlighted. The radius of the forward shock increases continuously with time, while the reverse shock is initially carried outwards before accelerating towards the centre (Fig. 3).

The forward shock is responsible for the processing and destruction of the dust grains already present in the CSM/ISM, while the reverse shock processes the dust freshly formed in the SN ejecta. This review focuses on this latter phenomenon.

3 Observational Evidence of Dust Processing and Destruction in SNRs

While there is increasing observational evidence that SNe form dust in their ejecta, as summarised in our companion paper (Sarangi et al. 2018), quantifying dust processing and destruction in SNRs is challenging. We present here observational examples of dust processing and destruction in the two young supernova remnants SN 1987A and Cas A, followed by a few more examples of older SNRs.

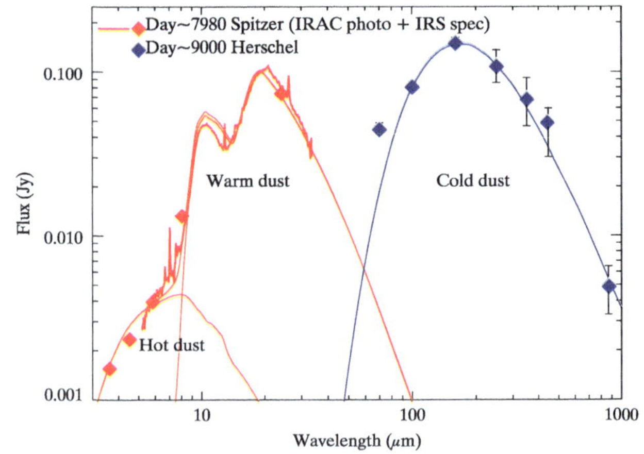

Fig. 4 Spectral energy distribution (SED) of SN 1987A, fitted with three discrete dust components: cold (~20 K), warm (~180 K) and hot (~400 K). The observational data are taken from Dwek et al. (2010) for day ~7980, and Matsuura et al. (2015) for day ~9000

3.1 SN 1987A

3.1.1 Ejecta Dust in SN 1987A—Newly Formed Dust and Possible Grain Growth?

SN 1987A is the first SN where dust formation was detected. The thermal emission of newly formed dust was detected starting from 615 days after the explosion (Danziger et al. 1989; Moseley et al. 1989; Wooden et al. 1993), with the optical light curve being extinguished by dust (Whitelock et al. 1989). Additionally, the detection of blue-shifted lines in the optical was suggested to be due to dust condensation within the ejecta (Lucy et al. 1989). The inferred dust mass was 10^{-4} $M_{\odot}$ at day 775, with a dust temperature of 300–400 K (Wooden et al. 1993).

About twenty years later, the HERSCHEL Space Observatory detected SN 1987A at far-infrared and submillimetre wavelengths (Fig. 4; Matsuura et al. 2011). The inferred dust mass was ~0.5 $M_{\odot}$ at days 8500–9090 (ignoring the current limit of ~0.25 $M_{\odot}$ of carbon predicted by nucleosynthesis models, Matsuura et al. 2015). This thermal dust emission was confirmed to be associated with the ejecta by spatially resolving the emission at submillimetre wavelengths (Fig. 5; Indebetouw et al. 2014; Zanardo et al. 2014). Radioactive decay of ^{44}Ti was suggested as the heating source of the ejecta dust (Matsuura et al. 2011).

After the detection of such a large amount of dust in the ejecta, the historical dust masses of SN 1987A have been revisited. There are currently two hypotheses about the time scale to acquire ~0.5 $M_{\odot}$ of dust. The first hypothesis is that such a large dust mass was already present in early days, but the emission was optically thick and the inferred dust mass was underestimated (Dwek and Arendt 2015). The second hypothesis is that the initial dust mass was indeed small (~10^{-3} $M_{\odot}$) and increased over 20 years via grain growth (Wesson et al. 2014; Bevan and Barlow 2016). In the future, far-infrared observations of newly exploded SNe performed in the first couple of years after explosion would help to disentangle these two hypotheses, because at these wavelengths dust emission tends to be optically thin, therefore possibly reflecting the actual mass of dust.

Although the reverse shock has been detected in SN 1987A, it is still just inside of the equatorial ring (France et al. 2010). The ejecta dust in SN 1987A is located much further inside, and has not yet encountered the reverse shock. Future monitoring will make possible to follow the propagation of the reverse shock inside the remnant and evaluate its efficiency at destroying dust.

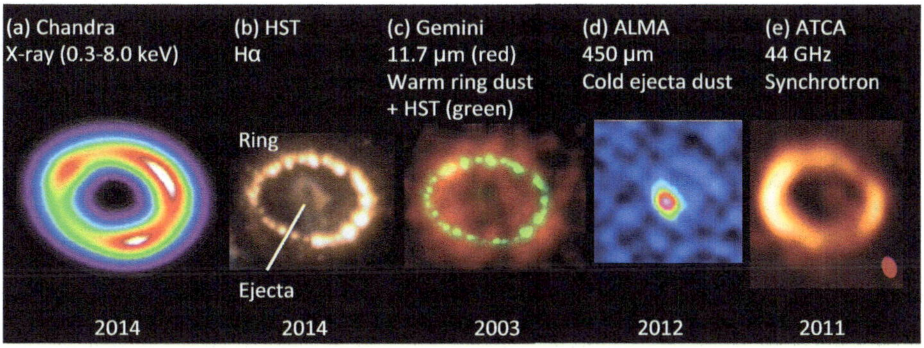

Fig. 5 Images of SN 1987A at different wavelengths. X-rays (**a**) and synchrotron radiation (**e**) trace the shocked gas in the ring. The HST image (**b**) shows the bright ring with faint ejecta in the centre. Warm dust has been identified in the ring (**c**), while the emission from the cold dust originates from the ejecta in the centre (**d**). Images from (**a**) to (**e**) are taken from Frank et al. (2016), Fransson et al. (2015), Bouchet et al. (2006), Indebetouw et al. (2014) and Zanardo et al. (2013), respectively

3.1.2 Dust in the Ring of SN 1987A—Collisional Heating by the Blast Wind

From the early days after explosion, the detection of asymmetries in polarisation observations has suggested the presence of a circumstellar nebula (e.g. Cropper et al. 1988). The HUBBLE Space Telescope has revealed that the circumstellar material is arranged in rings (Fig. 5; Jakobsen et al. 1991; Panagia et al. 1991). These rings are thought to be composed of the material lost from the progenitor, when the star was in the red-supergiant phase ~40,000 years ago (Arnett et al. 1989; McCray 1993).

The bright equatorial ring is another location within SN 1987A where dust emission has been detected. Using the mid-infrared instrument T-ReCS on the Gemini Telescope, Bouchet et al. (2004) found that dust emission at 10 μm is dominated by ring dust (Fig. 5). While the blast winds from the explosion collide with the ring, the collisional energy generates X-ray radiation (Fig. 5) and heats up the dust grains which had formed during the supergiant phase (Bouchet et al. 2006).

Since its launch in 2003, the SPITZER Space Telescope has been intensively used to investigate dust emission in SN 1987A. SPITZER data have allowed to identify two components. The first one is the warm (~180 K) dust, emitting in the range 8–35 μm and associated with the ring. In particular, from the data of the spectrometer on board SPITZER it has been possible to identify silicate dust features at 10 and 18 μm (Fig. 4; Bouchet et al. 2006). The second component is the excess emission detected from 3.5 μm up to 8 μm (Fig. 4). It has been suggested that this component is associated with hot dust in the ring, where small (radii < 0.03 μm) dust grains radiate at temperatures of ~400 K (Dwek et al. 2010).

SPITZER has been used to monitor the evolution of the warm and hot dust on a half-a-year basis. The infrared flux from 3.6 to 24 μm has continuously increased until SPITZER's helium ran out in 2009 (Dwek et al. 2010). Although the overall flux has increased, the shape of the silicate feature has remained the same. During that same period, the X-ray luminosity has also increased (Fig. 6; Dwek et al. 2010; Frank et al. 2016), as well as the optical brightness of the ring and the synchrotron radiation (Larsson et al. 2011; Zanardo et al. 2010; McCray and Fransson 2016). The interaction between the ejecta blast-wave and the ring generates energy, heating up the ring material and brightening it up at all wavelengths. More material from the ejecta has been plunging into the ring, increasing its brightness.

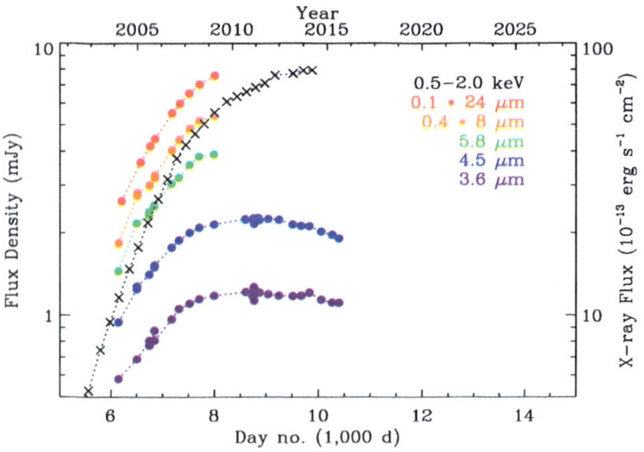

Fig. 6 Evolution of the infrared and X-ray fluxes of SN 1987A (Arendt et al. 2016). Both the infrared and X-ray fluxes have continuously increased until 2012. The infrared flux at 3.6 and 4.5 μm shows a peak followed by decline, while the X-ray radiation continues to increase

Dwek et al. (2008) have modelled the infrared emission of dust assuming that the grains are collisionally heated in the X-ray emitting plasma. Dust sputtering due to the fast moving ions in the plasma results in the erosion and possible destruction of the grains. The infrared luminosity increases slower than the X-ray luminosity, therefore Dwek et al. conclude that the dust grains are destroyed in the ring.

SPITZER continued to be used to monitor the infrared brightness employing the two bands still in operation after its helium ran out in 2009 (3.6 and 4.5 μm, Arendt et al. 2016). The infrared brightness kept increasing until 2010 (about day ~8500 after the explosion) and remained constant for two years, then started fading. The beginning of the decline was more or less coincident with the time when the HUBBLE Space Telescope captured indications that the outermost part of the blast wave had passed the bright ring (Fransson et al. 2015). In contrast to the infrared brightness, the X-ray luminosity continued to increase. Arendt et al. (2016) suggest that the time scale for the destruction of dust grains is shorter than the cooling time of the shocked gas: the bulk mass of dust has been reduced by shock destruction, decreasing the infrared brightness, while the X-ray flux reflects the temperature (but also the density and volume) of the shocked gas.

Dust grains in the ring will continue to interact with the blast wave, but at a lower rate than in the 2000s decade. While SPITZER will continue being used to monitor the flux of the ring, the launch of the JWST will allow us to study spatially resolved shock interaction during its mission.

3.2 Cassiopeia A—Forward Shock, Dust Processing by the Reverse Shock and Dust Emission from the Unshocked Centre

Cassiopeia A (Cas A) is the remnant of a Type IIb SN explosion which occurred approximately 335 years ago (Fesen et al. 2006; Krause et al. 2008). Estimates from X-ray observations indicate a rather high ejecta mass of 4 $M_\odot$ (Vink et al. 1996). CHANDRA observations of the SNR have revealed the location of the forward and reverse shocks (Fig. 7; Gotthelf et al. 2001; Hwang et al. 2004). The remnant has a filamentary structure with asymmetric shells. Maps of ionised lines show that elements from different nucleosynthesis processes are distributed in slightly different locations within the remnant (e.g. Isensee et al. 2012; Grefenstette et al. 2014; Milisavljevic and Fesen 2015).

Dust in Cas A has been found both in the reverse-shocked gas and in regions still unshocked. SPITZER observations at 19–23 μm have shown that mid-infrared dust emission is

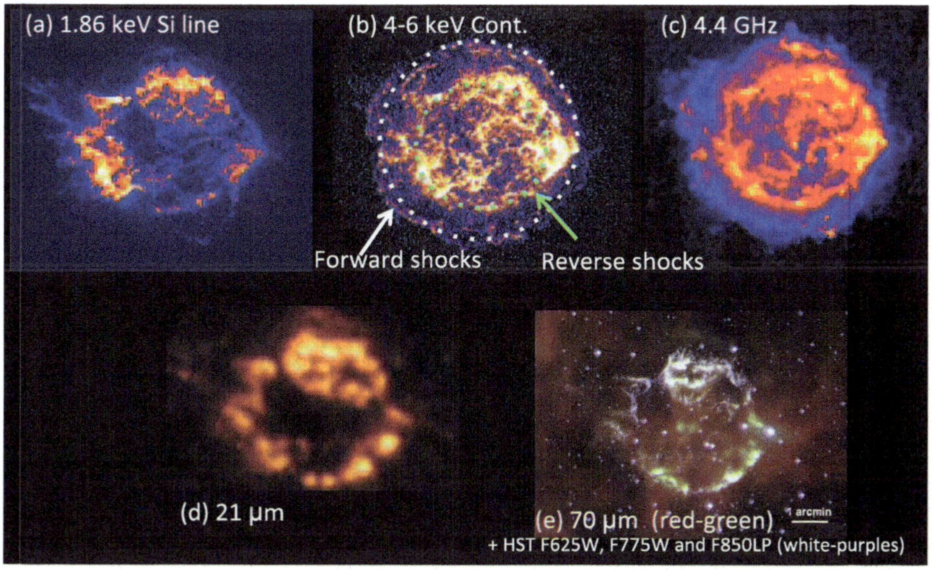

Fig. 7 Images of Cas A at various wavelengths. Panels (**a**) to (**c**) are taken from Gotthelf et al. (2001) and show the CHANDRA X-ray Kα silicon emission, the CHANDRA X-ray continuum with the location of the forward and reverse shocks highlighted, and a radio map at 4.4 GHz from the VLA, epoch 1997, respectively. The warm dust is traced by the SPITZER 21 μm emission (panel (**d**)—Rho et al. 2008), while colder dust is traced by the HERSCHEL 70 μm emission (panel (**e**)—De Looze et al. 2017)

mainly coincident with the [Ar II] emitting region (Ennis et al. 2006; Rho et al. 2008), and that this also corresponds to the reverse-shocked region (Fig. 7d). Dust emission at 70 μm has been also detected. The temperature of the dust has been estimated between 50–500 K, with an estimated mass of 0.020–0.054 $M_\odot$.

Hines et al. (2004) reported about SPITZER MIPS imaging of Cas A at 24 and 70 μm. From the total MIPS fluxes combined with IRAS, ISO and MSX observations, they estimated in Cas A a total dust mass up to 3×10^{-3} $M_\odot$.

Barlow et al. (2010) analysed HERSCHEL images of Cas A taken in five bands between 70 and 500 μm. They found that while warm (∼80 K) dust grains indeed reside in the reverse-shocked region, there is a cold (∼25 K) dust component located in the unshocked region in the centre of the system (green dots in the centre of Cas A in Fig. 7e). Their estimated dust mass is 0.075 $M_\odot$.

To determine the dust mass in Cas A, Bevan et al. (2017) adopted the method proposed by Lucy et al. (1989), which is based on photons scattering and absorption by dust in the expanding ejecta. Using the Monte Carlo line transfer code DAMOCLES (Bevan and Barlow 2016), Bevan et al. (2017) found that the most likely dust mass given by their modelling of the integrated optical spectrum of Cas A (Milisavljevic and Fesen 2013) is ∼1.1 $M_\odot$.

De Looze et al. (2017) revisited the analysis of dust mass and temperature in Cas A, using SPITZER and HERSCHEL images, together with PLANCK and HERSCHEL spectra. Their analysis using spatially resolved maps involves careful removal of ISM lines and ISM dust emission from the far-infrared images. The ISM dust model from Jones et al. (2013, 2017) was used to remove the contribution from the ISM dust. They concluded that the data are best-fitted by 0.3–0.5 $M_\odot$ of silicate grains (with a temperature of ∼30 K) or 0.4–0.6 $M_\odot$ of an equal mixture of silicates and carbonaceous grains. Their dust map shows a larger dust mass in the unshocked region than in the reverse-shocked region, suggesting that 70% of the dust formed in the SN could have been destroyed by the reverse shock. Different dust temperatures in the reverse-shocked region and in the unshocked region clearly indicate that dust grains are experiencing collisional heating in the reverse-shocked region, with potential dust destruction via sputtering and shattering.

In the case of Cas A, the evaluation and subtraction of the ISM dust along the line of sight affects the final estimate of the SN dust mass. That remains challenging in many Galactic SNRs with heavy contamination from ISM dust along the line of sight.

3.3 Sagittarius A East—a 10,000 Year old SNR with Surviving Dust?

Sagittarius A East is a SNR remnant close to the Galactic centre, with an estimated age of $\sim$10,000 years. Such an old age implies that the reverse shock should have reached the centre of the remnant. Lau et al. (2015) have detected infrared dust emission from the centre of the SNR at 5.8–70 μm. The location of dust emission is coincident with that of the iron Kα line, which traces shocked gas. While the fraction of small grains appears to be low, Lau et al. (2015) suggest that the excess far-infrared emission requires the presence of large dust grains ($\sim$0.04 μm) which could have survived the passage of the reverse shock.

4 Estimating Dust Destruction from Observations

4.1 Dust Depletion of Elements and Dust Destruction

Lines from ionised and neutral species detected in the UV, optical and X-ray domains are commonly used to measure elemental abundances. The discrepancies between the measured atomic abundances in the ISM and the solar abundances of refractory elements are generally attributed to the depletion of such refractory elements into dust grains in the ISM (e.g. Savage and Sembach 1996). Elemental abundances and dust depletion can be estimated in SNRs, and any spatial variation of the dust depletion rates across shocks can be due to dust destruction.

Vancura et al. (1994) showed that the X-ray spectra of shock waves are substantially modified when the effects of grain destruction are taken into account. Their theoretical results were reasonably in agreement with observations of the Cygnus Loop. Seab and Shull (1983) presented theoretical and observational evidence that both the depletion of heavy elements and the UV extinction curves are modified by the processing of interstellar dust grains due to SN blast waves propagating into the ISM. Raymond et al. (2013) measured the carbon abundance within the Cygnus Loop SNR, using UV spectra obtained by the HUBBLE Space Telescope. They found that the C IV to He II ratio changes from the shock front to the post shock region. They argue that carbon atoms are sputtered from dust grains, enhancing the carbon abundance in the post shocked region.

4.2 Modelling the Infrared Dust Emission

4.2.1 The Modified Blackbody Analysis—Advantages and Limitations

The Modified Blackbody Analysis This technique is widely used to estimate dust mass and temperature from the infrared emission of dust grains in stars and in the ISM. It can be also used to estimate the dust mass present in SNe and SNRs.

In the optically thin case, the flux density F_λ at the wavelength λ can be represented by a modified blackbody:

$$F_\lambda = 4\pi M_d \frac{\kappa_\lambda B_\lambda(T_d)}{4\pi D^2}, \tag{1}$$

where M_d is the dust mass, D is the distance to the object, $B_\lambda(T_d)$ is the Planck function and T_d is the dust temperature (Hildebrand 1983). The dust mass absorption coefficient κ_λ is expressed as

$$\kappa_{\lambda,a} = 3Q_\lambda/4\rho a, \tag{2}$$

where ρ is the mass density of the dust grains, Q_λ is the dust emissivity at the wavelength λ and a is the grain size. In the Rayleigh limit, i.e. grain size a much smaller than the emitting wavelength λ, the coefficient κ_λ for spherical grains can be simply expressed as a power-law

$$\kappa \propto \lambda^{-\beta}. \tag{3}$$

Thus, the flux F_λ at a given dust temperature, T_d, becomes independent from the grain size. These simple equations can provide dust masses from a few measured flux data points at infrared wavelengths.

Advantages and Limitations The modified blackbody analysis has both advantages and limitations. The advantage is that this method is very simple and easy to use. As far as all the grains have the same temperature, the equation does not depend on the grain size in the wavelength range where $a \ll \lambda$. One limitation is that the assumption of all grains having the same temperature might not be applicable in some cases. The dust absorption coefficients at UV and optical wavelengths depend on the size of the grains, and small grains can reach higher temperatures. In case of radiatively heated dust, if the incoming radiation has the spectrum of the interstellar radiation field, the dependence is approximately $T_d \propto a^{-0.06}$ for $\beta = 1$ (Tielens 2010; Draine 2011). The majority of dust in SNRs is collisionally heated in the shocked gas. This case is discussed in Sect. 4.3.

Can the Modified Blackbody Analysis be Used to Estimate Dust Destruction? The answer to this question depends on the degree of dust destruction and on the temperature and mass of both the SNR and ISM dust. In fact the definite answer is that this method should be cautiously used to estimate dust destruction, because the emission from the hot SNR dust can hide the emission from the cold ISM dust.

Often, SNRs are surrounded by interstellar clouds, therefore the infrared emission coming from a SNR is contaminated by the interstellar dust emission along the line of sight. The proper estimate of the amount of interstellar dust is crucial to correctly measure the amount of SNR dust (Sect. 3.2). The evaluation of the emission coming from the ISM is not a simple task. The modified blackbody analysis is often used for this purpose, but it requires particular attention because the results can be misleading, as demonstrated by the example below.

Dust in SNRs often has a slightly higher temperature than interstellar dust. The interstellar dust temperature has been found to be typically 15–25 K (Gordon et al. 2014), while the supernova dust located in shocked regions tends to have a higher temperature of 20–120 K (e.g. Sandstrom et al. 2009; Lakicevic et al. 2015; De Looze et al. 2017). Let's consider two dust components: warm dust from a SNR and cold interstellar dust along the line of sight of the remnant (Fig. 8 and Table 8). We add these two components as if they were both optically thin in the infrared, fit the summed emission using a modified blackbody and test whether we can recover the input dust masses and temperatures.

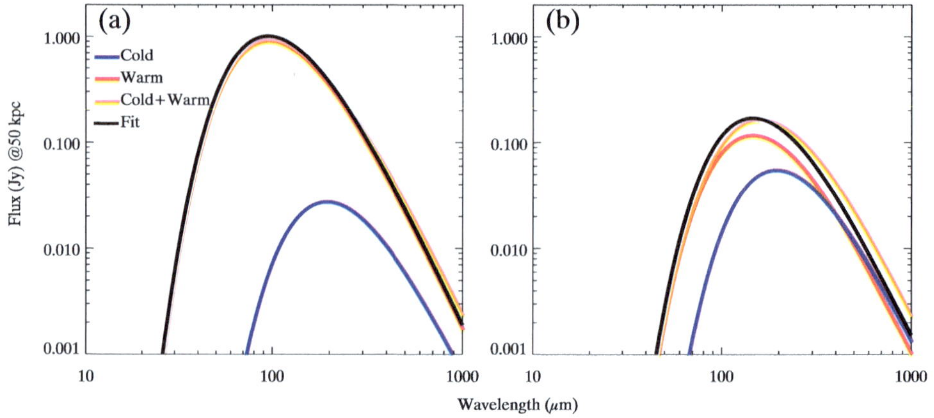

Fig. 8 Example of modified blackbody fits of two dust components: warm dust from a SNR (*red curve*) and cold interstellar dust along the line of sight of the remnant (*blue curve*). The two components are added together (*yellow curve*) and the summed emission is fitted with a modified blackbody (*black curve*) to recover the input dust masses and temperatures. In panel (**a**) the two components have an equal mass (1 $M_\odot$) and temperatures of 15 and 30 K respectively. The total emission is dominated by the warm dust, which hides the cold dust. The fitting yields a dust mass of 1.1 $M_\odot$ and a dust temperature of 29.9 K, almost reflecting the warm component only. In panel (**b**) the temperatures of the warm and cold dust are 20 and 15 K, and their masses are 1 and 2 $M_\odot$ respectively. Fitting of the total emission with a modified blackbody yields 1.5 $M_\odot$, which is only half of the total input dust mass (3 $M_\odot$). The input and output parameters are summarised in Table 1

Table 1 Input and output parameters of the modified blackbodies used in Fig. 8

	Warm		Cold		Fitted results	
	T_{warm} (K)	M_{warm} ($M_\odot$)	T_{cold} (K)	M_{cold} ($M_\odot$)	T_{fit} (K)	M_{fit} ($M_\odot$)
Example (a)	30	1	15	1	29.9	1.1
Example (b)	20	1	15	2	19.7	1.5

Panel (a) in Fig. 8 illustrates a case where the cold and warm dust components have an equal mass (1 $M_\odot$) and temperatures of 15 and 30 K, respectively (Table 8). We adopt for the grains the emissivity of silicates (Ossenkopf et al. 1992). As shown in Fig. 8a, the total emission is dominated by the warm dust component, which hides the cold dust. The χ-square fitting of the sum of the two emission components yields a dust mass of 1.1 $M_\odot$ and a dust temperature of 29.9 K (Table 1), almost reflecting the warm component only. In the direction of the SNR, the mass of interstellar dust appears reduced with respect to its actual value.

Panel (b) in Fig. 8 shows what happens if the two components have similar luminosities. The temperatures of the warm and cold dust are 20 and 15 K, and their masses are 1 and 2 $M_\odot$ respectively (Table 1). Fitting the sum of these two components with a modified blackbody yields 1.5 $M_\odot$, which is only half of the total input dust mass (3 $M_\odot$).

The above analysis shows that, when trying to find cold ISM dust behind hot SNRs, depending on the temperatures and column densities of both components, the inferred dust masses, particularly the cold ISM dust mass, can be distorted. The use of modified blackbody fitting to ISM and SNRs dust emission needs to be very cautious, and does not provide conclusive evidence of SNRs destroying ISM dust.

4.3 Modelling the Infrared Emission of Collisionally Heated Dust

As mentioned in Sect. 4.2.1, dust grains of different sizes have different temperatures. The calculation of the dust temperature gradient requires balancing the heating and cooling rates of the grains. This review article focuses on dust processing and destruction in SNRs, where dust grains are mainly heated via collisions with fast moving particles, predominantly electrons. Infrared emission is the main cooling channel for the hot dust. In this section, we briefly discuss the modelling of the infrared emission from collisionally heated dust, which has been studied intensively by Draine (1981), Dwek (1986, 1987), Dwek et al. (2008) and summarised by Draine (2011).

The collisional heating rate of a dust grain of radius a is given by

$$H = \pi a^2 \sum_j n_j v_j E^j_{\text{dep}}, \tag{4}$$

where the index j runs over the different constituents of the plasma (ions and electrons) embedding the dust, n_j is the number density of the jth plasma component, v_j is its thermal velocity and E_j is its thermally-averaged energy deposition in the grains (Dwek et al. 2008). It is assumed that ions and electrons have the same temperature, which implies $v_e \gg v_{\text{ions}}$. Therefore, the dust heating rate is dominated by collisions with electrons. For simplicity, we set $E^e_{\text{dep}} \equiv E_{\text{dep}}$. Dwek et al. (2008) consider two cases: if most of the electrons are stopped by the grains, on average E_{dep} is equal to the thermal energy of the electrons, i.e. $E_{\text{dep}} \propto T_e$, where T_e is the electron temperature. The electron velocity is given by $\sqrt{8kT_e/\pi m_e}$, where m_e is the electron mass, therefore

$$H \sim a^2 n_e T_e^{3/2}. \tag{5}$$

If most of the electrons traverse entirely the grains, E_{dep} is proportional to the electron stopping power in the solid, $\rho^{-1}(\mathrm{d}E/\mathrm{d}x)$, where ρ is the mass density of the material. In the energy regime considered here, $\mathrm{d}E/\mathrm{d}x \propto E^{-1/2}$, so that $E \propto T_e^{-1/2} a$ and

$$H \sim a^3 n_e. \tag{6}$$

The cooling of the dust is dominated by its own emission. The radiative cooling rate is given by:

$$L = 4\pi \sigma_d \int_0^\infty Q(\lambda) B(T_d, \lambda) d\lambda, \tag{7}$$

where σ_d is the geometrical cross section of the grains, $Q(\lambda)$ is the dust absorption efficiency at wavelength λ, B is the Planck function and T_d is the dust temperature. The average of $Q(\lambda)$ over the Planck function is given by the following expression:

$$\langle Q(T_d) \rangle = \frac{\pi}{\sigma T_d^4} \int_0^\infty Q(\lambda) B(T_d, \lambda) d\lambda, \tag{8}$$

where σ is the Stefan-Boltzmann constant. For $a \ll \lambda$, $Q(\lambda) \propto \lambda^{-\beta}$, thus $\langle Q \rangle \propto a T_d^\beta$ for spherical grains. The cooling rate becomes

$$L = 4\pi a^2 \sigma T_d^4 \langle Q \rangle \approx 4\pi\, a^3 \sigma T_d^{4+\beta}. \tag{9}$$

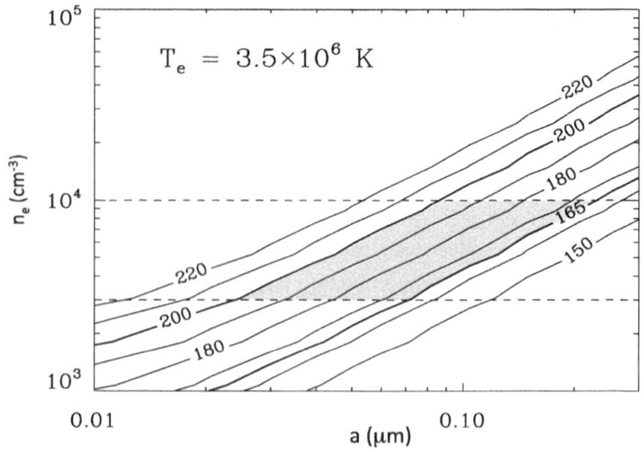

Fig. 9 Contour plot showing the dust temperature dependence on grain size (a) and electron density (n_e) for the given electron temperature, T_e (Dwek et al. 2008). The *dashed horizontal lines* indicate the range of electron densities in the ring of SN 1987A, estimated from the analysis of ionised emission. Within these values of n_e, the range of observed temperatures (between 165 and 200 K) limits the viable grain sizes between 0.023 and 0.22 μm (*gray area*)

In case of thermal equilibrium, $H = L$ and the dust temperature can be written as

$$T_d \propto \left(\frac{n_e}{a}\right)^{\gamma} T_e^{3\gamma/2} \tag{10}$$

for electrons stopped in a grain, and

$$T_d \propto n_e^{\gamma} \tag{11}$$

for electrons which traverse the grain, where $\gamma = 1/(4 + \beta)$ (Dwek et al. 2008). The values of β are typically between 1 and 2 ($\sim$2 for silicates and $\sim$1 for amorphous carbon; Mennella et al. 1998). Equation (11) shows that, when the electrons are not stopped inside the grain, the dust temperature becomes independent from the grain size a. In the optically thin case, the flux density F_λ at wavelength λ is then given by Eq. (1).

Figure 9 shows the modelled dust temperature as a function of the electron density and grain size. For the same electron temperature, smaller grains can reach higher temperatures than the larger ones. The figure also shows that a combined estimate of dust temperature (from infrared emission) and electron density (from X-ray or optical measurements) can potentially constrain the range of dust grain sizes. For the dust in the circumstellar ring of SN 1987A, Fig. 9 suggests grain sizes in the range 0.023–0.22 μm.

X-rays come from the hot shock-heated plasma, while the infrared luminosity reflects the energy emitted by dust grains collisionally heated in the shocked gas. Dwek et al. (2008) introduced the infrared-to-X-ray luminosity ratio as a tracer of the heating and cooling balance of dust grains. This will be discussed in Sect. 4.6. As already mentioned, dust heating is dominated by collisions with electrons. Dwek et al. (2008) suggested that the infrared-to-X-ray ratio, which shows some dependence on grain size, is also a good indicator of the electron temperature of the plasma.

4.4 Modelling the Infrared Emission from Dust with a Modified Size Distribution

Dust grains embedded in a shocked gas are affected by different destructive processes (Sect. 5.6). Two of the most important are erosion due to collisions with gas particles (sputtering) and fragmentation due to collisions with other grains (shattering). Grain-grain collisions are less frequent than gas-grain collisions because of the low dust density, but each collision can be more efficient in destroying the collision partners. Sputtering and shattering modify the initial size distribution of the dust, completely destroying the smallest grains and reducing the size of the bigger ones via progressive erosion or fragmentation. The result is a size distribution shifted towards smaller dust sizes with respect to the initial one (see e.g. Williams and Temim 2016, for a review).

Infrared emission from grains with a modified size distribution can be calculated assuming collisional heating of the dust (e.g. Dwek and Arendt 1992, Sect. 4.3). Using the method from Borkowski et al. (2006), the required input parameters are the electron and proton temperatures, the proton density, n_p, and the shock sputtering age. The optical spectra of non-radiative supernova remnants are dominated by Balmer lines, which are used to determine the shock parameters, in particular the electron/proton temperature ratio and the shock velocity (Ghavamian et al. 2007). The sputtering age is equivalent to the ionisation timescale, and for Sedov dynamics reaches a maximum at around $1/3 n_p t$, where t is the true age of the SNR (Borkowski et al. 2001). Knowing the dust sputtering rates (e.g. Tielens et al. 1994) and the sputtering age, it is possible to calculate the size distributions of the grains which are progressively modified by processing in the hot gas, and compute the corresponding emission spectra using the equations in Sect. 4.3.

4.5 Estimates of Dust Destruction Rates in Specific SNRs

The approach described in Sect. 4.4 can be used to model the infrared emission from dust whose size distribution has been modified by shocks. The results have been compared with the SPITZER 24 over 70 μm flux ratio in various SNRs. Borkowski et al. (2006b) estimated that 35–40% of dust has been destroyed in two LMC SNRs, DEM L71 and 0548–70.4, Sankrit et al. (2010) applied the same method to the Cygnus Loop, suggesting that 35% of the dust has been destroyed by shocks over 1350 years. Blair et al. (2007) analysed the same infrared ratio in the type Ia SNR Kepler, finding a dust destruction rate of 85%, though later Williams et al. (2012) determined that only 10% of dust could have been destroyed in some parts of this SNR. The same method applied to LMC core-collapse SNRs, suggests that 27–50% of the dust has been destroyed (Williams et al. 2006). Using SPITZER 24, 70 and 160 μm images of the Pup A SNR, Arendt et al. (2010) estimated that 25% of dust grains could have been destroyed. It should be noted that all these works do not specify which kind of shocks are responsible for dust destruction.

4.6 Is the Infrared-over-X-Ray Luminosity Ratio Universal?

Dust grains residing in a hot, X-ray emitting gas are mainly heated by collisions with electrons and cool down through infrared emission (Sect. 4.3). A correlation between X-ray and infrared emission could therefore be expected. Dwek (1987) tested this hypothesis first on Cas A and the Cygnus loop and later on seven more Galatic SNRs (Dwek et al. 1987). Koo et al. (2016) expanded this kind of study to 20 Galactic SNRs, and Seok et al. (2013, 2015) to SNRs in the Large Magellanic Cloud.

Koo et al. (2016) find that the measured infrared and X-ray luminosities of the 20 Galactic SNRs under examination do not follow the theoretical trend. The adopted infrared and X-ray luminosities are similar to those of Dwek (1987), but the more recent model of Weingartner and Draine (2001) for interstellar dust is used instead. Some SNRs are much more brighter than predicted in the X-ray with respect to the infrared, with discrepancies larger than a factor of 100. Their initial model considers a hot, dusty plasma in thermal and collisional equilibrium, with dust having the MRN interstellar grain size distribution (Mathis

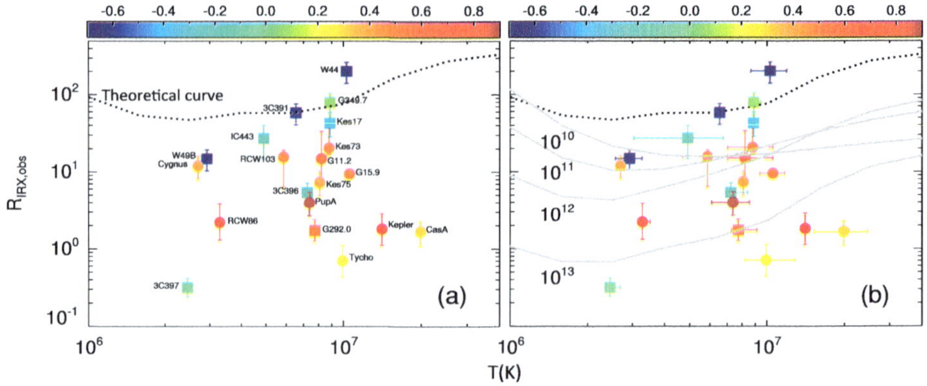

Fig. 10 Infrared-over-X-ray luminosity ratio, R_{IRX}, as a function of the electron temperature of the X-ray plasma (Koo et al. 2016). The *dotted line* on panel (**a**) shows the theoretically predicted ratio for a plasma in thermal and collisional equilibrium with dust having the MRN interstellar grain size distribution. Panel (**b**) includes the ratios resulting from an out-of-equilibrium gas where the grain size distribution is modified by sputtering (*grey lines*). The curves show R_{IRX} at different times, given by the ionisation timescale τ_{ion}, after the gas has been shocked, with $\tau_{\mathrm{ion}} = 10^{10}, 10^{11}, 10^{12}$ and $10^{13}\ \mathrm{s\,cm^{-3}}$. The colour of the SNRs symbols indicates the value of the coefficient of correlation between the morphology of the infrared and X-ray emissions. The coefficient ranges from -0.7 to 0.9 as in the *colour bar*

et al. 1977). The calculations are extended to consider a plasma where dust grains are continuously injected and sputtered in the hot gas. The theoretical IR-to-X-ray ratio curves corresponding to this out-of-equilibrium gas with modified grain size distributions cover a wider range of values and are able to reproduce the majority of the measurements (Fig. 10).

Koo et al. (2016) have carefully examined the properties of individual SNRs. Low infrared luminosity SNRs tend to have a known additional X-ray emission component, such as the ejecta, on top of the X-ray radiation due to the expanding remnant ploughing the surrounding ISM. On the other hand, SNRs with high infrared luminosities tend to have radiative shocks, and the infrared luminosity is largely due to dust heated by the shock radiation. The study of LMC SNRs by Seok et al. (2013, 2015) suggests that the lower dust-to-gas ratio in this galaxy may affect the infrared and X-ray luminosities. These studies show that modelling dust emission and dust destruction requires a careful examination of the SNRs properties.

5 Dust Processing by the Reverse Shock: One Problem, Different Approaches

Studying the survival of freshly formed dust in SNRs requires some key ingredients. It is necessary to have a physically motivated picture of the remnants: morphology, composition, temperature, density structure of the ejecta (e.g. clumpiness), ongoing radiative processes, in short everything that can affect the properties of the ejecta. These properties impact directly the processing of the dust. For instance, the composition of the ejecta determines which species will bombard and erode the dust, while the temperature determines which mechanisms will be more effective in processing the grains. The physical description of the SNRs must be included into reliable model(s) for the dynamical evolution of the remnants, i.e. the evolution of radius and velocity of the forward and reverse shocks. Well founded models capable of reproducing observations are necessary not only to properly evaluate current dust processing but also to make realistic predictions. The dynamical models will have to be coupled with the description of the dust (size, composition, physical and chemical properties) and of the different processes capable of eroding, fragmenting or even destroying dust grains while they reside inside supernova remnants.

It has been our choice to focus this section on the works which actually treat the processing of dust by the reverse shock, presenting them at a level of detail sufficient to appreciate the different approaches used by the authors. We do not include therefore the many papers only mentioning the possible effect of the reverse shock in modifying the dust. The studies that we have considered are briefly introduced below. In Sects. 5.1 to 5.6 we discuss how the different authors include in their models the key ingredients listed at the beginning of this section. In Sect. 5.7 we summarise the results in terms of the amount of dust mass surviving the processing by the reverse shock, and show how these findings compare with observations. It is interesting to note that, to our best knowledge, after the study of Itoh (1985) it has been necessary to wait twenty years to have dust survival in SNRs investigated again, with a renewed and this time persistent interest.

Itoh (1985) estimates the amount of dust erosion in both the blast-shocked circumstellar medium and the reverse-shocked ejecta, adopting self-similar solutions for the expansion of the SNR. The results are applied to the remnants of SN 1979c and SN 1980k.

Dwek (2005) discusses the destruction of SN-condensed dust by the reverse shock assuming homogenous ejecta expanding into a uniform medium, but emphasises that in reality, observations show that the ejecta are clumpy and that the dust resides primarily in these clumps. This is the case for instance for the remnant of Cas A, where dust has been detected in fast moving knots (Lagage et al. 1996; Arendt et al. 1999). When the reverse shock propagates inside the clumps, its velocity is reduced with respect to the value in the more tenuous medium. The presence of clumpy ejecta has therefore important consequences on the survival of the dust.

Bianchi and Schneider (2007) revise the dust formation model of Todini and Ferrara (2001) based on classical nucleation theory (CNT) and assess the role of the reverse shock in modifying the size and mass of the newly synthesised grains, making predictions on dust properties from formation up to few hundreds of years later, i.e., only considering the non-radiative phase of supernova remnants.

Nozawa et al. (2007) study the evolution of dust formed inside primordial type II supernovae (SNe II) and pair-instability supernovae (PI SNe). They follow the dynamics and destruction of the dust, together with the time evolution of gas temperature and density, until 10^5–10^6 years after explosion, well extending into the radiative phase. This provides a better idea of how much dust is actually injected into the ISM. An important point addressed by this work is how the thickness of the H envelope of the progenitor star affects the survival of the dust.

Nozawa et al. (2010) adapt the dust evolution model developed by Nozawa et al. (2007) to study the formation and processing of dust in hydrogen/helium poor CCSNe such as Type Ib/Ic and Type IIb.

Nath et al. (2008) adopt an analytical approach which treats directly the effect of reverse shocks traveling through ejecta with a power-law mass distribution, instead of a uniform distribution as in previous works. While various important processes such as dust dynamics are not included, this formalism allows to make a clear separation between destruction occurring within the reverse shock itself, and the processing taking place in the region between the reverse and forward shocks.

Silvia et al. (2010, 2012) tackle a different aspect of the problem of dust destruction inside supernova remnants considering the processing of dust when this is located inside overdense clumps, instead of being dispersed into a homogeneous medium. This aspect is particularly important because, as previously mentioned, SN ejecta are indeed clumpy (e.g. Fesen et al. 2006; Hammell and Fesen 2008) and observations have shown that the dust resides in such dense clumps. This is typically the case of Cas A. To study this problem, Silvia et al. perform numerical simulations on an individual ejecta clump containing dust and encountering a planar reverse shock.

Bocchio et al. (2016) study the survival of freshly formed dust inside four well-known core-collapse SNRs: Cas A and the Crab Nebula (in the Milky Way) and SN 1987A and N49 (in the Large Magellanic Cloud). The formation of dust and the evolution of SNRs are described as in Bianchi and Schneider (2007), with an upgraded molecular network and an updated sputtering treatment.

Biscaro and Cherchneff (2014, 2016) approach the problem of the formation and survival of molecules and dust in Cas A via a rigorous study of the chemistry of the ejecta of type IIb supernovae, which led to the Cas A supernova remnant, including in particular the chemistry of the reverse shock. Biscaro and Cherchneff (2014) focuses on the formation and processing of molecules and dust clusters synthesised in the gas-phase, and it is discussed in our companion paper. Biscaro and Cherchneff (2016) deals with the condensation of dust clusters, the grain size distributions expected from type IIb SNe and the processing of dust in the ejecta clumps and in the inter-clump medium.

Micelotta et al. (2016) tackle the question of dust survival in SNRs after the passage of the reverse shock, with special application to the clumpy ejecta of Cas A. They adopt a global approach which combines a realistic description of the remnant and its evolution with a robust treatment of dust sputtering, first in the clumps and later on into the smooth inter-clump medium. In particular, a combination of analytical methods and Monte Carlo simulations allows to follow dust sputtering "on-the-fly", taking into account simultaneously the deceleration and erosion of the grains as a function of their size and position inside the clumps.

Depending on their initial size, the newly synthesised grains will be progressively eroded and possibly completely destroyed due to collisions with the shocked gas. As previously mentioned, this phenomenon is known as sputtering (e.g. Tielens et al. 1994, and references therein) and consists on the progressive removal of atoms or molecules from the surface of the grains, induced by impacts with the particles populating the gas. According to the velocity distribution of the incoming projectiles, sputtering is classified as kinetic (or inertial) when the collision velocity is given by the relative motion of the grains through the gas, and thermal when the velocity of the projectiles is due to their thermal motion and is determined by the temperature of the shocked gas. Dust destruction via sputtering is discussed in Sect. 5.6.

5.1 Modelling the Dynamical Evolution of Supernova Remnants

The velocity of the reverse shock determines the velocity distributions of the gas particles which will impinge on dust grains, progressively eroding them. The velocity of the reverse shock is related to the velocity of the forward shock, and they can both be calculated from dynamical models for the expansion of SN ejecta. These models typically depend on the energy of the explosion, the mass and initial density profile of the ejecta, and on the density profile of the circumstellar medium the ejecta are expanding into.

Itoh (1985) adopts the self-similar solutions from Chevalier (1982) for ejecta-dominated (ED—see Sect. 2.1) supernova remnants having ejecta with a power-law density distribution

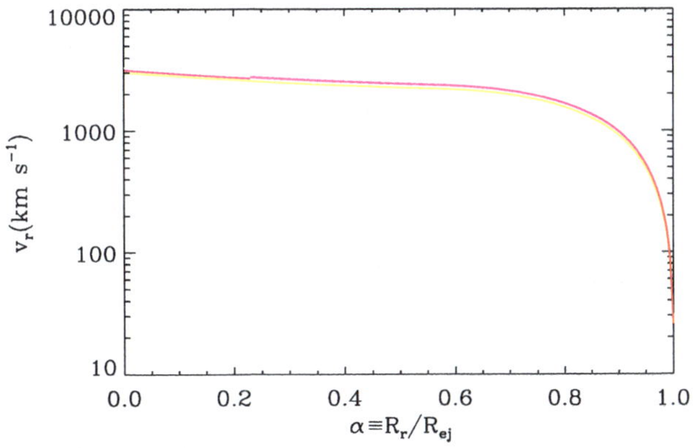

Fig. 11 Velocity of the reverse shock as a function of the parameter α for homogeneous SN ejecta expanding into a uniform medium. The value $\alpha = 1$ corresponds to the outermost ejecta layer, while $\alpha = 0$ to the innermost one. Figure reproduced from Dwek (2005)

(characterised by the index $n > 5$) expanding into a circumstellar medium which also has a power-law density distribution (with index $s < 3$).

Many of the works discussed in Sect. 5 are based on the analytical models of Truelove and McKee (1999) for the evolution of a young SNR transitioning between the ED and Sedov-Taylor (ST—see Sect. 2.1) phases. Truelove and McKee (1999) highlight the fact that the ED stage is strongly dependent on the ejecta parameters, and provide analytical expressions that smoothly merge the blast-wave and reverse shock solutions between the ED and ST stages. These expressions are particularly suited to describe the dynamical evolution of young remnants like Cas A, which are currently in an intermediate step between the ED and ST phases.

In Dwek (2005), the evolution of the reverse shock is provided by the solutions from Truelove and McKee (1999) for the specific case of homogeneous ejecta expanding into a uniform medium. The ejecta are characterised using the parameter $\alpha \equiv R_r/R_{ej}$, where R_r is the radius of the reverse shock and R_{ej} is the outer radius of the ejecta. With this parametrisation, $\alpha = 1$ corresponds to the outermost ejecta layer, and $\alpha = 0$ to the innermost one. The velocity v_r of the reverse shock as a function of the parameter α is reproduced in Fig. 11.

Also Bianchi and Schneider (2007) and later on Bocchio et al. (2016) describe the evolution of supernova remnants using the analytical solutions from Truelove and McKee (1999) for homogeneous ejecta expanding into a uniform medium. The ejecta are divided into spherical shells and those not yet visited by the reverse shock are considered to expand homologously. Three different values are adopted for the density of the interstellar medium where the remnants are expanding into. The strong adiabatic shocks under consideration are described using the classical Rankine-Hugoniot jump conditions. For these latter we think that the reader may find interesting to refer to the original publications (Rankine 1870; Hugoniot 1887, 1889). As Bianchi and Schneider (2007) emphasise, their approximation for the SNR evolution (density and temperature) is very crude, nevertheless it agrees with some 1D hydrodynamical models and simulations (van der Swaluw et al. 2003; Del Zanna et al. 2003).

To describe the evolution of the forward and reverse shocks in the supernova remnant, Nath et al. (2008) adopt a slightly more sophisticated approach, using again the self-similar solutions from Truelove and McKee (1999) but for ejecta with a core-envelope density structure, where the power-law envelope is characterised by the index n. The ejecta expand into a uniform ambient medium and different values of n are considered (which lead to different functional forms for the solutions—Truelove and McKee 1999): $n < 3$, $n = 4$ and $n > 5$.

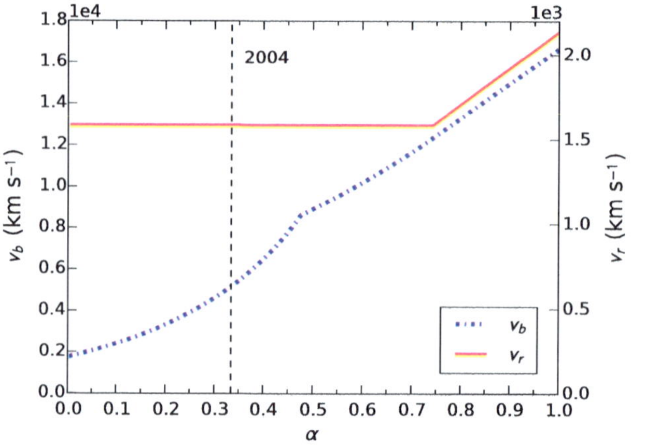

Fig. 12 Velocity of the blast-wave (v_b, *left* y-axis) and reverse shock (v_r, *right* y-axis) as a function of α, calculated for Cas A assuming inhomogeneous ejecta expanding into a non-uniform ambient medium. The curve for v_b gives the velocity of the blast-wave shock when the reverse shock hits a layer α of the ejecta. The reverse shock velocity is calculated in the frame of the unshocked ejecta ahead of it. The *vertical line* indicates the value $\alpha = 0.33$, which corresponds to the ejecta layer encountering the reverse shock in 2004. Figure reproduced from Micelotta et al. (2016)

The ejecta are divided into a series of shells identified by the same parameter α as in Dwek (2005), but here called w. It is assumed that the ejecta in a given shell evolve adiabatically until they are hit by the reverse shock. At that point, density and temperature jump to their post-shock values before continuing to evolve adiabatically, i.e. $\rho \propto t^{-3}$ and $T \propto t^{-2}$ for a monoatomic gas.

Biscaro and Cherchneff (2014, 2016) consider simple homologous expansion of the ejecta into a homogeneous medium, and the same approach is adopted by Nozawa et al. (2007) to study primordial type II supernovae (SNe II) and pair-instability supernovae (PI SNe), for which therefore the composition of the ambient medium is taken as primordial. Nozawa et al. (2010) consider Type IIb SNe with application to the specific case of Cas A, and adopt for the circumstellar medium not only a uniform profile but also a power-law density profile with exponent $s = 2$, produced by a steady stellar wind during the evolution of the progenitor.

To describe the dynamical evolution of Cas A, Micelotta et al. (2016) start from the seminal work of Truelove and McKee (1999) and its further development by Laming and Hwang (2003), but generalize their treatment to derive analytical expressions for the evolution of inhomogeneous ejecta (uniform core + power-law envelop with exponent $n = 9$) expanding into a non-uniform ambient medium (power-law with exponent $s = 2$—pre-supernova steady stellar wind). This configuration represents indeed a closer match to the physical situation of Cas A and the adopted parametrisation allows to match the radii and velocities of the forward and reverse shocks and the pre-shock density of the smooth ejecta. The ejecta layers are characterised by the same parameter α as in Dwek (2005). Figure 12 shows the velocities of the forward and reverse shocks as a function of α, calculated with the equations derived by Micelotta et al. (2016) for Cas A. The comparison with Fig. 11 reveals how a different density structure for the ejecta and the ambient medium modifies the behaviour of the reverse shock velocity. In the homogeneous ejecta—uniform ambient medium case (Fig. 11) the reverse shock starts at a very low velocity in the outermost ejecta layer ($\alpha = 1$) before rising steeply to its maximum value in the innermost layer ($\alpha = 0$). For inhomogeneous ejecta expanding into a power-law medium (Fig. 12) the situation is different: the velocity of the reverse shock is at maximum for $\alpha = 1$ and decreases to reach a constant value of $\sim$1600 km s^{-1}.

Finally, the results from the simulations in Silvia et al. (2010, 2012) are not included into an evolutionary model of the reverse shock.

5.2 The Physical Properties of SN Ejecta

There is a complex interplay between the physical properties of SN ejecta and the processing of dust grains within them. We already mentioned that the composition of the ejecta determines which species will bombard and erode dust grains. We have shown in Sect. 5.1 that different density profiles for the ejecta and the surrounding medium produce different temporal profiles for the reverse shock velocity. The proper evaluation of dust survival in SNRs requires a knowledge as accurate as possible of the conditions where dust processing takes place. Some of the ejecta properties can be derived from observations, while SN models are necessary for some others, as shown below.

For the parameters of SN 1979c and SN 1980k, Itoh (1985) adopts estimates based on, or consistent with, the analysis of Dwek (1983). Bianchi and Schneider (2007) use the physical properties derived from the SN models of Woosley and Weaver (1995). These include the metallicity and mass of the progenitor star, which determine the initial composition of the ejecta, and the mass of the ejecta and the kinetic energy of the explosion, where these latter control the dynamics of the ejecta. The initial temperature and density of the ejecta are chosen to match those of SN 1987A. Dust grains collide with H, He and O in the ejecta, taken as the most abundant elements. However, this assumption does not fully apply to remnants like Cas A, whose hydrogen content is negligible (e.g. Nozawa et al. 2007). For the specific case of Cas A, Bianchi and Schneider (2007) found that all the available observational constraints (at the time of publication) could be matched by a 15–25 $M_\odot$ progenitor which lost its hydrogen envelope. They apply their evolutionary model for the ejecta to a 12 $M_\odot$ progenitor star, but have to neglect the hydrogen mass. The adopted ejecta mass is $\sim$3 $M_\odot$.

For primordial SNe II and PI SNe, Nozawa et al. (2007) use the initial conditions for the density and velocity structure of the ejecta from the hydrodynamical models for Population III SNe by Umeda and Nomoto (2002). An important assumption is that SNe II and PI SNe have retained their thick hydrogen envelopes, because no mass loss seems possible in these massive metal-free stars. The best explosion model for each supernova studied by Bocchio et al. (2016) is selected comparing the available physical properties derived from observations with the output from the FRANEC stellar evolutionary code (Limongi and Chieffi 2006). For the shocked gas in all the considered remnants, Bocchio et al. (2016) adopt a composition of H^+, He^{++} and O^+, taken as the most abundant elements. They do not provide a description of the ionisation state of the gas, under the hypothesis that the very high temperatures considered will imply full ionisation for the gas. These assumption might not be completely valid in all the remnants under examination. Oxygen is the most abundant element in the smooth ejecta of Cas A, but Ne, Mg, Si, S, Ar, and Fe are also present (Hwang and Laming 2012) while H and He are negligible. The results from the photo-ionization and shock modelling code Mappings III (Allen et al. 2008) indicate that the dominant ionisation state for oxygen is +6 or higher, but full ionisation for the other elements is unlikely (Micelotta et al. 2016). For all remnants, the temperature of the shocked gas is calculated assuming equilibration between ions and electrons. However, this is not the case for Tycho (Yamaguchi et al. 2014) or Cas A (Micelotta et al. 2016, see Sect. 5.4).

Biscaro and Cherchneff (2014, 2016) consider for Cas A a model with stratified ejecta, microscopic mixing in each layer and no leakage between zones. The elemental composition is given by the model of Rauscher et al. (2002) for a 19 $M_\odot$ SN progenitor. Number density and temperature profiles are taken from the explosion models for type IIb SNe of Nozawa et al. (2010) which assume homologous expansion for the ejecta gas. All the parameters adopted by Micelotta et al. (2016) for Cas A or derived from their dynamical evolution model are in agreement with current estimates and/or observations and are explicitly reported in their Table 1. Both Biscaro and Cherchneff (2014, 2016) and Micelotta et al. (2016) model the ejecta as overdense clouds embedded into a tenuous and smoother medium, in agreement with observations (e.g. Fesen et al. 2001; Rho et al. 2009a, 2012; Wallström et al. 2013). It is assumed that the clouds have a composition of pure oxygen (Docenko and Sunyaev 2010; Chevalier and Kirshner 1979). For the smooth inter-clump medium Micelotta et al. (2016) adopt the composition determined by Hwang and Laming (2012) from X-ray observations, which consists of O, Ne, Mg, Si, S, Ar, and Fe, while H and He are negligible. The inclusion of such heavier elements implies higher sputtering rates with respect to the case when much lighter projectiles are considered. The properties of the ejecta clouds and the importance of considering a clumpy medium are discussed in Sect. 5.3.

5.3 Clumpy Ejecta

SN ejecta are not homogeneous but they show a variety of structures, ranging from dense clouds to filaments. As already mentioned, this is the case for Cas A, where dense clumps become optically visible due to heating by the reverse shock and fade out progressively. The dust residing inside the clumps is processed by the reverse shock. Biscaro and Cherchneff (2014) have shown that the presence of warm dust in Cas A "detected in the IR" seems to require high-density clumps in the ejecta as main sources of molecules and dust. Indeed, the low-density stratified ejecta resulting from explosion models of type IIb SNe which do not include over-dense regions are incapable of synthesising large amounts of molecules and are almost dust-free.

The presence of clumpy ejecta has a strong impact on the survival of dust because it changes the physical conditions under which processing occurs. To study this problem, Silvia et al. focus on the specific idealised case of an individual ejecta clump encountering a planar reverse shock, performing three-dimensional (3D) cloud-crushing simulations. The analogue problem of ISM clouds hit by SN-driven shocks has been already treated using hydrodynamic simulations (Bedogni and Woodward 1990; Stone and Norman 1992; Klein et al. 1994; Mac Low et al. 1994; Orlando et al. 2005; Patnaude and Fesen 2005; Nakamura et al. 2006). Here, the parameter space is adjusted to the new physical conditions, although it should be noted that the simulations are not included into an evolutionary model of the reverse shock. The simulations are performed using the cosmological, Eulerian adaptive mesh refinement (AMR), hydrodynamics + N-body code, Enzo (Bryan and Norman 1997; Norman and Bryan 1999; O'Shea et al. 2005). When the reverse shock with velocity $v_{\rm shock}$ encounters a cloud with constant overdensity χ, its propagation velocity inside the cloud is reduced by a factor of $\chi^{1/2}$. The simulations follow the dynamical fragmentation of the cloud due to the passage of the shock, whose timescale is given by the cloud-crushing time $t_{\rm cc} = \chi^{1/2} r_{\rm cloud}/v_{\rm shock}$, where $r_{\rm cloud}$ is the radius of the cloud (Klein et al. 1994). In fact, $t_{\rm cc}$ is the time required by the reverse shock to cross half of the cloud diameter. The calculations are performed for different values of $\chi^{1/2}$ and $v_{\rm shock}$ and consider the effect of gas cooling as well (see Sect. 5.4). An example of the results is shown in Fig. 13, which compares the evolution of an ejecta clump with and without cooling in the case of $\chi = 1000$

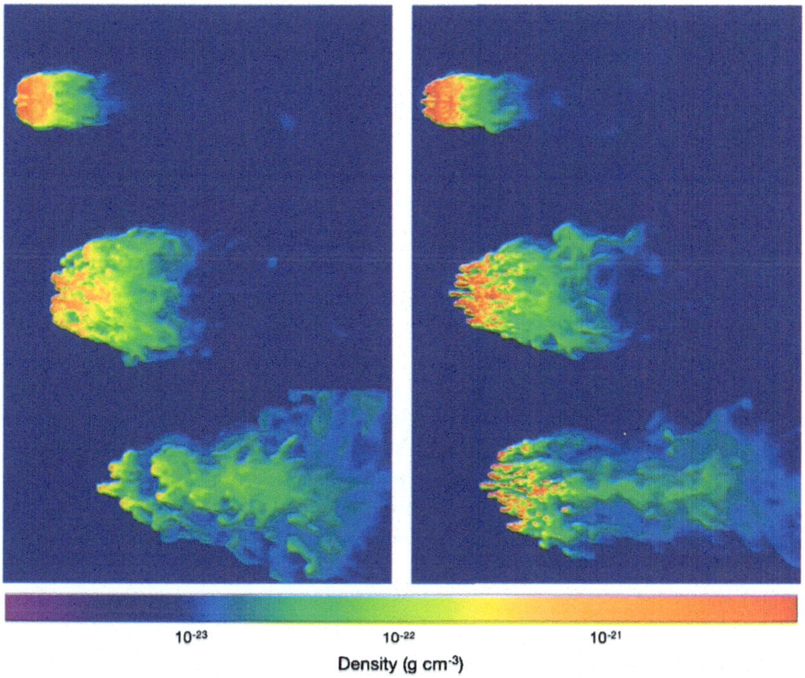

Fig. 13 Evolution of an ejecta clump with density contrast $\chi = 1000$ crushed by a shock with velocity $v_{\rm shock} = 1000\ {\rm km\,s^{-1}}$. The figure depicts the density projections (x–z plane) for a simulation with cooling off (*left panel*) and cooling on. Time progresses from top to bottom and the three snapshots have been taken at $t = 1.8t_{\rm cc}, 2.4t_{\rm cc}, 3.0t_{\rm cc}$. Figure reproduced from Silvia et al. (2010)

and $v_{\rm shock} = 1000\ {\rm km\,s^{-1}}$. It can be noted that while the simulation progresses, more and more material gets shredded by the constant inflow. When cooling is turned on, cold and dense nodules form in the fragmenting cloud. These features appear to be amplifications of overdensities created in the clump.

Biscaro and Cherchneff (2014, 2016) consider dense ejecta clumps which are currently being processed in Cas A. For the velocity of the reverse shock inside the clump, they explore the range 35–200 $\rm km\,s^{-1}$ (consistent with the results of Docenko and Sunyaev 2010) which can be obtained adopting $\chi = 100$–1000 and velocities of the reverse shock between 1000 and 2000 $\rm km\,s^{-1}$ (compatible with the value of $\sim$2000 $\rm km\,s^{-1}$ from Morse et al. 2004). The time variation of the gas parameters of the clumps shocked by the reverse shock is derived from the model of Borkowski and Shull (1990) and the results of Docenko and Sunyaev (2010), combined with the equation for homologous expansion of the ejecta from Nozawa et al. (2010). The lifetime of the cloud against shock disruption is given by $3 \times \tau_{\rm cc}$ (Klein et al. 1994; Silvia et al. 2010), where Biscaro and Cherchneff (2016) adopt for $\tau_{\rm cc}$ a value of 100 years (given by a clump radius of 10^{16} cm, a density contrast of 1000 and a velocity of the reverse shock of 1000 $\rm km\,s^{-1}$, Silvia et al. 2010). It should be noted, however, that this cloud lifetime does not agree with the optical lifetime of $\sim$30 years deduced from optical observations for the Fast Moving Knots (FMKs) in Cas A (Kamper and van den Bergh 1976).

Biscaro and Cherchneff (2016) assume that dust is processed via inertial sputtering in the clumps and it is instantaneously released into the smooth ejecta after the cloud has been disrupted. However, this approach does not take into account the fact that the residence time of the dust in the clumps can be shortened by the thermal evaporation of the clouds (Cowie and McKee 1977) and by the capability of the grains to escape from the cloud due to their ballistic velocities. These phenomena have been investigated by Micelotta et al. (2016) and are discussed in Sect. 5.6.

For the properties of the ejecta clouds Micelotta et al. (2016) adopt a fixed set of values capable of reproducing observations. The assumed pre-shock density for the clouds is $n_{\rm cloud} = 100\ {\rm cm}^{-3}$ with a density contrast $\chi = 100$, which allow to reproduce the optical spectra of the fast moving knots (FMKs) in Cas A (Sutherland and Dopita 1995). The resulting pre-shock density of the smooth ejecta, $n_{\rm smooth} = n_{\rm cloud}/\chi = 1.0\ {\rm cm}^{-3}$, is consistent with the ambient density of 0.1–10 $\rm cm^{-3}$ estimated by Morse et al. (2004). The diameter of a typical cloud is taken as 1.5×10^{16} cm $\simeq 0.005$ pc, within the observed range of $(1\text{–}5) \times 10^{16}$ cm (Fesen et al. 2011). The passage of the reverse shock inside the clouds heats them up so that they become optically visible before and/or while getting fragmented. This optical lifetime is defined as the time required by the reverse shock to cross the cloud and is therefore given by twice the cloud crushing time. With their calculated value of $\sim$1600 $\rm km\,s^{-1}$ for the current velocity of the reverse shock in the smooth ejecta of Cas A, Micelotta et al. (2016) obtain $2t_{\rm cc} = 30$ yr, in agreement with the determinations from Kamper and van den Bergh (1976).

5.4 Ejecta Cooling

The type and level of processing of dust in SNRs depend on the temperature of the ejecta embedding the grains. The temperature of the immediate post-shock gas can be calculated using the classical expression derived from the Rankine-Hugoniot jump conditions:

$$T = \frac{3}{16}\frac{\mu}{k_b}v_s^2, \tag{12}$$

where μ is the mean molecular mass per particle and v_s is the velocity of the shock. The temperature depends therefore not only on the velocity of the shock but also on the mass and ionisation state of the gas particles (included in the term μ). Equation (12) has been used (implicitly or explicitly) in all the papers discussed in Sect. 5. The equation assumes full electron-ion temperature equilibration, however the validity of this assumption depends on the velocity of the shocks considered. As pointed out by Micelotta et al. (2016), for the reverse shock in Cas A ($v_s \sim 1600\ {\rm km\,s^{-1}}$) the results of van Adelsberg et al. (2008) would indicate that the hypothesis of temperature equilibration may not be fully justified.

The typical temperature of the post-shock X-ray emitting gas in the smooth ejecta of SNRs like Cas A is of the order of 10^7–10^8 K. The value of the immediate post-shock temperature during the propagation of the reverse shock is determined by the model adopted for the evolution of the reverse shock itself coupled with Eq. (12). The ejecta at any given location of the remnant are assumed to evolve adiabatically until they are hit by the reverse shock. After the gas gets shocked, the temperature jumps to its post-shock values before continuing to evolve adiabatically. For the high temperatures under consideration, gas radiative cooling is generally negligible. This has been shown explicitly by Micelotta et al. (2016) for Cas A. The temperature evolution of the shocked smooth ejecta is calculated using Mappings III under non-equilibrium ionization (NEI) conditions. The results show that the temperature remains constant at $\sim 10^8$ K for more than 10^4 years before dropping abruptly. This is longer than the time frame for dust evolution considered in their work, and it holds for the majority of the studies considered here. In addition to atomic and inverse Compton cooling, Nozawa et al. (2007) include thermal cooling from the collisionally heated dust in the post-shock gas, showing that this has little effect on the destruction efficiency of the dust.

If the ejecta contain dense clumps, these latter will be shocked and heated up as well by the reverse shock propagating through them, but the post-shock temperature will be lower because of the reduced speed of the shock (the shock velocity inside the clumps is reduced by a factor of $\sqrt{\chi}$, where χ is the ratio between the density of the cloud and the density of the smooth medium outside). Silvia et al. (2010) investigate the effects on dust survival of including radiative cooling in a gas with half-solar metallicity. Cooling is described using the analytical formula of Sarazin (1987), which approximates the cooling rate as a function of temperature for a gas with half-solar metallicity in ionisation equilibrium. Silvia et al. (2012), considers the more realistic case of metal-enriched ejecta knots including the cooling of a gas with metallicity ranging from $Z = Z_{\odot}$ to $Z = 100\ Z_{\odot}$. This is an important point to address because the ejecta are indeed expected to have a high metal content, which would change their cooling properties and also provide heavier particles which are more efficient in sputtering the dust. In this case, cooling is calculated using the photoionisation code Cloudy (Ferland et al. 1998) with the assumptions that (i) the gas is in ionisation equilibrium, and (ii) both ions and electrons have the same temperature. These assumptions may not be fully valid in some supernova remnants like Cas A (Micelotta et al. 2016).

For the post-shock temperature of the oxygen-rich dense clumps in Cas A, Biscaro and Cherchneff (2014, 2016) adopt the values from Borkowski and Shull (1990) and Docenko and Sunyaev (2010). For the calculation of sputtering, the specific value of 1500 K from Docenko and Sunyaev (2010) is used.

To calculate the immediate post-shock temperature in the ejecta clouds of Cas A, Micelotta et al. (2016) as well use Eq. (12), assuming that the pre-shock ionisation state of oxygen is +2. The assumption of temperature equilibration between ions and electrons implicit in Eq. (12) is justified in the present case because the equilibration length scale is of the order of about (2–4) $\times 10^{13}$ cm (Itoh 1981, J. Raymond, private communication) corresponding to less than 0.27% of the diameter of clumps. Such a quick temperature equilibration is in agreement with the findings from van Adelsberg et al. (2008) and Ghavamian et al. (2007). The resulting post-shock temperature is $T_{\rm cloud} \sim 3 \times 10^6$ K, consistent with Sutherland and Dopita (1995) and Borkowski and Shull (1990). Micelotta et al. (2016) calculate explicitly the post-shock cooling of the dense clumps, using the functions computed by Sutherland and Dopita (1995, see their Fig. 2) which differ from the equilibrium curves often used (e.g. by Silvia et al. 2012). The work of Sutherland and Dopita (1995) appears indeed as the only available estimate of the cooling function of an oxygen-rich shocked gas. The calculations performed by Micelotta et al. (2016) show that cooling is very rapid due to the extreme metallicity of the gas. On a timescale of six months the temperature drops from $\sim 3 \times 10^6$ K to 300 K. The thickness of the layer beyond the shock front having a temperature sufficiently high for efficient thermal sputtering ($\sim 10^5$ K) represents only 1.4% of the diameter of the clump. A very small amount of dust populates at each time such a thin layer, which allows to neglect thermal sputtering in the ejecta clumps.

5.5 The Initial Grain Size Distribution and Composition of the Newly Formed Dust

The initial composition and size distribution of the dust synthesised in SN ejecta are two of the key input parameters in any model of dust processing inside SNRs. Such knowledge

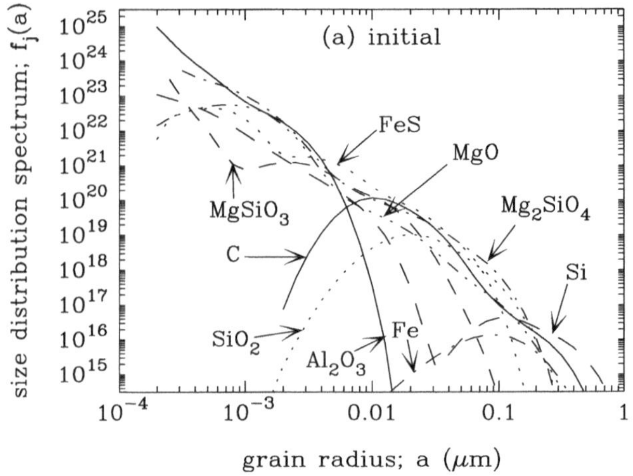

Fig. 14 Size distributions of the different dust species synthesised in SN ejecta (standard model). Figure reproduced from Nozawa et al. (2007)

relies heavily on theoretical models of dust formation, which can produce very different results even for similar initial conditions. The various scenarios for dust formation in SN ejecta are discussed in our companion paper (Sarangi et al. 2018). We summarise here the choices made in the studies we are discussing, in the perspective of evaluating the impact of these parameters in the resulting dust survival rate.

Nozawa et al. (2007) assume that dust condenses in the He core, where the metal-rich cooling gas provides the appropriate conditions for grain nucleation and grow according to classical nucleation theory (CNT—see e.g. Kalikmanov 2013, for a recent review). The resulting dust composition and size distributions depend on the elemental abundance inside the He core (Kozasa et al. 1989). Dust formation is based on the Nozawa et al. (2003) model, whose calculated distributions for dust synthesised in a CCSN with a progenitor mass of 20 $M_\odot$ are also those adopted by Silvia et al. (2010, 2012). Nozawa et al. (2003) consider two cases for the elemental composition of the ejecta: (i) Unmixed ejecta which retain their original onion-like structure, where different grains will be produced in the different layers: carbon grains in the C-rich He layer; Al_2O_3 (alumina), Mg_2SiO_4 (forsterite) and MgO grains in the O–Mg–Si layer; Al_2O_3, $MgSiO_3$ (enstatite) and SiO_2 (silica) grains in the O–Si–Mg layer; Si and FeS grains in the Si–S–Fe layer; and Fe grains in the innermost Fe–Ni core.; (ii) Mixed ejecta where oxides (Al_2O_3 and Fe_3O_4—magnetite) and silicates ($MgSiO_3$, Mg_2SiO_4 and SiO_2) form in uniformly mixed ejecta with C/O < 1. As an example, we reproduce in Fig. 14 the size distribution of each dust species expected from the "standard model" in Nozawa et al. (2007) i.e. the one with progenitor mass $M_{\rm pr} = 20\ M_\odot$, initial hydrogen ambient density $n_{\rm H,0} = 1\ {\rm cm}^{-3}$ and unmixed grain model inside the He core. The same approach is used by Nozawa et al. (2010), who find that the average radius of the produced dust is smaller than 0.01 μm, much less than the size of grains produced by SNe II-P with massive hydrogen envelopes.

CNT is also used by Bianchi and Schneider (2007) to form amorphous carbon (AC), iron, Al_2O_3, Fe_3O_4, $MgSiO_3$, Mg_2SiO_4 and SiO_2 with the expected log-normal distribution. The same code used in Bianchi and Schneider (2007) but with an upgraded molecular network from Marassi et al. (2015) is utilised by Bocchio et al. (2016). For the size distribution of dust grains condensed in SN ejecta, Nath et al. (2008) refer to the work of Todini and Ferrara (2001) and Nozawa et al. (2003) and adopt two power-laws $dn/da \propto a^\alpha$, where a is the radius of the grain and $\alpha = -2.5$ for smaller grains and $\alpha = -3.5$ for larger grains.

A simplified composition of graphite and silicate grains is considered (as in the interstellar medium, see the review from Draine 2003, and the references therein) and the same size range $10^{-7} - 3 \times 10^{-5}$ cm is assumed for all grains, although an upper limit of 10^{-6} cm is also considered.

Because of its relative simplicity, CNT has been widely used to model the formation of dust grains in astrophysical environments (starting from e.g. Yamamoto and Hasegawa 1977; Draine and Salpeter 1977) including SN ejecta (e.g. Kozasa et al. 1989; Todini and Ferrara 2001; Nozawa et al. 2003; Schneider et al. 2004; Nozawa et al. 2007; Bianchi and Schneider 2007), however Biscaro and Cherchneff (2014, 2016) emphasise the fact that this formalism is likely inadequate for this purpose because the assumed gas equilibrium conditions do not apply to a shocked gas in an environment controlled by radioactivity (Donn and Nuth 1985; Cherchneff and Dwek 2009). In addition, the CNT approach does not discriminate between possible dust chemical compositions, likely leading to a variety of condensates which have not been identified in SN emission spectra or pre-solar grains. Instead of CNT, Biscaro and Cherchneff (2014) use a chemical kinetic approach under non-equilibrium conditions to determine the masses and types of molecules and dust clusters synthesised by the SNe, and feed the results into the model developed by Sarangi and Cherchneff (2015) for grain growth via coalescence and coagulation of these dust clusters. This procedure is used to calculate the grain size distributions and masses expected from the SN progenitor for four types of dust: silicates in the form of forsterite (Mg_2SiO_4), alumina (Al_2O_3), carbon, and silicon carbide (SiC). The preferred chemical kinetic treatment provides a more realistic description of the environmental conditions where dust forms and a more complete approach to dust formation which includes both the gas phase (molecules formation) and the solid phase (dust formation via condensation of dust clusters).

Due to the uncertainties in the initial grain size distribution of SN-condensed dust, Micelotta et al. (2016) adopt as a test case the classical MRN power-law expression with exponent -3.5 typically found in the diffuse ISM (Mathis et al. 1977), although this may not fully apply to SN dust. A simplified composition of silicates ($MgSiO_3$) and amorphous carbon (AC) is assumed. The MRN distribution has also been the choice of Itoh (1985).

5.6 Dust Destruction in SNRs: Thermal and Inertial Sputtering

Dust grains progressively embedded in the reverse-shocked gas are affected by a variety of destructive processes resulting from the collisions with gas particles and other dust grains. However, not all these processes have the same importance in modifying the initial size distribution of dust under the conditions found in supernova remnants. The four main processes to consider are the following: (i) sputtering, i.e. the ejection of atoms from the grain surface due to grain collisions with high-velocity atoms and ions, with consequent erosion of the target dust particles; (ii) sublimation, i.e. the direct transition from the solid to the gas phase due to collisional heating in the high-temperature gas; (iii) shattering, i.e. fragmentation of the dust particles into smaller fragments due to grain-grain collisions; (iv) partial or complete vaporisation of the grains, i.e. rapid destruction with return of the dust constituents to the gas phase due to the intense heat generated in grain-grain collisions.

The relative contribution of these four processes has been evaluated by Bocchio et al. (2016). Sublimation is treated in the same way and gives the same results as in Bianchi and Schneider (2007), i.e. a negligible contribution to dust destruction for all the SNRs considered in their work. Shattering and vaporisation are evaluated using the method from Tielens et al. (1994) and Jones et al. (1996) and lead as well to a minor contribution to dust processing with respect to sputtering, because of the low dust density which prevents frequent grain-grain collisions, as also pointed out by Biscaro and Cherchneff (2016) and Bianchi and Schneider (2007). The main destruction agent for dust in SNRs is therefore sputtering, defined *kinetic* (or *inertial* or *non-thermal*) when the collision velocities are determined by the relative motion between the grains and the gas, and it is defined *thermal* when the collision velocities arise from the thermal motion of the gas. In this case, the erosion rate depends on the gas temperature.

The classical formalism for sputtering in solids (Matsunami 1981; Andersen and Bay 1981; Sigmund 1981; Bohdansky 1984) has been widely used to describe the sputtering of cosmic dust. Itoh (1985) considers thermal sputtering in the hot post-shock gas adopting a constant sputtering rate of 10^{-6} $\mu m\, cm^3\, yr^{-1}$ (Draine and Salpeter 1979a). This value represents a reasonable approximation (within 30%) of the sputtering rate of graphite, silicate, iron, and $H_2 0$ ice bombarded by hydrogen, helium and oxygen with relative abundances of 1, 0.1 and 0.001 respectively. All the other theoretical works discussed in Sect. 5 adopt the sputtering yields from Tielens et al. (1994) and/or Nozawa et al. (2006). The sputtering yield, Y, is defined as the number of atoms/molecules ejected from a grain per incident projectile of a given composition and depends on the energy of the projectiles. In particular, Bocchio et al. (2016) use a revised version of such formalism (Bocchio et al. 2014) where thermal and inertial velocities are not treated separately but their combined effect is estimated using a skewed Maxwellian distribution, as originally implemented by Shull (1978) and later on by Guillet (2008) and Bocchio et al. (2014). The sputtering yields are corrected to take into account the fact that sputtering is size-dependent because the targets are not semi-infinite slabs of material but particles with a finite radius (Jurac et al. 1998; Serra Díaz-Cano and Jones 2008; Bocchio et al. 2014). The mass of (sub)nano-sized species like PAHs is estimated negligible, thus the specific dissociation of these particles (Micelotta et al. 2010a,b) has not been implemented. However, this hypothesis still needs to be fully verified.

The dust erosion rate strongly depends on the dynamics of the dust. When passing through the reverse shock, grains of different sizes will have different velocities due to the effect of drag forces. These differential velocities are responsible for kinetic sputtering. Itoh (1985) estimates that the relative motion between dust and gas gets quickly and strongly reduced via direct collisions, dominating over Coulomb collisions. Magnetic forces, even if present, would not increase relative motions because of the reduction of the magnetic field due to the expansion of the remnant. Thermal sputtering is therefore the dominant dust erosion mechanism. Nozawa et al. (2007, 2010), Bocchio et al. (2016) and Biscaro and Cherchneff (2016) evaluate the contribution of both thermal and inertial sputtering. They treat explicitly the dynamics of dust using the classical expressions from Draine and Salpeter (1979a) and McKee et al. (1987). Betatron acceleration of the dust due to the effect of magnetic fields on charged grains is estimated to be unimportant (Bocchio et al. 2016). Bianchi and Schneider (2007) also consider both thermal and inertial sputtering but without treating dust dynamics. Gas drag and grain charge are not included and all grains are kept at the same initial velocity. Nath et al. (2008) do not include dust dynamics and consider only thermal sputtering in the hot post-shock gas. In the numerical simulations of Silvia et al. (2010, 2012) dust is assumed to be fully coupled with the flow of gas in the cloud. This implies operating in the limit of purely thermal sputtering.

The presence of overdense clumps strongly modifies the dynamics and processing timescale of the dust residing in there. Micelotta et al. (2016) study the dynamics of the dust grains located inside the cold ejecta clouds of Cas A using the formalism from Baines et al. (1965), Draine and Salpeter (1979a) and McKee et al. (1987). The grains acquire a relative velocity with respect to the gas due to the cloud shock propagating inside the cloud itself and generated by the interaction between the clump and the reverse shock. Gas-grain collisions

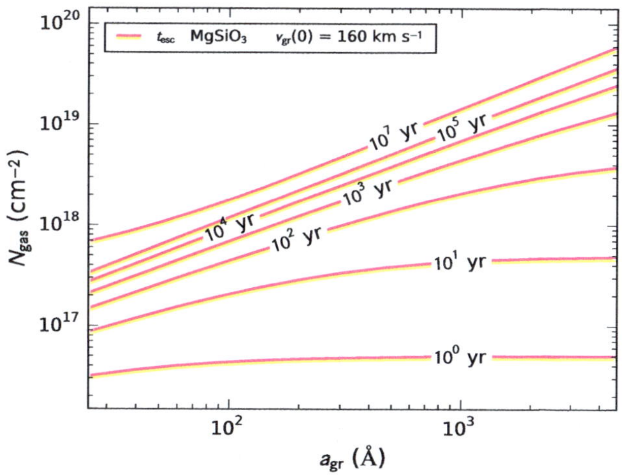

Fig. 15 Time t_{esc} required by a silicate dust grain with radius a_{gr} and initial velocity $v_{gr}(0) = 160\ \mathrm{km\,s^{-1}}$ to traverse a column density of gas N_{gas}. Figure reproduced from Micelotta et al. (2016)

reduce asymptotically to zero such relative velocity while eroding progressively the grains via kinetic sputtering. Figure 15 shows the time t_{esc} required by a silicate grain of a given size to traverse a column density of gas N_{gas}. The contour plot depicts t_{esc} as a function of the column density of gas traversed and dust grains radius. Large grains sitting close to the edge of the clump may be able to escape before suffering substantial sputtering, while smaller grains located in the inner regions are likely eroded and/or stopped before getting close to the surface.

The level of destruction due to inertial sputtering depends on how much time the grains spend inside the clumps. As mentioned in Sect. 5.3, there are three main phenomena affecting the residence time of the grains in the clouds (Micelotta et al. 2016). The first is the possibility for the grains to escape from the cloud due to their ballistic velocities, and the corresponding timescale is given by t_{esc}, discussed in the previous paragraph. The second is the dynamical fragmentation of the clouds due to the passage of the reverse shock. Following Klein et al. (1994), the time required for dynamical fragmentation, t_{dest}, is taken as $t_{dest} = 3.5 t_{cc}$ where t_{cc} is the cloud crushing time already discussed. With this choice of parameters, Micelotta et al. (2016) calculate $t_{cc} = 15$ yr and $t_{dest} = 52.5$ yr. This is much shorter than the value of 300 years adopted by Biscaro and Cherchneff (2016). It is assumed that after the time t_{dest} the ejecta cloud is fragmented and dispersed, releasing into the smooth ejecta the fresh dust still residing in there. The third competing process is the thermal evaporation of the clouds (Cowie and McKee 1977). The evaporation timescale, t_{sat}, is calculated using the results from Dalton and Balbus (1993) for the case of highly saturated conduction appropriate for the Cas A knots.

Figure 16 shows the above timescales for silicate grains as a function of the diameter of the clouds. The evaporation time is calculated for two temperatures of the smooth ejecta surrounding the clouds: $T = 10^7$ K (consistent with the electron temperature derived from X-ray observations Hwang and Laming 2009) and $T = 10^8$ K (derived from Eq. (12)). The two curves for the escape time t_{esc} are calculated for two grain sizes: 2500 Å (upper limit of the MRN grain size distribution) and 1000 Å (an intermediate value). Because the cloud diameter is the maximum distance that a dust grain would have to cross to escape, the plotted value of t_{esc} is the largest possible. Both t_{esc} and t_{dest} are calculated assuming for the grains an initial velocity of 160 $\mathrm{km\,s^{-1}}$, which is the lowest velocity reached by the reverse shock inside the ejecta clumps of Cas A (see Fig. 12). It follows therefore that the plotted t_{dest} is

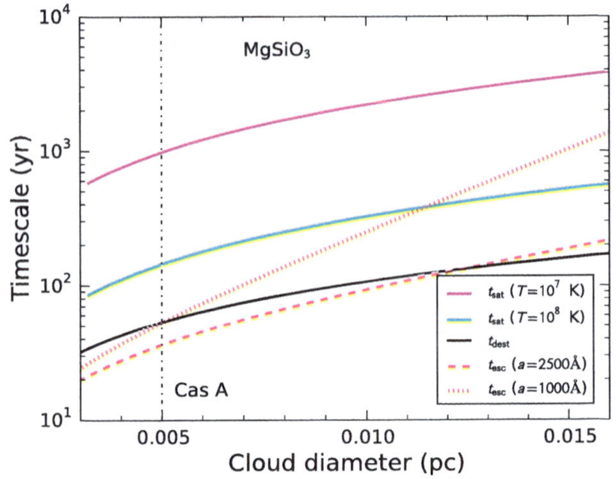

Fig. 16 Comparison between cloud destruction time t_{dest}, cloud evaporation time t_{sat} (calculated for two temperatures of the smooth ejecta) and dust escape time t_{esc} (calculated for two grain sizes) as a function of the diameter of the clump. For the calculation of both t_{esc} and t_{dest}, the initial velocity of the grains is $v_{gr}(0) = 160\ km\,s^{-1}$. The *vertical line* indicates the diameter adopted for the Cas A clumps. Figure reproduced from Micelotta et al. (2016)

also an upper limit. The inspection of the figure and the considerations above clearly show that in Cas A the trade is between the escape time and the destruction time. If the grains are sufficiently big, they will escape from the clouds before these latter get disrupted and they will start experiencing thermal sputtering in the hot smooth ejecta.

The initial size of the grains and their position inside the cloud determine whether they will be injected into the smooth ejecta before the cloud gets fragmented and dispersed. Such fugitive grains will thus start to be eroded by thermal sputtering in the smooth ejecta while the remaining dust population is still trapped inside the cloud. Micelotta et al. (2016) couple explicitly dust erosion and dust deceleration and implement their analytical expressions into a Monte Carlo simulation which allows to follow "on the fly" the journey of each particle through kinetic sputtering (in the clumps) and thermal sputtering (in the smooth ejecta) and determine the resulting dust size and mass.

Another important factor influencing the dust erosion rate is the location of the grains inside the remnants (Nozawa et al. 2007; Bianchi and Schneider 2007; Biscaro and Cherchneff 2014, 2016; Bocchio et al. 2016; Micelotta et al. 2016). The rates of deceleration and erosion depend on the chemical composition and size of the dust but also on the temperature and density of the shocked gas. All these may vary across the remnant and during its evolution. The compositions of the dust and of the projectiles that will be impinging on it depend on the composition of the ejecta layer where dust forms and on the level of mixing of the ejecta (Nozawa et al. 2007; Bianchi and Schneider 2007; Biscaro and Cherchneff 2014, 2016). As an example, Fe dust, which is formed in the innermost He core, will be crossing O-rich gas, experiencing efficient sputtering, while for C-dust in the outermost part of the He core, the possibility of a quick escape together with the smaller sputtering yield of carbon with respect to other species will result in a higher chance to survive (Nozawa et al. 2007). In the external layers, erosion is mostly due to He^+ and will affect carbon and SiC which form in there, while silicates and allumina form in the oxygen-rich zone and are mostly eroded by heavier O^+ and Mg^+ projectiles (Biscaro and Cherchneff 2016). Because large grains can travel longer distances in the shocked gas before being stopped, if they are located at the edges of the remnants they can more easily escape and avoid further erosion. The presence of ejecta clouds where the dust initially resides introduces an additional level of complexity. Because of the expansion of the supernova remnant, the density of the smooth ejecta changes while inertial sputtering is occurring inside the clouds. Therefore, dust is progressively injected into a medium whose density changes continuously, which implies different erosion rates (Micelotta et al. 2016).

5.7 Mass and Size Distribution of the Surviving Dust: Theoretical Results and Comparison with Observations

One of the main goals of all theoretical studies about dust processing in SNRs is to provide an estimate of the amount of surviving dust. Such estimate can be given as an *absolute value* or as a *fractional value*. This latter option depends exclusively on dust processing, with all the assumptions made to compute it, of course. The former option requires in addition an estimate of the amount of dust initially produced by the SN, which is still one of the most uncertain parameters in this field. In Sect. 5.7.1 we summarise the theoretical results obtained by the different studies under consideration and in Sect. 5.7.2 we show how the theoretical results for specific SNRs compare with observations. We would like to emphasise the fact that the different approaches, models, parameters and parameters' values chosen by the different authors prevent a direct global comparison between the results.

5.7.1 Theoretical Results for Dust Survival in SNRs

The calculations from Itoh (1985) already indicate that sputtering in the ejecta can heavily modify the amount of dust synthesised in SNe.

Dwek (2005) uses the ratio τ_{sput}/t versus α to identify where in the remnant dust sputtering takes place efficiently. The quantity τ_{sput} is the sputtering lifetime and t is the expansion age of the ejecta. As shown in Fig. 17, destruction occurs in the layers $0 < \alpha < 1$ where a combination of gas temperature (given by the reverse shock velocity) and ejecta density gives $\tau_{sput}/t < 1$. In the outer ejecta layer ($\alpha = 1$) the velocity of the reverse shock is zero, which implies a very low gas temperature and therefore projectiles which are not energetic enough to trigger the sputtering process. In the inner layer ($\alpha = 0$), the reverse shock velocity reaches its maximum value, but the corresponding high gas temperature does not compensate the very low density of projectiles. For smaller particles the condition $\tau_{sput}/t < 1$ is fulfilled over a wider range of α values than for the larger particles.

Similar conclusions about the location of efficient dust sputtering in SNRs are also reached by Bianchi and Schneider (2007) and Nath et al. (2008), which is not surprising because they also refer to the fully homogenous case of the Truelove and McKee (1999) model for the evolution of the supernova remnant. However, this trend also appears for ejecta density profiles with $n = 2, 4$ (Nath et al. 2008), even if in this case the velocity is a peaked function (Truelove and McKee 1999).

The calculated size distributions resulting from processing reveal that, for all the dust species considered, small grains (radius below a few 10^{-6} cm) are much more eroded that the bigger ones. These latter can survive sputtering but their distribution is shifted towards smaller sizes. The size distributions recovered by Nozawa et al. (2007) show an abrupt cut at small radii (Fig. 18), while those in Bianchi and Schneider (2007) flatten at the small-sizes end. According to Nozawa et al. (2007), this difference has to be ascribed to the fact that Bianchi and Schneider (2007) do not consider the dynamics of the dust: the grains are trapped and confined at their initial location where they are sputtered until the density drop prevents further erosion. In the case considered by Nozawa et al. (2007), even small grains with initial radius lower than 0.05 μm are able to penetrate into the hot plasma between the

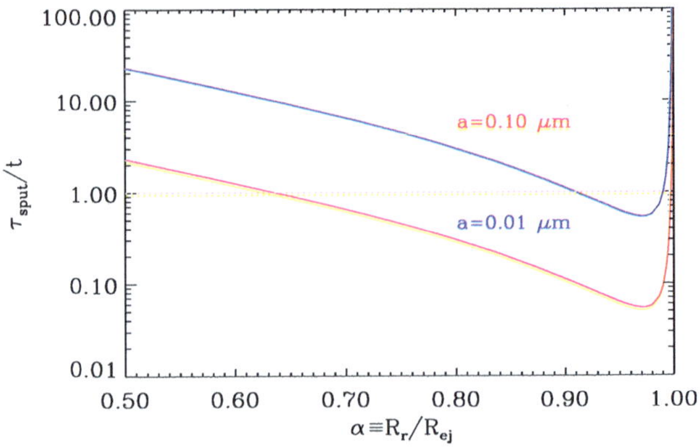

Fig. 17 Ratio τ_{sput}/t as a function of α calculated for grains with radius $a = 0.1, 0.01$ μm embedded in smooth oxygen-rich ejecta. Destruction occurs when $\tau_{sput}/t < 1$. Figure reproduced from Dwek (2005)

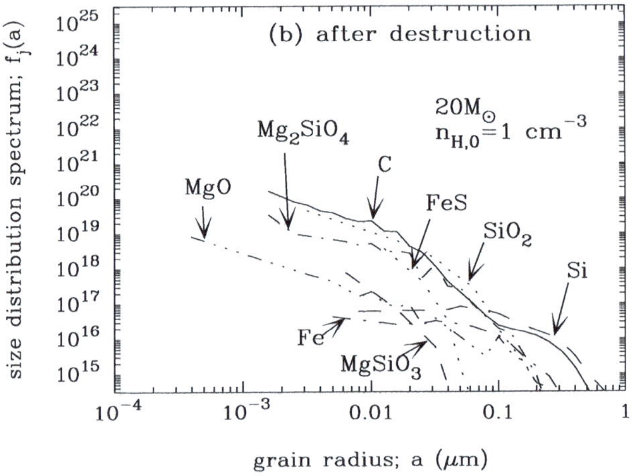

Fig. 18 Size distributions of the surviving dust resulting from processing in SNRs. Figure reproduced from Nozawa et al. (2007)

reverse and forward shocks, where they are completely eroded by thermal sputtering in the relatively high-density gas.

The amount of surviving dust depends in a complicated way on the approach used for the calculations and on the adopted initial conditions. It is strongly influenced by the composition and size distribution of the freshly formed dust (Bianchi and Schneider 2007; Nozawa et al. 2007, 2010; Bocchio et al. 2016; Biscaro and Cherchneff 2016). Because these latter differ from model to model, it is difficult to establish the relation between the destruction efficiency and the mass of the progenitor for each dust species (Nozawa et al. 2007). SNe exploding in a denser medium produce a reverse shock which travels faster through the ejecta and encounters denser gas, producing more sputtering. As a result, the mass of surviving dust decreases with increasing ambient density (Bianchi and Schneider 2007; Nozawa et al. 2007, Fig. 19 and Fig. 20a respectively) while the shape of the size distributions is almost unaffected (Bianchi and Schneider 2007). The amount of surviving dust is not impacted substantially by the metallicity of the progenitor (Bianchi and Schneider 2007) but it is reduced for higher explosion energies (Nozawa et al. 2007; Bocchio et al. 2016), in mixed SN ejecta (Nozawa et al. 2007) and when smaller sticking coefficients are considered for dust formation. This is because the resulting grains are smaller, therefore more prone to dissociation (Bianchi and Schneider 2007). Nath et al. (2008) find that the *fractions* of surviving dust

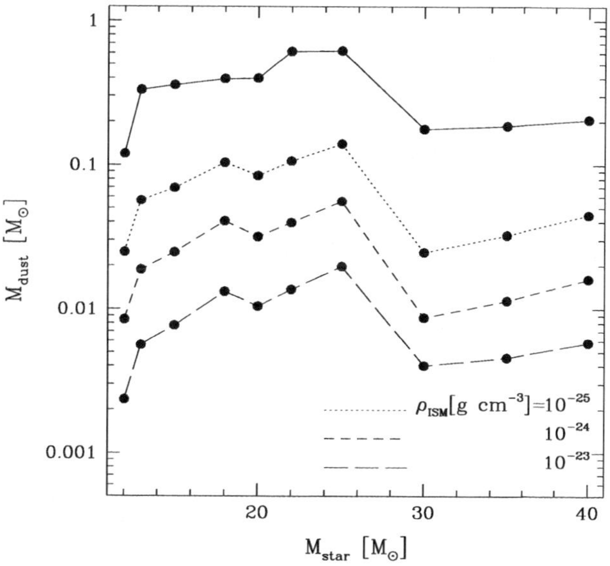

Fig. 19 Mass of dust surviving the processing by the reverse shock, as a function of the mass of the progenitor star, calculated for three densities of the ambient medium. The *solid line* represents the initial dust mass. Figure reproduced from Bianchi and Schneider (2007)

mass have an intricate dependency on the explosion energy and ejecta mass, but they do not depend on the density of the ambient medium.

An important point addressed by Nozawa et al. (2007) and Nozawa et al. (2010) is the effect of the thickness of the H envelope. In particular, the detailed study from Nozawa et al. (2010) on Type IIb SNe shows that the much thinner hydrogen envelope and higher expansion velocity of the He core in Type IIb SNRs with respect to Type II-P SNRs implies that the reverse shock reaches the He core at earlier times. Sputtering occurs in a denser medium and the increased collision rate between dust particles and gas projectiles leads to a more efficient erosion of the grains. This fact, combined with the smaller sizes of the synthesised grains, results in a complete destruction for all grains (and no injection into the ISM) if the ambient gas density exceeds 0.1 cm^{-3} (see Fig. 20b). Nozawa et al. (2010) conclude that, if the circumstellar medium is roughly spherical, low-mass envelope SNe cannot inject a significant amount of dust into the interstellar medium.

Bianchi and Schneider (2007) and Nozawa et al. (2007) provide the *absolute mass* of surviving dust with respect to the initial values calculated from their SN and dust coagulation models (Figs. 19–20b). Conversion to fractional values leads to various ranges, depending on the adopted parameters: between 2% and 20% of the initial value for Bianchi and Schneider (2007) and between 0 and 80% for Nozawa et al. (2007). Nath et al. (2008) calculate fractional values (Fig. 21) and with their power-law initial grain size distributions they find that between 80% and 99% of the mass of graphite and silicate grains is preserved if the ejecta have a shallow density profile ($n < 5$). If the profile is steeper, the inner shells may contain a large fraction of mass which will remain untouched. This suggests that a considerable fraction of the dust synthesised in SNe could reside in cold (still unshocked) ejecta. The fractions of surviving dust mass are larger than those reported by Bianchi and Schneider (2007) and Nozawa et al. (2007). These latter in particular include in their calculation the significant sputtering occurring in the hot gas located between the reverse and forward shocks, which is not considered in the analytical treatment of Nath et al. (2008). The density of such gas, and therefore the sputtering rate, strongly depends on the density of the ambient medium. Another point to consider is that the adopted rates for thermal sputtering from Tielens et al. (1994) are valid for a gas with solar composition, while SN ejecta are expected to

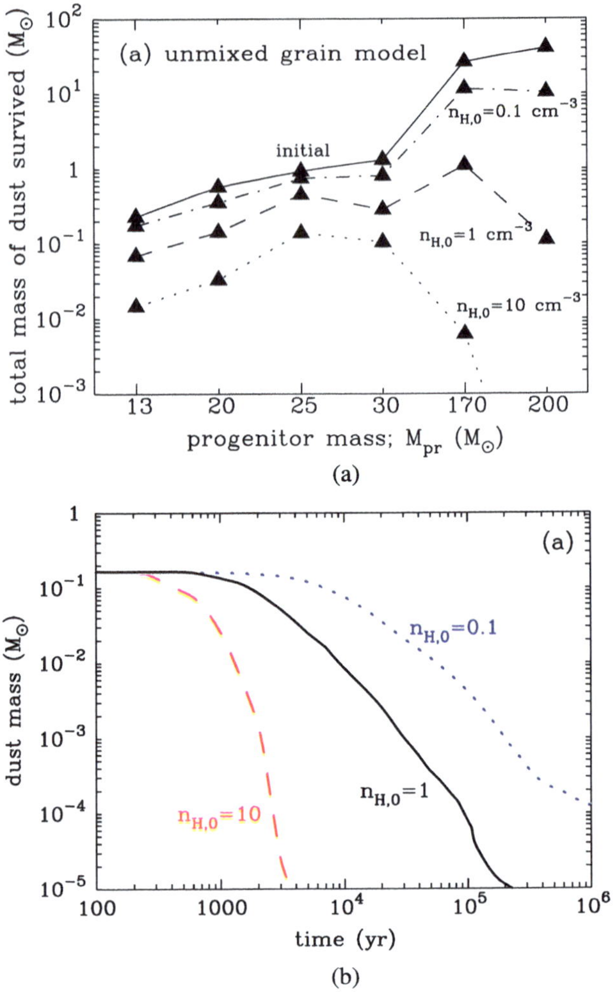

Fig. 20 (**a**)—Total mass of surviving dust as a function of the mass of the progenitor star, calculated for different densities of the ambient medium, $n_{H,0}$, for the unmixed grain model. The *solid line* represents the initial total mass of the formed dust. Figure reproduced from Nozawa et al. (2007). (**b**)—Temporal evolution of the total mass of dust inside a Type IIb SNR expanding into a homogeneous circumstellar medium having different densities. Figure reproduced from Nozawa et al. (2010)

be rich in metals, whose sputtering yields are much higher than those of hydrogen and helium. The power-laws adopted by Nath et al. (2008) extend over a wider range of sizes than the peaked distributions from Bianchi and Schneider (2007) which include smaller grains more prone to dissociation ($\leq$0.01 μm for silicates and $\leq$0.05 μm for graphite). These two factors would increase the fraction of mass sputtered away.

The simulations of Silvia et al. (2010) show that the relative velocity between the clumps and the reverse shock combined with the density of the clumps has a strong impact on the survival rate of the dust residing in the clouds, resulting in a ~50% additional mass loss (with respect to the low-velocity case) for high-density clumps ($\chi = 1000$) crossed by the highest velocity shock considered (5000 km s^{-1}). The total amount of surviving dust varies with the grain species and the adopted values of the parameters. As an example, Fig. 22 reproduces the *fractional* dust mass as a function of time for the case $\chi = 1000$ when the sputtering rates from Nozawa et al. (2006) are used. For high-velocity shocks impacting a high-density cloud, the percentage of surviving dust mass of some key species

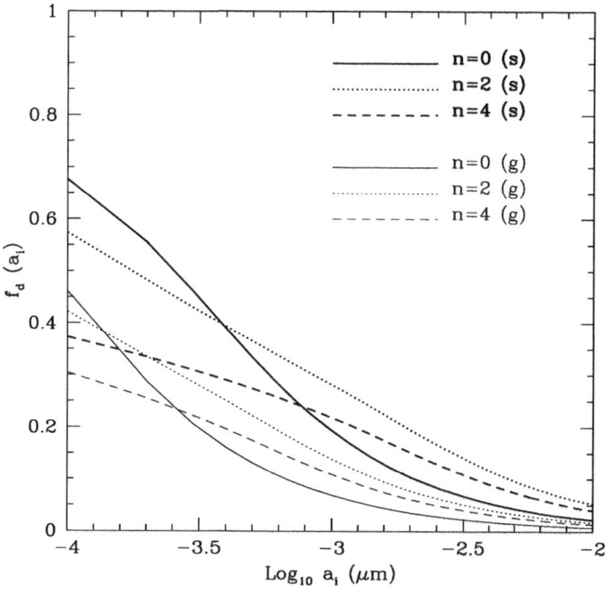

Fig. 21 Resulting fraction of dust mass sputtered by the reverse shock as a function of the initial radius of the grain, for graphite (*thin lines–g*) and silicate (*thick lines—s*) dust. The calculations have been performed for three ejecta mass profiles: $n = 0, 2, 4$ and the resulting fractions have been weighted by the mass of the shell where the grains are formed. Figure reproduced from Nath et al. (2008)

is the following: C—62%, SiO_2—20%, Fe—80%. Switching on cooling in the gas with half-solar metallicity considered in the study has little effect on the final amount of surviving dust.

In the higher metallicity gas investigated by Silvia et al. (2012) the abundances of hydrogen and helium are kept at their solar values, while for the other metals the abundances are increased by a factor of 10 and 100 with respect to solar. The results are shown in Fig. 23. The degree of destruction spans the entire parameter space, ranging from almost negligible destruction (Fe and Si, high-metallicity gas, low-velocity shock) to almost complete destruction (Al_2O_3, FeS, and $MgSiO_3$, high-metallicity gas, high-velocity shock). There is a clear correlation between shock velocity and amount of destroyed dust mass, and this latter increases sharply at early times when the shock velocity transitions from 1000 to 3000 km s^{-1}. For the key species C, SiO_2 and Fe, the simulations for a high-metallicity gas result in a wide range of surviving dust mass: 44–96%, 7–95%, and 24–99% respectively. The modifications introduced by Silvia et al. (2012), i.e. cooling functions and erosion rates more appropriate for high metallicities, a higher shock velocity considered, suppressed cooling in the medium surrounding the cloud, do not invalidate the results from Silvia et al. (2010). Indeed, the differences do not become significant until the abundances of metals are taken as 100 times their solar values.

Marassi et al. (2015) use the same code as in Bianchi and Schneider (2007) but with an upgraded molecular network to evaluate dust survival in the remnants from *standard* and *faint* Population III (Pop III) CCSNe, finding that 3–50% and 10–80% of the initially formed amount of dust survives, respectively.

Bocchio et al. (2016) study the evolution of dust mass in four supernova remnants: SN 1987A, Cas A, the Crab Nebula and N49. The results, given in terms of *absolute* masses of surviving dust, are compared with those from Bianchi and Schneider (2007), where the dynamics of the gas and the effect of the forward shock were not included, and with the observed dust masses (Fig. 24). Except for N49, the results from the models appear to be consistent with observations, but this should take into account the very large error bars of the datapoints. Between 10^3 and 10^4 years after explosion, Bocchio et al. (2016) predict a

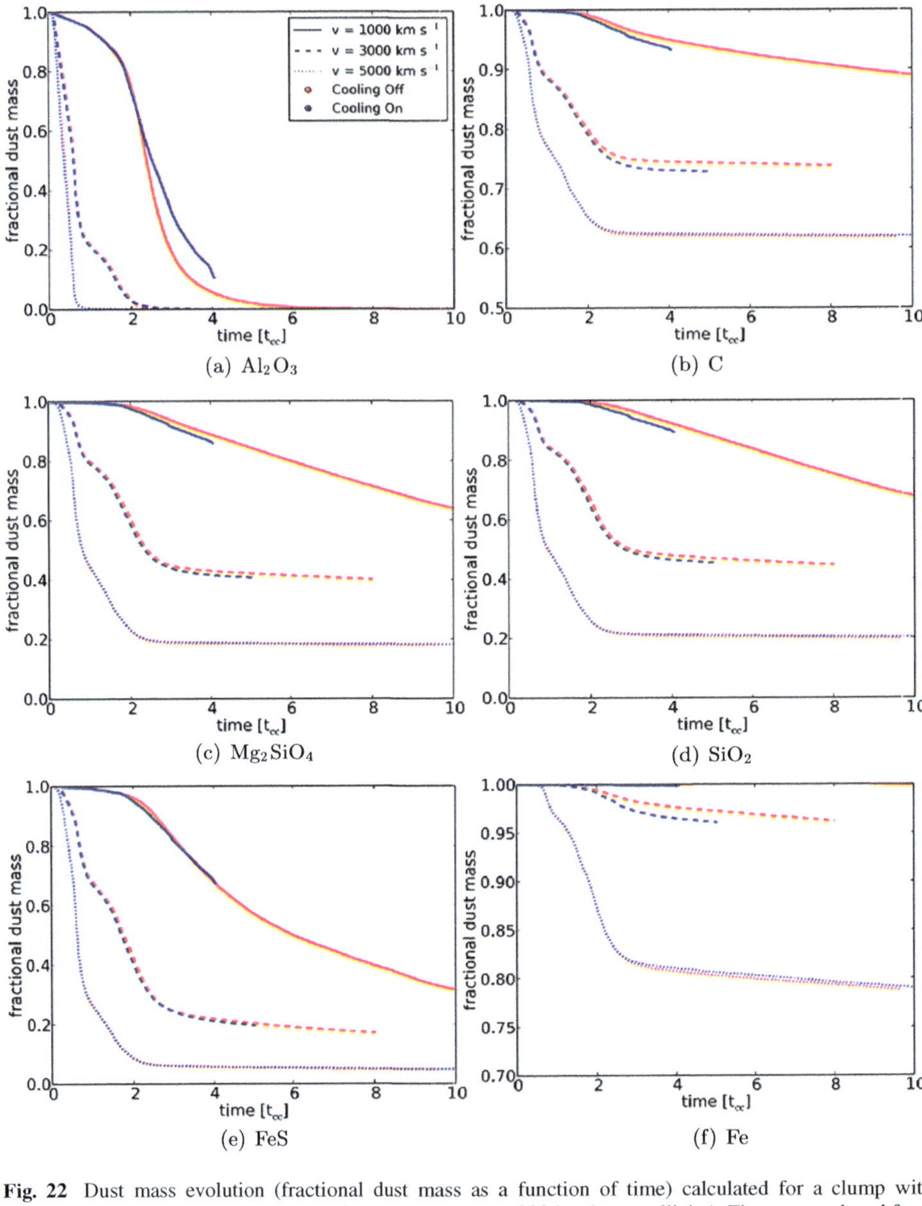

(a) Al_2O_3 (b) C

(c) Mg_2SiO_4 (d) SiO_2

(e) FeS (f) Fe

Fig. 22 Dust mass evolution (fractional dust mass as a function of time) calculated for a clump with $\chi = 1000$ adopting the sputtering rates from Nozawa et al. (2006, solar metallicity). Figure reproduced from Silvia et al. (2010)

larger mass of surviving dust than Bianchi and Schneider (2007), because in this time interval the dust decouples from the gas and travels in the region between the forward and reverse shocks, where the physical conditions are milder. For $t \gtrsim 10^4$ the situation is reversed because most of the grains have travelled beyond the forward shock, where dust erosion proceeds efficiently. Globally, the final masses estimated by Bocchio et al. (2016) are ~5–50% of those derived by Bianchi and Schneider (2007).

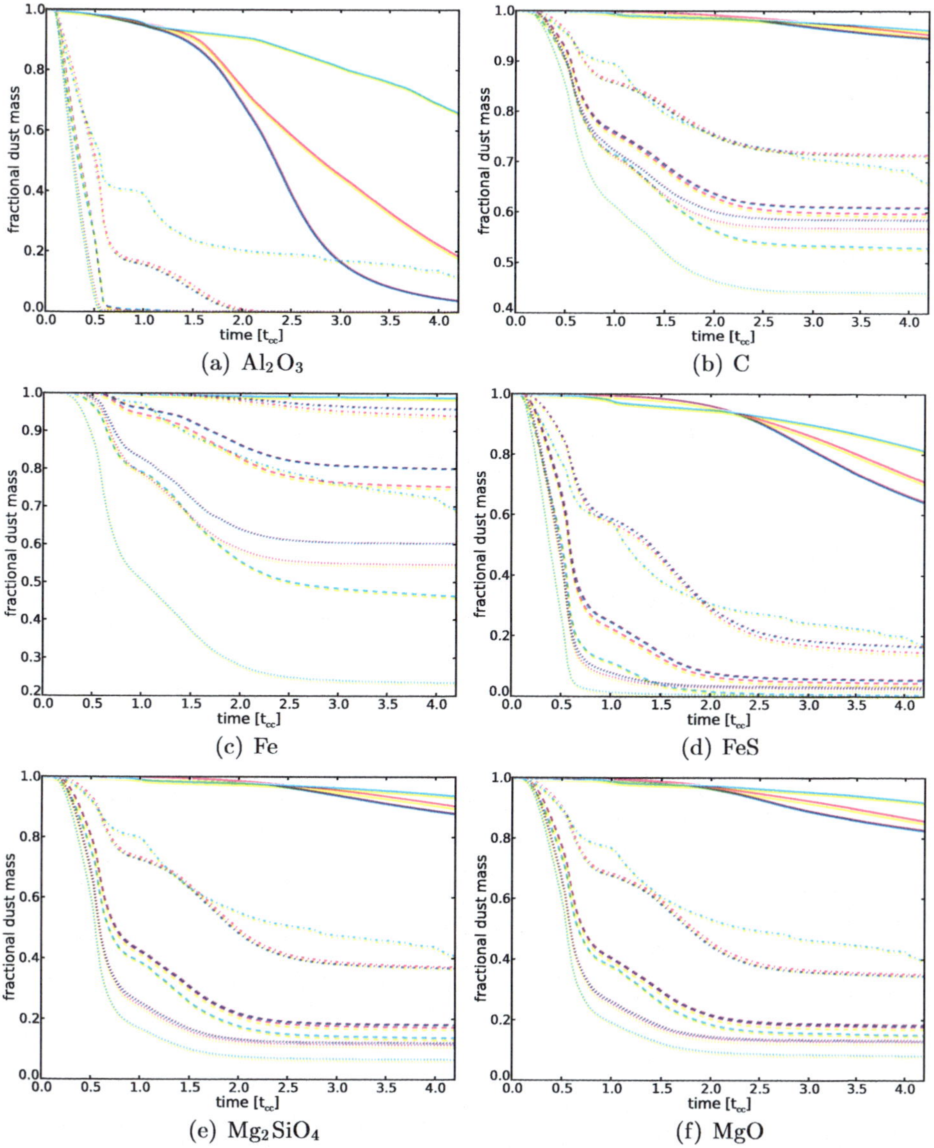

Fig. 23 Dust mass evolution (fractional dust mass as a function of time) calculated in the high-metallicity case. For each dust species, the curves show the results for four shock velocities: 10^3 km s^{-1}—*solid*, 3×10^3 km s^{-1}—*dot-dashed*, 5×10^3 km s^{-1}—*dashed*, and 10^4 km s^{-1}—*dotted*. Each colour corresponds to a different metallicity: $Z = 1\ Z_\odot$ (*blue*), $10\ Z_\odot$ (*red*), $100\ Z_\odot$ (*green*). Figure reproduced from Silvia et al. (2012)

Biscaro and Cherchneff (2016) study the processing of dust grains by the reverse shock first inside the clumps and after in the inter-clump medium. They consider two over-density factors: 200, appropriate for Cas A, and 2000, typical of the homogeneous ejecta of Type II-P SNe (Sarangi and Cherchneff 2013, 2015). For Cas A, Biscaro and Cherchneff (2016) assume that thermal sputtering starts at $t_{\rm thermal} = 340 + 3 \times \tau_{\rm cc}$ yr, where 340 years is the age

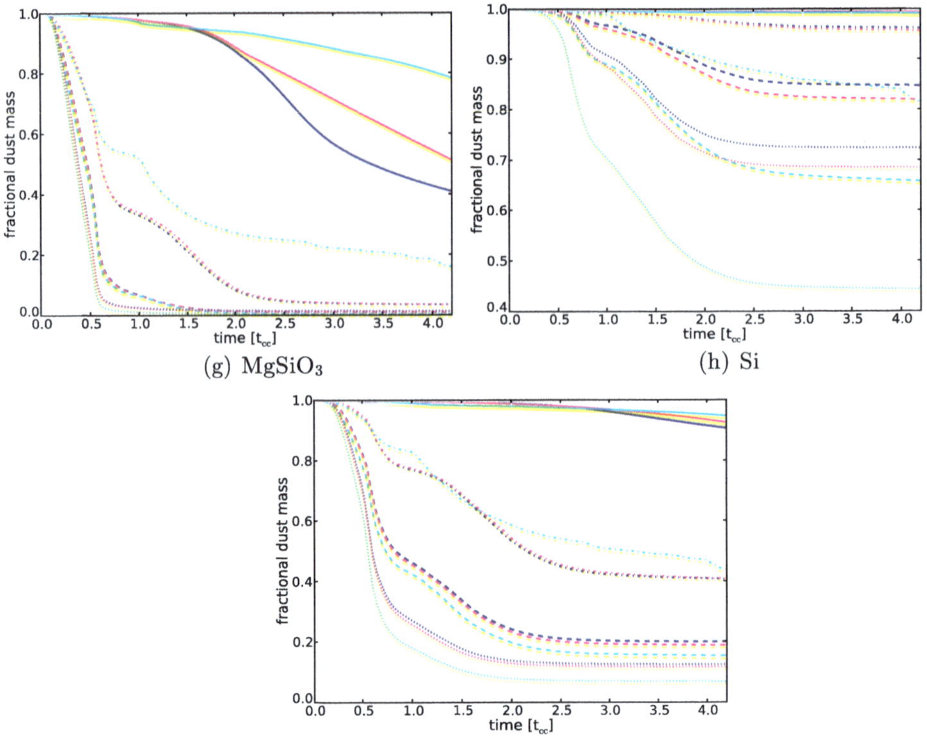

(g) $MgSiO_3$

(h) Si

(i) SiO_2

Fig. 23 (*Continued*)

of the remnant and $3 \times \tau_{cc}$ is the lifetime of a clump against shock disruption (Klein et al. 1994; Silvia et al. 2010). As already mentioned, a value of 100 years is taken for τ_{cc}, which implies that the lifetime of the cloud is comparable to the age of the remnant and that thermal sputtering should not be currently happening. This is in contrast with Micelotta et al. (2016), which recover for the cloud a much shorter lifetime, in agreement with optical observations and which implies that thermal sputtering already started and is currently ongoing in Cas A. Dust is eroded in the clumps by non-thermal sputtering for $\sim 3\tau_{cc} = 300$ yr, after which the resulting grain size distribution is injected into the hot inter-clump medium where it is further processed via thermal sputtering for ~4000 years, until the sputtering rate becomes negligible. During this time, it is assumed that the temperature of the gas remains in the initial range 10^6–10^8 K. As an example, Fig. 25 shows the resulting size distributions for Cas A for allumina processed by the heavy O^+ and Mg^+ ions in the oxygen-rich core.

Micelotta et al. (2016) calculate dust processing in Cas A over the entire remnant and during its evolution, using the appropriate profiles derived for quantities like the reverse shock velocity and the ejecta density. The results of the Monte Carlo simulation in terms of the total *fraction* of surviving dust (processed + still unprocessed) as a function of α are reported in Fig. 26 for both carbon and silicate dust. The curves show the effect of kinetic sputtering alone and in combination with thermal sputtering. The reduced fraction of surviving silicate grains is due to the combined effect of their slightly higher sputtering yield, the higher average mass of the sputtered particles (23 amu versus 12 amu for AC), and the longer exposure to thermal sputtering due to the fact that a larger fraction of silicate grains

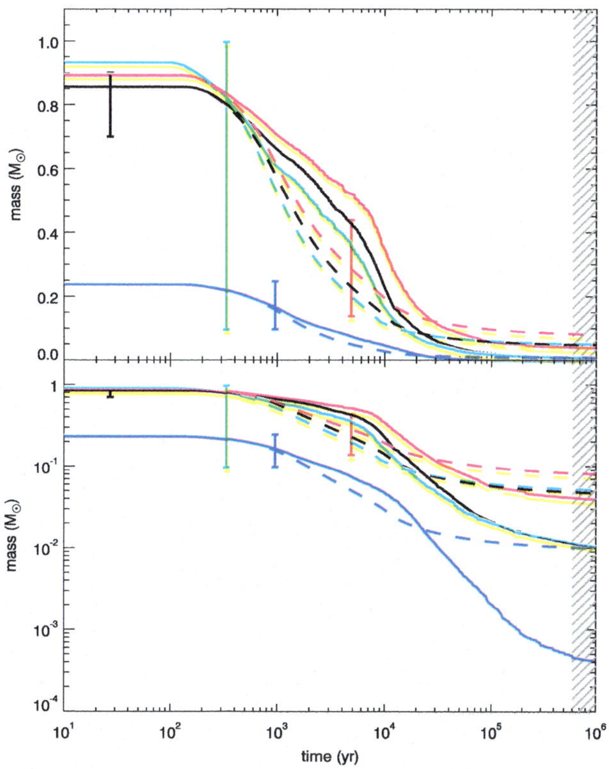

Fig. 24 Temporal evolution of dust mass in SN 1987A (*black*), Cas A (*green*), the Crab Nebula (*blue*) and N49 (*red*). The results obtained from the GRASH_Rev code (*solid lines*) are compared with the output from the Bianchi and Schneider (2007) model calculated using the same set of initial conditions (*dashed lines*). The theoretical curves are overlaid with the observed datapoints. The dust mass is shown on a linear scale in the top panel and on a logarithmic scale in the bottom one. The hatched area shows the time interval when the processing of dust becomes negligible. Figure reproduced from Bocchio et al. (2016)

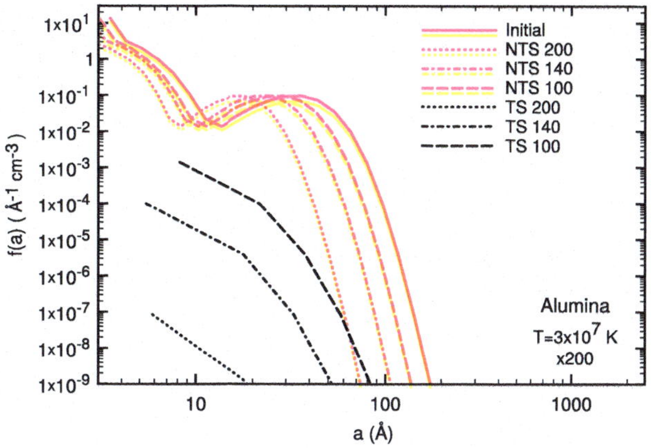

Fig. 25 Calculated grain size distributions for Cas A ($\chi = 200$) for alumina (*red*) processed by non-thermal sputtering (NTS) in the oxygen-rich clumps and by thermal sputtering (TS) in the inter-clump medium (with temperature $T = 3 \times 10^7$ K). Three shock velocities in the clump have been considered: 200, 140 and 100 km s^{-1}. The *solid line* represents the initial distribution. Figure reproduced from Biscaro and Cherchneff (2016)

tends to escape the ejecta clouds before disruption. The impact of a higher temperature for the smooth ejecta becomes noticeable in the inner layers of Cas A ($\alpha \lesssim 0.55$). When the reverse shock reaches the centre of the remnant, the surviving fraction is reduced to 15.9% for Am C and 11.8% for $MgSiO_3$ ($T = 10^7$ K), 12.3% for Am C and 8.7% for $MgSiO_3$ ($T = 10^8$ K). Kinetic sputtering alone is able to destroy 20% of the carbon and 40% of the silicate grains.

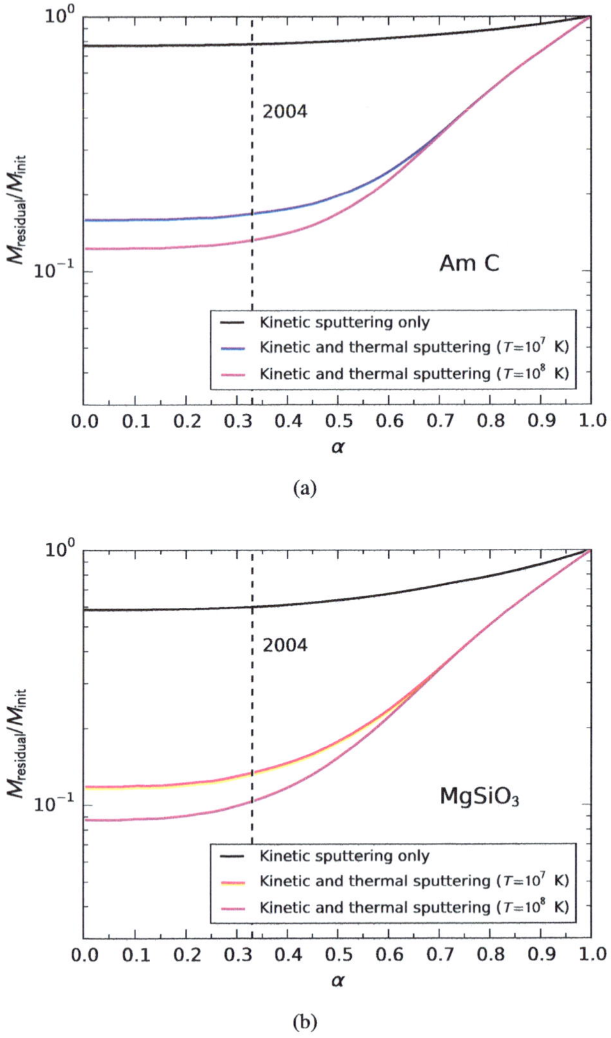

Fig. 26 Total mass fraction of surviving dust in Cas A as a function of α. The *top black curve* gives the effect of kinetic sputtering alone, the other two show the combined effect of kinetic plus thermal sputtering, this latter evaluated for smooth ejecta having $T = 10^7$ K and $T = 10^8$ K. Figure reproduced from Micelotta et al. (2016)

5.7.2 Comparison with Observations

The theoretical results from Itoh (1985) applied to SN 1979c and SN 1980k indicate that dust sputtering in the ejecta of these remnants may be important, although the lack of observations did not allow to validate such findings.

For the specific case of Cas A, the model adopted by Bianchi and Schneider (2007) predicts the formation of $\sim$0.1 $M_\odot$ of dust, reduced to $\approx$0.05 $M_\odot$ after $\sim$325 yr (current age of Cas A) and to $\sim$7% of this initial value when the reverse shock reaches the centre of the remnant. These results have been updated in the follow-up paper from Bocchio et al. (2016), where the initial amount of dust is set to $\sim$0.9 $M_\odot$ and little dust destruction is currently predicted for Cas A, with only 0.06 $M_\odot$ of dust heated in the post-shock gas and 0.77 $M_\odot$ unshocked and therefore cold. These values have to be compared with the results from fitting observations from polarimetry ($\sim$1 $M_\odot$—cold dust, Dunne et al. 2009) and SPITZER (0.02–0.054 $M_\odot$, Rho et al. 2008; Arendt et al. 2014) and with the recent estimates of (0.3–0.5) $M_\odot$ and $\sim$1.1 $M_\odot$ from De Looze et al. (2017) and Bevan et al. (2017) respectively.

Nozawa et al. (2010) use the output from their dust evolution simulations to recover the spectral energy distributions (SEDs) expected from shock-heated dust inside SNRs. The calculated SEDs are compared with the synchrotron-subtracted IR flux densities of Cas A in the window 8–100 μm taken from the photometric observations tabulated in Hines et al. (2004). The datapoints are reasonably fitted considering 0.008 $M_\odot$ of shocked dust and 0.072 $M_\odot$ of unshocked dust, where the newly formed grains are eroded and heated in a SNR expanding into a power-law medium with an initial hydrogen density of 120 cm^{-3}. The recovered dust masses have to be compared with more recent estimates of $\sim$1.1 $M_\odot$ and (0.3–0.5) $M_\odot$ (at $\sim$30 K) by Bevan et al. (2017) and De Looze et al. (2017) respectively (see Sect. 3.2). The sputtering calculations from Nozawa et al. (2010) indicate that all the dust produced in Cas A should be destroyed before reaching the ISM. However, their treatment does not consider the formation of an aspherical circumstellar medium (Nomoto et al. 1995, 1996) nor the clumpiness and asymmetry of the ejecta of Cas A (Willingale et al. 2002; Ennis et al. 2006; Smith et al. 2009). These factors may increase the chances of the grains to survive processing and being injected into the ISM.

The results from both Silvia et al. (2010) and Silvia et al. (2012) are qualitatively in agreement with the observation of Si dust in Cas A (Rho et al. 2008), and with the lack of Al_2O_3 and the presence of large amorphous carbon grains in SNR 1E0102-7219 (Sandstrom et al. 2009).

In terms of the amount of dust which should have formed in SN 1987A, Matsuura et al. (2015) report 0.8 $M_\odot$ of dust in the ejecta (0.5 $M_\odot$ of silicates and 0.3 $M_\odot$ of amorphous carbon). If the current limit of $\sim$0.25 $M_\odot$ of carbon predicted by nucleosynthesis models is ignored, then the observed SED can be fitted with 0.5 $M_\odot$ of amorphous carbon. Bocchio et al. (2016) predict the formation of $\sim$0.85 $M_\odot$ of dust in SN 1987A and estimate that this represents, and matches, current observations, because the reverse shock did not have time to process the ejecta yet.

For the Crab Nebula, the simulations from Bocchio et al. (2016) predict a current total dust mass of 0.17 $M_\odot$, almost equally distributed between carbon grains and silicates. These values have to be compared with the estimates deduced from observations. The HERSCHEL data on the Crab Nebula (51–670 μm, Gomez et al. 2012) have been fitted with two-temperature blackbodies using carbonaceous and silicate grains having either high or low temperature (55–63 K and 28–33 K respectively). The total inferred masses are 0.12 $M_\odot$ of carbonaceous dust and 0.25 $M_\odot$ of silicates. Using the same data, Temim and Dwek (2013) model the heating of dust grains by the pulsar wind nebula adopting a continuous distribution of grain sizes which leads to a continuous temperature gradient. The resulting total dust masses are 0.019 $M_\odot$ and 0.13 $M_\odot$, assuming a composition of pure amorphous carbon and pure silicate grains respectively. These values are lower than previous estimates and result from the continuous temperature distribution but also from the use of different dust emissivities. This is indeed the case for carbon dust, whose mass is not significantly modified by adopting a distribution of temperatures. Owen and Barlow (2015) constructed radiative transfer models used to fit the SEDs defined by SPITZER and HERSCHEL observations of the Crab Nebula. The results from the fits favour a clumpy mass distribution with (0.18–0.27) $M_\odot$ of amorphous carbon dust. If mixed dust chemistry models are considered, the resulting dust masses are (0.11–0.13) $M_\odot$ for amorphous carbon and (0.39–0.47) $M_\odot$ for silicates.

Observations of N49 indicate masses of warm dust of the order of 0.1 $M_\odot$ (carbonaceous) and 0.4 $M_\odot$ (silicates), taking these two species as the main constituents (Otsuka et al.

2010). The cold dust is difficult to detect because of contamination from the parent cloud. The fit has been done using a single dust temperature, which could lead to large uncertainties (Mattsson et al. 2015). The simulations predict a mass of warm (shocked) dust of $\sim$0.48 $M_\odot$ dominated by silicates, $\sim$15% higher than the dust mass deduced from observations.

Bocchio et al. (2016) conclude that between 10 and 40% of the synthesised dust has been destroyed up to now by the passage of the reverse shock but they do not provide the fraction of destroyed dust for each single SNR they consider (except for SN 1987A for which this fraction is zero). From Fig. 24 we derive that the fraction of destroyed dust is $\sim$10% for Cas A, $\sim$30% for the Crab Nebula and $\sim$40% for N49. For this latter remnant, the high fraction is consistent with its older age ($\sim$4800 years). The very low destroyed dust fraction in Cas A ($\sim$335 years old) has to be compared with the 40–70% range estimated by Biscaro and Cherchneff (2016) and with the 83–90% range (depending on dust composition and gas temperature) found by Micelotta et al. (2016). The significant destruction fraction calculated for the Crab Nebula (960 years old) is in contrast with the fact that there is no currently visible reverse shock in the remnant (Hester 2008, review). Indeed, the formation of the reverse shock in remnants like the Crab Nebula is expected to occur on a timescale of $\sim 10^4$ years (Blondin et al. 2001). At the end of the simulations (after 10^6 years) Bocchio et al. (2016) find that only 1–8% of the dust mass currently observed is expected to contribute to the enrichment of the ISM.

Biscaro and Cherchneff (2016) find that the gas and shock conditions characterising Cas A result in the complete destruction of silicates and the survival of a small fraction of allumina, SiC and carbon grains. For the total initial dust mass before sputtering, Biscaro and Cherchneff (2016) assume a value of 2.7×10^{-2} $M_\odot$. This should correspond to the dust mass synthesised by the SN and is much lower than the currently observed dust in Cas A estimated by Arendt et al. (2014), Bevan et al. (2017) and De Looze et al. (2017) and which includes the cold component. Biscaro and Cherchneff (2016) estimate that between 30% and 60% of their initial dust mass should have survived non-thermal sputtering and be present today in the remnant. This is globally consistent with the values derived from mid-infrared observations for the warm dust (see Fig. 27). After $\sim$4000 years of processing, the surviving fraction is reduced to 6–11% of the initial value.

For $\chi = 2000$, the fraction of surviving dust mass is larger, between 14% and 45%. Non-thermal sputtering is important and reduces the dust mass to 20–47% of its initial value before thermal sputtering starts. Biscaro and Cherchneff (2016) also study a specific sub-case where a Type II-P SN with a progenitor mass of 19 $M_\odot$ is taken as a surrogate for SN 1987A (with the caveat that SN 1987A is a Peculiar Type II-P supernova). Their results indicate that non-thermal sputtering in the clumps is unimportant and that even after $\sim$4000 years of thermal sputtering in the inter-clump medium, erosion is negligible.

Biscaro and Cherchneff (2016) point out that kinetic models produce smaller amounts of dust with different size distributions and chemical compositions than those based on CNT. This has to be taken into account, because sputtering is indeed very sensitive to dust size and composition. The formation and survival of dust in SNe is also strongly influenced by the clumpiness of the ejecta. Biscaro and Cherchneff conclude that Type II-b SNe may not be significant dust contributors, while SNe with dense and clumpy ejecta and their remnants could enrich the ISM with a significant amount of dust.

To compare their results with observations, Micelotta et al. (2016) adopt a different approach. The calculated current fractions of surviving dust in Cas A ($\alpha = 0.33$) are 16.9% and 13.3% for Am C, and of 13.4% and 10.4% for $MgSiO_3$ (for a gas temperature of 10^7 K and 10^8 K respectively). If the total amount of dust (warm + cold) currently observed in Cas A is $\sim$0.1 $M_\odot$ (Arendt et al. 2014) and represents the surviving fraction, the corresponding

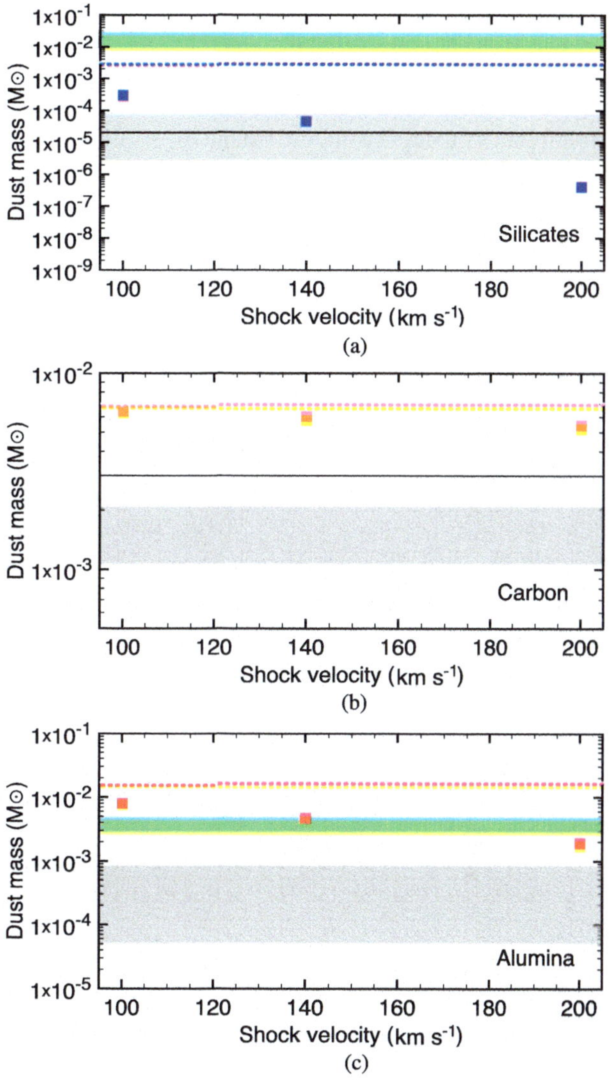

Fig. 27 Calculated masses of carbon, silicates and alumina in Cas A (*solid squares*) compared with the masses derived from SPITZER observations (Rho et al. 2008; Arendt et al. 2014, *grey shaded area* and *green shaded area* respectively). The mass of dust of each species initially synthesised by the supernova is represented by the *dotted lines*. The *solid black line* indicates the dust masses derived for SNR 1E0102.2-7219 in the Small Magellanic Cloud (SPITZER data, Sandstrom et al. 2009). Figure reproduced from Biscaro and Cherchneff (2016)

amount of SN produced dust is about 0.8–1 $M_\odot$, consistent with theoretical predictions from CNT-based dust coagulation models in CCSNe ejecta (0.1–1 $M_\odot$). If the estimate of ∼0.5 $M_\odot$ from De Looze et al. (2017) is adopted, the corresponding amount of fresh dust would be around 5 $M_\odot$. However, the survival rates from Micelotta et al. (2016) should be considered as a lower limit, because they do not include phenomena that could favor the survival of dust, as the possibility for the grains to escape the shell of shocked ejecta thanks to their residual ballistic velocities.

6 The Role of SNRs as a Dust Source and Sink

The impact between the blast wave (or forward shock) generated by the explosion of a high-mass star as a SN, and the circumstellar material results in the formation of a reverse shock which processes and destroys the dust newly formed in the ejecta. At the same time, the forward shock propagating into the ISM processes and destroys the preexisting ambient dust. While this review focuses on the processing and survival of dust in SNRs, in this section we briefly discuss some implications of such processing (Sect. 6.1) and the role of forward shock processing in determining the lifetime of dust in the ISM (Sect. 6.2).

6.1 Dust Destruction by the Reverse Shock

As shown in Sect. 5.7.1, the calculated fraction of surviving dust ranges from almost 0 to 100%. Determining the dust survival rate is crucial to establish whether the dust produced in SN ejecta will be eventually integrated into the ISM, and thus whether SNe can be an important dust source in the ISM. The results from the theoretical models that we have discussed are summarised in Table 2. We would like to stress again that, while a direct comparison between them is not really feasible, such a summary table allows to appreciate the variety of such results and how they evolved with time.

Explaining the amount of dust detected in high-redshift galaxies with SN sources would require the release of about 0.1 $M_\odot$ per SN (Nozawa et al. 2006; Dwek and Cherchneff 2011). If dust destruction is close to 100%, SNe could not be sufficient to provide the required 0.1 $M_\odot$ each. However, if each SN could form 0.2–0.5 $M_\odot$ of dust (Gall et al. 2011; Matsuura 2017; Sarangi et al. 2018) and nearly half of it can survive, SNe could be an important dust supplier in the ISM.

6.2 The Lifetime of Dust in the ISM: Destruction by Expanding SNRs

Dust residing in the ISM is heavily processed and destroyed by the expanding blast waves generated by SN explosions. The dust destruction rate increases with increasing shock velocities. Table 2 summarises some relevant estimations of the fraction of surviving dust resulting from SN shock processing in the ISM.

These results have been used to estimate the lifetime of dust in the ISM, $t_{\rm ISM}$, as described in e.g. McKee and Ostriker (1977), McKee et al. (1987), McKee (1989), Draine (2009):

$$t_{\rm ISM} = \frac{M_{\rm ISM} t_{\rm SN}}{\epsilon M_s}, \tag{13}$$

where $M_{\rm ISM}$ is the total mass of dust and gas residing in the ISM of a single galaxy, $t_{\rm SN}$ is the effective interval between SNe in a galaxy, M_s is the mass of gas shocked by a SNR in the Sedov-Taylor stage to at least 100 $\rm km\,s^{-1}$ and ϵ is the dust destruction efficiency. In the Sedov-Taylor phase, the shock speed drops continuously. Assuming that the initial explosion energy, E_0, of a SN is converted into kinetic energy (Salpeter 1977; Draine 2009), and that the final shock velocity for which dust can be destroyed is 100–200 $\rm km\,s^{-1}$, we have that M_s would be around 1000–3000 $M_\odot$. The explosion energy of a SN is typically around 10^{51} ergs. It is not well determined yet how much of its energy will be converted into kinetic energy, and how much will be carried away by neutrinos. Nevertheless, 10^{51} ergs provides the correct order of magnitude.

The typical time scale for dust formation and injection from asymptotic giant branch (AGB) stars is $\sim 2\times10^9$ years, much longer than the calculated dust lifetimes in the ISM (except for Slavin et al. 2015, see Table 3) thus, dust should not be detectable, unless there is an additional dust source. These estimates are based upon the assumption that SNe form little dust. Recent theoretical models have shown that a reasonable fraction of the ejecta dust could survive the processing by the reverse shock. This could potentially speed up the

Table 2 Theoretically calculated fractions of surviving dust mass, F_{srv} resulting from processing by the reverse shock (in the SNR cavity) and by the forward shock (in the ISM)

Work	F_{srv} %	Remarks
Reverse shock		
Bianchi and Schneider (2007)	2–20	
Nozawa et al. (2007)	0–80	Primordial SNRs
Nath et al. (2008)	80–99	Graphite and silicates, shallow ejecta profile ($n < 5$)
Silvia et al. (2010)	62	C, high-velocity shocks, high-density clouds
	20	SiO_2
	80	Fe
Silvia et al. (2012)	44–96	C
	7–95	SiO_2
	24–99	Fe
Marassi et al. (2015)	3–50	*Standard* Population III CCSNe
	10–80	*Faint* Population III CCSNe
Bocchio et al. (2016)	90	Cas A, current
	70	Crab Nebula, but no reverse shock detected yet
	60	N49
	1–8	Global, end of processing (after 10^6 yr)
Biscaro and Cherchneff (2016)	30–60	Cas A ($\chi = 200$), current
	6–11	Cas A, end of processing (after 4000 yr)
	14–45	Dense clumps ($\chi = 2000$)
Micelotta et al. (2016)	13–17	Amorphous carbon, Cas A, current
	10–13	$MgSiO_3$
	12–16	Amorphous carbon, Cas A, after 8000 yr
	9–12	$MgSiO_3$
Forward shock		
Draine and Salpeter (1979b)	70–85	Graphite, $v_s = 100$ km s^{-1}, $a_{gr} = 0.1$ µm
	40–60	Silicates
Seab and Shull (1983)	85	Graphite, $B = 1$ µm, $v_s = 100$ km s^{-1}, $n_0 = 10$ cm^{-3}
	50	Silicates
McKee et al. (1987)	62–75	Silicates, $B = 1$ µG, $v_s = 100$ km s^{-1}, $n_0 = 0.25$ cm^{-3}
	84–90	$B = 3$ µG
Jones et al. (1994)	81	Graphite, $B = 1$ µG, $v_s = 100$ km s^{-1}, $n_0 = 0.25$ cm^{-3}
	89	Graphite, $B = 3$ µG
	62	Silicates, $B = 1$ µG
	78	Silicates, $B = 3$ µG
Jones et al. (1996)	93	Graphite, $B = 3$ µG, $v_s = 100$ km s^{-1}, $n_0 = 0.25$ cm^{-3}
	82	Silicates
Serra Díaz-Cano and Jones (2008)	69	a-C:H, $B = 3$ µG, $v_s = 100$ km s^{-1}, $n_0 = 0.25$ cm^{-3}
Bocchio et al. (2014)	9	a-C:H, $B = 3$ µG, $v_s = 100$ km s^{-1}, $n_0 = 0.25$ cm^{-3}
	71	Silicates

Notes—χ: density contrast between the ejecta clumps and the smooth ejecta, v_s: shock velocity, a_{gr}: initial dust grain radius, B: pre-shock magnetic field intensity, n_0: pre-shock ambient density, a-C:H: hydrogenated amorphous carbon

Table 3 Estimated dust lifetime against processing by supernova shocks in the ISM

Work	Dust lifetime (yrs)	Note
Salpeter (1977)	$\sim 7 \times 10^8$	Silicates
Barlow (1978)	2×10^8	Ice grains
	5×10^8	$MgSiO_3$
	2×10^9	Graphite
Jones et al. (1994)	4×10^8	Graphite
	2.2×10^8	Silicates
Jones et al. (1996)	6×10^8	Graphite
	4×10^8	Silicates
Serra Díaz-Cano and Jones (2008)	2×10^8	a-C:H
Micelotta et al. (2010a)	$(1.4–4) \times 10^8$	PAHs
Jones and Nuth (2011a)	$(0.3–10) \times 10^8$	Silicates
	$(0.2–5.1) \times 10^8$	a-C:H
Bocchio et al. (2014)	$(6.2 \pm 5.7) \times 10^7$	a-C:H
	$(3.1 \pm 2.7) \times 10^8$	Silicates
Temim et al. (2015)	$(2.2 \pm 1.3) \times 10^7$	Silicates, LMC
	$(2.0 \pm 1.7) \times 10^7$	Carbon dust, LMC
	$(5.4 \pm 3.2) \times 10^7$	Silicates, SMC
	$(7.2 \pm 4.3) \times 10^7$	Carbon dust, SMC
Slavin et al. (2015)	3.2×10^9	Carbon dust
	2.0×10^9	Silicates

dust injection time scale from SNe. In the past few years there has been more evidence that core-collapse SNe can form a substantial amount of dust (from 10^{-2} to nearly 1 $M_\odot$; Matsuura 2017; Sarangi et al. 2018). Even if 20 % of this dust survives the passage of the reverse shock, it will not be sufficient to compensate for the high destruction rate in the ISM. Jones and Nuth (2011b) and Bocchio et al. (2014) have carefully examined the uncertainties involved in dust destruction calculations, finding that they can get to 90%. This implies that the estimates of dust destruction will substantially improve when better models and more refined parameters' values will be available (Table 3).

7 Summary and Conclusions

This paper is the second of a series of two aiming to review the topic of dust in supernovae and supernova remnants. In Paper I (Sarangi et al. 2018) we have summarised the scenarios for dust formation in supernova ejecta and the observations of dust formation in core-collapse supernovae. Here in Paper II, we have focused on the theoretical models aiming to describe the processing of dust *inside* supernova remnants and to quantify the amount of surviving grains. We have also discussed the observational evidence of dust processing and destruction in supernova remnants and summarised some of the methods adopted to quantify dust destruction from observations.

In young SN 1987A the reverse shock did not reach the ejecta yet, but there is evidence of interaction between the blast wave shock and the circumstellar material, arranged in series of rings, where dust grains should undergo shock processing and heating. In Cas A (age ∼335 years), dust has been detected in both the shocked and unshocked regions inside the cavity of the remnant. The temperature of the dust in these two regions is different, which would indicate that dust is collisionally heated and potentially destroyed by sputtering in the shocked gas. In the 10,000 years old Sagittarius A East SNR, where the reverse shock should have reached the centre of the remnant, dust emission has been detected in the central region, coincident with shocked gas. The excess far-infrared emission suggests the presence of large grains which could have survived shock processing.

Different methods can be used to derive the amount of emitting dust and to make estimates of the fraction which will eventually survive processing. These include fitting measured flux data points at infrared wavelengths, and modelling the infrared emission from dust whose size distribution has been modified by sputtering and shattering in a hot shocked gas, and which has been collisionally heated by the same gas. Both methods have their own advantages and limitations and they should always carefully take into account the properties of every specific SNR under examination.

From a theoretical point of view, the different studies presented in this review clearly show that modelling the processing of dust in SNRs is challenging. The ideal model should be based on a realistic description of the remnant under consideration and include an accurate description of the dust and of all processes potentially capable of affecting it. In reality, our knowledge of these fundamental ingredients is affected by large uncertainties. SNRs are complex objects, often inhomogenous and exhibiting a variety of physical conditions. Dust composition and size distribution are subject of debate while dust processing depends on the environmental conditions, which vary within the remnant.

Many of the works presented here are built on a common ground: take an evolutionary model for the SNRs (typically young ones) and combine it with a description of dust and dust processing. However, it is revealing and instructing to compare the different implementations of this basic recipe and the specific aspects of the problem which have been investigated.

For the dynamics of the remnant, in particular the evolution of the reverse shock velocity, which directly affects the processing of dust via sputtering, Itoh (1985) adopts the self-similar solutions from Chevalier (1982) for ejecta-dominated supernova remnants having ejecta with a power-law density distribution (characterised by the index $n > 5$) expanding into a circumstellar medium which also has a power-law density distribution (with index $s < 3$). Dwek (2005) uses the analytical expressions of Truelove and McKee (1999) but for the simplest case: homogeneous ejecta expanding into a uniform medium. The solutions from Truelove and McKee (1999) are particularly suited for young SNRs (like Cas A) which are transitioning between the ejecta-dominated and Sedov-Taylor phases, because they allow a smooth merging of the solutions corresponding to the two stages. The simplest of solutions have been used by Bianchi and Schneider (2007), Marassi et al. (2015) and Bocchio et al. (2016), while Nath et al. (2008) include the core-envelope density structure of the ejecta from Truelove and McKee (1999), considering power-law envelopes with $n < 3$, $n = 4$ and $n > 5$. Nozawa et al. (2007) and Biscaro and Cherchneff (2014, 2016) consider simple homologous expansion of homogeneous ejecta expanding into a uniform medium, while Nozawa et al. (2010) consider the case of a power-law ambient medium (exponent $s = 2$). Following the work of Truelove and McKee (1999) and Laming and Hwang (2003), Micelotta et al. (2016) develop for Cas A the solutions corresponding to core-envelope ejecta expanding into a power-law medium, while Silvia et al. (2010, 2012) do not include their results into an evolutionary model for the reverse shock. Dwek (2005) also points out that SN ejecta are clumpy and that this will affect dust processing. The hydrodynamics + N-body simulations of Silvia et al. (2010, 2012) focus on the behaviour of a single dust-containing over-dense clump impacted by a planar shock, in terms of cloud fragmentation and dust sputtering. Biscaro and Cherchneff (2014, 2016) and Micelotta et al. (2016) include the treatment of dusty ejecta clumps into a supernova model.

The adopted or calculated initial grain size distributions of the newly formed dust are based of the CNT formalism (Bianchi and Schneider 2007; Nozawa et al. 2007; Nath et al. 2008; Nozawa et al. 2010; Silvia et al. 2010, 2012; Marassi et al. 2015; Bocchio et al. 2016) or on a chemical kinetic approach (Biscaro and Cherchneff 2014, 2016). Itoh (1985) and Micelotta et al. (2016) adopt as a test case the MRN interstellar grain size distribution. Dust composition includes silicates and carbon grains, but it can be more or less sophisticated depending on the model used. Sputtering, both thermal and inertial, is the main destructive process for dust inside supernova remnants. It is treated using the classical formalism for sputtering in solids (Matsunami 1981; Andersen and Bay 1981; Sigmund 1981; Bohdansky 1984) adopting the sputtering yields from Draine and Salpeter (1979a), Tielens et al. (1994) and/or Nozawa et al. (2006). In addition, Bocchio et al. (2016) estimate the combined effect of thermal and inertial sputtering and introduce a correction to the sputtering yields to take into account the finite size of the grains. Dust dynamics is included in the work of Itoh (1985), Nozawa et al. (2007), Nozawa et al. (2010), Bocchio et al. (2016), Biscaro and Cherchneff (2016) and Micelotta et al. (2016). This latter couples explicitly dust erosion and dust slowing-down in the ejecta clouds, to calculate the time required by a grain to escape from cloud depending on its size and position. These analytical expressions implemented into a Monte Carlo simulation allow to follow dust sputtering "on-the-fly". Cooling of the ejecta has been taken into account by Nozawa et al. (2007), by Silvia et al. (2010, 2012) who compared dust sputtering in a dense cloud with cooling both off and on, and by Micelotta et al. (2016) who calculated explicitly for Cas A the cooling of both the smooth ejecta and the over-dense clumps, using for these latter the cooling function for an oxygen-rich shocked gas from Sutherland and Dopita (1995).

The results of these different works in terms of the fraction of surviving dust mass span a wide range of values. Bianchi and Schneider (2007) find that after (4–8) $\times$ 10^4 years of processing, the fraction of surviving dust mass is between 2% and 20% (depending on the adopted parameters). These values are reduced to 1–8% using the updated model presented by Bocchio et al. (2016), which extend the simulations up to 10^6 years after explosion (into the radiative phase of supernova remnant evolution). Using an updated version of the code from Bianchi and Schneider (2007), Marassi et al. (2015) find that 3–50% and 10–80% of the dust initially formed in standard and faint Pop III CCSNe, respectively, survives the passage of the reverse shock. Nozawa et al. (2007) as well extend their calculations to the radiative stage, finding that 0-80% of the dust mass survives processing, while Nath et al. (2008) recover a surviving fraction larger than 80% considering the ED and ST phases only. The numerical simulations of Silvia et al. (2010, 2012) predict that between 7 and 99% of the initial dust mass will survive, these results however are not included into a supernova remnant evolutionary model and cannot be directly compared with other works. Biscaro and Cherchneff (2016) find that after 4000 years of processing, the fraction of surviving dust will be 6% to 11% in Cas A and between 14 and 45% in Type II-P supernovae. For Cas A, the results of Micelotta et al. (2016) indicate that $\sim$8000 years after explosion (when the reverse shock reaches the centre of the remnant) the survival fractions range from 9% to 16%, depending on dust composition and temperature of the smooth ejecta.

This variety of results reflects the different approaches used (analytical or numerical) and the different physical descriptions and conditions adopted for the supernova remnants, the ambient medium and the dust grains. Apart from the N49 remnant investigated by Bocchio et al. (2016), the theoretical amounts of surviving dust are consistent with the dust masses derived from observations, and/or with theoretical predictions about the amount of dust which should have been initially synthesised by supernovae. This with the caveat that the dust mass estimate from photometric and spectroscopic data and the theoretical evaluation of dust formation and processing are affected by major uncertainties. In the specific case of the Crab Nebula, Bocchio et al. (2016) find that the amount of observed dust is consistent with their model which predicts that $\sim$30% of the dust has been already destroyed by the reverse shock. However, no reverse shock has been detected so far in the Crab Nebula (960 years old) and it is not expected to form before $\sim 10^4$ years. This illustrates the importance of developing realistic models for each specific remnant under examination.

Determining the net amount of SN-condensed dust that is ultimately able to reach the ISM is crucial to determine whether SNe can indeed supply the large dust masses detected in high-redshift galaxies, especially if alternative routes like dust reformation and growth in the ISM are not viable (e.g. Ferrara et al. 2016). While this goal has not been reached yet, the increasing interest in facing the problem in its whole complexity puts us on the right track.

Acknowledgements We would like to thank the referee for the careful reading and the many useful suggestions. We also deeply acknowledge the support from the International Space Science Institute in Bern, Switzerland and their successful endeavour to organise the Supernovae workshop. We are grateful to Isabelle Cherchneff for her help at the beginning of this project. E. R. M. wishes to acknowledge the support of Academy of Finland grant 1285769 and M. M. acknowledges support from an STFC Ernest Rutherford fellowship (ST/L003597/1).

References

M.G. Allen, B.A. Groves, M.A. Dopita, R.S. Sutherland, L.J. Kewley, The MAPPINGS III library of fast radiative shock models. Astrophys. J. Suppl. Ser. **178**, 20–55 (2008). https://doi.org/10.1086/589652. 0805.0204.

H.H. Andersen, H.L. Bay, *Sputtering Yield Measurements* (1981), p. 145. https://doi.org/10.1007/3540105212_9

R.G. Arendt, E. Dwek, S.H. Moseley, Newly synthesized elements and pristine dust in the Cassiopeia A supernova remnant. Astrophys. J. **521**, :234–245 (1999). https://doi.org/10.1086/307545. astro-ph/9901042

R.G. Arendt, E. Dwek, W.P. Blair, P. Ghavamian, U. Hwang, K.S. Long, R. Petre, J. Rho, P.F. Winkler, Spitzer observations of dust destruction in the puppis a supernova remnant. Astrophys. J. **725**(1), 585–597 (2010)

R.G. Arendt, E. Dwek, G. Kober, J. Rho, U. Hwang, Interstellar and ejecta dust in the Cas A supernova remnant. Astrophys. J. **786**, 55 (2014). https://doi.org/10.1088/0004-637X/786/1/55. 1403.3008

R.G. Arendt, E. Dwek, P. Bouchet, I.J. Danziger, K.A. Frank, R.D. Gehrz, S. Park, C.E. Woodward, Infrared continuum and line evolution of the equatorial ring around SN 1987A Astron. J. **151**(3), 62 (2016)

W.D. Arnett, J.N. Bahcall, R.P. Kirshner, S.E. Woosley, Supernova 1987A. in: *Annual Review of Astronomy and Astrophysics*, vol. 27 Palo Alto 27 (1989), p. 629 (A90-29983 12-90)

M.J. Baines, I.P. Williams, A.S. Asebiomo, Resistance to the motion of a small sphere moving through a gas. Mon. Not. R. Astron. Soc. **130**, 63 (1965)

M.J. Barlow, The destruction and growth of dust grains in interstellar space. I—Destruction by sputtering. Mon. Not. R. Astron. Soc. **183**(3), 367–395 (1978)

M.J. Barlow, O. Krause, B.M. Swinbank, B. Sibthorpe, M.A. Besel, R. Wesson, R.J. Ivison, L. Dunne, W.K. Gear, H.L. Gomez née Morgan, P.C. Hargrave, T. Henning, S.J. Leeks, T.L. Lim, G. Olofsson, E.T. Polehampton, A Herschel PACS and SPIRE study of the dust content of the Cassiopeia A supernova remnant. Astron. Astrophys. **518**, L138 (2010)

M.J. Barlow, O. Krause, B.M. Swinyard, B. Sibthorpe, M.A. Besel, R. Wesson, R.J. Ivison, L. Dunne, W.K. Gear, H.L. Gomez, P.C. Hargrave, T. Henning, S.J. Leeks, T.L. Lim, G. Olofsson, E.T. Polehampton, A Herschel PACS and SPIRE study of the dust content of the Cassiopeia A supernova remnant. Astron. Astrophys. **518**, L138 (2010b). https://doi.org/10.1051/0004-6361/201014585. 1005.2688

R. Bedogni, P.R. Woodward, Shock wave interactions with interstellar clouds. Astron. Astrophys. **231**, 481–498 (1990)
F. Bertoldi, C.L. Carilli, P. Cox, X. Fan, M.A. Strauss, A. Beelen, A. Omont, R. Zylka, Dust emission from the most distant quasars. Astron. Astrophys. **406**, L55–L58 (2003). https://doi.org/10.1051/0004-6361:20030710. astro-ph/0305116
A. Bevan, M.J. Barlow, Modelling supernova line profile asymmetries to determine ejecta dust masses: SN 1987A from days 714 to 3604. Mon. Not. R. Astron. Soc. **456**(2), 1269–1293 (2016)
A. Bevan, M.J. Barlow, D. Milisavljevic, Dust masses for SN 1980K, SN1993J and Cassiopeia A from red-blue emission line asymmetries. Mon. Not. R. Astron. Soc. **465**, 4044–4056 (2017). https://doi.org/10.1093/mnras/stw2985. 1611.05006
S. Bianchi, R. Schneider, Dust formation and survival in supernova ejecta. Mon. Not. R. Astron. Soc. **378**, 973–982 (2007). https://doi.org/10.1111/j.1365-2966.2007.11829.x. 0704.0586
C. Biscaro, I. Cherchneff, Molecules and dust in Cassiopeia A. I. Synthesis in the supernova phase and processing by the reverse shock in the clumpy remnant. Astron. Astrophys. **564**, A25 (2014). https://doi.org/10.1051/0004-6361/201322932. 1401.5594
C. Biscaro, I. Cherchneff, Molecules and dust in Cassiopeia A. II. Dust sputtering and diagnosis of supernova dust survival in remnants. Astron. Astrophys. **589**, A132 (2016). https://doi.org/10.1051/0004-6361/201527769. 1511.05487
W.P. Blair, P. Ghavamian, K.S. Long, B.J. Williams, K.J. Borkowski, S.P. Reynolds, R. Sankrit, Spitzer Space Telescope observations of Kepler's supernova remnant: a detailed look at the circumstellar dust component. Astrophys. J. **662**(2), 998–1013 (2007)
J.M. Blondin, R.A. Chevalier, D.M. Frierson, Pulsar wind nebulae in evolved supernova remnants. Astrophys. J. **563**, 806–815 (2001). https://doi.org/10.1086/324042. astro-ph/0107076
M. Bocchio, A.P. Jones, J.D. Slavin, A re-evaluation of dust processing in supernova shock waves. Astron. Astrophys. **570**, A32 (2014). https://doi.org/10.1051/0004-6361/201424368
M. Bocchio, S. Marassi, R. Schneider, S. Bianchi, M. Limongi, A. Chieffi, Dust grains from the heart of supernovae. Astron. Astrophys. **587**, A157 (2016). https://doi.org/10.1051/0004-6361/201527432. 1601.06770
J. Bohdansky, A universal relation for the sputtering yield of monatomic solids at normal ion incidence. Nucl. Instrum. Methods Phys. Res. B **2**, 587–591 (1984). https://doi.org/10.1016/0168-583X(84)90271-4
K.J. Borkowski, J.M. Shull, Pure-oxygen radiative shocks with electron thermal conduction. Astrophys. J. **348**, 169–185 (1990). https://doi.org/10.1086/168225
K.J. Borkowski, W.J. Lyerly, S.P. Reynolds, Supernova remnants in the Sedov expansion phase: thermal X-ray emission. Astrophys. J. **548**, 820–835 (2001). https://doi.org/10.1086/319011. astro-ph/0008066
K.J. Borkowski, B.J. Williams, S.P. Reynolds, W.P. Blair, P. Ghavamian, R. Sankrit, S.P. Hendrick, K.S. Long, J.C. Raymond, R.C. Smith, S. Points, P.F. Winkler, Dust destruction in Type Ia supernova remnants in the Large Magellanic Cloud. Astrophys. J. Lett. **642**, L141–L144 (2006). https://doi.org/10.1086/504472. astro-ph/0602313
K.J. Borkowski, B.J. Williams, S.P. Reynolds, W.P. Blair, P. Ghavamian, R. Sankrit, S.P. Hendrick, K.S. Long, J.C. Raymond, R.C. Smith, S. Points, P.F. Winkler, Dust destruction in Type Ia supernova remnants in the Large Magellanic Cloud. Astrophys. J. **642**(2), L141–L144 (2006b)
P. Bouchet, J.M. De Buizer, N.B. Suntzeff, I.J. Danziger, T.L. Hayward, C.M. Telesco, C. Packham, High-resolution mid-infrared imaging of SN 1987A. Astrophys. J. **611**, 394–398 (2004). https://doi.org/10.1086/421936. astro-ph/0312240
P. Bouchet, E. Dwek, J. Danziger, SN 1987A after 18 years: mid-infrared Gemini and Spitzer observations of the remnant. Astrophys. J. **650**, 212 (2006)
G.L. Bryan, M.L. Norman, Simulating X-ray clusters with adaptive mesh refinement, in *Computational Astrophysics; 12th Kingston Meeting on Theoretical Astrophysics*, ed. by D.A. Clarke, M.J. West. Astronomical Society of the Pacific Conference Series, vol. 123 (1997), p. 363. astro-ph/9710186
I. Cherchneff, E. Dwek, The chemistry of population III supernova ejecta. I. Formation of molecules in the early universe. Astrophys. J. **703**, 642–661 (2009). https://doi.org/10.1088/0004-637X/703/1/642. 0907.3621
R.A. Chevalier, Self-similar solutions for the interaction of stellar ejecta with an external medium. Astrophys. J. **258**, 790–797 (1982). https://doi.org/10.1086/160126
R.A. Chevalier, R.P. Kirshner, Abundance inhomogeneities in the Cassiopeia A supernova remnant. Astrophys. J. **233**, 154–162 (1979). https://doi.org/10.1086/157377
D.D. Clayton, L.R. Nittler, Astrophysics with presolar stardust. Annu. Rev. Astron. Astrophys. **42**, 39–78 (2004). https://doi.org/10.1146/annurev.astro.42.053102.134022
L.L. Cowie, C.F. McKee, The evaporation of spherical clouds in a hot gas. I—Classical and saturated mass loss rates. Astrophys. J. **211**, 135–146 (1977). https://doi.org/10.1086/154911

M. Cropper, J. Bailey, J. McCowage, R.D. Cannon, W.J. Couch, Spectropolarimetry of SN 1987A—observations up to 1987 July 8. Mon. Not. R. Astron. Soc. **231**, 695 (1988)

W.W. Dalton, S.A. Balbus, A flux-limited treatment for the conductive evaporation of spherical interstellar gas clouds. Astrophys. J. **404**, 625–635 (1993). https://doi.org/10.1086/172316

I.J. Danziger, C. Gouiffes, P. Bouchet, L.B. Lucy, Supernova 1987A in the Large Magellanic Cloud, IAU Circ 4746 (1989)

I. De Looze, M.J. Barlow, B.M. Swinyard, J. Rho, H.L. Gomez, M. Matsuura, R. Wesson, The dust mass in Cassiopeia A from a spatially resolved Herschel analysis. Mon. Not. R. Astron. Soc. **465**(3), 3309–3342 (2017)

L. Del Zanna, N. Bucciantini, P. Londrillo, An efficient shock-capturing central-type scheme for multidimensional relativistic flows. II. Magnetohydrodynamics. Astron. Astrophys. **400**, 397–413 (2003). https://doi.org/10.1051/0004-6361:20021641. astro-ph/0210618

D. Docenko, R.A. Sunyaev, Fine-structure infrared lines from the Cassiopeia A knots. Astron. Astrophys. **509**, A59 (2010). https://doi.org/10.1051/0004-6361/200810366. 0806.1801

B. Donn, J.A. Nuth, Does nucleation theory apply to the formation of refractory circumstellar grains? Astrophys. J. **288**, 187–190 (1985). https://doi.org/10.1086/162779

B.T. Draine, Infrared emission from dust in shocked gas. Astrophys. J. **245**, 880 (1981)

B.T. Draine, Interstellar dust grains. Annu. Rev. Astron. Astrophys. **41**, 241–289 (2003). https://doi.org/10.1146/annurev.astro.41.011802.094840. astro-ph/0304489

B.T. Draine, Interstellar dust models and evolutionary implications. Cosmic dust—near and far. Astron. Soc. Pac. Conf. Ser. **414**, 453 (2009)

B.T. Draine, *Physics of the Interstellar and Intergalactic Medium* (Princeton University Press, Princeton, 2011)

B.T. Draine, E.E. Salpeter, Time-dependent nucleation theory. J. Chem. Phys. **67**, 2230–2235 (1977). https://doi.org/10.1063/1.435116

B.T. Draine, E.E. Salpeter, Destruction mechanisms for interstellar dust. Astrophys. J. **231**, 77–94 (1979a). https://doi.org/10.1086/157165

B.T. Draine, E.E. Salpeter, On the physics of dust grains in hot gas. Astrophys. J. **231**, 438–455 (1979b). https://doi.org/10.1086/157206

L. Dunne, S.J. Maddox, R.J. Ivison, L. Rudnick, T.A. Delaney, B.C. Matthews, C.M. Crowe, H.L. Gomez, S.A. Eales, S. Dye, Cassiopeia A: dust factory revealed via submillimetre polarimetry. Mon. Not. R. Astron. Soc. **394**, 1307–1316 (2009). https://doi.org/10.1111/j.1365-2966.2009.14453.x. 0809.0887

E. Dwek, The infrared echo of a type II supernova with a circumstellar dust shell—applications to SN 1979c and SN 1980k. Astrophys. J. **274**, 175–183 (1983). https://doi.org/10.1086/161435

E. Dwek, Temperature fluctuations and infrared emission from dust particles in a hot gas. Astrophys. J. **302**, 363–370 (1986)

E. Dwek, The infrared diagnostic of a dusty plasma with applications to supernova remnants. Astrophys. J. **322**, 812 (1987)

E. Dwek, Interstellar dust: what is it, how does it evolve, and what are its observational consequences? in *The Spectral Energy Distributions of Gas-Rich Galaxies: Confronting Models with Data*, ed. by C.C. Popescu, R.J. Tuffs. American Institute of Physics Conference Series, vol. 761 (2005), pp. 103–122. https://doi.org/10.1063/1.1913921. astro-ph/0412344

E. Dwek, R.G. Arendt, Dust-gas interactions and the infrared emission from hot astrophysical plasmas. Annu. Rev. Astron. Astrophys. **30**, 11–50 (1992). https://doi.org/10.1146/annurev.aa.30.090192.000303

E. Dwek, R.G. Arendt, The evolution of dust mass in the ejecta of SN1987A. Astrophys. J. **810**(1), 75 (2015)

E. Dwek, I. Cherchneff, The origin of dust in the early universe: probing the star formation history of galaxies by their dust content. Astrophys. J. **727**, 63 (2011). https://doi.org/10.1088/0004-637X/727/2/63. 1011.1303

E. Dwek, R. Petre, A. Szymkowiak, W.L. Rice, IRAS observations of supernova remnants—a comparison between their infrared and X-ray cooling rates. Astrophys. J. Lett. **320**, L27–L33 (1987). https://doi.org/10.1086/184971

E. Dwek, S.H. Moseley, W. Glaccum, J.R. Graham, R.F. Loewenstein, R.F. Silverberg, R.K. Smith, Dust and gas contributions to the energy output of SN 1987A on day 1153. Astrophys. J. Lett. **389**, L21–L24 (1992). https://doi.org/10.1086/186339

E. Dwek, R.G. Arendt, P. Bouchet, D.N. Burrows, P. Challis, I.J. Danziger, J.M. De Buizer, R.D. Gehrz, R.P. Kirshner, R.A. McCray, S. Park, E.F. Polomski, C.E. Woodward, Infrared and X-ray evidence for circumstellar grain destruction by the blast wave of supernova 1987A. Astrophys. J. **676**, 1029 (2008)

E. Dwek, R.G. Arendt, P. Bouchet, D.N. Burrows, P. Challis, I.J. Danziger, J.M. De Buizer, R.D. Gehrz, S. Park, E.F. Polomski, J.D. Slavin, C.E. Woodward, Five years of mid-infrared evolution of the remnant of SN 1987A: the encounter between the blast wave and the dusty equatorial ring. Astrophys. J. **722**(1), 425–434 (2010)

J.A. Ennis, L. Rudnick, W.T. Reach, J.D. Smith, J. Rho, T. DeLaney, H. Gomez, T. Kozasa, Spitzer IRAC images and sample spectra of Cassiopeia A's explosion. Astrophys. J. **652**, 376–386 (2006). https://doi.org/10.1086/508142. astro-ph/0610838

G.J. Ferland, K.T. Korista, D.A. Verner, J.W. Ferguson, J.B. Kingdon, E.M. Verner, CLOUDY 90: numerical simulation of plasmas and their spectra. Publ. Astron. Soc. Pac. **110**, 761–778 (1998). https://doi.org/10.1086/316190

A. Ferrara, S. Viti, C. Ceccarelli, The problematic growth of dust in high-redshift galaxies. Mon. Not. R. Astron. Soc. **463**, L112–L116 (2016). https://doi.org/10.1093/mnrasl/slw165. 1606.07214

R.A. Fesen, J.A. Morse, R.A. Chevalier, K.J. Borkowski, C.L. Gerardy, S.S. Lawrence, S. van den Bergh, Hubble space telescope WFPC2 imaging of Cassiopeia A. Astron. J. **122**, 2644–2661 (2001). https://doi.org/10.1086/323539. astro-ph/0108193

R.A. Fesen, M.C. Hammell, J. Morse, R.A. Chevalier, K.J. Borkowski, M.A. Dopita, C.L. Gerardy, S.S. Lawrence, J.C. Raymond, S. van den Bergh, Discovery of outlying high-velocity oxygen-rich ejecta in Cassiopeia A. Astrophys. J. **636**, 859–872 (2006). https://doi.org/10.1086/498092. astro-ph/0509067

R.A. Fesen, J.A. Zastrow, M.C. Hammell, J.M. Shull, D.W. Silvia, Ejecta knot flickering, mass ablation, and fragmentation in Cassiopeia A. Astrophys. J. **736**, 109 (2011). https://doi.org/10.1088/0004-637X/736/2/109. 1105.3970.

K. France, R.A. McCray, K. Heng, R.P. Kirshner, P. Challis, P. Bouchet, A. Crotts, E. Dwek, C. Fransson, P.M. Garnavich, J. Larsson, S.S. Lawrence, P. Lundqvist, N. Panagia, C.S.J. Pun, N. Smith, J. Sollerman, G. Sonneborn, J.T. Stocke, L. Wang, J.C. Wheeler, Observing supernova 1987A with the refurbished Hubble space telescope. Science **329**, 1624 (2010)

K.A. Frank, S.A. Zhekov, S. Park, R.A. McCray, E. Dwek, D.N. Burrows, Chandra observes the end of an era in SN 1987A. Astrophys. J. **829**(1), 40 (2016)

C. Fransson, J. Larsson, K. Migotto, D. Pesce, P. Challis, R.A. Chevalier, K. France, R.P. Kirshner, B. Leibundgut, P. Lundqvist, R.A. McCray, J. Spyromilio, F. Taddia, A. Jerkstrand, S. Mattila, N. Smith, J. Sollerman, J.C. Wheeler, A. Crotts, P. Garnavich, K. Heng, S.S. Lawrence, N. Panagia, C.S.J. Pun, G. Sonneborn, B.E.K. Sugerman, The destruction of the circumstellar ring of SN 1987A. Astrophys. J. **806**(1), L19 (2015)

B.M. Gaensler, P.O. Slane, The evolution and structure of pulsar wind nebulae. Annu. Rev. Astron. Astrophys. **44**, 17–47 (2006). https://doi.org/10.1146/annurev.astro.44.051905.092528. astro-ph/0601081

C. Gall, J. Hjorth, A.C. Andersen, Production of dust by massive stars at high redshift. Annu. Rev. Astron. Astrophys. **19**, 43 (2011). https://doi.org/10.1007/s00159-011-0043-7. 1108.0403

P. Ghavamian, J.M. Laming, C.E. Rakowski, A physical relationship between electron-proton temperature equilibration and Mach number in fast collisionless shocks. Astrophys. J. Lett. **654**, L69–L72 (2007). https://doi.org/10.1086/510740. astro-ph/0611306.

P. Ghavamian, J.C. Raymond, W.P. Blair, K.S. Long, A. Tappe, S. Park, P.F. Winkler, Spitzer spectroscopy of the galactic supernova remnant G292.0+1.8: structure and composition of the oxygen-rich ejecta. Astrophys. J. **696**, 1307–1318 (2009). https://doi.org/10.1088/0004-637X/696/2/1307. 0902.2804

H.L. Gomez, O. Krause, M.J. Barlow, B.M. Swinyard, P.J. Owen, C.J.R. Clark, M. Matsuura, E.L. Gomez, J. Rho, M.A. Besel, J. Bouwman, W.K. Gear, T. Henning, R.J. Ivison, E.T. Polehampton, B. Sibthorpe, A cool dust factory in the Crab nebula: a Herschel study of the filaments. Astrophys. J. **760**, 96 (2012). https://doi.org/10.1088/0004-637X/760/1/96. 1209.5677

K.D. Gordon, J. Roman-Duval, C. Bot, M. Meixner, B. Babler, J.P. Bernard, A. Bolatto, M.L. Boyer, G.C. Clayton, C. Engelbracht, Y. Fukui, M. Galametz, F. Galliano, S. Hony, A. Hughes, R. Indebetouw, F.P. Israel, K. Jameson, A. Kawamura, V. Lebouteiller, A. Li, S.C. Madden, M. Matsuura, K. Misselt, E. Montiel, K. Okumura, T. Onishi, P. Panuzzo, D. Paradis, M. Rubio, K.M. Sandstrom, M. Sauvage, J. Seale, M. Sewiło, K. Tchernyshyov, R. Skibba, Dust and gas in the magellanic clouds from the heritage Herschelkey project. I. Dust properties and insights into the origin of the submillimeter excess emission. Astrophys. J. **797**(2), 85 (2014)

E.V. Gotthelf, B. Koralesky, L. Rudnick, T.W. Jones, U. Hwang, R. Petre, Chandra detection of the forward and reverse shocks in Cassiopeia A. Astrophys. J. **552**(1), L39–L43 (2001)

B.W. Grefenstette, F.A. Harrison, S.E. Boggs, S.P. Reynolds, C.L. Fryer, K.K. Madsen, D.R. Wik, A. Zoglauer, C.I. Ellinger, D.M. Alexander, H. An, D. Barret, F.E. Christensen, W.W. Craig, K. Forster, P. Giommi, C.J. Hailey, A. Hornstrup, V.M. Kaspi, T. Kitaguchi, J.E. Koglin, P.H. Mao, H. Miyasaka, K. Mori, M. Perri, M.J. Pivovaroff, S. Puccetti, V. Rana, D. Stern, N.J. Westergaard, W.W. Zhang, Asymmetries in core-collapse supernovae from maps of radioactive 44Ti in Cassiopeia A. Nature **506**(7488), 339–342 (2014)

V. Guillet, Dust evolution in interstellar shocks. Theses, Université Paris Sud—Paris XI (2008). https://tel.archives-ouvertes.fr/tel-00332738

M.C. Hammell, R.A. Fesen, A catalog of outer ejecta knots in the Cassiopeia A supernova remnant. Astrophys. J. Suppl. Ser. **179**, 195 (2008). https://doi.org/10.1086/591528

J.J. Hester, The Crab nebula: an astrophysical chimera. Annu. Rev. Astron. Astrophys. **46**, 127–155 (2008). https://doi.org/10.1146/annurev.astro.45.051806.110608
R.H. Hildebrand, The determination of cloud masses and dust characteristics from submillimetre thermal emission. Q. J. R. Astron. Soc. **24**, 267 (1983)
D.C. Hines, G.H. Rieke, K.D. Gordon, J. Rho, K.A. Misselt, C.E. Woodward, M.W. Werner, O. Krause, W.B. Latter, C.W. Engelbracht, E. Egami, D.M. Kelly, J. Muzerolle, J.A. Stansberry, K.Y.L. Su, J.E. Morrison, E.T. Young, A. Noriega-Crespo, D.L. Padgett, R.D. Gehrz, E. Polomski, J.W. Beeman, E.E. Haller, Imaging of the supernova remnant Cassiopeia A with the Multiband Imaging Photometer for Spitzer (MIPS). Astrophys. J. Suppl. Ser. **154**, 290–295 (2004). https://doi.org/10.1086/422583
H. Hugoniot, Mémoire sur la propagation des mouvements dans les corps et spécialement dans les gaz parfaits (première partie). J. Éc. Polytech. **57**, 3–97 (1887)
H. Hugoniot, Mémoire sur la propagation des mouvements dans les corps et spécialement dans les gaz parfaits (deuxième partie). J. Éc. Polytech. **58**, 1–125 (1889)
U. Hwang, J.M. Laming, The circumstellar medium of Cassiopeia A inferred from the outer ejecta knot properties. Astrophys. J. **703**, 883–893 (2009). https://doi.org/10.1088/0004-637X/703/1/883. 0907.5177
U. Hwang, J.M. Laming, A Chandra X-ray survey of ejecta in the Cassiopeia A supernova remnant. Astrophys. J. **746**, 130 (2012). https://doi.org/10.1088/0004-637X/746/2/130. 1111.7316
U. Hwang, J.M. Laming, C. Badenes, F. Berendse, J.M. Blondin, D. Cioffi, T.A. DeLaney, D. Dewey, R. Fesen, K.A. Flanagan, C.L. Fryer, P. Ghavamian, J.P. Hughes, J.A. Morse, P.P. Plucinsky, R. Petre, M. Pohl, L. Rudnick, R. Sankrit, P.O. Slane, R.K. Smith, J. Vink, J.S. Warren, A million second Chandra view of Cassiopeia A. Astrophys. J. **615**(2), L117–L120 (2004)
R. Indebetouw, M. Matsuura, E. Dwek, G. Zanardo, M.J. Barlow, M. Baes, P. Bouchet, D.N. Burrows, R. Chevalier, G.C. Clayton, C. Fransson, B. Gaensler, R. Kirshner, M. Lakićević, K.S. Long, P. Lundqvist, I. Martí-Vidal, J. Marcaide, R. McCray, M. Meixner, C.Y. Ng, S. Park, G. Sonneborn, L. Staveley-Smith, C. Vlahakis, J. van Loon, Dust production and particle acceleration in supernova 1987A revealed with ALMA. Astrophys. J. Lett. **782**, L2 (2014). https://doi.org/10.1088/2041-8205/782/1/L2. 1312.4086
K. Isensee, G. Olmschenk, L. Rudnick, T.A. DeLaney, J. Rho, J.D. Smith, W.T. Reach, T. Kozasa, H.L. Gomez, Nucleosynthetic layers in the shocked ejecta of Cassiopeia A. Astrophys. J. **757**(2), 126 (2012)
H. Itoh, Shockwave model for the optical emission from oxygen-rich supernova ejecta—part two—precursor region. Publ. Astron. Soc. Jpn. **33**, 521 (1981)
H. Itoh, The destruction of dust grains in collisions of a supernova with a circumstellar medium. Mon. Not. R. Astron. Soc. **212**, 309–323 (1985). https://doi.org/10.1093/mnras/212.2.309
P. Jakobsen, R. Albrecht, C. Barbieri, J.C. Blades, A. Boksenberg, P. Crane, J.M. Deharveng, M.J. Disney, T.M. Kamperman, I.R. King, F. Macchetto, C.D. Mackay, F. Paresce, G. Weigelt, D. Baxter, P. Greenfield, R. Jedrzejewski, A. Nota, W.B. Sparks, R.P. Kirshner, N. Panagia, First results from the faint object camera—SN 1987A. Astrophys. J. **369**, L63 (1991)
A.P. Jones, J.A. Nuth, Dust destruction in the ISM: a re-evaluation of dust lifetimes. Astron. Astrophys. **530**, A44 (2011a). https://doi.org/10.1051/0004-6361/201014440
A.P. Jones, J.A. Nuth III., Dust destruction in the ISM: a re-evaluation of dust lifetimes. Astron. Astrophys. **530**, A44 (2011b)
A.P. Jones, A.G.G.M. Tielens, D.J. Hollenbach, C.F. McKee, Grain destruction in shocks in the interstellar medium. Astrophys. J. **433**, 797–810 (1994). https://doi.org/10.1086/174689
A.P. Jones, A.G.G.M. Tielens, D.J. Hollenbach, Grain shattering in shocks: the interstellar grain size distribution. Astrophys. J. **469**, 740 (1996). https://doi.org/10.1086/177823
A.P. Jones, L. Fanciullo, M. Köhler, L. Verstraete, V. Guillet, M. Bocchio, N. Ysard, The evolution of amorphous hydrocarbons in the ISM: dust modelling from a new vantage point. Astron. Astrophys. **558**, A62 (2013)
A.P. Jones, M. Köhler, N. Ysard, M. Bocchio, L. Verstraete, The global dust modelling framework THEMIS. Astron. Astrophys. **602**, A46 (2017)
S. Jurac, R.E. Johnson, B. Donn, Monte Carlo calculations of the sputtering of grains: enhanced sputtering of small grains. Astrophys. J. **503**, 247–252 (1998). https://doi.org/10.1086/305994
V.I. Kalikmanov (ed.), *Nucleation Theory*. Lecture Notes in Physics, vol. 860 (Springer, Berlin, 2013). https://doi.org/10.1007/978-90-481-3643-8
K. Kamper, S. van den Bergh, Optical studies of Cassiopeia A. V—A definitive study of proper motions. Astrophys. J. Suppl. Ser. **32**, 351–366 (1976). https://doi.org/10.1086/190400
R.I. Klein, C.F. McKee, P. Colella, On the hydrodynamic interaction of shock waves with interstellar clouds. 1: nonradiative shocks in small clouds. Astrophys. J. **420**, 213–236 (1994). https://doi.org/10.1086/173554
B.C.C. Koo, J.J. Lee, I.G. Jeong, J.Y. Seok, H.J. Kim, Infrared supernova remnants and their infrared-to-X-ray flux ratios. Astrophys. J. **821**(1), 20 (2016)

T. Kozasa, H. Hasegawa, K. Nomoto, Formation of dust grains in the ejecta of SN 1987A. Astrophys. J. **344**, 325–331 (1989). https://doi.org/10.1086/167801
O. Krause, S.M. Birkmann, T. Usuda, T. Hattori, M. Goto, G.H. Rieke, K.A. Misselt, The Cassiopeia A supernova was of Type IIb. Science **320**(5880), 1195–1197 (2008)
P.O. Lagage, A. Claret, J. Ballet, F. Boulanger, C.J. Cesarsky, D. Cesarsky, C. Fransson, A. Pollock, Dust formation in the Cassiopeia A supernova. Astron. Astrophys. **315**, L273–L276 (1996)
M. Lakicevic, M. van Loon J.T. Meixner, K.D. Gordon, C. Bot, J. Roman-Duval, B. Babler, A. Bolatto, C. Engelbracht, M. Filipović, S. Hony, R. Indebetouw, K. Misselt, E. Montiel, K. Okumura, P. Panuzzo, F. Patat, M. Sauvage, J. Seale, G. Sonneborn, D. temim t Urošević, G. Zanardo, The influence of supernova remnants on the interstellar medium in the Large Magellanic Cloud seen at 20–600 μm, wavelengths Astrophys. J. **799**(1), 50 (2015)
J.M. Laming, U. Hwang, On the determination of ejecta structure and explosion asymmetry from the X-ray knots of Cassiopeia A. Astrophys. J. **597**, 347–361 (2003). https://doi.org/10.1086/378268. arXiv:astro-ph/0306119
J. Larsson, C. Fransson, G. Östlin, P. Groningsson, A. Jerkstrand, C. Kozma, J. Sollerman, P. Challis, R.P. Kirshner, R.A. Chevalier, K. Heng, R.A. McCray, N.B. Suntzeff, P. Bouchet, A. Crotts, I.J. Danziger, E. Dwek, K. France, P.M. Garnavich, S.S. Lawrence, B. Leibundgut, P. Lundqvist, N. Panagia, C.S.J. Pun, N. Smith, G. Sonneborn, L. Wang, J.C. Wheeler, X-ray illumination of the ejecta of supernova 1987A. Nature **474**(7352), 484–486 (2011)
R.M. Lau, T.L. Herter, M.R. Morris, Z. Li, J.D. Adams, Old supernova dust factory revealed at the Galactic center. Science **348**(6233), 413–418 (2015)
M. Limongi, A. Chieffi, The nucleosynthesis of ^{26}Al and ^{60}Fe in solar metallicity stars extending in mass from 11 to 120 M_{solar}: the hydrostatic and explosive contributions. Astrophys. J. **647**, 483–500 (2006). https://doi.org/10.1086/505164. astro-ph/0604297
L.B. Lucy, I.J. Danziger, C. Gouiffes, P. Bouchet, Dust condensation in the ejecta of SN 1987 A, in *Structure and Dynamics of the Interstellar Medium* vol. 350 (1989), p. 164
M.M. Mac Low, C.F. McKee, R.I. Klein, J.M. Stone, M.L. Norman, Shock interactions with magnetized interstellar clouds. 1: Steady shocks hitting nonradiative clouds. Astrophys. J. **433**, 757–777 (1994). https://doi.org/10.1086/174685
S. Marassi, R. Schneider, M. Limongi, A. Chieffi, M. Bocchio, S. Bianchi, The metal and dust yields of the first massive stars. Mon. Not. R. Astron. Soc. **454**, 4250–4266 (2015). https://doi.org/10.1093/mnras/stv2267. 1509.08923
J.S. Mathis, W. Rumpl, K.H. Nordsieck, The size distribution of interstellar grains. Astrophys. J. **217**, 425–433 (1977). https://doi.org/10.1086/155591
N. Matsunami, A semiempirical formula for the energy dependence of the sputtering yield. Radiat. Eff. **57**(1–2), 15–21 (1981). https://doi.org/10.1080/01422448008218676
M. Matsuura, Dust and molecular formation in supernovae, in *Handbook of Supernovae*, ed. by W.A. Athem M. Paul (Springer, Berlin, 2017)
M. Matsuura, E. Dwek, M. Meixner, M. Otsuka, B. Babler, M.J. Barlow, J. Roman-Duval, C. Engelbracht, K.M. Sandstrom, M. Lakicevic, G. van Loon J.T. Sonneborn, G.C. Clayton, K.S. Long, P. Lundqvist, T. Nozawa, K.D. Gordon, S. Hony, P. Panuzzo, K. Okumura, K.A. Misselt, E. Montiel, M. Sauvage, Herschel detects a massive dust reservoir in supernova 1987A. Science **333**(6047), 1258–1261 (2011)
M. Matsuura, E. Dwek, M.J. Barlow, B. Babler, M. Baes, M. Meixner, J. Cernicharo, G.C. Clayton, L. Dunne, C. Fransson, J. Fritz, W. Gear, H.L. Gomez, M.A.T. Groenewegen, R. Indebetouw, R.J. Ivison, A. Jerkstrand, V. Lebouteiller, T.L. Lim, P. Lundqvist, C.P. Pearson, J. Roman-Duval, P. Royer, L. Staveley-Smith, A stubbornly large mass of cold dust in the ejecta of supernova 1987A. Astrophys. J. **800**, 50 (2015). https://doi.org/10.1088/0004-637X/800/1/50. 1411.7381
L. Mattsson, H.L. Gomez, A.C. Andersen, M. Matsuura, From flux to dust mass: does the grain-temperature distribution matter for estimates of cold dust masses in supernova remnants? Mon. Not. R. Astron. Soc. **449**, 4079–4090 (2015). https://doi.org/10.1093/mnras/stv487. 1504.02664
R. McCray, Supernova 1987A revisited. Annu. Rev. Astron. Astrophys. **31**, 175–216 (1993). https://doi.org/10.1146/annurev.aa.31.090193.001135
R.A. McCray, Supernova 1987A revisited. Annu. Rev. Astron. Astrophys. **31**, 175 (1993)
R.A. McCray, C. Fransson, The remnant of supernova 1987A. Annu. Rev. Astron. Astrophys. **54**(1), 19–52 (2016)
C. McKee, Dust destruction in the interstellar medium, in *Proceedings of the International Astronomical Union*, vol. 135 (1989), p. 431
C.F. McKee, J.P. Ostriker, A theory of the interstellar medium—three components regulated by supernova explosions in an inhomogeneous substrate. Astrophys. J. **218**, 148 (1977)

C.F. McKee, D.J. Hollenbach, G.C. Seab, A.G.G.M. Tielens, The structure of the time-dependent interstellar shocks and grain destruction in the interstellar medium. Astrophys. J. **318**, 674–701 (1987). https://doi.org/10.1086/165403

V. Mennella, J.R. Brucato, L. Colangeli, P. Palumbo, A. Rotundi, E. Bussoletti, Temperature dependence of the absorption coefficient of cosmic analog grains in the wavelength range 20 microns to 2 millimeters. Astrophys. J. **496**(2), 1058–1066 (1998)

E.R. Micelotta, A.P. Jones, A.G.G.M. Tielens, Polycyclic aromatic hydrocarbon processing in a hot gas. Astron. Astrophys. **510**, A37 (2010a). https://doi.org/10.1051/0004-6361/200911683. 0912.1595

E.R. Micelotta, A.P. Jones, A.G.G.M. Tielens, Polycyclic aromatic hydrocarbon processing in interstellar shocks. Astron. Astrophys. **510**, A36 (2010b). https://doi.org/10.1051/0004-6361/200911682. 0910.2461

E.R. Micelotta, E. Dwek, J.D. Slavin, Dust destruction by the reverse shock in the Cassiopeia A supernova remnant. Astron. Astrophys. **590**, A65 (2016). https://doi.org/10.1051/0004-6361/201527350. 1602.02754

D. Milisavljevic, R.A. Fesen, A detailed kinematic map of Cassiopeia A's optical main shell and outer high-velocity ejecta. Astrophys. J. **772**, 134 (2013). https://doi.org/10.1088/0004-637X/772/2/134. 1306.2310

D. Milisavljevic, R.A. Fesen, The bubble-like interior of the core-collapse supernova remnant Cassiopeia A. Science **347**(6221), 526–530 (2015)

J.A. Morse, R.A. Fesen, R.A. Chevalier, K.J. Borkowski, C.L. Gerardy, S.S. Lawrence, S. van den Bergh, Location of the optical reverse shock in the Cassiopeia A supernova remnant. Astrophys. J. **614**, 727–736 (2004). https://doi.org/10.1086/423709

S.H. Moseley, E. Dwek, W. Glaccum, J.R. Graham, R.F. Loewenstein, Far-infrared observations of thermal dust emission from supernova 1987A. Nature **340**, 697–699 (1989). https://doi.org/10.1038/340697a0

F. Nakamura, C.F. McKee, R.I. Klein, R.T. Fisher, On the hydrodynamic interaction of shock waves with interstellar clouds. II. The effect of smooth cloud boundaries on cloud destruction and cloud turbulence. Astrophys. J. Suppl. Ser. **164**, 477–505 (2006). https://doi.org/10.1086/501530. astro-ph/0511016

B.B. Nath, T. Laskar, J.M. Shull, Dust sputtering by reverse shocks in supernova remnants. Astrophys. J. **682**, 1055–1064 (2008). https://doi.org/10.1086/589224. 0804.3472

K. Nomoto, K. Iwamoto, T. Suzuki, O.R. Pols, H. Yamaoka, M. Hashimoto, P. Hoflich, E.P.J. van den Heuvel, The origin of Type Ib-Ic-IIb-IIL supernovae and binary star evolution, in *Compact Stars in Binaries, IAU Symposium*, vol. 165, ed. by J. van Paradijs, E.P.J. van den Heuvel, E. Kuulkers (1996), p. p 119

K.I. Nomoto, K. Iwamoto, T. Suzuki, The evolution and explosion of massive binary stars and Type Ib-Ic-IIb-IIL supernovae. Phys. Rep. **256**, 173–191 (1995). https://doi.org/10.1016/0370-1573(94)00107-E

M.L. Norman, G.L. Bryan, Cosmological adaptive mesh refinement^{CD}, in *Numerical Astrophysics, Astrophys. Space Sci. Library*, vol. 240, ed. by S.M. Miyama, K. Tomisaka, T. Hanawa (1999), p. 19. https://doi.org/10.1007/978-94-011-4780-4_3. astro-ph/9807121

T. Nozawa, T. Kozasa, H. Umeda, K. Maeda, K. Nomoto, Dust in the early universe: dust formation in the ejecta of population III supernovae. Astrophys. J. **598**, 785–803 (2003). https://doi.org/10.1086/379011. astro-ph/0307108

T. Nozawa, T. Kozasa, A. Habe, Dust destruction in the high-velocity shocks driven by supernovae in the early universe. Astrophys. J. **648**, 435–451 (2006). https://doi.org/10.1086/505639. astro-ph/0605193

T. Nozawa, T. Kozasa, A. Habe, E. Dwek, H. Umeda, N. Tominaga, K. Maeda, K. Nomoto, Evolution of dust in primordial supernova remnants: can dust grains formed in the ejecta survive and be injected into the early interstellar medium? Astrophys. J. **666**, 955–966 (2007). https://doi.org/10.1086/520621. 0706.0383

T. Nozawa, T. Kozasa, N. Tominaga, K. Maeda, H. Umeda, K. Nomoto, O. Krause, Formation and evolution of dust in type IIb supernovae with application to the Cassiopeia A supernova remnant. Astrophys. J. **713**, 356–373 (2010). https://doi.org/10.1088/0004-637X/713/1/356. 0909.4145

S. Orlando, G. Peres, F. Reale, F. Bocchino, R. Rosner, T. Plewa, A. Siegel, Crushing of interstellar gas clouds in supernova remnants. I. The role of thermal conduction and radiative losses. Astron. Astrophys. **444**, 505–519 (2005). https://doi.org/10.1051/0004-6361:20052896. astro-ph/0508638

B.W. O'Shea, G.L. Bryan, J. Bordner, M.L. Norman, T. Abel, R. Harkness, A. Kritsuk, Introducing Enzo, an AMR cosmology application, in *Adaptive Mesh Refinement—Theory and Applications*. Lecture Notes in Computational Science and Engineering, vol. 41 (Springer, Berlin, 2005)

V. Ossenkopf, T. Henning, J.S. Mathis, Constraints on cosmic silicates. Astron. Astrophys. **261**, 567 (1992)

M. Otsuka, J.T. van Loon, K.S. Long, M. Meixner, M. Matsuura, W.T. Reach, J. Roman-Duval, K. Gordon, M. Sauvage, S. Hony, K. Misselt, C. Engelbracht, P. Panuzzo, K. Okumura, P.M. Woods, F. Kemper, G.C. Sloan, Dust in the bright supernova remnant N49 in the LMC. Astron. Astrophys. **518**, L139 (2010). https://doi.org/10.1051/0004-6361/201014642. 1005.2787

P.J. Owen, M.J. Barlow, The dust and gas content of the Crab nebula. Astrophys. J. **801**, 141 (2015). https://doi.org/10.1088/004-637X/801/2/141. 1501.01510
T. Padmanabhan, *Theoretical Astrophysics*, vol. 2. Stars and Stellar Systems (Springer, Berlin, 2001). https://doi.org/10.2277/0521562414
N. Panagia, R. Gilmozzi, F. Macchetto, H.M. Adorf, R.P. Kirshner, Properties of the SN 1987A circumstellar ring and the distance to the Large Magellanic Cloud. Astrophys. J. **380**, L23–L26 (1991)
D.J. Patnaude, R.A. Fesen, Model simulations of a shock-cloud interaction in the Cygnus Loop. Astrophys. J. **633**, 240–247 (2005). https://doi.org/10.1086/452627. astro-ph/0507330
W.J.M. Rankine, On the thermodynamic theory of waves of finite longitudinal disturbance. Philos. Trans. R. Soc. Lond. **160**, 277–288 (1870). https://doi.org/10.1098/rstl.1870.0015. http://rstl.royalsocietypublishing.org/content/160/277.short. http://rstl.royalsocietypublishing.org/content/160/277.full.pdf+html
T. Rauscher, A. Heger, R.D. Hoffman, S.E. Woosley, Nucleosynthesis in massive stars with improved nuclear and stellar physics. Astrophys. J. **576**, 323–348 (2002). https://doi.org/10.1086/341728. astro-ph/0112478
J.C. Raymond, P. Ghavamian, B.J. Williams, W.P. Blair, K.J. Borkowski, T.J. Gaetz, R. Sankrit, Grain destruction in A supernova remnant shock wave. Astrophys. J. **778**(2), 161 (2013)
S.P. Reynolds, Supernova remnants at high energy. Annu. Rev. Astron. Astrophys. **46**, 89–126 (2008). https://doi.org/10.1146/annurev.astro.46.060407.145237
J. Rho, T. Kozasa, W.T. Reach, J.D. Smith, L. Rudnick, T. DeLaney, J.A. Ennis, H. Gomez, A. Tappe, Freshly formed dust in the Cassiopeia A supernova remnant as revealed by the Spitzer Space Telescope. Astrophys. J. **673**, 271–282 (2008). https://doi.org/10.1086/523835. 0709.2880
J. Rho, T.H. Jarrett, W.T. Reach, H. Gomez, M. Andersen, Carbon monoxide in the Cassiopeia A supernova remnant. Astrophys. J. Lett. **693**, L39–L43 (2009). https://doi.org/10.1088/0004-637X/693/1/L39. 0901.2308
J. Rho, W.T. Reach, A. Tappe, U. Hwang, J.D. Slavin, T. Kozasa, L. Dunne, Spitzer observations of the young core-collapse supernova remnant 1E0102-72.3: infrared ejecta emission and dust formation. Astrophys. J. **700**, 579–596 (2009b). https://doi.org/10.1088/0004-637X/700/1/579
J. Rho, W.T. Reach, A. Tappe, L. Rudnick, T. Kozasa, U. Hwang, M. Andersen, H. Gomez, T. Delaney, L. Dunne, J. Slavin, Dust formation observed in young supernova remnants with Spitzer, in *Cosmic Dust—Near and Far*, ed. by T. Henning, E. Grün, J. Steinacker. Astronomical Society of the Pacific Conference Series, vol. 414 (2009c), p. p 22. 0901.1699
J. Rho, T. Onaka, J. Cami, W.T. Reach, Spectroscopic detection of carbon monoxide in the young supernova remnant Cassiopeia A. Astrophys. J. Lett. **747**, L6 (2012). https://doi.org/10.1088/2041-8205/747/1/L6. 1202.4540
E.E. Salpeter, Formation and destruction of dust grains. Annu. Rev. Astron. Astrophys. **15**(1), 267–293 (1977)
K.M. Sandstrom, A.D. Bolatto, S. Stanimirović, J.T. van Loon, J.D.T. Smith, Measuring dust production in the small Magellanic Cloud core-collapse supernova remnant 1E 0102.2-7219. Astrophys. J. **696**, 2138–2154 (2009). https://doi.org/10.1088/0004-637X/696/2/2138. 0810.2803
R. Sankrit, B.J. Williams, K.J. Borkowski, T.J. Gaetz, J.C. Raymond, W.P. Blair, P. Ghavamian, K.S. Long, S.P. Reynolds, Dust destruction in a non-radiative shock in the Cygnus Loop supernova remnant. Astrophys. J. **712**(2), 1092–1099 (2010)
A. Sarangi, I. Cherchneff, The chemically controlled synthesis of dust in Type II-P supernovae. Astrophys. J. **776**, 107 (2013). https://doi.org/10.1088/0004-637X/776/2/107. 1309.5887
A. Sarangi, I. Cherchneff, Condensation of dust in the ejecta of Type II-P supernovae. Astron. Astrophys. **575**, A95 (2015). https://doi.org/10.1051/0004-6361/201424969. 1412.5522
A. Sarangi, M. Matsuura, E. Micelotta, Dust in supernovae and supernova remnants I: formation scenarios. Space Sci. Rev. (2018). https://doi.org/10.1007/s11214-018-0484-7
C.L. Sarazin, R.E. White III, Steady state cooling flow models for normal elliptical galaxies. Astrophys. J. **320**, 32–48 (1987). https://doi.org/10.1086/165522
B.D. Savage, K.R. Sembach, Interstellar abundances from absorption-line observations with the Hubble space telescope. Annu. Rev. Astron. Astrophys. **34**(1), 279–329 (1996)
R. Schneider, A. Ferrara, R. Salvaterra, Dust formation in very massive primordial supernovae. Mon. Not. R. Astron. Soc. **351**, 1379–1386 (2004). https://doi.org/10.1111/j.1365-2966.2004.07876.x. astro-ph/0307087
C.G. Seab, J.M. Shull, Shock processing of interstellar grains. Astrophys. J. **275**, 652–660 (1983). https://doi.org/10.1086/161563
L.I. Sedov, *Similarity and Dimensional Methods in Mechanics* (1959)
J.Y. Seok, B.C.C. Koo, T. Onaka, A survey of infrared supernova remnants in the Large Magellanic Cloud. Astrophys. J. **779**(2), 134 (2013)

J.Y. Seok, B.C. Koo, H. Hirashita, Dust cooling in supernova remnants in the Large Magellanic Cloud. Astrophys. J. **807**, 100 (2015). https://doi.org/10.1088/0004-637X/807/1/100. 1506.07926
L. Serra Díaz-Cano, A.P. Jones, Carbonaceous dust in interstellar shock waves: hydrogenated amorphous carbon (a-C:H) vs. graphite. Astron. Astrophys. **492**, 127–133 (2008). https://doi.org/10.1051/0004-6361:200810622
J.M. Shull, Disruption and sputtering of grains in intermediate-velocity interstellar clouds. Astrophys. J. **226**, 858–862 (1978). https://doi.org/10.1086/156666
P. Sigmund, *Sputtering by Particle Bombardment, vol. 1*. Topics in Applied Physics, vol. 47 (Springer, Berlin, 1981)
D.W. Silvia, B.D. Smith, J.M. Shull, Numerical simulations of supernova dust destruction. I. Cloud-crushing and post-processed grain sputtering. Astrophys. J. **715**, 1575–1590 (2010). https://doi.org/10.1088/0004-637X/715/2/1575. 1001.4793
D.W. Silvia, B.D. Smith, J.M. Shull, Numerical simulations of supernova dust destruction. II. Metal-enriched ejecta knots. Astrophys. J. **748**, 12 (2012). https://doi.org/10.1088/0004-637X/748/1/12. 1111.0302
J.D. Slavin, E. Dwek, A.P. Jones, Destruction of interstellar dust in evolving supernova remnant shock waves. Astrophys. J. **803**, 7 (2015). https://doi.org/10.1088/0004-637X/803/1/7. 1502.00929
J.D.T. Smith, L. Rudnick, T. Delaney, J. Rho, H. Gomez, T. Kozasa, W. Reach, K. Isensee, Spitzer spectral mapping of supernova remnant Cassiopeia A. Astrophys. J. **693**, 713–721 (2009). https://doi.org/10.1088/0004-637X/693/1/713. 0810.3014
J.M. Stone, M.L. Norman, The three-dimensional interaction of a supernova remnant with an interstellar cloud. Astrophys. J. Lett. **390**, L17–L19 (1992). https://doi.org/10.1086/186361
R.S. Sutherland, M.A. Dopita, Young oxygen-rich supernova remnants. 2: an oxygen-rich emission mechanism. Astrophys. J. **439**, 381–398 (1995). https://doi.org/10.1086/175181
G. Taylor, The formation of a blast wave by a very intense explosion. I. Theoretical discussion. Proc. R. Soc. Lond. Ser. A **201**, 159–174 (1950). https://doi.org/10.1098/rspa.1950.0049
T. Temim, E. Dwek, The importance of physical models for deriving dust masses and grain size distributions in supernova ejecta. I. Radiatively heated dust in the Crab nebula. Astrophys. J. **774**, 8 (2013). https://doi.org/10.1088/0004-637X/774/1/8. 1302.5452
T. Temim, P. Slane, S.P. Reynolds, J.C. Raymond, K.J. Borkowski, Deep Chandra observations of the Crab-like pulsar wind nebula G54.1+0.3 and Spitzer spectroscopy of the associated infrared shell. Astrophys. J. **710**, 309–324 (2010). https://doi.org/10.1088/0004-637X/710/1/309. 0912.4538
T. Temim, E. Dwek, K. Tchernyshyov, M.L. Boyer, M. Meixner, C. Gall, J. Roman-Duval, Dust destruction rates and lifetimes in the Magellanic Clouds. Astrophys. J. **799**, 158 (2015). https://doi.org/10.1088/0004-637X/799/2/158. 1411.4574
T. Temim, E. Dwek, R.G. Arendt, K.J. Borkowski, S.P. Reynolds, P. Slane, J.D. Gelfand, J.C. Raymond, A massive shell of supernova-formed dust in SNR G54.1+0.3. Astrophys. J. **836**, 129 (2017). https://doi.org/10.3847/1538-4357/836/1/129. 1701.01117
A.G.G.M. Tielens, *The Physics and Chemistry of the Interstellar Medium* (Cambridge University Press, Cambridge, 2010)
A.G.G.M. Tielens, C.F. McKee, C.G. Seab, D.J. Hollenbach, The physics of grain-grain collisions and gas-grain sputtering in interstellar shocks. Astrophys. J. **431**, 321–340 (1994). https://doi.org/10.1086/174488
P. Todini, A. Ferrara, Dust formation in primordial Type II supernovae. Mon. Not. R. Astron. Soc. **325**, 726–736 (2001). https://doi.org/10.1046/j.1365-8711.2001.04486.x. astro-ph/0009176
J.K. Truelove, C.F. McKee, Evolution of nonradiative supernova remnants. Astrophys. J. Suppl. Ser. **120**, 299–326 (1999). https://doi.org/10.1086/313176
H. Umeda, K. Nomoto, Nucleosynthesis of zinc and iron peak elements in population III Type II supernovae: comparison with abundances of very metal poor halo stars. Astrophys. J. **565**, 385–404 (2002). https://doi.org/10.1086/323946. astro-ph/0103241
M. van Adelsberg, K. Heng, R. McCray, J.C. Raymond, Spatial structure and collisionless electron heating in Balmer-dominated shocks. Astrophys. J. **689**, 1089–1104 (2008). https://doi.org/10.1086/592680. 0803.2521
E. van der Swaluw, A. Achterberg, Y.A. Gallant, T.P. Downes, R. Keppens, Interaction of high-velocity pulsars with supernova remnant shells. Astron. Astrophys. **397**, 913–920 (2003). https://doi.org/10.1051/0004-6361:20021488. astro-ph/0202232
O. Vancura, J.C. Raymond, E. Dwek, W.P. Blair, K.S. Long, S. Foster, A study of X-ray and infrared emissions from dusty nonradiative shock waves. Astrophys. J. **431**, 188–200 (1994). https://doi.org/10.1086/174477
J. Vink, J.S. Kaastra, J.A.M. Bleeker, A new mass estimate and puzzling abundances of SNR Cassiopeia A. Astron. Astrophys. **307**, L41–L44 (1996)

S.H.J. Wallström, C. Biscaro, F. Salgado, J.H. Black, I. Cherchneff, S. Muller, O. Berné, J. Rho, A.G.G.M. Tielens, CO rotational line emission from a dense knot in Cassiopeia A. Evidence for active post-reverse-shock chemistry. Astron. Astrophys. **558**, L2 (2013). https://doi.org/10.1051/0004-6361/201322576. 1309.4229

D. Watson, L. Christensen, K.K. Knudsen, J. Richard, A. Gallazzi, M.J. Michałowski, A dusty, normal galaxy in the epoch of reionization. Nature **519**, 327–330 (2015). https://doi.org/10.1038/nature14164. 1503.00002

K.W. Weiler, R.A. Sramek, Supernovae and supernova remnants (1988). Annu. Rev. Astron. Astrophys. 26:295–341, https://doi.org/10.1146/annurev.aa.26.090188.001455

J.C. Weingartner, B.T. Draine, Dust grain-size distributions and extinction in the Milky Way, Large Magellanic Cloud, and Small Magellanic Cloud. Astrophys. J. **548**, 296 (2001)

R. Wesson, M.J. Barlow, M. Matsuura, B. Ercolano, The timing and location of dust formation in the remnant of SN 1987A. Mon. Not. R. Astron. Soc. **446**(2), 2089–2101 (2014)

P.A. Whitelock, R.M. Catchpole, J.W. Menzies, M.W. Feast, S.E. Woosley, D. Alen, F. van Wyk, F. Marang, C.D. Laney, H. Winkler, K. Sekiguchi, L.A. Balona, B.S. Carter, J.H. Spencer Jones, J.D. Laing, T.L. Evans, A.P. Fairall, D.A.H. Buckley, I.S. Glass, M.V. Penston, L.N. da Costa, S.A. Bell, C. Hellier, M. Shara, A.F.J. Moffat, Spectroscopic and photometric observations of SN1987A. VI—Days 617 to 792. Mon. Not. R. Astron. Soc. **240**, 7–24 (1989)

B.J. Williams, T. Temim, Infrared emission from supernova remnants: formation and destruction of dust (2016). https://doi.org/10.1007/978-3-319-20794-0_94-1

B.J. Williams, K.J. Borkowski, S.P. Reynolds, W.P. Blair, P. Ghavamian, S.P. Hendrick, K.S. Long, S. Points, J.C. Raymond, R. Sankrit, R.C. Smith, P.F. Winkler, Dust destruction in fast shocks of core-collapse supernova remnants in the Large Magellanic Cloud. Astrophys. J. **652**(1), L33–L36 (2006)

B.J. Williams, K.J. Borkowski, S.P. Reynolds, P. Ghavamian, W.P. Blair, K.S. Long, R. Sankrit, Dust in a type Ia supernova progenitor: Spitzer spectroscopy of Kepler's supernova remnant. Astrophys. J. **755**(1), 3 (2012)

R. Willingale, J.A.M. Bleeker, K.J. bparticevan der Heyden, J.S. Kaastra, J. Vink, X-ray spectral imaging and Doppler mapping of Cassiopeia A. Astron. Astrophys. **381**, 1039–1048 (2002). https://doi.org/10.1051/0004-6361:20011614. astro-ph/0107270

J. Wilms, Supernova Remnants. X-ray Astronomy II, University of Erlangen-Nuremberg 4 (2012). http://pulsar.sternwarte.uni-erlangen.de/wilms/teach/xray2_0809/xray2chap4toc.html

A. Wongwathanarat, E. Müller, H.T. Janka, Three-dimensional simulations of core-collapse supernovae: from shock revival to shock breakout. Astron. Astrophys. **577**, A48 (2015). https://doi.org/10.1051/0004-6361/201425025. 1409.5431

D.H. Wooden, Observational evidence for mixing and dust condensation in core-collapse supernovae, in *American Institute of Physics Conference Series*, ed. by E.K. Zinner, T.J. Bernatowicz. American Institute of Physics Conference Series, vol. 402 (1997), pp. 317–376. https://doi.org/10.1063/1.53315

D.H. Wooden, D.M. Rank, J.D. Bregman, F.C. Witteborn, A.G.G.M. Tielens, M. Cohen, P.A. Pinto, T.S. Axelrod, Airborne spectrophotometry of SN 1987A from 1.7 to 12.6 microns—time history of the dust continuum and line emission. Astron. Astrophys. Suppl. Ser. **88**, 477 (1993). (ISSN 0067-0049)

S.E. Woosley, T.A. Weaver, The evolution and explosion of massive stars. II. Explosive hydrodynamics and nucleosynthesis. Astrophys. J. Suppl. Ser. **101**, 181 (1995). https://doi.org/10.1086/192237

H. Yamaguchi, K.A. Eriksen, C. Badenes, J.P. Hughes, N.S. Brickhouse, A.R. Foster, D.J. Patnaude, R. Petre, P.O. Slane, R.K. Smith, New evidence for efficient collisionless heating of electrons at the reverse shock of a young supernova remnant. Astrophys. J. **780**, 136 (2014). https://doi.org/10.1088/0004-637X/780/2/136. 1310.8355

T. Yamamoto, H. Hasegawa, Grain formation through nucleation process in astrophysical environment. Prog. Theor. Phys. **58**, 816–828 (1977). https://doi.org/10.1143/PTP.58.816

G. Zanardo, L. Staveley-Smith, L. Ball, B.M. Gaensler, M.J. Kesteven, R.N. Manchester, C.Y. Ng, A.K. Tzioumis, T.M. Potter, Multifrequency radio measurements of supernova 1987A over 22 years. Astrophys. J. **710**(2), 1515–1529 (2010)

G. Zanardo, L. Staveley-Smith, C.Y. Ng, B.M. Gaensler, T.M. Potter, R.N. Manchester, A.K. Tzioumis, High-resolution radio observations of the remnant of SN 1987A at high frequencies. Astrophys. J. **767**(2), 98 (2013)

G. Zanardo, L. Staveley-Smith, R. Indebetouw, R.A. Chevalier, M. Matsuura, B.M. Gaensler, M.J. Barlow, C. Fransson, R.N. Manchester, M. Baes, J.R. Kamenetzky, M. Lakićević, P. Lundqvist, J.M. Marcaide, I. Marti-Vidal, M. Meixner, C.Y. Ng, S. Park, G. Sonneborn, J. Spyromilio, J.T. van Loon, Spectral and morphological analysis of the remnant of supernova 1987A with ALMA and ATCA Astrophys. J. **796**(2), 82 (2014)

E. Zinner, Stardust in the laboratory. Publ. Astron. Soc. Aust. **25**, 7–17 (2008). https://doi.org/10.1071/AS07039

Space Sci Rev (2018) 214:41
https://doi.org/10.1007/s11214-018-0479-4

Cosmic Ray Production in Supernovae

A.M. Bykov[1,2,3] · D.C. Ellison[4] · A. Marcowith[5] ·
S.M. Osipov[1]

Received: 4 December 2017 / Accepted: 24 January 2018 / Published online: 30 January 2018

Abstract We give a brief review of the origin and acceleration of cosmic rays (CRs), emphasizing the production of CRs at different stages of supernova evolution by the first-order Fermi shock acceleration mechanism. We suggest that supernovae with trans-relativistic outflows, despite being rather rare, may accelerate CRs to energies above 10^{18} eV over the first year of their evolution. Supernovae in young compact clusters of massive stars, and interaction powered superluminous supernovae, may accelerate CRs well above the PeV regime. We discuss the acceleration of the bulk of the galactic CRs in isolated supernova remnants and re-acceleration of escaped CRs by the multiple shocks present in superbubbles produced by associations of OB stars. The effects of magnetic field amplification by CR driven instabilities, as well as superdiffusive CR transport, are discussed for nonthermal radiation produced by nonlinear shocks of all speeds including trans-relativistic ones.

Keywords Supernovae · Cosmic rays · Supernova remnants · Interstellar medium · Magnetic turbulence

---

Supernovae
Edited by Andrei Bykov, Roger Chevalier, John Raymond, Friedrich-Karl Thielemann, Maurizio Falanga and Rudolf von Steiger

---

✉ A.M. Bykov
byk@astro.ioffe.ru

D.C. Ellison
don_ellison@ncsu.edu

1 Ioffe Institute, 194021, St. Petersburg, Russia

2 St. Petersburg Polytechnic University, St. Petersburg, Russia

3 International Space Science Institute, Bern, Switzerland

4 Department of Physics, North Carolina State University, Raleigh, NC 27695-8202, USA

5 Laboratoire Univers et Particules de Montpellier CNRS/Universite de Montpellier, Place E. Bataillon, 34095 Montpellier, France

1 Introduction

Based on early inconclusive evidence, Baade and Zwicky (1934) suggested that cosmic rays (CRs) are produced by supernovae (SNe). Since then it has become clear that SNe are the most likely source of CRs, at least for those with energies below the so-called "knee" at $\sim 10^{15}$ eV, for four main reasons. First, and most important, SNe are the only known galactic source with sufficient energy to power CRs (e.g., Ginzburg and Syrovatskii 1964; Axford 1981; Berezinskii et al. 1990; Hillas 2005; Reynolds 2008; Lingenfelter 2017). Even for supernova remnants (SNRs) the acceleration of CRs must be highly efficient. Second, the source composition of the bulk of CR material is primarily well-mixed interstellar medium (ISM) material (including those elements preferentially locked in dust) with no more than a 20% contribution from fresh core-collapse SN ejecta material (i.e., Meyer et al. 1997; Ellison et al. 1997; Meyer and Ellison 1999; Binns et al. 2014). This requires a source that injects *old* (i.e., not freshly synthesized) material from the entire galaxy including material from low-mass stars that never explode. No source other than SNRs has a galactic filling factor large enough to do this. Third, there is now a wealth of observational evidence from SNRs showing non-thermal radiation emitted by ultra-relativistic electrons and ions (e.g., Helder et al. 2012; Ackermann et al. 2013a; Blasi 2013; Amato 2014; Blandford et al. 2014; Slane et al. 2015a). The *in situ* production of CRs is observed in SNRs! Fourth, the collisionless shocks associated with SNRs can utilize the first-order Fermi shock acceleration mechanism to simultaneously accelerate various ion species with high efficiency (e.g., Ahn et al. 2010). Perhaps even more important, for the same set of parameters, first-order Fermi shock acceleration accelerates electrons and ions with similar spectral shapes, as observed in CRs and solar energetic particles (e.g., Ellison and Ramaty 1985; Boyle et al. 2008). No other mechanism as naturally produces similar spectral shapes for leptons and hadrons.

The power required to maintain the observed energy density of CRs (dominated by nuclei in the GeV range) is $\sim 2 \times 10^{41}$ $\mathrm{erg\,s^{-1}}$ (e.g., Berezinskii et al. 1990). This is estimated by assuming some CR propagation model with the CR escape length from the galaxy derived from the observed secondary to primary nuclei ratio (e.g., Ptuskin 2012; Aharonian et al. 2012). If in the Milky Way, there is one SN every 30–100 yr with an average kinetic energy of 2×10^{51} erg, an efficiency of 10–30% is required to power CRs, i.e., $> 10\%$ of the ram kinetic energy of the SNR blast wave must be put into relativistic particles of the GeV energy regime.

If acceleration efficiencies are this large, nonlinear feedback effects from CR production will influence the shock dynamics and the resultant CR spectral shape (e.g., Berezhko and Ellison 1999). Consequences for the dynamics include the fact that the test-particle relation between the postshock temperature, T_2, and the shock speed, u_0, often assumed for strong shocks, will be modified, i.e., $(1/\mu m_H)(kT_2/u_0^2 \neq 3/16)$ (e.g., Decourchelle et al. 2000; Hughes et al. 2000). Here μm_H is the mean particle mass. Likewise, the Sedov solutions relating the SNR age to the shock radius and speed will also be modified (e.g., Ellison et al. 2007).

Recent detailed observations of the energy spectrum, composition, and anisotropy of the angular distribution of CRs up to TeV energies have been made by spacecraft and balloon instruments, e.g., *ACE*, *PAMELA*, *AMS2*, *CREAM*, *SuperTIGER*, and *Fermi*. For higher energies, ground-based experiments are required and these include *KASCADE-Grande*, *TUNKA*, and *LOFAR* up to a few hundred PeV, while the *Telescope Array* and the *Pierre Auger* observatories provided detailed CR observations to EeV energies. The *LOFAR* array measures radio emission produced by relativistic leptons created in high-energy CR air showers. This relatively new technique has a large duty cycle and can determine the atmospheric depth of shower maximum, and thus CR energy, with high resolution (Buitink et al. 2016). The high statistics of the above mentioned measurements have allowed a closer look into the spectral features which have long been used to understand the sources of CRs, their propagation in the Galaxy and beyond, and the assumed transition from a galactic source to extragalactic sources of CRs above $\sim 10^{18}$ eV (e.g., Bergman and Belz 2007; Troitsky 2013; Lipari 2017).

A particularly interesting feature has been the spectral hardening of protons and helium at rigidities above ~ 230 GV reported by *PAMELA* and *AMS2*.[1] This hardening is likely fairly broad and may extend beyond the rigidity range of both *PAMELA* and *AMS2*. The combined fit of *AMS2* and *CREAM* data made by Lipari (2017) indicates a very broad feature that extends from 200 GeV to 2 TeV with a spectral index difference of ~ 0.2 (see Malkov 2017).

Other important features in the all particle spectrum are the so-called "knee" at 10^{15-16} eV and the "ankle" in the EeV range. The energy band between the knee and the ankle is crucial for understanding the transition between galactic and extragalactic CRs (e.g., Bergman and Belz 2007; Aloisio et al. 2007). Regardless of the energy range, and despite decades of portraying the all particle spectrum observed at Earth as a simple power law, improved statistics are beginning to show that irregularities and substructures are present in the spectrum (e.g., Apel et al. 2013; Berezhnev et al. 2012).

Recent high resolution measurements by the *LOFAR* telescope below ~ 1 EeV (Buitink et al. 2016) favor a CR composition with a light-mass fraction (protons and helium nuclei) of about 80%. A light-mass fraction this high is surprising and may show the existence of an extragalactic component at energies between the knee and the ankle, or a previously unsuspected light-mass galactic CR component at these energies (e.g., Thoudam et al. 2016).

Another important measurement is the anisotropy. This is below a few percent for CR energies $\lesssim 1$ EeV. The data thus far does not allow clear conclusions at higher energies but recent observations have shown a large-scale anisotropy with an amplitude of $\sim 10^{-3}$, along with small-scale structure (angular size 10°–30°) at about the 10^{-4} level (i.e., Abeysekara et al. 2017). It is still not possible to constrain particular source populations, or specific nearby sources, from the anisotropy observations.

The direct observation of CRs can only be done at the Earth since the charged CRs meander through the irregular galactic magnetic field and virtually all directional information from the source is lost. Source information can be obtained, however, by observing the non-thermal radiation CRs produce in specific objects such as SNRs. Multi-wavelength observations of SNRs give detailed information on the CR acceleration process. Diffuse galactic emission on the other hand gives direct information on the distribution of CRs throughout the galaxy.

Recent observations of non-thermal radiation from young (< 1000 yr) and middle-aged SNRs have proven that particles are accelerated to above 10 TeV by the shocks in these objects. Synchrotron radiation from relativistic leptons is detected from radio to X-rays, and the high angular resolution observations by *VLA* and *Chandra* of young SNRs like Tycho's SNR, Cas A, SN1006, and RX J1713.7-3946 provide compelling evidence for magnetic field amplification (MFA) directly associated with the acceleration process at the remnant forward shock (e.g., Reynolds 2008; Vink 2012; Helder et al. 2012).

The Fermi acceleration process (i.e., Fermi 1949, 1954), and its particular realization called diffusive shock acceleration (DSA) (Axford et al. 1977; Krymskii 1977; Bell 1978;

[1] Rigidity is defined as pc/Ze where p is the particle momentum, c is the speed of light, e is the electronic charge, Z is the total charge number of the particle, and the units are volts.

Blandford and Ostriker 1978), provides a plausible way to efficiently convert the kinetic energy released in a supernova shock to a wide energy spectrum of accelerated particles (e.g., Blandford and Eichler 1987; Jones and Ellison 1991; Malkov and Drury 2001; Schure et al. 2012). In fact, Fermi acceleration can occur at any collisionless shock and there is direct observational confirmation of efficient particle acceleration at the quasi-parallel Earth bow shock (i.e., Ellison et al. 1990).[2]

The forward shocks in most galactic SNRs are observed to have speeds below $\sim 10^4$ km s^{-1}. While these shocks are believed to be capable of producing the bulk of galactic CRs to energies approaching the knee via Fermi acceleration uncertainties remain. It is still uncertain how particles escape from the accelerator without experiencing strong adiabatic losses, and the maximum CR energy produced in observed SNRs does not extend into the knee. Particle escape depends on the galactic environment at the late stages of SNR evolution which may be in the warm ISM or in a superbubble produced by clustered SNe (e.g., Mac Low and McCray 1988; Heiles 1990; Cox 2005). As for the maximum CR energy a given shock can produce, $E_{\rm max}$, Lagage and Cesarsky (1983) estimated $E_{\rm max}$ for Fermi acceleration at the forward shock of an isolated SNR over its lifetime as $E_{\rm max} \sim 10^5\ Z\ B_{-6}$ GeV, where Z is the ion charge number and B_{-6} is the ISM magnetic field measured in μG (see also Hillas 2005).

The γ-ray observations of SNRs W44 and IC 443 provide clear evidence for pion production in a SNR due to TeV CR proton interactions (Giuliani et al. 2011; Ackermann et al. 2013a). The evidence for PeV CRs is less certain. All of the SNRs identified so far show breaks in their γ-ray spectra well below 100 TeV (e.g., Funk 2015). These remnants have been observed with the currently operating ground-based Cherenkov telescopes *H.E.S.S*, *MAGIC* and *VERITAS*.

We note that there is an unidentified diffuse H.E.S.S. source in the Galactic Center (GC) region (i.e., Abramowski et al. 2016) with no clear spectral break or cut-off until tens of TeV and this may well be a PeV CR accelerator. There are a number of possible interpretations of this γ-ray source. The Pevatron can be associated with the massive black hole Sgr A* (e.g., Aharonian and Neronov 2005; Fujita et al. 2017; Guo et al. 2017) which may produce a blast wave after a tidal disruption event to accelerate particles (e.g., Liu et al. 2016). An alternative explanation is that a supernova exploded close to the GC and interacted with either the fast wind from the GC or with a wind from its parent compact cluster of massive stars. We will discuss such a "colliding shock flow system" later in this paper and give estimates of the maximum CR energies achievable at very different evolution stages and for different types of SNe, as well as the possible role superbubbles play (Bykov 2001; Parizot et al. 2004; Ferrand and Marcowith 2010; Ackermann et al. 2011; Bykov 2014; Ohm 2016).

The review lay out is as follows. In Sect. 2 we detail the observational status of particle acceleration in SNe and in SNRs. In Sect. 3 we discuss CR acceleration beyond the PeV regime by trans-relativistic SNe or in superluminous SNe. In Sect. 4 we describe the different ingredients relevant for the investigation of particle acceleration in fast SN shocks as well as the expected radiation spectra produced at fast shocks. Section 5 discusses other galactic sources of PeV-EeV CRs: superbubbles and SNe in clusters. Section 6 addresses future ob-

[2] We note that while the terms "Fermi acceleration" and "diffusive shock acceleration" are often used interchangeably there is an important difference when relativistic shocks are discussed. Particle transport in relativistic flows need not be diffusive. In fact, non-diffusive behavior in steep density gradients, as near the sharp subshock transition and at an upstream free escape boundary, is critically important for all aspects of particle acceleration and MFA. Henceforth we use Fermi acceleration to include all cases including those where diffusive behavior does dominate.

servational facilities which should result in a substantial improvement in our understanding of CR acceleration in SNe.

2 Observational Status

In this section we discuss some observations of SNe. In Sect. 2.1 we consider mainly radio observations of extragalactic SNe where the SN explosion and its immediate aftermath can be observed. In Sect. 2.2 we consider evidence for MFA from galactic SNRs where detailed observations can be made from radio to γ-rays.

2.1 Radio Supernovæ

Today about 200 extragalactic SNe have been detected in radio, but only seven are sufficiently close to show well-resolved light curves at radio wavebands. These are SN 1979C (SN IIL), SN 1986J (SN IIn), SN 1987A (SN IIpec), SN 1993J (SN IIb), SN1996cr (SN IIn), SN 2008iz, and SN 2011dh (SN IIb) (Bartel et al. 2017, and references therein), where the type of the core-collapse SN has been added in parenthesis when available. Among the objects in this list we discard SN 1987A and SN 1996cr. The former shows peculiar behavior difficult to account for in the following simplified models and the latter does not have enough data to constrain the shock dynamics and magnetic fields. Considering type Ib/c SNe, we have selected the following objects on the basis of the amount of radio data available: SN 1983N, SN 1994I, SN 2003L (Weiler et al. 1986; Soderberg et al. 2005), and discuss the case of SN 2009bb since it is a relativistic Ibc SN (Soderberg et al. 2010) (see also Sect. 3.1).

Radio emission is due to synchrotron radiation from electrons accelerated by turbulent magnetic fields which develop in a region including both reverse and forward shocks. However, the exact nature of these turbulent magnetic fields, and the exact region of acceleration, remain elusive (Björnsson and Keshavarzi 2017). There are different locations where such turbulent magnetic fields can develop: at the contact discontinuity where Rayleigh-Taylor instabilities develop with fingers of the ejecta stretching into the shocked circumstellar medium (CSM), and/or at the forward or reverse shock where existing turbulence can be amplified and the acceleration of CRs in high-speed shocks can drive new MFA. Modeling of the radio lightcurves in different wavebands requires accounting for a number of processes. These include synchrotron self-absorption (SSA), free-free absorption by the ambient thermal plasma (an internal effect) or by CSM matter (an external effect), and possibly plasma processes like the Razin-Tsytovich effect (Fransson and Björnsson 1998).

Radio observations are important for understanding many aspects of particle acceleration. First, the spectral turnover produced by SSA leads to an estimate of the magnetic field intensity of the synchrotron emitting zone. Second, the synchrotron spectral index provides a constraint on the electron distribution function and then on the acceleration process. Third, radio images are used to derive the SN shell dynamics, time evolution of the shock radius and its velocity, both quantities mandatory for any microphysical calculation of particle acceleration efficiency (e.g., Tatischeff 2009). The shell radius and speed can also be compared to a self-similar expansion model (e.g., Chevalier 1982).

The properties of the radio emission depend on the SN type: type Ib/c SNe show steep spectral indices ($\alpha > 1$, with a radio flux scaling as $S_\nu \propto \nu^{-\alpha}$), and have similar flux peaking before optical maximum at a wavelength around 6 cm, while other type II SNe show flatter spectra ($\alpha < 1$) with a wider range of radio luminosities usually peaking at 6 cm significantly after optical maximum (e.g., Weiler et al. 2002). Björnsson and Keshavarzi (2017) suggested

Table 1 Magnetic field and shock radius evolution in a set of type II SNe. The spectral index α corresponds to the optically thin synchrotron spectrum. Estimations for SN 1979C and SN 1986J are based on a SSA model. Magnetic field strengths are derived at day 5 and are compatible with the strength of equipartition magnetic field given in Martí-Vidal et al. (2011). The value for n is obtained from the solution of propagation in a wind with constant mass loss rate with $s = 2$ (see Eq. (1)). SN 1993J: Fransson and Björnsson (1998) use an expansion index $m = 0.74$ at $t > 100$ days, Martí-Vidal et al. (2011) present a long term radio survey where the expansion law index at frequencies above 1.7 GHz varies from $m \leq 0.925 \pm 0.016$ before $t_{\rm br} = 360 \pm 50$ days to $m \geq 0.87 \pm 0.02$ after. SN2008iz: Kimani et al. (2016) derive the equipartition magnetic field strength, the lower and upper solutions depend if protons are not or are accounted in the estimation. References: Weiler et al. (1991), Marcaide et al. (2009), Martí-Vidal et al. (2011) for SN 1979C, Weiler et al. (1990), Bietenholz et al. (2010), Martí-Vidal et al. (2011) for SN 1986J, Fransson and Björnsson (1998), Tatischeff (2009) for SN 1993J, Kimani et al. (2016) for SN 2008iz, Horesh et al. (2013), Krauss et al. (2012), Yadav et al. (2016) for SN 2011dh

Name	α	t_0 (days)	$([B_0\ (\mathrm{G})], n)$	$(R_0\ (\mathrm{cm}), m)$
SN 1979C	$0.74^{+0.05}_{-0.08}$	5	$([20\text{–}30], -1.00)$	$(8.68(e14), 0.91 \pm 0.09)$
SN 1986J	$0.67^{+0.04}_{-0.08}$	5	$([30\text{–}50], -1.00)$	$(3.18(e15), 0.69 \pm 0.03)$
SN 1993J	1.00	10	$([25.5], -0.93 \pm 0.08)$	$(1.9(e15), 1.00)$
...	0.90	100	$([2.4 \pm 1.0], -1.16 \pm 0.20)$	$(1.6(e16), 0.829 \pm 0.005)$
SN 2008iz	1.00	100	$([0.2\text{–}1.5], -1.00)$	$(2.1(e16), 0.86 \pm 0.02)$
SN 2011dh	1.15	4	$([5.9], -1.00 \pm 0.12)$	$(5.0(e14), 1.14 \pm 0.24)$
...	0.95	15	$([1.1], -1.00)$	$(3.1(e15), 0.87 \pm 0.07)$

that the radio emission in type Ib/c SNe comes from a narrow region in the vicinity of the forward shock, while the radio emission region of type IIb SNe (e.g. SN 1993J) is wider due to the effect of the Rayleigh–Taylor instability.

In Tables 1 and 2 we summarize the main properties of the above SNe from the available data deduced from VLBI observations. In addition to α, we use the following notation. The magnetic field strength is characterized by its amplitude B_0 at a reference time t_0 to be specified and the spectral index $n > 0$ of its time variation: $B(t) = B_0(t/t_0)^{-n}$. The VLBI shell radius is characterized by its radius at t_0, R_0 and by the index $m > 0$ of its time variation: $R_{\rm sh} = R_0(t/t_0)^m$. The shell velocity can then be deduced by $V_{\rm sh} = dR_{\rm sh}/dt$. Notice that the time evolution of the magnetic field is proportional to the energy density in the shock as described by Martí-Vidal et al. (2011)

$$B \propto \rho V_{\rm sh}^2 \propto t^{m\frac{2-s}{2}-1}. \tag{1}$$

Here the mass density profile of the pre-SN stellar wind is assumed to be $\rho \propto r^{-s}$. In the case of a shock propagating in a wind profile with a constant mass loss rate, that is with $s = 2$, we find from Eq. (1) an index for the magnetic field dependence of $n = 1$.

What is obvious from these tables is that the magnetic field quite early in the SN evolution is, in general, far in excess of critical magnetic field strengths obtained by balancing magnetic and wind kinetic energies (e.g., ud-Doula and Owocki 2002). The critical magnetic field is given by $B_{\rm cr}^2/8\pi = \rho v_{\rm w}^2/2$, where $v_{\rm w}$ is the stellar wind speed of the massive star. A general value in Gauss is

$$B_{\rm w,eq,G}(t) \simeq \left[\frac{2.5 \times 10^{13}}{R_0}\right] \dot{M}_{-5}^{1/2}\, V_{\rm w,10}^{1/2} \left(\frac{t}{t_0}\right)^{-ms/2}, \tag{2}$$

where the stellar mass loss rate is in units of $10^{-5} M_{\odot}/\mathrm{yr}$ and the wind speed is in units of $10\ \mathrm{km/s}$. For the case of a red supergiant wind where the pre-SN mass loss rate $\dot{M}_{-5} \sim 1$,

Table 2 Magnetic field and shock radius evolution in a set of type Ib/Ic SNe. In the case of SN 1983N, the magnetic field strength dependence with time is assumed, Slysh (1992) only derives estimates of an upstream magnetic field strength of 0.9 G which is multiplied by 4 here. In the case of SN 1994I, the fitted parameters of the model derived by Alexander et al. (2015) do not show a simple power-law time dependence for both B and R. References: Weiler et al. (1986), Slysh (1992) for SN 1983N, Weiler et al. (2011), Alexander et al. (2015) for SN 1994I, Soderberg et al. (2005) for SN 2003L, Soderberg et al. (2010), Chakraborti and Ray (2011) for SN 2009bb

Name	α	t_0 (days)	$(B_0$ (G), $n)$	$(R_0$ (cm), $m)$
SN 1983N	1.03 ± 0.06	13	$(3.6, -1.00?)$	$(2.3(e15), 0.81)$
SN 1994I	1.22	10.125	$(2.3, -)$	$(2.39(e15), -)$
SN 2003L	1.1	10	$(4.5, -1.00)$	$(4.30(e15), 0.96)$
SN 2009bb	1.0	20	$(0.6, -1.00)$	$(3.20(e16), 1.00)$

$V_{w,10} \sim 1$, and $R_0 \sim 10^{15}$ cm, field strengths of the order of 25 mG are obtained, far below those shown in the tables. It requires $V_w > 1000$ km/s and high mass loss rates to find $B_{eq} \sim$ Gauss, which may be the case in some WR winds. In general, however, amplification factors of two to three orders of magnitude above B_{eq} are deduced from radio observations.

Recently, Kundu et al. (2017) discussed possible constraints on the circumstellar medium from the radio non-detection of two Type Ia supernovae, SN 2011fe and SN 2014J. They found a very low-density medium around both the SNe assuming that about 20% of the shock bulk energy was shared equally between electrons and magnetic fields. Note that the non-linear DSA modeling by Bykov et al. (2014) predicted about 10% efficiency of non-adiabatic magnetic field amplification by CR-driven instabilities (see Fig. 6 below), while the electron acceleration efficiency is usually lower.

As already mentioned, the physical processes producing magnetic field amplification are not fully understood. The possibility that CR driven plasma instabilities can produce this amplification is discussed below and we refer to Tatischeff (2009) and Marcowith et al. (2014) for additional work. Before providing more details on the CR instabilities we now briefly review the observational evidence that such instabilities develop in more evolved SNRs.

2.2 Supernova Remnants

Supernova remnants are known sites of particle acceleration as they produce non-thermal radiation from radio to γ-rays (see Marcowith et al. 2016, for a review). Associated with this particle acceleration is clear evidence of magnetic field amplification in all historical SNRs making it likely that the two processes are tightly linked (e.g., Parizot et al. 2006).

The most direct evidence for MFA comes from high angular resolution observations of nonthermal X-rays. The *Chandra* X-ray satellite has detected in historical SNRs, in the hard 4–6 keV band, thin filaments of typical size 1–10% of the remnant radius. Such X-ray emission can hardly be interpreted as Bremsstrahlung radiation from relativistic electrons as it would required a very high ambient ISM density (Ballet 2006). Hence, this radiation is interpreted as synchrotron emission produced by multi-TeV electrons. The size of the filaments puts some constraints on the post-shock magnetic field intensity: electrons downstream from the shock are transported either diffusively or are advected with the flow but can radiate X-rays only within a synchrotron loss timescale. Since both the diffusion coefficient and the synchrotron loss time depend on the magnetic field intensity the radiation zone is limited. Typical field strengths between 50 (in older SNR like SN 1006) and 500 μG (in younger SNRs like Cassiopeia A) are deduced by this technique (Parizot et al. 2006). In fact, since *Chandra* has a finite point spread function, the deduced magnetic field intensity is only a lower limit in some cases.

Lately, *Chandra* has detected regularly spaced stripes in the X-ray synchrotron emission from Tycho's SNR (Eriksen et al. 2011). This pattern and spacing has been interpreted by Bykov et al. (2011) as a signature of magnetic turbulence produced by the short-scale Bell instability driven by PeV CR protons. However, recent observations by VERITAS and Fermi (i.e., Archambault et al. 2017) suggest that the proton spectrum may be cutting off well below PeV energies. If the stripes are, in fact, the result of CR-driven turbulence, the turbulence may result from the long-wavelength mirror instability driven by CRs undergoing super-diffusion (i.e., *Lévy* flight; see Sect. 3.1) (Bykov et al. 2017a). The turbulence resulting from the mirror instability will have characteristic wavelengths long compared to the CR gyroradius, in contrast to the shorter wavelength turbulence produced by the Bell instability.

An important observational perspective would be to identify filaments at other wavelengths to constrain the MFA process. The improved sensitivity at 1 GHz of the *Square Kilometer Array* (SKA) (see Sect. 6.1), with less than an arcsecond angular resolution, could bring some strong constraints on radio filaments and radio precursors (e.g., Achterberg et al. 1994).

Supernova remnants are also strong γ-ray emitters. This γ-ray emission can be of leptonic origin, i.e., inverse-Compton (IC) or bremsstrahlung radiation in dense environments, or hadronic, i.e., pion-decay. A trend seems to appear as a function of the remnant age (Acero et al. 2015, 2016). Young (with an age < 1000 yr) SNRs usually show hard spectra with a photon index close to 2 whereas older SNRs exhibit soft spectra beyond GeV energies. The γ-ray spectra of young SNRs can be interpreted within the framework of one-zone models as being produced by IC radiation. But from X-ray observations we know that in SNRs where MFA is occurring (see above), a one zone description of acceleration and cooling is likely a poor approximation since the magnetic field will vary behind the forward shock (Marcowith and Casse 2010). Also, if SNRs propagate into the perturbed ISM or CSM, density inhomogeneities can boost γ-ray emission from pion-decay forcing multi-zone models (Gabici and Aharonian 2014).

In the few young core-collapse SNRs which are γ-ray emitters (e.g., Cassiopeia A, RX J0852.0-4622), the γ-ray spectrum seems to cut off around 10 TeV implying a CR spectrum cutting off below ~ 100 TeV. For the moment there are no hints of a source of PeV CRs which would then produce pion-decay γ-rays of ~ 300 TeV. Older SNRs show a cut off below one TeV. One possibility is that the Pevatron phase is very short after the SN outburst and ends before an age of ~ 1000 yr. This question is treated in the next section.

3 CR Acceleration Beyond PeV

Observations of the spectrum, composition, and anisotropy of CRs below the knee were briefly mentioned in Sect. 1 (see also Hillas 2005). An important question that is yet to have a definitive answer for higher energy CRs is where the transition between the galactic and extragalactic CR components occurs. It is even possible that the spectral breaks that have been observed below PeV energies, combined with the light-mass CR component between 10^{17}–10^{18} eV suggested by *LOFAR* observations, may indicate the presence of different types of galactic sources which have not been previously considered.

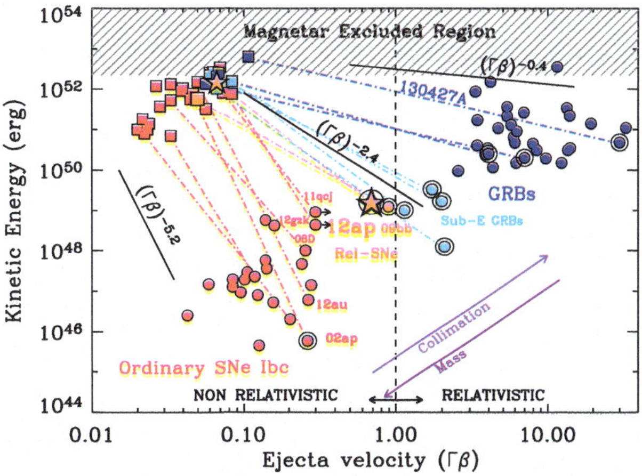

Fig. 1 The kinetic energy of the ejecta as a function of the spatial component of the ejecta four velocity $\Gamma\beta$ is shown for distinct classes of core collapsed events: ordinary type Ibc SNe (*red*), GRBs (*blue*), sub-energetic GRBs (*light-blue*), and relativistic supernovae SN 2009bb and SN 2012ap (*orange*). Adapted from Margutti et al. (2014)

While providing a sufficient energy budget is the most important constraint for sources of GeV-TeV CRs, the energy content in CRs beyond the knee is considerably less. This opens the possibility these CRs may be produced by low-power sources or by rare powerful events. Of course SNe are still among the possible sources of high-energy galactic CRs and Sveshnikova (2003) discussed the role the diversity in supernova explosion energies might have on CR spectral features around the knee. In particular, PeV CRs may be produced by a subset of type Ibc SNe with trans-relativistic shocks (e.g., Budnik et al. 2008; Chakraborti et al. 2011; Ellison et al. 2013), or by superluminous SNe (e.g., Murase et al. 2011, 2014; Zirakashvili and Ptuskin 2016), or by SNe in young compact stellar clusters (Bykov 2014; Bykov et al. 2017). We shall now briefly discuss these sources.

3.1 CR Acceleration in Trans-relativistic Supernovae

Soon after the cosmological nature of γ-ray bursts was established an unusually radio bright supernova SN1998bw was associated with GRB980425. This SN showed clear evidence for a blast wave moving at relativistic speeds (e.g., Kulkarni et al. 1998). The mass in the relativistic ejecta of SN1998bw was estimated as $M_{\rm ej} \sim 10^{-5} M_{\odot}$, with the energy in the relativistic outflow $\gtrsim 10^{49}$ erg. There are also indications for the existence of relativistic SNe without a detected GRB. The rare subclass of type Ibc SNe is distinguished from ordinary SNe by the presence of ejecta moving at mildly relativistic speeds (e.g., Soderberg et al. 2006). In Fig. 1 (adapted from Margutti et al. 2014) the kinetic energy of the ejecta as a function of the spatial component of the ejecta four velocity is shown for distinct classes of core collapsed events: ordinary type Ibc SNe (red), GRBs (blue), sub-E GRBs (i.e., sub-energetic; light-blue), and the relativistic supernovae SN 2009bb and SN 2012ap (orange). The events lying near the $\Gamma\beta = 1$ line represent a subset of SNe with mildly relativistic ejecta speeds and other properties intermediate between ordinary SNe and GRBs.

A prototype of a relativistic SNe without a detected GRB is SN 2009bb (Soderberg et al. 2010; Margutti et al. 2014). In this remnant, radio observations were used to probe the fastest ejecta as the expanding relativistic blast wave propagated through the fast wind of the progenitor star. In the case of SN 2009bb, Soderberg et al. (2010) estimated that the ejecta had a minimum energy of $(1.3 \pm 0.1) \times 10^{49}$ erg, with a Lorentz factor $\Gamma \simeq 1.3$ (see Fig. 1), and were propagating in the wind of the progenitor star with an estimated mass loss rate of $\dot{\rm M} = (2 \pm 0.2) \times 10^{-6}\ M_{\odot}\ {\rm yr}^{-1}$.

A simple estimate of the deceleration length, R_s, of a SN shock assuming the ejected mass, M_{ej}, equals the amount of swept up mass in a progenitor wind with speed v_w and mass loss rate $\dot{M}$ is

$$R_s \approx 10^{17}\left[\frac{M_{ej}}{10^{-5} M_\odot}\right]\left[\frac{v_w}{3\times 10^3\ \mathrm{km\,s^{-1}}}\right]\left[\frac{10^{-6} M_\odot\ \mathrm{yr}^{-1}}{\dot{M}}\right]\ \mathrm{cm}. \tag{3}$$

The stellar wind mass-loss rate of $10^{-6} M_\odot\ \mathrm{yr}^{-1}$, and the wind terminal speed of $3000\ \mathrm{km\,s^{-1}}$, are consistent with predictions for the He main sequence stars which were presented recently by Vink (2017), while the Wolf-Rayet type stars collected by Nugis and Lamers (2000) demonstrated somewhat higher mass-loss rates and lower wind velocities. The values and configurations of the magnetic fields in the winds of the potential progenitors are not very certain (e.g., Walder et al. 2012), while the analysis of radio observations of SN2009bb (Chakraborti et al. 2011) revealed a magnetic field of ~ 0.1 G at the estimated radius of $\sim 10^{17}$ cm (see Table 2 for more details).

It is important to note that in a relativistic SN, both the shock speed and magnetic field in the wind of the progenitor star are expected to be much higher than in the ISM. This will strongly reduce the acceleration time of particles subject to Fermi acceleration and may make it possible to accelerate nuclei well beyond PeV energies in a few months.

Budnik et al. (2008) estimated the conditions for Fermi acceleration in trans-relativistic SNe and concluded that the estimated rate and energy production by these sources are high enough to allow them to power CRs with energies up to 10^{18} eV. They assumed that the magnetic field is amplified by the shock both in the upstream and downstream regions to values close to equipartition. Moreover, Chakraborti et al. (2011) argued that relativistic SNe similar to SN 2009bb could be the sources of ultra-high-energy-cosmic-rays (UHECRs) with energies beyond the Greisen-Zatsepin-Kuzmin limit (i.e., $\gtrsim 10^{20}$ eV) if they have magnetic fields above ~ 0.1 G at radii beyond $\sim 10^{17}$ cm.

Of course relativistic SNe are scarce. The Chakraborti et al. (2011) conjecture was made assuming that the type Ibc SN rate is $\sim 1.7 \times 10^4\ \mathrm{Gpc^{-3}\,yr^{-1}}$ and the fraction of these which are mildly relativistic is $\sim 0.7\%$ (the uncertainty is large, however, $0.7^{+1.6}_{-0.6}\%$, or (0.1–2.3)%). If the fraction 0.7% is right then one would expect a few trans-relativistic SNe occurring within a distance of a few hundred Mpc every year, and one every 5×10^4 yr in the Milky Way. A crucial point in these CR energy estimates is the magnetic field strength. Strong MFA is simply assumed but this must be justified by self-consistent models for MFA in nonlinear Fermi acceleration in trans-relativistic shocks.

Relativistic SNe may provide a distinct component of CRs beyond PeV energies. Supernova statistics in a volume-limited sample based on nearby SN rates from the Lick Observatory Supernova Search (Li et al. 2011) found that the most abundant type of SNe are type II (57% of the total), while type Ibc comprised about 19%, a fraction similar to that of SNe Ia which is estimated as 24%.

If indeed relativistic SNe are just 0.7% of all type Ibc SNe, as suggested by Soderberg et al. (2010) and Chakraborti et al. (2011), than their rate in the Galaxy is about one every 5×10^4 yr. The diffusion coefficient D required for the high energy CRs to be confined in the Galactic halo of size R for 5×10^4 yr can be estimated to be $D \sim 2 \times 10^{31} (R/3\ \mathrm{kpc})^2\ \mathrm{cm^2\,s^{-1}}$. Note that the gyroradius of a proton of energy 10^{18} eV in the typical interstellar magnetic field is $\sim 10^{21}$ cm providing a minimum value for $D \sim 10^{31}\ \mathrm{cm^2\,s^{-1}}$. The minimum value was derived assuming Bohm diffusion where the CR mean-free-path is equal to the gyroradius, however, the diffusion coefficient in the halo is expected to be well above the minimal value. This value, and the 2% upper limit on the anisotropy of $\sim 10^{18}$ eV CRs (see Ptuskin 2012, for a discussion), requires an extended CR halo of size ~ 10 kpc if the relativistic SNe are to be the dominant source of CRs up to $\sim 10^{18}$ eV. The CR-driven galactic winds (e.g., Farber et al. 2017; Recchia et al. 2017, and references therein) may affect the scale size R of the CR confinement region. The power required to maintain this high-energy galactic CR component (assuming it is quasi-steady) is about 10^{36} erg s^{-1} meaning trans-relativistic shocks would need to place $\sim 10\%$ of the mildly relativistic outflow kinetic energy into CRs extending to 10^{18} eV. It has also been suggested by Loeb and Waxman (2006) that relativistic SNe may produce high-energy CRs in starburst galaxies and be a source of high-energy neutrinos.

The UHECRs with energies well above 10^{18} eV are most probably of extragalactic origin. The main source candidates are γ-ray bursts and powerful radio-galaxies. Galactic magnetars may also contribute. In all cases, high power and relativistic outflows are required to produce the most extreme CR energies (e.g., Waxman 1995; Blasi et al. 2000; Arons 2003; Lemoine 2013; Asano and Mészáros 2016).

To address MFA with high CR acceleration efficiency in trans-relativistic shocks we generalize a Monte Carlo technique that has been used previously to study nonlinear Fermi acceleration in shocks of all Lorentz factors but without a self-consistent calculation of MFA in relativistic flows (Ellison et al. 2013; Warren et al. 2015, 2017). This previous model assumed a given structure for the magnetic turbulence which was not directly determined by the CR distribution. While neglecting MFA in relativistic shocks might be acceptable if Fermi acceleration is inefficient, the efficiency estimate given above suggests that Fermi acceleration is efficient enough in trans-relativistic shocks to produce CR pressure gradients and CR currents strong enough to drive instabilities responsible for MFA in a way similar to what is believed to occur in non-relativistic shocks (e.g., Bell and Lucek 2001; Bell 2004; Bell et al. 2013; Schure et al. 2012; Bykov et al. 2013; Marcowith et al. 2016; Bykov et al. 2017a; van Marle et al. 2017).

The non-relativistic Monte Carlo simulation self-consistently finds a steady-state, planar-shock solution for the nonlinear shock structure including (1) injection of particles from the thermal background, (2) modification of the precursor flow by energetic particle and magnetic field pressure, (3) calculation of the magnetic turbulence (i.e., MFA) due to the anisotropic distribution of the accelerated particles, (4) turbulence cascade, (5) a determination of the momentum and space dependent scattering mean free path from the self-generated turbulence, (6) a self-consistent calculation of the scattering center speed[3] (which may model an effect of a mean electric field in the shock upstream) without reference to the Alfvén velocity, and (7) particle escape at an upstream free escape boundary. Most recently (i.e., Bykov et al. 2014, 2017a), super-diffusion (also called *Lévy*-walk or *Lévy*-flight propagation) has been included within the full nonlinear code structure.

In contrast to PIC simulations, where the magnetic turbulence is determined directly from particle motion using Maxwell's equations, the Monte Carlo code calculates the resonant and non-resonant instabilities by coupling analytic descriptions of the growth rates with the anisotropic particle transport simulated by the code. In Fig. 2 we show preliminary results from our trans-relativistic code.

The figure shows spectra of magnetic fluctuations for Lorentz factors between 1.5 and 4 assuming no turbulence cascade along the initial magnetic field (cf., Lithwick and Goldreich 2001). The turbulence was calculated self-consistently with the CR distributions shown in Fig. 3. It is critical to note that, for clarity, Figs. 2 and 3 only show spectra for the downstream region (the particle distributions are calculated in the shock-rest frame). There is no strict

[3] See Fiorito et al. (1990) for an early calculation of the scattering center speed using Monte Carlo methods.

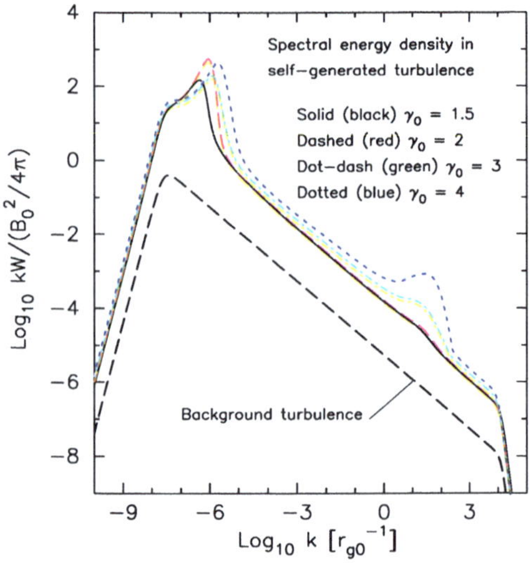

Fig. 2 Preliminary results showing spectral energy densities of the CR-driven magnetic fluctuations measured in the downstream region of trans-relativistic shocks with indicated Lorentz factors, γ_0. The *black dashed curve* is the background turbulence. There is no turbulence cascade in this model (Bykov et al. 2017b)

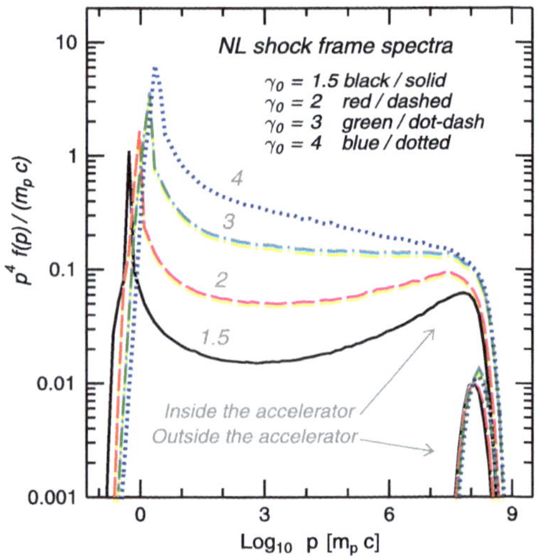

Fig. 3 Shown are proton phase-space distributions measured in the shock rest frame (preliminary results from Bykov et al. 2017b). These nonlinear, steady-state, Monte Carlo simulations were performed for the trans-relativistic shock Lorentz factors, γ_0, indicated. The amplified magnetic turbulence determined self-consistently with these spectra is shown in Fig. 2

one-to-one correspondence between the CRs and the turbulence in these plots. All spectra are calculated throughout the shock precursor with the convection of turbulence and CRs taken into account. The turbulence shown in Fig. 2 is the net result, in the downstream region, of production, convection, and cascading, if present.

The effect of strong magnetic field amplification is clearly seen in Fig. 2. The self-generated turbulence stands more than an order of magnitude above the background level (dashed black curve) for all wavenumbers and Lorentz factors. In addition, there is much stronger amplification of long-wavelength turbulence (i.e., $k \sim 10^{-6}\, r_{g0}^{-1}$) produced by CRs that escape at the upstream FEB. The quasi-thermal CRs also produce enhanced turbulence for $k \sim 10$–$100\, r_{g0}^{-1}$.[4] While not shown, we have performed these calculations with

[4]Particle-particle interactions, i.e., Coulomb collisions, are not explicitly modeled in the Monte Carlo simulation so true thermal distributions are not produced. Therefore, we use the term "quasi-thermal" to describe

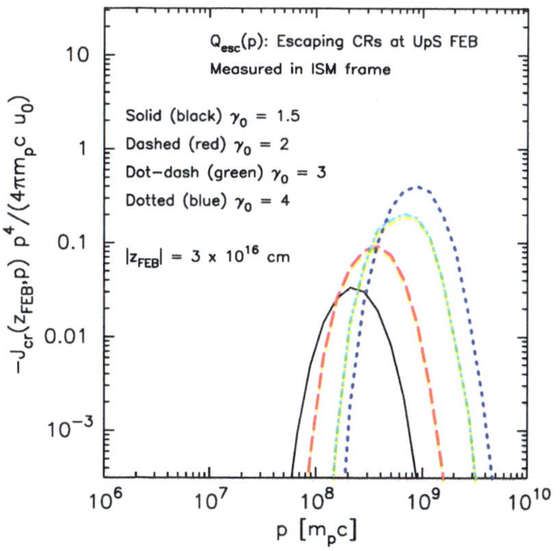

Fig. 4 Shown are distributions of protons escaping at the upstream free escape boundary (FEB) (i.e., Eq. (4)), as measured in the far upstream (i.e., the ISM) rest frame. The trans-relativistic shock Lorentz factors are indicated. These are preliminary results from Bykov et al. (2017b)

Kolmogorov-type cascade and found little difference in the particle spectra or the total value of the amplified magnetic field. The background turbulent magnetic field in the wind of the progenitor star (dashed line in Fig. 2), integrated over the spectrum, had a magnitude $B_{\rm turb} \sim 0.01$ G for these simulations. This is considerably less than the shock ram pressure and the amplified turbulent magnetic field in the shock downstream which is ~ 0.2 G. The wind of the WR type progenitor star with the magnetic field value ~ 0.01 G at R $\sim 10^{17}$ cm is magnetized if the mass-loss rate is $\sim 10^{-6} M_{\odot}$ yr^{-1}.

We find, even for $\gamma_0 = 4$, that escaping CRs are important. In Fig. 4 we show distributions of escaping CRs as seen by a far upstream observer. These spectra are determined by

$$Q_{\rm esc}(p) = -\frac{J_{\rm cr}(z_{\rm FEB}, p)p^4}{4\pi m_p c u_0}, \tag{4}$$

where u_0 is the shock speed and $J_{\rm cr}(z_{\rm FEB}, p)$ is the CR current, both measured in the far upstream rest frame. Here $z_{\rm FEB}$ is the position of the upstream FEB and for the plots shown, $z_{\rm FEB}$ was 3×10^{16} cm from the subshock at $z = 0$.

3.2 CR Acceleration in Interaction Powered Luminous Supernovae

Superluminous SNe with a peak luminosity in the optical-UV of $\sim 10^{44}$ erg s^{-1} were recently recognized to be a broad class of events with the rate estimated to be $\sim 10^{-4}$ of the core-collapse SN rate (e.g., Branch and Wheeler 2017). The detected events were both hydrogen rich and hydrogen poor, while some of them were apparently dominated by radioactive ^{56}Ni decay (e.g., Gal-Yam 2012). A superluminous SN event may occur in the case of a dense enough circumstellar medium around the SN which provides a radiation-dominated shock propagating through the envelope (e.g., Falk and Arnett 1977; Chugai and Danziger 1994; Chevalier and Irwin 2011; Chevalier 2012; Blinnikov 2016b,a; Sorokina et al. 2016). Pre-supernova activity of a massive progenitor star during some evolutionary

the peaks at thermal energies seen in Fig. 3. These peaks have essentially the same mean and total energy as the equivalent Maxwell-Boltzmann distributions.

phases may have a very large mass-loss rate in the form of a dense slow wind that produces dense circumstellar envelopes (Smith 2017; Morozova et al. 2017). Multiple envelopes may be formed if outbursts occur at different progenitor star evolutionary stages prior to the SN event.

Assuming that $\sim 10\%$ of the shock ram pressure can be converted to the fluctuating magnetic field (cf. Fig. 6) and using the minimal Bohm diffusion coefficient of CRs in the amplified field, Zirakashvili and Ptuskin (2016) estimated the maximal energy of protons accelerated by the Fermi mechanism in type IIn SNe to be

$$\frac{E_{\max}}{80\ \mathrm{PeV}} \le \frac{E_{\mathrm{SN}}}{10^{52}\ \mathrm{erg}}\left(\frac{M_{\mathrm{ej}}}{10\ M_{\odot}}\right)^{-1}\left(\frac{v_{\mathrm{w}}}{100\ \mathrm{km\,s^{-1}}}\right)^{-0.5}\left(\frac{\dot{\mathrm{M}}}{10^{-2}\ M_{\odot}\ \mathrm{yr}^{-1}}\right)^{0.5}. \tag{5}$$

This implies that hypernovae with ejected kinetic energies of $E_{\mathrm{SN}} \sim 10^{52}$ erg can accelerate protons beyond PeV energies. To reproduce the observed light curves of the superluminous SN 2006gy, assuming that the supernova is powered by the collision of supernova ejecta with a dense circumstellar medium, Moriya et al. (2013) derived an ejecta mass $\lesssim 15 M_{\odot}$ and an explosion energy $\gtrsim 4 \times 10^{51}$ erg. Moriya and Maeda (2014), based on the observed rise times and peak luminosities in type IIn SNe, found substantial diversity in their progenitor wind densities, SN ejecta energies, and ejecta masses.

The interactions of the accelerated CRs with the dense circumstellar envelope of the interaction powered SNe would result in rich multi-wavelength nonthermal emission, as well as high-energy neutrinos in the energy range of the *IceCube Observatory* (e.g., Katz et al. 2011; Murase et al. 2014; Zirakashvili and Ptuskin 2016; Petropoulou et al. 2017). The pp collisions between the accelerated CRs and the dense circumstellar envelope would efficiently produce γ-rays and neutrinos. Moreover, Murase et al. (2014) argued that secondary electrons and positrons, which are copiously produced in the inelastic pp collisions, radiate efficiently and would produce high-frequency synchrotron radio emission which could be observed by the *Jansky Very Large Array* and the *Atacama Large Millimeter/submillimetre Array* from type IIn SNe at Gpc distances. Petropoulou et al. (2017) estimated the diffuse neutrino emission from the SN IIn to be $\sim 10\%$ of the neutrino flux above 60 TeV observed by the *IceCube Observatory* (Aartsen et al. 2014). They also concluded that to produce the observed neutrino flux, the high-energy neutrino sources associated with type IIn events should comprise $\sim 4\%$ of all core collapse SNe and have a high-energy proton acceleration efficiency $> 20\%$.

4 Hydrodynamic Models of Evolving SNRs

An effective way to model the dynamic nature of SNRs is to couple the hydrodynamic evolution of the remnant with the CR production. A number of different models have been presented along these lines and we refer the reader to the original papers for details (e.g., Berezhko et al. 1996; Ellison et al. 2007; Kang et al. 2009; Telezhinsky et al. 2011; Ferrand et al. 2012; Lee et al. 2012; Brose et al. 2016). Supernovae and first-order Fermi shock acceleration have been actively studied for several decades and any reasonably consistent model of an evolving SNR producing CRs with an efficiency above 5 or 10% will be complex with a number of parameters and assumptions. In simplest terms, the different models combine various physical processes in a more or less self-consistent fashion. The important physical processes and model assumptions include:

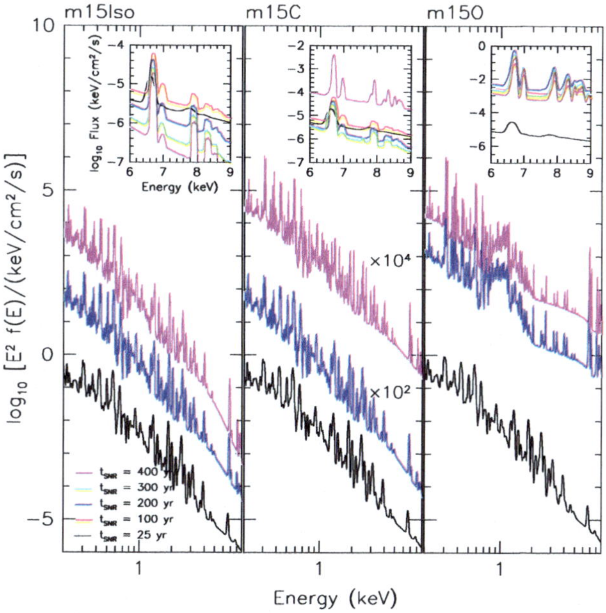

Fig. 5 Integrated X-ray spectra for three different end-to-end (i.e., pre-main sequence, through explosion, into the remnant phase) core-collapse SN models. In all panels X-ray spectra are plotted for different ages, as indicated by *color*. In the main panels, the spectra at 200 and 400 years are scaled relative to the spectra at 25 years. The *inserts* show absolutely normalized emission around the Fe-K line. This work shows that X-ray emission, even 100's of years after the explosion, differs noticeably depending on the progenitor mass-loss history and other details of the explosion. This is Fig. 8 from Patnaude et al. (2017) and this paper should be seen for details

4.1 Remnant Geometry

Most models assume the SNR is spherically symmetric but multi-dimensional models (e.g., Ferrand et al. 2012) have been presented. Spherically symmetric models can include dense shells of material and mimic in an approximate fashion dense clumps of circumstellar material (e.g., Ellison and Bykov 2011). Multi-dimensional models are essential for modeling the irregular nature of many remnants and particularly those, such as SN1006, with clear asymmetries in emission (e.g., Cassam-Chenaï et al. 2008). Besides resulting from irregularities in the CSM, asymmetries, such as that seen in SN1006, may result from a varying shock geometry, as determined by the ambient magnetic field direction. Most theories of Fermi acceleration suggest that quasi-parallel shocks, those where the ambient magnetic field direction is nearly parallel to the shock normal, inject and accelerate CRs more efficiently than quasi-perpendicular shocks (e.g., Ellison et al. 1995). Observations and modeling of SN1006 (e.g., Rothenflug et al. 2004; Cassam-Chenaï et al. 2008) support this view.

4.2 Ejecta Profile

The density of material ejected from the SN explosion is typically assumed to have a power law (e.g., Chevalier 1982) or exponential spatial distribution in radius (e.g., Dwarkadas and Chevalier 1998). A power law distribution, as assumed in Eq. (1), is amenable to self-similar solutions (e.g., Decourchelle et al. 2000) while an exponential distribution may be more appropriate for type Ia thermonuclear SNe. Of course, the ejecta distributions from actual SNe are likely to be more complex, as discussed in the next section.

4.3 Circumstellar Material

The ejecta and outer blast wave will interact with the circumstellar medium adding another level of complexity (e.g., Raymond 2018). While type Ia SNe may explode in a uniform medium, core-collapse SNe are likely to explode in a more complex environment produced by pre-SN stellar winds (i.e., wind-blown bubbles) (e.g., Dwarkadas 2005), nearby dense molecular clouds, and/or colliding plasmas from nearby SNe and SNRs (e.g., Bykov et al. 2017). If the SN blast wave interacts with dense external material, strong γ-ray emission from proton-proton interactions is expected as in SNRs W44, IC 443, and 3C 391 (see e.g. Tavani et al. 2010; Uchiyama et al. 2010; Abdo et al. 2010a,b; Ackermann et al. 2013b; Slane et al. 2015b). Recent work by Patnaude et al. (2017) has included ejecta profiles obtained from stellar evolution codes of core-collapse SNe and demonstrated that the mass-loss history leaves an imprint on X-ray emission lasting days to years after the collapse (see Fig. 5).

4.4 Production of a Forward and Reverse Shock Pair

In young SNRs the reverse shock will co-exist with the forward shock for a period of time and both will heat the thermal plasma to X-ray emitting temperatures and simultaneously produce CRs. Except for the extreme remnant limb, the observed line-of-sight will pass through material with very different parameters and shock histories. While single component models are often used, care must be taken to ensure the multi-component nature of the remnant is accurately accounted for. Interpreting the integrated multi-component emission in terms of simple power laws, as is often done, may be misleading.

4.5 Effect of CR Acceleration on Remnant Dynamics, Plasma Heating, and Thermal X-ray Production

If CRs are produced efficiently (i.e., $> 10\%$ of the shock bulk flow kinetic energy is placed in relativistic particles), the backpressure of CRs will influence the remnant hydrodynamics since relativistic particles produce less pressure for a given energy density than non-relativistic ones. The energy placed in CRs comes from the thermal plasma so the shocked temperature is less than expected for test-particle acceleration. Furthermore, the efficient production of CRs can result in an increase in the shock compression ratio from standard Rankine-Hugoniot values. The change in temperature and density of the shocked plasma will modify the X-ray line emission. This effect has been studied extensively with a code coupling the remnant hydrodynamics with efficient CR production (see Ellison et al. 2012; Patnaude et al. 2017, and references therein). The effect of CR production influences the non-equilibrium ionization X-ray emission in observable ways.

4.6 Magnetic Field Amplification (MFA)

The self-generation of magnetic turbulence is necessary for first-order Fermi shock acceleration to produce CRs to high energies. Beyond that, as shown in Sect. 2.1, it has become clear from observations of young SNRs that not only must turbulence be generated, it must be generated far more efficiently then envisioned when the Fermi mechanism was first proposed (e.g., Parizot et al. 2006). This MFA is strongly nonlinear and is generally added to hydro models in some approximate way. Obtaining self-consistent descriptions of MFA from both resonant and non-resonant instabilities is an active area of current research, as discussed in Sect. 3.1.

In Fig. 6 we show Fig. 11 from Bykov et al. (2014) giving the pressure in magnetic turbulence versus shock speed u_0, as obtained with Monte Carlo techniques including the resonant CR-streaming instability and two non-resonant CR-current instabilities (i.e., Bell

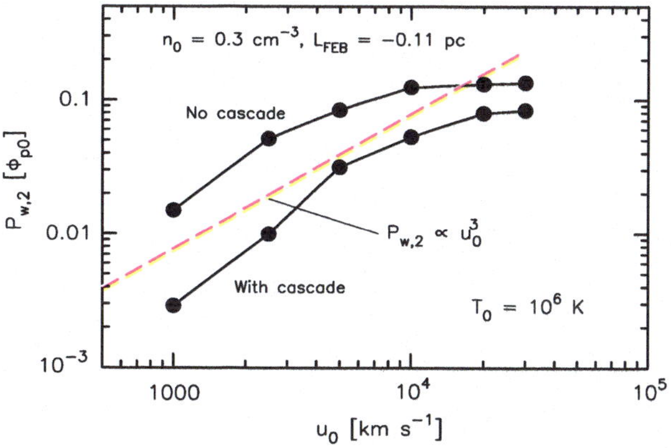

Fig. 6 The downstream pressure in magnetic fluctuations in units of the far upstream ram pressure vs. shock speed u_0 for various shock models simulated with a non-linear Monte Carlo model (see Bykov et al. 2014, for details). Models with and without turbulence cascade are included as indicated. The *dashed* (*red*) *line* approximates the behavior for low shock speeds. At higher speeds the turbulence saturates at the 10–15% level

2005; Bykov et al. 2013). The Monte Carlo results show that the efficiency of MFA, as defined by the pressure in turbulence, saturates at $\sim$ 10–15% of the far upstream bulk flow ram pressure.

4.7 Cosmic Ray Escape in Fermi Acceleration

For the Fermi mechanism to work, particles must be confined to the shock by magnetic turbulence. The scale of this confinement is $\sim D(p)/u_{sh}$, where $D(p)$ is the diffusion coefficient of a CR with momentum p and u_{sh} is the shock velocity.[5] Due to the self-generated turbulence, this diffusion coefficient is typically much smaller than that of the quiet ISM (e.g., Bell 1978; Lagage and Cesarsky 1983; Bell 2004), i.e., CR driven instabilities can provide amplification of the seed interstellar turbulence and, in some energy ranges, reduce the diffusion coefficient by orders of magnitude.

However, regardless of how efficient the MFA is, as long as the shock age is long compared to the acceleration time, there will always be some CR energy, determined by shock geometry, above which the accelerated particles can no longer generate enough turbulence to confine themselves to the shock. These high-energy particles will escape the accelerator while lower-energy CRs remain confined.[6] If Fermi acceleration is efficient, the modified compression ratio for the highest energy CRs can become greater than four (see Berezhko and Ellison 1999) and the escaping energy flux can be a significant fraction of the total shock energy flux (see Caprioli et al. 2009; Drury 2011; Ellison and Bykov 2011; Kang 2013; Malkov et al. 2013, for further discussion).

A consequence of escape is that, if SN shocks are producing galactic CRs by Fermi acceleration, an outside observer would see at each particular moment just a relatively narrow spectrum of the escaped particles centered at some maximum momentum $p_{max}(t)$. However, since p_{max} evolves with the SNR expansion, the time integrated CR spectrum can be an extended power law (e.g., Ptuskin and Zirakashvili 2005).

Basic Fermi acceleration assumes that fast particle transport in the shock precursor is described by standard diffusion due to particle scattering by magnetic turbulence. However,

---

[5] Note that the far upstream shock speed is written as either u_0 or u_{sh}.

[6] This assertion depends only on a diffusion coefficient which is an increasing function of CR momentum and the fact that all real shocks are finite in extent. In this case, at some p, $D(p)/u_{sh} \sim R_{sh}$, where R_{sh} is the shock radius, and the CR can no longer be confined independent of any plasma physics details (see Drury 2011, for a discussion of effects of the background ISM field).

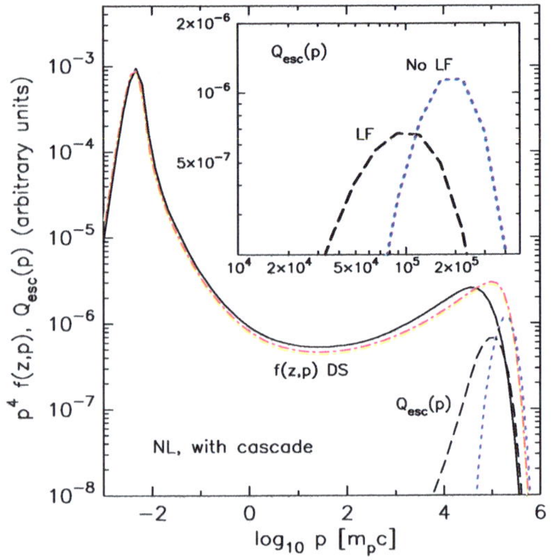

Fig. 7 Shown are proton phase-space distributions simulated with a nonlinear Monte Carlo model of superdiffusive shock acceleration (Bykov et al. 2017a). Downstream (DS) spectra, as well as the distributions of particles escaping the upstream FEB, are plotted. In the *main panel*, the *black curves* are with superdiffusion and the *red* and *blue curves* are with normal diffusion. The *insert* shows the escaping fluxes with superdiffusion (labeled 'LF') and with normal diffusion (labeled 'No LF'). All spectra are absolutely normalized relative to each other

the transport of energetic particles in the presence of intermittent (i.e., non-Gaussian) magnetic turbulence produced by anisotropic CR distributions can be different from standard diffusion (see Zimbardo et al. 2015, for a recent review). This superdiffusive propagation (also called *Lévy*-walk or *Lévy*-flight) modifies the spectrum of escaping CRs. A nonlinear Monte Carlo model of Fermi acceleration with MFA, including superdiffusive transport in the shock precursor, has been presented by Bykov et al. (2017a). In Fig. 7 we illustrate the effect of superdiffusive particle transport on the shape of the spectrum of escaping CRs with superdiffusion (labeled 'LF') and without (labeled 'No LF'). The escaping CR spectrum is clearly broadened when superdiffusion is taken into account. This broadening will influence the pion-decay emission and measurements of γ-ray spectra produced by CRs escaping a SNR adjacent to a molecular cloud can be used to constrain superdiffusive transport models.

4.8 Broad-Band Continuum Radiation

The shock-heated plasma and relativistic CRs produce radiation that must be described if particular SNRs are to be modeled. Since the thermal X-ray line emission is coupled to the CR production through nonlinear Fermi acceleration, the broad-band continuum emission must be determined consistently with the thermal emission. The continuum processes that must be considered are synchrotron, inverse-Compton (IC), and bremsstrahlung from relativistic electrons, and pion-decay from relativistic ions interacting with the background material. Fitting the broad-band emission with a single set of parameters can strongly constraint models. This is particularly true for synchrotron because of the large lever arm between radio (from $\sim$ GeV electrons) and X-ray synchrotron emission (from 10–100 TeV electrons). Care must be taken of course to ensure the observations from different instruments with different fields of view, energy resolutions, etc. are compatible.

In Fig. 8 we show a fit obtained by Ellison et al. (2012) to the broadband emission observed from SNR RX J1713.7-3946. The various emission processes are indicated and for this remnant IC emission dominates over pion-decay (p-p) at TeV energies. In other SNRs, e.g., Tycho, W44, and IC 443, pion-decay is seen to dominate. As shown by Castro et al. (2012), there are also cases, such as CTB 109, where IC and pion-decay contribute almost equally to the γ-ray emission.

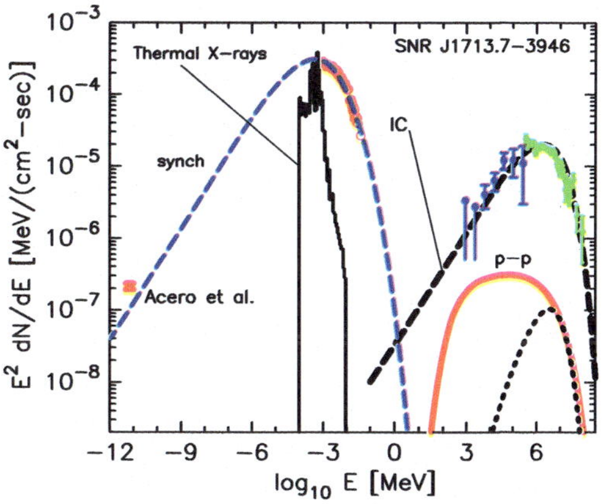

Fig. 8 Fit to SNR J1713 continuum and X-ray emission line observations. The different emission processes are indicated except the *dotted black curve* which is p-p emission from escaping CRs. This figure is adapted from Fig. 2 in Ellison et al. (2012) and that paper should be seen for details and references to the observations

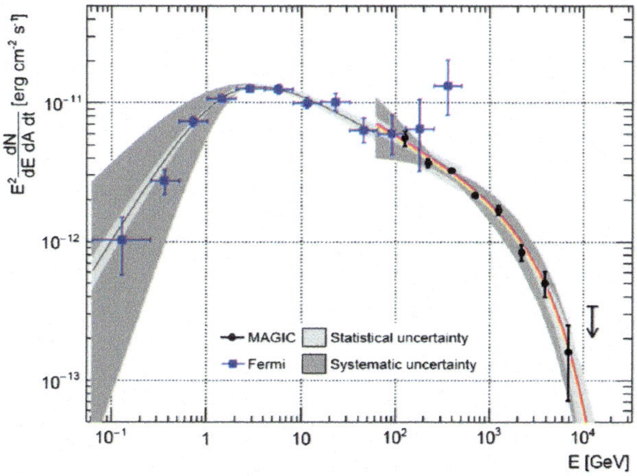

Fig. 9 Fig. 1 from Ahnen et al. (2017) ("A cut-off in the TeV gamma-ray spectrum of the SNR Cassiopeia A") showing the spectral energy distribution from the Cassiopeia A SNR. As indicated, the *black dots* are from the MAGIC telescopes while the *blue squares* were measured by the *Fermi* spacecraft. The cutoff at low energies is an indication of the kinematic threshold for pion production, while the break at high energies, if this emission is from pion-decay, indicates a break in the energy spectrum of hadrons produced by the SNR blast wave (see Funk 2015)

First-order Fermi shock acceleration naturally puts more energy into hadrons than leptons. This is because Fermi acceleration taps the mechanical energy of the shock with near-elastic scatterings between the converging upstream and downstream flows rather than electromagnetic energy—heavy particles get more energy than light ones. All consistent models of Fermi acceleration show this. On the other hand, leptons radiate far more efficiently than hadrons. In any particular SNR, environmental factors, most importantly the density and magnetic field of the circumstellar medium, determine which of the two competing processes, IC or pion-decay, dominate.

Supernovae that explode in a dense environment, such as W44 and IC 443, are likely to show strong pion-decay emission. If the remnant is in a low density region, such as SNR RX J1713.7-3946, IC is likely to dominate. While specific features in the photon spectrum in a limited energy band (such the low-energy turnover shown in Fig. 9 for Cas A) can point to one mechanism over another, broadband fits from radio to γ-rays, which depend on both leptonic and hadronic emission, produce stronger constraints. This is particularly true when thermal X-rays are coupled to CR ion production, as in the fit shown in Fig. 8. It is important to emphasize that, in non-relativistic shocks at least, even if IC emission dominates, all consistent theories of Fermi acceleration show the underlying shock putting a large majority (e.g., 99%) of energy into hadrons.

5 Supernovae in Clusters and Clustered Supernovae

Massive stars which are the progenitors of core-collapse SNe are typically born in dense cores of molecular clouds (e.g., Lada and Lada 2003; Krumholz 2017) and therefore are spatially and temporarily correlated. This correlation is important for global models of the interstellar medium (e.g., Heiles 1990; Cox 2005) and influences many aspects of CR physics.

Observations have revealed massive stars in both dense young compact clusters, like Westerlund 1 with a mass close to $10^5\ M_\odot$ (e.g., Clark et al. 2008), and in unbound OB associations with stellar volume densities below 100 stars pc^{-3}, such as Cygnus OB2 (e.g., Wright et al. 2014).

5.1 Cosmic Rays in Superbubbles

Powerful stellar winds and supernovae in OB associations had been predicted to be sources of CRs (e.g., Cesarsky and Montmerle 1983; Bykov and Toptygin 1990; Bykov 2001; Lingenfelter 2017). During the early period of imaging atmospheric Cherenkov telescopes, *HEGRA* detected an excess of TeV emission with a hard spectrum spatially coincident with the Cyg OB2 region (Aharonian et al. 2002). The authors discussed the possibility that this γ-ray emission originated from TeV particles accelerated by multiple young massive stars and SNe in the Cyg OB2 region.

More recently, an extended (50–200 pc wide) γ-ray source was discovered with the *Fermi Large Area Telescope* in the Cygnus X region, a giant complex of molecular clouds and star-forming regions located at an estimated distance of ~ 1.4 kpc (Rygl et al. 2012). The source (dubbed the Cygnus cocoon) was identified by Ackermann et al. (2011) as a possible superbubble filled with freshly accelerated CRs. The hard emission from the Cygnus cocoon extends to 100 GeV with a flux of $(5.8 \pm 0.9) \times 10^{-8}\ \mathrm{ph\,cm^{-2}\,s^{-1}}$ in the 1–100 GeV range. At a distance of 1.4 kpc, this corresponds to a γ-ray luminosity of $(9 \pm 2) \times 10^{34}\ \mathrm{erg\,s^{-1}}$. This γ-ray luminosity is below 1 percent of the kinetic power of the stellar winds in Cyg OB2, the rich OB star association located in the direction of the Cygnus cocoon.

Recently, a second extended *Fermi LAT* γ-ray source with a hard spectrum in the GeV range, possibly associated with the star forming region G25.0+0.0, was reported by Katsuta et al. (2017). The authors estimate the γ-ray luminosity of G25.0+0.0 to be about 10 times larger than that of the Cygnus cocoon for otherwise similar parameters. This implies much higher efficiencies of particle acceleration and/or radiation in the G25.0+0.0 OB association. Given the total inelastic cross section of proton-proton interaction to be about 30 mb above a few GeV, the proton cooling time would be $\sim 3 \times 10^7/n$ yr, where n is the ambient density in $\mathrm{cm^{-3}}$. If before being released into the surrounded dense shell, GeV-TeV CRs are confined in a superbubble filled predominantly with a hot tenuous plasma, the radiative efficiency of the superbubble will be low even if the particle acceleration efficiency is high.

The age of an OB association is important for estimating the kinetic power available from SNe events. Using a population synthesis approach Martin et al. (2010) estimated the mechanical luminosity over the first 3 Myr of the life of Cyg OB2 to be $\sim 4 \times 10^{38}$ erg s^{-1}. Cygnus OB2 is part of the Cygnus X region and is a rich OB association with many hundreds of OB stars (e.g., Knödlseder 2000; Wright et al. 2015). Wright et al. (2014) analyzed a selected sample of *Chandra* X-ray observations of young stars and concluded that Cyg OB2 formed as a highly substructured, unbound association with a low volume density (≤ 100 stars pc^{-3}). They found no signs of dynamical evolution of Cyg OB2 which is inconsistent with the idea that all stars form in dense, compact clusters. On the other hand, Comerón et al. (2016) investigated the past star formation history of Cygnus OB2 using the red supergiants detectable in the near infrared as a probe to trace massive stars with initial masses between 7 and 40 $M_{\odot}$. They concluded that Cygnus OB2 has a history of star formation extending into the past for at least 20 Myr. This age is such that star formation started long before the latest star formation burst which produced the dense aggregate of O-type stars currently dominating the appearance of Cyg OB2.

Multiple SNe and powerful winds of early-type stars have been suggested as favorable sites of CR acceleration (see Bykov 2014, for a review). Extended superbubbles, filled with hot X-ray emitting gas, can be created by multiple clustered SNe over a time scale of $\gtrsim 10^7$ years. The bubbles will contain an ensemble of MHD shocks which may be able to strongly amplify the turbulent magnetic fields within the superbubble leading to a high efficiency (i.e., $\gtrsim 10\%$) for converting kinetic power to freshly re-accelerated CRs. This may result in a substantial temporal evolution of the CR spectra over times on the order of 10 million years. Nonlinear modeling of this process (i.e., Bykov 2001) predicted the time asymptotic CR spectrum to be a power law of index close to 2 in MeV-TeV range where the energy independent turbulent diffusion dominates the CR propagation inside the superbubble. This is consistent with the spectrum of γ-rays observed in the Cygnus cocoon if the main radiation mechanism is pion production from CR hadron interactions.

The Fermi shock acceleration mechanism assumes that CRs are confined relatively close to the shock due to a small diffusion coefficient. This strong scattering is likely provided by the turbulence produced by CR-driven instabilities, i.e., strong magnetic field amplification (e.g., Schure et al. 2012). This strong turbulence will also increase the maximum energy CRs can obtain in the shock. Observations show that the maximum particle energies in some young isolated SNRs with shock velocities $\gtrsim 1{,}000$ km s^{-1} are typically above a TeV.

The bulk of CRs observed at Earth have GeV energies and these must escape from the SNR at late stages when the CR-generated turbulence has weakened. For core-collapse SNe, this escape is expected to occur in a superbubble. These CRs are expected to be confined and re-accelerated by multiple shocks in the superbubble (Bykov 2001; Parizot et al. 2004; Ferrand and Marcowith 2010; Bykov 2014). Recently, the *Advanced Composition Explorer CRIS* instrument discovered ^{60}Fe nuclei in CRs in the energy range 195–500 MeV per nucleon (Binns et al. 2016). The short ~ 2.6 Myr lifetime of ^{60}Fe suggests these nuclei originated in nearby clusters of massive stars.

It is worth noting that the observed γ-ray emission from the starburst galaxies NGC 253, NGC 1068, NGC 4945 and M82 (see Ohm 2012, 2016) show spectra with photon indexes ~ 2. This is consistent with that observed in both the Cygnus cocoon and G25.0+0.0 but much flatter than that of the Milky Way. This may indicate that superbubble type objects are the dominant CR sources in starburst galaxies or, alternatively, that CR transport in starburst galaxies is energy independent at least up to TeV energies. Furthermore, since superbubbles are adjacent to their parent molecular complexes, the enhanced density of low-energy CRs they contain may provide ionization and heating of the dark molecular clouds. The CR

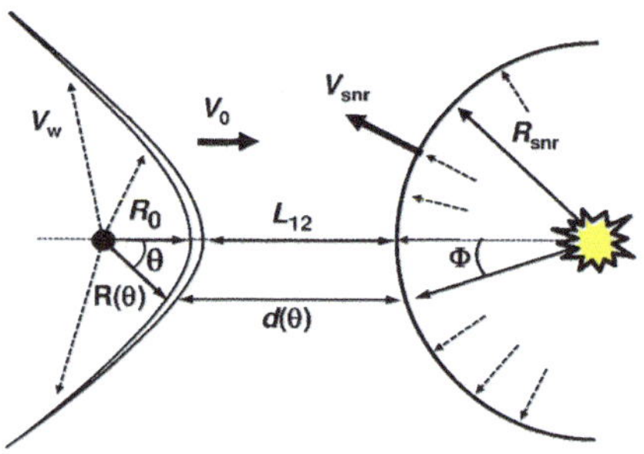

Fig. 10 A sketch illustrating the colliding shock flow geometry where the star cluster wind is impacted by a supernova shock. See Bykov et al. (2015) for details

energy deposition may keep the gas temperature at ~ 10 K and provide the gas ionization fraction $\sim 10^{-7}$ required by current models of the molecular chemistry in dark clouds (e.g., Grenier et al. 2015).

High-energy neutrinos produced by inelastic nuclei collisions in the Cygnus X region may have a large enough flux to be detected with the *IceCube Observatory*, as estimated by Yoast-Hull et al. (2017) using a single-zone model of CR interactions with the molecular gas. The planned observations of the Cygnus region with the high sensitivity and good angular resolution *Cherenkov Telescope Array* have the potential to provide imaging and spectra of the region between a few tens of GeV up to ~ 100 TeV (Weinstein et al. 2015). Highly informative MeV to GeV observations of the region can be performed with the planned *e-ASTROGAM* mission (De Angelis et al. 2017).

5.2 Cosmic Ray Acceleration by Supernovae in Compact Clusters

Compact clusters of young massive stars are sites of SN explosions. Contrary to loose OB associations, young stellar clusters have large core star densities $\gtrsim 10^3 M_\odot\,\mathrm{pc}^{-3}$, with a total cluster mass $> 10^4 M_\odot$ within a virial radius ~ 1 pc (e.g., Portegies Zwart et al. 2010). The total mechanical power of the OB stars in such a cluster may exceed 5×10^{38} erg s^{-1} and will likely launch a powerful cluster wind, as modeled by Chevalier and Clegg (1985) for starburst-galaxy nuclei.

A supernova in such a cluster (see sketch in Fig. 10) may be an efficient CR accelerator. As shown by Bykov (2014) and Bykov et al. (2015), a SN blast wave, colliding with a fast wind from a compact cluster of young stars, allows the acceleration of protons to energies well above the standard limits of Fermi acceleration in an isolated SN. The proton spectrum in such a wind-SN PeVatron accelerator is hard with a large flux in the high-energy-end of the spectrum producing copious γ-rays and neutrinos in inelastic nuclear collisions. In Fig. 11 we illustrate the model predictions for the γ-ray (dashed line) and neutrino spectra (solid line) produced by a SN in a cluster similar to the galactic Westerlund 1 cluster (see Bykov et al. 2015, for details). It is estimated that the Westerlund 1 cluster may accelerate protons to $\gtrsim 40$ PeV and result in enough neutrino production to account for a few events detected by the *Ice Cube Observatory* from the inner Milky Way direction.

6 Prospects for Future Observational Facilities

6.1 The Square Kilometer Array

The *Square Kilometer Array* (SKA, http://skatelescope.org/) is the next high sensitivity radio telescope scheduled for 2020 to be located in South Africa and Australia. The SKA will have

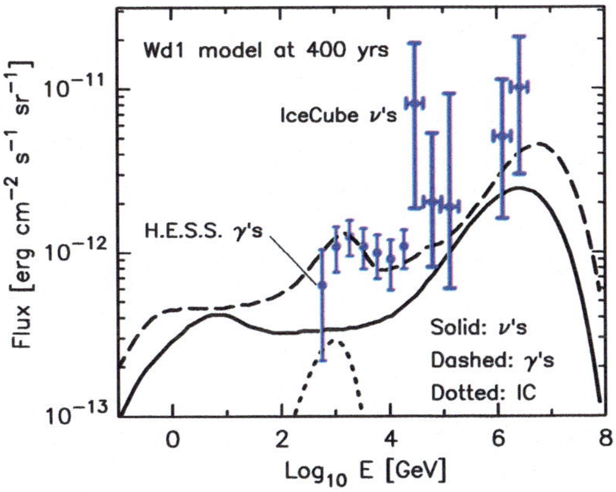

Fig. 11 Model predictions of γ-rays (*dashed curve*) and neutrinos (*solid curve*) from proton-proton interactions calculated in a colliding shock flow source of age $\sim$ 400 yrs as given by Bykov et al. (2015). The *dotted curve* is the IC emission from primary and secondary electrons accelerated directly in this source. The extreme upward curvature in the neutrino spectrum above $\sim$ 10 TeV reflects the transition from CR acceleration in the single SNR shock for low-energy particles to the more efficient acceleration for high-energy particles as they scatter back and forth between the SNR shock and the cluster wind. The data points for the H.E.S.S. source, and the five *Ice Cube* events, are presented just to illustrate the model predictions

an effective surface of 1 km^2 with unprecedented sensitivities of 3.36 and 0.75 $\mu J/\sqrt{Hz}$ for continuum emission at two wavebands 0.05–0.35 GHz (SKA-Low) and 0.35–14 GHz (SKA-Mid), respectively. The SKA-Mid will have an angular resolution of 0.25 arcsecond at 1 GHz. As we have emphasized above, radio observations are essential for understanding many aspects of SN physics and the CSM and the improved sensitivity of the SKA should permit a breakthrough in SNe radio observations.

6.1.1 Radio Observations of Type Ia Supernova Explosions

As of yet, no type Ia SN explosion has been observed in the radio band. Hence, the high sensitivity of the SKA may lead to this discovery as well as the ability to follow in time the radio emission of such objects. The SKA-mid configuration should reach flux sensitivities below the upper limits obtained by the VLA for the most luminous type Ia SNe like SN 2014J or SN 2011fe.[7] A SNIa radio detection will help to determine if the stellar system prior to the explosion is double or single degenerate. Radio emission is expected from a single degenerate case as a result of the interaction with the CSM deposited by the companion star (Wang et al. 2015).

6.1.2 Radio Observations of Core-Collapse Supernova Explosions

The term core-collapse supernova (CCSN) includes a wide variety of sub-types, the names of which convey little or no information on the underlying properties of the object. For example, type IIP SNe (about half of the SNe), type IIL, IIb, Ib/c, or IIn (see Smith 2014, for details). At present, about 50 CCSN explosions have been detected and followed in time at radio wavebands (Weiler et al. 2002; Perez-Torres et al. 2015). The SKA, with its improved

[7]The most recent anticipated instrumental performance for SKA1 is given in https://astronomers.skatelescope.org/wp-content/uploads/2017/10/SKA-TEL-SKO-0000818-01_SKA1_Science_Perform.pdf.

sensitivity at different wavelengths and combined pointed and survey modes, is expected to detect several thousand CCSNe (Perez-Torres et al. 2015) up to a redshift $z \sim 0.25$ for the most intense type IIn SNe. With the advent of the SKA the community will be able to start statistical analysis of CCSN properties depending of their subclass type.

The SKA will also be able to monitor the closest and most intense events as it was already done with the VLA for emblematic objects like SN1987A and SN1993J. Time evolution of the radio spectrum is of particular importance to understand shock dynamics, magnetic field generation, particle acceleration in different CSM environments. These issues are intimately related to the problem of the origin of cosmic rays and make these observations particularly relevant to trigger multi-wavelength observation campaigns involving high-energy instruments like the X-ray satellites XMM-Newton or Chandra and γ-ray telescopes like the future Cherenkov Telescope Array (see next sections).

6.2 The Large Synoptic Survey Telescope

The Large Synoptic Survey Telescope (LSST) (https://www.lsst.org/) is a wide-field survey telescope with an 8.4-meter primary mirror currently under construction in Chile with a first light expected in 2019. It is sensitive in the 320–1050 nm band and will scan the entire visible (mostly southern) sky every few nights. The LSST is an important telescope for SN optical monitoring and should obtain photosphere emission lightcurves and act as a trigger for other multi-wave length observatories. The LSST survey is designed to detect more than 100,000 core-collapse SNe, 200,000 type Ia SNe and 20,000 luminous SNe per year.

6.3 X-ray Facilities

By the time the SKA comes on line, a number of telescopes very relevant for SN studies should be operating (Chandra et al. 2015). These include the Cherenkov Telescope Array (CTA) (scheduled for 2019, see Sect. 6.4), Chandra, NuSTAR, and hopefully Swift. XMM-Newton has a planned end date in December 2018. A planned X-ray telescope Athena+ (www.the-athena-x-ray-observatory.eu/) should be operational in 2028 when both SKA and CTA are active. There are several scientific synergies between SKA and CTA, as made clear from topics discussed in some recent joined meetings.[8] Athena+ will be sensitive in the energy band 0.2–12 keV and will have an angular resolution of 5 arcsecond at an energy < 8 keV. Sub-arcsecond angular resolution X-ray imaging combined with a sensitive high resolution spectrometry of the proposed *Lynx X-ray Observatory* (https://wwwastro.msfc.nasa.gov/lynx/) are very promising for the deep studies of both SNe and SNRs.

Until very recently no type Ia SN explosion has been confirmed in X-rays, but Bochenek et al. (2018) have reported the detection in Chandra data of SN 2012c, a type Ia-CSM object, i.e., a type Ia SN surrounded by very dense material. On the other hand more than 60 core-collapse SNe have been detected in X-rays (see Ross and Dwarkadas 2017, and the SNaX database at http://kronos.uchicago.edu/snax/).

The thermal X-ray continuum (i.e., bremsstrahlung radiation) and the thermal X-ray line emission of the shocked and heated CSM and ejecta material behind the forward and reverse shocks respectively, are tracers of the total density of the material into which the shocks are expanding as well as the shocked temperature. The evolution of the X-ray lightcurve can be used to understand the density structure of the material, and hence the mass-loss from

[8] See, e.g., symposium 15 at EWASS 2017, http://eas.unige.ch/EWASS2017/.

the progenitor star (see Sect. 4.3 and the discussion in the referenced papers). The nonthermal radio and X-ray synchrotron emission from relativistic electrons generated by the forward and reverse shocks, can strongly constrain the shock acceleration process including the electron acceleration efficiency and the magnetic field strength in the acceleration zone. As described in Sect. 4.8, modeling the thermal and nonthermal radiation consistently can constrain the overall Fermi acceleration efficiency and the electron-to-proton ratio.

6.4 The Cherenkov Telescope Array

The *Cherenkov Telescope Array* (CTA, www.cta-observatory.org/) is a γ-ray observatory complex scheduled for completion in 2019. It will be located in the Canary islands and in Chile. The northern site will focus on low- and mid-energy ranges from 20 GeV to 20 TeV. The southern site will also include some large telescopes sensitive in the high-energy band up to 100 TeV. The expected γ-ray sensitivity is on the order of 0.2% Crab in 50 hours of observation at 1 TeV. At energies above 100 TeV, LHAASO (located in Tibet) is expected to reach a sensitivity of 0.01 Crab in 1 year (Di Sciascio 2016).

Extending the energy range of sensitive γ-ray telescopes to 10 TeV and beyond is critical for testing CR acceleration at fast shocks. As mentioned in Sect. 4.8, inverse-Compton from leptons competes with pion-decay emission from hadrons at energies up to $\sim$ 10 TeV, and which process dominates depends on often poorly constrained environmental factors. For higher electron energies however, IC emission drops due to the Klein-Nishina effect. Gamma-rays detected well above 10 TeV will almost certainly be of hadronic origin. Such a discovery would constitute a breakthrough in CR physics proving that the hadronic CR component can be accelerated quite early in the lifetime of a SNR (Schure and Bell 2013). From calculations of CR production performed in Marcowith et al. (2014) using a nonlinear Fermi acceleration model including MFA, the horizon for γ-ray detection is typically 10–20 Mpc.

The detection rate of extragalactic SNe in γ-rays is uncertain since no γ-ray signal has yet been detected from such objects. Upper limits been reported (i.e., Simoni et al. 2017) and the expected number of SNe to be observed by CTA is strongly model dependent and highly uncertain. However, CTA, together with SKA, LSST and some X-ray facilities will be active at the same time. This highlights the importance of using multi-messenger radio, optical and X-ray wave bands as triggers for γ-ray monitoring for luminous sources close enough not to be absorbed by the cosmic microwave background.

7 Summary

Apart from the limited number of observed astrophysical neutrinos, and the small amount of interstellar dust so far collected, CRs are the only source of mass from beyond the solar system. Because of this, CRs convey information not obtainable from photons, and understanding their origin is of fundamental importance for astrophysics.

As we emphasize in this review, supernovae, both extragalactic, where the explosion and its aftereffect can often be witnessed, and galactic remnants, which can be studied in detail from radio to γ-rays, offer the best way to study the *in situ* acceleration of CRs. While other sources contribute to the CRs observed at Earth at some level, SNRs are clearly the source of the bulk of CRs with energies below the knee at $\sim 10^{15}$ eV (Sect. 1). As for the acceleration mechanism, first-order Fermi shock acceleration is the most highly developed theoretically, and has by far the most observational confirmation, of any proposed mechanism. While other mechanisms (e.g., those occurring in pulsars) may be important at some level, Fermi acceleration in collisionless shocks is most likely the prime accelerator of CRs and we concentrate on some of the current problems facing nonlinear models.

The origin of CRs with energies beyond the knee is far more uncertain than those below the knee. We discuss the possibilities of CR acceleration well beyond PeV energies by galactic sources associated with the infrequent but powerful relativistic supernovae which can accelerate CRs up to EeV (Sect. 3). In addition, we show how superluminous interaction powered supernovae, and normal core-collapsed SNe in clusters of young massive stars, can efficiently accelerate CRs to above PeV. At some energy above 10^{18} eV CRs are almost certainly of extragalactic origin and we mention some possible acceleration scenarios for such high energies. A critical piece of the puzzle is the transition between galactic and extragalactic CRs and understanding the production of galactic CRs in SNRs is essential for modeling this transition region.

While some aspects of CR origin and Fermi acceleration remain uncertain, we have concentrated on magnetic field amplification, Fermi acceleration in trans-relativistic shocks, acceleration in star clusters from multiple shocks, and prospects for future multi-messenger observations from radio to γ-ray energies.

Acknowledgements A.M.B., D.C.E. and A.M. thank the staff of ISSI for their generous hospitality and assistance. The authors thank the referees for the constructive comments. A.M. Bykov and S.M. Osipov were supported by the RSF grant 16-12-10225. Some of the modeling was performed at the "Tornado" subsystem of the St. Petersburg Polytechnic University supercomputing center. A.M. Bykov thanks R.A. Chevalier and J.C. Raymond for discussions, R. Margutti for the Fig. 1, and M.A. Grekov for his support with computations.

References

M.G. Aartsen, M. Ackermann, J. Adams et al., Observation of high-energy astrophysical neutrinos in three years of IceCube data. Phys. Rev. Lett. **113**(10), 101101 (2014). https://doi.org/10.1103/PhysRevLett.113.101101

A.A. Abdo, M. Ackermann, M. Ajello et al., Gamma-ray emission from the shell of supernova remnant W44 revealed by the Fermi LAT. Science **327**, 1103 (2010a)

A.A. Abdo, M. Ackermann, M. Ajello et al., Observation of supernova remnant IC 443 with the Fermi Large Area Telescope. Astrophys. J. **712**, 459–468 (2010b)

A.U. Abeysekara, A. Albert, R. Alfaro et al., HAWC contributions to the 35th International Cosmic Ray Conference (ICRC2017). arXiv:1708.02572 (2017)

A. Abramowski, F. Aharonian et al. (HESS Collaboration), Acceleration of petaelectronvolt protons in the Galactic Centre. Nature **531**, 476–479 (2016). https://doi.org/10.1038/nature17147

F. Acero, M. Lemoine-Goumard, M. Renaud et al., Study of TeV shell supernova remnants at gamma-ray energies. Astron. Astrophys. **580**, 74 (2015). https://doi.org/10.1051/0004-6361/201525932

F. Acero, M. Ackermann, M. Ajello et al., The first Fermi LAT Supernova Remnant Catalog. Astrophys. J. Suppl. Ser. **224**, 8 (2016). https://doi.org/10.3847/0067-0049/224/1/8

A. Achterberg, R.D. Blandford, S.P. Reynolds, Evidence for enhanced MHD turbulence outside sharp-rimmed supernova remnants. Astron. Astrophys. **281**, 220–230 (1994)

M. Ackermann, M. Ajello, A. Allafort et al., A cocoon of freshly accelerated cosmic rays detected by Fermi in the Cygnus superbubble. Science **334**, 1103 (2011). https://doi.org/10.1126/science.1210311

M. Ackermann, M. Ajello, A. Allafort et al., Detection of the characteristic pion-decay signature in supernova remnants. Science **339**, 807–811 (2013a). https://doi.org/10.1126/science.1231160

M. Ackermann, M. Ajello, A. Allafort et al., Detection of the characteristic pion-decay signature in supernova remnants. Science **339**, 807–811 (2013b). https://doi.org/10.1126/science.1231160

F. Aharonian, A. Neronov, High-energy gamma rays from the massive black hole in the Galactic Center. Astrophys. J. **619**, 306–313 (2005). https://doi.org/10.1086/426426

F. Aharonian, A. Akhperjanian, M. Beilicke et al., An unidentified TeV source in the vicinity of Cygnus OB2. Astron. Astrophys. **393**, 37–40 (2002). https://doi.org/10.1051/0004-6361:20021171

F. Aharonian, A. Bykov, E. Parizot et al., Cosmic rays in galactic and extragalactic magnetic fields. Space Sci. Rev. **166**, 97–132 (2012)

H.S. Ahn, P. Allison, M.G. Bagliesi et al., Discrepant hardening observed in cosmic-ray elemental spectra. Astrophys. J. **714**, 89–93 (2010). https://doi.org/10.1088/2041-8205/714/1/L89

M.L. Ahnen, S. Ansoldi, L.A. Antonelli et al., A cut-off in the TeV gamma-ray spectrum of the SNR Cassiopeia A. Mon. Not. R. Astron. Soc. **472**, 2956–2962 (2017). https://doi.org/10.1093/mnras/stx2079

K.D. Alexander, A.M. Soderberg, L.B. Chomiuk, A new model for the radio emission from SN 1994I and an associated search for radio transients in M51. Astrophys. J. **806**, 106 (2015). https://doi.org/10.1088/0004-637X/806/1/106

R. Aloisio, V. Berezinsky, P. Blasi et al., A dip in the UHECR spectrum and the transition from galactic o extragalactic cosmic rays. Astropart. Phys. **27**, 76–91 (2007). https://doi.org/10.1016/j.astropartphys.2006.09.004

E. Amato, The origin of galactic cosmic rays. Int. J. Mod. Phys. D **23**, 30013 (2014). https://doi.org/10.1142/S0218271814300134

W.D. Apel, J.C. Arteaga-Velàzquez, K. Bekk et al., Ankle-like feature in the energy spectrum of light elements of cosmic rays observed with KASCADE-Grande. Phys. Rev. D **87**(8), 081101 (2013). https://doi.org/10.1103/PhysRevD.87.081101

S. Archambault, A. Archer, W. Benbow et al., Gamma-ray observations of Tycho's supernova remnant with VERITAS and Fermi. Astrophys. J. **836**, 23 (2017). https://doi.org/10.3847/1538-4357/836/1/23

J. Arons, Magnetars in the Metagalaxy: an origin for ultra-high-energy cosmic rays in the nearby Universe. Astrophys. J. **589**, 871–892 (2003). https://doi.org/10.1086/374776

K. Asano, P. Mészáros, Ultrahigh-energy cosmic ray production by turbulence in gamma-ray burst jets and cosmogenic neutrinos. Phys. Rev. D **94**(2), 023005 (2016). https://doi.org/10.1103/PhysRevD.94.023005

W.I. Axford, The acceleration of galactic cosmic rays, in *Origin of Cosmic Rays*, ed. by G. Setti, G. Spada, A.W. Wolfendale. IAU Symposium, vol. 94 (1981), pp. 339–358

W.I. Axford, E. Leer, G. Skadron, The acceleration of cosmic rays by shock waves, in *Proc. 15th ICRC*, Plovdiv, vol. 11 (1977), p. 132

W. Baade, F. Zwicky, Remarks on super-novae and cosmic rays. Phys. Rev. **46**, 76–77 (1934). https://doi.org/10.1103/PhysRev.46.76.2

J. Ballet, X-ray synchrotron emission from supernova remnants. Adv. Space Res. **37**, 1902–1908 (2006). https://doi.org/10.1016/j.asr.2005.03.047

N. Bartel, B. Karimi, M.F. Bietenholz, VLBI of supernovae and gamma-ray bursts. Astron. Rep. **61**, 299–306 (2017). https://doi.org/10.1134/S1063772917040011

A.R. Bell, The acceleration of cosmic rays in shock fronts. I. Mon. Not. R. Astron. Soc. **182**, 147–156 (1978)

A.R. Bell, Turbulent amplification of magnetic field and diffusive shock acceleration of cosmic rays. Mon. Not. R. Astron. Soc. **353**, 550–558 (2004)

A.R. Bell, The interaction of cosmic rays and magnetized plasma. Mon. Not. R. Astron. Soc. **358**, 181–187 (2005)

A.R. Bell, S.G. Lucek, Cosmic ray acceleration to very high energy through the non-linear amplification by cosmic rays of the seed magnetic field. Mon. Not. R. Astron. Soc. **321**, 433–438 (2001). https://doi.org/10.1046/j.1365-8711.2001.04063.x

A.R. Bell, K.M. Schure, B. Reville et al., Cosmic-ray acceleration and escape from supernova remnants. Mon. Not. R. Astron. Soc. **431**, 415–429 (2013)

E.G. Berezhko, D.C. Ellison, A simple model of nonlinear diffusive shock acceleration. Astrophys. J. **526**, 385 (1999)

E.G. Berezhko, V.K. Elshin, L.T. Ksenofontov, Cosmic ray acceleration in supernova remnants. Sov. Phys. JETP **82**, 1–21 (1996)

S.F. Berezhnev, D. Besson, N.M. Budnev et al., The Tunka-133 EAS Cherenkov light array: status of 2011. Nucl. Instrum. Methods Phys. Res., Sect. A, Accel. Spectrom. Detect. Assoc. Equip. **692**, 98–105 (2012). https://doi.org/10.1016/j.nima.2011.12.091

V.S. Berezinskii, S.V. Bulanov, V.A. Dogiel et al., *Astrophysics of Cosmic Rays* (North-Holland, Amsterdam, 1990)

D.R. Bergman, J.W. Belz, Cosmic rays: the second knee and beyond. J. Phys. G, Nucl. Part. Phys. **34**, 359–400 (2007). https://doi.org/10.1088/0954-3899/34/10/R01

M.F. Bietenholz, N. Bartel, M.P. Rupen, Supernova 1986J Very Long Baseline interferometry. II. The evolution of the shell and the central source. Astrophys. J. **712**, 1057–1069 (2010). https://doi.org/10.1088/0004-637X/712/2/1057

W.R. Binns, E.R. Christian, A.C. Cummings et al., Constraints on galactic cosmic-ray origins from elemental composition measurements, in *APS April Meeting Abstracts* (2014). https://doi.org/10.1103/BAPS.2014.APRIL.E9.4

W.R. Binns, M.H. Israel, E.R. Christian et al., Observation of the ^{60}Fe nucleosynthesis-clock isotope in galactic cosmic rays. Science **352**, 677–680 (2016). https://doi.org/10.1126/science.aad6004
C.-I. Björnsson, S.T. Keshavarzi, Inhomogeneities and the modeling of radio supernovae. Astrophys. J. **841**, 12 (2017). https://doi.org/10.3847/1538-4357/aa6cad
R. Blandford, D. Eichler, Particle acceleration at astrophysical shocks: a theory of cosmic ray origin. Phys. Rep. **154**, 1–75 (1987)
R.D. Blandford, J.P. Ostriker, Particle acceleration by astrophysical shocks. Astrophys. J. **221**, 29–32 (1978)
R. Blandford, P. Simeon, Y. Yuan, Cosmic ray origins: an introduction. Nucl. Phys. B, Proc. Suppl. **256**, 9–22 (2014). https://doi.org/10.1016/j.nuclphysbps.2014.10.002
P. Blasi, The origin of galactic cosmic rays. Astron. Astrophys. Rev. **21**, 70 (2013). https://doi.org/10.1007/s00159-013-0070-7
P. Blasi, R.I. Epstein, A.V. Olinto, Ultra-high-energy cosmic rays from young neutron star winds. Astrophys. J. **533**, 123–126 (2000). https://doi.org/10.1086/312626
S. Blinnikov, Radiative shock waves and their role in solving puzzles of Superluminous Supernovae. arXiv:1611.00513 (2016a)
S.I. Blinnikov, Light produced by shocks and shocks produced by light: superluminous supernovae and GRB afterglows. Astron. Astrophys. Trans. **29**, 129–142 (2016b)
C.D. Bochenek, V.V. Dwarkadas, J.M. Silverman et al., X-ray emission from SN 2012ca: a Type Ia-CSM supernova explosion in a dense surrounding medium. Mon. Not. R. Astron. Soc. **473**, 336–344 (2018). https://doi.org/10.1093/mnras/stx2029
P.J. Boyle, F. Gahbauer, C. Höppner et al., Cosmic ray composition at high energies: the TRACER project. Adv. Space Res. **42**, 409–416 (2008). https://doi.org/10.1016/j.asr.2007.03.063
D. Branch, J.C. Wheeler, *Supernova Explosions* (Springer, Berlin, 2017). https://doi.org/10.1007/978-3-662-55054-0
R. Brose, I. Telezhinsky, M. Pohl, Transport of magnetic turbulence in supernova remnants. Astron. Astrophys. **593**, 20 (2016). https://doi.org/10.1051/0004-6361/201527345
R. Budnik, B. Katz, A. MacFadyen et al., Cosmic rays from transrelativistic supernovae. Astrophys. J. **673**, 928–933 (2008)
S. Buitink, A. Corstanje, H. Falcke et al., A large light-mass component of cosmic rays at 10^{17}–$10^{17.5}$ electronvolts from radio observations. Nature **531**, 70–73 (2016). https://doi.org/10.1038/nature16976
A.M. Bykov, Particle acceleration and nonthermal phenomena in superbubbles. Space Sci. Rev. **99**, 317–326 (2001)
A.M. Bykov, Nonthermal particles and photons in starburst regions and superbubbles. Astron. Astrophys. Rev. **22**, 77 (2014). https://doi.org/10.1007/s00159-014-0077-8
A.M. Bykov, I.N. Toptygin, Theory of charge particle acceleration by a shock wave ensemble in a turbulent medium. Sov. Phys. JETP **71**, 702–708 (1990)
A.M. Bykov, D.C. Ellison, S.M. Osipov et al., X-ray stripes in Tycho's supernova remnant: synchrotron footprints of a nonlinear cosmic-ray-driven instability. Astrophys. J. **735**, 40 (2011). https://doi.org/10.1088/2041-8205/735/2/L40
A.M. Bykov, A. Brandenburg, M.A. Malkov et al., Microphysics of cosmic ray driven plasma instabilities. Space Sci. Rev. **178**, 201–232 (2013). https://doi.org/10.1007/s11214-013-9988-3
A.M. Bykov, D.C. Ellison, S.M. Osipov et al., Magnetic field amplification in nonlinear diffusive shock acceleration including resonant and non-resonant cosmic-ray driven instabilities. Astrophys. J. **789**, 137 (2014). https://doi.org/10.1088/0004-637X/789/2/137
A.M. Bykov, D.C. Ellison, P.E. Gladilin et al., Ultrahard spectra of PeV neutrinos from supernovae in compact star clusters. Mon. Not. R. Astron. Soc. **453**, 113–121 (2015). https://doi.org/10.1093/mnras/stv1606
A.M. Bykov, D.C. Ellison, P.E. Gladilin et al., Supernovae in compact star clusters as sources of high-energy cosmic rays and neutrinos. Adv. Space Res. (2017). https://doi.org/10.1016/j.asr.2017.05.043. 1706.01135.
A.M. Bykov, D.C. Ellison, S.M. Osipov, Nonlinear Monte Carlo model of superdiffusive shock acceleration with magnetic field amplification. Phys. Rev. E **95**(3), 033207 (2017a). https://doi.org/10.1103/PhysRevE.95.033207
A.M. Bykov, S.M. Osipov, D.C. Ellison, Magnetic field amplification in nonlinear trans-relativistic shocks. (2017b in preparation)
D. Caprioli, P. Blasi, E. Amato, On the escape of particles from cosmic ray modified shocks. Mon. Not. R. Astron. Soc. **396**, 2065–2073 (2009)
G. Cassam-Chenaï, J.P. Hughes, E.M. Reynoso et al., Morphological evidence for azimuthal variations of the cosmic-ray ion acceleration at the blast wave of SN 1006. Astrophys. J. **680**, 1180–1197 (2008)
D. Castro, P. Slane, D.C. Ellison et al., Fermi-LAT observations and a broadband study of supernova remnant CTB 109. Astrophys. J. **756**, 88 (2012). https://doi.org/10.1088/0004-637X/756/1/88

C.J. Cesarsky, T. Montmerle, Gamma rays from active regions in the galaxy—the possible contribution of stellar winds. Space Sci. Rev. **36**, 173–193 (1983). https://doi.org/10.1007/BF00167503
S. Chakraborti, A. Ray, Baryon loaded relativistic blast waves in supernovae. Astrophys. J. **729**, 57 (2011). https://doi.org/10.1088/0004-637X/729/1/57
S. Chakraborti, A. Ray, A.M. Soderberg et al., Ultra-high-energy cosmic ray acceleration in engine-driven relativistic supernovae. Nat. Commun. **2**, 175 (2011)
P. Chandra, R.A. Chevalier, N. Chugai et al., X-ray and radio emission from type IIn supernova SN 2010jl. Astrophys. J. **810**, 32 (2015). https://doi.org/10.1088/0004-637X/810/1/32
R.A. Chevalier, Self-similar solutions for the interaction of stellar ejecta with an external medium. Astrophys. J. **258**, 790–797 (1982)
R.A. Chevalier, Common envelope evolution leading to supernovae with dense interaction. Astrophys. J. **752**, 2 (2012). https://doi.org/10.1088/2041-8205/752/1/L2
R.A. Chevalier, A.W. Clegg, Wind from a starburst galaxy nucleus. Nature **317**, 44 (1985). https://doi.org/10.1038/317044a0
R.A. Chevalier, C.M. Irwin, Shock breakout in dense mass loss: luminous supernovae. Astrophys. J. **729**, 6 (2011). https://doi.org/10.1088/2041-8205/729/1/L6
N.N. Chugai, I.J. Danziger, Supernova 1988Z—low-mass ejecta colliding with the clumpy wind. Mon. Not. R. Astron. Soc. **268**, 173 (1994). https://doi.org/10.1093/mnras/268.1.173
J.S. Clark, M.P. Muno, I. Negueruela et al., Unveiling the X-ray point source population of the young massive cluster Westerlund 1. Astron. Astrophys. **477**, 147–163 (2008). https://doi.org/10.1051/0004-6361:20077186
F. Comerón, A.A. Djupvik, N. Schneider et al., Red supergiants and the past of Cygnus OB2. Astron. Astrophys. **586**, 46 (2016). https://doi.org/10.1051/0004-6361/201527517
D.P. Cox, The three-phase interstellar medium revisited. Annu. Rev. Astron. Astrophys. **43**, 337–385 (2005). https://doi.org/10.1146/annurev.astro.43.072103.150615
A. De Angelis, V. Tatischeff, M. Tavani et al., The e-ASTROGAM mission. Exploring the extreme Universe with gamma rays in the MeV–GeV range. Exp. Astron. **44**, 25–82 (2017). https://doi.org/10.1007/s10686-017-9533-6
A. Decourchelle, D.C. Ellison, J. Ballet, Thermal X-ray emission and cosmic-ray production in young supernova remnants. Astrophys. J. **543**, 57 (2000)
G. Di Sciascio (LHAASO Collaboration), The LHAASO experiment: from gamma-ray astronomy to cosmic rays. Nucl. Part. Phys. Proc. **279**, 166–173 (2016). https://doi.org/10.1016/j.nuclphysbps.2016.10.024
L.O. Drury, Escaping the accelerator: how, when and in what numbers do cosmic rays get out of supernova remnants? Mon. Not. R. Astron. Soc. **415**, 1807–1814 (2011). https://doi.org/10.1111/j.1365-2966.2011.18824.x
V.V. Dwarkadas, The evolution of supernovae in circumstellar wind-blown bubbles. I. Introduction and one-dimensional calculations. Astrophys. J. **630**, 892–910 (2005)
V.V. Dwarkadas, R.A. Chevalier, Interaction of type IA supernovae with their surroundings. Astrophys. J. **497**, 807 (1998)
D.C. Ellison, A.M. Bykov, Gamma-ray emission of accelerated particles escaping a supernova remnant in a molecular cloud. Astrophys. J. **731**, 87 (2011)
D.C. Ellison, R. Ramaty, Shock acceleration of electrons and ions in solar flares. Astrophys. J. **298**, 400–408 (1985)
D.C. Ellison, E. Moebius, G. Paschmann, Particle injection and acceleration at earth's bow shock—comparison of upstream and downstream events. Astrophys. J. **352**, 376–394 (1990)
D.C. Ellison, M.G. Baring, F.C. Jones, Acceleration rates and injection efficiencies in oblique shocks. Astrophys. J. **453**, 873 (1995)
D.C. Ellison, L.O. Drury, J. Meyer, Galactic cosmic rays from supernova remnants. II. Shock acceleration of gas and dust. Astrophys. J. **487**, 197 (1997)
D.C. Ellison, D.J. Patnaude, P. Slane et al., Particle acceleration in supernova remnants and the production of thermal and nonthermal radiation. Astrophys. J. **661**, 879–891 (2007)
D.C. Ellison, P. Slane, D.J. Patnaude et al., Core-collapse model of broadband emission from SNR RX J1713.7-3946 with thermal X-rays and gamma rays from escaping cosmic rays. Astrophys. J. **744**, 39 (2012)
D.C. Ellison, D.C. Warren, A.M. Bykov, Monte Carlo simulations of nonlinear particle acceleration in parallel trans-relativistic shocks. Astrophys. J. **776**, 46 (2013). https://doi.org/10.1088/0004-637X/776/1/46
K.A. Eriksen, J.P. Hughes, C. Badenes et al., Evidence for particle acceleration to the knee of the cosmic ray spectrum in Tycho's supernova remnant. Astrophys. J. **728**, 28 (2011). https://doi.org/10.1088/2041-8205/728/2/L28
S.W. Falk, W.D. Arnett, Radiation dynamics, envelope ejection, and supernova light curves. Astrophys. J. Suppl. Ser. **33**, 515–562 (1977)

R. Farber, M. Ruszkowski, H.-Y.K. Yang et al., Impact of cosmic ray transport on galactic winds. arXiv:1707.04579 (2017)
E. Fermi, On the origin of the cosmic radiation. Phys. Rev. **75**, 1169–1174 (1949). https://doi.org/10.1103/PhysRev.75.1169
E. Fermi, Galactic magnetic fields and the origin of cosmic radiation. Astrophys. J. **119**, 1 (1954). https://doi.org/10.1086/145789
G. Ferrand, A. Marcowith, On the shape of the spectrum of cosmic rays accelerated inside superbubbles. Astron. Astrophys. **510**, 101 (2010). https://doi.org/10.1051/0004-6361/200913520
G. Ferrand, A. Decourchelle, S. Safi-Harb, Three-dimensional simulations of the thermal X-ray emission from young supernova remnants including efficient particle acceleration. Astrophys. J. **760**, 34 (2012). https://doi.org/10.1088/0004-637X/760/1/34
R.B. Fiorito, D. Eichler, D.C. Ellison, A study of the phase velocity and growth of waves at parallel shocks. Astrophys. J. **364**, 582–589 (1990). https://doi.org/10.1086/169441
C. Fransson, C.-I. Björnsson, Radio emission and particle acceleration in SN 1993J. Astrophys. J. **509**, 861–878 (1998). https://doi.org/10.1086/306531
Y. Fujita, K. Murase, S.S. Kimura, Sagittarius A* as an origin of the Galactic PeV cosmic rays? J. Cosmol. Astropart. Phys. **4**, 037 (2017). https://doi.org/10.1088/1475-7516/2017/04/037
S. Funk, Ground- and space-based gamma-ray astronomy. Annu. Rev. Nucl. Part. Sci. **65**, 245–277 (2015). https://doi.org/10.1146/annurev-nucl-102014-022036
S. Gabici, F.A. Aharonian, Hadronic gamma-rays from RX J1713.7-3946? Mon. Not. R. Astron. Soc. **445**, 70–73 (2014). https://doi.org/10.1093/mnrasl/slu132
A. Gal-Yam, Luminous supernovae. Science **337**, 927 (2012). https://doi.org/10.1126/science.1203601
V.L. Ginzburg, S.I. Syrovatskii, *The Origin of Cosmic Rays* (Macmillan Co., New York, 1964)
A. Giuliani, M. Cardillo, M. Tavani et al., Neutral pion emission from accelerated protons in the supernova remnant W44. Astrophys. J. **742**, 30 (2011). https://doi.org/10.1088/2041-8205/742/2/L30
I.A. Grenier, J.H. Black, A.W. Strong, The nine lives of cosmic rays in galaxies. Annu. Rev. Astron. Astrophys. **53**, 199–246 (2015). https://doi.org/10.1146/annurev-astro-082214-122457
Y.-Q. Guo, Z. Tian, Z. Wang et al., The galactic center: a petaelectronvolt cosmic-ray acceleration factory. Astrophys. J. **836**, 233 (2017). https://doi.org/10.3847/1538-4357/aa5f58
C. Heiles, Clustered supernovae versus the gaseous disk and halo. Astrophys. J. **354**, 483–491 (1990). https://doi.org/10.1086/168709
E.A. Helder, J. Vink, A.M. Bykov et al., Observational signatures of particle acceleration in supernova remnants. Space Sci. Rev. **173**, 369–431 (2012). https://doi.org/10.1007/s11214-012-9919-8
A.M. Hillas, Can diffusive shock acceleration in supernova remnants account for high-energy galactic cosmic rays? J. Phys. G, Nucl. Part. Phys. **31**, 95 (2005)
A. Horesh, C. Stockdale, D.B. Fox et al., An early and comprehensive millimetre and centimetre wave and X-ray study of SN 2011dh: a non-equipartition blast wave expanding into a massive stellar wind. Mon. Not. R. Astron. Soc. **436**, 1258–1267 (2013). https://doi.org/10.1093/mnras/stt1645
J.P. Hughes, C.E. Rakowski, A. Decourchelle, Electron heating and cosmic rays at a supernova shock from Chandra X-ray observations of 1E 0102.2-7219. Astrophys. J. **543**, 61–65 (2000)
F.C. Jones, D.C. Ellison, The plasma physics of shock acceleration. Space Sci. Rev. **58**, 259–346 (1991)
H. Kang, Effects of wave-particle interactions on diffusive shock acceleration at supernova remnants. J. Korean Astron. Soc. **46**, 49–63 (2013)
H. Kang, D. Ryu, T.W. Jones, Self-similar evolution of cosmic-ray modified shocks: the cosmic-ray spectrum. Astrophys. J. **695**, 1273–1288 (2009)
J. Katsuta, Y. Uchiyama, S. Funk, Extended gamma-ray emission from the G25.0+0.0 region: a star-forming region powered by the newly found OB association? Astrophys. J. **839**, 129 (2017). https://doi.org/10.3847/1538-4357/aa6aa3
B. Katz, N. Sapir, E. Waxman, X-rays, gamma-rays and neutrinos from collisionless shocks in supernova wind breakouts. arXiv:1106.1898 (2011)
N. Kimani, K. Sendlinger, A. Brunthaler et al., Radio evolution of supernova SN 2008iz in M 82. Astron. Astrophys. **593**, 18 (2016). https://doi.org/10.1051/0004-6361/201628800
J. Knödlseder, Cygnus OB2—a young globular cluster in the Milky Way. Astron. Astrophys. **360**, 539–548 (2000)
M.I. Krauss, A.M. Soderberg, L. Chomiuk et al., Expanded Very Large Array observations of the radio evolution of SN 2011dh. Astrophys. J. **750**, 40 (2012). https://doi.org/10.1088/2041-8205/750/2/L40
M. Krumholz, Star Formation (2017)
G.F. Krymskii, A regular mechanism for the acceleration of charged particles on the front of a shock wave. Dokl. Akad. Nauk SSSR **234**, 1306–1308 (1977)

S.R. Kulkarni, D.A. Frail, M.H. Wieringa et al., Radio emission from the unusual supernova 1998bw and its association with the γ-ray burst of 25 April 1998. Nature **395**, 663–669 (1998). https://doi.org/10.1038/27139
E. Kundu, P. Lundqvist, M.A. Pérez-Torres et al., Constraining magnetic field amplification in SN shocks using radio observations of SNe 2011fe and 2014J. Astrophys. J. **842**, 17 (2017). https://doi.org/10.3847/1538-4357/aa704c
C.J. Lada, E.A. Lada, Embedded clusters in molecular clouds. Annu. Rev. Astron. Astrophys. **41**, 57–115 (2003). https://doi.org/10.1146/annurev.astro.41.011802.094844
P.O. Lagage, C.J. Cesarsky, The maximum energy of cosmic rays accelerated by supernova shocks. Astron. Astrophys. **125**, 249–257 (1983)
S.-H. Lee, D.C. Ellison, S. Nagataki, A generalized model of nonlinear diffusive shock acceleration coupled to an evolving supernova remnant. Astrophys. J. **750**, 156 (2012)
M. Lemoine, Acceleration and propagation of ultrahigh energy cosmic rays. J. Phys. Conf. Ser. **409**, 012007 (2013). https://doi.org/10.1088/1742-6596/409/1/012007
W. Li, J. Leaman, R. Chornock et al., Nearby supernova rates from the Lick Observatory Supernova Search—II. The observed luminosity functions and fractions of supernovae in a complete sample. Mon. Not. R. Astron. Soc. **412**, 1441–1472 (2011). https://doi.org/10.1111/j.1365-2966.2011.18160.x
R.E. Lingenfelter, Cosmic rays from supernova remnants and superbubbles. Adv. Space Res. (2017). https://doi.org/10.1016/j.asr.2017.04.006
P. Lipari, Spectral features in the cosmic ray fluxes. arXiv:1707.02504 (2017)
Y. Lithwick, P. Goldreich, Compressible magnetohydrodynamic turbulence in interstellar plasmas. Astrophys. J. **562**, 279–296 (2001). https://doi.org/10.1086/323470
R.-Y. Liu, X.-Y. Wang, A. Prosekin et al., Modeling the gamma-ray emission in the Galactic Center with a fading cosmic-ray accelerator. Astrophys. J. **833**, 200 (2016). https://doi.org/10.3847/1538-4357/833/2/200
A. Loeb, E. Waxman, The cumulative background of high energy neutrinos from starburst galaxies. J. Cosmol. Astropart. Phys. **5**, 003 (2006). https://doi.org/10.1088/1475-7516/2006/05/003
M.-M. Mac Low, R. McCray, Superbubbles in disk galaxies. Astrophys. J. **324**, 776–785 (1988). https://doi.org/10.1086/165936
M. Malkov, Newly-discovered anomalies in galactic cosmic rays: time for exotic scenarios? arXiv:1703.05772 (2017)
M.A. Malkov, L.O. Drury, Nonlinear theory of diffusive acceleration of particles by shock waves. Rep. Prog. Phys. **64**, 429–481 (2001)
M.A. Malkov, P.H. Diamond, R.Z. Sagdeev et al., Analytic solution for self-regulated collective escape of cosmic rays from their acceleration sites. Astrophys. J. **768**, 73 (2013). https://doi.org/10.1088/0004-637X/768/1/73
J.M. Marcaide, I. Martí-Vidal, M.A. Perez-Torres et al., 1.6 GHz VLBI observations of SN 1979C: almost-free expansion. Astron. Astrophys. **503**, 869–872 (2009). https://doi.org/10.1051/0004-6361/200912485
A. Marcowith, F. Casse, Postshock turbulence and diffusive shock acceleration in young supernova remnants. Astron. Astrophys. **515**, 90 (2010). https://doi.org/10.1051/0004-6361/200913022
A. Marcowith, M. Renaud, V. Dwarkadas et al., Cosmic-ray acceleration and gamma-ray signals from radio supernovæ. Nucl. Phys. B, Proc. Suppl. **256**, 94–100 (2014). https://doi.org/10.1016/j.nuclphysbps.2014.10.011
A. Marcowith, A. Bret, A. Bykov et al., The microphysics of collisionless shock waves. Rep. Prog. Phys. **79**(4), 046901 (2016). https://doi.org/10.1088/0034-4885/79/4/046901
R. Margutti, D. Milisavljevic, A.M. Soderberg et al., Relativistic supernovae have shorter-lived central engines or more extended progenitors: the case of SN 2012ap. Astrophys. J. **797**, 107 (2014). https://doi.org/10.1088/0004-637X/797/2/107
I. Martí-Vidal, J.M. Marcaide, A. Alberdi et al., Radio emission of SN1993J: the complete picture. II. Simultaneous fit of expansion and radio light curves. Astron. Astrophys. **526**, 143 (2011). https://doi.org/10.1051/0004-6361/201014517
P. Martin, J. Knödlseder, G. Meynet et al., Predicted gamma-ray line emission from the Cygnus complex. Astron. Astrophys. **511**, 86 (2010). https://doi.org/10.1051/0004-6361/200912864
J.P. Meyer, D.C. Ellison, The origin of present day cosmic rays: fresh SN ejecta or interstellar medium material? I. Cosmic ray composition and SN nucleosynthesis. A conflict with the early Galactic evolution of Be? arXiv:astro-ph/9905037 (1999)
J. Meyer, L.O. Drury, D.C. Ellison, Galactic cosmic rays from supernova remnants. I. A cosmic-ray composition controlled by volatility and mass-to-charge ratio. Astrophys. J. **487**, 182 (1997)
T.J. Moriya, K. Maeda, Constraining physical properties of type IIn supernovae through rise times and peak luminosities. Astrophys. J. **790**, 16 (2014). https://doi.org/10.1088/2041-8205/790/2/L16

T.J. Moriya, S.I. Blinnikov, N. Tominaga et al., Light-curve modelling of superluminous supernova 2006gy: collision between supernova ejecta and a dense circumstellar medium. Mon. Not. R. Astron. Soc. **428**, 1020–1035 (2013). https://doi.org/10.1093/mnras/sts075

V. Morozova, A.L. Piro, S. Valenti, Unifying type II supernova light curves with dense circumstellar material. Astrophys. J. **838**, 28 (2017). https://doi.org/10.3847/1538-4357/aa6251

K. Murase, T.A. Thompson, B.C. Lacki et al., New class of high-energy transients from crashes of supernova ejecta with massive circumstellar material shells. Phys. Rev. D **84**(4), 043003 (2011). https://doi.org/10.1103/PhysRevD.84.043003

K. Murase, T.A. Thompson, E.O. Ofek, Probing cosmic ray ion acceleration with radio-submm and gamma-ray emission from interaction-powered supernovae. Mon. Not. R. Astron. Soc. **440**, 2528–2543 (2014). https://doi.org/10.1093/mnras/stu384

T. Nugis, H.J.G.L.M. Lamers, Mass-loss rates of Wolf-Rayet stars as a function of stellar parameters. Astron. Astrophys. **360**, 227–244 (2000)

S. Ohm, Gamma-rays from starburst galaxies. AIP Conf. Proc. **1505**, 64–71 (2012). https://doi.org/10.1063/1.4772221

S. Ohm, Starburst galaxies as seen by gamma-ray telescopes. C. R. Phys. **17**, 585–593 (2016). https://doi.org/10.1016/j.crhy.2016.04.003

E. Parizot, A. Marcowith, E. van der Swaluw et al., Superbubbles and energetic particles in the Galaxy. I. Collective effects of particle acceleration. Astron. Astrophys. **424**, 747–760 (2004). https://doi.org/10.1051/0004-6361:20041269

E. Parizot, A. Marcowith, J. Ballet et al., Observational constraints on energetic particle diffusion in young supernova remnants: amplified magnetic field and maximum energy. Astron. Astrophys. **453**, 387–395 (2006). https://doi.org/10.1051/0004-6361:20064985

D.J. Patnaude, S.-H. Lee, P.O. Slane et al., The impact of progenitor mass loss on the dynamical and spectral evolution of supernova remnants. arXiv:1708.04984 (2017)

M. Perez-Torres, A. Alberdi, R.J. Beswick et al., Core-collapse and Type Ia supernovae with the SKA, in *Advancing Astrophysics with the Square Kilometre Array (AASKA14)* (2015), p. 60

M. Petropoulou, S. Coenders, G. Vasilopoulos et al., Point-source and diffuse high-energy neutrino emission from Type IIn supernovae. Mon. Not. R. Astron. Soc. **470**, 1881–1893 (2017). https://doi.org/10.1093/mnras/stx1251

S.F. Portegies Zwart, S.L.W. McMillan, M. Gieles, Young massive star clusters. Annu. Rev. Astron. Astrophys. **48**, 431–493 (2010). https://doi.org/10.1146/annurev-astro-081309-130834

V. Ptuskin, Propagation of galactic cosmic rays. Astropart. Phys. **39**, 44–51 (2012). https://doi.org/10.1016/j.astropartphys.2011.11.004

V.S. Ptuskin, V.N. Zirakashvili, On the spectrum of high-energy cosmic rays produced by supernova remnants in the presence of strong cosmic-ray streaming instability and wave dissipation. Astron. Astrophys. **429**, 755–765 (2005)

J.C. Raymond, Shock waves in supernova ejecta. Space Sci. Rev. **214**, 27 (2018). https://doi.org/10.1007/s11214-017-0461-6

S. Recchia, P. Blasi, G. Morlino, Cosmic ray-driven winds in the Galactic environment and the cosmic ray spectrum. Mon. Not. R. Astron. Soc. **470**, 865–881 (2017). https://doi.org/10.1093/mnras/stx1214

S.P. Reynolds, Supernova remnants at high energy. Annu. Rev. Astron. Astrophys. **46**, 89–126 (2008). https://doi.org/10.1146/annurev.astro.46.060407.145237

M. Ross, V.V. Dwarkadas, SNaX: a database of supernova X-ray light curves. Astron. J. **153**, 246 (2017). https://doi.org/10.3847/1538-3881/aa6d50

R. Rothenflug, J. Ballet, G. Dubner et al., Geometry of the non-thermal emission in SN 1006. Azimuthal variations of cosmic-ray acceleration. Astron. Astrophys. **425**, 121–131 (2004)

K.L.J. Rygl, A. Brunthaler, A. Sanna et al., Parallaxes and proper motions of interstellar masers toward the Cygnus X star-forming complex. I. Membership of the Cygnus X region. Astron. Astrophys. **539**, 79 (2012). https://doi.org/10.1051/0004-6361/201118211

K.M. Schure, A.R. Bell, Cosmic ray acceleration in young supernova remnants. Mon. Not. R. Astron. Soc. **435**, 1174–1185 (2013). https://doi.org/10.1093/mnras/stt1371

K.M. Schure, A.R. Bell, L. O'C Drury et al., Diffusive shock acceleration and magnetic field amplification. Space Sci. Rev. **173**, 491–519 (2012)

R. Simoni, N. Maxted, M. Renaud et al., Upper limits on gamma-ray emission from supernovae serendipitously observed with H.E.S.S, in *Supernova 1987A:30 Years Later—Cosmic Rays and Nuclei from Supernovae and Their Aftermaths*, ed. by A. Marcowith, M. Renaud, G. Dubner et al.. IAU Symposium, vol. 331 (2017), pp. 325–328. https://doi.org/10.1017/S1743921317004628

P. Slane, S.-H. Lee, D.C. Ellison et al., Erratum: "A CR-hydro-NEI model of the structure and broadband emission from Tycho's supernova remnant". Astrophys. J. **799**, 238 (2015a). https://doi.org/10.1088/0004-637X/799/2/238

P. Slane, A. Bykov, D.C. Ellison et al., Supernova remnants interacting with molecular clouds: X-ray and gamma-ray signatures. Space Sci. Rev. **188**, 187–210 (2015b). https://doi.org/10.1007/s11214-014-0062-6

V.I. Slysh, Radio supernovae and particle acceleration. Astron. Astrophys. Trans. **1**, 171–193 (1992). https://doi.org/10.1080/10556799208260465

N. Smith, Mass loss: its effect on the evolution and fate of high-mass stars. Annu. Rev. Astron. Astrophys. **52**, 487–528 (2014). https://doi.org/10.1146/annurev-astro-081913-040025

N. Smith, Luminous blue variables and the fates of very massive stars. Philos. Trans. R. Soc. Lond. Ser. A **375**, 20160268 (2017). https://doi.org/10.1098/rsta.2016.0268

A.M. Soderberg, S.R. Kulkarni, E. Berger et al., The radio and X-ray-luminous Type Ibc supernova 2003L. Astrophys. J. **621**, 908–920 (2005). https://doi.org/10.1086/427649

A.M. Soderberg, S.R. Kulkarni, E. Nakar et al., Relativistic ejecta from X-ray flash XRF 060218 and the rate of cosmic explosions. Nature **442**, 1014–1017 (2006)

A.M. Soderberg, S. Chakraborti, G. Pignata et al., A relativistic type Ibc supernova without a detected γ-ray burst. Nature **463**, 513–515 (2010)

E. Sorokina, S. Blinnikov, K. Nomoto et al., Type I superluminous supernovae as explosions inside non-hydrogen circumstellar envelopes. Astrophys. J. **829**, 17 (2016). https://doi.org/10.3847/0004-637X/829/1/17

L.G. Sveshnikova, The knee in the Galactic cosmic ray spectrum and variety in Supernovae. Astron. Astrophys. **409**, 799–807 (2003). https://doi.org/10.1051/0004-6361:20030909

V. Tatischeff, Radio emission and nonlinear diffusive shock acceleration of cosmic rays in the supernova SN 1993J. Astron. Astrophys. **499**, 191–213 (2009). https://doi.org/10.1051/0004-6361/200811511

M. Tavani, A. Giuliani, A.W. Chen et al., Direct evidence for hadronic cosmic-ray acceleration in the supernova remnant IC 443. Astrophys. J. **710**, 151–155 (2010). https://doi.org/10.1088/2041-8205/710/2/L151

I. Telezhinsky, V.V. Dwarkadas, M. Pohl, Time-dependent escape of cosmic rays from supernova remnants, and their interaction with dense media. arXiv:1112.3194 (2011)

S. Thoudam et al., Cosmic-ray energy spectrum and composition up to the ankle: the case for a second Galactic component. Astron. Astrophys. **595**, 33 (2016). https://doi.org/10.1051/0004-6361/201628894

S.V. Troitsky, Cosmic particles with energies above 10^{19} eV: a brief summary of results. Phys. Usp. **56**, 304–310 (2013). https://doi.org/10.3367/UFNe.0183.201303i.0323

Y. Uchiyama, R.D. Blandford, S. Funk et al., Gamma-ray emission from crushed clouds in supernova remnants. Astrophys. J. **723**, 122–126 (2010)

A. ud-Doula, S.P. Owocki, Dynamical simulations of magnetically channeled line-driven stellar winds. I. Isothermal, nonrotating, radially driven flow. Astrophys. J. **576**, 413–428 (2002). https://doi.org/10.1086/341543

A.J. van Marle, F. Casse, A. Marcowith, On magnetic field amplification and particle acceleration near non-relativistic astrophysical shocks: particles in MHD Cells simulations. arXiv:1709.08482 (2017)

J. Vink, Supernova remnants: the X-ray perspective. Astron. Astrophys. Rev. **20**, 49 (2012). https://doi.org/10.1007/s00159-011-0049-1

J.S. Vink, Winds from stripped low-mass Helium stars and Wolf-Rayet stars. arXiv:1710.02010 (2017)

R. Walder, D. Folini, G. Meynet, Magnetic fields in massive stars, their winds, and their nebulae. Space Sci. Rev. **166**, 145–185 (2012). https://doi.org/10.1007/s11214-011-9771-2

L. Wang, X. Cui, H. Zhu et al., Investigations of supernovae and supernova remnants in the era of SKA, in *Advancing Astrophysics with the Square Kilometre Array (AASKA14)* (2015), p. 64

D.C. Warren, D.C. Ellison, A.M. Bykov et al., Electron and ion acceleration in relativistic shocks with applications to GRB afterglows. Mon. Not. R. Astron. Soc. **452**, 431–443 (2015). https://doi.org/10.1093/mnras/stv1304

D.C. Warren, D.C. Ellison, M.V. Barkov et al., Nonlinear particle acceleration and thermal particles in GRB afterglows. Astrophys. J. **835**, 248 (2017). https://doi.org/10.3847/1538-4357/aa56c3

E. Waxman, Cosmological gamma-ray bursts and the highest energy cosmic rays. Phys. Rev. Lett. **75**, 386–389 (1995). https://doi.org/10.1103/PhysRevLett.75.386

K.W. Weiler, R.A. Sramek, N. Panagia et al., Radio supernovae. Astrophys. J. **301**, 790–812 (1986). https://doi.org/10.1086/163944

K.W. Weiler, N. Panagia, R.A. Sramek, Radio emission from supernovae. II—SN 1986J: a different kind of type II. Astrophys. J. **364**, 611–625 (1990). https://doi.org/10.1086/169444

K.W. Weiler, S.D. van Dyk, J.L. Discenna et al., The 10 year radio light curves for SN 1979C. Astrophys. J. **380**, 161–166 (1991). https://doi.org/10.1086/170571

K.W. Weiler, N. Panagia, M.J. Montes et al., Radio emission from supernovae and gamma-ray bursters. Annu. Rev. Astron. Astrophys. **40**, 387–438 (2002). https://doi.org/10.1146/annurev.astro.40.060401.093744

K.W. Weiler, N. Panagia, C. Stockdale et al., Radio emission from SN 1994I in NGC 5194 (M 51): the best-studied Type Ib/c radio supernova. Astrophys. J. **740**, 79 (2011). https://doi.org/10.1088/0004-637X/740/2/79

A. Weinstein, E. Aliu, S. Casanova et al., Creating a high-resolution picture of Cygnus with the Cherenkov Telescope Array. arXiv:1509.02189 (2015)

N.J. Wright, R.J. Parker, S.P. Goodwin et al., Constraints on massive star formation: Cygnus OB2 was always an association. Mon. Not. R. Astron. Soc. **438**, 639–646 (2014). https://doi.org/10.1093/mnras/stt2232

N.J. Wright, J.E. Drew, M. Mohr-Smith, The massive star population of Cygnus OB2. Mon. Not. R. Astron. Soc. **449**, 741–760 (2015). https://doi.org/10.1093/mnras/stv323

N. Yadav, A. Ray, S. Chakraborti, Low-frequency radio observations of SN 2011dh and the evolution of its post-shock plasma properties. Mon. Not. R. Astron. Soc. **459**, 595–602 (2016). https://doi.org/10.1093/mnras/stw594

T.M. Yoast-Hull, J.S. Gallagher, F. Halzen et al., Gamma-ray puzzle in Cygnus X: implications for high-energy neutrinos. Phys. Rev. D **96**(4), 043011 (2017). https://doi.org/10.1103/PhysRevD.96.043011

G. Zimbardo, E. Amato, A. Bovet et al., Superdiffusive transport in laboratory and astrophysical plasmas. J. Plasma Phys. **81**(6), 495810601 (2015). https://doi.org/10.1017/S0022377815001117

V.N. Zirakashvili, V.S. Ptuskin, Type IIn supernovae as sources of high energy astrophysical neutrinos. Astropart. Phys. **78**, 28–34 (2016). https://doi.org/10.1016/j.astropartphys.2016.02.004

Physics & Material science

Printed by Printforce, the Netherlands